T0239718

Diffractive Optics
and Nanophotonics

Diffractive Optics and Nanophotonics

Edited by
V. A. Soifer

CISP

CRC Press
Taylor & Francis Group
Boca Raton London New York

CRC Press is an imprint of the
Taylor & Francis Group, an **informa** business

CRC Press
Taylor & Francis Group
6000 Broken Sound Parkway NW, Suite 300
Boca Raton, FL 33487-2742

First issued in paperback 2020

© 2017 by CISP
CRC Press is an imprint of Taylor & Francis Group, an Informa business

No claim to original U.S. Government works

ISBN 13: 978-0-367-57310-2 (pbk)
ISBN 13: 978-1-4987-5447-7 (hbk)

This book contains information obtained from authentic and highly regarded sources. Reasonable efforts have been made to publish reliable data and information, but the author and publisher cannot assume responsibility for the validity of all materials or the consequences of their use. The authors and publishers have attempted to trace the copyright holders of all material reproduced in this publication and apologize to copyright holders if permission to publish in this form has not been obtained. If any copyright material has not been acknowledged please write and let us know so we may rectify in any future reprint.

Except as permitted under U.S. Copyright Law, no part of this book may be reprinted, reproduced, transmitted, or utilized in any form by any electronic, mechanical, or other means, now known or hereafter invented, including photocopying, microfilming, and recording, or in any information storage or retrieval system, without written permission from the publishers.

For permission to photocopy or use material electronically from this work, please access www.copyright.com (http://www.copyright.com/) or contact the Copyright Clearance Center, Inc. (CCC), 222 Rosewood Drive, Danvers, MA 01923, 978-750-8400. CCC is a not-for-profit organization that provides licenses and registration for a variety of users. For organizations that have been granted a photocopy license by the CCC, a separate system of payment has been arranged.

Trademark Notice: Product or corporate names may be trademarks or registered trademarks, and are used only for identification and explanation without intent to infringe.

Visit the Taylor & Francis Web site at
http://www.taylorandfrancis.com

and the CRC Press Web site at
http://www.crcpress.com

Contents

Introduction   **xii**

**1.   The Fourier modal method and its use in plasmonics and
         the theory of resonant diffraction gratings**   **1**
1.1.   The Fourier modal method of solving the diffraction problems   3
1.1.1.   The equation of a plane wave   3
1.1.2.   The Fourier modal method for two-dimensional periodic
         structures   7
1.1.2.1. The geometry of the structure and formulation of the problem   8
1.1.2.2. Description of the field above and below the structure   9
1.1.2.3. The system of differential equations for the description of
         the field in the layer   10
1.1.2.4. Representation of the field inside the layer   21
1.1.2.5. 'Stitching' of the electromagnetic field at the boundaries of
         layers   23
1.1.2.6. The scattering matrix algorithm   26
1.1.2.7. Calculation of the field distribution   30
1.1.2.8. Intensities of the diffraction orders   31
1.1.2.9. Numerical example   33
1.1.3.   The Fourier modal method for three-dimensional
         periodic structures   33
1.1.4.   The Fourier modal method for two-dimensional non-
         periodic structures   40
1.1.4.1. The geometry of the structure and formulation of the problem   40
1.1.4.2. Perfectly matched absorbing layers   41
1.1.4.3. Solution of the diffraction problem   52
1.1.4.4 The energy characteristics of the reflected and transmitted
         light   54
1.1.4.5. Numerical example   55
1.2.   Methods for calculating eigenmodes of periodic diffractive
         structures   56
1.2.1.   Calculation of modes based on the calculation of the poles
         of the scattering matrix   56
1.2.1.1. Resonant representation of the scattering matrix   57
1.2.1.2. Calculation of the poles of the scattering matrix   60

1.2.1.3. Numerical examples   67
1.2.2.   Calculation of events based on calculating eigenvalues of the
         transfer matrix   73
1.2.2.1. Calculation of modes   73
1.2.2.2. Calculation of the field distribution   75
1.2.2.3. Numerical example   78
1.3.   Resonant diffraction gratings for transformation of optical
         pulses   78
1.3.1.   The integration and differentiation of optical pulses.
         Wood's anomalies   79
1.3.2.   Types of gratings for integration and differentiation of
         pulses   82
1.3.3.   Fractional integration and differentiation of optical pulses.
         Rayleigh–Wood anomalies   84
1.3.4.   Examples of calculation of pulse differentiation   86
1.3.4.1. Calculation of the grating for differentiating pulses   86
1.3.4.2. Calculation of the grating for fractional differentiation of
         pulses   87
1.3.5.   Example of calculation of the integrating gratings   88
1.3.5.1. Calculation of the grating for the integration of pulses   88
1.3.5.2. Calculation of the grating for the fractional integration of
         pulses   89
1.4.   Diffractive optical elements for controlling the propagation of
         surface plasmon–polaritons   91
1.4.1.   The dispersion relation and the properties of surface
         plasmon–polaritons   92
1.4.1.1. Dispersion relations   92
1.4.1.2. The properties of surface plasmon–polaritons   101
1.4.1.3. Plasmonic waveguides   104
1.4.1.4. Excitation of surface plasmon–polaritons   106
1.4.2.   Phase modulation of surface plasmon–polaritons by
         dielectric structures   107
1.4.3.   Suppression of parasitic scattering of surface plasmon–
         polaritons   112
1.4.3.1. Theoretical analysis of matching of transverse profiles of
         the field of the surface plasmon–polariton and the plasmonic
         mode of the dielectric/dielectric/metal waveguide   112
1.4.3.2. Suppression of parasitic scattering during the passage of
         surface plasmon–polaritons through two-layer dielectric
         structures   118
1.4.4.   Design of the lens for focusing the surface plasmon–
         polaritons   122
1.4.4.1. Integral representations of the field in the form of the
         spectrum of surface plasmon–polaritons   122
1.4.4.2. Diffraction lenses for focusing the surface plasmon–

polaritons   125
1.4.4.3. The calculation of diffractive lenses for surface plasmon
         polaritons with parasitic scattering suppression   128
1.4.5.   Calculation of reflective Bragg gratings for surface
         plasmon–polaritons   131
1.5.   Generation of interference patterns of evanescent
         electromagnetic waves by means of diffraction gratings   132
1.5.1.   Theoretical description of the interference patterns of
         evanescent diffraction orders of gratings with one-dimensional
         and two-dimensional periodicity   133
1.5.1.1. One-dimensional interference patterns   133
1.5.1.2. Two-dimensional interference patterns   137
1.5.2.   Formation of interference patterns of evanescent
         electromagnetic waves in the metal–dielectric diffraction
         gratings   146
1.5.2.1. Generation of one-dimensional interference patterns   146
1.5.2.2. Controlling the period of one-dimensional interference
         patterns by changing the parameters of the incident wave   152
1.5.2.3. Generation of two-dimensional interference patterns and
         control of their shape by changing the parameters of the
         incident wave   157
1.5.3.   Generation of interference patterns of evanescent
         electromagnetic waves in dielectric diffraction gratings   163
1.5.3.1. Generation of one-dimensional interference patterns   166
1.5.3.2. Generation of two-dimensional interference patterns   173

2.   Components of photonic crystals   182
2.1   One-dimensional and two-dimensional photonic crystals   182
2.1.1   Photonic bandgaps   182
2.1.2.   Diffraction of a plane wave on a photonic crystal with no
         defects   185
2.1.3.   The propagation of light in a photonic crystal waveguide   186
2.1.4.   Photonic crystal collimators   188
2.2.   The two-dimensional photonic crystal gradient Mikaelian lens   189
2.2.1.   The modal solution for a gradient secant medium   191
2.2.2.   Photonic crystal gradient lenses   193
2.2.3.   Photonic crystal lens for coupling two waveguides   198
2.3.   Simulation of three-dimensional nanophotonic devices for
         input of radiation in a planar waveguide   212
2.3.1.   The two-dimensional simulation of the input of light in
         a planar waveguide using a grating   213
2.3.2.   Three-dimensional modelling Mikaelian photonic crystal lens
         for coupling two waveguides   216
2.3.3.   Three-dimensional modeling of the entire nanophotonic
         device   218

2.4.   Photonic crystal fibres   219
2.4.1.   Calculation of modes of photonic-crystal fibres by the
          method of matched sinusoidal modes (MSM method)   225
2.4.2.   Method of matched sinusoidal modes in the vector case   239
2.4.3.   Krylov method for solving non-linear eigenvalue problems   246
2.4.4.   Calculation of stepped fibre modes   251
2.4.5.   Calculation of modes of photonic crystal fibres   256
2.4.6.   Calculation of modes using Fimmwave   259

3.   **Sharp focusing of light with microoptics components**   **267**
3.1.   Numerical and experimental methods for studying sharp
          focusing of light   267
3.1.1.   BOR-FDTD method   267
3.1.2.   Richards–Wolf vector formula   284
3.1.3.   Scanning near-field optical microscopy   291
3.2.   Axicon   292
3.2.1.   Sharp focusing of light with radial polarization
          using a microaxicon   293
3.2.2.   Diffractive logarithmic microaxicon   303
3.2.3.   Binary axicons with period 4, 6 and 8 μm   312
3.2.4.   Binary axicon with a period of 800 nm   327
3.3.   Fresnel zone plate   330
3.3.1.   Comparison with the Richards–Wolf formulas   331
3.3.2.   Symmetry of the intensity and power flow of the
          subwavelength focal spot   336
3.4.   Focusing light with gradient lenses   348
3.4.1.   Mechanism of superresolution in a planar hyperbolic secant
          lens   348
3.4.2.   Gradient elements of micro-optics for achieving
          superresolution   358
3.4.3.   Construction of an enlarged image with superresolution using
          planar Mikaelian lenses   373
3.4.4.   Hyperbolic secant lens with a slit for subwavelength
          focusing of light   387
3.4.5.   Sharp focusing of radially polarized light with a 3D
          hyperbolic secant lens   393
3.4.6.   Optimizing the parameters of the planar binary lens for the
          visible radiation range   396
3.5.   Formation of a photon nanojet by a sphere   402
3.5.1.   Numerical simulation of passage of continuous radiation
          through the microsphere   404
3.5.2.   Numerical simulation of passage of pulsed radiation
          through a microsphere   410
3.5.3.   Experiments with focusing of light by a microsphere   413

4.   Introduction   421
4.1.   Formation of vortex laser beams by using singular optics
        elements   423
4.1.1. Theoretical analysis of the diffraction of plane and Gaussian
        waves on the spiral phase plate   424
4.1.2. Theoretical analysis of the diffraction of plane and Gaussian
        waves on a spiral axicon   432
4.1.3. Numerical modeling of the diffraction of different beams on
        elements of singular optics   437
4.1.4. Conclusions   441
4.2.   Vector representation of the field in the focal region for the
        vortex transmission function   443
4.2.1. Uniformly polarized beams (linear and circular polarization) 447
4.2.1.1. Linear x-polarization   447
4.2.1.2. Circular polarization   455
4.2.2. Cylindrical vector beams (radial and azimuthal polarization) 456
4.2.2.1. The radial polarization   456
4.2.2.2. The azimuthal polarization   461
4.2.3. Generalized vortex polarization   464
4.2.4. Conclusions   467
4.3.   Application of axicons in high-aperture focusing systems 469
4.3.1. Adding the axicon to the lens: the paraxial scalar model 471
4.3.2. Apodization of the short-focus lens by the axicon:
        the non-paraxial vector model in the Debye approximation 475
4.3.3. Using vortex axicons for spatial redistribution of components
        of the electric field in the focal region   483
4.3.4. Conclusions   486
4.4.   Control of contributions of components of the vector electric
        field at the focus of the high-aperture lens using binary
        phase structures   487
4.4.1. Maximizing the longitudinal component for linearly
        polarized radiation   488
4.4.2. Increasing efficiency by focusing the radially polarized
        radiation   494
4.4.3. Overcoming of the diffraction limit for azimuthal polarization
        by the transverse components of the electric field 496
4.4.4. Conclusions   496
4.5.   Minimizing the size of the light or shadow focal spot with
        controlled growth of the side lobes   498
4.5.1. Scalar diffraction limit: theoretical analysis   499
4.5.2. Optimization of the transmission function of the focusing
        system   502
4.5.3. Minimization of the light spot for radially polarized
        radiation   503

4.5.4.   Formation of light rings of the subwavelength radius
         azimuthal polarization of laser radiation   509
4.5.5.   Conclusions   512

5.   **Optical trapping and manipulation of micro- and
         nano-objects   519**
5.1.   Calculation of the force acting on the micro-object by a
         focused laser beam   522
5.1.1.   Electromagnetic force for the three-dimensional case   522
5.1.2.   Electromagnetic force for the two-dimensional case   524
5.1.3.   Calculation of force for a plane wave   525
5.1.4.   Calculation of force for a non-paraxial Gaussian beam   528
5.1.5.   Calculation of forces for the refractive index of the object
         smaller than the refractive index of the medium   534
5.2.   Methods for calculating the torque acting on a micro-object
         by a focused laser beam   535
5.2.1.   The orbital angular momentum in cylindrical microparticles 537
5.2.2.   The results of numerical simulation of the torque   538
5.1.7.   A geometrical optics method for calculating the force acting
         by light on a microscopic object   544
5.1.8.   Comparison of results of calculations by geometrical
         optics and electromagnetic methods   549
5.2.   Rotation of micro-objects in a Bessel beam   551
5.2.1.   Rotation of micro-objects in Bessel beams   551
5.2.2.   Optical rotation using a multiorder spiral phase plate   565
5.2.3.   Rotation of microscopic objects in a vortex light ring formed
         by an axicon   568
5.2.4.   Optical rotation in a double light ring   569
5.2.5.   Formation of the DOE with a liquid-crystal display   572
5.2.5.   Rotation of micro-objects by means of hypergeometric
         beams and beams that do not have the orbital angular
         momentum using the spatial light modulator   575
5.2.6.   Quantitative investigation of rotation of micro-objects in light
         beams with an orbital angular momentum   586
5.3.   Rotating microparticles in light beams designed and formed to
         obtain the maximum torque   594
5.3.1.   Formation of the light beams, consisting of several light
         rings   594
5.3.2.   Formation of the superposition of high-order optical vortices 604
5.3.3.   Experimental formation of vortex beams superpositions   606
5.4.   Light beams specially formed for linear displacement and
         positioning of micro-objects   607
5.4.1.   Encoding of amplitude by the local phase jump method   608
5.4.2.   Modification of the method of encoding the amplitude
         for the calculation of light fields with a complex structure 611

5.4.3.	Positioning of micro-objects using focusators	612
5.4.4.	Focusators to move micro-objects along predetermined paths	619
5.4.5.	Focusators for sorting and filtering micro-objects	624
5.4.6.	Focusators to filter certain micro-objects	626
5.4.8.	The linear movement of micro-objects in superposition of two vortex beams	637
5.5.	Formation of arrays of light 'bottles' with the DOE	642
5.5.1.	Bessel light beams and their remarkable properties	644
5.5.2.	Multimode Bessel beams	645
5.5.3.	Formation of light 'bottles' through the use of composite Bessel beams of 0th order	647
5.5.4.	Algorithm for increasing the uniformity of light traps formed on the basis of the gradient procedure	652
5.5.5.	Formation of arrays of light 'bottles'	660
5.5.6.	Formation of arrays of hollow light beams	669
5.5.7.	Experimental formation of arrays of light 'bottles' with the binary DOE	675
5.5.8.	Capture of transparent micro-objects in the system of light 'bottle'	679
5.5.9.	Capture and displacement of metallic tin microparticles with the shape close to spherical	686
5.5.10.	The deposition and positioning of micro-objects using arrays of hollow light beams	687
5.6.	The light beams generated by the DOE for non-damaging capture of biological micro-objects	688
5.6.1.	Modification of the Gaussian beam to optimize the power characteristics of the optical trap	691
5.6.2	Measurements of the energy efficiency of the DOE forming beams–crescents	695
5.6.3.	Experiments with optical manipulation of cells of *Saccharomyces cerevisiae*	697
Conclusions		**711**
Index		**714**

Introduction

The phenomenon of diffraction, which was originally seen as a limiting factor in optics, is now a fundamental basis for the creation of a new component base and advanced information technology.

The development of diffractive optics and nanophotonics devices is based on a computer solution of direct and inverse problems of diffraction theory, based on Maxwell's equations. Among the numerical methods for solving Maxwell's equations most widely used are: the finite-difference time-domain method, the finite element method and the Fourier modal method, and also approximate methods for calculating diffraction integrals.

The book is devoted to modern achievements of diffractive optics, focused on the development of new components and devices for nanophotonics, and devices and information technologies based on them.

The first chapter describes the Fourier modal method, designed for the numerical solution of Maxwell's equations, as well as some of its applications in problems of calculation of diffractive gratings with the resonance properties and plasmon optics components.

The Fourier modal method (or rigorous-coupled wave analysis) has a wide range of applications. In the standard formulation the method is used to solve the problems of diffraction of a monochromatic plane wave on diffraction gratings. Introduction of the light beam in a plane-wave basis allows to use the method for modelling the diffraction of optical pulses. Using the so-called perfectly matched layers in combination with artificial periodization enables the method to be used efficiently to solve the problems of the diffraction of light waves by non-periodic structures. In this chapter the Fourier modal method is considered for solving the diffraction of a plane wave in the two-dimensional and three-dimensional diffractive gratings, as well as in the case of non-periodic structures. The implementation of the method is based on the numerical-stable approach, known as the scattering matrix method.

The resonance features in the spectra of periodic diffraction structures are studied using the methods developed by the authors for calculating the scattering matrix poles. These methods take into account the form of the matrix in the vicinity of the resonances associated with the excitation of eigenmodes in the lattice and, compared with the known methods, have better convergence.

The Fourier modal method and scattering matrix formalism are applied to the calculation of diffractive gratings with the resonance properties for the conversion of optical pulses. The chapter proposes a theoretical model of resonant gratings performing operations of differentiation and integration of the optical pulse envelope and the results of the calculation and research of diffractive gratings for the differentiation and integration of picosecond pulses are presented.

The non-periodic variant of the Fourier modal method is used in the problem of calculating the diffractive optical elements for controlling the propagation of surface plasmon-polaritons. The principle of operation is based on the phase modulation of surface plasmon-polaritons by dielectric steps with changing height and length and located on the surface of the metal.

The Fourier modal method is also applied to the task of calculating the diffractive gratings forming, in the near-field, interference patterns of evanescent electromagnetic waves and, in particular plasmon modes. The chapter provides a theoretical description and a number of numerical examples of calculation of the gratings forming interference patterns of evanescent electromagnetic waves and plasmon modes with a substantially subwavelength period and demonstrates the ability to control the type and period of the interference patterns of damped waves due to changes in the parameters of the incident radiation.

The practical use of the results of the first chapter includes systems for optical computing and ultra-fast optical information processing, the creation of high-performance components plasmon optics, the contact lithography systems and the systems for optical trapping and manipulation of nanoscale objects.

The second chapter deals with the nanophotonics components based on photonic crystals: the gradient planar photonic crystal (PC) lens and photonic crystal fibers. The ultra-compact nanophotonic device is described for effectively connecting two-dimensional waveguides of different widths using the PC-lens. It is shown that the PC-lens focuses the light into a small focal spot directly behind the lens whose size is substantially smaller than the scalar diffraction limit. The simulation was performed using a finite difference solution of Maxwell's equations.

The second chapter also describes the method of calculation of optical fibres with the PC cladding. In this waveguide the light propagates within the core not due to the effect of total internal reflection from the core–cladding interface, and by reflection from a multilayer Bragg mirror formed by the system of periodically spaced holes around the core. Calculation of spatial modes in PC-fibres is based on partitioning the inhomogeneous fibre cross-section into a set of rectangular cells, each with a set of known spatial sinusoidal modes. Further cross-linking of all modes is carried out at the interfaces of all cells. The PC waveguides differ from the step and gradient waveguides by that they allow the modes to be localized within the core, that is all the modes propagate inside the core and almost do not penetrate into the cladding thus increasing the diameter of the mode localized within the core. In addition, the PC waveguides with a hollow core help to avoid chromatic dispersion in the fibre and transmit light with higher power. A short pulse of light passing through in a PC waveguide of finite size is transformed at the output to white light due to non-linear dispersion. Sections of the PC waveguides are used as filters, white light sources, and non-linear optics for second harmonic generation.

The third chapter discusses the focusing of laser radiation. The concept of the diffraction limit was established in the 19th century: $d_{min} = \lambda/(2n)$, where λ is the wavelength of light in vacuum, n is the refractive index of the medium. The third chapter shows that using diffractive micro-optics components, focusing the light near their surface it is possible to overcome the diffraction limit. Attention is given to the sharp focusing of laser light using micro-components such as the axicon, the zone plate, the binary and gradient planar microlens, microspheres. Focusing light near the surface of the micro-components allows to overcome the diffraction limit as a result of the presence of surface waves and the influence of the refractive index of the material of the focusing element. Simulation of focusing the laser beam is carried out by the approximate Richards–Wolf vector method and the finite difference solution of Maxwell's equations.

Reducing the size of the focal spot and overcoming the diffraction limit is an urgent task in the near-field microscopy, optical micromanipulation, contact photolithography, increasing the density of recording information on an optical disc, and coupling planar waveguides of different widths.

The fourth chapter describes the focusing of singular vortex laser beams. At the point of singularity the intensity of the light field is zero, and the phase is not defined. There are abrupt phase changes in the vicinity of this point. Singularities in light fields can appear as they pass through randomly inhomogeneous and non-linear media. It is also possible to excite vortex fields in laser resonators and multimode optical fibres. The most effective method of forming the vortex laser beams is to use spiral diffractive optical elements, including spiral phase plates and spiral axicons. The fourth chapter discusses the formation of vortex beams represented as a superposition of Bessel, Laguerre–Gauss, Hermite–Gauss, etc. modes. When focusing the vortex beams attention is paid to the combination of different types of polarization and phase singularities which lead to overcoming the diffraction limit of the far-field diffraction zone.

Main applications of vortex laser beams are sharp focusing of laser light, manipulation of microscopic objects and multiplexing the channels of information transmission.

In the fifth chapter we consider the problem of optical trapping, rotation, moving, positioning of micro-objects through the use of diffractive optical elements. Micro-objects are rotated by light beams with an orbital angular momentum. Considerable attention is paid to the methods of calculating the forces acting on the micro-objects in light fields. The problem of creating the torque in micromechanical systems using light beams has a fairly long history. In a number of studied the problem of rotation is considered in conjunction with other tasks: sorting, moving, positioning, etc. It should be noted that in all the above cases the focus is primarily on the manufacturing technology of micromechanics elements and no attempts are made to improve the light beams. At the same time, the calculation and application of diffractive optical elements, forming the vortex light beams for a specific form of the micromechanical component can improve the transmission efficiency of the torque in micromechanical systems.

This chapter discusses two methods of calculating the diffractive optical elements for forming light fields with a given amplitude-phase distribution. One of them is based on calculating a focusator forming a light field with a predetermined phase gradient along the contour. Another method uses the superposition of zero-order Bessel beams to form light traps in the form of hollow beams for opaque microscopic objects. The results of experiments on optical trapping and relocation of micro-objects are presented. The chapter examines the possibility of using light beams to move the biological micro-objects.

The book has been written by experts of the Image Processing Systems Institute, Russian Academy of Sciences. In the first chapter, sections 1.1 and 1.2 were written by D.A. Bykov, E.A. Bezus and L.L. Doskolovich, section 1.3 by D.A. Bykov, L.L. Doskolovich and V.A. Soifer, sections 1.4 and 1.5 by E.A. Bezus and L.L. Doskolovich. The second chapter was written by V.V. Kotlyar, A.A. Kovalev, A.G. Nalimov and V.A. Soifer. The third chapter – by V.V. Kotlyar, A.A. Kovalev, A.G. Nalimov and S.S. Stafeev. The fourth chapter was written by S.N. Khonina and the fifth chapter by R.V. Skidanov, A.P. Porfir'ev and V.A. Soifer.

1

The Fourier modal method and its use in plasmonics and the theory of resonant diffraction gratings

This chapter describes the Fourier modal method (also called rigorous coupled-wave analysis) designed for the numerical solution of Maxwell's equations, as well as some of its applications in problems of diffractive nanophotonics.

The Fourier modal method has a wide range of applications. The formulation of the method is used for the solution of the problem of diffraction of a monochromatic plane wave on diffraction gratings. The representation of a light beam in the plane wave basis allows to use the method for the simulation of the diffraction of spatio-temporal optical signals (optical pulses). The use of the so-called perfectly matched absorbing layers in combination with artificial periodization enables the method to be used to solve the problems of the diffraction of light waves by non-periodic structures.

In section 1.1 the Fourier modal method is considered in the standard formulation for solving the problem of diffraction of a plane wave on two-dimensional and three-dimensional multilayered periodic diffraction structures and for the case of non-periodic structures. The considered implementation of the method is based on a numerically stable approach, known as the scattering matrix method. The method is described in detail and was used by the authors to create original computer programs for the solution of problems of diffraction on structures of various types. The readers can safely use the formulas given in section 1.1 when creating their own programs for electromagnetic modelling.

The interaction of light with a periodic structure, in particular with metal–dielectric structures, is the subject of intense research. In such structures there is a wide range of extraordinary (resonant) optical effects, including extraordinary transmission, total absorption of incident radiation, resonant changes of the transmission and reflection coefficients, the formation of regions with a high degree of localization of energy due to interference of the evanescent and plasmon waves. These effects are usually associated with the excitation in a periodic structure of quasi-guided and plasmonic eigenmodes. The modes of the structure correspond to the poles of the scattering matrix. Section 1.2 presents a number of methods proposed by the authors for calculating the poles of the scattering matrix. These methods take into account the representation of scattering matrix in the vicinity of resonances and, compared with the known techniques, have better convergence.

In section 1.3 the method and the scattering matrix technique are applied to the calculation of diffraction gratings with the resonant properties to transform optical pulses. The first part of this section describes in a common approach a class of transformations of optical pulses realized by resonant diffraction gratings, and proposes a theoretical model of resonant gratings performing operations of differentiation and integration of the optical pulse envelope. The model is based on the resonant representations of the complex amplitude of the zeroth diffraction order in the vicinity of the frequencies corresponding to the waveguide resonances (Wood's anomalies) and in the vicinity of the frequency of the Rayleigh–Wood's anomalies associated with the emergence of propagating diffraction orders. The section also presents the results of calculations and studies of diffraction gratings to perform differentiation and integration of picosecond pulses (Wood's anomalies), as well as the operations of fractional differentiation and integration (Rayleigh–Wood anomalies).

The possible applications of the results of this section include optical computing and ultrafast optical information processing.

In section 1.4 the aperiodic Fourier modal method is used for the design of diffractive optical elements (DOE) to control the propagation of surface plasmon–polaritons (SPP). The principle of operation of the conventional DOE is based on the phase modulation of the input wave field by a diffraction microrelief of variable height.

A similar approach can be used to create optical elements for the SPP. Attention is given to the mechanisms of phase modulation of the SPP and plasmonic modes of thin metal films using dielectric ridges of variable height and fixed length (or variable height and length), located on the surface of the metal. The section also describes the method of suppressing parasitic scattering in diffraction of SPP (plasmonic modes) on a dielectric ridge, based on the use of the structure of two isotropic dielectric layers on the metal surface. This two-layer structure enables phase modulation of the SPP (plasmonic modes) while reducing the energy loss in the parasitic scattering by an order of magnitude compared to single-layer steps. The section also presents the results of calculation and research of diffractive lenses for focusing the SPP based on different phase modulation methods.

The main practical application of the results of this section is the creation of high-performance elements of plasmon optics: lens, Bragg gratings, plasmonic crystals.

In section 1.5 the Fourier modal method is applied to the problem of calculating diffraction gratings which generate interference patterns of evanescent electromagnetic waves in the near field (especially of plasmonic modes). The first part of the section provides a theoretical description of the interference patterns of evanescent diffraction orders for gratings with one- and two-dimensional periodicity. This is followed by a series of numerical examples of the gratings lattices, forming interference patterns of evanescent electromagnetic waves and plasmonic modes with a substantially subwavelength period, and the ability to control the type and period of the interference patterns due to changes in the parameters of the incident radiation is demonstrated.

The practical use of the results of this section includes contact nanolithography systems and systems for optical trapping and manipulation of nanoscale objects.

1.1. The Fourier modal method of solving the diffraction problems

1.1.1. The equation of a plane wave

This subsection is auxiliary. It considers the derivation of the general equation of a plane wave in an isotropic medium used in the following description of the Fourier modal method.

We write the Maxwell equations and material equations in the Gaussian system of units:

$$\begin{cases} \nabla \times \mathbf{E} = -\dfrac{1}{c}\dfrac{\partial}{\partial t}\mathbf{B}, \\[2mm] \nabla \times \mathbf{H} = \dfrac{1}{c}\dfrac{\partial}{\partial t}\mathbf{D}, \\[2mm] \mathbf{B} = \mu\,\mathbf{H}, \\[2mm] \mathbf{D} = \varepsilon\,\mathbf{E}. \end{cases} \tag{1.1}$$

Equations (1.1) are written in the absence of charges, in this case ε is the complex permittivity. For a monochromatic field

$$\begin{aligned} \mathbf{E}(x,y,z,t) &= \mathbf{E}(x,y,z)\exp(-i\omega t), \\ \mathbf{H}(x,y,z,t) &= \mathbf{H}(x,y,z)\exp(-i\omega t), \end{aligned} \tag{1.2}$$

system (1.1) takes the form:

$$\begin{cases} \nabla \times \mathbf{E} = ik_0\mu\mathbf{H}, \\ \nabla \times \mathbf{H} = -ik_0\varepsilon\mathbf{E}, \end{cases} \tag{1.3}$$

wherein $k_0 = 2\pi/\lambda$, λ is the wavelength. Expanding the rotor operator we get

$$\nabla \times \mathbf{E} = \begin{bmatrix} \dfrac{\partial E_z}{\partial y} - \dfrac{\partial E_y}{\partial z} \\[2mm] \dfrac{\partial E_x}{\partial z} - \dfrac{\partial E_z}{\partial x} \\[2mm] \dfrac{\partial E_y}{\partial x} - \dfrac{\partial E_x}{\partial y} \end{bmatrix} = ik_0\mu \begin{bmatrix} H_x \\ H_y \\ H_z \end{bmatrix}, \quad \nabla \times \mathbf{H} = \begin{bmatrix} \dfrac{\partial H_z}{\partial y} - \dfrac{\partial H_y}{\partial z} \\[2mm] \dfrac{\partial H_x}{\partial z} - \dfrac{\partial H_z}{\partial x} \\[2mm] \dfrac{\partial H_y}{\partial x} - \dfrac{\partial H_x}{\partial y} \end{bmatrix} = -ik_0\varepsilon \begin{bmatrix} E_x \\ E_y \\ E_z \end{bmatrix}. \tag{1.4}$$

To obtain the plane wave equation we will seek the solution of (1.4) in the form

$$\Phi(x,y,z) = \Phi\exp\big(ik_0(\alpha x + \beta y + \gamma z)\big), \tag{1.5}$$

where $\Phi = \begin{bmatrix} E_x & E_y & E_z & H_x & H_y & H_z \end{bmatrix}^{\mathrm{T}}$ is the column vector of the components of the field. The values α, β, γ in (1.5) define the direction of propagation of the plane wave. Substituting (1.5) into

(1.4), we obtain

$$ik_0 \begin{bmatrix} \beta E_z - \gamma E_y \\ \gamma E_x - \alpha E_z \\ \alpha E_y - \beta E_x \end{bmatrix} = ik_0 \mu \begin{bmatrix} H_x \\ H_y \\ H_z \end{bmatrix}, \quad ik_0 \begin{bmatrix} \beta H_z - \gamma H_y \\ \gamma H_x - \alpha H_z \\ \alpha H_y - \beta H_x \end{bmatrix} = -ik_0 \varepsilon \begin{bmatrix} E_x \\ E_y \\ E_z \end{bmatrix}. \quad (1.6)$$

We represent the equation (1.6) in the matrix form:

$$\begin{bmatrix} 0 & -\gamma & \beta & -\mu & 0 & 0 \\ \gamma & 0 & -\alpha & 0 & -\mu & 0 \\ -\beta & \alpha & 0 & 0 & 0 & -\mu \\ \varepsilon & 0 & 0 & 0 & -\gamma & \beta \\ 0 & \varepsilon & 0 & \gamma & 0 & -\alpha \\ 0 & 0 & \varepsilon & -\beta & \alpha & 0 \end{bmatrix} \Phi = \mathbf{A} \cdot \Phi = 0. \quad (1.7)$$

Direct calculation shows that the determinant of the system (1.7) has the form:

$$\det \mathbf{A} = \varepsilon \mu \left(\alpha^2 + \beta^2 + \gamma^2 - \varepsilon \mu \right)^2. \quad (1.8)$$

The sought non-trivial solution of the system (1.7) exists when the determinant (1.8) is zero. If $\mu = 1$ the determinant of (1.8) vanishes provided

$$\alpha^2 + \beta^2 + \gamma^2 = \varepsilon. \quad (1.9)$$

We write down explicitly the solution of (1.6). Under the condition (1.9) the rank of the system (1.7) is equal to four, so to write the solution the values of the amplitudes E_z and H_z are fixed. We introduce the so-called E- and H-waves. For the E-wave $E_z \neq 0$, $H_z = 0$, and for the H-wave $H_z \neq 0$, $E_z = 0$. We represent the desired solution as a superposition of the E- and H-waves in the form

$$\Phi = \frac{1}{\sqrt{\alpha^2 + \beta^2}} \begin{bmatrix} -\alpha\gamma \\ -\beta\gamma \\ \alpha^2 + \beta^2 \\ \beta\varepsilon \\ -\alpha\varepsilon \\ 0 \end{bmatrix} \frac{A_E}{\sqrt{\varepsilon}} + \frac{1}{\sqrt{\alpha^2 + \beta^2}} \begin{bmatrix} -\beta \\ \alpha \\ 0 \\ -\alpha\gamma \\ -\beta\gamma \\ \alpha^2 + \beta^2 \end{bmatrix} A_H, \quad (1.10)$$

where A_E and A_H are the the amplitudes of the E- and H-waves, respectively.

We write the components of the vector $\mathbf{p} = (\alpha, \beta, \gamma)$, representing the propagation direction of the wave, by means of the angles θ and φ of the spherical coordinate system:

$$\alpha = \sqrt{\varepsilon}\,\cos\varphi\sin\theta, \ \ \beta = \sqrt{\varepsilon}\,\sin\varphi\sin\theta, \ \ \gamma = \sqrt{\varepsilon}\,\cos\theta, \quad (1.11)$$

where θ is the angle between the vector $\mathbf{p}$ and the axis Oz, φ is the angle between the plane of incidence and the plane xOz. For these angles the following relations are fulfilled:

$$\sqrt{\alpha^2 + \beta^2} = \sqrt{\varepsilon}\,\sin\theta, \ \frac{\alpha}{\sqrt{\alpha^2 + \beta^2}} = \cos\varphi, \ \frac{\beta}{\sqrt{\alpha^2 + \beta^2}} = \sin\varphi. \ (1.12)$$

We write separately y- and x-components of the electric and magnetic fields:

$$\bar{\Phi} = \begin{bmatrix} E_y \\ E_x \\ H_y \\ H_x \end{bmatrix} = \begin{bmatrix} -\cos\theta\sin\varphi \\ -\cos\theta\cos\varphi \\ -\sqrt{\varepsilon}\,\cos\varphi \\ \sqrt{\varepsilon}\,\sin\varphi \end{bmatrix} A_E + \begin{bmatrix} \cos\varphi \\ -\sin\varphi \\ -\sqrt{\varepsilon}\,\cos\theta\sin\varphi \\ -\sqrt{\varepsilon}\,\cos\theta\cos\varphi \end{bmatrix} A_H$$

$$= \begin{bmatrix} \mathbf{R} & 0 \\ 0 & \mathbf{R} \end{bmatrix} \begin{bmatrix} 0 & 1 \\ -\gamma/\sqrt{\varepsilon} & 0 \\ -\sqrt{\varepsilon} & 0 \\ 0 & -\gamma \end{bmatrix} \begin{bmatrix} A_E \\ A_H \end{bmatrix},$$

$$(1.13)$$

where $\mathbf{R}$ is the rotation matrix of the following form:

$$\mathbf{R} = \begin{bmatrix} \cos\varphi & \sin\varphi \\ -\sin\varphi & \cos\varphi \end{bmatrix}. \qquad (1.14)$$

To describe the polarization of the wave, we introduce the angle ψ – the angle between the vector $\mathbf{E}$ and the incidence plane. The incidence plane is the plane containing the direction of the wave propagation vector $\mathbf{p}$ and the axis Oz. In this case

$$A_H = A\sin\psi,$$
$$A_E = A\cos\psi. \qquad (1.15)$$

Note that in the solution of the diffraction problem the case φ = 0 corresponds to the so-called planar diffraction. In this case the E-waves are waves with TM-polarization and H-waves the waves with TE-polarization.

Consider the time-averaged Umov–Poynting vector $\mathbf{S} = \frac{1}{2}\mathrm{Re}(\mathbf{E}\times\mathbf{H}^*)$. We compute the z-component of the vector, which corresponds to the density of the energy flux through the plane xOy. For the E-wave

$$2S_z = \mathrm{Re}\left(E_x H_y^* - E_y H_x^*\right) = |A_E|^2\, \mathrm{Re}\left(\sqrt{\varepsilon}^{\,*}\right)\cdot\cos\theta = |A_E|^2\,\mathrm{Re}\sqrt{\varepsilon}\cdot\cos\theta,$$
(1.16)

and similarly for the H-wave:

$$2S_z = \mathrm{Re}\left(E_x H_y^* - E_y H_x^*\right) = |A_H|^2\, \mathrm{Re}\left(\sqrt{\varepsilon}^{\,*}\right)\cdot\cos\theta = |A_H|^2\,\mathrm{Re}\sqrt{\varepsilon}\cdot\cos\theta,$$
(1.17)

The expressions (1.16) and (1.17) describe the energy flux through the plane *xOy*. These expressions are important for the control of the energy conservation law in solving the problem of diffraction on diffraction structures of a lossless material.

1.1.2. The Fourier modal method for two-dimensional periodic structures

The Fourier modal method focuses on the numerical solution of Maxwell's equations for the case of periodic structures, consisting of a set of 'binary layers' [1–13]. In each layer, dielectric permittivity (magnetic permeability) of the material of the structure is independent of the variable z and the axis Oz is perpendicular to the structure. In the method the electromagnetic field in the areas above and below the structure is given as a superposition of plane waves (diffraction orders). The function of dielectric permittivity and magnetic permeability of the material in each layer of the structure are represented as segments of the Fourier series, and the components of the electromagnetic field are written as an expansion in the Fourier modes basis. The calculation of the Fourier modes is reduced to eigenvalue problems. Sequential imposition of the condition of equality of the tangential components of the electromagnetic field at the boundaries of the layers reduces the determination of the amplitudes of the reflected and transmitted diffraction orders to solving a system of linear equations [1–13].

Here the method is considered for the two-dimensional diffraction gratings (gratings with one-dimensional periodicity). In the two-dimensional case, the material properties of the structure are constant along one of the axes. The structure materials are in a general case defined by tensors of permittivity and permeability.

1.1.2.1. The geometry of the structure and formulation of the problem

Consider a diffraction grating with a period d along the axis x (Fig. 1.1). For gratings with a continuous profile (dashed line in Fig. 1.1) the method approximates the grating profile by a set of binary layers. We assume that the diffraction grating is made up of L binary layers (Fig. 1.1). The permittivity ε and permeability μ in the layers depend only on the variable x. Layer boundaries are the lines $z = d_l$, the l-th layer is located in the area $z_l < z < z_{l-1}$.

Over the considered structure in the region $z > z_0 = 0$ there is a homogeneous dielectric with a refractive index $n_U = \sqrt{\varepsilon_U}$. Under the structure there is a homogeneous dielectric with a refractive index $n_D = \sqrt{\varepsilon_D}$.

A plane monochromatic wave is incident on the top of the grating (wavelength λ, wave number $k_0 = \frac{2\pi}{\lambda}$), the direction of which is defined by the angles θ and φ in the spherical coordinate system

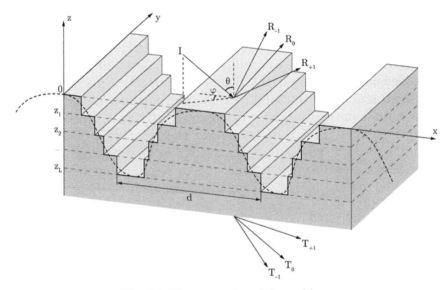

Fig. 1.1. The geeometry of the problem.

(Fig. 1.1). The polarization of the wave is determined by the angle ψ between the incidence plane and the vector **E** (see. (1.15)).

The problem of light diffraction by a periodic structure is solved by calculating the intensities or complex amplitudes of diffraction orders. The diffraction orders are the reflected and transmitted plane waves arising in the diffraction of the incident wave on the structure. There are reflected orders with amplitudes R_i, $i = 0, \pm 1, \pm 2, \ldots$ and transmitted orders with amplitudes T_i, $i = 0, \pm 1, \pm 2, \ldots$ (Fig. 1.1). The diffraction orders can also be divided into evanescent and propagating. The amplitude of the evanescent orders decreases exponentially away from the gratings.

1.1.2.2. Description of the field above and below the structure

The field above and below the structure is written as a superposition of plane waves (Rayleigh expansion). According to (1.5), (1.10), the plane wave is represented by a vector of six components:

$$\Phi = \begin{bmatrix} E_x & E_y & E_z & H_x & H_y & H_z \end{bmatrix}^T. \tag{1.18}$$

The equation of the incident wave can be represented in the form

$$\Phi^{inc}(x, y, z) = \Phi^{inc}(\psi) \exp\left(i\left(k_{x,0}x + k_y y - k_{z,U,0}z\right)\right), \tag{1.19}$$

where $\Phi^{inc}(\psi)$ denotes the dependence of the incident wave on polarization (see (1.15)). The constants $k_{x,0} = k_0 n_U \sin\theta \cos\varphi$, $k_{x,0} = k_0 n_U \sin\theta \cos\varphi$, $k_y = k_0 n_U \sin\theta \sin\varphi$, $k_{z,U,0} = \sqrt{(k_0 n_U)^2 - k_{x,0}^2 - k_y^2}$ in (1.19) are determined through the angles θ and φ defining the direction of the incident wave.

The field above the grating corresponds to the superposition of the incident wave and the reflected diffraction orders:

$$\Phi^U(x, y, z) = \Phi^{inc}(x, y, z) + \sum_m \Phi_m^R (R_m) \exp\left(i(k_{x,m}x + k_y y + k_{z,U,m}z)\right),$$

$$\tag{1.20}$$

where $\Phi(R) = \Phi_E R_E + \Phi_H R_H$. Expression $\Phi(R)$ corresponds to the representation (1.10) of the reflected wave as a sum of E- and H-waves with complex amplitudes R_E and R_H respectively.

The field below the grating is expressed similarly as a superposition of transmitted waves:

$$\Phi^D(x, y, z) = \sum_m \Phi_m^T (T_m) \exp\left(i\left(k_{x,m}x + k_y y - k_{z,D,m}(z - z_L)\right)\right). \tag{1.21}$$

The directions of the orders in (1.20) (1.21) are given by propagation constants $k_{x,m}$, k_y, $k_{z,p,m}$. The propagation constants have the form

$$k_{x,m} = k_0 \left(n_\mathrm{U} \sin\theta \cos\varphi + m\frac{\lambda}{d} \right),$$

$$k_y = k_0 n_\mathrm{U} \sin\theta \sin\varphi, \qquad (1.22)$$

$$k_{z,p,m} = \sqrt{\left(k_0 n_p\right)^2 - k_{x,m}^2 - k_y^2},$$

where p is taken as 'U' for the reflected orders (the field above the grid) and 'D' – for the transmitted orders (the field under the grating). The expression for $k_{x,m}$ follows from the Floquet–Bloch theorem [14, 15]. From a mathematical point of view the form of $k_{x,m}$ provides the implementation of the so-called quasi-periodicity condition

$$\Phi^p \left(x+d, y, z \right) = \Phi^p \left(x, y, z \right) \exp\left(i k_{x,0} d \right), \; p = \mathrm{U}, \mathrm{D}. \qquad (1.23)$$

According to (1.23), the amplitude of the field does not change with shifts by a period. The waves with real $k_{z,p,m}$ are propagating, those with imaginary one – evanescent.

1.1.2.3. The system of differential equations for the description of the field in the layer

We turn now to the field within the particular l-th layer. For simplicity, we omit the index l.

The field in each layer is described by the Maxwell's equations for the monochromatic field in the form

$$\begin{cases} \nabla \times \mathbf{E} = i k_0 \ddot{\mu} \mathbf{H}, \\ \nabla \times \mathbf{H} = -i k_0 \ddot{\varepsilon} \mathbf{E}, \end{cases} \qquad (1.24)$$

where $\ddot{\varepsilon}$, $\ddot{\mu}$ are tensors in the most general case:

$$\ddot{\varepsilon} = \begin{bmatrix} \varepsilon_{1,1} & \varepsilon_{1,2} & \varepsilon_{1,3} \\ \varepsilon_{2,1} & \varepsilon_{2,2} & \varepsilon_{2,3} \\ \varepsilon_{3,1} & \varepsilon_{3,2} & \varepsilon_{3,3} \end{bmatrix}, \quad \ddot{\mu} = \begin{bmatrix} \mu_{1,1} & \mu_{1,2} & \mu_{1,3} \\ \mu_{2,1} & \mu_{2,2} & \mu_{2,3} \\ \mu_{3,1} & \mu_{3,2} & \mu_{3,3} \end{bmatrix}. \qquad (1.25)$$

The tensor components depend only on the coordinate x. Expanding the rotor operator (1.24), we obtain:

$$
\begin{cases}
\dfrac{\partial E_z}{\partial y} - \dfrac{\partial E_y}{\partial z} = ik_0(\mu_{1,1}H_x + \mu_{1,2}H_y + \mu_{1,3}H_z), \\[2mm]
\dfrac{\partial E_x}{\partial z} - \dfrac{\partial E_z}{\partial x} = ik_0(\mu_{2,1}H_x + \mu_{2,2}H_y + \mu_{2,3}H_z), \\[2mm]
\dfrac{\partial E_y}{\partial x} - \dfrac{\partial E_x}{\partial y} = ik_0(\mu_{3,1}H_x + \mu_{3,2}H_y + \mu_{3,3}H_z), \\[2mm]
\dfrac{\partial H_z}{\partial y} - \dfrac{\partial H_y}{\partial z} = -ik_0(\varepsilon_{1,1}E_x + \varepsilon_{1,2}E_y + \varepsilon_{1,3}E_z), \\[2mm]
\dfrac{\partial H_x}{\partial z} - \dfrac{\partial H_z}{\partial x} = -ik_0(\varepsilon_{2,1}E_x + \varepsilon_{2,2}E_y + \varepsilon_{2,3}E_z), \\[2mm]
\dfrac{\partial H_y}{\partial x} - \dfrac{\partial H_x}{\partial y} = -ik_0(\varepsilon_{3,1}E_x + \varepsilon_{3,2}E_y + \varepsilon_{3,3}E_z).
\end{cases}
\tag{1.26}
$$

We represent the components of the electric and magnetic fields in the form of Fourier series with respect to variable x:

$$
\begin{cases}
E_x = \sum_j S_{x,j}(z)\exp\!\left(i(k_{x,j}x + k_y y)\right), \\[2mm]
E_y = \sum_j S_{y,j}(z)\exp\!\left(i(k_{x,j}x + k_y y)\right), \\[2mm]
E_z = \sum_j S_{z,j}(z)\exp\!\left(i(k_{x,j}x + k_y y)\right), \\[2mm]
H_x = -i\sum_j U_{x,j}(z)\exp\!\left(i(k_{x,j}x + k_y y)\right), \\[2mm]
H_y = -i\sum_j U_{y,j}(z)\exp\!\left(i(k_{x,j}x + k_y y)\right), \\[2mm]
H_z = -i\sum_j U_{z,j}(z)\exp\!\left(i(k_{x,j}x + k_y y)\right).
\end{cases}
\tag{1.27}
$$

The representations (1.27) are written with the quasi-periodicity of the field components with respect to variable x taken into account. We restrict ourselves to a finite number of terms in the expansion (1.27), corresponding to $-N \le j \le N$. From variables $S_{x,j}$, $S_{y,j}$, $S_{z,j}$, $U_{x,j}$, $U_{y,j}$, $U_{z,j}$, we form column vectors, $\mathbf{S}_x$, $\mathbf{S}_y$, $\mathbf{S}_z$, $\mathbf{U}_x$, $\mathbf{U}_y$, $\mathbf{U}_z$, containing $2N + 1$ elements each.

Consider the expansion of the product of two functions $\varepsilon_a(x)E_b(x,y,z)$ into the Fourier series:

$$\varepsilon_a(x)E_b(x) = \left(\sum_m e_{a,m}\exp\left(i\frac{2\pi}{d}mx\right)\right)\cdot\left(\sum_l S_{b,l}(z)\exp\left(i\left(k_{x,l}x+k_yy\right)\right)\right)$$

$$= [j = l+m] = \sum_j\left(\sum_l e_{a,j-l}S_{b,l}(z)\right)\exp\left(i\left(k_{x,j}x+k_yy\right)\right). \tag{1.28}$$

Retaining a finite number of terms corresponding to $-N \le j \le N, -N \le l \le N$ in (1.28), we obtain:

$$[\varepsilon_a(x)E_b(x)] = [\![\varepsilon_a(x)]\!]S_b = [\![\varepsilon_a(x)]\!][E_b(x)], \tag{1.29}$$

where the square brackets denote the vectors composed from the Fourier coefficients of the expansion of functions $\varepsilon_a(x)E_b(x)$ and $E_b(x)$, and $[\![\varepsilon_a(x)]\!]$ is the Toeplitz matrix of the Fourier coefficients, which has the following form:

$$\mathbf{Q} = \begin{bmatrix} e_0 & e_{-1} & e_{-2} & \cdots & \cdots & e_{-2N} \\ e_1 & e_0 & e_{-1} & e_{-2} & \ddots & e_{1-2N} \\ e_2 & e_1 & e_0 & e_{-1} & \ddots & \vdots \\ \vdots & e_2 & e_1 & \ddots & \ddots & \vdots \\ \vdots & \ddots & \ddots & \ddots & \ddots & e_{-1} \\ e_{2N} & e_{2N-1} & \cdots & \cdots & e_1 & e_0 \end{bmatrix}. \tag{1.30}$$

Substituting (1.27) into (1.26) and equating coefficients of the same Fourier harmonics, we obtain:

$$\begin{cases} ik_0\mathbf{K}_y\mathbf{S}_z - \dfrac{d\mathbf{S}_y}{dz} &= k_0(\mathbf{M}_{1,1}\mathbf{U}_x + \mathbf{M}_{1,2}\mathbf{U}_y + \mathbf{M}_{1,3}\mathbf{U}_z), \\[2mm] \dfrac{d\mathbf{S}_x}{dz} - ik_0\mathbf{K}_x\mathbf{S}_z &= k_0(\mathbf{M}_{2,1}\mathbf{U}_x + \mathbf{M}_{2,2}\mathbf{U}_y + \mathbf{M}_{2,3}\mathbf{U}_z), \\[2mm] ik_0\mathbf{K}_x\mathbf{S}_y - ik_0\mathbf{K}_y\mathbf{S}_x &= k_0(\mathbf{M}_{3,1}\mathbf{U}_x + \mathbf{M}_{3,2}\mathbf{U}_y + \mathbf{M}_{3,3}\mathbf{U}_z), \\[2mm] ik_0\mathbf{K}_y\mathbf{U}_z - \dfrac{d\mathbf{U}_y}{dz} &= k_0(\mathbf{E}_{1,1}\mathbf{S}_x + \mathbf{E}_{1,2}\mathbf{S}_y + \mathbf{E}_{1,3}\mathbf{S}_z), \\[2mm] \dfrac{d\mathbf{U}_x}{dz} - ik_0\mathbf{K}_x\mathbf{U}_z &= k_0(\mathbf{E}_{2,1}\mathbf{S}_x + \mathbf{E}_{2,2}\mathbf{S}_y + \mathbf{E}_{2,3}\mathbf{S}_z), \\[2mm] ik_0\mathbf{K}_x\mathbf{U}_y - ik_0\mathbf{K}_y\mathbf{U}_x &= k_0(\mathbf{E}_{3,1}\mathbf{S}_x + \mathbf{E}_{3,2}\mathbf{S}_y + \mathbf{E}_{3,3}\mathbf{S}_z), \end{cases} \tag{1.31}$$

where $\mathbf{K}_x = \operatorname{diag} \dfrac{k_{x,j}}{k_0}$, $\mathbf{K}_y = \operatorname{diag} \dfrac{k_y}{k_0}$ – diagonal matrices, $\mathbf{E}_{i,j} = \left[\!\left[\varepsilon_{i,j}\right]\!\right]$ and $\mathbf{M}_{i,j} = \left[\!\left[\mu_{i,j}\right]\!\right]$ are the Toeplitz matrices of the form (1.30), made up of the expansion coefficients in the Fourier series $\varepsilon_{i,j}(x)$ and $\mu_{i,j}(x)$.

We perform a change of variables $z' = k_0 z$ and transform the system (1.31) to the form:

$$
\begin{cases}
-\dfrac{d\mathbf{S}_y}{dz'} = \mathbf{M}_{1,1}\mathbf{U}_x + \mathbf{M}_{1,2}\mathbf{U}_y + \mathbf{M}_{1,3}\mathbf{U}_z - i\mathbf{K}_y\mathbf{S}_z, \\[2mm]
-\dfrac{d\mathbf{S}_x}{dz'} = -\mathbf{M}_{2,1}\mathbf{U}_x - \mathbf{M}_{2,2}\mathbf{U}_y - \mathbf{M}_{2,3}\mathbf{U}_z - i\mathbf{K}_x\mathbf{S}_z, \\[2mm]
-\dfrac{d\mathbf{U}_y}{dz'} = \mathbf{E}_{1,1}\mathbf{S}_x + \mathbf{E}_{1,2}\mathbf{S}_y + \mathbf{E}_{1,3}\mathbf{S}_z - i\mathbf{K}_y\mathbf{U}_z, \\[2mm]
-\dfrac{d\mathbf{U}_x}{dz'} = -\mathbf{E}_{2,1}\mathbf{S}_x - \mathbf{E}_{2,2}\mathbf{S}_y - \mathbf{E}_{2,3}\mathbf{S}_z - i\mathbf{K}_x\mathbf{U}_z, \\[2mm]
\mathbf{E}_{3,3}\mathbf{S}_z = i\mathbf{K}_x\mathbf{U}_y - i\mathbf{K}_y\mathbf{U}_x - \mathbf{E}_{3,1}\mathbf{S}_x - \mathbf{E}_{3,2}\mathbf{S}_y, \\[2mm]
\mathbf{M}_{3,3}\mathbf{U}_z = i\mathbf{K}_x\mathbf{S}_y - i\mathbf{K}_y\mathbf{S}_x - \mathbf{M}_{3,1}\mathbf{U}_x - \mathbf{M}_{3,2}\mathbf{U}_y.
\end{cases}
\tag{1.32}
$$

We rewrite the first four equations of the system in matrix form:

$$
\frac{d}{dz'}
\begin{bmatrix}
\mathbf{S}_y \\ \mathbf{S}_x \\ \mathbf{U}_y \\ \mathbf{U}_x
\end{bmatrix}
= \mathbf{Y}
\begin{bmatrix}
\mathbf{S}_y \\ \mathbf{S}_x \\ \mathbf{U}_y \\ \mathbf{U}_x
\end{bmatrix}
-
\begin{bmatrix}
-i\mathbf{K}_y & \mathbf{M}_{1,3} \\
-i\mathbf{K}_x & -\mathbf{M}_{2,3} \\
\mathbf{E}_{1,3} & -i\mathbf{K}_y \\
-\mathbf{E}_{2,3} & -i\mathbf{K}_x
\end{bmatrix}
\begin{bmatrix}
\mathbf{S}_z \\ \mathbf{U}_z
\end{bmatrix},
\tag{1.33}
$$

where

$$
\mathbf{Y} = -
\begin{bmatrix}
\mathbf{0} & \mathbf{0} & \mathbf{M}_{1,2} & \mathbf{M}_{1,1} \\
\mathbf{0} & \mathbf{0} & -\mathbf{M}_{2,2} & -\mathbf{M}_{2,1} \\
\mathbf{E}_{1,2} & \mathbf{E}_{1,1} & \mathbf{0} & \mathbf{0} \\
-\mathbf{E}_{2,2} & -\mathbf{E}_{2,1} & \mathbf{0} & \mathbf{0}
\end{bmatrix}.
\tag{1.34}
$$

From the last two equations of (1.32) we express S_z and U_z and substitute the resulting expressions into (1.33). As a result, we obtain a system of linear differential equations with respect to x- and y-Fourier components of the fields:

where

$$\frac{d}{dz'}\begin{bmatrix} \mathbf{S}_y \\ \mathbf{S}_x \\ \mathbf{U}_y \\ \mathbf{U}_x \end{bmatrix} = \mathbf{A}\begin{bmatrix} \mathbf{S}_y \\ \mathbf{S}_x \\ \mathbf{U}_y \\ \mathbf{U}_x \end{bmatrix}, \quad (1.35)$$

where

$$\mathbf{A} = \mathbf{Y} - \begin{bmatrix} i\mathbf{K}_y & -\mathbf{M}_{1,3} \\ i\mathbf{K}_x & \mathbf{M}_{2,3} \\ -\mathbf{E}_{1,3} & i\mathbf{K}_y \\ \mathbf{E}_{2,3} & i\mathbf{K}_x \end{bmatrix}\begin{bmatrix} \mathbf{E}_{3,3}^{-1} & 0 \\ 0 & \mathbf{M}_{3,3}^{-1} \end{bmatrix}\begin{bmatrix} \mathbf{E}_{3,2} & \mathbf{E}_{3,1} & -i\mathbf{K}_x & i\mathbf{K}_y \\ -i\mathbf{K}_x & i\mathbf{K}_y & \mathbf{M}_{3,2} & \mathbf{M}_{3,1} \end{bmatrix}.$$

$$(1.36)$$

Thus, we have obtained a system of linear differential equations of the first order for the vectors $\mathbf{S}_x$, $\mathbf{S}_y$, $\mathbf{U}_x$, $\mathbf{U}_y$. Note that the matrix $\mathbf{A}$ has the dimensions $4(2N+1) \times 4(2N+1)$.

Correct rules of the Fourier expansion of the product of functions

The derivation of the system of differential equations (1.35) was based on the representation in (1.26) of the component of electromagnetic fields and of the components of permittivity and permeability tensors in the form of segments of the Fourier series. The representation of the products of the functions in (1.26) by the Fourier series has its own peculiarities. The used formulas (1.28)–(1.30) have limited applicavility.

Consider two periodic functions

$$f(x) = \sum_m f_m \exp(iKmx), \quad g(x) = \sum_m g_m \exp(iKmx), \quad K = \frac{2\pi}{d} \quad (1.37)$$

and the expansion of their product into the Fourier series

$$h(x) = f(x)g(x) = \sum_m h_m \exp(iKmx). \qquad (1.38)$$

The Fourier coefficients of the product are the values

$$h_j = \sum_{m=-N}^{N} f_{j-m} g_m \qquad (1.39)$$

obtained by direct multiplication of the series (1.37), (1.38). Equation (1.39) for calculating the Fourier coefficients is called the Laurent rule. In matrix notation (1.39) can be represented as [5]

$$[h] = [\![f]\!][g], \qquad (1.40)$$

where, as in (1.29), the square brackets denote the vectors composed from the Fourier coefficients of the expansion of the functions, and $[\![f]\!]$ is the Toeplitz matrix of the Fourier coefficients.

As was shown in [5, 6], the use of formula (1.40) is correct if there is no value x for which the functions $f(x)$ and $g(x)$ show a discontinuity at the same time. Using the Laurent rule (1.40) for the product of the functions having identical points of discontinuity leads to poor convergence of the Fourier series at the points of discontinuity.

If the functions $f(x)$ and $g(x)$ are discontinuous at the same time, but the function $h(x) = f(x)g(x)$ is continuous, it is correct to use the so-called inverse Laurent rules [5]:

$$[h] = \left[\!\!\left[\frac{1}{f}\right]\!\!\right]^{-1}[g]. \qquad (1.41)$$

In the products $\mu_{i,1}(x) \cdot H_x(x,y,z)$ and $\varepsilon_{i,1}(x) \cdot E_x(x,y,z)$ in (1.26), both expanded functions have discontinuities at the same points x, corresponding to vertical interfaces of the media in the layers. Accordingly, it is erroneous to use the Laurent rule (1.28)–(1.30) when writing (1.31). Errors in the use of the Laurent rule are significant when working with diffraction gratings of conductive materials. In particular, the erroneous use of the Laurent rule leads to slow convergence of the solutions for binary metal gratings in the case of TM-polarization [5–10]. In this case we define the convergence as the stabilization of the results of the calculation of the amplitudes of diffraction orders with increasing number of the Fourier harmonics N in the components of the field (1.27).

Continuous components at the vertical boundaries between the media are the tangential components E_y, E_z, H_y, H_z and the normal components of the electric displacement and magnetic induction D_x, B_x. We express discontinuous components E_x, H_x through the continuous field components

$$E_x = \frac{1}{\varepsilon_{11}} D_x - \frac{\varepsilon_{12}}{\varepsilon_{11}} E_y - \frac{\varepsilon_{13}}{\varepsilon_{11}} E_z,$$

$$H_x = \frac{1}{\mu_{11}} B_x - \frac{\mu_{12}}{\mu_{11}} H_y - \frac{\mu_{13}}{\mu_{11}} H_z \tag{1.42}$$

and substitute into Maxwell's equations (1.26). The result is:

$$
\begin{cases}
\dfrac{\partial E_z}{\partial y} - \dfrac{\partial E_y}{\partial z} = ik_0 B_x, \\[2mm]
\dfrac{\partial E_x}{\partial z} - \dfrac{\partial E_z}{\partial x} = ik_0\left(\dfrac{\alpha_{2,1}}{\alpha_{1,1}} B_x + \left(\alpha_{2,2} - \alpha_{2,1}\dfrac{\alpha_{1,2}}{\alpha_{1,1}} \right) H_y + \left(\alpha_{2,3} - \alpha_{2,1}\dfrac{\alpha_{1,3}}{\alpha_{1,1}} \right) H_z \right), \\[2mm]
\dfrac{\partial E_y}{\partial x} - \dfrac{\partial E_x}{\partial y} = ik_0\left(\dfrac{\alpha_{3,1}}{\alpha_{1,1}} B_x + \left(\alpha_{3,2} - \alpha_{3,1}\dfrac{\alpha_{1,2}}{\alpha_{1,1}} \right) H_y + \left(\alpha_{3,3} - \alpha_{3,1}\dfrac{\alpha_{1,3}}{\alpha_{1,1}} \right) H_z \right), \\[2mm]
\dfrac{\partial H_z}{\partial y} - \dfrac{\partial H_y}{\partial z} = -ik_0 D_x, \\[2mm]
\dfrac{\partial H_x}{\partial z} - \dfrac{\partial H_z}{\partial x} = -ik_0\left(\dfrac{\varepsilon_{2,1}}{\varepsilon_{1,1}} D_x + \left(\varepsilon_{2,2} - \varepsilon_{2,1}\dfrac{\varepsilon_{1,2}}{\varepsilon_{1,1}} \right) E_y + \left(\varepsilon_{2,3} - \varepsilon_{2,1}\dfrac{\varepsilon_{1,3}}{\varepsilon_{1,1}} \right) E_z \right), \\[2mm]
\dfrac{\partial H_y}{\partial x} - \dfrac{\partial H_x}{\partial y} = -ik_0\left(\dfrac{\varepsilon_{3,1}}{\varepsilon_{1,1}} D_x + \left(\varepsilon_{3,2} - \varepsilon_{3,1}\dfrac{\varepsilon_{1,2}}{\varepsilon_{1,1}} \right) E_y + \left(\varepsilon_{3,3} - \varepsilon_{3,1}\dfrac{\varepsilon_{1,3}}{\varepsilon_{1,1}} \right) E_z \right).
\end{cases}
\tag{1.43}
$$

Equations (1.43) do not contain products of functions simultaneously suffering a discontinuity. Therefore, in the transition to the space–frequency domain, we can use the direct Laurent rule (1.29), (1.30). According to (1.42), the vectors of the Fourier coefficients of the functions D_x, B_x have the form

$$[D_x] = \left[\left[\frac{1}{\varepsilon_{11}}\right]\right]^{-1}\left([E_x] + \left[\frac{\varepsilon_{12}}{\varepsilon_{11}}\right][E_y] + \left[\frac{\varepsilon_{13}}{\varepsilon_{11}}\right][E_z]\right),$$

$$[B_x] = \left[\left[\frac{1}{\mu_{11}}\right]\right]^{-1}\left([H_x] + \left[\frac{\mu_{12}}{\mu_{11}}\right][H_y] + \left[\frac{\mu_{13}}{\mu_{11}}\right][H_z]\right).$$

(1.44)

Performing in (1.43) the transition to the space–frequency domain and transforming the result, we also obtain the system of differential equations (1.35)–(1.34) where the matrices $\mathbf{M}_{i,j}$ and $\mathbf{E}_{i,j}$ have the form

$$\mathbf{M}_{1,1} = \left[\left[\frac{1}{\varpi_{11}}\right]\right]^{-1}, \quad \mathbf{M}_{1,2} = \left[\left[\frac{1}{\varpi_{11}}\right]\right]^{-1}\left[\frac{\varpi_{12}}{\varpi_{11}}\right], \quad \mathbf{M}_{1,3} = \left[\left[\frac{1}{\varpi_{11}}\right]\right]^{-1}\left[\frac{\varpi_{13}}{\varpi_{11}}\right],$$

$$\mathbf{M}_{2,1} = \left[\frac{\varpi_{21}}{\varpi_{11}}\right]\left[\left[\frac{1}{\varpi_{11}}\right]\right]^{-1}, \quad \mathbf{M}_{2,2} = \left[\varpi_{22} - \frac{\varpi_{12}\varpi_{21}}{\varpi_{11}}\right] + \left[\frac{\varpi_{21}}{\varpi_{11}}\right]\left[\left[\frac{1}{\varpi_{11}}\right]\right]^{-1}\left[\frac{\varpi_{12}}{\varpi_{11}}\right],$$

$$\mathbf{M}_{2,3} = \left[\varpi_{23} - \frac{\varpi_{13}\varpi_{21}}{\varpi_{11}}\right] + \left[\frac{\varpi_{21}}{\varpi_{11}}\right]\left[\left[\frac{1}{\varpi_{11}}\right]\right]^{-1}\left[\frac{\varpi_{13}}{\varpi_{11}}\right],$$

$$\mathbf{M}_{3,1} = \left[\frac{\varpi_{31}}{\varpi_{11}}\right]\left[\left[\frac{1}{\varpi_{11}}\right]\right]^{-1}, \quad \mathbf{M}_{3,2} = \left[\varpi_{32} - \frac{\varpi_{31}\varpi_{12}}{\varpi_{11}}\right] + \left[\frac{\varpi_{31}}{\mu_{11}}\right]\left[\left[\frac{1}{\mu_{11}}\right]\right]^{-1}\left[\frac{\mu_{12}}{\mu_{11}}\right],$$

$$\mathbf{M}_{3,3} = \left[\mu_{33} - \frac{\mu_{31}\mu_{13}}{\mu_{11}}\right] + \left[\frac{\mu_{31}}{\mu_{11}}\right]\left[\left[\frac{1}{\mu_{11}}\right]\right]^{-1}\left[\frac{\mu_{13}}{\mu_{11}}\right],$$

$$\mathbf{E}_{1,1} = \left[\left[\frac{1}{\varepsilon_{11}}\right]\right]^{-1}, \quad \mathbf{E}_{1,2} = \left[\left[\frac{1}{\varepsilon_{11}}\right]\right]^{-1}\left[\frac{\varepsilon_{12}}{\varepsilon_{11}}\right], \quad \mathbf{E}_{1,3} = \left[\left[\frac{1}{\varepsilon_{11}}\right]\right]^{-1}\left[\frac{\varepsilon_{13}}{\varepsilon_{11}}\right],$$

$$\mathbf{E}_{2,1} = \left[\frac{\varepsilon_{21}}{\varepsilon_{11}}\right]\left[\left[\frac{1}{\varepsilon_{11}}\right]\right]^{-1}, \quad \mathbf{E}_{2,2} = \left[\varepsilon_{22} - \frac{\varepsilon_{12}\varepsilon_{21}}{\varepsilon_{11}}\right] + \left[\frac{\varepsilon_{21}}{\varepsilon_{11}}\right]\left[\left[\frac{1}{\varepsilon_{11}}\right]\right]^{-1}\left[\frac{\varepsilon_{12}}{\varepsilon_{11}}\right],$$

$$\mathbf{E}_{2,3} = \left[\varepsilon_{23} - \frac{\varepsilon_{13}\varepsilon_{21}}{\varepsilon_{11}}\right] + \left[\frac{\varepsilon_{21}}{\varepsilon_{11}}\right]\left[\left[\frac{1}{\varepsilon_{11}}\right]\right]^{-1}\left[\frac{\varepsilon_{13}}{\varepsilon_{11}}\right],$$

$$\mathbf{E}_{3,1} = \left[\frac{\varepsilon_{31}}{\varepsilon_{11}}\right]\left[\left[\frac{1}{\varepsilon_{11}}\right]\right]^{-1}, \quad \mathbf{E}_{3,2} = \left[\varepsilon_{32} - \frac{\varepsilon_{31}\varepsilon_{12}}{\varepsilon_{11}}\right] + \left[\frac{\varepsilon_{31}}{\varepsilon_{11}}\right]\left[\left[\frac{1}{\varepsilon_{11}}\right]\right]^{-1}\left[\frac{\varepsilon_{12}}{\varepsilon_{11}}\right],$$

$$\mathbf{E}_{3,3} = \left[\varepsilon_{33} - \frac{\varepsilon_{31}\varepsilon_{13}}{\varepsilon_{11}}\right] + \left[\frac{\varepsilon_{31}}{\varepsilon_{11}}\right]\left[\left[\frac{1}{\varepsilon_{11}}\right]\right]^{-1}\left[\frac{\varepsilon_{13}}{\varepsilon_{11}}\right].$$

(1.45)

The system of differential equations (1.35) with the matrices $\mathbf{M}_{i,j}$, $\mathbf{E}_{i,j}$ in the form (1.45) will be called a system, obtained using the correct rules of the Fourier expansions of the product of the functions.

The form of the matrix of the system for different permittivity tensors

We consider some special kinds of tensors of permittivity and permeability and the corresponding matrices of the system of differential equations (1.35). The systems of differential equations are presented for the case of correct rules of the Fourier expansions. Consider the case of an isotropic material. For this material, $\mu = 1$ and ε is a scalar. In this case, the matrices $\mathbf{E}_{i,j}$, $\mathbf{M}_{i,j}$ take the following form:

$$\mathbf{E}_{1,1} = \mathbf{E}^* = \left[\frac{1}{\varepsilon}\right]^{-1}, \ \mathbf{E}_{2,2} = \mathbf{E}_{3,3} = \mathbf{E} = [\![\varepsilon]\!],$$

$$\mathbf{E}_{1,2} = \mathbf{E}_{1,3} = \mathbf{E}_{2,1} = \mathbf{E}_{2,3} = \mathbf{E}_{3,1} = \mathbf{E}_{3,2} = \mathbf{0},$$

$$\mathbf{M}_{1,1} = \mathbf{M}_{2,2} = \mathbf{M}_{3,3} = \mathbf{I},$$

$$\mathbf{M}_{1,2} = \mathbf{M}_{1,3} = \mathbf{M}_{2,1} = \mathbf{M}_{2,3} = \mathbf{M}_{3,1} = \mathbf{M}_{3,2} = \mathbf{0},$$

$$(1.46)$$

where $\mathbf{I}$ is the identity matrix with the dimensions $(2N+1)\times(2N+1)$. Substituting (1.46) into (1.36)–(1.34), we obtain the matrix of the system of differential equations in the form

$$\mathbf{A} = -\begin{bmatrix} \mathbf{0} & \mathbf{0} & \mathbf{K}_y\mathbf{E}^{-1}\mathbf{K}_x & \mathbf{I}-\mathbf{K}_y\mathbf{E}^{-1}\mathbf{K}_y \\ \mathbf{0} & \mathbf{0} & \mathbf{K}_x\mathbf{E}^{-1}\mathbf{K}_x-\mathbf{I} & -\mathbf{K}_x\mathbf{E}^{-1}\mathbf{K}_y \\ \mathbf{K}_y\mathbf{K}_x & \mathbf{E}^*-\mathbf{K}_y^2 & \mathbf{0} & \mathbf{0} \\ \mathbf{K}_x^2-\mathbf{E} & -\mathbf{K}_x\mathbf{K}_y & \mathbf{0} & \mathbf{0} \end{bmatrix}. \quad (1.47)$$

Consider the case of a planar incidence when $k_y = 0$ ($\varphi = 0$) in (1.22) and the direction vector of the incident wave lies in the plane xOz. In this case, $\mathbf{K}_y = \mathbf{0}$ (1.47) and the system of differential equations (1.35) splits into two independent systems

$$\frac{d}{dz'}\begin{bmatrix} \mathbf{S}_y \\ \mathbf{U}_x \end{bmatrix} = \mathbf{A}^{\text{TE}}\begin{bmatrix} \mathbf{S}_y \\ \mathbf{U}_x \end{bmatrix}, \ \mathbf{A}^{\text{TE}} = -\begin{bmatrix} \mathbf{0} & \mathbf{I} \\ \mathbf{K}_x^2-\mathbf{E} & \mathbf{0} \end{bmatrix},$$

$$\frac{d}{dz'}\begin{bmatrix} \mathbf{U}_y \\ \mathbf{S}_x \end{bmatrix} = \mathbf{A}^{\text{TM}}\begin{bmatrix} \mathbf{U}_y \\ \mathbf{S}_x \end{bmatrix}, \ \mathbf{A}^{\text{TM}} = -\begin{bmatrix} \mathbf{0} & \mathbf{E}^* \\ \mathbf{K}_x\mathbf{E}^{-1}\mathbf{K}_x-\mathbf{I} & \mathbf{0} \end{bmatrix}. \quad (1.48)$$

This result in the case of a planar incidence reduces the solution of the problem of diffraction to two independent problems of diffraction of the waves with TM- and TE-polarization.

Currently, there is considerable interest in the structures comprising layers of a magnetic material. In such structures the optical properties of the materials can be modified by an external magnetic field. This allows to effectively control the amplitude and phase of the diffraction orders with the help of an external magnetic field. For magnetic materials the dielectric constant is given by the tensor [16, 17]:

$$
\ddot\varepsilon = \begin{bmatrix} \varepsilon & ig\cos\theta_M & -ig\sin\theta_M\sin\varphi_M \\ -ig\cos\theta_M & \varepsilon & ig\sin\theta_M\cos\varphi_M \\ ig\sin\theta_M\sin\varphi_M & -ig\sin\theta_M\cos\varphi_M & \varepsilon \end{bmatrix}, \quad (1.49)
$$

where ε is the main dielectric constant of the medium, g is the modulus of the gyration vector of the medium proportional to the magnetization [16, 17], θ_M and φ_M are the spherical coordinates, describing the direction of the magnetization vector. In the optical frequency range $\mu = 1$.

We consider three basic cases corresponding to the direction of the magnetization vector along the three coordinate axes.

For the *polar geometry of magnetization* (magnetization vector is perpendicular to the plane of the layers of the structure and directed along the axis Oz) $\theta_M = 0$ and the tensor (1.49) takes the form:

$$
\ddot\varepsilon = \begin{bmatrix} \varepsilon & ig & 0 \\ -ig & \varepsilon & 0 \\ 0 & 0 & \varepsilon \end{bmatrix}. \quad (1.50)
$$

In this case

$$
\mathbf{A} = -\begin{bmatrix} 0 & 0 & \mathbf{K}_y\mathbf{E}^{-1}\mathbf{K}_x & \mathbf{I}-\mathbf{K}_y\mathbf{E}^{-1}\mathbf{K}_y \\ 0 & 0 & \mathbf{K}_x\mathbf{E}^{-1}\mathbf{K}_x-\mathbf{I} & -\mathbf{K}_x\mathbf{E}^{-1}\mathbf{K}_y \\ i\mathbf{E}^*\mathbf{H}+\mathbf{K}_y\mathbf{K}_x & \mathbf{E}^*-\mathbf{K}_y^2 & 0 & 0 \\ \mathbf{K}_x^2-\mathbf{E}_2 & i\mathbf{HE}^*-\mathbf{K}_x\mathbf{K}_y & 0 & 0 \end{bmatrix}, \quad (1.51)
$$

where

$$\mathbf{E}^* = \left[\left[\frac{1}{\varepsilon}\right]\right]^{-1}, \quad \mathbf{E} = [[\varepsilon]], \quad \mathbf{H} = \left[\left[\frac{g}{\varepsilon}\right]\right], \quad \mathbf{E}_2 = \left[\left[\varepsilon - \frac{g^2}{\varepsilon}\right]\right] + \mathbf{H}\mathbf{E}^*\mathbf{H}$$

In the case of the meridional geometry of magnetization (the magnetization vector is parallel to the plane of the layers and directed along the axis Ox) $\theta_M = \dfrac{\pi}{2}$, $\varphi_M = 0$ and the tensor (1.49) takes the form

$$\ddot{\varepsilon} = \begin{bmatrix} \varepsilon & 0 & 0 \\ 0 & \varepsilon & ig \\ 0 & -ig & \varepsilon \end{bmatrix}. \tag{1.52}$$

In the case of (1.52) the matrix of the system takes the form

$$A = -\begin{bmatrix} \mathbf{K}_y\mathbf{E}^{-1}\mathbf{G} & 0 & \mathbf{K}_y\mathbf{E}^{-1}\mathbf{K}_x & \mathbf{I} - \mathbf{K}_y\mathbf{E}^{-1}\mathbf{K}_y \\ \mathbf{K}_x\mathbf{E}^{-1}\mathbf{G} & 0 & \mathbf{K}_x\mathbf{E}^{-1}\mathbf{K}_x - \mathbf{I} & -\mathbf{K}_x\mathbf{E}^{-1}\mathbf{K}_y \\ \mathbf{K}_y\mathbf{K}_x & \mathbf{E}^* - \mathbf{K}_y^2 & 0 & 0 \\ \mathbf{G}\mathbf{E}^{-1}\mathbf{G} + \mathbf{K}_x^2 - \mathbf{E} & -\mathbf{K}_x\mathbf{K}_y & \mathbf{G}\mathbf{E}^{-1}\mathbf{K}_x & -\mathbf{G}\mathbf{E}^{-1}\mathbf{K}_y \end{bmatrix},$$

$$\tag{1.53}$$

where $\mathbf{G} = [[g]]$.

For the *equatorial geometry of magnetization* (the magnetization vector is parallel to the plane of layers and the direction of the axis Oy) $\theta_M = \varphi_M = \dfrac{\pi}{2}$. In this case from (1.49) we obtain:

$$\ddot{\varepsilon} = \begin{bmatrix} \varepsilon & 0 & ig \\ 0 & \varepsilon & 0 \\ -ig & 0 & \varepsilon \end{bmatrix}. \tag{1.54}$$

For the tensor (1.54) the matrix of the system takes the form

$$A = -\begin{bmatrix} 0 & \mathbf{K}_y\mathbf{E}_2^{-1}\mathbf{H}\mathbf{E}^* & \mathbf{K}_y\mathbf{E}_2^{-1}\mathbf{K}_x & \mathbf{I} - \mathbf{K}_y\mathbf{E}_2^{-1}\mathbf{K}_y \\ 0 & \mathbf{K}_x\mathbf{E}_2^{-1}\mathbf{H}\mathbf{E}^* & \mathbf{K}_x\mathbf{E}_2^{-1}\mathbf{K}_x - \mathbf{I} & -\mathbf{K}_x\mathbf{E}_2^{-1}\mathbf{K}_y \\ \mathbf{K}_y\mathbf{K}_x & \mathbf{E}^* - \mathbf{K}_y^2 - \mathbf{E}^*\mathbf{H}\mathbf{E}_2^{-1}\mathbf{H}\mathbf{E}^* & -\mathbf{E}^*\mathbf{H}\mathbf{E}_2^{-1}\mathbf{K}_x & \mathbf{E}^*\mathbf{H}\mathbf{E}_2^{-1}\mathbf{K}_y \\ \mathbf{K}_x^2 - \mathbf{E} & -\mathbf{K}_x\mathbf{K}_y & 0 & 0 \end{bmatrix}$$

$$\tag{1.55}$$

1.1.2.4. Representation of the field inside the layer

For direct representation of the field in the layer, we consider the eigendecomposition of the matrix $\mathbf{A}$:

$$\mathbf{A} = \mathbf{W\Lambda W}^{-1}, \tag{1.56}$$

wherein $\mathbf{\Lambda} = \underset{i}{\mathrm{diag}}\, \lambda_i$ is the diagonal matrix of eigenvalues of the matrix $\mathbf{A}$, and $\mathbf{W}$ is the matrix of eigenvectors. Then the solution of system of differential equations (1.35) can be written as

$$\begin{bmatrix} \mathbf{S}_y \\ \mathbf{S}_x \\ \mathbf{U}_y \\ \mathbf{U}_x \end{bmatrix} = \mathbf{W}\exp\left(\mathbf{\Lambda}\,z'\right)\mathbf{C}' = \mathbf{W}\exp\left(\mathbf{\Lambda}\,k_0 z\right)\mathbf{C}' = \sum_i c_i'\mathbf{w}_i \exp\left(\lambda_i k_0 z\right). \tag{1.57}$$

We split the last sum into two depending on the sign of the real part λ_i and write the expression in matrix notation:

$$\begin{aligned}
\begin{bmatrix} \mathbf{S}_y \\ \mathbf{S}_x \\ \mathbf{U}_y \\ \mathbf{U}_x \end{bmatrix} &= \sum_i c_i'\mathbf{w}_i \exp\left(\lambda_i k_0 z\right) = \sum_{i:\mathrm{Re}\lambda_i<0} c_i\mathbf{w}_i \exp\left(\lambda_i k_0(z-z_l)\right) + \\
&\quad \sum_{i:\mathrm{Re}\lambda_i>0} c_i\mathbf{w}_i \exp\left(\lambda_i k_0(z-z_{l-1})\right) = \\
&= \mathbf{W}^{(-)}\exp\left(\mathbf{\Lambda}^{(-)}k_0(z-z_l)\right)\mathbf{C}^{(-)} + \\
&\quad \mathbf{W}^{(+)}\exp\left(\mathbf{\Lambda}^{(+)}k_0(z-z_{l-1})\right)\mathbf{C}^{(+)},
\end{aligned} \tag{1.58}$$

where $\mathbf{\Lambda}^{(+)}$ and $\mathbf{\Lambda}^{(-)}$ are the diagonal matrices of the eigenvalues whose the real parts are positive and negative, respectively, $\mathbf{W}^{(+)}$ and $\mathbf{W}^{(-)}$ are the corresponding eigenvector matrices, $\mathbf{C}^{(-)}$, $\mathbf{C}^{(+)}$ are the vectors of arbitrary constants. The representation (1.58) is suitable for numerical calculations. The exponent in (1.58) always has a negative real part. This ensures that there is no numerical overflow.

The procedure for calculating the eigenvectors and eigenvalues can in some cases be significantly speeded up taking into account the specific form of the matrix $\mathbf{A}$ [4]. In particular, the matrix $\mathbf{A}$ in (1.47), (1.48) and (1.51) has the following block structure:

$$A = \begin{bmatrix} 0 & A_{12} \\ A_{21} & 0 \end{bmatrix}. \tag{1.59}$$

We write for A the matrices of eigenvectors and eigenvalues in the form

$$W = \begin{bmatrix} W_{11} & W_{12} \\ W_{21} & W_{22} \end{bmatrix}, \quad A = \begin{bmatrix} \Lambda_{11} & 0 \\ 0 & \Lambda_{22} \end{bmatrix} \tag{1.60}$$

Since $A = W \Lambda W^{-1}$, then

$$\begin{bmatrix} 0 & A_{12} \\ A_{21} & 0 \end{bmatrix} \cdot \begin{bmatrix} W_{11} & W_{12} \\ W_{21} & W_{22} \end{bmatrix} = \begin{bmatrix} W_{11} & W_{12} \\ W_{21} & W_{22} \end{bmatrix} \cdot \begin{bmatrix} \Lambda_{11} & 0 \\ 0 & \Lambda_{22} \end{bmatrix}$$

$$\begin{bmatrix} A_{12}W_{21} & A_{12}W_{22} \\ A_{21}W_{11} & A_{21}W_{12} \end{bmatrix} = \begin{bmatrix} W_{11}\Lambda_{11} & W_{12}\Lambda_{22} \\ W_{21}\Lambda_{11} & W_{22}\Lambda_{22} \end{bmatrix}$$

and we have:

$$A_{12}A_{21}W_{11} = A_{12}W_{21}\Lambda_{11} = W_{11}\Lambda_{11}\Lambda_{11},$$
$$A_{12}A_{21}W_{12} = A_{12}W_{22}\Lambda_{22} = W_{12}\Lambda_{22}\Lambda_{22}. \tag{1.61}$$

We introduce the matrix $B = A_{12}A_{21}$ and write (1.61) in the form

$$B \cdot W_{11} = W_{11}\Lambda_{11}^2, \quad B \cdot W_{12} = W_{12}\Lambda_{22}^2. \tag{1.62}$$

According to (1.62) $W_{11}, \Lambda' = \Lambda_{11}^2, W_{12}, \Lambda_{22}^2$ are the matrices of the eigenvectors and the diagonal matrices of the eigenvalues of the same matrix B. Therefore

$$W_{11} = W_{12}, \quad \Lambda_{22} = -\Lambda_{11} \text{ and } W_{21} = A_{21}W_{11}\Lambda_{11}^{-1}, \quad W_{22} = -W_{21}. \tag{1.63}$$

The relations (1.63) determine the eigenvalues and eigenvectors of the matrix A by the eigenvalues and eigenvectors of the matrix B half their size in the form:

$$W = \begin{bmatrix} W_{11} & W_{11} \\ A_{21}W_{11}\sqrt{\Lambda'}^{-1} & -A_{21}W_{11}\sqrt{\Lambda'}^{-1} \end{bmatrix}, \quad \Lambda = \begin{bmatrix} \sqrt{\Lambda'} & 0 \\ 0 & -\sqrt{\Lambda'} \end{bmatrix}. \tag{1.64}$$

In [4] it is pointed out that halving the dimensions of the eigenvalue problem is equivalent to transforming the system (1.35) of $4(2N+1)$ first-order differential equations to a system $2(2N+1)$ of second-order differential equations.

1.1.2.5. 'Stitching' of the electromagnetic field at the boundaries of layers

The general representation of the field in the layer was described above. To obtain a solution that satisfies the Maxwell's equations, it is necessary to equate the tangential components of the fields at the boundaries of the layers. Equating the tangential field components is equivalent to equating the functions (1.58) corresponding to the Fourier coefficients at each fixed z. We write preliminary solutions of (1.58) on the upper and lower boundaries of all the layers. For convenience, we assume that the matrix of eigenvectors $\mathbf{W}$ is given by:

$$\mathbf{W} = \begin{bmatrix} \mathbf{W}^{(-)} & \mathbf{W}^{(+)} \end{bmatrix}. \tag{1.65}$$

In addition, we introduce a vector of unknown constants $\mathbf{C}$ as follows:

$$\mathbf{C} = \begin{bmatrix} \mathbf{C}^{(-)} \\ \mathbf{C}^{(+)} \end{bmatrix}. \tag{1.66}$$

In view of this notation the solution of (1.58) on the upper and lower boundaries of the layer has the form

$$\begin{bmatrix} \mathbf{S}_y(z_{l-1}) \\ \mathbf{S}_x(z_{l-1}) \\ \mathbf{U}_y(z_{l-1}) \\ \mathbf{U}_x(z_{l-1}) \end{bmatrix} = \mathbf{W}^{(-)}\mathbf{X}^{(-)}\mathbf{C}^{(-)} + \mathbf{W}^{(+)}\mathbf{C}^{(+)} = \mathbf{W}\begin{bmatrix} \mathbf{X}^{(-)} & \mathbf{0} \\ \mathbf{0} & \mathbf{I} \end{bmatrix}\mathbf{C} = \mathbf{NC},$$
$$\tag{1.67}$$

$$\begin{bmatrix} \mathbf{S}_y(z_l) \\ \mathbf{S}_x(z_l) \\ \mathbf{U}_y(z_l) \\ \mathbf{U}_x(z_l) \end{bmatrix} = \mathbf{W}^{(-)}\mathbf{C}^{(-)} + \mathbf{W}^{(+)}\mathbf{X}^{(+)}\mathbf{C}^{(+)} = \mathbf{W}\begin{bmatrix} \mathbf{I} & \mathbf{0} \\ \mathbf{0} & \mathbf{X}^{(+)} \end{bmatrix}\mathbf{C} = \mathbf{MC}, \tag{1.68}$$

where

$$\mathbf{X}^{(+)} = \exp\left(\mathbf{\Lambda}^{(+)}k_0\left(z_l - z_{l-1}\right)\right), \quad \mathbf{X}^{(-)} = \exp\left(-\mathbf{\Lambda}^{(-)}k_0\left(z_l - z_{l-1}\right)\right). \tag{1.69}$$

Equating the tangential components at the interface of the adjacent layers, we obtain the equations

$$\mathbf{M}_{l-1}\mathbf{C}_{l-1} = \mathbf{N}_l\mathbf{C}_l, \qquad l = 2,\ldots,L, \qquad (1.70)$$

where the index $l = 2$ corresponds to the condition for the lower boundary of the upper layer, $l = L$ for the upper boundary of the lower layer.

The relations (1.70) must be supplemented by the conditions of the equality of the tangential components of the field above the structure of (1.20) and at the upper boundary of the 1st layer, as well as the field below the structure (1.21) and at the lower boundary of the L-th layer. Given that the components in the field in the region of the layers are represented by segments of the Fourier series with the dimension $2N + 1$, in representations of the field above and under the grating (1.20)–(1.22) we should also take $2N + 1$ waves at $-N \leq i \leq N$.

By adding these relations for the upper and lower boundaries of the diffractive structure, we obtain the following system of linear equations:

$$\begin{cases} \mathbf{D} + \mathbf{P}^{(U)}\mathbf{R} = \mathbf{N}_1\mathbf{C}_1, \\ \mathbf{M}_{l-1}\mathbf{C}_{l-1} = \mathbf{N}_l\mathbf{C}_l, \ l = 2,\ldots,L, \\ \mathbf{M}_L\mathbf{C}_L = \mathbf{P}^{(D)}\mathbf{T}, \end{cases} \qquad (1.71)$$

where $\mathbf{R}$ and $\mathbf{T}$ are the vectors of complex amplitudes of the reflected and transmitted orders, respectively. These vectors are of the form:

$$\mathbf{R} = \begin{bmatrix} \mathbf{R}_E \\ \mathbf{R}_H \end{bmatrix}, \quad \mathbf{T} = \begin{bmatrix} \mathbf{T}_E \\ \mathbf{T}_H \end{bmatrix}, \qquad (1.72)$$

where $\mathbf{R}_E$ and $\mathbf{T}_E$ are the vectors of complex amplitudes of the E-waves, $\mathbf{R}_H$ and $\mathbf{T}_H$ are the vectors of complex amplitudes of the H-waves.

The vector $\mathbf{D}$ in (1.71) represents the incident wave and has the form (see (1.13) for $\theta \to \pi - \theta$):

$$\mathbf{D} = \begin{bmatrix} \cos\theta\sin\varphi\cdot\delta_i \\ \cos\theta\cos\varphi\cdot\delta_i \\ -in_U\cos\varphi\cdot\delta_i \\ in_U\sin\varphi\cdot\delta_i \end{bmatrix}\cos\psi + \begin{bmatrix} \cos\varphi\cdot\delta_i \\ -\sin\varphi\cdot\delta_i \\ in_U\cos\theta\sin\varphi\cdot\delta_i \\ in_U\cos\theta\cos\varphi\cdot\delta_i \end{bmatrix}\sin\psi, \qquad (1.73)$$

where $\boldsymbol{\delta}_i$ is the column vector in which only one element, standing in the middle, is nonzero and is equal to one. Column vector $\boldsymbol{\delta}_i$ has the dimension $2N + 1$ and vector $\mathbf{D}$ the dimension $4(2N + 1)$.

The matrices $\mathbf{P}^{(U)}$ and $\mathbf{P}^{(D)}$ (1.71) correspond to the reflected and transmitted orders, respectively, and have the form

$$
\mathbf{P}^{(p)} = \begin{bmatrix} -\cos\boldsymbol{\Theta}^{(p)}\sin\boldsymbol{\Phi} & \cos\boldsymbol{\Phi} \\ -\cos\boldsymbol{\Theta}^{(p)}\cos\boldsymbol{\Phi} & -\sin\boldsymbol{\Phi} \\ -in_p\cos\boldsymbol{\Phi} & -in_p\cos\boldsymbol{\Theta}^{(p)}\sin\boldsymbol{\Phi} \\ in_p\sin\boldsymbol{\Phi} & -in_p\cos\boldsymbol{\Theta}^{(p)}\cos\boldsymbol{\Phi} \end{bmatrix}, \quad p = U, D, \quad (1.74)
$$

where $\boldsymbol{\Phi}$, $\boldsymbol{\Theta}^{(P)}$ are the diagonal matrices of the angles determining the direction of the scattered diffraction orders. They satisfy the following relations:

$$
\sin\boldsymbol{\Phi} = \operatorname*{diag}_i \frac{k_y}{\sqrt{k_{x,i}^2 + k_y^2}}, \quad \cos\boldsymbol{\Phi} = \operatorname*{diag}_i \frac{k_{x,i}}{\sqrt{k_{x,i}^2 + k_y^2}}, \quad (1.75)
$$

$$
\cos\boldsymbol{\Theta}^{(U)} = \operatorname*{diag}_i \frac{k_{z,U,i}}{k_0 n_U}, \quad \cos\boldsymbol{\Theta}^{(D)} = \operatorname*{diag}_i \frac{-k_{z,D,i}}{k_0 n_D}. \quad (1.76)
$$

The expressions (1.73)–(1.74) follow directly from the general formulas (1.12) and (1.13) for a plane wave with the form of propagation constants of the orders (1.22) taken into account.

According to (1.71), the solution of the diffraction problem is reduced to solving a system of linear equations. The sequential expression of the coefficients $\mathbf{C}_{l-1}$ in the layer with the index $(l-1)$ through the coefficients $\mathbf{C}_{l-1}$ in the l-th layer allows to reduce the system (1.71) to a system of equations for the coefficients $\mathbf{R}$ and $\mathbf{T}$. Indeed, from the last two equations in (1.71), we obtain

$$
\mathbf{M}_{L-1}\mathbf{C}_{L-1} = \mathbf{N}_L\mathbf{M}_L^{-1}\mathbf{P}^{(D)}\mathbf{T}. \quad (1.77)
$$

By substituting (1.77) into the equation with index $l = L-1$ in (1.71) we have

$$
\mathbf{M}_{L-2}\mathbf{C}_{L-2} = \mathbf{N}_{L-1}\mathbf{M}_{L-1}^{-1}\mathbf{N}_L\mathbf{M}_L^{-1}\mathbf{P}^{(D)}\mathbf{T}. \quad (1.78)
$$

Continuing this process to the equation with the index $l = 2$, we obtain

$$\mathbf{M}_1\mathbf{C}_1 = \left(\prod_{l=2}^{L}\mathbf{N}_l\mathbf{M}_l^{-1}\right)\mathbf{P}^{(D)}\mathbf{T}. \tag{1.79}$$

Finally, substituting (1.79) in the first equation (1.71), we obtain the desired system of linear equations for the coefficients **R** and **T** in the form of:

$$\mathbf{D} + \mathbf{P}^{(U)}\mathbf{R} = \left(\prod_{l=1}^{L}\mathbf{N}_l\mathbf{M}_l^{-1}\right)\mathbf{P}^{(D)}\mathbf{T}. \tag{1.80}$$

Note that writing the system in this form allows to calculate the vectors **R** and **T**, but the calculation directly from (1.80) can lead to a numerical instability of the problem [3]. There are several numerically stable approaches [3, 18] to calculating the vectors **R** and **T**. We consider the so-called method of the scattering matrix [18].

1.1.2.6. The scattering matrix algorithm

Consider the Rayleigh expansion (1.20), (1.21) of the field above the structure, but instead of one incident wave (corresponding to the zeroth diffraction order), we consider a set of incident waves (respective orders $-N,...,N$). Then, the propagation constant can be described by the following expressions:

$$k_{x,m} = k_{x,0} + k_0 \cdot m\frac{\lambda}{d}, \quad m = -N,...,N; \tag{1.81}$$

$$k_{z,m}^{R} = \sqrt{k_0^2\varepsilon_U - k_{x,m}^2}; \tag{1.82}$$

$$k_{z,m}^{I} = -\sqrt{k_0^2\varepsilon_U - k_{x,m}^2}, \tag{1.83}$$

where $k_{z,m}^{R}$ are the propagation constants corresponding to the reflected orders; $k_{z,m}^{I}$ are the propagation constants of the waves incident on the structure from above.

Also, consider a set of waves incident on the structure from the substrate and write similar relations for the orders under the structure:

$$k_{z,m}^{T} = -\sqrt{k_0^2\varepsilon_D - k_{x,m}^2}; \tag{1.84}$$

$$k_{z,m}^{J} = \sqrt{k_0^2\varepsilon_D - k_{x,m}^2}, \tag{1.85}$$

where $k_{z,m}^T$ are the propagation constants corresponding to the transmitted orders; $k_{z,m}^J$ are the propagation constants of the waves incident on the structure from the substrate side.

In this case, when in addition to the $2N +1$ reflected and $2N +1$ transmitted diffraction orders there are also $2N +1$ waves incident from above and $2N +1$ waves incident from below, the calculation of complex amplitudes of diffraction orders consists in the determination of the S-matrix satisfying the equation

$$\begin{bmatrix} \mathbf{T} \\ \mathbf{R} \end{bmatrix} = \mathbf{S} \begin{bmatrix} \mathbf{I}_1 \\ \mathbf{J}_{L+1} \end{bmatrix}, \tag{1.86}$$

where $\mathbf{R}$ and $\mathbf{T}$ are the vectors of the complex amplitudes of the reflected and transmitted diffraction orders, and $\mathbf{I}_1$ and $\mathbf{J}_{L+1}$ are the vectors of the complex amplitudes of the waves incident on the structure from the top and the bottom, respectively. The matrix $\mathbf{S}$ in (1.86) is called the *scattering matrix*. The scattering matrix $\mathbf{S}$ is completely determined by the geometry of the structure, the optical properties of materials and the parameters of the incident radiation.

The difference between $\mathbf{S}$ and the matrix of the system in the previously obtained expression (1.80) is that now it is necessary to consider not one incident wave but a set of waves incident on the structure from both the top and the substrate side. For the incident waves instead of (1.73) we will use the following expression at $p = \mathrm{U}$:

$$\mathbf{D}^{(p)} = \begin{bmatrix} \cos\Theta^{(p)}\sin\Phi & \cos\Phi \\ \cos\Theta^{(p)}\cos\Phi & -\sin\Phi \\ -in_p\cos\Phi & in_p\cos\Theta^{(p)}\sin\Phi \\ in_p\sin\Phi & in_p\cos\Theta^{(p)}\cos\Phi \end{bmatrix}. \tag{1.87}$$

When $p = \mathrm{D}$ the expression (1.87) describes the Fourier coefficients of the tangential components of the fields $\mathbf{E}$ and $\mathbf{H}$ corresponding to the waves incident on the structure from the 'bottom'.

The system (1.71) will now take the form:

$$\begin{cases} \mathbf{D}^{(U)}\mathbf{I}_1 + \mathbf{P}^{(U)}\mathbf{R} = \mathbf{N}_1\mathbf{C}_1, \\ \mathbf{M}_{l-1}\mathbf{C}_{l-1} = \mathbf{N}_l\mathbf{C}_l, \qquad l = 2,\dots,L, \\ \mathbf{M}_L\mathbf{C}_L = \mathbf{D}^{(D)}\mathbf{J}_{L+1} + \mathbf{P}^{(D)}\mathbf{T}, \end{cases} \qquad (1.88)$$

We introduce the notation

$$\begin{aligned} \mathbf{I}_l &= \mathbf{X}_{l-1}^{(+)}\mathbf{C}_{l-1}^{(+)}; & \mathbf{J}_l &= \mathbf{X}_l^{(-)}\mathbf{C}_l^{(-)}; \\ \mathbf{R}_l &= \mathbf{C}_{l-1}^{(-)}; & \mathbf{T}_l &= \mathbf{C}_l^{(+)}, \end{aligned} \qquad (1.89)$$

having the meaning of the complex amplitudes of the incident and scattered orders in each layer (see Fig. 1.2). Furthermore, to write system (1.88) in the most compact form, we introduce the notation

$$\begin{aligned} \mathbf{C}_0^{(-)} &= \mathbf{R}_1 = \mathbf{R}; & \mathbf{C}_{L+1}^{(+)} &= \mathbf{T}_{L+1} = \mathbf{T}; \\ \mathbf{X}_0^{(+)}\mathbf{C}_0^{(+)} &= \mathbf{I}_1; & \mathbf{X}_{L+1}^{(-)}\mathbf{C}_{L+1}^{(-)} &= \mathbf{J}_{L+1}; \\ \mathbf{W}_0^{(-)} &= \mathbf{P}^{(U)}; & \mathbf{W}_{L+1}^{(-)} &= \mathbf{D}^{(D)}; \\ \mathbf{W}_0^{(+)} &= \mathbf{D}^{(U)}; & \mathbf{W}_{L+1}^{(+)} &= \mathbf{P}^{(D)}. \end{aligned} \qquad (1.90)$$

Then the system (1.88) can be written as

$$\left\{ \mathbf{W}_{l-1}^{(-)}\mathbf{R}_l + \mathbf{W}_{l-1}^{(+)}\mathbf{I}_l = \mathbf{W}_l^{(-)}\mathbf{J}_l + \mathbf{W}_l^{(+)}\mathbf{T}_l, \qquad l = 1,\dots,L+1. \qquad (1.91)\right.$$

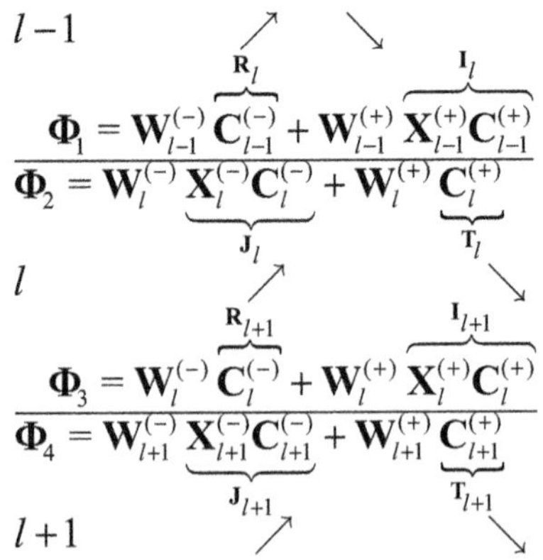

$l-1$

$$\mathbf{\Phi}_1 = \mathbf{W}_{l-1}^{(-)}\overbrace{\mathbf{C}_{l-1}^{(-)}}^{\mathbf{R}_l} + \mathbf{W}_{l-1}^{(+)}\overbrace{\mathbf{X}_{l-1}^{(+)}\mathbf{C}_{l-1}^{(+)}}^{\mathbf{I}_l}$$
$$\mathbf{\Phi}_2 = \mathbf{W}_l^{(-)}\underbrace{\mathbf{X}_l^{(-)}\mathbf{C}_l^{(-)}}_{\mathbf{J}_l} + \mathbf{W}_l^{(+)}\underbrace{\mathbf{C}_l^{(+)}}_{\mathbf{T}_l}$$

l

$$\mathbf{\Phi}_3 = \mathbf{W}_l^{(-)}\overbrace{\mathbf{C}_l^{(-)}}^{\mathbf{R}_{l+1}} + \mathbf{W}_l^{(+)}\overbrace{\mathbf{X}_l^{(+)}\mathbf{C}_l^{(+)}}^{\mathbf{I}_{l+1}}$$
$$\mathbf{\Phi}_4 = \mathbf{W}_{l+1}^{(-)}\underbrace{\mathbf{X}_{l+1}^{(-)}\mathbf{C}_{l+1}^{(-)}}_{\mathbf{J}_{l+1}} + \mathbf{W}_{l+1}^{(+)}\underbrace{\mathbf{C}_{l+1}^{(+)}}_{\mathbf{T}_{l+1}}$$

$l+1$

Fig. 1.2. Representation of fields at the bottom border of the $l-1$-th layer ($\mathbf{\Phi}_1$), at the upper ($\mathbf{\Phi}_2$) and lower ($\mathbf{\Phi}_3$) borders of the l-th layer and the upper border of the $l+1$-th layer ($\mathbf{\Phi}_4$).

We consider the numerically stable method for finding the scattering matrix of the multilayer structure [18]. We introduce the notation $\mathbf{S}^{(l)}$ for the scattering matrix, which connects the Fourier components of the field at the lower boundary of the structure and the lower boundary of the l-th layer

$$\begin{bmatrix} \mathbf{S}_{1,1}^{(l)} & \mathbf{S}_{1,2}^{(l)} \\ \mathbf{S}_{2,1}^{(l)} & \mathbf{S}_{2,2}^{(l)} \end{bmatrix} \begin{bmatrix} \mathbf{I}_{l+1} \\ \mathbf{J}_{l+1} \end{bmatrix} = \begin{bmatrix} \mathbf{T}_{L+1} \\ \mathbf{R}_{l+1} \end{bmatrix} \tag{1.92}$$

Also, we introduce the scattering matrix $\tilde{\mathbf{S}}^{(l)}$ linking the Fourier components of the field at the lower boundary of the structure and the upper boundary of lth layer:

$$\begin{bmatrix} \tilde{\mathbf{S}}_{1,1}^{(l)} & \tilde{\mathbf{S}}_{1,2}^{(l)} \\ \tilde{\mathbf{S}}_{2,1}^{(l)} & \tilde{\mathbf{S}}_{2,2}^{(l)} \end{bmatrix} \begin{bmatrix} \mathbf{T}_{l} \\ \mathbf{J}_{l+1} \end{bmatrix} = \begin{bmatrix} \mathbf{T}_{L+1} \\ \mathbf{J}_{l} \end{bmatrix}. \tag{1.93}$$

Consider the sequential method of constructing the scattering matrix structure, starting from the matrix $\tilde{\mathbf{S}}^{(L+1)} = \mathbf{I}$, where $\mathbf{I}$ is the identity matrix. We consistently find the matrices $\mathbf{S}^{(l)}$ and $\tilde{\mathbf{S}}^{(l)}$ in the following sequence:

$$\mathbf{I} = \tilde{\mathbf{S}}^{(L+1)} \rightarrow \mathbf{S}^{(L)} \rightarrow \tilde{\mathbf{S}}^{(L)} \rightarrow \mathbf{S}^{(L-1)} \rightarrow \cdots \rightarrow \tilde{\mathbf{S}}^{(1)} \rightarrow \mathbf{S}^{(0)} = \mathbf{S}. \tag{1.94}$$

Equalities (1.89) make it easy to find the matrix $\tilde{\mathbf{S}}^{(l)}$, knowing the matrix $\mathbf{S}^{(l)}$:

$$\tilde{\mathbf{S}}^{(l)} = \begin{bmatrix} \mathbf{I} & \mathbf{0} \\ \mathbf{0} & \mathbf{X}_{l}^{(-)} \end{bmatrix} \mathbf{S}^{(l)} \begin{bmatrix} \mathbf{X}_{l}^{(+)} & \mathbf{0} \\ \mathbf{0} & \mathbf{I} \end{bmatrix}. \tag{1.95}$$

We find the numerically stable equations for calculating the matrix $\mathbf{S}^{(l-1)}$ on the basis of the matrix $\tilde{\mathbf{S}}^{(l)}$. To do this, we write the equation of the system (1.91) with the number $l+1$ in the following matrix form:

$$\begin{bmatrix} \mathbf{W}_{l}^{(+)} & -\mathbf{W}_{l+1}^{(-)} \end{bmatrix} \begin{bmatrix} \mathbf{I}_{l+1} \\ \mathbf{J}_{l+1} \end{bmatrix} = \begin{bmatrix} \mathbf{W}_{l+1}^{(+)} & -\mathbf{W}_{l}^{(-)} \end{bmatrix} \begin{bmatrix} \mathbf{T}_{l+1} \\ \mathbf{R}_{l+1} \end{bmatrix} \tag{1.96}$$

We introduce the notation $\mathbf{H}^{(l)}$ by rewriting the expression (1.96) as follows:

$$\underbrace{\left[\mathbf{W}_{l+1}^{(+)} \quad -\mathbf{W}_l^{(-)} \right]^{-1} \left[\mathbf{W}_l^{(+)} \quad -\mathbf{W}_{l+1}^{(-)} \right]}_{\mathbf{H}^{(l)}} \left[\begin{matrix} \mathbf{I}_{l+1} \\ \mathbf{J}_{l+1} \end{matrix} \right] = \left[\begin{matrix} \mathbf{T}_{l+1} \\ \mathbf{R}_{l+1} \end{matrix} \right] \qquad (1.97)$$

Matrix $\mathbf{H}^{(l)}$ has the meaning of the scattering matrix linking the field at the lower boundary of the l-th layer and the upper boundary of the $l+1$-th layer. On the basis of expressions (1.92), (1.93), (1.97) it can be shown that the matrix $\mathbf{S}^{(l)}$ is calculated as

$$\mathbf{S}^{(l)} = \mathbf{H}^{(l)} \otimes \tilde{\mathbf{S}}^{(l+1)}, \qquad (1.98)$$

where the associative operation "$\otimes$" is defined as follows [18]:

$$\left[\begin{matrix} \mathbf{H}_{1,1} & \mathbf{H}_{1,2} \\ \mathbf{H}_{2,1} & \mathbf{H}_{2,2} \end{matrix} \right] \otimes \left[\begin{matrix} \mathbf{S}_{1,1} & \mathbf{S}_{1,2} \\ \mathbf{S}_{2,1} & \mathbf{S}_{2,2} \end{matrix} \right] =$$

$$= \left[\begin{matrix} \mathbf{S}_{1,1}(\mathbf{I} - \mathbf{H}_{1,2}\mathbf{S}_{2,1})^{-1}\mathbf{H}_{1,1} & \mathbf{S}_{1,2} + \mathbf{S}_{1,1}\mathbf{H}_{1,2}(\mathbf{I} - \mathbf{S}_{2,1}\mathbf{H}_{1,2})^{-1}\mathbf{S}_{2,2} \\ \mathbf{H}_{2,1} + \mathbf{H}_{2,2}\mathbf{S}_{2,1}(\mathbf{I} - \mathbf{H}_{1,2}\mathbf{S}_{2,1})^{-1}\mathbf{H}_{1,1} & \mathbf{H}_{2,2}(\mathbf{I} - \mathbf{S}_{2,1}\mathbf{H}_{1,2})^{-1}\mathbf{S}_{2,2} \end{matrix} \right].$$

$$(1.99)$$

Note that in (1.95), (1.97), (1.99) there is no inversion of the ill-conditioned matrices, therefore, this method allows to find the scattering matrix of the structure $\mathbf{S}$ and to avoid problems with numerical stability.

If a single plane wave of unit intensity is incident on the structure from above, complex transmission $\mathbf{T}$ and reflection $\mathbf{R}$ coefficients are found from (1.86):

$$\left[\begin{matrix} \mathbf{T} \\ \mathbf{R} \end{matrix} \right] = \mathbf{S} \left[\begin{matrix} 1 \cdot \delta_i \\ 0 \cdot \delta_i \end{matrix} \right]. \qquad (1.100)$$

1.1.2.7. Calculation of the field distribution
In the calculation of the field distribution in the structure there may be the same problems with numerical stability as in the calculation of the complex amplitudes of diffraction orders. To construct the distribution of the field on the basis of expressions (1.27) and (1.58), it is necessary to know the values of the vectors $\mathbf{C}^{(-)}$, $\mathbf{C}^{(+)}$. Consider an iterative procedure for numerically-stable calculation of these vectors [19].

We assume that the vector $\mathbf{R}$ of the complex amplitudes of the reflected orders and matrices $\mathbf{S}^{(l)}$, $\tilde{\mathbf{S}}^{(l)}$ is already calculated by the method described in the previous section.

We construct a recursive procedure for computing $C_l^{(+)}$. From (1.97) at $l = 0$ we obtain the expression $C_1^{(+)}$ in the form:

$$C_1^{(+)} = T_1 = H_{1,1}^{(0)}I_1 + H_{1,2}^{(0)}J_1 = H_{1,1}^{(0)}I_1 + H_{1,2}^{(0)}\left(H_{2,2}^{(0)}\right)^{-1}\left(R - H_{2,1}^{(0)}I_1\right).$$

(1.101)

Now, assuming that $C_l^{(+)}$ is known we find expression $C_{l+1}^{(+)} = T_{l+1}$. From equation (1.93), taking into account $J_{L+1} = 0$, we obtain

$$T_{l+1} = C_{l+1}^{(+)} = \left(\tilde{S}_{1,1}^{(l+1)}\right)^{-1}T_{L+1} = \left(\tilde{S}_{1,1}^{(l+1)}\right)^{-1}T.$$

(1.102)

We write the matrix $S_{1,1}^{(l)}$ equation of (1.98) (1.99) in the form:

$$S_{1,1}^{(l)} = \tilde{S}_{1,1}^{(l+1)}\left(I - H_{1,2}^{(l)}\tilde{S}_{2,1}^{(l+1)}\right)^{-1}H_{1,1}^{(l)}$$

(1.103)

We express the matrix $\tilde{S}_{1,1}^{(l)}$ from formula (1.95):

$$\tilde{S}_{1,1}^{(l)} = S_{1,1}^{(l)}X_l^{(+)}.$$

(1.104)

We express from equations (1.103) the matrix $\left(\tilde{S}_{1,1}^{(l+1)}\right)^{-1}$, and using the equation (1.104) we exclude matrix $S_{1,1}^{(l)}$:

$$\left(\tilde{S}_{1,1}^{(l+1)}\right)^{-1} = \left(I - H_{1,2}^{(l)}\tilde{S}_{2,1}^{(l+1)}\right)^{-1}H_{1,1}^{(l)}\left(S_{1,1}^{(l)}\right)^{-1} = \left(I - H_{1,2}^{(l)}\tilde{S}_{2,1}^{(l+1)}\right)^{-1}H_{1,1}^{(l)}X_l^{(+)}\left(\tilde{S}_{1,1}^{(l)}\right)^{-1}.$$

(1.105)

Multiplying the left and right hand sides of (1.105) on the right by T, with (1.102) taken into account, we finally obtain:

$$C_{l+1}^{(+)} = T_{l+1} = \left(I - H_{1,2}^{(l)}\tilde{S}_{2,1}^{(l+1)}\right)^{-1}H_{1,1}^{(l)}X_l^{(+)}C_l^{(+)}.$$

(1.106)

The recurrence relation (1.106) with the formula (1.101) allows to find $C_l^{(+)}$ for all l. We now obtain a formula for finding $C_l^{(-)} = R_{l+1}$. From formula (1.92) at $J_{L+1} = 0$ we have:

$$C_l^{(-)} = R_{l+1} = S_{2,1}^{(l)}I_{l+1} = S_{2,1}^{(l)}X_l^{(+)}C_l^{(+)}.$$

(1.107)

1.1.2.8. Intensities of the diffraction orders
In the analysis of the field away from the grating researchers are usually not interested in complex amplitudes (1.72) and pay attention to the intensities of the reflected and transmitted propagating

diffraction orders. The propagating orders are determined by the real values $k_{z,p,m}$ in (1.22). The intensities of the diffraction orders are defined as the flux of the Umov–Poynting vector through the plane z = const, normalized to the correspondng flux of the incident wave [15]. Taking into account the expressions (1.16) and (1.17) the intensity of the orders can be found from the following expressions:

$$\mathbf{I}^R = \left| \frac{n_U \operatorname{Re}\left(\cos\Theta^{(U)}\right)}{n_U \cos\theta} \right| \left(|\mathbf{R}_E|^2 + |\mathbf{R}_H|^2 \right); \qquad (1.108)$$

$$\mathbf{I}^T = \left| \frac{n_D \operatorname{Re}\left(\cos\Theta^{(D)}\right)}{n_U \cos\theta} \right| \left(|\mathbf{T}_E|^2 + |\mathbf{T}_H|^2 \right), \qquad (1.109)$$

where the diagonal matrices $\cos\Theta^{(U)}$, $\cos\Theta^{(D)}$ are defined in (1.76), and squaring of vectors $\mathbf{R}_E$, $\mathbf{R}_H$, $\mathbf{T}_E$, $\mathbf{T}_H$ is performed element wise. For evanescent diffraction orders $\operatorname{Re}(\cos\Theta^{(p)}) = 0$, p = U, D, therefore their intensities are equal to zero.

In general, propagating diffraction orders are plane waves with elliptical polarization. Indeed, each diffraction order corresponds to the superposition of the E- and H-waves. Let E_E, E_H be the complex amplitude of the electric field at the E- and H-waves. Note that the electric field vectors in the E- and H-waves are perpendicular to each other and perpendicular to the direction of wave propagation. The addition of perpendicular oscillations results in the formation of an elliptically polarized wave. The polarization ellipse is characterized by two parameters: the angle ϕ of the major axis of the polarization ellipse and the ellipticity parameter χ [20]. The ellipticity parameter characterizes the ratio of the lengths of the a, b axes of the polarization ellipse in the form $\operatorname{tg}\chi = a/b$. The parameters ϕ and χ are determined through complex amplitudes E_E, E_H as [20]

$$tg(2\phi) = \frac{2\operatorname{Re}\left(E_E/E_H\right)}{1 - |E_E/E_H|^2},$$

$$\sin(2\chi) = \frac{2\operatorname{Im}\left(E_E/E_H\right)}{1 + |E_E/E_H|^2}. \qquad (1.110)$$

In conclusion, let us make a few remarks about the choice of the parameter N that determines the length of the segments of the Fourier series approximating the components of the electric and magnetic fields in the grating. For a given N the number of calculated orders is equal to $2N+1$, from $-N$ to $+N$. The parameter N must be greater than the number of propagating orders. If the grating consists only of dielectrics (all refractive indices are real numbers), and the parameter N satisfies this condition, the energy conservation law in the following form should be satisfied:

$$\sum I^R + \sum I^T = 1 \tag{1.111}$$

If the grating contains absorbing materials, the sum (1.111) should be less than one. In general, N is selected in computational experiment on the basis of the condition of stabilizing the intensities of the orders.

1.1.2.9. Numerical example
Consider the example of calculation of the intensities of diffraction orders of a binary grating. The grating geometry is shown in Fig. 1.3. The grating parameters are shown in the Fig. caption.

In the calculations the parameter N that determines the length of the segments of the Fourier series was $N = 20$. The calculations were carried out for the TM-polarized normally incident wave. The results of calculation of the reflection and transmission spectra of the grating (the intensities of diffraction orders) are shown in Fig. 1.4.

1.1.3. The Fourier modal method for three-dimensional periodic structures

Consider the described method in the case of three-dimensional periodic diffraction structures. The z-axis is perpendicular to the plane in which the diffraction grating is positioned. The functions of permittivity and magnetic permeability in the grating region are assumed to be periodic with respect to the variables x, y with periods d_x and d_y respectively. As in the two-dimensional case, we assume

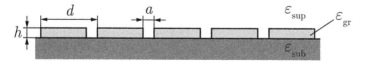

Fig. 1.3. The geometry of the binary grating (Parameters: $d = 1000$ nm, $a = 200$ nm, $h = 200$ nm, $\varepsilon_{sup} = 1$, $\varepsilon_{gr} = 4$, $\varepsilon_{sub} = 2.25$).

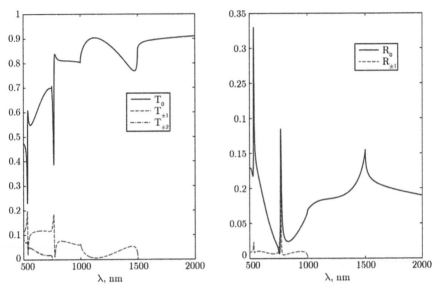

Fig. 1.4. The intensities of the reflected and transmitted diffraction orders.

that the diffraction grating is made up of L binary layers and the permittivity and permeability in each layer are independent of the variable z.

The method of solving the problem of diffraction in the three-dimensional case is similar to the two-dimensional case considered previously. The main details of the three-dimensional problem are given below.

Upon the plane wave diffraction on a three-dimensional diffraction grating, a set of reflected and transmitted diffraction orders is formed. In this case the field above and below the structure is as follows:

$$\Phi^U(x,y,z) = \Phi^{\text{inc}}(x,y,z) + \sum_n \sum_m \Phi^R_{n,m}(R_{n,m})\exp\left(i(k_{x,n}x + k_{y,m}y + k_{z,U,n,m}z)\right),$$

$$(1.112)$$

$$\Phi^D(x,y,z) = \sum_n \sum_m \Phi^T_{n,m}(T_{n,m})\exp\left(i\left(k_{x,n}x + k_{y,m}y - k_{z,D,n,m}(z - z_L)\right)\right),$$

$$(1.113)$$

where $\Phi^{\text{inc}}(x,y,z)$ is the incident wave. The incident wave is assumed to be given in the form (1.19). The propagation constants of the diffraction orders with numbers (n,m) are described as follows:

$$k_{x,n} = k_0 \left(n_U \sin\theta \cos\varphi + n\frac{\lambda}{d_x} \right),$$

$$k_{y,m} = k_0 \left(n_U \sin\theta \sin\varphi + m\frac{\lambda}{d_y} \right), \qquad (1.114)$$

$$k_{z,p,n,m} = \sqrt{\left(k_0 n_p\right)^2 - k_{x,n}^2 - k_{y,m}^2},$$

where as p, 'U' is taken for the reflected orders (the fields above the grating) and 'D' – for the transmitted ones (the field below the grating). The form of the propagation constants ensures that the two-dimensional quasi-periodicity condition is fulfilled:

$$\Phi^p(x+d_x, y+d_y, z) = \Phi^p(x,y,z)\exp\left(ik_{x,0}d_x + ik_{y,0}d_y\right), \quad p = U, D. \qquad (1.115)$$

According to (1.115), the amplitude of the field does not change upon a shift along the x and y axes by integer multiples of the corresponding periods. The waves with the real $k_{z,p,n,m}$ are propagating, those with the imaginary value are evanescent.

The electromagnetic field in each layer, as in the two-dimensional case, is described by the basic Maxwell's equations for the monochromatic field in the form (1.24)–(1.27). We represent the components of the electric and magnetic fields in the form of a two-dimensional Fourier series in the variables x, y:

$$\begin{cases}
E_x &= \sum_n\sum_m S_{x,n,m}(z)\exp\left(i\left(k_{x,n}x + k_{y,m}y\right)\right), \\[4pt]
E_y &= \sum_n\sum_m S_{y,n,m}(z)\exp\left(i\left(k_{x,n}x + k_{y,m}y\right)\right), \\[4pt]
E_z &= \sum_n\sum_m S_{z,n,m}(z)\exp\left(i\left(k_{x,n}x + k_{y,m}y\right)\right), \\[4pt]
H_x &= -i\sum_n\sum_m U_{x,n,m}(z)\exp\left(i\left(k_{x,n}x + k_{y,m}y\right)\right), \\[4pt]
H_y &= -i\sum_n\sum_m U_{y,n,m}(z)\exp\left(i\left(k_{x,n}x + k_{y,m}y\right)\right), \\[4pt]
H_z &= -i\sum_n\sum_m U_{z,n,m}(z)\exp\left(i\left(k_{x,n}x + k_{y,m}y\right)\right).
\end{cases} \qquad (1.116)$$

The equations (1.116) are written with the quasi-periodicity of the field components in the variables x, y taken into account. We restrict ourselves to a finite number of terms in the expansions

(1.116), corresponding to $-N_x \le n \le N_x$, $-N_y \le m \le N_y$. Substituting the expansion (1.116) into (1.26) and equating the coefficients of the same Fourier harmonics, we obtain a system of differential equations in the form

$$
\begin{cases}
ik_0 \mathbf{K}_y \mathbf{S}_z - \dfrac{d\mathbf{S}_y}{dz} & = k_0 (\mathbf{M}_{1,1}\mathbf{U}_x + \mathbf{M}_{1,2}\mathbf{U}_y + \mathbf{M}_{1,3}\mathbf{U}_z), \\[2mm]
\dfrac{d\mathbf{S}_x}{dz} - ik_0 \mathbf{K}_x \mathbf{S}_z & = k_0 (\mathbf{M}_{2,1}\mathbf{U}_x + \mathbf{M}_{2,2}\mathbf{U}_y + \mathbf{M}_{2,3}\mathbf{U}_z), \\[2mm]
ik_0 \mathbf{K}_x \mathbf{S}_y - ik_0 \mathbf{K}_y \mathbf{S}_x & = k_0 (\mathbf{M}_{3,1}\mathbf{U}_x + \mathbf{M}_{3,2}\mathbf{U}_y + \mathbf{M}_{3,3}\mathbf{U}_z), \\[2mm]
ik_0 \mathbf{K}_y \mathbf{U}_z - \dfrac{d\mathbf{U}_y}{dz} & = k_0 (\mathbf{E}_{1,1}\mathbf{S}_x + \mathbf{E}_{1,2}\mathbf{S}_y + \mathbf{E}_{1,3}\mathbf{S}_z), \\[2mm]
\dfrac{d\mathbf{U}_x}{dz} - ik_0 \mathbf{K}_x \mathbf{U}_z & = k_0 (\mathbf{E}_{2,1}\mathbf{S}_x + \mathbf{E}_{2,2}\mathbf{S}_y + \mathbf{E}_{2,3}\mathbf{S}_z), \\[2mm]
ik_0 \mathbf{K}_x \mathbf{U}_y - ik_0 \mathbf{K}_y \mathbf{U}_x & = k_0 (\mathbf{E}_{3,1}\mathbf{S}_x + \mathbf{E}_{3,2}\mathbf{S}_y + \mathbf{E}_{3,3}\mathbf{S}_z).
\end{cases}
\tag{1.117}
$$

The form of the resulting system is the same as that of the system (1.31) for the two-dimensional case. The difference is in the details of the representation of vectors and matrices in the system. The vectors $\mathbf{S}_x$, $\mathbf{S}_y$, $\mathbf{S}_z$, $\mathbf{U}_x$, $\mathbf{U}_y$, $\mathbf{U}_z$ in (1.117) are a row-wise one-dimensional representation of the matrices $S_{x,j,k}$, $S_{y,j,k}$, $S_{z,j,k}$, $U_{x,j,k}$, $U_{y,j,k}$, $U_{z,j,k}$, $-N_x \le j \le N_x$, $-N_y \le k \le N_y$. This means that the element of the vector $\mathbf{S}_x$ with the number

$$
l(i,j) = i(2N_y + 1) + j.
\tag{1.118}
$$

corresponds to the value $S_{x,i,j}$. The vectors introduced in this manner have the dimension $(2N_x + 1)(2N_y + 1)$ equal to the total number of the calculated diffraction orders. For example, the vector $\mathbf{S}_x$ has the form

$$
\mathbf{S}_x = \left[S_{x,-N_x,-N_y}, S_{x,-N_x,1-N_y}, \ldots, S_{x,-N_x,N_y}, S_{x,1-N_x,-N_y}, S_{x,1-N_x,1-N_y}, \ldots, S_{x,N_x,N_y} \right]^T
\tag{1.119}
$$

The matrices $\mathbf{K}_x$, $\mathbf{K}_y$, $\mathbf{E}_{i,j}$, $\mathbf{M}_{i,j}$ in (1.117) have the dimensions $(2N_x + 1)(2N_y + 1) \times (2N_x + 1)(2N_y + 1)$. The matrices $\mathbf{K}_x$ and $\mathbf{K}_y$ are defined by the following expressions:

$$
\begin{aligned}
K_{x,l(i,j),l(n,m)} &= k_{x,i}\delta_{i-n}\delta_{j-m}/k_0, \\
K_{y,l(i,j),l(n,m)} &= k_{y,j}\delta_{i-n}\delta_{j-m}/k_0,
\end{aligned}
\tag{1.120}
$$

where $-N_x \le i, n \le N_x$, $-N_y \le j, m \le N_y$.

Consider the preliminary form of the matrices $\mathbf{E}_{i,j}$, $\mathbf{M}_{i,j}$ obtained by using the direct Laurent rules for the expansion of the product of the function into a Fourier series. The matrices $\mathbf{E}_{i,j}$, $\mathbf{M}_{i,j}$ consist of the Fourier coefficients of the permittivity and permeability tensors, the structure of the matrices is the same and has the form

$$T_{l(i,j),l(n,m)} = e_{i-n,j-m},\qquad (1.121)$$

where $e_{i,j}$ are the Fourier coefficients $-N_x \le i, n \le N_x$, $-N_y \le j, m \le N_y$.

Since the form of the systems of differential equations in the two-dimensional and three-dimensional cases is identical, all subsequent changes are also identical. The system of differential equations for the vectors $\mathbf{S}_x$, $\mathbf{S}_y$, $\mathbf{U}_x$, $\mathbf{U}_y$, in (1.117) also has the form (1.35)–(1.34). In particular, for a grating of an isotropic material $\varepsilon = \varepsilon(x,y)$ and $\mu = 1$ are the scalars, and the matrix of the differential equation system (1.35) has the form

$$\mathbf{A} = -\begin{bmatrix} \mathbf{0} & \mathbf{0} & \mathbf{K}_y\mathbf{E}^{-1}\mathbf{K}_x & \mathbf{I}-\mathbf{K}_y\mathbf{E}^{-1}\mathbf{K}_y \\ \mathbf{0} & \mathbf{0} & \mathbf{K}_x\mathbf{E}^{-1}\mathbf{K}_x - \mathbf{I} & -\mathbf{K}_x\mathbf{E}^{-1}\mathbf{K}_y \\ \mathbf{K}_y\mathbf{K}_x & \mathbf{E}-\mathbf{K}_y^2 & \mathbf{0} & \mathbf{0} \\ \mathbf{K}_x^2-\mathbf{E} & -\mathbf{K}_x\mathbf{K}_y & \mathbf{0} & \mathbf{0} \end{bmatrix},\ (1.122)$$

where the matrix $\mathbf{E}$ has the form (1.120) and is composed of the Fourier coefficients of functions $\varepsilon(x,y)$. The formula (1.122) is obtained from the general expressions (1.36) and (1.34) with

$$\begin{aligned} \mathbf{E}_{1,1} &= \mathbf{E}_{2,2} = \mathbf{E}_{3,3} = \mathbf{E}, \\ \mathbf{E}_{1,2} &= \mathbf{E}_{1,3} = \mathbf{E}_{2,1} = \mathbf{E}_{2,3} = \mathbf{E}_{3,1} = \mathbf{E}_{3,2} = \mathbf{0}, \\ \mathbf{M}_{1,1} &= \mathbf{M}_{2,2} = \mathbf{M}_{3,3} = \mathbf{I}, \\ \mathbf{M}_{1,2} &= \mathbf{M}_{1,3} = \mathbf{M}_{2,1} = \mathbf{M}_{2,3} = \mathbf{M}_{3,1} = \mathbf{M}_{3,2} = \mathbf{0}. \end{aligned} \qquad (1.123)$$

Consider the transition to the spatial–frequency representation (1.117) using the correct rules of the Fourier expansion of the product of the functions. Derivation of the formulas is presented for the case of an isotropic material. In this case the system of Maxwell's equations (1.26) contains only the following three products: εE_z, εE_x, εE_y.

The tangential component E_z is continuous, so the product εE_z is expanded into a Fourier series using the Laurent rules (1.40). In this case, the corresponding matrix $E_{3,3}$ (1.117) has the form (1.121).

We assume that the boundary between media with different dielectric constants in each layer is parallel to the axes [6]. Consider the product $D_x = \varepsilon E_x$. The product D_x is continuous at the sections of the boundaries between two media, parallel to the axis Oy. Indeed, in these areas $D_x = \varepsilon E_x$ is a normal component of the electric displacement. The component E_x and the function of the dielectric constant ε are discontinuous at these borders. Thus, the product $D_x = \varepsilon E_x$ is continuous with respect to x for any fixed y.

Accordingly, for the expansion of $D_x = \varepsilon E_x$ into the Fourier series in the variable x we use the inverse Laurent rule (1.41):

$$d_{x,i}(y,z) = \sum_n \varepsilon_{x\,i,n}(y) S_{x,n}(y,z), \tag{1.124}$$

where $d_{x,i}(y,z)$, $S_{x,n}(y,z)$ are the Fourier coefficients of the functions $D_x = \varepsilon E_x$, and $\varepsilon_{x\,i,n}(y)$ are the elements of the Toeplitz matrix $[\![1/\varepsilon_x]\!]^{-1}$ formed from the Fourier coefficients with respect to the variable x of the function $1/\varepsilon(x,y)$. In the sections of the border between the media parallel to the axis Ox, component E_x is tangential and therefore continuous in the variable y. The Fourier coefficients $S_{x,n}(y,z)$ of the function E_x will also be continuous. Therefore, the expansion in (1.124) of the terms $\varepsilon_{x\,i,n}(y)\,S_{x,n}(y,z)$ into the Fourier series in the variable y is performed using the Laurent rule (1.40):

$$d_{x,i,j}(z) = \sum_{n,m} \varepsilon_{x\,i,n,j-m} S_{x,n,m}(z), \tag{1.125}$$

where $\varepsilon_{x\,i,n,\,j-m}$ is the Fourier coefficient with the number $(j-m)$ of the function $\varepsilon_{x\,i,n}(y)$.

Repeating similar reasoning for $D_y = \varepsilon E_y$, we obtain

$$d_{y,j}(x,z) = \sum_m \varepsilon_{y\,j,m}(x) S_{y,m}(x,z), \tag{1.126}$$

$$d_{y,i,j}(z) = \sum_{n,m} \varepsilon_{y\,i-n,\,j,m} S_{y,n,m}(z), \tag{1.127}$$

where $\varepsilon_{y\,j,m}(x)$ are the elements of the Toeplitz matrix $[\![1/\varepsilon_y]\!]^{-1}$ formed from the Fourier coefficients in the variable y of the functions $1/\varepsilon(x,y)$, and $\varepsilon_{y\,i-n,\,j,m}$ is the Fourier coefficient with the number $(i-n)$ of the function $\varepsilon_{y\,j,m}$.

The equations (1.125) and (1.127) were obtained with the use of correct rules of expansion into a Fourier series of the products $D_x = \varepsilon E_x$, $D_y = \varepsilon E_y$. Accordingly, in transition from the system (1.26) to the spatial–frequency representation (1.117), the matrices $\mathbf{E}_{1,1}$, $\mathbf{E}_{2,2}$ will have the form

$$
\begin{aligned}
E_{1,1\,l(i,n),l(j,m)} &= \varepsilon_{x\,i,n,j-m}, \\
E_{2,2\,l(i,n),l(j,m)} &= \varepsilon_{y\,i-n,j,m},
\end{aligned}
\tag{1.128}
$$

where $l(i,j)$ is defined in (1.118), $-N_x \le i,n \le N_x$, $-N_y \le j,m \le N_y$. The matrices (1.128) have the dimensions $(2N_x + 1)(2N_y + 1) \times (2N_x + 1)(2N_y + 1)$.

As a result, the matrix $\mathbf{A}$ of the system of of the linear differential equations takes the form:

$$
\mathbf{A} = -
\begin{bmatrix}
\mathbf{0} & \mathbf{0} & \mathbf{K}_y \mathbf{E}^{-1} \mathbf{K}_x & \mathbf{I} - \mathbf{K}_y \mathbf{E}^{-1} \mathbf{K}_y \\
\mathbf{0} & \mathbf{0} & \mathbf{K}_x \mathbf{E}^{-1} \mathbf{K}_x - \mathbf{I} & -\mathbf{K}_x \mathbf{E}^{-1} \mathbf{K}_y \\
\mathbf{K}_y \mathbf{K}_x & \mathbf{E}_{1,1} - \mathbf{K}_y^2 & \mathbf{0} & \mathbf{0} \\
\mathbf{K}_x^2 - \mathbf{E}_{2,2} & -\mathbf{K}_x \mathbf{K}_y & \mathbf{0} & \mathbf{0}
\end{bmatrix}.
\tag{1.129}
$$

A detailed description of the correct rules of expansion into the Fourier series for the general case of the tensors of permittivity and permeability can be found in [8, 9].

Subsequent operations of the 'stitching' of the solutions at the layer boundaries and the numerically stable implementation of calculation of the matrix of the system of linear equations for the amplitudes of the diffraction orders are also the same. The matrices $\mathbf{E},\mathbf{F}$ in the systems of linear equations, representing the tangential field components (1.112), (1.113) at the upper and lower boundaries of the grating, also have the form of (1.74), where

$$
\sin \mathbf{\Phi} = \operatorname*{diag}_{l(i,j)} \frac{k_{y,j}}{\sqrt{k_{x,i}^2 + k_{y,j}^2}}, \quad
\cos \mathbf{\Phi} = \operatorname*{diag}_{l(i,j)} \frac{k_{x,i}}{\sqrt{k_{x,i}^2 + k_{y,j}^2}},
$$
$$
\cos \mathbf{\Phi}^{(U)} = \operatorname*{diag}_{l(i,j)} \frac{k_{z,U,i,j}}{k_0 n_U}, \quad
\cos \mathbf{\Phi}^{(D)} = \operatorname*{diag}_{l(i,j)} \frac{-k_{z,D,i,j}}{k_0 n_D}.
\tag{1.130}
$$

Here $\operatorname*{diag}_{l(i,j)} f(i,j)$ denotes a diagonal matrix composed of elements $f(i,j)$ arranged in the ascending order of the magnitude of $l(i,j)$.

1.1.4. The Fourier modal method for two-dimensional non-periodic structures

1.1.4.1. The geometry of the structure and formulation of the problem

The Fourier modal method can be adapted to simulate the diffraction of the waveguide and plasmonic modes at inhomogeneities of the waveguide. Consider the method for the case when the geometry of the considered structure is independent of the coordinate y. In addition, we assume that, as before, the structure can be divided into regions and in each region $z_l < z < z_{l-1}$ the material parameters depend only on the coordinate x (Fig. 1.5). To apply the Fourier modal method we introduce the artificial periodization along the coordinate x, while for the elimination of the interaction between adjacent periods (i.e., to ensure vanishing of the electromagnetic field on the borders of the period) special absorbing layers are added to the boundaries of the period. These layers can be represented by multi-layered gradient absorbers or perfectly matched absorbing layers (PML) can be used [21, 22]. The gradient absorbing layers are layers in which the real part of the permittivity equals the permittivity of the adjacent medium, while the imaginary part, which characterizes absorption, increases while approaching the boundary of the period. A disadvantage of the gradient absorbing layers is non-zero reflection of incident radiation on them which increases substantially with increasing incidence angle. In some cases, this fact can substantially affect the solution of the problem with artificial periodization, decreasing the accuracy of the solution with respect to the solution of the original non-periodic problem. Reflections at the boundaries with absorbing layers are eliminated using perfectly matched absorbing layers described in the next section.

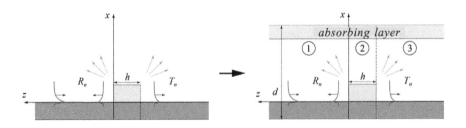

Fig. 1.5. The geometry of the problem of diffraction of surface plasmon polaritons in the aperiodic structure.

Unlike the standard version of the Fourier modal method, intended for the solution of the problem of diffraction on periodic structures, the result of the application of non-periodic modification will not be a set of intensities of diffraction orders but a set of the amplitudes of the reflected and transmitted modes.

1.1.4.2. Perfectly matched absorbing layers

Perfectly matched absorbing layers as anisotropic materials
Consider perfectly matched absorbing layers represented by a layer of anisotropic materials added to the boundaries of the period of the structure. The property of the perfectly matched absorbing layers to absorb incident radiation without back-reflection in this case is achieved by a special choice of the permittivity and permeability tensors.

To determine the type of tensor we consider initially an overview of the field of a plane electromagnetic waves in an anisotropic medium, described by the diagonal tensors of dielectric permittivity and magnetic permeability:

$$[\varepsilon] = \begin{pmatrix} \varepsilon_x & 0 & 0 \\ 0 & \varepsilon_y & 0 \\ 0 & 0 & \varepsilon_z \end{pmatrix}, \ [\mu] = \begin{pmatrix} \mu_x & 0 & 0 \\ 0 & \mu_y & 0 \\ 0 & 0 & \mu_z \end{pmatrix}. \tag{1.131}$$

Since the properties of the medium do not depend on the variable z, the electric and magnetic fields have the form

$$\begin{aligned} \mathbf{E}(x,y,z) &= \mathbf{E}(x,y)\exp(ik_0\gamma z), \\ \mathbf{H}(x,y,z) &= \mathbf{H}(x,y)\exp(ik_0\gamma z). \end{aligned} \tag{1.132}$$

Substituting (1.131) into (1.3), we obtain:

$$\begin{aligned} \partial_y H_z - ik_0\gamma H_y &= -ik_0\varepsilon_x E_x, & \partial_y E_z - ik_0\gamma E_y &= ik_0\mu_x H_x, \\ -\partial_x H_z + ik_0\gamma H_x &= -ik_0\varepsilon_y E_y, & -\partial_x E_z + ik_0\gamma E_x &= ik_0\mu_y H_y, \\ \partial_x H_y - \partial_y H_x &= -ik_0\varepsilon_z E_z, & \partial_x E_y - \partial_y E_x &= ik_0\mu_z H_z, \end{aligned} \tag{1.133}$$

where $\partial_x f = \partial f / \partial x$. Using (1.133), we represent the tangential components by the components E_z, H_z in the form

$$E_x = \frac{-1}{ik_0\left(\varepsilon_x - \gamma^2/\mu_y\right)}\left(\partial_y H_z + \frac{\gamma}{\mu_y}\cdot\partial_x E_z\right),$$

$$E_y = \frac{1}{ik_0\left(\varepsilon_y - \gamma^2/\mu_x\right)}\left(\partial_x H_z - \frac{\gamma}{\mu_x}\cdot\partial_y E_z\right),$$

$$H_x = \frac{1}{ik_0\left(\mu_x - \gamma^2/\varepsilon_y\right)}\left(\partial_y E_z - \frac{\gamma}{\varepsilon_y}\cdot\partial_x H_z\right),$$

$$H_y = \frac{-1}{ik_0\left(\mu_y - \gamma^2/\varepsilon_x\right)}\left(\partial_x E_z + \frac{\gamma}{\varepsilon_x}\cdot\partial_y H_z\right),$$

(1.134)

where E_z, H_z satisfy the equations

$$\frac{1}{\left(\mu_y - \gamma^2/\varepsilon_x\right)}\left(\frac{\partial^2 E_z}{\partial x^2} + \frac{\gamma}{\varepsilon_x}\cdot\frac{\partial^2 H_z}{\partial x\partial y}\right) +$$

$$+ \frac{1}{\left(\mu_x - \gamma^2/\varepsilon_y\right)}\left(\frac{\partial^2 E_z}{\partial y^2} - \frac{\gamma}{\varepsilon_y}\cdot\frac{\partial^2 H_z}{\partial y\partial x}\right) = -k_0^2\varepsilon_z E_z,$$

$$\frac{1}{\left(\varepsilon_y - \gamma^2/\mu_x\right)}\left(\frac{\partial^2 H_z}{\partial x^2} - \frac{\gamma}{\mu_x}\cdot\frac{\partial^2 E_z}{\partial x\partial y}\right) +$$

$$+ \frac{1}{\left(\varepsilon_x - \gamma^2/\mu_y\right)}\left(\frac{\partial^2 H_z}{\partial y^2} + \frac{\gamma}{\mu_y}\cdot\frac{\partial^2 E_z}{\partial y\partial x}\right) = -k_0^2\mu_z H_z.$$

(1.135)

For an uniaxial anisotropic medium of the type

$$[\varepsilon] = \begin{pmatrix} \varepsilon & 0 & 0 \\ 0 & \varepsilon & 0 \\ 0 & 0 & \varepsilon_z \end{pmatrix}, \quad [\mu] = \begin{pmatrix} \mu & 0 & 0 \\ 0 & \mu & 0 \\ 0 & 0 & \mu_z \end{pmatrix}$$

(1.136)

we obtain from (1.135) that E_z, H_z satisfy the Helmholtz equation

$$\frac{\partial^2 E_z}{\partial x^2} + \frac{\partial^2 E_z}{\partial y^2} = -k_0^2\varepsilon_z\left(\mu - \gamma^2/\varepsilon\right)E_z,$$

$$\frac{\partial^2 H_z}{\partial x^2} + \frac{\partial^2 H_z}{\partial y^2} = -k_0^2\mu_z\left(\varepsilon - \gamma^2/\mu\right)H_z.$$

(1.137)

Solving (1.137) by separation of variables, we obtain

$$E_z(x,y,z) = \exp\left(ik_0\sqrt{\varepsilon_z\mu}\,(\alpha x + \beta y) \pm ik_0\gamma z\right),$$
$$H_z(x,y,z) = \exp\left(ik_0\sqrt{\varepsilon_z\mu}\,(\alpha x + \beta y) \pm ik_0\gamma_1 z\right),$$

(1.138)

where

$$\gamma = \sqrt{\varepsilon\mu}\sqrt{1-\alpha^2-\beta^2}, \quad \gamma_1 = \mu\sqrt{\varepsilon/\mu - (\alpha^2+\beta^2)\varepsilon_z/\mu_z}. \quad (1.139)$$

Next, we consider the E- and H-type waves. For the E-wave $E_z \neq 0, H_z = 0$, and for the H-wave $H_z \neq 0, E_z = 0$. From (1.134) (1.138) for the E-waves we get

$$E_x = \frac{\mp\gamma}{ik_0(\varepsilon\mu - \gamma^2)} \cdot \partial_x E_z = \frac{\mp\alpha\gamma\sqrt{\varepsilon_z\mu}}{(\mu\varepsilon - \gamma^2)} \cdot \exp\left(ik_0\sqrt{\varepsilon_z\mu}\,(\alpha x + \beta y) \pm ik_0\gamma z\right),$$

$$E_y = \frac{\mp\gamma}{ik_0(\varepsilon\mu - \gamma^2)} \cdot \partial_y E_z = \frac{\mp\gamma\beta\sqrt{\varepsilon_z\mu}}{(\varepsilon\mu - \gamma^2)} \cdot \exp\left(ik_0\sqrt{\varepsilon_z\mu}\,(\alpha x + \beta y) \pm ik_0\gamma z\right),$$

$$H_x = \frac{\varepsilon}{ik_0(\mu\varepsilon - \gamma^2)}\partial_y E_z = \frac{\varepsilon\beta\sqrt{\varepsilon_z\mu}}{(\mu\varepsilon - \gamma^2)} \cdot \exp\left(ik_0\sqrt{\varepsilon_z\mu}\,(\alpha x + \beta y) \pm ik_0\gamma z\right),$$

$$H_y = \frac{-\varepsilon}{ik_0(\mu\varepsilon - \gamma^2)}\partial_x E_z = \frac{-\varepsilon\alpha\sqrt{\varepsilon_z\mu}}{(\mu\varepsilon - \gamma^2)} \cdot \exp\left(ik_0\sqrt{\varepsilon_z\mu}\,(\alpha x + \beta y) \pm ik_0\gamma z\right).$$

(1.140)

Since the solution of the diffraction problem requires the imposition of conditions of equality of the tangential components of the fields at the interface, it is convenient to introduce the following four-component vector of the tangential components:

$$\mathbf{E}_E = \begin{pmatrix} E_x \\ H_y \\ E_y \\ H_x \end{pmatrix} = \frac{1}{\sqrt{\gamma\varepsilon(\alpha^2+\beta^2)}}\begin{pmatrix} \mp\alpha\gamma \\ -\varepsilon\alpha \\ \mp\gamma\beta \\ \varepsilon\beta \end{pmatrix}\exp\left(ik_0\sqrt{\varepsilon_z\mu}\,(\alpha x + \beta y) \pm ik_0\gamma z\right)$$

(1.141)

Vector (1.141) is written with a normalizing factor chosen from the condition that the unit z-component of the Poynting vector $S_z = |E_x H_y - E_y H_x|$ is unity.

From (1.134) and (1.138) for the H-wave we get

$$E_x = \frac{-\mu}{ik_0\left(\varepsilon\mu - \gamma_1^2\right)}\partial_y H_z = \frac{-\mu\beta\sqrt{\varepsilon_z\mu}}{\left(\varepsilon\mu - \gamma_1^2\right)}\cdot\exp\left(ik_0\sqrt{\varepsilon_z\mu}\left(\alpha x + \beta y\right)\pm ik_0\gamma_1 z\right),$$

$$H_y = \frac{\mp\gamma_1}{ik_0\left(\mu - \gamma_1^2\right)}\cdot\partial_y H_z = \frac{\mp\gamma_1\beta\sqrt{\varepsilon_z\mu}}{\left(\mu - \gamma_1^2\right)}\cdot\exp\left(ik_0\sqrt{\varepsilon_z\mu}\left(\alpha x + \beta y\right)\pm ik_0\gamma_1 z\right),$$

$$E_y = \frac{\mu}{ik_0\left(\varepsilon\mu - \gamma_1^2\right)}\cdot\partial_x H_z = \frac{\mu\alpha\sqrt{\varepsilon_z\mu}}{\left(\varepsilon\mu - \gamma_1^2\right)}\cdot\exp\left(ik_0\sqrt{\varepsilon_z\mu}\left(\alpha x + \beta y\right)\pm ik_0\gamma_1 z\right),$$

$$H_x = \frac{\mp\gamma_1}{ik_0\left(\mu\varepsilon - \gamma_1^2\right)}\cdot\partial_x H_z = \frac{\mp\gamma_1\alpha\sqrt{\varepsilon_z\mu}}{\left(\mu\varepsilon - \gamma_1^2\right)}\cdot\exp\left(ik_0\sqrt{\varepsilon_z\mu}\left(\alpha x + \beta y\right)\pm ik_0\gamma_1 z\right).$$

$$(1.142)$$

Similarly to (1.141), we introduce the vector of the tangential components of the H-wave in the form

$$\mathbf{E}_H = \begin{pmatrix} E_x \\ H_y \\ E_y \\ H_x \end{pmatrix} = \frac{1}{\sqrt{\gamma_1\mu\left(\alpha^2 + \beta^2\right)}}\begin{pmatrix} -\mu\beta \\ \mp\gamma_1\beta \\ \mu\alpha \\ \mp\gamma_1\alpha \end{pmatrix}\exp\left(ik_0\sqrt{\varepsilon_z\mu}\left(\alpha x + \beta y\right)\pm ik_0\gamma_1 z\right)\cdot$$

$$(1.143)$$

Fresnel equations. We write the Fresnel equations for the interface between a homogeneous dielectric and a medium described by the tensor (1.136). Without the loss of generality, we assume that the interface is the plane $z = 0$. We also assume that in the region $z > 0$ there is a homogeneous dielectric, and at $z < 0$ – the anisotropic medium (1.136).

Suppose that a plane wave impinges on the interface from the side of the homogeneous dielectric. The plane wave is represented as a superposition of the E- and H-waves. In a general case, upon the reflection and refraction of the incident wave at the interface reflected and refracted waves will be formed as a superposition of the E- and H-waves.

Assuming in (1.141) and (1.143) $\varepsilon = \varepsilon_z = \varepsilon_0$, $\mu = \mu_z = \mu_0$, we write the E- and H-waves incident and reflected from the interface

$$\mathbf{E}_{inc,E} = \frac{1}{\sqrt{\gamma_0 \varepsilon_0 \left(\alpha_0^2 + \beta_0^2\right)}} \begin{pmatrix} \alpha_0 \gamma_0 \\ -\varepsilon_0 \alpha_0 \\ \gamma_0 \beta_0 \\ \varepsilon_0 \beta_0 \end{pmatrix} \exp\left(ik_0 \sqrt{\varepsilon_0 \mu_0}\left(\alpha_0 x + \beta_0 y\right) - ik_0 \gamma_0 z\right),$$

$$\mathbf{E}_{inc,H} = \frac{1}{\sqrt{\gamma_0 \mu_0 \left(\alpha_0^2 + \beta_0^2\right)}} \begin{pmatrix} -\mu_0 \beta_0 \\ \gamma_0 \beta_0 \\ \mu_0 \alpha_0 \\ \gamma_0 \alpha_0 \end{pmatrix} \exp\left(ik_0 \sqrt{\varepsilon_0 \mu_0}\left(\alpha_0 x + \beta_0 y\right) - ik_0 \gamma_0 z\right),$$

$$(1.144)$$

$$\mathbf{E}_{ref,E} = \frac{1}{\sqrt{\gamma_0 \varepsilon_0 \left(\alpha_0^2 + \beta_0^2\right)}} \begin{pmatrix} -\alpha_0 \gamma_0 \\ -\varepsilon_0 \alpha_0 \\ -\gamma_0 \beta_0 \\ \varepsilon_0 \beta_0 \end{pmatrix} \exp\left(ik_0 \sqrt{\varepsilon_0 \mu_0}\left(\alpha_0 x + \beta_0 y\right) + ik_0 \gamma_0 z\right),$$

$$\mathbf{E}_{ref,H} = \frac{1}{\sqrt{\gamma_0 \mu_0 \left(\alpha_0^2 + \beta_0^2\right)}} \begin{pmatrix} -\mu_0 \beta_0 \\ -\gamma_0 \beta_0 \\ \mu_0 \alpha_0 \\ -\gamma_0 \alpha_0 \end{pmatrix} \exp\left(ik_0 \sqrt{\varepsilon_0 \mu_0}\left(\alpha_0 x + \beta_0 y\right) + ik_0 \gamma_0 z\right),$$

$$(1.145)$$

where $\gamma_0 = \sqrt{\varepsilon_0 \mu_0}\sqrt{1 - \alpha_0^2 - \beta_0^2}$. 'inc' and 'ref' in the subscripts of the waves in (1.144) and (1.145) denote the incident and reflected waves, respectively.

If $z > 0$ the field is a superposition of the incident and reflected waves

$$\mathbf{E}_{z>0}(\mathbf{x}) = I_E \mathbf{E}_{inc,E}(\mathbf{x}) + I_H \mathbf{E}_{inc,H}(\mathbf{x}) + R_E \mathbf{E}_{ref,E}(\mathbf{x}) + R_H \mathbf{E}_{ref,H}(\mathbf{x}),$$

$$(1.146)$$

where $\mathbf{x} = (x, y, z)$, I_E, I_H are the coefficients of the E- and H-waves for the incident wave, and R_E, R_H are the reflection coefficients. The transmitted field $z < 0$ has the form:

$$\mathbf{E}_{z<0}(\mathbf{x}) = T_E \cdot \mathbf{E}_{tr,E}(\mathbf{x}) + T_H \cdot \mathbf{E}_{tr,H}(\mathbf{x}) , \qquad (1.147)$$

where T_E, T_H are the transmission coefficients,

$$\mathbf{E}_{tr,E} = \frac{1}{\sqrt{\gamma\varepsilon\left(\alpha^2 + \beta^2\right)}} \begin{pmatrix} \alpha\gamma \\ -\varepsilon\alpha \\ \gamma\beta \\ \varepsilon\beta \end{pmatrix} \exp\left(ik_0\sqrt{\varepsilon_z\mu}\left(\alpha x + \beta y\right) - ik_0\gamma z\right),$$

$$\mathbf{E}_{tr,H} = \frac{1}{\sqrt{\gamma_1\mu\left(\alpha^2 + \beta^2\right)}} \begin{pmatrix} -\mu\beta \\ \gamma_1\beta \\ \mu\alpha \\ \gamma_1\alpha \end{pmatrix} \exp\left(ik_0\sqrt{\varepsilon_z\mu}\left(\alpha x + \beta y\right) - ik_0\gamma_1 z\right),$$

$$(1.148)$$

where $\gamma = \sqrt{\varepsilon\mu}\sqrt{1 - \alpha^2 - \beta^2}$, $\gamma_1 = \mu\sqrt{\varepsilon/\mu - \left(\alpha^2 + \beta^2\right)\varepsilon_z/\mu_z}$.

To determine the reflection and transmission coefficients, we write the condition of equality of the tangential components at $z = 0$ ($\mu_0 = 1$):

$$\frac{I_E}{\sqrt{\gamma_0\varepsilon_0\left(\alpha_0^2 + \beta_0^2\right)}} \begin{pmatrix} \gamma_0\alpha_0 \\ -\varepsilon_0\alpha_0 \\ \gamma_0\beta_0 \\ \varepsilon_0\beta_0 \end{pmatrix} + \frac{I_H}{\sqrt{\gamma_0\left(\alpha_0^2 + \beta_0^2\right)}} \begin{pmatrix} -\beta_0 \\ \gamma_0\beta_0 \\ \alpha_0 \\ \gamma_0\alpha_0 \end{pmatrix} + \frac{R_E}{\sqrt{\gamma_0\varepsilon_0\left(\alpha_0^2 + \beta_0^2\right)}} \begin{pmatrix} -\gamma_0\alpha_0 \\ -\varepsilon_0\alpha_0 \\ -\gamma_0\beta_0 \\ \varepsilon_0\beta_0 \end{pmatrix} +$$

$$+ \frac{R_H}{\sqrt{\gamma_0\left(\alpha_0^2 + \beta_0^2\right)}} \begin{pmatrix} -\beta_0 \\ -\gamma_0\beta_0 \\ \alpha_0 \\ -\gamma_0\alpha_0 \end{pmatrix} = \frac{T_E}{\sqrt{\gamma\varepsilon\left(\alpha^2 + \beta^2\right)}} \begin{pmatrix} \alpha\gamma \\ -\varepsilon\alpha \\ \gamma\beta \\ \varepsilon\beta \end{pmatrix} + \frac{T_H}{\sqrt{\gamma_1\mu\left(\alpha^2 + \beta^2\right)}} \begin{pmatrix} -\mu\beta \\ \gamma_1\beta \\ \mu\alpha \\ \gamma_1\alpha \end{pmatrix}.$$

$$(1.149)$$

The refraction law should be satisfied in this case;

$$\alpha_0\sqrt{\varepsilon_0} = \alpha\sqrt{\varepsilon_z\mu}, \quad \beta_0\sqrt{\varepsilon_0} = \beta\sqrt{\varepsilon_z\mu} \qquad (1.150)$$

In the particular case

$$[\varepsilon] = \varepsilon_0 \begin{pmatrix} a & 0 & 0 \\ 0 & a & 0 \\ 0 & 0 & 1/a \end{pmatrix}, \quad [\mu] = \begin{pmatrix} a & 0 & 0 \\ 0 & a & 0 \\ 0 & 0 & 1/a \end{pmatrix} \qquad (1.151)$$

we obtain

$$\alpha = \alpha_0, \beta = \beta_0, \gamma = \gamma_1 = a\sqrt{\varepsilon_0}\sqrt{1-\alpha_0^2-\beta_0^2} = a\gamma_0. \qquad (1.152)$$

In this case, the equation (1.149) takes the form

$$
\frac{I_E}{\sqrt{\gamma_0\varepsilon_0\left(\alpha_0^2+\beta_0^2\right)}}\begin{pmatrix}\gamma_0\alpha_0\\-\varepsilon_0\alpha_0\\\gamma_0\beta_0\\\varepsilon_0\beta_0\end{pmatrix}+\frac{I_H}{\sqrt{\gamma_0\left(\alpha_0^2+\beta_0^2\right)}}\begin{pmatrix}-\beta_0\\\gamma_0\beta_0\\\alpha_0\\\gamma_0\alpha_0\end{pmatrix}+\frac{R_E}{\sqrt{\gamma_0\varepsilon_0\left(\alpha_0^2+\beta_0^2\right)}}\begin{pmatrix}-\gamma_0\alpha_0\\-\varepsilon_0\alpha_0\\-\gamma_0\beta_0\\\varepsilon_0\beta_0\end{pmatrix}+
$$

$$
+\frac{R_H}{\sqrt{\gamma_0\left(\alpha_0^2+\beta_0^2\right)}}\begin{pmatrix}-\beta_0\\-\gamma_0\beta_0\\\alpha_0\\-\gamma_0\alpha_0\end{pmatrix}=\frac{T_E}{\sqrt{\gamma_0\varepsilon_0\left(\alpha_0^2+\beta_0^2\right)}}\begin{pmatrix}\alpha_0\gamma_0\\-\varepsilon_0\alpha_0\\\gamma_0\beta_0\\\varepsilon_0\beta_0\end{pmatrix}+\frac{T_H}{\sqrt{\gamma_0\left(\alpha_0^2+\beta_0^2\right)}}\begin{pmatrix}-\beta_0\\\gamma_0\beta_0\\\alpha_0\\\gamma_0\alpha_0\end{pmatrix}.
$$

$$(1.153)$$

It is easy to see that the solution of (1.153) is as follows:

$$R_E = R_H = 0, \; T_H = I_H, \; T_E = I_E. \qquad (1.154)$$

According to (1.154), medium (1.151) does not reflect the incident waves. At complex $a = a' + ia''$ in (1.151) the transmitted waves (1.148) will be evansecent. The rate of decay is determined by the imaginary part a''. Thus, a layer of material (1.151) at complex a is a perfectly matched absorbing layer.

When $\beta_0 = 0$ in (1.144) the direction vector of the incident wave lies in the plane xOz. In the case of 'planar incidence' the E-wave corresponds to a wave with TM-polarization and the H-wave to a wave with TE-polarization.

For the incident wave with TM-polarization the medium (1.151) takes the form

$$[\varepsilon] = \varepsilon_0\begin{pmatrix}a & 0 & 0\\0 & a & 0\\0 & 0 & 1/a\end{pmatrix}, \; \mu = a \cdot \qquad (1.155)$$

Indeed, it is easy to show that under the condition (1.155), $\beta_0 = 0$ and $I_H = 0$ the solution of (1.149) has the form

$$R_E = R_H = 0, \ T_H = 0, \ T_E = I_E. \tag{1.156}$$

According to (1.156), the medium (1.155) does not reflect the incident waves.

Similarly, for the incident wave with TE-polarization the following condition is sufficient

$$\varepsilon = \varepsilon_0 a, \ [\mu] = \begin{pmatrix} a & 0 & 0 \\ 0 & a & 0 \\ 0 & 0 & 1/a \end{pmatrix}. \tag{1.157}$$

In the case of (1.157) the solution of (1.149) with $\beta_0 = 0$ and $I_H = 0$ has the form

$$R_E = R_H = 0, \ T_H = I_H, \ T_E = 0. \tag{1.158}$$

Perfectly matched absorbing layers as a complex coordinate transformation

We now introduce the perfectly matched absorbing layers as a complex coordinate transformation [23]. For simplicity, consider the case of diffraction of a TM-polarized wave (in particular, the plasmonic modes – the geometry of the structure is shown on the left in Fig. 1.6) on the structure composed of isotropic materials. In this case, the electric and magnetic fields have the form $\mathbf{E} = (E_x, 0, E_z)$, $\mathbf{H} = (0, H_y, 0)$. In each of the layers of the structure at $\mu = 1$ we can obtain the Helmholtz equation for the component H_y from Maxwell's equations (1.24):

$$\frac{\partial^2 H_y}{\partial z^2} + \varepsilon(x) \frac{\partial}{\partial x} \left[\frac{1}{\varepsilon(x)} \frac{\partial H_y}{\partial x} \right] + k_0^2 \varepsilon(x) H_y = 0. \tag{1.159}$$

Equation (1.159) is written in a general form and is valid for both the initial non-periodic problem and, for example, for the problem of diffraction of the TM-polarized wave on a diffraction structure with one-dimensional periodicity. Note that in solving the diffraction problem by the Fourier modal method, equation (1.159), which describes the field in the layer of the structure, is reduced to a second system of differential equations in (1.48).

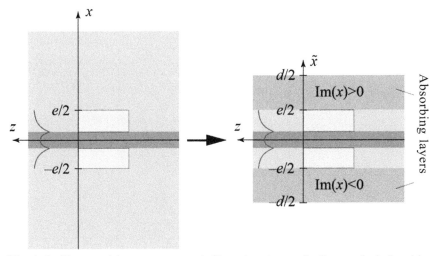

Fig. 1.6. The transition to a non-periodic task using perfectly matched absorbing layers in the form of coordinate transformations.

Suppose we are interested in the solution of the original aperiodic problem in the region $|x| > e/2$ in addition, the regions $|x| > e/2$ are homogeneous (Fig. 1.6). Note that because of physical reasons (absence of incident waves apart from the incident waveguide or plasmonic mode) the regions $|x| > e/2$ contain only the waves propagating from the structure (in the direction of increase $|x| > e/2$).

As indicated above, to solve a non-periodic problem by the Fourier modal method it is necessary to introduce artificial periodization with a certain period $d > e$ so as to eliminate the interaction between adjacent periods. To do this, it is necessary to ensure that the components of the electromagnetic field at the boundaries between periods are equal to zero. Consider the analytic continuation of the solution of equation (1.159) for the original problem in the variable x in the complex plane: $x = x' + ix''$. Note that the propagating waves are evanescent in this case. Indeed, for waves propagating in the positive direction of the x axis ($k_x > 0$), we get:

$$\exp(ik_x x) = \exp(ik_x x') \exp(-k_x x''). \tag{1.160}$$

Thus, these waves become evanescent when $x'' \to +\infty$. Similarly, the waves propagating in the negative direction of the axis x, decay when $x'' \to -\infty$. Thus, we consider the solution of (1.159) on the contour in the complex plane having the form shown in Fig. 1.7.

The horizontal dashed line shows the real axis, the vertical dotted lines – the area $x'/d \in \left[-e/(2d), e/(2d)\right]$.

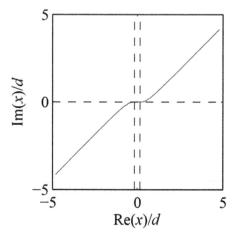

Fig. 1.7. The considered contour in the complex plane.

At $x' \in (-e/2, e/2)$ the contour lies on the real axis, away from the real axis $x'' \to \pm\infty$ and the imaginary part x' also tends to infinity, with the condition $x'x'' > 0$. This selection of the contour provides decay of the electromagnetic field with increasing distance from the real axis. To introduce artificial periodization, consider the transformation of coordinates $F(\tilde{x}) = x$ mapping the contour on the complex plane into the interval $\tilde{x} \in (-d/2, d/2)$ where $\tilde{x}$ is a new real coordinate. At the same time at $\tilde{x} \in (-e/2, e/2)$ the field will be the same as the solution of the original equation (1.159), while at $\tilde{x} \to \pm d/2$ the field components will tend to zero by the choice of the contour in the complex plane x. To move from equation (1.159) to a new differential equations depending on a variable $\tilde{x}$, we need to make the following change of the differential operator: $\partial/\partial x \to f \cdot \partial/\partial \tilde{x}$ where $f(\tilde{x}) = d\tilde{x}/dx = \left(dF(\tilde{x})/d\tilde{x}\right)^{-1}$. As a result, we obtain the differential equation

$$\frac{\partial^2 H_y}{\partial z^2} + \varepsilon(x) f \frac{\partial}{\partial \tilde{x}}\left[\frac{f}{\varepsilon(x)}\frac{\partial H_y}{\partial \tilde{x}}\right] + k_0^2 \varepsilon(x) H_y = 0. \qquad (1.161)$$

Equation (1.161) can be solved by the Fourier modal method since the task now allows artificial periodization (since the field at the boundaries of the period $\tilde{x} = \pm d/2$ tends to zero). Using the expansions (1.27), the solution of the differential equation (1.161), which describes a general representation of the field in the layer can be reduced to a system of differential equations

$$\frac{d}{dz'}\begin{bmatrix}\mathbf{U}_y \\ \mathbf{S}_x\end{bmatrix} = -\begin{bmatrix}\mathbf{0} & \mathbf{E}^* \\ \mathbf{F}_x\mathbf{K}_x\mathbf{E}^{-1}\mathbf{F}_x\mathbf{K}_x - \mathbf{I} & \mathbf{0}\end{bmatrix}\begin{bmatrix}\mathbf{U}_y \\ \mathbf{S}_x\end{bmatrix}, \qquad (1.162)$$

where $z' = k_0 z$, $\mathbf{F}_x$ is the Toeplitz matrix formed from the coefficients of expansion of the function f into the Fourier series. Comparing the second system of differential equations (1.48) with (1.162), one can see that the latter is derived from the first formal change $\mathbf{K}_x \to \mathbf{F}_x\mathbf{K}_x$. Note that in the general case, when the field in the layer is described by a system of differential equations (1.35) with the matrix (1.36), the perfectly matched absorbing layers can be introduced using the same formal change.

We now give as an example the type of function $f(\tilde{x})$ that describes the coordinate transformation [23]:

$$f(\tilde{x}) = \begin{cases} 1, |\tilde{x}| < e/2, \\ \left[1 - \gamma\sin^2\left(\pi\frac{|\tilde{x}| - e/2}{q}\right)\right]\cos^2\left(\pi\frac{|\tilde{x}| - e/2}{q}\right), \frac{e}{2} < |\tilde{x}| < \frac{d}{2}, \end{cases}$$

$$(1.163)$$

where $q = d - e$ is the total size of the region corresponding to the absorbing layers, γ is a complex parameter characterizing the absorbing layer (typical value $\gamma = 1/(1-i)$). For this function, the coefficients of the Fourier series can be found analytically and have the form

$$f_n = \delta_{n0} - \frac{q}{2d}(-1)^n\left[\left(1 + \frac{\gamma}{4}\right)\text{sinc}\left(\frac{nq}{d}\right) + \frac{1}{2}\text{sinc}\left(\frac{nq}{d} - 1\right) + \right.$$
$$\left. + \frac{1}{2}\text{sinc}\left(\frac{nq}{d} + 1\right) - \frac{\gamma}{8}\text{sinc}\left(\frac{nq}{d} - 2\right) - \frac{\gamma}{8}\text{sinc}\left(\frac{nq}{d} + 2\right)\right], \qquad (1.164)$$

where δ_{n0} is the Kronecker symbol, $\text{sinc}(x) = \sin(\pi x)/(\pi x)$. We note that the function $f(\tilde{x})$ corresponds to the contour in the complex plane

$$F(\tilde{x}) = \left\{ \begin{array}{l} x, |\tilde{x}| < e/2, \\ \dfrac{\tilde{x}}{|\tilde{x}|}\left[\dfrac{e}{2} + \dfrac{q}{\pi(1-\gamma)}\left\{ \text{tg}\left(\pi\dfrac{|\tilde{x}|-e/2}{q}\right) \\ -\dfrac{\gamma}{\sqrt{1-\gamma}}\text{arctg}\left[\sqrt{1-\gamma}\,\text{tg}\left(\pi\dfrac{|\tilde{x}|-e/2}{q}\right)\right]\right\}\right], \\ \dfrac{e}{2} < |\tilde{x}| < \dfrac{d}{2}. \end{array}\right.$$

(1.165)

The real and imaginary parts of the function $f(\tilde{x})$ at $e = d/3$, $\gamma = 1/(1-i)$ are shown in Fig. 1.8, the contour in the complex plane is shown in Fig. 1.7.

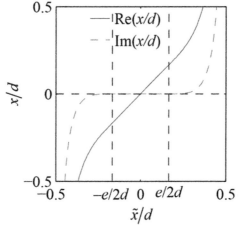

Fig. 1.8. The real and imaginary parts of the contour in the complex plane, depending on the value of the new real variable.

1.1.4.3. Solution of the diffraction problem

Stages of solving the difraction problem in this case coincide with the stages of solving the diffraction problem of a plane wave on a two-dimensional diffraction structure described in subsection 1.2. The differences lie in the form of the field above and below the structure (in regions 1 and 3 in Fig. 1.5). Since the regions above and below the structure in this case are not uniform, the electromagnetic field therein can not be represented in the form of Rayleigh expansions (1.20) and (1.21). Representation of the field in these areas can be obtained similarly to the representation of the field in the layer of the structure obtained by solving a system of differential equations of the form (1.35) (with the possible modification of the matrix of the system by introducing perfectly matched layers in the form of coordinate transformations). Thus, the field above and below the structure will be presented in the form of the Fourier series (1.27) and the expressions for the x- and y-Fourier components of the fields will be similar to (1.58):

$$\begin{bmatrix} \mathbf{S}_y^U \\ \mathbf{S}_x^U \\ \mathbf{U}_y^U \\ \mathbf{U}_x^U \end{bmatrix} = \mathbf{W}_{inc} \exp\left(\lambda_{inc} k_0 z\right) A_{inc} + \mathbf{W}_U^{(-)} \exp\left(\mathbf{\Lambda}_U^{(-)} k_0 z\right) \mathbf{R}, \quad (1.166)$$

$$\begin{bmatrix} \mathbf{S}_y^D \\ \mathbf{S}_x^D \\ \mathbf{U}_y^D \\ \mathbf{U}_x^D \end{bmatrix} = \mathbf{W}_D^{(+)} \exp\left[\mathbf{\Lambda}_D^{(+)} k_0 \left(z - z_L\right)\right] \mathbf{T}. \quad (1.167)$$

Similarly to the expressions (1.20), (1.21), the indices U and D correspond to the regions above and below the structure (reas 1 and 3 in Fig. 1.5). The first term in (1.166) describes the incident wave with an amplitude A_{inc}, with its eigenvalue λ_{inc} selected from $\mathbf{\Lambda}_U^{(+)}$ as closest one to the propagation constant of the incident mode calculated by solving the corresponding dispersion equation.

The system of linear equations (1.71), which describes the condition of equality of the tangential components of the electromagnetic field at the boundaries of layers of the structure, retains its form, however, the expressions for $\mathbf{D}$, $\mathbf{P}^{(U)}$ and $\mathbf{P}^{(D)}$, included in the equation, change. In this case,

$$\begin{aligned} \mathbf{D} &= \mathbf{W}_{inc} A_{inc}, \\ \mathbf{P}^{(U)} &= \mathbf{W}_U^{(-)}, \\ \mathbf{P}^{(D)} &= \mathbf{W}_D^{(+)}. \end{aligned} \quad (1.168)$$

As before, for obtaining a numerically-stable solution of equations (1.71) and finding the amplitudes of the transmitted and reflected waves we can use the scattering matrix algorithm, described in section 1.2.6. For the calculation of the distribution of the field we can use a stable procedure described in subsection 1.2.7.

1.1.4.4 The energy characteristics of the reflected and transmitted light

First we find the energy characteristics of the transmitted radiation. To do this, we calculate the flux of the z-component of the Umov–Poynting vector $S_z = \text{Re}(E_x H_y^* - E_y H_x^*)/2$ of the transmitted waves within a period through the segment $z = z_L$. We rewrite the representation of the field (1.167) for $z = z_L$, separating parts of the eigenvectors $\mathbf{W}_D^{(+)}$ corresponding to the various components of the electromagnetic field:

$$\begin{bmatrix} \mathbf{S}_y^D \\ \mathbf{S}_x^D \\ \mathbf{U}_y^D \\ \mathbf{U}_x^D \end{bmatrix} = \begin{bmatrix} \mathbf{W}_{D,\mathbf{S}_y}^{(+)} \\ \mathbf{W}_{D,\mathbf{S}_x}^{(+)} \\ \mathbf{W}_{D,\mathbf{U}_y}^{(+)} \\ \mathbf{W}_{D,\mathbf{U}_x}^{(+)} \end{bmatrix} \mathbf{T}. \tag{1.169}$$

Taking into account (1.169) in (1.27) and assuming for the sake of simplicity that $y = 0$, we obtain:

$$\begin{cases} E_x = \sum_k t_k \sum_j w_{j,k,S_x} \exp(ik_{x,j}x), \\ E_y = \sum_k t_k \sum_j w_{j,k,S_y} \exp(ik_{x,j}x), \\ H_x = -i\sum_k t_k \sum_j w_{j,k,U_x} \exp(ik_{x,j}x), \\ H_y = -i\sum_k t_k \sum_j w_{j,k,U_y} \exp(ik_{x,j}x), \end{cases} \tag{1.170}$$

where t_i are the elements of the vector $\mathbf{T}$, $w_{j,i,S_x}, w_{j,i,S_y}, w_{j,i,U_x}$, and w_{j,i,U_y} are the elements of vectors $\mathbf{W}_{D,\mathbf{S}_x}^{(+)}, \mathbf{W}_{D,\mathbf{S}_y}^{(+)}, \mathbf{W}_{D,\mathbf{U}_x}^{(+)}$, and $\mathbf{W}_{D,\mathbf{U}_y}^{(+)}$ respectively. Taking into account (1.170) in the expression for the z-component of the Umov–Poynting vector, we get:

$$S_d^D = 2\int_0^d S_z dx = \sum_{k,m,j,n} \int_0^d \text{Re}\left(\frac{it_k t_m^* \left[w_{j,k,S_x} w_{n,m,U_y}^* - w_{j,k,S_y} w_{n,m,U_x}^* \right]}{\exp\left[i\left(k_{x,j} - k_{x,n}\right)x \right]} \right) dx.$$

$$\tag{1.171}$$

Given the equality $\int_0^d \text{Re}\left(a\exp\left[i\left(k_{x,j} - k_{x,n}\right)x\right]\right)dx = \text{Re}(a)d\delta_{jn}$ from (1.171) we obtain:

$$S_d^D = d \sum_{k,m} \mathrm{Im}\left(t_k t_m^* \sum_j \left[w_{j,k,S_y} w_{j,m,U_x}^* - w_{j,k,S_x} w_{j,m,U_y}^* \right] \right). \quad (1.172)$$

From (1.172) it follows that, in contrast to the above-mentioned periodic structures, the flux of the Umov–Poynting vector in the period can not be represented as a sum of terms, each of which corresponds to a single wave. Note, however, that the waveguide and plasmonic modes, propagating in the original structure, the following condition of orthogonality is numerically satisfied (up to the order of 10^{-10}):

$$\sum_j \left[w_{j,k,S_y} w_{j,m,U_x}^* - w_{j,k,S_x} w_{j,m,U_y}^* \right] = 0, \ k \neq m, \quad (1.173)$$

where k is a fixed number, and m runs through all values from 1 to the number of waves corresponding to the dimensionality of the system of differential equations (1.35). Verification of fulfillment of this condition allows us to judge the correctness of the choice of the values of the period d for artificial periodization and the parameters of absorbing layers. But the waves, for which the orthogonality condition (1.173) is fulfilled, can be given the energy characteristics similar to the intensity of the diffraction orders and determined by expression

$$I_n^T = \frac{|t_n|^2 \left| \mathrm{Im}\left(\sum_j \left[w_{j,n,S_y} w_{j,n,U_x}^* - w_{j,n,S_x} w_{j,n,U_y}^* \right] \right) \right|}{|A_{inc}|^2 \left| \mathrm{Im}\left(\sum_j \left[w_{j,inc,S_y} w_{j,inc,U_x}^* - w_{j,inc,S_x} w_{j,inc,U_y}^* \right] \right) \right|}, \quad (1.174)$$

where $w_{j,inc,S_x}, w_{j,inc,S_y}, w_{j,inc,U_x}$, and w_{j,inc,U_y} are the elements of the vector $\mathbf{W}_{inc}$. The denominator of (1.174) gives the flux of the Umov–Poynting vector of the incident wave. Similarly, the energy characteristics of the reflected waves are calculated: the elements of the vector $\mathbf{T}$ (1.174) are replaced by the elements of the vector $\mathbf{R}$, and the elements of the eigenvectors from the matrix $\mathbf{W}_D^{(+)}$ are replaced by the corresponding values from the matrix $\mathbf{W}_D^{(-)}$.

1.1.4.5. Numerical example
Consider as an example the propagation of a TM-polarized

fundamental mode of the plane-parallel waveguide through a couple of grooves in it (Fig. 1.9) [21]. The refractive indices of the materials are shown in the Fig., the waveguide thickness is 300 nm, the width of the grooves and the step separating them are 150 nm. The wavelength of the radiation in free space is 975 nm.

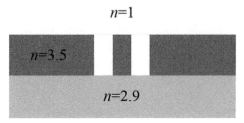

Fig. 1.9. The geometry of the diffraction problem.

Table 1.1 shows the values of the coefficients of reflection and transmission (energy characteristics of the reflected and transmitted modes) depending on the number of harmonics N (the total number of harmonics is $2N+1$). Simulation was carried out using perfectly matched absorbing layers in the form of coordinate transformations. The following simulation parameters were: $e = 900$ nm, $d = 3e$, $\gamma = 1/(1-i)$, the middle of the waveguide coincided with the middle of the period.

Table 1.1. Coefficients of reflection and transmission for the TM-polarized fundamental mode

N	I^R	I^T
50	0.356216	0.129970
100	0.355637	0.129675
150	0.355569	0.129619
200	0.355531	0.129604

1.2. Methods for calculating eigenmodes of periodic diffractive structures

1.2.1. Calculation of modes based on the calculation of the poles of the scattering matrix

The diffractive micro- and nanostructures with the resonant properties are of great interest for designing of modern elements of integrated optics and photonics [19,24–31]. The resonant properties are manifested in the abrupt change in the transmission and reflection spectra, and are usually associated with the excitation of the eigenmodes of the structure. The eigenmodes can be calculated as the poles of the scattering matrix [28,32]. Such an approach to the explanation of the optical properties of diffractive structures is widely used to describe the optical properties of diffraction gratings [28, 32], including those containing anisotropic [33] and gyrotropic [29] materials, photonic crystal structures [30, 31] and laser resonators [19].

Calculation of the poles of the scattering matrix is a computationally difficult task. The known papers propose several iterative methods for solving it. The simplest method is to calculate the poles of the determinant of the scattering matrix [30, 31] or the poles of its maximum eigenvalue [34]. In [26,27,35] the authors proposed a more sophisticated method based on the linearization of the scattering matrix inverse. To adequately describe the optical properties of the metal-dielectric diffraction gratings and plasmonic structures it is required to use the scattering matrix with high dimensions. Typically, the scattering matrix is calculated on the basis of thr Fourier modal method [2]. Despite the variety of approaches to improving the convergence of the method, the scattering matrix dimensions may amount to several hundreds [36]. Calculations of the determinant and the inverse scattering matrix for such large matrices carried out in [26, 27, 31, 34, 35] become numerically unstable [37], significantly limiting the area of applicability of the methods.

We should also highlight the method for calculating the frequency of modes, based on Cauchy's integral formula. This method is used in calculating the modes propagating in plane-parallel waveguides [38, 39] and photonic crystal structures [34,40,41]. The distinctive feature of this method is that it allows one to find all the poles of the scattering matrix, located in a given area [34,40,41]. However, in these studies the authors made a number of assumptions about the form of the scattering matrix which usually do not take place for the scattering matrix of a diffraction grating.

In this section, we compare the known iterative methods of calculating the scattering matrix poles and propose, in our opinion, a new method, which takes into account the form of the scattering matrix in the vicinity of resonances and has better convergence. In this section we also proposes a modification of the method based on Cauchy's integral formula which generalizes the methods of [24–26].

1.2.1.1. Resonant representation of the scattering matrix
Consider the scattering matrix of a periodic structure. Let $N' =$

2N +1 denote the number of Fourier harmonics considered in the representation of the field (1.27). In this case, the scattering matrix relates $2N'$ incident and $2N'$ scattered waves. Taking into account the two polarization states of the incident and scattered waves, the size of the scattering matrix is $4N' \times 4N'$ [28,2]. For the given geometry and grating materials the scattering matrix **S** is a function of frequency ω and the *x*-component of the wave vector of the incident wave with the number $m = 0$: $\mathbf{S} = \mathbf{S}(\omega, k_{x,0})$. In order to describe the resonances in the transmission or reflection spectrum of the grating we fix the direction of the incident wave, and consider the scattering matrix as a function of frequency ω: $\mathbf{S} = \mathbf{S}(\omega)$. The resonances of the structure correspond to the poles of the analytic continuation of $\mathbf{S}(\omega)$ [19]. The real part of the poles corresponds to the frequency of the incident wave that can excite a mode, while the imaginary part determines mode decay.

Consider the analytic continuation $\mathbf{S}(\omega)$, $\omega \in \mathbb{C}$ of the scattering matrix for the domain D of the complex ω-plane bounded by a closed curve Γ. We assume that the Rayleigh anomalies of the structure are located far from the frequency region of interest. In this case, the analytic continuation $\mathbf{S}(\omega)$ in the considered domain D is single-valued [42,43].

Suppose the analytic continuation of the scattering matrix in domain D has a simple pole at $\omega = \omega_p$. Next, we consider the case when the poles of the scattering matrix are simple. This assumption is valid for problems of electrodynamics and quantum mechanics [40]. In this case, it makes sense to determine the residue of the scattering matrix $\mathbf{S}(\omega)$:

$$\operatorname*{Res}_{\omega=\omega_p} \mathbf{S}(\omega) = \frac{1}{2\pi i} \oint_\gamma \mathbf{S}(\omega) d\omega, \qquad (1.175)$$

where the integration contour γ is chosen in such a way that it contains only one pole ω_p. Equation (1.175) should be understood as an *elementwise* operation. If the scattering matrix has only one pole in domain D, then the following representation holds:

$$\mathbf{S}(\omega) = \mathbf{A}(\omega) + \frac{\mathbf{B}}{\omega - \omega_p} \qquad (1.176)$$

where $\mathbf{B} = \underset{\omega=\omega_p}{\mathrm{Res}}\,\mathbf{S}(\omega)$, and the matrix-valued function $\mathbf{A}(\omega)$ has no poles in the domain D and is holomorphic in this area. In general, when domain D contains M poles, the expansion (1.176) takes the following form:

$$\mathbf{S}(\omega) = \mathbf{A}(\omega) + \sum_{m=1}^{M} \frac{\mathbf{B}_m}{\omega - \omega_p^{(m)}} \tag{1.177}$$

where $\mathbf{B}_m = \underset{\omega=\omega_p^{(m)}}{\mathrm{Res}}\,\mathbf{S}(\omega)$. The first term in (1.176) and (1.177) describes the non-resonant scattering, the second one – the resonant scattering.

We now consider the *properties of the matrix* $\mathbf{S}(\omega)$. Assume that $\mathbf{S}(\omega)$ has a pole at $\omega = \omega_p$, and the elements of the inverse matrix $\mathbf{S}^{-1}(\omega)$ do not have a pole at a given frequency. In this case, the kernel of the matrix $\mathbf{S}^{-1}(\omega_p)$ determines the non-trivial solutions of the homogeneous equation:

$$\mathbf{S}^{-1} \cdot \mathbf{\Psi}^{\mathrm{scatt}} = \mathbf{0}, \tag{1.178}$$

where, according to (1.86), $\mathbf{\Psi}^{\mathrm{scatt}} = \begin{bmatrix} \mathbf{T} \\ \mathbf{R} \end{bmatrix}$.

Thus, $\ker \mathbf{S}^{-1}(\omega_p)$ describes the distribution of the field that exists in the absence of incident waves, that is, the frequency ω_p corresponds to the frequency of the quasi-guided grating mode.

It is easy to show that $\mathrm{Im}\,\mathbf{B} = \ker\mathbf{S}^{-1}(\omega_p)$, so $\mathrm{rank}\,\mathbf{B} = \dim \ker \mathbf{S}^{-1}$ [40]. Usually $\mathrm{rank}\,\mathbf{B} = 1$, i.e. the same frequency matches only one mode. However, under certain parameters of the structure the frequencies of several different modes may coincide. In this case $\mathrm{rank}\,\mathbf{B} > 1$. Such resonances are called degenerate. Structures with degenerate resonances have interesting optical properties [24,25].

Suppose that in a general case $\mathrm{rank}\,\mathbf{B} > r$. In this case, for the matrix $\mathbf{B}$ we can write the rank factorization [44]

$$\mathbf{B} = \mathbf{LR}, \tag{1.179}$$

where $\mathbf{L} \in \mathbb{C}^{n \times r}$, $\mathbf{R} \in \mathbb{C}^{r \times n}$, $\mathrm{rank}\,\mathbf{L} = \mathrm{rank}\,\mathbf{R} = r$. Taking into account (1.179), the expression (1.177) can be written as

$$\mathbf{S} = \mathbf{A}(\omega) + \mathbf{L}\frac{1}{\omega - \omega_p}\mathbf{R}. \tag{1.180}$$

The general expansion (1.177) can be written as follows:

$$\mathbf{S} = \mathbf{A}(\omega) + \sum_{m=1}^{M} \mathbf{L}_m \frac{1}{\omega - \omega_p^{(m)}} \mathbf{R}_m = \mathbf{A}(\omega) + \mathbf{L}(\mathbf{I}\omega - \mathbf{\Omega}_p)^{-1}\mathbf{R}, \quad (1.181)$$

where $\mathbf{L},\mathbf{R}$ are the block matrices $\mathbf{L} = [\mathbf{L}_1, \mathbf{L}_2 ... \mathbf{L}_M]$, $\mathbf{R}^\mathrm{T} = [\mathbf{R}_1^\mathrm{T} \quad \mathbf{R}_2^\mathrm{T} \quad \cdots \quad \mathbf{R}_M^\mathrm{T}]$ and $\mathbf{\Omega}_p$ is a diagonal matrix composed of $\omega_p^{(m)}$, $m = 1...M$, with the number of repetitions of frequency $\omega_p^{(m)}$ equal to the rank of the corresponding matrix $\mathbf{B}_m$. Equations (1.180) and (1.181) determine the resonant representation of the scattering matrix, which will be used in the further derivation of numerical methods for the calculation of the scattering matrix poles.

1.2.1.2. Calculation of the poles of the scattering matrix

As shown above, the poles of the scattering matrix determine the eigenmodes of the structure. Consider the problem of calculating the poles of the scattering matrix $\mathbf{S}(\omega)$ in domain D. The easiest method is to move from the problem of calculating the poles of the matrix-valued function $\mathbf{S}(\omega)$ to the problem of finding the poles of a scalar function $\det\mathbf{S}(\omega)$, that is, to the numerical solution of the equation

$$1/\det\mathbf{S}(\omega) = 0. \quad (1.182)$$

This method allows us to calculate the poles for a small dimensional matrix $\mathbf{S}$. When a large dimensional matrix $\mathbf{S}(\omega)$ is considered the calculation of the determinant becomes numerically unstable. A more approach consists in solving the following equation [34]:

$$1/\max \mathrm{eig}\, \mathbf{S}(\omega) = 0. \quad (1.183)$$

where $\max \mathrm{eig}\, \mathbf{S}(\omega)$ is the eigenvalue of the matrix $\mathbf{S}(\omega)$ having the maximal modulus. To solve the equations (1.182) and (1.183) we can use iterative methods for finding the root of the non-linear equation, such as Newton's method, or a more general Householder's method [45]. For equation (1.183), this method can be written as the following iterative procedure:

$$\omega_{n+1} = \omega_n + p \frac{\dfrac{d^{p-1}}{d\omega^{p-1}}\max \mathrm{eig}\, \mathbf{S}(\omega)}{\dfrac{d^p}{d\omega^p}\max \mathrm{eig}\, \mathbf{S}(\omega)}\Bigg|_{=\omega_n} \quad (1.184)$$

where ω_n is an initial approximation of the pole. When $p = 1$ the equation (1.184) corresponds to Newton's method, when $p = 2$ to Halley's method. In solving the equations (1.182) and (1.183) the method (1.184) does not take into account the form of the scattering matrix (1.181), replacing it with a scalar integer (either with the determinant or with the maximum modulus eigenvalue $\mathbf{S}(\omega)$). This is the reason for the relatively slow convergence of the above methods. In the following sections we consider the iterative methods of calculation of the poles of the scattering matrix using the matrix expansion.

Calculation of the poles based on linearization of the scattering matrix inverse

Consider an iterative method of calculation of the poles of the scattering matrix proposed in [19,32,33,34]. Let ω_n be some initial approximation of the pole. We expand the matrix $\mathbf{S}^{-1}(\omega)$ into a Taylor series up to the first order term

$$\mathbf{S}^{-1}(\omega) = \mathbf{S}^{-1}(\omega_n) + \left.\frac{d\mathbf{S}^{-1}}{d\omega}\right|_{\omega_n} (\omega - \omega_n). \qquad (1.185)$$

Let ω_p be the pole of the scattering matrix, in this case there exists the vector $\mathbf{\Psi}^{\text{scatt}}$ which is a non-trivial solution of (1.178). Multiplying (1.185) by $\mathbf{\Psi}^{\text{scatt}}$ on the right side when $\omega = \omega_p$ we have:

$$\mathbf{S}^{-1}(\omega_n)\mathbf{\Psi}^{\text{scatt}} = (\omega_n - \omega_p)\left.\frac{d\mathbf{S}^{-1}}{d\omega}\right|_{\omega_n} \mathbf{\Psi}^{\text{scatt}}. \qquad (1.186)$$

Equation (1.186) is a generalized eigenvalue problem. Solving this problem, we obtain a set of eigenvalues $\lambda_k = \omega_n - \omega_p$. Choosing from the eigenvalues the number with a minimal modulus, we obtain the following iterative procedure:

$$\omega_{n+1} = \omega_n - \min \operatorname{eig}\left(\mathbf{S}^{-1}(\omega_n), \left.\frac{d\mathbf{S}^{-1}}{d\omega}\right|_{\omega_n}\right), \qquad (1.187)$$

where $\operatorname{eig}(\mathbf{F},\mathbf{G})$ denotes the vector composed of the eigenvalues for a generalized eigenvalue problem $\mathbf{F}\cdot\mathbf{X} = \lambda\mathbf{G}\cdot\mathbf{X}$. Selecting in (1.187) the eigenvalue with the minimum modulus means that in the next approximation of the pole ω_{n+1} we select the closest approximation to the original value ω_n.

We note a number of shortcomings of the method (1.187). The main drawback is the inability to carry out numerically stable computation of the matrix inverse to the scattering matrix for a large number of orders. [37]. Second, note that if in the vicinity of the pole of the scattering matrix $S(\omega)$ there is a pole of the matrix $S^{-1}(\omega)$, the Taylor series in (1.185) will not converge [46].

Solution of the system of matrix equations $A = LXR$, $B = LYR$
Before proceeding to the description of more advanced methods for calculating the poles of the scattering matrix, based on the resonance approximation, we consider the necessary mathematical apparatus. Let us assume we have matrices $A, B \in \mathbb{C}^{n \times n}$ that have the following form:

$$\begin{cases} A = LXR; \\ B = LYR, \end{cases} \qquad (1.188)$$

where $L \in \mathbb{C}^{n \times r}$, $R \in \mathbb{C}^{r \times n}$ are some unknown matrices; $X \in \mathbb{C}^{r \times r}$, $Y \in \mathbb{C}^{r \times r}$ are unknown diagonal matrices, with the value of r is also unknown. We assume that $\text{rank} A = r$. It is required to find the connection between the matrices X and Y. Note that systems of this kind occur in the problem of determining the poles of LTI-systems [47, 48].

Consider the method of solving the system (1.188), based on the singular value decomposition [49]. First, we find the rank r of the matrix A. To do this, we calculate its singular value decomposition $A = U\Sigma V^*$ (here V denotes the Hermitian conjugation). Rank r is determined by the number of non-zero singular numbers. In practice, the matrix A can be noisy, in this case r should be taken as the number of singular values greater than a certain threshold. Knowing the rank A, we can write compact singular value expansion:

$$A = U_r \Sigma_r V_r^* \qquad (1.189)$$

where $\Sigma_r \in \mathbb{R}^{r \times r}$ is a diagonal matrix composed of r largest singular values A; $U_r, V_r \in \mathbb{C}^{n \times r}$ are the matrices of the corresponding left and right singular vectors. In the case of the noised matrix the equation (1.189) holds approximately and it makes sense to approximate the matrix A with the matrix of rank r.

Note that the matrices U and V are unitary and hence $U_r^* U_r = V_r^* V_r = 1$. We multiply the equations of the system (1.188) on the left by U_r^* and on the right by V_r. We obtain

$$(U_r^*L)X(RV_r) = (U_r^*U_r)\Sigma_r(V_r^*V_r) = \Sigma_r;$$
$$(U_r^*L)Y(RV_r) = U_r^*BV_r. \tag{1.190}$$

Note rank Σ_r = rank A = rank U_r = rank V_r = r hence U_r^*L and RV_r are the invertible matrices from matrices $\mathbb{C}^{r \times r}$. Multiplying the second equation (1.190) by Σ_r^{-1} on the right, we get:

$$(U_r^*L)YX^{-1}(U_r^*L)^{-1} = U_r^*BV_r\Sigma_r^{-1}. \tag{1.191}$$

Equation (1.191) can be considered as eigendecomposition of the matrix on the right-hand side of the equation. Thus, we found an explicit expression for the matrix YX^{-1}:

$$YX^{-1} = \text{diag eig}(U_r^*BV_r\Sigma_r^{-1}). \tag{1.192}$$

Consider the important special case when the matrices A, B are the matrix of unit rank, that is $r = 1$. In this case, the vector U_1 is an eigenvector of the matrix A, and V_1 the eigenvector of matrix A^*. Consequently, the matrix A can be represented as

$$A = U_1 \frac{\text{max eig } A}{V_1^*U_1} V_1^*, \tag{1.193}$$

where max eig A is the unique non-zero eigenvalue of matrix A. Vectors U_1, V_1 are also eigenvectors of the matrices B and B^*, respectively. Therefore we have

$$B = U_1 \frac{\text{max eig } B}{V_1^*U_1} V_1^*. \tag{1.194}$$

By substituting (1.193) and (1.194) into (1.192), we obtain:

$$y/x = \frac{\text{max eig } B}{\text{max eig } A}. \tag{1.195}$$

Note that in practice the matrices A and B can be noised and therefore in formula (1.195) one should take eigenvalue of maximum modulus.

Calculation of the poles on the basis of the resonance approximation
Consider an iterative method of calculation of the poles of the scattering matrix, taking into account the form of the scattering matrix in the vicinity of resonances. Let there be some initial

approximation of the pole $\omega = \omega_n$. We write resonance approximation (1.181), neglecting the nonresonant term dependence on frequency:

$$\mathbf{S}(\omega) = \mathbf{A} + \mathbf{L}(\omega\mathbf{I} - \mathbf{\Omega}_p)^{-1}\mathbf{R}. \qquad (1.196)$$

We calculate the first two derivatives $\mathbf{S}(\omega)$:

$$\begin{aligned}\mathbf{S}'(\omega) &= -\mathbf{L}(\omega\mathbf{I} - \mathbf{\Omega}_p)^{-2}\mathbf{R},\\ \mathbf{S}''(\omega) &= 2\mathbf{L}(\omega\mathbf{I} - \mathbf{\Omega}_p)^{-3}\mathbf{R}.\end{aligned} \qquad (1.197)$$

We assume that $\operatorname{rank} \boldsymbol{L} = \operatorname{rank} \boldsymbol{R} = \Sigma_m \operatorname{rank} \boldsymbol{L}_m = \Sigma_m \operatorname{rank} \boldsymbol{R}_m$. This assumption suggests that the columns of the matrix $\mathbf{L}$ are linearly independent or, which is the same, the kernels of the matrices $\mathbf{S}^{-1}(\omega_p^{(m)})$, $m = 1 \ldots M$ are linearly independent. This means that the distribution of the scattered fields for different modes are linearly independent. This assumption is usually performed when the size of the scattering matrix is much larger than the number of modes $(\dim\mathbf{S} > \operatorname{rank}\mathbf{L})$.

Assuming in (1.197) $\omega = \omega_n$, we obtain a system of two matrix equations for the unknown diagonal matrix $\mathbf{\Omega}_p$. A method for solving systems of the form (1.197) is given in the preceding paragraph. Following this method (see (1.192)), the diagonal matrix $\mathbf{\Omega}_p$ is calculated according to the formula:

$$\mathbf{\Omega}_p = \omega_n\mathbf{I} + 2\operatorname{diag}\operatorname{eig}(\mathbf{U}_r^*\mathbf{S}'(\omega_n)\mathbf{V}_r\mathbf{\Sigma}_r^{-1}), \qquad (1.198)$$

where $\operatorname{diag}\operatorname{eig}\mathbf{F}$ is a diagonal matrix composed of the eigenvalues $\mathbf{F}$ and the matrix $\mathbf{U}_r, \mathbf{\Sigma}_r, \mathbf{V}_r$ obtained from the singular value compact decomposition of matrix $\mathbf{S}''(\omega_n) = \mathbf{U}_r\mathbf{\Sigma}_r\mathbf{V}_r^*$. Calculating the eigenvalues of the right-hand side of (1.198), we arrive at the following iterative procedure:

$$\omega_{n+1} = \omega_n + 2\min\operatorname{eig}\left(\mathbf{U}_r^*\mathbf{S}'(\omega_n)\mathbf{V}_r\mathbf{\Sigma}_r^{-1}\right). \qquad (1.199)$$

Similar to (1.187), the choice of the minimum-modulus eigenvalue (1.199) means that as the next approximation of the pole ω_{n+1} we select the closest approximation to the original ω_n.

The iterative procedure (1.199) suggests that in the vicinity of the initial approximation there are several poles. In practice, we can assume that in the neighborhood of ω_n there is only one pole corresponding to the non-degenerate resonance. In this case, using the ratio (1.195), iterative procedure takes the following simple form:

$$\omega_{n+1} = \omega_n + 2\frac{\max \text{eig}\, \mathbf{S}'(\omega_n)}{\max \text{eig}\, \mathbf{S}''(\omega_n)}. \qquad (1.200)$$

Unlike (1.187), the iterative methods (1.199) and (1.200) are based on the resonance approximation (1.196) and not on the linearization of the matrix $\mathbf{S}^{-1}(w)$. The results of numerical investigations in the following paragraphs show that the iterative methods (1.199) and (1.200) have a better convergence than (1.187). In addition, the advantage of the proposed approach is that it remains valid for the case of a large number of diffraction orders (large dimension of the matrix S) in the case when the poles of the scattering matrix and of the scattering matrix inverse.

Note that the method (1.187) has the meaning of Newton's method for matrix-valued functions, while the methods (1.199) and (1.200) can be considered as a generalization of Halley's matrix for solving equations of the form $1/f(x) = 0$. In general, we can write the matrix analogue of the Householder method [45], based on derivatives of the scattering matrix of order p and $p-1$ ($\mathbf{S}^{(p)}(\omega_n)$, $\mathbf{S}^{(p-1)}(\omega_n)$). In this case, the iterative procedure (1.199) is a special case ($p = 2$) of the following iterative procedure

$$\omega_{n+1} = \omega_n + p \min \text{eig}\left(\mathbf{U}_r^* \mathbf{S}^{(p-1)}(\omega_n)\mathbf{V}_r\mathbf{\Sigma}_r^{-1}\right), \qquad (1.201)$$

where the matrices $\mathbf{U}_r, \mathbf{\Sigma}_r, \mathbf{V}_r$ are determined by the compact singular value decomposition of the matrix $\mathbf{S}^{(p)}(\omega_n) = \mathbf{U}_r\mathbf{\Sigma}_r\mathbf{V}_r^*$. An analogue of the procedure (1.200) is written similarly:

$$\omega_{n+1} = \omega_n + p\frac{\max \text{eig}\, \mathbf{S}^{(p-1)}(\omega_n)}{\max \text{eig}\, \mathbf{S}^{(p)}(\omega_n)}. \qquad (1.202)$$

The last expression is reminiscent of the Householder method (1.184) for the solution of equation (1.183) with the exception that the operations of differentiation and calculation of the maximum eigennumber are swapped. At $p = 1$ the methods (1.201) and (1.202) will be the analogue of Newton's method, but, in contrast to (1.187), this method is based on calculating the scattering matrix $\mathbf{S}(\omega)$ and its derivative, not the matrix $\mathbf{S}^{-1}(\omega)$.

Calculation of the poles based on Cauchy's integral formula
The efficiency of the above methods (1.182), (1.183), (1.187), (1.199) and (1.200) depends significantly on the initial approximation of

the pole $\omega = \omega_n$. In addition, these methods allow one to find only one pole in the vicinity of the initial approximation. Note that the calculation methods of the poles by (1.187), (1.199) and (1.201) can use all the eigenvalues when computing these approximations. However, this approach does not guarantee that *all* the poles of the scattering matrix can be calculated in the region of interest D.

The following is an approach based on Cauchy's integral to calculate all the poles of the scattering matrix in a given area. This approach is widely used in the calculation of modes of multilayer slab waveguides [38, 39] and photonic waveguides [34,40].

Following [24,25,26], we calculate two contour integral of matrix-valued functions:

$$C_1 = \frac{1}{2\pi i} \oint_\Gamma S(\omega)d\omega,$$
$$C_2 = \frac{1}{2\pi i} \oint_\Gamma \omega S(\omega)d\omega,$$
(1.203)

where Γ is the boundary of region D. Using the representation (1.177) it is easy to get

$$C_1 = \sum_n B_n ;$$
$$C_2 = \sum_n \omega_p^{(n)} B_n ,$$
(1.204)

where $B_n = \operatorname*{Res}_{\omega = \omega_p^{(n)}} S(\omega)$. Note that in [40] several examples are used to demonstrate the possibility of finding all the poles, bounded by contour Γ, on the basis of the integrals (1.203). It should be noted that in [40] the matrices B_n are symmetric and the columns of the matrix L (1.181) are orthogonal. These assumptions generally do not hold for the scattering matrices of diffraction gratings [27].

Consider the method for solving equations (1.204) under more general assumptions. For this we write (1.204) in the notation of (1.181):

$$C_1 = LR;$$
$$C_2 = L\Omega_p R.$$
(1.205)

As before, we assume that the columns of the matrix L are linearly independent. This condition is more common than the orthogonality condition, used in [40]. The system of equations (1.205) can be solved for the unknown diagonal matrix Ω_p using the method described above. Following this method, the matrix $\mathbf{\Omega}_p$ is calculated by the formula

$$\mathbf{\Omega}_p = \operatorname{diag} \operatorname{eig}(\mathbf{U}_r^* \mathbf{C}_2 \mathbf{V}_r \mathbf{\Sigma}_r^{-1}). \qquad (1.206)$$

where the matrices $\mathbf{U}_r, \mathbf{\Sigma}_r, \mathbf{V}_r$ are obtained from the compact singular value decomposition of matrix $\mathbf{C}_1 = \mathbf{U}_r \mathbf{\Sigma}_r \mathbf{V}_r^*$.

Thus, calculating the integrals (1.203), one can find all the poles of the scattering matrix, located inside the integration contour. The accuracy of calculation of the poles is determined by the accuracy of numerical calculation of integrals (1.203). For calculation of poles with high precision one requires a large number of calculations of the scattering matrix [34, 40]. Therefore, the method based on Cauchy's integral formula should be applied only to calculate the initial approximations of the poles, while the exact values can be obtained using the iterative procedures discussed above.

1.2.1.3. Numerical examples
This section describes the calculation of the poles of the scattering matrix of the diffraction grating using the above methods. The scattering matrix of the grating was calculated by the Fourier modal method [2, 5, 18]. We note that the condition of analyticity of the scattering matrix requires proper implementation of the Fourier modal method for complex frequencies [28, 35].

Consider the calculation of eigenmodes of the diffraction grating corresponding to the periodic array of slits in a silver film, located on an dielectric waveguide layer (Fig. 1.10). We are interested in the modes which can be excited by a normally incident plane wave ($k_{x,0} = 0$). The grating parameters are shown in the caption to Fig. 1.10. The dielectric constant of silver was described by the Drude–Lorentz model [50], and the analytical expression for $\varepsilon_{Ag}(\omega)$ is considered as a function of complex frequency. Note that the analytical function $\varepsilon(\omega)$ is a prerequisite for the analyticity of $\mathbf{S}(\omega)$. If this condition is not met, then the convergence rate of the iterative methods significantly deteriorates, and the method based on Cauchy's integral theorem even stops working. Furthermore, we assume that the frequency of the mode of the grating is calculated correctly if max svd $\mathbf{S}(\omega) \geq 10^{10}$, where max svd is the maximum singular value of the scattering matrix. This inequality will be used as a criterion for stopping the iterative algorithm. Numerical calculations show that this condition for the given grating corresponds to the accuracy of determining the poles $|\Delta\lambda_p| < 10^{-8}$ nm. This accuracy is sufficient for practical applications, including the calculation of the field distribution of the modes.

Fig. 1.10. A silver grating on a waveguide layer (grating parameters: period $d = 1000$ nm, the grating height $h_{gr} = 50$ nm, the slit width $a = 200$ nm, the thickness of the waveguide layer $h = 800$ nm, the permittivity of the layer $\varepsilon = 5.5$).

Method based on Cauchy's integral formula
We find all the modes of the grating in the frequency range $\omega \in [1.0 \cdot 10^{15} \text{ s}^{-1}; 1.8 \cdot 10^{15} \text{ s}^{-1}]$. We shall consider only the high-Q modes ($|\text{Im } \omega_p| < 5 \cdot 10^{12} \text{ s}^{-1}$). In further calculations, we used the value $N' = 21$ for the number of Fourier harmonics.

To assess the mode frequencies, we calculated the contour integrals (1.203) using the trapezoidal method with for 500 points. Formula (1.206) was used to obtain 10 approximate values of the poles. These values are listed in the first column of Table 1.2. The second column shows the exact frequency of the poles, calculated using an iterative procedure (1.200). The third column contains the error of initial approximation. The table shows that the calculation of the poles based on the method of contour integration is sufficiently accurate.

For comparison, we note that if we calculate the initial approximations of the poles by the calculation of the scattering matrix on the grid consisting of 500×20 points (10 000 calculations of the scattering matrix), we are able to find only 9 poles. This is due to the fact that the poles ω_6, ω_7 are very close to each other. Thus, using the method based on Cauchy's integral formula is justified in this case.

Table 1.2. The poles, calculated by the formula (1.206), and their refinement by different iterative methods

	ω_{cauchy}	ω_{exact}	$|\omega_{exact} - \omega_{cauchy}|$	n_{iter} (1.200)	(1.187)	(1.182)
ω_1	$1.14246 \times 10^{15} -$ $3.35055 \times 10^{12}i$	$1.14247 \times 10^{15} -$ $3.34402 \times 10^{12}i$	7.8665×10^{9}	1	2	3
ω_2	$1.15554 \times 10^{15} -$ $6.85407 \times 10^{11}i$	$1.15551 \times 10^{15} -$ $6.86647 \times 10^{12}i$	3.40401×10^{10}	2	2	3
ω_3	$1.40431 \times 10^{15} -$ $1.87394 \times 10^{12}i$	$1.40428 \times 10^{15} -$ $1.87643 \times 10^{12}i$	3.45275×10^{10}	2	2	3
ω_4	$1.48168 \times 10^{15} -$ $3.86736 \times 10^{12}i$	$1.48168 \times 10^{15} -$ $3.86743 \times 10^{12}i$	2.35895×10^{10}	1	2	3
ω_5	$1.5047 \times 10^{15} -$ $1.11064 \times 10^{12}i$	$1.50469 \times 10^{15} -$ $1.11116 \times 10^{12}i$	1.52001×10^{10}	2	2	3
ω_6	$1.66425 \times 10^{15} -$ $7.11868 \times 10^{11}i$	$1.66425 \times 10^{15} -$ $7.12688 \times 10^{11}i$	3.4421×10^{9}	1	2	3
ω_7	$1.66551 \times 10^{15} -$ $1.0516 \times 10^{11}i$	$1.66545 \times 10^{15} -$ $1.06986 \times 10^{12}i$	5.64199×10^{10}	2	2	4
ω_8	$1.73171 \times 10^{15} -$ $3.70973 \times 10^{12}i$	$1.73171 \times 10^{15} -$ $3.7085 \times 10^{12}i$	1.23369×10^{9}	1	2	3
ω_9	$1.73768 \times 10^{15} -$ $2.4303 \times 10^{12}i$	$1.73768 \times 10^{15} -$ $2.43055 \times 10^{12}i$	2.84178×10^{10}	1	2	3
ω_{10}	$1.7766 \times 10^{15} -$ $2.79575 \times 10^{12}i$	$1.77659 \times 10^{15} -$ $2.79747 \times 10^{12}i$	6.09846×10^{9}	1	2	3

Comparison of the convergence rates of iterative methods

Consider the question of the convergence rate of iterative methods (1.187) and (1.200) and Newton's method for solving the equation (1.182). First, we consider the case where we know quite good initial approximations of the poles. In this case, we apply these methods to improve estimates of the poles specified in column 2 of Table 1.2. Columns 5, 6, 7 in Table 1.2 show the number of iterations in the calculation of the poles by the methods of (1.200) (1.187) and with the use of Newton's method for solving the equation (1.182). From Table 1.2 it follows that the slowest convergence is observed for Newton's method in equation (1.182). The proposed method (1.200) and the method (1.187) require approximately the same number of iterations.

Let us now consider the convergence of iterative methods in more detail when the initial approximations of the poles are unknown. In this case, the following numerical experiment was carried out. A random point was selected from the considered frequency range and the iterative procedure was started. Tables 1.3–1.5 give the average percentage of the initial approximations (n_{conv}) converged to the pole, the average number of iterations (n_{iter}), and the average number of calculations of the scattering matrix ($n_{s\text{-}calc}$).

Note that various methods require different numbers of scattering matrix calculations at each iteration. For example, Newton's method and the method of (1.187) calculate the first derivative of the scattering matrix and require two calculations of the scattering matrix at each iteration, while Halley's method and the proposed method (1.199), (1.200) calculate the second derivative and require three scattering matrix computations per iteration. Therefore, when comparing the methods it is necessary to pay attention to the value $n_{s\text{-}calc}$ which is the total number of calculations of the scattering matrix. At the same time, at each iteration the scattering matrix can be calculated independently in parallel. Therefore, in parallel implementation of the iterative methods the value n_{iter} is more important.

Table 1.3 was calculated for $N' = 21$. The table shows that among the scalar methods of solving the equations (1.182) and (1.183) fewer iterations are required when using Halley's method. This is because Halley's method has the order of convergence 3, while Newton'method has 2 [45]. Note also that Newton's method used for solving the equation (1.182) converges better than for the equation (1.183).

According to the number of calculations of the scattering matrix, the *matrix* methods (1.187), (1.199) and (1.200) have a better convergence compared with the scalar methods. The tables show that the proposed methods (1.199) and (1.200) require a smaller number of iterations of all the methods considered. In particular, the average number of iterations for the method (1.199) is almost 2 times smaller

Table 1.3. The convergence of iterative methods of the number of harmonics $N' = 21$ (84×84 scattering matrix); [a] – scalar method; [b] – matrix method

	n_{conv},%	n_{iter}	$n_{s\text{-}calc}$	n_{near}
Newton's method for (1.182)[a]	98.5	9.85	19.7	57.2
Newton's method for (1.183)[a]	94.9	7,08	14,16	42.4
Halley's method for (1.183)[a]	99.9	4.20	12.60	54.5
Method (1.187)[b]	99.7	5.48	10.96	63.9
Method (1.199)[b]	100	2.78	8.35	96.1
Method (1.200)[b]	100	3.40	10.2	64.6

Table 1.4. The convergence of iterative methods of the number of harmonics $N' = 61$ (244×244 scattering matrix)

	$n_{conv}, \%$	n_{iter}	n_{s-calc}	n_{near}
Newton's method for (1.182)	82.5	11.79	23.58	37.7
Newton's method for (1.183)	87.2	7.1	14.2	24.5
Halley's method for (1.183)	99.7	4.54	13.62	40.3
Method (1.187)	78	5.95	11.9	48.6
Method (1.199)	100	2.88	8.64	87.6
Method (1.200)	100	3.72	11.16	43.75

Table 1.5. The convergence of iterative methods of the number of harmonics $N' = 201$ (804×904 scattering matrix)

$n_{conv}, \%$	n_{iter}	n_{s-calc}	$n_{near}, \%$
0	–	–	–
94	6.83	13.66	26.5
99	4.85	14.55	40
0	–	–	–
100	3.86	11.58	75.5
100	3.94	11.82	45

than for the method (1.187). At the same time, the average number of calculations for the scattering matrix method (1.199) is only 1.31 times smaller than for the method (1.187). Comparing the methods (1.199) and (1.200), we note that the method of (1.200), built on the assumption of the presence of a single resonance in the vicinity of the initial approximation of the pole, requires more iterations than the general method (1200), on average 1.22 times.

We now consider the methods at a higher value of the harmonics. Table 1.4 shows that when $N' = 61$ the convergence of the method (1.187) and the Newton method for equation (1.182) is deteriorating: the first method find the pole only in 78% of cases, the second is 83% of cases. If $N' = 201$ (Table 1.5), the method (1.187) and Newton's method for (1.182) stop working. Newton's and Halley's methods for the solution of equation (1.183) converge, but require about 30% more iterations (the number of calculations of the scattering matrix) than the proposed methods (1.199) and (1.200). Note that methods (1.199) and (1.200) converge in 100% of cases.

In addition to the rate of convergence of another important characteristic of the iterative method is the shape of the attraction basin. Speaking about the calculation of the pole, the attraction basin of the pole ω_p is a set of points from which the iterative method converges to ω_p. Ideally, the method should converge to the nearest pole. In this case, for the variation of some grating parameter one can calculate the corresponding change in the eigenfrequency of the mode. Therefore, we can easily calculate the functional dependence of the frequency on some parameter (for example, one can calculate the dispersion curves by varying the wave number $k_{x,0}$). Unfortunately, Newton's method (like Halley' method) is characterized by fractal attraction basins [51].

Figure 1.11 represents the attraction domains for the region $\mathrm{Re}\omega_0 \in [1.64 \cdot 10^{15}\ \mathrm{s}^{-1};\ 1.75 \cdot 10^{15}\ \mathrm{s}^{-1}]$, $\mathrm{Im}\omega_0 \in [-5 \cdot 10^{12}\ \mathrm{s}^{-1};\ 5 \cdot 10^{12}\ \mathrm{s}^{-1}]$. We used a number of harmonics $N = 21$. In accordance with Fig. 1.11 the attraction basins of the proposed method (1.199) have a fairly regular form, while for the rest of the methods the attraction domains have a distinctive fractal shape.

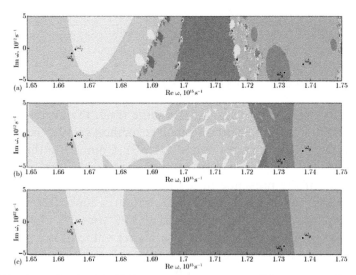

Fig. 1.11. Attraction basins (different colours represent different attraction poles): (a) Newton's method for (1.182);b the equation (1.187); (c) the equation (1.199).

As a numerical characteristic of the shape of the attraction basins we consider the percentage of the starting points for which the method converged to the nearest pole. This value (n_{near}), presented in the last column of the Tables 1.3–1.5, should be as close to 100% as possible. As shown in the Tables 1.3–1.5, the proposed method (1.199) in the majority of cases (70–90%) converges to the nearest pole. At the same time, for the other methods the value n_{near} lies in the range from 20% to 60%.

Tables 1.3–1.5 show that the performance of the methods considered decreases with increasing dimensions of the scattering matrix. Note that for all methods except, method (1.187), it is possible to use only part of the scattering matrix of the large dimension. For example, we can use the central part 84 × 84 in size of the large scattering matrix, calculated with the number of harmonics $N' = 201$. In this case, it is expected that the performance of the methods (1.199) and (1.200), as well as the Newton and Halley methods, to solve the equations (1.182) and (1.183) will be close to the values given in Table 1.3.

1.2.2. Calculation of events based on calculating eigenvalues of the transfer matrix

1.2.2.1. Calculation of modes

We now consider the method of calculation of eigenmodes of the structures based on the use of a modification of the Fourier modal method for non-periodic structures, described in subsection 1.1.4. Suppose that the structure in question is periodic along the z axis with the period d and consists of L regions (layers) within a period, in each of which the material parameters depend only on the coordinate x (Fig. 1.12). To calculate the eigenmodes of the structure using the Fourier modal method, the authors of [21,52] introduced artificial periodization along the x axis with a certain period d'. The interaction between adjacent periods in the x axis was prevented by adding special absorbing layers (for example, perfectly matched absorbing layers are discussed in section 1.1.4.2).

After the introduction of artificial periodization the structure can be regarded as a periodic structure along the axis x, which consists of an infinite number of layers along the axis z. In each of the layers at a selected frequency (wavelength in free space) the field can be represented in the form (1.58). Similar to the expression (1.71), the boundary conditions of the equality of the tangential components of the electromagnetic field on the boundaries of the layers can be written as

$$\mathbf{M}_{l-1}\mathbf{C}_{l-1} = \mathbf{N}_l\mathbf{C}_l, \; l = ..., -1, 0, 1, ..., L, ..., \tag{1.207}$$

where the vector $\mathbf{C}_l$ and the matrices $\mathbf{N}_l$ and $\mathbf{M}_l$ are given by (1.66), (1.67) and (1.68) respectively.

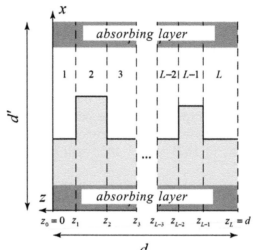

Fig. 1.12. The geometry of the structure.

From (1.207) we can obtain the relationship of the wave amplitudes in the layers with numbers 1 and $L + 1$:

$$\mathbf{C}_{L+1} = \left(\prod_{l=1}^{L} \mathbf{N}_{l+1}^{-1} \mathbf{M}_l \right) \mathbf{C}_1 = \mathbf{TC}_1, \tag{1.208}$$

where $\mathbf{T}$ is the so-called the transfer matrix. Since the structure is periodic in the z-axis direction, and its period is composed of L areas (layers) along this axis, $L + 1$-th layer coincides with the first layer. Consequently, the following equation is fulfilled for the eigenmodes of the structure :

$$\mathbf{C}_{L+1} = \beta \mathbf{C}_1, \tag{1.209}$$

where $\beta = \exp(ik_0 n_{eff} d)$, n_{eff} is the effective refractive index of the mode (normalized propagation constant). Taking into account (1.209) and (1.208), we obtain:

$$\mathbf{TC}_1 = \beta \mathbf{C}_1. \tag{1.210}$$

Thus, the problem of finding the propagation constants of modes is reduced to the problem of computing the eigenvalues of the transfer matrix. The solution of the eigenvalue problem (1.210) is a set of values β, from which we can determine n_{eff}. Representing n_{eff} as the sum of the real and imaginary parts $n'_{eff} + n''_{eff} i$, and we have:

$$\beta = \exp\left[ik_0 \left(n'_{eff} + n''_{eff} i \right) d \right] = \exp\left(ik_0 n'_{eff} d \right) \exp\left(-k_0 n''_{eff} d \right). \tag{1.211}$$

From (1.211) it follows immediately that:

$$n''_{eff} = -\frac{\ln|\beta|}{k_0 d}.$$ (1.212)

Also from (1.211) we can obtain the following expression :

$$n'_{eff} = \frac{2\pi l + \arg \beta}{k_0 d},$$ (1.213)

where l is an integer. When $l = 0$ the propagation constant of the mode $k_0 n'_{eff} d$ lies in the range $[-\pi/d, \pi/d]$, i.e. in the first Brillouin zone.

It should be noted that the calculation of the transfer matrix of from formula (1.208) can lead to a numerical instability similar to the numerical instability when calculating the matrix of the equation system (1.80). In both cases, the numerical instability is connected with the necessity of inverting the matrices N_l and M_l, which include, as a factor, the matrices $X_l^{(+)}$ and $X_l^{(-)}$ given by (1.69) and represent diagonal matrices with elements of the form $\exp\left(\lambda^{(+)}k_0\left(z_l - z_{l-1}\right)\right)$ and $\exp\left(-\lambda^{(-)}k_0\left(z_l - z_{l-1}\right)\right)$. All the real parts of the exponents are negative, thus with the inversion of these matrices can cause a numerical overflow. Below is a stable algorithm suggested by the authors [53] for calculating the propagation constants and the distribution of the electromagnetic field of the modes of the structure is presented.

1.2.2.2. Calculation of the field distribution

We introduce the scattering matrix $S_l^{(L)}$ linking the amplitudes of the waves on the left boundaries of the layers with numbers 1 and l:

$$\begin{pmatrix} C_l^{(+,L)} \\ C_1^{(-,L)} \end{pmatrix} = S_l^{(L)} \begin{pmatrix} C_1^{(+,L)} \\ C_l^{(-,L)} \end{pmatrix}.$$ (1.214)

Also, we introduce the matrix $S_l^{(R)}$ linking the wave amplitudes on the right boundaries of the layers with numbers l and $L + 1$:

$$\begin{pmatrix} C_{L+1}^{(+,R)} \\ C_l^{(-,R)} \end{pmatrix} = S_l^{(R)} \begin{pmatrix} C_l^{(+,R)} \\ C_{L+1}^{(-,R)} \end{pmatrix}.$$ (1.215)

According to (1.58), (1.67) and (1.68)

$$C_l^{(-,L)} = X_l^{(-)}C_l^{(-)}, \ C_l^{(+,L)} = C_l^{(+)},$$
$$C_l^{(-,R)} = C_l^{(-)}, \ C_l^{(+,R)} = X_l^{(+)}C_l^{(+)}. \tag{1.216}$$

Stable construction of the scattering matrix (1.214) and (1.215) is achieved similar to the process of constructing the scattering matrix as described above in section 1.2.6. We use the following block representation for the matrices $S_l^{(L)}$ and $S_l^{(R)}$:

$$S_l^{(L)} = \begin{pmatrix} T_{++,l}^{(L)} & R_{+-,l}^{(L)} \\ R_{-+,l}^{(L)} & T_{--,l}^{(L)} \end{pmatrix}, \ S_l^{(R)} = \begin{pmatrix} T_{++,l}^{(R)} & R_{+-,l}^{(R)} \\ R_{-+,l}^{(R)} & T_{--,l}^{(R)} \end{pmatrix}. \tag{1.217}$$

We write the equations (1.214) and (1.215) at $l = L +1$ in (1.214) and $l = 1$ in (1.215), considering the equalities (1.216):

$$\begin{pmatrix} I & 0 \\ 0 & X_l^{(-)} \end{pmatrix} \begin{pmatrix} C_{L+1}^{(+)} \\ C_l^{(-)} \end{pmatrix} = S_{L+1}^{(L)} \begin{pmatrix} I & 0 \\ 0 & X_{L+1}^{(-)} \end{pmatrix} \begin{pmatrix} C_1^{(+)} \\ C_{L+1}^{(-)} \end{pmatrix}, \tag{1.218}$$

$$\begin{pmatrix} X_{L+1}^{(+)} & 0 \\ 0 & I \end{pmatrix} \begin{pmatrix} C_{L+1}^{(+)} \\ C_l^{(-)} \end{pmatrix} = S_l^{(R)} \begin{pmatrix} X_l^{(+)} & 0 \\ 0 & I \end{pmatrix} \begin{pmatrix} C_1^{(+)} \\ C_{L+1}^{(-)} \end{pmatrix}. \tag{1.219}$$

We rewrite the first equation of (1.218), and the second equation of (1.219) using a block representation (1.217):

$$\begin{pmatrix} C_{L+1}^{(+)} \\ C_1^{(-)} \end{pmatrix} = \begin{pmatrix} T_{++,L+1}^{(L)} & R_{+-,L+1}^{(L)}X_{L+1}^{(-)} \\ R_{-+,1}^{(R)}X_1^{(+)} & T_{--,1}^{(R)} \end{pmatrix} \begin{pmatrix} C_1^{(+)} \\ C_{L+1}^{(-)} \end{pmatrix}. \tag{1.220}$$

We rewrite (1.220), transferring $C_{L+1}^{(-)}$ to the left side:

$$\begin{pmatrix} I & -R_{+-,L+1}^{(L)}X_{L+1}^{(-)} \\ 0 & -T_{--,1}^{(R)} \end{pmatrix} \begin{pmatrix} C_{L+1}^{(+)} \\ C_{L+1}^{(-)} \end{pmatrix} = \begin{pmatrix} T_{++,L+1}^{(L)} & 0 \\ R_{-+,1}^{(R)}X_1^{(+)} & -I \end{pmatrix} \begin{pmatrix} C_1^{(+)} \\ C_1^{(-)} \end{pmatrix}. \tag{1.221}$$

Taking into account (1.209) in (1.221), we obtain a generalized eigenvalue problem for finding β values:

$$\begin{pmatrix} \mathbf{T}^{(L)}_{++,L+1} & \mathbf{0} \\ \mathbf{R}^{(R)}_{-+,1}\mathbf{X}^{(+)}_1 & -\mathbf{I} \end{pmatrix}\begin{pmatrix} \mathbf{C}^{(+)}_1 \\ \mathbf{C}^{(-)}_1 \end{pmatrix} = \beta\begin{pmatrix} \mathbf{I} & -\mathbf{R}^{(L)}_{+-,L+1}\mathbf{X}^{(-)}_{L+1} \\ \mathbf{0} & -\mathbf{T}^{(R)}_{--,1} \end{pmatrix}\begin{pmatrix} \mathbf{C}^{(+)}_1 \\ \mathbf{C}^{(-)}_1 \end{pmatrix}. \quad (1.222)$$

Since scattering matrices $\mathbf{S}^{(L)}_l$ and $\mathbf{S}^{(R)}_l$ are calculated using a stable algorithm, construction and solution of the eigenvalue problem (1.222) does not entail numerical overflows.

The solution of the generalized eigenvalue problem (1.222) in addition to a set of values β yields a set of characteristic vectors $\mathbf{C}^{(\pm)}_1$ describing the electromagnetic field of the eigenmodes in the first layer of the structure. To calculate the wave amplitudes in an arbitrary layer with number l, we write expressions (1.214) and (1.215) with (1.216) and (1.217) taken into account:

$$\begin{pmatrix} \mathbf{I} & \mathbf{0} \\ \mathbf{0} & \mathbf{X}^{(-)}_1 \end{pmatrix}\begin{pmatrix} \mathbf{C}^{(+)}_l \\ \mathbf{C}^{(-)}_1 \end{pmatrix} = \begin{pmatrix} \mathbf{T}^{(L)}_{++,l} & \mathbf{R}^{(L)}_{+-,l}\mathbf{X}^{(-)}_l \\ \mathbf{R}^{(L)}_{-+,l} & \mathbf{T}^{(L)}_{--,l}\mathbf{X}^{(-)}_l \end{pmatrix}\begin{pmatrix} \mathbf{C}^{(+)}_1 \\ \mathbf{C}^{(-)}_l \end{pmatrix}, \quad (1.223)$$

$$\begin{pmatrix} \mathbf{X}^{(+)}_{L+1} & \mathbf{0} \\ \mathbf{0} & \mathbf{I} \end{pmatrix}\begin{pmatrix} \mathbf{C}^{(+)}_{L+1} \\ \mathbf{C}^{(-)}_l \end{pmatrix} = \begin{pmatrix} \mathbf{T}^{(R)}_{++,l}\mathbf{X}^{(+)}_l & \mathbf{R}^{(R)}_{+-,l} \\ \mathbf{R}^{(R)}_{-+,l}\mathbf{X}^{(+)}_l & \mathbf{T}^{(R)}_{--,l} \end{pmatrix}\begin{pmatrix} \mathbf{C}^{(+)}_l \\ \mathbf{C}^{(-)}_{L+1} \end{pmatrix}. \quad (1.224)$$

Substituting the second equation of (1.224) into the first equation of (1.223), and transforming the result, we obtain:

$$\mathbf{C}^{(+)}_l = \left(\mathbf{I} - \mathbf{R}^{(L)}_{+-,l}\mathbf{X}^{(-)}_l\mathbf{R}^{(R)}_{-+,l}\mathbf{X}^{(+)}_l\right)^{-1}\left(\mathbf{T}^{(L)}_{++,l}\mathbf{C}^{(+)}_1 + \mathbf{R}^{(L)}_{+-,l}\mathbf{X}^{(-)}_l\mathbf{T}^{(R)}_{--,l}\mathbf{C}^{(-)}_{L+1}\right). \quad (1.225)$$

Substituting the first equation of (1.223) into the second equation of (1.224), we have:

$$\mathbf{C}^{(-)}_l = \left(\mathbf{I} - \mathbf{R}^{(R)}_{-+,l}\mathbf{X}^{(+)}_l\mathbf{R}^{(L)}_{+-,l}\mathbf{X}^{(-)}_l\right)^{-1}\left(\mathbf{R}^{(R)}_{-+,l}\mathbf{X}^{(-)}_l\mathbf{T}^{(L)}_{++,l}\mathbf{C}^{(+)}_1 + \mathbf{T}^{(R)}_{--,l}\mathbf{C}^{(-)}_{L+1}\right). \quad (1.226)$$

Given the equality $\mathbf{C}^{(-)}_{L+1} = \mathbf{C}^{(-)}_1\beta$ and combining (1.225) and (1.226), we finally obtain:

$$\begin{pmatrix} \mathbf{C}^{(+)}_l \\ \mathbf{C}^{(-)}_l \end{pmatrix} = \begin{pmatrix} \mathbf{I} - \mathbf{R}^{(L)}_{+-,l}\mathbf{X}^{(-)}_l\mathbf{R}^{(R)}_{-+,l}\mathbf{X}^{(+)}_l & \mathbf{0} \\ \mathbf{0} & \mathbf{I} - \mathbf{R}^{(R)}_{-+,l}\mathbf{X}^{(+)}_l\mathbf{R}^{(L)}_{+-,l}\mathbf{X}^{(-)}_l \end{pmatrix}^{-1} \cdot$$
$$\cdot \begin{pmatrix} \mathbf{T}^{(L)}_{++,l} & \mathbf{R}^{(L)}_{+-,l}\mathbf{X}^{(-)}_l\mathbf{T}^{(R)}_{--,l} \\ \mathbf{R}^{(R)}_{-+,l}\mathbf{X}^{(-)}_l\mathbf{T}^{(L)}_{++,l} & \mathbf{T}^{(R)}_{--,l} \end{pmatrix}\begin{pmatrix} \mathbf{C}^{(+)}_1 \\ \mathbf{C}^{(-)}_1\beta \end{pmatrix}. \quad (1.227)$$

The expressions (1.227) for the wave amplitudes $\mathbf{C}^{(\pm)}_l$ are stable because they do not contain inversions of the potentially ill-posed matrices containing the matrices $\mathbf{X}_l^{(\pm)}$ as factors. The calculated amplitudes may be used in the description of the field (1.58) for stable calculations of the field in the structure.

1.2.2.3. Numerical example

As an example, consider the calculation of the modes of a diffraction grating for forming high-frequency interference patterns of evanescent waves and discussed below in section 1.5.2.1 (Figs. 1.57–1.60). Calculations are carried out using the stable algorithm (expressions (1.222) and (1.227)) described above in subsection 1.2.2.2. Fig. 1.13 shows the distribution of the intensity of the electric field corresponding to the interference of two eigenmodes of the structure propagating in opposite directions with the propagation constants closest to zero (respectively excited at normal incidence). The distributions in Figures 1.13a and b have been calculated with the number of harmonics $N = 200$ and $N = 400$ respectively (the total number of harmonics is $2N + 1$). In this case the numerical instability is not observed, and for a sufficiently large number of Fourier harmonics the calculated field distribution converges and is close to the field distribution in the structure at resonance (Fig. 1.59).

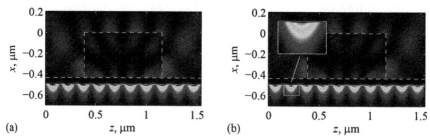

Fig. 1.13. Distribution of electric field intensity $|\mathbf{E}|^2$ corresponding to the interference of two eigenmodes of the considered structure propagating in opposite directions and calculated using the proposed stable algorithm at $N = 200$ (a) and $N = 400$ (b).

1.3. Resonant diffraction gratings for transformation of optical pulses

The interaction of light with the resonant gratings is the subject of intense research. The resonant gratings are characterised by a wide range of extraordinary (resonance) optical effects, including extraordinary optical transmission, the total absorption of incident radiation, sharp (resonant) changes of the transmission and reflection spectrum [54–59]. The resonance characteristics in the transmission spectrum (reflection) are associated with the excitation of the quasi-waveguide modes [28,60] and the emergence of new propagating diffraction orders (Rayleigh–Wood anomalies) [42,43].

Of great interest is the diffraction of optical pulses by the resonant gratings. While diffracting by the resonant grating there can occur strong changes in the shape of the pulse envelope [61, 62]. This makes the resonant gratings an effective tool for performing the specified transformations of the pulse envelope. The basic operations of conversion of the pulses are differentiation and integration of the pulse envelope. Currently, the integration and differentiation of optical pulses are carried out using various options for Bragg gratings [63–68]. The longitudinal dimensions of Bragg gratings for the integration (differentiation) of picosecond pulses range from a few millimeters to several centimeters.

In this section we discuss possible transformations of the optical pulse that can be performed by the diffraction gratings in the vicinity of the waveguide resonance frequencies and Rayleigh–Wood anomalies. The form of the performed transformations is determined by the general form of the complex transmission coefficient of the diffraction grating in the vicinity of the frequency of these anomalies. It is shown that the diffraction gratings in the vicinity of the resonance frequency waveguide may perform differentiation and integration operations (with an exponential weight function). In the vicinity of the frequency of the Rayleigh–Wood anomalies the diffraction gratings allow one to calculate the fractional derivative and the integral of the order of 1/2.

1.3.1. The integration and differentiation of optical pulses. Wood's anomalies

Consider an optical pulse with a central frequency of ω_0 and the envelope $P(t)$, propagating in the positive direction on the axis Oz. The field of the pulse is:

$$E(z,t) = \exp\left(ik(\omega_0)z - i\omega_0 t\right)P_{inc}\left(t - z/v_g\right) =$$
$$= \int_{-\infty}^{\infty} F(\omega - \omega_0)\exp\left(ik(\omega)z - i\omega t\right)d\omega, \tag{1.228}$$

where the function $E(z,t)$ represents, depending on the polarization, the x- or y-component of the electric field, $k(\omega) = \sqrt{\varepsilon}\,\omega/c$ is the wave number, $v_g = c/\sqrt{\varepsilon}$ is the group velocity, ε is the dielectric constant of the medium, $F(\omega)$ is the spectrum of the pulse envelope. Note that the function $E(z,t)$ represents the analytical signal [20].

Suppose that the pulse is normally incident on the diffraction grating (Fig. 1.14). The envelope of the transmitted pulse in the 0-th diffraction order is as follows:

$$P_{\mathrm{tr}}(t) = \int_{-\infty}^{\infty} T(\omega + \omega_0)F(\omega)\exp(-i\omega t)\,d\omega \qquad (1.229)$$

where $T(\omega)$ is the integrated transmittance (the complex amplitude of the 0-th transmitted diffraction order) as a function of frequency. According to (1.229), the conversion of the incident pulse envelope by the diffraction grating in the 0-th diffraction order corresponds to the passage of the signal $P(t)$ through a linear system with the transfer function (TF) $H(\omega) = T(\omega + \omega_0)$.

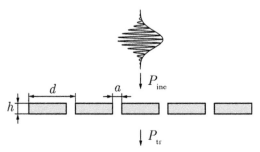

Fig. 1.14. The geometry of the structure under consideration: incident pulse; binary metal diffraction grating.

Consider the transformation of the envelope of the incident pulse, which can be performed by the diffraction grating in the vicinity of the waveguide resonance frequency corresponding to the excitation of the quasi-guided modes in the grating. For this we use the resonant scattering matrix representation (1.180). Consider the case where in the vicinity of the central frequency of the incident pulse ω_0 the grating has a resonance frequency corresponding to the mode with frequency ω_p. Consider a scattering matrix element corresponding to the scattering of the normally incident plane wave into the zeroth transmitted diffraction order. From the approximation (1.180) in the vicinity of the waveguide resonance frequencies

(Wood's anomalies) we obtain the following approximation for the transmission coefficient:

$$T(\omega) \approx a + \frac{b}{\omega - \omega_p} = a \frac{\omega - \omega_z}{\omega - \omega_p},$$

$$(1.230)$$

wherein a is the non-resonant transmission coefficient while $b/(\omega - \omega_p)$ describes resonant transmission. The value b can be regarded as the coupling coefficient between the mode and the incident light. It should be noted that each transmission zero ω_p corresponds to complex pole $\omega_z = \omega_p - b/a$. Further, the transmittance $T(\omega)$ is considered as a function of the *real* frequency.

By substituting equations (1.230) into (1.229), we can obtain an approximate expression for the envelope of the transmitted pulse. The general form of the transformation of the envelope of the incident pulse, which can be performed by the diffraction grating in the vicinity of the frequencies of Wood's anomalies can be described by the TF $H(\omega) = T(\omega - \omega_0)$, where $T(\omega)$ has the form (1.230). We show that the diffraction grating with TF of the above form allows one to perform integration and differentiation of the pulse envelope. The integration of the signal is performed by the LTI system with the TF $H_{int}(\omega) = -1/i\omega$. Due to the pole at $\omega = 0$ the specified TF is unfeasible. In practice, we can obtain only the approximate evaluation of the integral via a linear system with the following TF:

$$\tilde{H}_{int}(\omega) \sim \frac{b}{\omega + i/\tau},$$

$$(1.231)$$

where constant b determines the amplitude of the transmitted signal, $\tau \geq 0$ is a time constant, which determines the accuracy of the integration. The TF of the diffraction grating $H(\omega) = T(\omega + \omega_0)$ takes the form (1.231) when $a = 0$ and $\omega_0 = \mathrm{Re}\,\omega_p$ in (1.230). Here $\tau = i/(\omega_0 - \omega_p) = 1/(-\mathrm{Im}\,\omega_p)$. Note that under the terms of causality $\mathrm{Im}\,\omega_p \leq 0$. The pulse response of the system with the TF (1.231) has the form $h(t) = ib\exp(-t/\tau)\theta(t)$ (see [69]), where $\theta(t)$ is the Heaviside function. The transmitted signal is given by convolution $P_{tr}(t) = P_{inc}(t) * h(t)$. Thus, we have found that a diffraction grating allows to integrate the input signal with an exponential weight:

$$P_{tr}(t) = -ib \int_0^{+\infty} P_{inc}(t - T)\exp(-T/\tau)dT = -ib \int_{-\infty}^t P_{inc}(T)\exp(-(t - T)/\tau)dT.$$

$$(1.232)$$

Consider now the differentiation operation. The differentiation of the signal is performed by a linear system with the TF $H_{\text{diff}}(\omega) = -i\omega$. The differentiation operation is unstable with respect to high-frequency noise. Nevertheless, the differentiation can be carried out using the system with the following TF:

$$\tilde{H}_{\text{diff}}(\omega) = \tilde{a}\omega\frac{1}{\omega\tau + i}, \cdot \qquad (1.233)$$

where $\tilde{a} = a\tau$ determines the amplitude of the transmitted signal. The last factor in (1.233) corresponds to the integrator and plays the role of the regularizing low-pass filter. TF of the diffraction grating (1.230) takes the form (1.233) at $\omega_0 = \omega_z \in \mathbb{R}$ where ω_z is the complex-valued zero of the function $T(\omega)$. At the same time $\tau = i/(\omega_0 - \omega_p)$ is the integrated analogue of the time constant.

Thus, it can be said that the differentiation is carried out at the zeros of the function $T(\omega)$ and integration at the poles. TF of the ideal differentiator is obtained from the TF (1.233) with $|\tau| \to 0$ (at an infinite distance of the pole from the central frequency of the pulse ω_0). In contrast, the TF of the ideal integrator is obtained from the TF (1.231) at $\tau \to \infty$ (when the pole ω_p tends to the central frequency of the pulse ω_0).

From the energy conservation law $|T(\omega_0)|^2 \leq 1$ and the equation (1.230) it is easy to obtain the relation between the time constant τ and the amplitude $|b|$ of the signal at the output of the integrator:

$$|b| \cdot \tau \leq 1. \qquad (1.234)$$

Similarly, from the condition $|T(+\infty)|^2 \leq 1$ the relationship between τ and the amplitude of the signal $|\tilde{a}| = |T'(\omega_0)| = |a\tau|$ at the output of the differentiator is given by the expression:

$$|\tilde{a}||\tau|^{-1} \leq 1. \qquad (1.235)$$

From (1.234) and (1.235) it follows that the ideal integration $(\tau \to \infty)$ and the ideal differentiation $(\tau \to 0)$ are not possible since they correspond to the zero amplitudes (energy) of the transmitted signals.

1.3.2. Types of gratings for integration and differentiation of pulses

The analysis in the previous section refers to the diffraction grating of the general type, without any assumptions about the grating lattice (dielectric or metal), its symmetry, or the period to the wavelength ratio. Now consider the question of what kind of diffraction gratings is most suitable for integration and differentiation. The type of the function $T(\omega)$ and parameters a, b, ω_p (1.230) depend on the geometric parameters of the grating (the period, width and height of the slits, etc.). By the proper selection of these parameters we can be obtained TF functions in the form (1.231), (1.233), and manage the relationship between the quality of integration (differentiation) and the energy of the transmitted signal. Defining the configuration of the diffraction grating and the calculation of its parameters from the conditions for obtaining a given function $H(\omega)$ is a complex task that is usually solved using optimization techniques.

Suppose that the geometrical parameters of the grating having a TF of the (1.231) and (1.233) type were calculated. Because of technological errors the transmission function $T(\omega)$ of the produced grating will be different from the designed one. This will lead to some changes in the function $T(\omega)$. In some cases, the manufacture error can be compensated by changing the central frequency of the incident pulse [65–67]. Indeed, the errors in the parameters of the integrating gratings will lead to shift of the pole that can be compensated by changing the central frequency of the incident pulse. Harder is the case with differentiating gratings. Errors in the manufacture of the differentiating grating could lead to the fact that the zero of the transmission function $T(\omega)$ of the grating becomes complex. In this case, the grating will not perform differentiation under any central frequency. In the *subwavelength dielectric* grating the function of complex transmission will always have a *real* zero in the vicinity of the resonance [70]. Therefore, for the differentiation of optical pulses we should use dielectric gratings operating in transmission. Such differentiating gratings were considered in [71,72]. To differentiate in reflection we should use subwavelength dielectric gratings with a horizontal plane of symmetry. In [70] it is shown that such symmetric gratings can have zero reflection. In particular, this condition is satisfied by the Bragg gratings with a defect commonly used for differentiation [66, 68]. It should be noted that the dielectric grating can simultaneously differentiate in transmission and integrate in reflection with the same time constant.

When integrating optical pulses there is no requirement for the function $T(\omega)$ to be actually zero, so the grating may be made of conductive materials. To integrate, the value a in (1.230) must be equal to zero. This means that the integrating grating must have high transmittance at the resonance frequency $\omega_0 = \mathrm{Re}\,\omega_p$, at low background transmission. Such a transmission spectrum is characteristic of the metal gratings, with the effect of extraordinary optical transmission [54, 55]. This effect consists of the appearance of sharp peaks in the transmission spectrum of the grating. This effect is usually observed for subwavelength metal gratings supporting plasmonic modes. Note that these plasmonic gratings can integrate only TM-polarized pulses.

1.3.3. Fractional integration and differentiation of optical pulses. Rayleigh–Wood anomalies

Equation (1.230) is written on the assumption that the analytic continuation of $T(\omega)$ is a single-valued function in the vicinity of ω_p. This condition is fulfilled far from the so-called Rayleigh anomalies that are associated with the appearance of additional propagating diffraction orders. The Rayleigh frequencies are associated with the emergence of the $\pm m$-th diffraction orders, defined as $\omega_R = \dfrac{2\pi}{d}\dfrac{c}{\sqrt{\varepsilon}}\cdot m$. In the vicinity of the Rayleigh frequency the function $T(\omega)$ has specific breaks that are not described by the equation (7). The break of the function $T(\omega)$ is related to the fact that the point $\omega = \omega_R$ is a branch point [42,43]. The presence of a break allows to impletement transformation other than those that are described by equations (1.231)–(1.233).

Consider the complex transmission T as a function of the z-component of the wave vector $k_z = \sqrt{\varepsilon}\sqrt{\omega^2 - \omega_R^2}\big/c$ of the $\pm m$-th diffraction orders. In this case, the function $T(k_z)$ will be single-valued in the vinicity of frequency ω_R [61]. Single-valued is also the function T of the argument $\xi = \sqrt{\omega - \omega_R}$ since $k_z(\xi)$ is the single-valued function in the vicinity of frequency ω_R. In this case, the equations analogous to equations (1.176), (1.178) and (1.230) will hold for the function $\tilde{T}(\xi)$ (and the corresponding scattering matrix $\tilde{S}(\xi)$). Thus, in the vicinity of the Rayleigh frequency, we can use the following approximation for the transmission coefficient:

$$T(\omega) = \tilde{T}(\xi) \approx a + \frac{b}{\xi - \xi_p} = a\frac{\xi - \xi_z}{\xi - \xi_p}. \qquad (1.236)$$

where $\xi = \sqrt{\omega - \omega_R}$, and the coefficients a, b describe the resonant and non-resonant transmission. ξ_p is the complex pole, and ξ_z is the complex zero of the function $\tilde{T}(\xi)$.

Consider the conversion implemented by the grating when the pulse central frequency coincides with the Rayleigh frequency ($\omega_0 = \omega_R$). The corresponding TF $H(\omega) = T(\omega + \omega_0) = \tilde{T}(\sqrt{\omega})$ at $a = 0$ has the form:

$$H(\omega) \approx b \cdot \frac{1}{\sqrt{\omega} - \xi_p}. \tag{1.237}$$

The pulse characteristic of the system with TF (1.237) can be obtained in the form of:

$$h(t) = \frac{1}{2\pi} \int_{-\infty}^{+\infty} b \frac{e^{-i\omega t}}{\sqrt{\omega} - \xi_p} d\omega = b \left[\frac{1}{\sqrt{i\pi t}} - i\xi_p \cdot w\left(-\xi_p \sqrt{it}\right) \right] \cdot \theta(t), \tag{1.238}$$

where $\theta(t)$ is the Heaviside function, $w(x) = e^{-x^2} \operatorname{erfc}(-ix)$ is the complex error function (or Faddeeva function) [73], erfc is the complementary error function. The derivation for the integral (1.238) is given in [74]. According to (1.238), the envelope of the pulse passing through the diffraction grating has the form:

$$P_{tr}(t) = b \int_0^{+\infty} P_{inc}(t - T) \cdot \left[\frac{1}{\sqrt{i\pi T}} - i\xi_p \cdot w\left(-\xi_p \sqrt{iT}\right) \right] dT. \tag{1.239}$$

Equation (1.239), as well as the expression (1.232) describes the integration of the incident pulse, but with another decay function. We represent (1.239) as

$$P_{tr}(t) = b\sqrt{-i} \cdot \frac{1}{\Gamma(1/2)} \int_{-\infty}^{t} (t - T)^{-1/2} P_{inc}(T) dT - ib\xi_p \cdot \int_{-\infty}^{t} w\left(-\xi_p \sqrt{i(t - T)}\right) \cdot P_{inc}(T) dT.$$

$$\tag{1.240}$$

Equation (1.240) can be used to consider the transmitted signal as the fractional integral of the order of 1/2 of the input signal. In this case, the second term in (1.240) describes an additive error of calculating the fractional integral. By analogy with (1.234), from the energy conservation law $\left|\tilde{T}(0)\right|^2 \leq 1$ it is easy to obtain the following relation between the amplitude of the transmitted signal and the accuracy of the integration $\left|\xi_p\right|^{-1}$:

$$|b| \cdot \left|\xi_p\right|^{-1} \leq 1. \tag{1.241}$$

When $\xi_z = 0$, where $\xi_z = \xi_p - b/a$ is the complex zero of the function $\tilde{T}(\xi)$ in (1.236), the TF of the grating becomes:

$$H(\omega) = a\frac{\sqrt{\omega}}{\sqrt{\omega} - \xi_p} = a\xi_p^{-1}\sqrt{\omega}\frac{1}{\sqrt{\omega} \cdot \xi_p^{-1} - 1}. \qquad (1.242)$$

where the coefficient $\tilde{a} = a\xi_p^{-1}$ determines the amplitude of the transmitted signal. TF (1.242) describes the fractional differentiation of order 1/2. According to (1.242), the calculation of the half-derivative using the resonant grating can be represented as an exact calculation of the half-derivative followed by smoothing the signal by fractional integration. The TF of the ideal fractional differentiator is obtained from the TF (1.242) at $\left|\xi_p^{-1}\right| \to 0$. By analogy with (1.235) from the condition $\left|T(+\infty)\right|^2 \le 1$ we can write the following relation between the amplitude of the signal $\left|\tilde{a}\right| = \left|a\xi_p^{-1}\right|$ at the output of the fractional differentiator and the value $\left|\xi_p\right|$ (analogue of the time constant):

$$\left|\tilde{a}\right| \cdot \left|\xi_p\right| \le 1. \qquad (1.243)$$

1.3.4. Examples of calculation of pulse differentiation

1.3.4.1. Calculation of the grating for differentiating pulses
As an example, calculations were carried out of a resonant diffraction grating (inset to Fig. 1.15) for differentiating TM-polarized pulses with a central frequency $\omega_0 = 1.2153\cdot10^{15}$ s^{-1} ($\lambda_p \approx 1550$ nm). The grating parameters are shown in the legend to Fig. 1. These parameters were obtained by minimizing the constant τ provided $T(\omega_0) = 0$. To calculate the time constant τ at each iteration we built a Padé approximant of order [1/1] of the function $T(\omega_0)$ in the vicinity of the frequency ω_0. Fig. 1.15a shows the envelope of the incident pulse. Fig. 1.15b shows the modulus of the envelope of the pulse transmitted through the grating and the modulus of the analytically calculated derivative. Calculation of the transmitted pulse was conducted by the Fourier modal method (see section 1.1.2). The quality of differentiation was assessed using the sampling correlation coefficient. For the graphs in Fig. 1.15b this coefficient is 0.995, indicating the high-precision calculation of the derivative. To confirm that the differentiation occurs at resonance conditions, the complex frequencies of the eigenmodes of the grating were carried out by the scattering matrix method (see section 1.2.1). According to

calculations, the grating includes a quasi-waveguide TM mode with a complex frequency $\omega_p = 1.2226 \cdot 10^{15} - 2.6347 \cdot 10^{13} \, i \, s^{-1}$, the real part of which is close to the central frequency $\omega_0 = 1.2153 \cdot 10^{15} \, s^{-1}$.

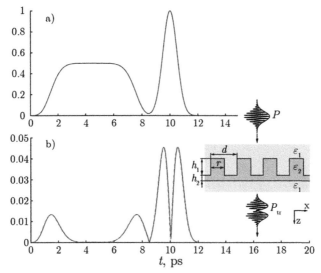

Fig. 1.15. (a) The envelope of the incident pulse. (b) modulus of the amplitude of the transmitted signal (solid line) and the modulus of the analytically calculated derivative (dashed line) (inset). The geometry of the resonant diffraction grating (period $d = 1010$ nm, height of steps $h_1 = 620$ nm, width of steps $r = 530$ nm, thickness of the layer $h_2 = 210$ nm, $\varepsilon = 2.1$, $\varepsilon = 5.5$).

1.3.4.2. Calculation of the grating for fractional differentiation of pulses

A diffraction grating was designed (inset to Fig. 1.16) to calculate the fractional derivative of order 1/2. The grating parameters are given in the caption to Fig. 1.16. These parameters were obtained by minimizing the value $|\xi_p|$ provided when $T(\omega) = 0$. To calculate the time constant ξ_p at each iteration we built a Padé approximant of order [1/1] of the function $T(\xi)$ in the vicinity of the frequency ω_0. Fig. 1.16 shows the modulus of the envelope of the transmitted signal, and the analytically calculated fractional derivative. The incident signal is the pulse shown in Fig. 1.15a.

In comparison with the previous example (Fig. 1.15b), the quality of the differentiation in Fig. 1.16 deteriorated, the correlation coefficient is 0.961. At the same time, according to said relation between the quality of differentiation and the power of the transmitted signal, the amplitude of the transmitted pulse in Fig. 1.16 is an order of magnitude higher.

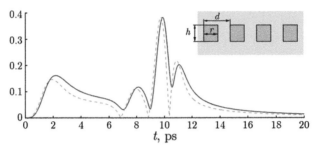

Fig. 1.16. The modulus of the amplitude of the transmitted signal (solid line) and the modulus of the analytically calculated fractional derivative (dotted line). The geometry of the resonance differentiating structure (period $d = \lambda_0/\sqrt{\varepsilon_1}$, height of the grating $h = 650$ nm, width of the step $r = 570$ nm).

1.3.5. Example of calculation of the integrating gratings

1.3.5.1. Calculation of the grating for the integration of pulses

As an example, a diffraction grating in the form of a periodic system of slits in a silver film was calculated (Fig. 1.10) and used for the integration of TM-polarized pulsed with the central frequency $\omega_0 = 1.2153\cdot10^{15}$ s^{-1} ($\lambda_p \approx 1550$ nm). The permittivity of silver was expressed by the Drude–Lorentz model [50]. The dielectric constants of the materials above the grating, under the grating and in the slits of the grating were $\varepsilon = 1$. The calculated grating parameters are shown in the legend to Fig. 1.10. The values of these parameters were obtained by optimization according to the criteria $b \to \max$, $\tau \to \max$ under the conditions $a = 0$, $\mathrm{Re}\,\omega_p = \omega_0$. To calculate the criteria b, τ, at each iteration a Padé approximant of order [1/1] of function $T(\omega)$ in the vicinity of frequency ω_0 was constructed. The function $T(\omega)$ was calculated by the Fourier modal method [2, 18, 5] (see section 1.1.2).

The transmission spectrum of the calculated grating is shown in Fig. 1.17. To explain the resonance peak in Fig. 1.17, the complex frequencies eigenmodes of the grating were calculated as the poles of the scattering matrix (see section 1.2.1). The following value of the complex frequency of the mode was obtained $\omega_p = (1.215\cdot10^{15} - 3.276\cdot10^{11}\,i)$ s^{-1}, corresponding to the resonance in Fig. 1.17. The model spectrum $I(\omega) \sim \left| b/(\omega - \omega_p) \right|^2$ for the estimated value of ω_p is shown in Fig. 1.17a by a dashed line and agrees well with the calculated spectrum. Evaluation of ω_p allows one to assess the width of the resonance at half intensity $\Delta\omega = -2\mathrm{Im}\,\omega_p = 6.552\cdot10^{11}$ s^{-1} ($\Delta\lambda = 0.836$ nm) and the time constant $\tau = 3.05$ ps.

Note that the thickness of the calculated grating is 90 nm. For comparison, the characteristic size of the Bragg grating, carrying out integration of the pulse enveloped with duration of a few picoseconds, is a few millimeters [63,64].

Fig. 1.17b presents the envelope of the incident Gaussian pulse (width FWHM 0.912 ps) and the calculated envelopes of the pulse passing through the grating. The envelope of the transmitted pulse is in good agreement with the integral (1.232) from the envelope of the incident pulse with an exponential weighting function $\exp(-t/\tau)$, where $\tau \approx 3.05$ ps. The integral (1.232) is shown in Fig. 1.17b by the dashed line.

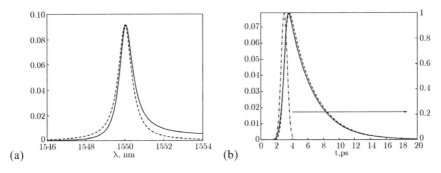

(a)

(b)

Fig. 1.17. (a) The transmission spectrum of the integrating grating (the period $d =$ 1540 nm, height $h = 80$ nm, width of slits $a = 440$ nm) and the model spectrum (dotted line). (b) The envelope of the incident pulse (dash-dot curve; axis on the right), the pulse envelope passing transmitted through the structure (solid line, axis of the left); The integral analytically calculated using the formula (1.232) (dotted line, axis of the left).

1.3.5.2. Calculation of the grating for the fractional integration of pulses

This subsection is an example of calculating the grating to compute the fractional integral. The transmission spectrum of the calculated lattice is shown in Fig. 1.18a, the grating parameters are shown in the Fig. caption. The grating has a period $d = \lambda_0 = 1550$ nm, so the Rayleigh–Wood anomalies in this grating are related to the ± 1-th diffraction orders. The scattering matrix method was used to calculate the scattering matrix pole $\mathbf{S}(\xi)$: $\xi_p = (-2.154 \cdot 10^5 + 11.025 \cdot 10^4\ i)$ $\text{s}^{-1/2}$ ($k_z = (-8.738 \cdot 10^{-3} + 4.1571\ i \cdot 10^{-4})\ \omega_0/c$). The model spectrum

$$I(\omega) \sim \left| a\frac{\sqrt{\omega - \omega_0}}{\sqrt{\omega - \omega_0} - \xi_p} \right|^2 \quad \text{at the estimated value of } \xi_p \text{ is shown in}$$

Fig. 1.18a by a dashed line. From (1.236) it is easy to obtain the following estimate for the resonance width at intensity half-maximum:

$$\Delta\omega_{RW} = \left(\sqrt{2\left(\mathrm{Im}\,\xi_p\right)^2 + \left(\mathrm{Re}\,\xi_p\right)^2} + \mathrm{Im}\,\xi_p\right)^2 + \left(\sqrt{2\left(\mathrm{Re}\,\xi_p\right)^2 + \left(\mathrm{Im}\,\xi_p\right)^2} + \mathrm{Re}\,\xi_p\right)^2.$$

$$(1.244)$$

For the given ξ_p $\Delta\omega_{RW} = 5.913 \cdot 10^{10}$ s^{-1}, which is significantly less than the minimum possible width of plasmon resonances corresponding to Wood's anomalies [75]. In Fig. 1.18b the solid line shows the envelope of the pulse passing through the diffraction grating. The incident pulse is the Gaussian pulse in Fig. 1.17b. The dotted line in Fig. 1.18b shows the integral (1.239). The difference in the continuous and dotted lines in Fig. 1.18b is explained by a difference of the transmission spectrum in comparison with the model spectrum in Fig. 1.18a. The dot-dash line in Fig. 1.18b shows the first term of the formula (1.240), corresponding to the analytically calculated fractional integral of the order of 1/2 of the envelope of the incident pulse. The dotted line indicates the second term in (1.240), calculated based on the calculated value ξ_p of the structure. The analytically calculated values from equation (1.240) are shown with a scale factor, ensuring the equality of the maximum values of the moduli of the analytical semi-integral and the transmitted signal.

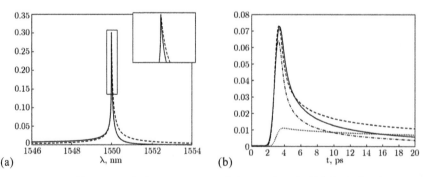

Fig. 1.18. (a) The transmission spectrum of the grating (solid line) for fractional integration (period $d = 1550$ nm, height $h = 280$ nm, width of slits $a = 60$ nm) and the model spectrum (dotted line). (b) The envelope of the pulse passing through the structure (solid line); Fractional integral of the order of 1/2 (dash-dotted line); calculation error of the fractional integral (the modulus of the second term of the formula (1.240)) (dotted line); The integral calculated by the formula (1.240) (dotted line).

Note different lengths of the transmitted pulses in Figs. 1.17 and 1.18. The pulse transmitted through the grating for fractional integration (Fig. 1.18) is longer. In the initial time interval (4 ps $< t <$ 6 ps) the pulse decays faster than the pulse in Fig. 1.17. Different duration and nature of decay of the pulses are explained by the different kind of the impulse response (weighting functions) in (1.232), (1.239). The impulse response of the integrating grating decreases as $e^{-t/\tau}$. It can be shown that the modulus of the pulse characteristic (1.238) of the grating for fractional integration at $t \to \infty$ decreases as a function of $t^{-3/2}$. The slower decay of the impulse response (1.238) leads to slower decay of the transmitted pulse [42].

1.4. Diffractive optical elements for controlling the propagation of surface plasmon–polaritons

This section describes the calculation of diffractive optical elements for controlling the propagation of surface plasmon–polaritons (SPP) – electromagnetic waves propagating in the metal–dielectric structures. In subsection 1.4.1, the derivation of the dispersion relation of the surface plasmon polaritons propagating at the interface between a metal and an insulator is considered and its characteristics are described. Subsection 1.4.2 presents the results of studies of the mechanisms of phase modulation of SPP using dielectric ridges of varying length and height used in subsection 1.4.4 for the calculation of diffractive lenses for focusing the SPP. Subsection 1.4.3 is devoted to the description of the method proposed by the authors for suppressing parasitic scattering in elements for SPP and plasmonic modes using two-layer dielectric structures. The calculation of lenses for the SPP in subsection 1.4.4 is similar to the calculation of diffractive optical elements within the scalar diffraction theory, where the propagation of the incident wave through an element is described by phase modulation of the input wave field [12, 76–78]. The phase shift at each point is calculated by solving the model problem of propagation of the incident light wave through the dielectric plate. The plate thickness is equal to the thickness of the optical element at a given point. As with the description of the light wave propagating through a diffractive optical element in the scalar diffraction theory, it is proposed to describe the passage of the SPP through the diffraction structure using the solution of the problem of diffraction of SPP on a dielectric ridge on the propagation surface. Changes in the amplitude and phase of the SPP at each point at the exit plane of the structures are considered the same as upon transmission of SPP through the ridge, in which the length and height match those of the dielectric microrelief at this point. Subsection 1.4.5 reviews the calculation of the dielectric Bragg reflective gratings for the SPP.

1.4.1. The dispersion relation and the properties of surface plasmon–polaritons

Surface plasmon–polaritons (SPP) are electromagnetic waves propagating along the interface between the metal and the dielectric [79]. In this subsection we derive the dispersion relation of surface plasmon– polaritons and examine the properties of the SPP.

1.4.1.1. Dispersion relations

We consider derivation of the dispersion relation of surface plasmon– polaritons on a flat interface between two semi-infinite isotropic media. The magnetic permeability of the medium is set equal to one. Let the interface be the plane $z = 0$, with the media with dielectric constants ε_d and ε_m corresponding to the regions $z > 0$ and $z < 0$ (Fig. 1.19). We assume that $\varepsilon_d \in \mathbb{R}$, $\varepsilon_m \in \mathbb{C}$.

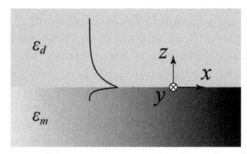

Fig. 1.19. The geometry of the structure supporting the surface plasmon–polaritons.

We seek a solution in the form

$$\mathbf{E}^{(l)}(x,y,z,t) = \mathbf{E}_0^{(l)} \exp\left[-i\left(\omega t - k_x x - k_y y - k_z^{(l)} z\right)\right],$$
$$\mathbf{H}^{(l)}(x,y,z,t) = \mathbf{H}_0^{(l)} \exp\left[-i\left(\omega t - k_x x - k_y y - k_z^{(l)} z\right)\right],$$

(1.245)

where

$$k_z^{(l)} = \pm\sqrt{k_0^2 \varepsilon_l - k_x^2 - k_y^2},$$

(1.246)

$l = d$ at $z > 0$ and $l = m$ at $z < 0$. The sign of the root in (1.246) will be selected from the following physical considerations. Without the loss of generality, we assume that the surface plasmon–polariton propagates along the x axis and in this case $k = 0$. Substituting (1.245) into (1.1), we obtain a system of linear equations for the amplitudes of the components of the electric and magnetic fields:

$$\begin{cases} k_z H_{y0}^{(l)} = k_0 \varepsilon_l E_{x0}^{(l)}, \\ k_x H_{z0}^{(l)} - k_z H_{x0}^{(l)} = k_0 \varepsilon_l E_{y0}^{(l)}, \\ -k_x H_{y0}^{(l)} = k_0 \varepsilon_l E_{z0}^{(l)}, \\ -k_z E_{y0}^{(l)} = k_0 H_{x0}^{(l)}, \\ k_z E_{x0}^{(l)} - k_x E_{z0}^{(l)} = k_0 H_{y0}^{(l)}, \\ k_x E_{y0}^{(l)} = k_0 H_{z0}^{(l)}, \end{cases} \qquad (1.247)$$

$l = d,m$. The system (1.247) is divided into two independent systems: the system

$$\begin{cases} k_z H_{y0}^{(l)} = k_0 \varepsilon_l E_{x0}^{(l)}, \\ -k_x H_{y0}^{(l)} = k_0 \varepsilon_l E_{z0}^{(l)}, \\ k_z E_{x0}^{(l)} - k_x E_{z0}^{(l)} = k_0 H_{y0}^{(l)}, \end{cases} \qquad (1.248)$$

having the amplitudes of the components H_y, E_x and E_z, and the system

$$\begin{cases} -k_z E_{y0}^{(l)} = k_0 H_{x0}^{(l)}, \\ k_x E_{y0}^{(l)} = k_0 H_{z0}^{(l)}, \\ k_x H_{z0}^{(l)} - k_z H_{x0}^{(l)} = k_0 \varepsilon_l E_{y0}^{(l)}, \end{cases} \qquad (1.249)$$

containing the amplitudes of the components E_y, H_x and H_z. The rank of the matrices of the systems (1.248) and (1.249) is equal to two, so we fix the amplitudes of the components E_y and H_y and introduce the notation $e_l = E_{y0}^{(l)}$, $h_l = H_{y0}^{(l)}$. Expressing the amplitudes $E_{x0}^{(l)}$, $E_{z0}^{(l)}$, $H_{x0}^{(l)}$, $H_{z0}^{(l)}$, through $E_{y0}^{(l)}$ and $H_{y0}^{(l)}$, we obtain a representation of the components of the electric and magnetic fields:

$$\mathbf{E}^{(l)}(x,z,t) = \left(\frac{k_z^{(l)} h_l}{k_0 \varepsilon_l}, e_l, -\frac{k_x h_l}{k_0 \varepsilon_l} \right) \exp\left[-i\left(\omega t - k_x x - k_z^{(l)} z \right) \right],$$

$$\mathbf{H}^{(l)}(x,z,t) = \left(-\frac{k_z^{(l)} e_l}{k_0}, h_l, \frac{k_x e_l}{k_0} \right) \exp\left[-i\left(\omega t - k_x x - k_z^{(l)} z \right) \right]. \tag{1.250}$$

We now write the condition of equality of the tangential components of the electric and magnetic fields at the interface at $z = 0$:

$$\begin{cases} E_y^{(m)}(x,0,t) = E_y^{(d)}(x,0,t); \\ H_y^{(m)}(x,0,t) = H_y^{(d)}(x,0,t); \\ E_x^{(m)}(x,0,t) = E_x^{(d)}(x,0,t); \\ H_x^{(m)}(x,0,t) = H_x^{(d)}(x,0,t). \end{cases} \tag{1.251}$$

Substituting (1.250) into the first two equations of the system (1.251), we obtain:

$$\begin{cases} e_m = e_d, \\ h_m = h_d. \end{cases} \tag{1.252}$$

Substituting (1.250) into the second two equations of the system (1.251), taking into account (1.252) and omitting the indices d and m of the values, e_d, h_d and e_m, h_m we have:

$$\begin{cases} \left(\dfrac{k_z^{(d)}}{\varepsilon_d} - \dfrac{k_z^{(m)}}{\varepsilon_m} \right) h = 0, \\ \left(k_z^{(d)} - k_z^{(m)} \right) e = 0. \end{cases} \tag{1.253}$$

The system of equations (1.253) has a nontrivial solution when

$$\left(\frac{k_z^{(d)}}{\varepsilon_d} - \frac{k_z^{(m)}}{\varepsilon_m} \right) \left(k_z^{(d)} - k_z^{(m)} \right) = 0. \tag{1.254}$$

In the case of media with different dielectric constants $k_z^{(d)} \neq k_z^{(m)}$ and equality (1.254) will be satisfied when

$$\frac{k_z^{(d)}}{\varepsilon_d} = \frac{k_z^{(m)}}{\varepsilon_m}. \qquad (1.255)$$

Taking into account the relation (1.246), we obtain that (1.255) holds for

$$k_x = \pm k_0 \sqrt{\frac{\varepsilon_m \varepsilon_d}{\varepsilon_m + \varepsilon_d}}. \qquad (1.256)$$

Symbol $\pm$ in equation (1.256) corresponds to the two possible directions of propagation of the SPP (in positive and negative directions of the x axis).

A non-trivial solution of (1.253) at the same time will have the form

$$\begin{cases} e = 0, \\ h \neq 0. \end{cases} \qquad (1.257)$$

Taking into account (1.257) we write down the expressions for the components of the electric and magnetic fields for the case when the SPP is propagating along the x axis:

$$\mathbf{E}^{(l)}(x,z,t) = h\left(\frac{k_z^{(l)}}{k_0 \varepsilon_l}, 0, -\frac{k_x}{k_0 \varepsilon_l}\right) \exp\left[-i\left(\omega t - k_x x - k_z^{(l)} z\right)\right],$$

$$\mathbf{H}^{(l)}(x,z,t) = h(0,1,0) \exp\left[-i\left(\omega t - k_x x - k_z^{(l)} z\right)\right]. \qquad (1.258)$$

Expressions (1.258) show that surface plasmon–polaritons propagating along the x axis are TM-polarized.

Note that the value k_x is complex, since the permittivity of a metal is complex. At the same time, according to the physical considerations the SPP must decay upon propagation. We write the expression for the y-component of the magnetic field, assuming for the sake of simplicity that $h = 1$:

$$H_y^{(l)}(x,z,t) = \exp\left[-i\left(\omega t - k_x x - k_z^{(l)} z\right)\right] = \exp\left[-i\left(\omega t - k_x' x - k_z^{(l)} z\right)\right] \exp(-k_x'' x),$$

$$(1.259)$$

where $k_x = k_x' + i k_x''$. Consider first the case $k_x' > 0$. The decay conditions of the wave at $x \to +\infty$: $\left|H_y^{(l)}(x,z,t)\right| \to 0$. From (1.259) it follows that in this case the inequality $k_x'' > 0$ holds. Consider also the wave propagating in the negative direction of the Ox axis and combining the two cases, we obtain the condition

$$k_x' k_x'' > 0. \tag{1.260}$$

Let us verify condition (1.260) for the values of the propagation constant k_x given by (1.256). From (1.256), it follows that

$$k_x' k_x'' = \frac{1}{2} k_0^2 \frac{\varepsilon_d^2 \varepsilon_m''}{\left(\varepsilon_m' + \varepsilon_d\right)^2 + \left(\varepsilon_m''\right)^2}, \tag{1.261}$$

where ε_m' and ε_m'' are the real and imaginary parts of the dielectric constant of the metal, respectively. Equation (1.261) shows that the sign of the product $k_x' k_x''$ is determined by the sign of the imaginary part of the dielectric constant of the metal. Given that the imaginary part of the absorbing material is positive, we see that the condition (1.260) is satisfied automatically.

We now write down the expression for the components $k_z^{(l)}$ substituting the dispersion relation (1.256) into (1.246):

$$k_z^{(l)} = \pm k_0 \sqrt{\frac{\varepsilon_l^2}{\varepsilon_m + \varepsilon_d}}, \tag{1.262}$$

where $l = d,m$. The sign of the root in (1.262) is determined from the condition of wave decay with increasing distance from the interface. Substituting $k_z^{(l)} = k_z'^{(l)} + i k_z''^{(l)}$ the expression (1.259), we write into the expression for the y-component of the magnetic field:

$$H_y^{(l)}\left(x,z,t\right) = \exp\left[-i\left(\omega t - k_x x - k_z^{(l)} z\right)\right] = \exp\left[-i\left(\omega t - k_x x - k_z'^{(l)} z\right)\right] \exp\left(-k_z''^{(l)} z\right). \tag{1.263}$$

From the condition $\left|H_y^{(l)}\left(x,z,t\right)\right| \to 0$ we get at $z \to \pm\infty$:

$$
\begin{aligned}
k_z''^{(d)} &> 0, \\
k_z''^{(m)} &< 0.
\end{aligned} \tag{1.264}
$$

We now find the relationships between the signs of the real and imaginary parts of values $k_z^{(l)}$. For convenience, we set

$$k_d = k_z^{(d)}, \; k_m = k_z^{(m)}. \tag{1.265}$$

For $k_d = k_d' + i k_d''$ it can be shown that

$$k_d' k_d'' = -\frac{1}{2} k_0^2 \frac{\varepsilon_d^2 \varepsilon_m''}{\left(\varepsilon_m' + \varepsilon_d\right)^2 + \left(\varepsilon_m''\right)^2}. \tag{1.266}$$

Note that the right-hand side of (1.266) differs from the right-hand side of (1.261) only in sign. Considering in the expression (1.266) the positive imaginary part of the dielectric constant of the metal, we find that

$$k'_d k''_d < 0. \tag{1.267}$$

Combining the first inequality of (1.264) with (1.267) and using the notation (1.265), we obtain:

$$k'_d < 0. \tag{1.268}$$

For $k_m = k'_m + ik''_m$ it can be shown that

$$k'_m k''_m = \frac{1}{2} k_0^2 \frac{\varepsilon''_m \left(|\varepsilon_m|^2 + 2\varepsilon'_m \varepsilon_d \right)}{\left(\varepsilon'_m + \varepsilon_d \right)^2 + \left(\varepsilon''_m \right)^2}. \tag{1.269}$$

When $\varepsilon'_m > -|\varepsilon_m|^2 / 2\varepsilon_d$ (this condition is satisfied, in particular for the air/gold and air/silver wavelengths of visible and infrared range)

$$k'_m k''_m > 0. \tag{1.270}$$

Considering the second inequality of (1.264) and taking into account the notation (1.265), when the inequality (1.270) is satisfied we have:

$$k'_m < 0. \tag{1.271}$$

For convenience we rewrite the expression for the y-component of the magnetic field (1.263), introducing the notation $\kappa_l = +\sqrt{k_x^2 - k_0^2 \varepsilon_l}$, $l = m, d$ ($\kappa_l = \kappa'_l + i\kappa''_l$):

$$H_y(x,z,t) = \begin{cases} \exp\left[-i(\omega t - k_x x)\right] \exp(-\kappa_d z), & z > 0, \\ \exp\left[-i(\omega t - k_x x)\right] \exp(\kappa_m z), & z < 0. \end{cases} \tag{1.272}$$

Similarly, from (1.267) we can show that

$$\kappa''_d > 0. \tag{1.273}$$

When the condition $\varepsilon'_m > -|\varepsilon_m|^2 / 2\varepsilon_d$ is satisfied,

$$\kappa''_m < 0. \tag{1.274}$$

We now find the value of the time-averaged Umov–Poynting vector given by [80]:

$$\overline{\mathbf{S}} = \frac{\mathrm{Re}\left[\mathbf{E} \times \mathbf{H}^*\right]}{2}, \tag{1.275}$$

where * denotes complex conjugation. Rewrite the representation of the field (1.258), separating in the exponent the real and imaginary parts of the quantities k_x and $k_z^{(l)}$:

$$\mathbf{E}^{(l)}(x,z,t) = h\left(\frac{k_z^{(l)}}{k_0 \varepsilon_l}, 0, -\frac{k_x}{k_0 \varepsilon_l}\right)\exp\left[-i\left(\omega t - k_x' x - k_z'^{(l)} z\right)\right]\exp\left[-\left(k_x'' x + k_z''^{(l)} z\right)\right],$$

$$\mathbf{H}^{(l)}(x,z,t) = h(0,1,0)\exp\left[-i\left(\omega t - k_x' x - k_z'^{(l)} z\right)\right]\exp\left[-\left(k_x'' x + k_z''^{(l)} z\right)\right]. \tag{1.276}$$

From this

$$\overline{\mathbf{S}}^{(l)}(x,z) = \frac{|h|^2}{2k_0}\left(\mathrm{Re}\left[\frac{k_x}{\varepsilon_l}\right], 0, \mathrm{Re}\left[\frac{k_z^{(l)}}{\varepsilon_l}\right]\right)\exp\left[-2\left(k_x'' x + k_z''^{(l)} z\right)\right]. \tag{1.277}$$

We rewrite the expression (1.277) with the notation (1.265):

$$\overline{\mathbf{S}}(x,z) = \begin{cases} \dfrac{|h|^2}{2k_0}\left(\dfrac{k_x'}{\varepsilon_d}, 0, \dfrac{k_d'}{\varepsilon_d}\right)\exp\left[-2\left(k_x'' x + k_d'' z\right)\right], & z > 0, \\[4mm] \dfrac{|h|^2}{2k_0}\left(\dfrac{k_x' \varepsilon_m' + k_x'' \varepsilon_m''}{|\varepsilon_m|^2}, 0, \dfrac{k_m' \varepsilon_m' + k_m'' \varepsilon_m''}{|\varepsilon_m|^2}\right)\exp\left[-2\left(k_x'' x + k_m'' z\right)\right], & z < 0. \end{cases} \tag{1.278}$$

We calculate the flux of the Umov–Poynting vector through the area of the unit width (along the coordinate y), located in the plane $x = 0$:

$$\int \overline{\mathbf{S}} \cdot d\mathbf{S} = \int \overline{\mathbf{S}} \cdot \mathbf{n} dS = \int \overline{S}_x dS = \int_0^1 dy\left[\int_0^{+\infty} \overline{S}_x^{(d)}(0,z)dz + \int_{-\infty}^0 \overline{S}_x^{(m)}(0,z)dz\right]. \tag{1.279}$$

Substituting (1.277) into (1.279), and taking into account (1.264), we obtain:

$$\int \overline{\mathbf{S}} \cdot d\mathbf{S} = \frac{|h|^2}{4k_0}\left(\mathrm{Re}\left[\frac{k_x}{\varepsilon_d}\right]\frac{1}{|k_z''^{(d)}|} + \mathrm{Re}\left[\frac{k_x}{\varepsilon_m}\right]\frac{1}{|k_z''^{(m)}|}\right). \tag{1.280}$$

In the notation of (1.265), we rewrite expression (1.280):

$$\int \overline{\mathbf{S}} \cdot d\mathbf{S} = \frac{|h|^2}{4k_0} \left(\frac{k_x'}{\varepsilon_d |k_d''|} + \frac{k_x' \varepsilon_m' + k_x'' \varepsilon_m''}{|\varepsilon_m|^2 |k_m''|} \right). \tag{1.281}$$

Consider this example: a surface plasmon–polariton with the wavelength in free space of 600 nm, propagating at the interface between silver [81] ($\varepsilon_m = -16.07 + 0.43i$) and a dielectric with the dielectric constant equal to 2.56. The normalized propagation constant of the SPP in this case is equal to $1.7449 + 0.0045i$. Fig. 1.20 shows the transverse distribution of the absolute value component H_y and the time-averaged Umov–Poynting vector $\overline{S}_x$. Also shown are the directions of the Umov–Poynting vector in the dielectric ($\overline{\mathbf{S}}_d$) and the metal ($\overline{\mathbf{S}}_m$) and the signs of the real parts of the wave vector components of the SPP k_x', k_d', and k_m' (shown as the vectors with directions coinciding with the positive or negative direction of the respective axes). Fig. 1.20 shows that the Umov–Poynting vector in both media is directed at a small angle to the Ox axis. We define the fractions of the energy carried in both the metal and the dielectric as

$$W_d = \frac{\left| \int\limits_{0}^{+\infty} \overline{S}_x (0,z) dz \right|}{\left| \int\limits_{0}^{+\infty} \overline{S}_x (0,z) dz \right| + \left| \int\limits_{-\infty}^{0} \overline{S}_x (0,z) dz \right|},$$

$$W_m = \frac{\left| \int\limits_{-\infty}^{0} \overline{S}_x (0,z) dz \right|}{\left| \int\limits_{0}^{+\infty} \overline{S}_x (0,z) dz \right| + \left| \int\limits_{-\infty}^{0} \overline{S}_x (0,z) dz \right|}. \tag{1.282}$$

For this example, $W_d \approx 0.975$, $W_m \approx 0.975$, i.e. most of the energy is carried in the dielectric. Note that the value $\int_{-\infty}^{0} \overline{S}_x (0,z) dz < 0$, therefore $\int_0^{+\infty} \overline{S}_x (0,z) dz > \int_{-\infty}^{+\infty} \overline{S}_x (0,z) dz$.

In general, considering $k_y \neq 0$ in expression (1.245) and further, it is possible to obtain an analogue of (1.256) for the SPP propagating in an arbitrary direction :

$$k_x^2 + k_y^2 = k_0^2 \frac{\varepsilon_m \varepsilon_d}{\varepsilon_m + \varepsilon_d}. \tag{1.283}$$

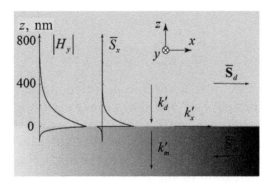

Fig. 1.20. The transverse distributions of the absolute value of the y-component of the magnetic field and the x-component of the Umov–Poynting vector of the SPP propagating along the OX axis.

Equation (1.283) is the dispersion relation of the surface plasmon–polariton. For values k_x and k_y, satisfying the relation (1.283), we obtain an analogue of the expressions (1.258) and it can be shown that, in general, the vector **H** of the SPP lies in the plane of the boundary between two media, and is perpendicular to the direction of propagation, and the vector **E** lies in the plane perpendicular to the plane of the interface of the media and parallel to the direction of propagation.

For convenience we rewrite the dispersion relation (1.283) in the form

$$k_{spp} = k_0 \sqrt{\frac{\varepsilon_m \varepsilon_d}{\varepsilon_m + \varepsilon_d}}. \tag{1.284}$$

The value k_{spp} thus denotes the component of the wave vector lying in the plane Oxy (the propagation constant of the surface plasmon–polariton). Taking into account (1.284), k_x and k_y values value can be written as

$$k_x = k_{spp} \alpha_0,$$
$$k_y = k_{spp} \beta_0, \tag{1.285}$$

where α_0, β_0 are the components of the unit vector defining the direction of propagation of the SPP. Note that due to the fact that $\varepsilon'_m < 0$ the inequality $\left| \sqrt{\varepsilon_m / (\varepsilon_m + \varepsilon_d)} \right| > 1$ is satisfied and it follows that

$$\left| k_{spp} \right| > k_0 \sqrt{\varepsilon_d}. \tag{1.286}$$

From (1.286) it follows that the SPP can not be excited by a plane wave incident from the dielectric medium with dielectric constant

ε_d, and it is necessary to use special structures such as diffraction gratings or the Kretschmann configuration.

1.4.1.2. The properties of surface plasmon–polaritons

We consider a number of quantities characterizing the surface plasmon–polaritons, such as the wavelength, the length of propagation, the depth of penetration into the dielectric and metal media [82]. As an illustration, we discuss the dependences of these values on the wavelength in free space for surface plasmon–polaritons propagating at the interface between the dielectric with a refractive index $n_d = 1.6$ and silver. As in the example discussed above, the values of the dielectric constant of silver were determined by spline interpolation based on the tabulated data shown in [81] (Fig. 1.21). When $\lambda \geq 365$ nm, $\varepsilon'_m + \varepsilon_d < 0$ and SPP may exist at the interface.

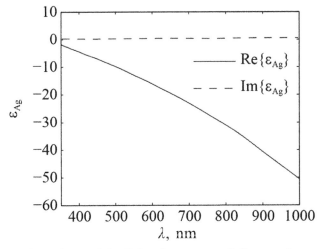

Fig. 1.21. The dependence of the dielectric constant of silver on the wavelength in free space (the real part – solid line, the imaginary part – dashed line).

The wavelength of the surface plasmon polariton

The wavelength of the SPP is determined from the expression [82]

$$\lambda_{spp} = \frac{2\pi}{k'_{spp}}, \tag{1.287}$$

where $k'_{spp} = \mathrm{Re}\{k_{spp}\}$. When the condition

$$|\varepsilon'_m + \varepsilon_d| \gg |\varepsilon''_m| \tag{1.288}$$

is satisfied, the following approximate equality applies

$$k'_{spp} \approx k_0 \sqrt{\frac{\varepsilon'_m \varepsilon_d}{\varepsilon'_m + \varepsilon_d}}. \tag{1.289}$$

In view of (1.289)

$$\lambda_{spp} \approx \lambda \sqrt{\frac{\varepsilon'_m + \varepsilon_d}{\varepsilon'_m \varepsilon_d}}. \tag{1.290}$$

Thus, $\lambda_{spp}/\lambda \approx \sqrt{1/\varepsilon_d + 1/\varepsilon'_m}$. From (1.286) it follows that $\lambda_{spp} < \lambda/\sqrt{\varepsilon_d}$. This fact is the basis for a variety of potential applications of surface plasmon–polaritons, in particular, use in contact photolithography discussed below.

Figure 1.22 shows the dependence of the wavelength of the SPP on the free-space wavelength of light, calculated using the formula (1.287). For comparison, also the wavelength of the light propagating in the dielectric is shown. From Fig. 1.22 it is clear that while the wavelength approaches the so-called the resonance value, which ensures the equality $\varepsilon'_m = -\varepsilon_d$, the ratio λ/λ_{spp} increases. However, as will be shown below, the propagation length of the SPP decreases significantly at the same time.

The propagation length of the surface plasmon–polariton
We define the length of the surface plasmon–polariton δ_{spp} as the

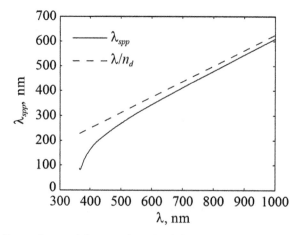

Fig. 1.22. Dependence of the wavelength of the SPP (solid line) and the wavelength in the dielectric (dotted line) on the free-space wavelength.

distance at which the intensity decreases e-fold compared with the initial value. In accordance with this definition

$$\delta_{spp} = \frac{1}{2k''_{spp}}.$$ (1.291)

When the condition (1.288) is satisfied, the following approximate equality holds

$$k''_{spp} \approx k_0 \frac{\varepsilon''_m}{2(\varepsilon'_m)^2} \left(\frac{\varepsilon'_m \varepsilon_d}{\varepsilon'_m + \varepsilon_d} \right)^{\frac{3}{2}}.$$ (1.292)

Taking into account (1.292) in (1.291), we obtain an approximate expression for the propagation length of the SPP:

$$\delta_{spp} \approx \lambda \frac{(\varepsilon'_m)^2}{2\pi \varepsilon''_m} \left(\frac{\varepsilon'_m + \varepsilon_d}{\varepsilon'_m \varepsilon_d} \right)^{\frac{3}{2}}.$$ (1.293)

If the condition $|\varepsilon'_m| \gg |\varepsilon_d|$ is fulfilled, we can write a simpler approximate expression for δ_{spp}:

$$\delta_{spp} \approx \lambda \frac{(\varepsilon'_m)^2}{2\pi \varepsilon''_m \varepsilon_d^{3/2}}.$$ (1.294)

Equation (1.294) confirms the obvious assumption that in order to obtain the larger length propagation of SPP the real and imaginary parts of the dielectric constant of the metal should be larger and small, respectively, i.e. it is necessary to use a metal with low losses.

Figure 1.23 shows the dependence of the propagation length of SPP (1.291) on the free-space wavelength for the investigated pair of materials. The graph shows that when the wavelength approaches the resonance wavelength the propagation length tends to zero. This fact limits the range of wavelengths that allow the practical application of the SPP.

The penetration depth of surface plasmon polaritons in metal and dielectric

Determine the depth of penetration of the SPP in the medium as the distance at which the amplitude of the wave decreases e times:

$$\delta_l = \frac{1}{|\mathrm{Im}\{k_l\}|},$$ (1.295)

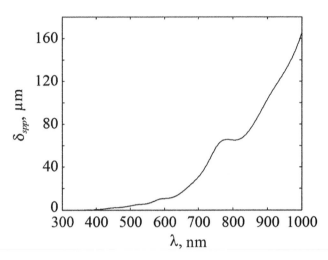

Fig. 1.23. The dependence of the propagation length of the SPP on the wavelength in free space.

where $l \in \{m,d\}$, k_m and k_d are defined by (1.265). When the condition (1.288) is satisfied the following approximate expression holds:

$$\delta_d \approx \frac{1}{k_0}\left|\frac{\varepsilon'_m + \varepsilon_d}{\varepsilon_d^2}\right|^{\frac{1}{2}}, \qquad (1.296)$$

$$\delta_m \approx \frac{1}{k_0}\left|\frac{\varepsilon'_m + \varepsilon_d}{\left(\varepsilon'_m\right)^2}\right|^{\frac{1}{2}}. \qquad (1.297)$$

It should be noted that the values δ_d and δ_m are of practical importance, since they allow to determine the minimum thickness of the layers required for the excitation of the SPP.

Figures 1.24 and 1.25 show the dependence of the penetration depth of the SPP to the dielectric and the metal, respectively, of the wavelength of the incident light calculated by the expression (1.295). We see that at wavelengths far from the resonance value, the penetration depth of the SPP in the metal is close to a constant value of 22 nm.

1.4.1.3. Plasmonic waveguides
The simplest geometry of the plasmonic waveguide is the metal/insulator (dielectric) interface discussed above. The plasmonic modes may also exist in the 'dielectric/metal/dielectric' systems (in thin metal films, in which case they have a large propagation

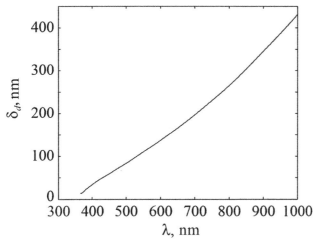

Fig. 1.24. The dependence of the penetration depth of the SPP into the dielectric on the wavelength in free space.

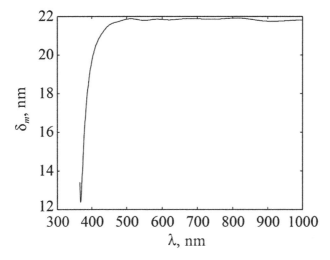

Fig. 1.25. Dependence of the penetration depth of SPP into metal on the wavelength in free space.

length) [83,84], the 'metal/dielectric/metal' structures (in the gaps between the metal layers in this case the energy of the mode may be localized in the subwavelength region) [85], as well as metal–dielectric multilayer structures [86]. In the first two cases, the plasmonic modes are described by the dispersion relation for the TM-polarized plane-parallel waveguide modes [87]. In this section, the propagation constants of the modes are determined using a stable iterative procedure described in [88].

There are also plasmonic waveguides in which the mode is confined in two dimensions, in particular, metal strips in a dielectric medium [89], dielectric structures on the surface of the metal [90] and others. However, this section discusses the structures in which the excited plasmonic modes are limited in one dimension.

1.4.1.4. Excitation of surface plasmon–polaritons

In accordance with (1.286) the excitation of the SPP requires the use of special structures. We consider the Kretschmann configuration [79], which is a dielectric prism with a metal film on the base (Fig. 1.26). For the wave incident on the prism with a refractive index n_1 to excite a SPP on the interface of the metal film and the medium with a refractive index n_2 it is necessary to satisfy the condition $n_1 > \mathrm{Re}\left\{\sqrt{\varepsilon_m \varepsilon_2 / (\varepsilon_m + \varepsilon_2)}\right\}$ where $\varepsilon_2 = n_2^2$. Then at a certain angle of incidence θ_{spp} an SPP is excited at the bottom interface and the angle θ_{spp} is close to the angle at which the condition of equality of the x-components of the wave vectors of the incident and surface waves is satisfied:

$$\sin\left(\theta_{spp}\right) \approx \frac{1}{n_1} \mathrm{Re}\left\{\sqrt{\frac{\varepsilon_m \varepsilon_2}{\varepsilon_m + \varepsilon_2}}\right\}. \qquad (1.298)$$

The equality in equation (1.298) is approximate because it does not account for the thickness of the metal film. More accurate analytical estimates of angle θ_{spp} are given in [91]. It is also possible to calculate analytically the leaky modes in the metal film [84], but note that for a sufficiently large thickness of a metal film several times greater than the penetration depth of SPP in the metal δ_m, the approximate equality (1.298) is satisfied with high accuracy.

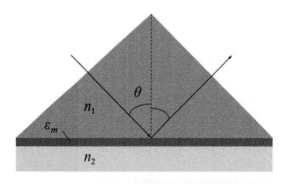

Fig. 1.26. The Kretschmann configura.

In this chapter we consider the excitation of SPP (or, more precisely, plasmonic modes) using diffraction gratings. This process will be discussed in detail in subsection 1.5.2.

1.4.2. Phase modulation of surface plasmon–polaritons by dielectric structures

As mentioned above, for the design of the diffraction structures to control the propagation of SPP it is necessary to solve the problem of diffraction of SPP on a dielectric rectangular step. The geometry of the model problem is shown in Fig. 1.27. SPP in medium 1 is normally incident on the rectangular ridge from the left, and a transmitted SPP forms on the right in medium 3. It should be noted that the geometry of the model problem is independent of the coordinate y (i.e. we solve the problem of diffraction of the SPP with a plane wavefront at an infinite cylinder with a rectangular cross-section), but for simplicity we use the term 'dielectric ridge'.

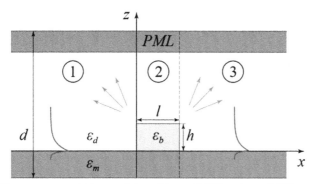

Fig. 1.27. The geometry of the model problem of diffraction of SPP on a dielectric ridge.

The model problem of diffraction of SPP on a dielectric ridge is solved using a modification of the Fourier modal method for non-periodic structures, discussed in section 1.4.

Figure 1.28 shows the calculated dependence of the modulus and phase of the complex transmission coefficient of the SPP T_{spp} on the length l and the height h of the dielectric ridge. The calculation was performed using the following parameters: $\lambda = 550$ nm, $\varepsilon_m = -13.686 + 0.444i$, $\varepsilon_d = 1$, $\varepsilon_b = 2.25$. The value ε_m corresponds to the permittivity of silver at this wavelength. Note that the maximum length of the ridges l_{max} in Fig. 1.28 was chosen from the condition $\Delta\varphi = \left(k_{spp}^b - k_{spp}\right)l_{max} = 4\pi$. This condition provides a range of the

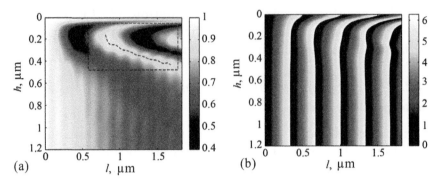

Fig. 1.28. Modulus $\left|T_{spp}(l,h)\right|$ (a) and phase $\varphi(l,h)=\arg\{T_{spp}(l,h)\}$ (b) of the complex transmittance coefficient of SPP depending on the length and height of the dielectric ridge.

phase differences $[0,4\pi]$ between the SPP which has passed through the ridge with dielectric constant ε_b and the SPP distributed in the intial medium.

Note that as shown in Fig. 1.28b the dependence of the phase of the transmitted SPP on the length of the ridge is close to linear for each fixed value of the ridge height. This kind of dependence suggests that when the SPP passes through the ridge most of the energy is transferred by the plasmonic mode propagating in it. To confirm this assumption, we calculate the propagation constant k_{mode} of the plasmonic mode of the 'dielectric/dielectric/metal' type with the dielectric constants ε_d, ε_d, ε_m, at different values of the thickness of the core layer. To find the values of the propagation constants the dispersion relation for TM-polarized modes of a planar waveguide is solved numerically. The dependence of the real part of the propagation constants of the plasmonic modes, normalized by the wave number, is shown in Fig. 1.29 (a). The dashed lines show the values of the real parts of the normalized propagation constants of the SPP propagating at the interfaces $\varepsilon_d - \varepsilon_m$ and $\varepsilon_b - \varepsilon_m$. Fig. 1.29b shows the phase of the plasmonic mode of the dielectric ridge $\varphi_{mode}(l) = \mathrm{Re}\{k_{mode}\}l$ reduced to the interval $[0,2\pi]$ in propagation over the distance l depending on the ridge height.

Figures 1.28b and 1.29b show that the calculated values of the phase obtained by numerical simulation are close to the theoretically calculated values of the phase of the plasmonic modes in the area of the ridge, indicating that the assumption to move the main part of the energy within the dielectric ridge by the plasmonic mode is correct.

Fig. 1.28b also shows that there are several methods of phase modulation of the SPP using dielectric ridges. In particular, at

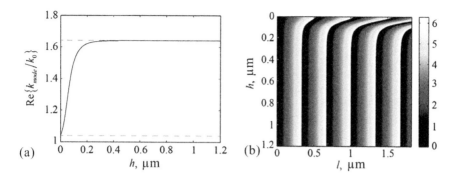

Fig. 1.29. The dependence of the real part of the normalized propagation constant of the plasmonic modes in the ridges on the ridge height (solid line) and the real part of the normalized propagation constants of the SPP, propagating at the $\varepsilon_d - \varepsilon_m$ (lower dashed line) and $\varepsilon_b - \varepsilon_m$ (upper dotted line) interfaces (a) and the dependence of the phase of the plasmonic modes in the ridge upon propagation over distance l on the height of the ridge (b).

sufficiently large height of the ridge the phase of the transmitted SPP depends only on its length. This is consistent with Fig. 1.29a which shows that with increasing height the propagation constant of the plasmonic mode in the ridge approaches the propagation constant of the SPP at the interface of materials with dielectric constants ε_b and ε_m. The depth of penetration of the SPP into the material of the ridge at the parameters is $\delta_d = 0.13$ μm. Recall that the penetration depth is defined as the distance at which the amplitude of the wave decreases by e times and is calculated according to formulas (1.295) and (1.296). When $h > 3\delta_d$ the phase is well described by the expression

$$\varphi(l,h) = k_0 \, \mathrm{Re}\left\{ \sqrt{\frac{\varepsilon_b \varepsilon_m}{\varepsilon_b + \varepsilon_m}} \right\} l = \mathrm{Re}\left\{ k_{spp}^b \right\} l \qquad (1.299)$$

where k_{spp}^b is the propagation constant of the SPP at the interface of the metal and the dielectric with permittivity ε_b. Equation (1.299) is similar to the formula of geometrical optics used for the phase shift of a plane wave through a layer with a thickness l. We introduce the function $\Delta(l,h) = \arg\left\{ T_{spp}(l,h) \right\} - \mathrm{Re}\left\{ k_{spp}^b \right\} l$ that characterizes the error of formula (1.299) for a fixed value of h. The maximum value of errors $\Delta(l,h)$ at $h = 1$ μm is less than $\pi/17$. When calculating the optical elements a similar error in the phase function in most cases is not essential. The modulus of the transmittance coefficient at $h = 1$ μm is higher than 0.7.

Fig. 1.30. Dielectric diffractive structure for the SPP with varying length and a fixed height.

The linear dependence of the phase shift on the length of the dielectric ridge allows one to create the specified phase distributions by changing the length of the ridge. Thus, for the conversion and focusing of the SPP we can use dielectric diffractive structures of variable length and a fixed height (see Fig. 1.30). According to (1.299), in the case of large height the length of the microrelief of the element for forming the required phase function $\varphi(y)$ has the form

$$l(y) = \frac{\varphi(y)}{\mathrm{Re}\left\{k_{spp}^b - k_{spp}\right\}}.$$
(1.300)

Fig. 1.28 also shows that it is possible to change the phase of the transmitted SPP by changing the height of the ridge with a fixed length $l = 1055$ nm. In particular, Fig. 1.31 shows the dependence of the modulus and phase of the complex transmission coefficient of the transmitted SPP on the ridge height at a fixed length $l = 1055$ nm. Fig. 1.31 shows the possibility of phase modulation in the range $[0, 2\pi]$ when changing the ridge height between 0 and 180 nm. The modulus of the transmittance coefficient is more than 0.7. Thus, for the transformation and focusing of the SPP we can also use the dielectric structures of variable height and a fixed length (Fig. 1.32).

A generalization of the described methods of phase modulation of the SPP is the use of structures of varying length and height (Fig. 1.33). At the same time, as shown in Fig. 1.28, the set of points (l, h) in the space of the geometrical parameters of the ridge, providing the possibility of forming an arbitrary phase distribution in the output plane of the element is not unique. The criterion for the selection of such a set can be, for example, maximization of the average value of the transmittance of the SPP $|T_{spp}(l, h)|$. More

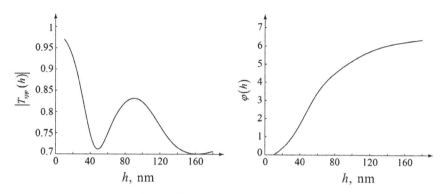

Fig. 1.31. Dependences $\left|T_{spp}(h)\right|$ (a) and $\varphi(h)$ (b) on the height of the ridge at the ridge length $l = 1055$ nm.

Fig. 1.32. Dielectric diffractive structure for the SPP with varying height and a fixed length.

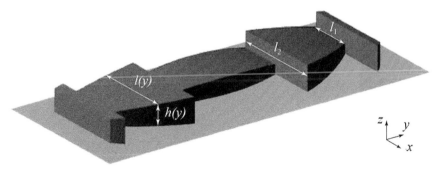

Fig. 1.33. Dielectric diffractive structure for the SPP with varying height and length.

details of the optimization criterion are described below in subsection 1.4.4. Maximization of the the observed parameters was carried out in region D, marked by a dotted rectangle in Fig. 128a. The optimal

(l,h) values are found on the curve shown by the dashed line in Fig. 1.28a. This curve passes through the maxima of the modulus of the transmittance coefficient. This modulus is higher than 0.83 which is 10% higher than in the case of structures in which one of the geometric parameters is fixed.

Calculation of the diffractive lenses for focusing the SPP on the basis on the studied techniques of phase modulation is given below in subsection 1.4.4.

1.4.3. Suppression of parasitic scattering of surface plasmon–polaritons

1.4.3.1. Theoretical analysis of matching of transverse profiles of the field of the surface plasmon–polariton and the plasmonic mode of the dielectric/dielectric/metal waveguide

According to the expression for the transverse component of the wave vector of the SPP (1.262), the propagation of the SPP through the interface of the element situated on the propagation surface in the region $0 < x < l$ and made of a material with dielectric constant $\varepsilon_1 \neq \varepsilon_d$, results in a change in the transverse profile of the field of SPP that leads to parasitic scattering. Thus, to reduce the scattering it is necessary that the transverse profile of the plasmonic mode in the element region was close to the profile of the field of the incident SPP. To suppress the scattering the authors proposed to use two-layer dielectric structures of isotropic materials (geometry of the structure in the case of passing through the interface at infinite thickness of the layer with dielectric constant ε_2 is shown in Fig. 1.34) [92–94].

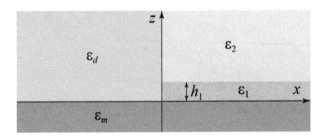

Fig. 1.34. The interface between a plasmonic metal/dielectric waveguide and a dielectric/dielectric/metal waveguide.

We consider the case when the thickness of the upper dielectric layer h_2 is large enough so that the plasmonic modes in the region of the structure (at $0 < x < l$) can be described by the dispersion relation of the TM-polarized modes of a planar waveguide [87]

$$\tanh\left(\gamma_1 h_1\right) = -\frac{\gamma_1 \varepsilon_1 \left(\varepsilon_2 \gamma_m + \varepsilon_m \gamma_2\right)}{\varepsilon_2 \varepsilon_m \gamma_1^2 + \varepsilon_1^2 \gamma_2 \gamma_m}, \tag{1.301}$$

where $\gamma_j^2 = k_{IIM}^2 - k_0^2 \varepsilon_j$, $j = m,1,2$, k_{IIM} is the plasmonic mode propagation constant in the structure to the right of the interface.

Let us find the relation between the dielectric constants ε_d, ε_1 and ε_2 in which the transverse profiles of the of plasmonic modes in the region $z > h_1$ can be matched. For ease of analysis, we neglect losses in the metal and assume that ε_m is a real number. Taking into account (1.301) the matching condition of the profiles of the mode field can be written as

$$\kappa_d = \gamma_2. \tag{1.302}$$

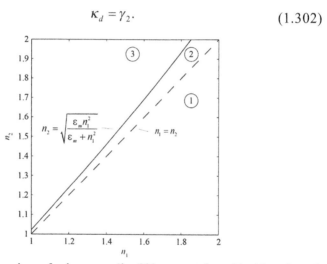

Fig. 1.35. Different regions of values ε_1, ε_2 ($\lambda = 800$ nm, metal – gold without losses).

Since most of the SPP energy is carried in the dielectric [95], then if (1.302) is satisfied at $h_1 \ll \delta_a$, where δ_a – the SPP penetration depth (1.295), one can expect a substantial reduction of parasitic scattering. Note that as part of the energy is concentrated outside the area $h_1 < z < h_1 + h_2$, a set of parameters ε_1, ε_2, h_1 ensuring the fulfillment of the condition (1.302) may not be optimal.

We consider separately the different regions of values $\varepsilon_2 = n_2^2$, $\varepsilon_1 = n_1^2$ (Fig. 1.35).

Consider first the case where the values of ε_1, ε_2 lie in region 1, i.e. when $\varepsilon_1 > \varepsilon_2$. In this case, the plasmonic mode in the structure exists at all values h_1 [96]. When the thickness changes from 0 to $+\infty$ the value k_{IIM} increases continuously from the propagation constant of the plasmon at the interface $\varepsilon_m - \varepsilon_2$ to the propagation constant of the plasmon at the interface $\varepsilon_m - \varepsilon_1$. When $h_1 = 0$ $\gamma_2 = k_0\sqrt{-\varepsilon_2^2/(\varepsilon_m + \varepsilon_2)}$, at $h_1 \to +\infty$ $\gamma_2^2 \to k_0^2(\varepsilon_m\varepsilon_1/(\varepsilon_m + \varepsilon_1) - \varepsilon_2)$ and, taking this into account consequently $\gamma_2 \to k_0\sqrt{[\varepsilon_m(\varepsilon_1 - \varepsilon_2) - \varepsilon_1\varepsilon_2]/(\varepsilon_m + \varepsilon_1)}$. Thus, the profile of the mode field can be matched when

$$k_0\sqrt{\frac{-\varepsilon_2^2}{\varepsilon_m + \varepsilon_2}} \le \kappa_d \le k_0\sqrt{\frac{\varepsilon_m(\varepsilon_1 - \varepsilon_2) - \varepsilon_1\varepsilon_2}{\varepsilon_m + \varepsilon_1}}. \tag{1.303}$$

As the value $\sqrt{-\varepsilon_2^2/\varepsilon_m + \varepsilon_2}$ increases with increasing ε_2, the left part of the double inequality (1.303) shows that the the profile of the modes can be matched when

$$\varepsilon_d \ge \varepsilon_2. \tag{1.304}$$

We transform the right side of the double inequality (1.303):

$$\sqrt{\frac{-\varepsilon_d^2}{\varepsilon_m + \varepsilon_d}} \le \sqrt{\frac{\varepsilon_m(\varepsilon_1 - \varepsilon_2) - \varepsilon_1\varepsilon_2}{\varepsilon_m + \varepsilon_1}}. \tag{1.305}$$

From (1.305), it follows that

$$\varepsilon_d^2 + A\varepsilon_d + A\varepsilon_m \le 0, \tag{1.306}$$

where $A = [\varepsilon_m(\varepsilon_1 - \varepsilon_2) - \varepsilon_1\varepsilon_2]/(\varepsilon_m + \varepsilon_1)$. Note that under the conditions $\varepsilon_m < 0$, $|\varepsilon_m| > \varepsilon_a$ the inequality $A > 0$ is satisfied. Because ε_d is an essentially positive quantity, the inequality (1.306) is satisfied for

$$0 < \varepsilon_d \le \frac{-A + \sqrt{A^2 - 4A\varepsilon_m}}{2}. \tag{1.307}$$

Introducing the notation

$$\varepsilon_{max} = \frac{\sqrt{A^2 - 4A\varepsilon_m} - A}{2}, \tag{1.308}$$

we have:

$$\varepsilon_2 \le \varepsilon_d \le \varepsilon_{max}. \tag{1.309}$$

It is easy to show that $\varepsilon_{max} > \varepsilon_1$. Indeed, with $\varepsilon_d = \varepsilon_{max}$

$$\kappa_d = \gamma_2 = k_0\sqrt{\frac{\varepsilon_m\varepsilon_1}{\varepsilon_m + \varepsilon_1} - \varepsilon_2} > k_0\sqrt{\frac{\varepsilon_m\varepsilon_1}{\varepsilon_m + \varepsilon_1} - \varepsilon_1}. \tag{1.310}$$

The right-hand side of the inequality (1.310) is the transverse component of the wave vector of the plasmon propagating at the interface $\varepsilon_m - \varepsilon_1$. As this value at a fixed metal permittivity increases with the permittivity of the dielectric, we obtain from (1.310) the required inequality $\varepsilon_{max} > \varepsilon_1$. It should be noted that the second inequality in (1.309) is more formal, because for large values ε_d the matching of the field profile in the region $z > h_1$ will take place at large h_1. The latter does not allow to expect a significant reduction of scattering in this case.

Now consider the case when the values ε_1, ε_2 lie in region 2 (Fig. 1.35). Analyzing the dispersion relation (1.301), one can show that for the values that satisfy the inequality $\varepsilon_1 < \varepsilon_2 < \varepsilon_m \varepsilon_1 / (\varepsilon_m + \varepsilon_1)$ the plasmonic mode in the structure also exists for any values of h_1. When changing h_1 from 0 to $+\infty$ the value k_{IIM} decreases continuously from the propagation constant of the plasmon at the interface $\varepsilon_m - \varepsilon_2$ to the propagation constant of the plasmon at the interface $\varepsilon_m - \varepsilon_1$. At $h_1 = 0$ $\gamma_2 = k_0 \sqrt{-\varepsilon_2^2 / (\varepsilon_m + \varepsilon_2)}$ and at $h_1 \to +\infty$ $\gamma_2 \to k_0 \sqrt{[\varepsilon_m(\varepsilon_1 - \varepsilon_2) - \varepsilon_1 \varepsilon_2] / (\varepsilon_m + \varepsilon_1)}$. In this case, the profile of the mode field can be matched at

$$k_0 \sqrt{\frac{\varepsilon_m(\varepsilon_1 - \varepsilon_2) - \varepsilon_1 \varepsilon_2}{\varepsilon_m + \varepsilon_1}} \leq \gamma_d \leq k_0 \sqrt{\frac{-\varepsilon_2^2}{\varepsilon_m + \varepsilon_2}}. \tag{1.311}$$

Transforming the inequality (1.311) similar to inequality (1.303), we find that (1.302) can be satisfied at

$$\varepsilon_{min} \leq \varepsilon_d \leq \varepsilon_2, \tag{1.312}$$

where

$$\varepsilon_{min} = \frac{\sqrt{A^2 - 4A\varepsilon_m} - A}{2}. \tag{1.313}$$

Note that the expression (1.313) coincides with the expression (1.308) obtained for the region 1. Similarly to the previous case, we can show that for the region 2 $\varepsilon_{min} < \varepsilon_1$.

Let us now consider the region 3 (Fig. 1.35), that is the case when $\varepsilon_2 > \varepsilon_m \varepsilon_1 / (\varepsilon_m + \varepsilon_1)$. In this case, the waveguide modes exist in the structure only when $h_1 \in [0, h_1^{max})$ [97]. When $h_1 = 0$ the mode is a plasmon propagating at the interface $\varepsilon_m - \varepsilon_2$. With increase of h_1 the propagation constant of the mode begins to decrease and at $h_1 = h_1^{max}$

reaches the value $k_0\sqrt{\varepsilon_2}$, that is at $h_1 > h_1^{max}$ the mode ceases to be a waveguide mode and becomes a leaky mode. The value h_1^{max} can be found from (1.301) and is

$$h_1^{max} = \tanh^{-1}\left[-\frac{\varepsilon_1\sqrt{\varepsilon_2 - \varepsilon_m}}{\varepsilon_m\sqrt{\varepsilon_2 - \varepsilon_1}}\right]\bigg/\left(k_0\sqrt{\varepsilon_2 - \varepsilon_1}\right). \qquad (1.314)$$

Note that when h_1 changes from 0 to the value h_1^{max} the value γ_2 decreases from $k_0\sqrt{-\varepsilon_2^2/(\varepsilon_m + \varepsilon_2)}$ to 0, and the field profiles of the modes can be matched when the following condition is satisfied:

$$\varepsilon_d < \varepsilon_2 \qquad (1.315)$$

The results obtained for the three regions are combined in Table 1.6. Tables 1.7 and 1.8 show the corresponding values of h_1 and the values $l_{2\pi}$ corresponding to the length of the structure in which the phase shift $\Delta\varphi = |k_{SPP} - k_{IIM}|l$ is 2π.

Table 1.6. The boundaries of the range of possible values of ε_d for different vales of ε_1, ε_2.

	Region 1: $\varepsilon_1 > \varepsilon_2$	Region 2: $\varepsilon_1 < \varepsilon_2 < \varepsilon_1\varepsilon_m/(\varepsilon_1 + \varepsilon_m)$	Region 3: $\varepsilon_2 > \varepsilon_1\varepsilon_m/(\varepsilon_1 + \varepsilon_m)$
ε_d^{min}	ε_2	$\dfrac{\sqrt{A^2 - 4A\varepsilon_m} - A}{2} < \varepsilon_1$	1
ε_d^{max}	$\dfrac{\sqrt{A^2 - 4A\varepsilon_m} - A}{2} < \varepsilon_1$	ε_2	ε_2

Table 1.7. The boundaries of the range of possible values of h_1 for different values of ε_1, ε_2.

	Region 1: $\varepsilon_1 > \varepsilon_2$	Region 2: $\varepsilon_1 < \varepsilon_2 < \varepsilon_1\varepsilon_m/(\varepsilon_1 + \varepsilon_m)$	Region 3: $\varepsilon_2 > \varepsilon_1\varepsilon_m/(\varepsilon_1 + \varepsilon_m)$
h_1 at $\varepsilon_d = \varepsilon_d^{min}$	0	∞	h_1^{max}
h_1 at $\varepsilon_d = \varepsilon_d^{max}$	∞	0	0

Table 1.8. The boundaries of the range of possible values of $l_{2\pi}$ for different ratios between ε_1, ε_2.

	Region 1: $\varepsilon_1 > \varepsilon_2$	Region 2: $\varepsilon_1 < \varepsilon_2 < \varepsilon_1\varepsilon_m/(\varepsilon_1 + \varepsilon_m)$	Region 3: $\varepsilon_2 > \varepsilon_1\varepsilon_m/$ $(\varepsilon_1 + \varepsilon_m)$
$l_{2\pi}$ at $\varepsilon_d = \varepsilon_d^{\min}$	∞	$l'_{2\pi} = \lambda_0 \Big/ \left\| \sqrt{\dfrac{\varepsilon_m\varepsilon_d}{\varepsilon_m + \varepsilon_d}} - \sqrt{\dfrac{\varepsilon_m\varepsilon_1}{\varepsilon_m + \varepsilon_1}} \right\|$	—
$l_{2\pi}$ at $\varepsilon_d = \varepsilon_d^{\max}$	$l'_{2\pi}$	∞	∞

Until now we solved the problem of finding possible ε_d for the given, ε_1, ε_2. Using these inequalities, we solve another problem: to find those ε_1, ε_2 at which the field profile of the plasmonic mode can be matched with the profile of the field of incident SPP at a fixed value ε_d. As before, we consider separately the three regions in which ε_1, ε_2 may lie and just write out inequalities for ε_1, ε_2.

For region 1 (Fig. 1.35) the field profile of the plasmonic mode and of the incident SPP can be matched at

$$\varepsilon_2 < \varepsilon_d,$$

$$\varepsilon_2 < \frac{\varepsilon_m\varepsilon_1}{\varepsilon_m + \varepsilon_1} + \frac{\varepsilon_d^2}{\varepsilon_m + \varepsilon_d}. \tag{1.316}$$

It can be shown that the second inequality in (1.316) ensures that the condition $\varepsilon_d > \varepsilon_{\max}$ is fulfilled, where $\varepsilon_{\max}$ is given by (1.308).

For the region 2 the field profiles of the plasmonic mode and the incident SPP can be matched at

$$\varepsilon_2 > \varepsilon_d,$$

$$\varepsilon_2 > \frac{\varepsilon_m\varepsilon_1}{\varepsilon_m + \varepsilon_1} + \frac{\varepsilon_d^2}{\varepsilon_m + \varepsilon_d}. \tag{1.317}$$

Similar to (1.316), the second inequality in (1.317) results in fulfilling the condition $\varepsilon_d > \varepsilon_{\min}$.

For region 3 matching is possible when

$$\varepsilon_2 > \varepsilon_d. \tag{1.318}$$

In practice, the dielectric constants of the materials are usually given. For the given ω, ε_1, ε_2, ε_m, ε_d, which ensure the existence of the solution (1.302), the corresponding value of h_1 can be analytically calculated from (1.301) at

$$\gamma_1 = k_0\sqrt{\varepsilon_m\varepsilon_d/(\varepsilon_m+\varepsilon_d)+\varepsilon_2-\varepsilon_d-\varepsilon_1},$$

$$\gamma_2 = k_0\sqrt{-\varepsilon_d^2/(\varepsilon_m+\varepsilon_d)},\qquad\qquad\qquad (1.319)$$

$$\gamma_m = k_0\sqrt{\varepsilon_m\varepsilon_d/(\varepsilon_m+\varepsilon_d)+\varepsilon_2-\varepsilon_d-\varepsilon_m}.$$

Expressions (1.319) were obtained from (1.302) and the expression $\gamma_j^2 = k_{IIM}^2 - k_0^2\varepsilon_j$. Note that when ε_m corresponds to the permittivity of the real metal and $\mathrm{Im}(\varepsilon_m) \neq 0$, the value of h_1 calculated from (1.301) will be a complex number. In the case of small losses the real part of this value can serve as an estimate of the optimum thickness of the first layer.

1.4.3.2. Suppression of parasitic scattering during the passage of surface plasmon–polaritons through two-layer dielectric structures

Consider the example of the scattering suppression at normal incidence of the SPP on the structure in Fig. 1.36a at $\lambda = 800$ nm. The following values were selected for the dielectric constants of the materials: $\varepsilon_m = -24.2$ (corresponds to the real part of the dielectric constant of gold), $\varepsilon_d = \varepsilon_1 = 1.96$, $\varepsilon_2 = 4$. In this case,

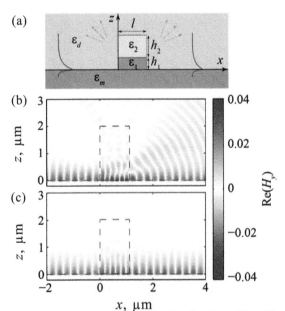

Fig. 1.36. Geometry of the structure (a) and the distribution of Re (H_y) corresponding to the SPP propagation through structure with $l = 1.1$ μm at $h_1 = 0$ (b) and $h_1 = 16$ μm (c).

the structure corresponds to a rectangular block embedded in the dielectric cladding of the plasmon waveguide at height h_1. For these parameters, the value h_1 calculated using the expressions (1.301) and (1.319) is 14 nm. Unlike Fig. 1.34, the structure in Fig. 1.36a has a final thickness h_2 of the upper layer and a finite length. The selected value $h_2 = 2$ μm is 14 times greater than the penetration depth of the SPP propagating at the $\varepsilon_m - \varepsilon_d$ interface. So, the above described model of a planar waveguide is applicable. Note that for the selected parameters the plasmonic modes in the region of the structure correspond to the modes of the 'conductor/gap/dielectric' waveguide considered in [97].

As before, the diffraction of SPP on the structure under consideration was simulated using the Fourier modal method in the formulation of [2, 5], adapted to the case of non-periodic structures [21] and described in subsection 1.1.4.

Figure 1.37 shows the calculated dependence of the transmission coefficient T (Fig. 1.37a) and the phase shift $\Delta\varphi = |k_{spp} - k_{IIM}| l$ of the transmitted SPP (Fig. 1.37b) on the height h_1 and length of the structure l. The solid line shows the thickness at which the minimum scattering loss (minimum value $\int_0^{l_{max}} S dl$ where $l_{max} = 1.8$ μm, $S = 1 - T - R$, where R is the reflectance coefficient of SPP). Figure 1.37 shows that this value (16 nm) is close to the calculated value of 14 nm, shown by the dotted line. Figure 1.37 shows that there may be effective phase modulation of the SPP by changing

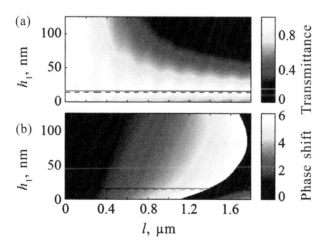

Fig. 1.37. SPP transmittance coefficient (a) and phase shift (b) vs. h_1 and the length of the structure l. The solid lines show the optimum value $h_1 = 16$ nm, the dashed lines – the calculated value $h_1 = 14$ nm.

the length of the structure at a high transmittance coefficient. Let $l_{2\pi}$ be the length of the structure in which $\Delta\varphi = 2\pi$. Note that since the propagation constant k_{IIM} depends on the thickness of the first layer $l_{2\pi}$ also depends on h_1. According to Fig. 1.37, $l_{2\pi} < 1.8$ μm when $h_1 < 120$ nm.

Figure 1.38 shows the dependence of the transmission T and scattering S on the normalized length of the structure $l/l_{2\pi}$ at $h_1 = 0$, the design value $h_1 = 14$ nm and the optimal value $h_1 = 16$ nm. The value $l_{2\pi}$ is 1.11 μm at $h_1 = 0$, 1.37 μm to 1.4 μm, $h_1 = 14$ nm and at $h_1 = 16$ nm. Figure 1.38 shows that the transition to the two-layer structure under consideration reduces the parasitic scattering losses by an order of magnitude: maximum scattering is reduced from 33.1% at $h_1 = 0$ to 1.6% at $h_1 = 14$ nm at 1.2% at $h_1 = 16$ nm (by 27 times); the average value of the scattering $\bar{S} = \int_0^{l_{2\pi}} Sdl / \int_0^{l_{2\pi}} dl$ is reduced from 16.6% at $h_1 = 0$ to 1.2% at $h_1 = 14$ nm at 0.8% at $h_1 = 16$ nm. Figures 1.36b 1.36c show the distribution $Re(H_y)$ in passage of SPP through the structure with the length $l = 1.11$ μm and $h_1 = 16$ nm correspondingly. Scattering losses in this case are§ 33.1% and 0.8%, respectively. Note that the value h_1 at which scattering is suppressed rather weakly depends on the wavelength. So, for this example the maximum scattering losses for $h_1 = 16$ nm is 2.1% at $\lambda = 700$ nm and 0.9% at $\lambda = 900$ nm $l \in [0, l_{2\pi})$.

Similar calculations were also carried out for $\varepsilon_2 = 1$ and $h_2 \gg 1$ μm (this case corresponds to a groove formed in the dielectric

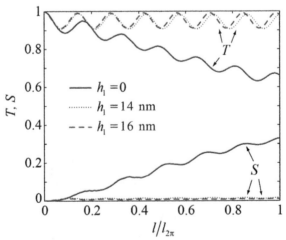

Fig. 1.38. Transmittance of SPP (T) and scattering losses (S) at $h_1 = 0$ (solid lines), $h_1 = 14$ nm (dotted line) and $h_1 = 16$ nm (dashed line) as a function of the normalized length of the structure $l/l_{2\pi}$.

cladding of the plasmonic waveguide). In this case, the optimum value of h_1 is 53 nm, and the maximum scattering is reduced from 14.1% to 1.9%. The average value of scattering is reduced from 7.9% at $h_1 = 0$ to 1.4% at $h_1 = 53$ nm.

We now consider the example of the scattering suppression in the structure of the real (lossy) metal. The dielectric constants of the materials in the example were: $\varepsilon_m = -24.2 + 1.442\,i$ (Au), $\varepsilon_d = 1$, $\varepsilon_1 = 2.1$, $\varepsilon_2 = 2.89$. With these parameters the value of h_1 calculated using the expressions (1.301) and (1.319) is 53 nm. The selected value $h_2 = 1.5$ μm is 2.5 times the penetration depth of the incident SPP into the dielectric.

Figure 1.39 shows the dependence of the transmittance T and scattering losses S on the normalized length of the structure $l/l_{2\pi}$ at $h_1 = 0$, and the calculated value of $h_1 = 53$ nm found by modelling the optimal value $h_1 = 65$ nm at which the minimum of the total losses due to scattering $\int_0^{l_{2\pi}} S\,dl$. For the given h_1 the values of $l_{2\pi}$ are 1.06 μm, 1.16 μm and 1.17 μm, respectively. Fig. 1.39 shows that the transition to a two-layer structure under consideration reduces the parasitic scattering losses as follows: maximum scattering is reduced from 29.5% at $h_1 = 0$ to under 3% at $h_1 = 53$ nm and 2.5% at $h_1 = 65$ nm (by 12 times); the average value of the scattering loss $\bar{S} = \int_0^{l_{2\pi}} S\,dl \Big/ \int_0^{l_{2\pi}} dl$ is reduced from 14.2% at $h_1 = 0$ to 1.9% at $h_1 = 53$ nm and 1.5% at $h_1 = 65$ nm.

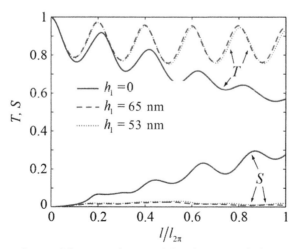

Fig. 1.39. Dependence of the transmittance of the plasmon–polariton T and scattering losses S on the normalized length of the structure $l/l_{2\pi}$ at $h_1 = 0$ (solid lines), $h_1 = 53$ nm (dotted line) and $h_1 = 65$ nm (dashed line).

As in the previous example, the values h_1 at which the scattering suppression of the scattering takes place vary slowly with wavelength. Figure 1.40 shows the dependence of the scattering losses and optimum thickness $h_{1,opt}$ on wavelength in the transmission of the plasmon–polariton through the interface between the standard 'metal/ dielectric' plasmonic waveguide and the considered structure (inset in Fig. 1.40). Figure 1.40 shows that in the wavelength range of 700–900 nm the optimal values of $h_{1,opt}$ are in the range of 58–63 nm, while the maximum scattering at the optimum height is 3.2%. This result shows the possibility of using the structure under consideration for creating elements for converting plasmonic pulses, The study of the propagation of these pulses is currently an urgent problem [98].

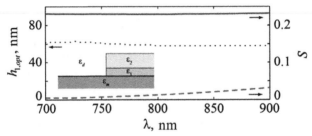

Fig. 1.40. Dependences of the scattering losses S at $h_1 = 0$ (solid line) and the optimal value at $h_1 = h_{1,opt}$ (dashed line) and of the optimum thickness $h_{1,opt}$ (dotted line) on the wavelength upon transmission of the plasmon–polariton through the interface between the standard 'dielectric/metal' plasmonic waveguide and the studied 'dielectric/dielectric/metal' waveguide (inset).

Figure 1.41 shows the distribution of $|\mathrm{Re}(H_y)|$ corresponding to the transmission of a plasmonic Gaussian pulse with a duration of 10 fs through the dielectric ridge ($h_1 = 0$, Fig. 1.41a) and the considered the structure ($h_1 = 65$ nm, Fig. 1.41b) with a length of 950 nm. Permittivities ε_1, ε_2, ε_d, were assumed constant, the dispersion of the dielectric constant of the metal was taken into account by interpolation of tabulated data [81]. The energy losses in parasitic scattering were thus reduced from 29% to 3%.

1.4.4. Design of the lens for focusing the surface plasmon–polaritons

1.4.4.1. Integral representations of the field in the form of the spectrum of surface plasmon–polaritons

The scalar diffraction theory uses widely the representation of the diffracted field in the form of the angular spectrum of plane waves

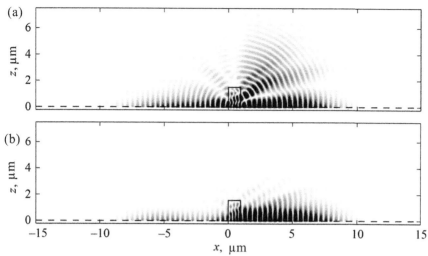

Fig. 1.41. Distribution of value of $|\text{Re}(H_y)|$ upon transmission of the plasmonic Gaussian pulse of 10 fs duration and a central wavelength of 800 nm through a structure with (a) $h_1 = 0$ and (b) $h_1 = 65$ nm.

and the diffraction Kirchhoff integral [77,78]. By analogy with the scalar theory these representations can also be used to describe the diffraction of the SPP. The only difference is that the wavelength of the radiation is the length of the SPP. In particular, the authors of [99] describe the integral representation of the electromagnetic field on the propagation surface through the angular spectrum of the SPP. In [100] this representation is used for the calculation of lenses to focus the SPP and for the simulation of SPP propagation. In [101] the lens for SPP is calculated and SPP diffraction is described using the Fresnel–Kirchhoff diffraction integral. For correctness, we give a brief derivation of these integral representations for the SPP.

In accordance with (1.245), the SPP field with the propagation direction $\mathbf{p} = \alpha_0, \beta_0, |\mathbf{p}| = 1$ in the plane $z = 0$ has the form

$$
\begin{aligned}
\mathbf{E}^{(l)}(x,y,z,t) &= \mathbf{E}_0^{(l)} \exp(-i\omega t)\exp\left[ik_0(\alpha x + \beta y)\right] = \\
&= \mathbf{E}_0^{(l)} \exp(-i\omega t)\exp\left[ik_{spp}(\alpha_0 x + \beta_0 y)\right], \\
\mathbf{H}^{(l)}(x,y,z,t) &= \mathbf{H}_0^{(l)} \exp(-i\omega t)\exp\left[ik_0(\alpha x + \beta y)\right] = \\
&= \mathbf{H}_0^{(l)} \exp(-i\omega t)\exp\left[ik_{spp}(\alpha_0 x + \beta_0 y)\right],
\end{aligned}
\tag{1.320}
$$

where the index $l = m,d$ refers to the medium, k_{spp} is the propagation constant of the SPP (1.284). In accordance with (1.283), constants α, β in (1.320) satisfy the dispersion relation

$$k_0^2 \left(\alpha^2 + \beta^2 \right) = k_{spp}^2. \qquad (1.321)$$

The selected optical axis is the Ox axis. In this case, the field is efficiently described by the component H_y, which is similar to the expression (1.272) and has the form

$$H_y \left(x, y, z, t \right) = \begin{cases} H_y^{(d)} \left(x, y, z, t \right), & z > 0, \\ H_y^{(m)} \left(x, y, z, t \right), & z < 0, \end{cases} \qquad (1.322)$$

where the quantities $H_y^{(d)}$ and $H_y^{(m)}$ at the unit amplitude have the form

$$H_y^{(d)} \left(x, y, z, t \right) = \exp(-i\omega t) \exp\left[ik_0 \left(\alpha x + \beta y \right) \right] \exp(-\kappa_d z),$$
$$H_y^{(m)} \left(x, y, z, t \right) = \exp(-i\omega t) \exp\left[ik_0 \left(\alpha x + \beta y \right) \right] \exp(\kappa_m z). \qquad (1.323)$$

Usually, the term SPP refers to a wave in the form (1.320), in which the propagation constants α, β are defined by the unit vector $\mathbf{p}$ of the direction of SPP in the form

$$k_0 \left(\alpha, \beta \right) = k_{spp} \left(\alpha_0, \beta_0 \right) = k_{spp} \mathbf{p}. \qquad (1.324)$$

In this case α, β are complex numbers, since the dielectric constant of the metal ε_m is complex and k_{spp} is also complex. In general, the constants α, β must only satisfy the dispersion equation (1.321). We will use the SPP (1.323) at

$$k_0 \alpha = \sqrt{k_{spp}^2 - k_0^2 \beta^2}, \ \beta \in \mathbb{R}. \qquad (1.325)$$

At the real values of β component H_y is bounded with respect to the variable y and the interface $z = 0$ has the form

$$H_y \left(x, y, 0, t \right) = \exp(-i\omega t) \exp(ik_0 \beta y) \exp\left(ik_0 \sqrt{k_{spp}^2 - k_0^2 \beta^2} \, x \right). \qquad (1.326)$$

Equation (1.326) allows to obtain integral representations for the function H_y at the interface $z = 0$ analogous to the field through the angular spectrum of plane waves and the Kirchhoff integral, which are widely used in the scalar diffraction theory [77,78]. Indeed, let us write the general solution at $z = 0$ in the form of a superposition of the SPP (we also omit the time dependence):

$$H_y \left(x, y \right) = \int_{-\infty}^{+\infty} I(\beta) \exp(ik_0 \beta y) \exp\left(i\sqrt{k_{spp}^2 - k_0^2 \beta^2} \, x \right) d\beta. \qquad (1.327)$$

This function $I(\beta)$ is defined by the values of the field at $x = 0$ at:

$$I(\beta) = \frac{k_0}{2\pi} \int_{-\infty}^{+\infty} H_y(0,y) \exp(-ik_0 \beta y) dy. \tag{1.328}$$

Equation (1.327) is completely identical to the integral representation of the field through the angular spectrum of plane waves used in the scalar diffraction theory [77,78]. Equation (1.327) can be used to write the Kirchhoff integral for SPP in the form

$$H_y(x,y) = \int_{-\infty}^{+\infty} H_y(0,u) G(x, y-u) du, \tag{1.329}$$

where

$$G(x,y) = \frac{k_0}{2\pi} \int_{-\infty}^{+\infty} \exp\left(i\sqrt{k_{spp}^2 - k_0^2 \beta^2}\, x\right) \exp(ik_0 \beta y) d\beta =$$

$$= \frac{ik_{spp} x}{2\sqrt{x^2 + y^2}} H_1^1\left(k_{spp}\sqrt{x^2 + y^2}\right), \tag{1.330}$$

$H_1^1(x)$ being the Hankel function of the first kind, the first order [73].

1.4.4.2. Diffraction lenses for focusing the surface plasmon–polaritons

Consider as an example the calculation of lenses to focus the SPP using the phase modulation techniques [102,103] discussed in section 1.4.2. According to [12], the phase function of the diffraction lens with focus f has the form

$$\varphi(y) = \mathrm{mod}_{2\pi}\left(-\mathrm{Re}(k_{spp})\sqrt{y^2 + f^2} + \varphi_0\right), \tag{1.331}$$

where φ_0 is an arbitrary constant. Figure 1.42a shows the function of the length of the microrelief of the lens with a varying length and a fixed height (Fig. 1.30) and the absolute value of the SPP transmission coefficient (modulus of the transmittance of SPP), calculated with the focal length $f = 8\lambda_{spp}$ and the lens aperture $2a = 10\lambda_{spp}$, where λ_{spp} is the SPP wavelength (1.287). The length of the microrelief is normalized by the SPP wavelength. The design of length was performed using the formulas (1.300) and (1.331). The height of the lens is 1 μm. The values of other parameters are specified in subsection 1.4.2. Figure 1.42b shows the distribution of $|H_y(x,y)|$ generated by a lens and calculated according to formulas (1.329) and (1.330). Fig.

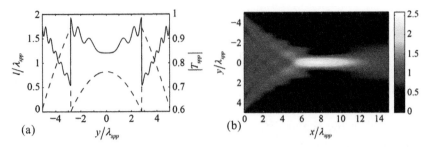

Fig. 1.42. The function of the length of the lens, normalized by the wavelength of the SPP (dotted line) and the absolute value of the transmission coefficient (solid line) within the aperture of the lens (a) the generated $|H_y(x,y)|$ distribution (b).

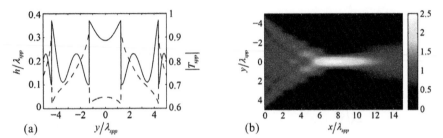

Fig. 1.43. The function of the height of the lens, normalized by the wavelength of the SPP (dotted line) and the absolute value of the transmission coefficient (solid line) within the aperture of the lens (a) the generated $|H_y(x,y)|$ distribution (b).

1.42b shows the focus on the axis on the line $x = f = 8\lambda_{spp}$.

Figure 1.43a shows the functions of the height of the microrelief of the lens and the absolute value of the transmission coefficient for the lens, designed for a fixed length of the structure l = 1055 nm (Fig. 1.32). The parameters of the lens coincide with the previous example. The resultant distribution of $|H_y(x,y)|$ is shown in Fig. 1.43b which also shows the focusing to the point. The distributions in Figs. 1.42b 1.43 andb have similar structures. Note that according to Figs. 1.31 and 1.43 the methods of phase modulation on the basis of changes in the height allow focusing the SPP using structures with subwavelength height.

The energy efficiency of the lens (the fraction of energy in the wave transmitted through the lens) can be estimated by the formula [100]

$$T_e = 100 \times \int_{-\infty}^{+\infty} \frac{|I(\beta)|^2 \, \mathrm{Re}\left(\sqrt{k_{spp}^2 - k_0^2 \beta^2}\right)}{\mathrm{Re}(k_{spp})} \, d\beta \ (\%). \qquad (1.332)$$

The efficiency is 68.5% for the lens shown in Fig. 1.42 and 63.9% for the lens shown in Fig. 1.43. Note that the value of the constant φ_0 for both lenses was chosen from the condition of maximizing the energy efficiency (1.332). It should be noted that despite the somewhat lower energy efficiency of the lens with a fixed length and variable height, its advantages are the flat surfaces of the element (such a lens is an analogue of the lens with the gradient refractive index in conventional optics) and a height substantially smaller than the wavelength.

The highest efficiency of the lens can be achieved by modulating the SPP by simultaneously changing the length and height of the structure. Indeed, we assume that the lens is located in the area $-L \leq x \leq 0$, where L is the maximum length of the lens. For every value of y the height h and length l of the structure can be determined from the condition of the maximum modulus of the transmittance

$$\left|T_{spp}\left(l,h;y\right)\right|\exp\left[-\text{Im}\left(k_{spp}\right)\left(L-l\right)\right]\rightarrow\max \qquad (1.333)$$

where h, l are determined by the conditions of formation of a given phase

$$\text{mod}_{2\pi}\left[\varphi\left(l,h\right)+\text{Re}\left\{k_{spp}\right\}\left(L-l\right)\right]=\varphi\left(y\right). \qquad (1.334)$$

Equation (1.334) suggests that a predetermined phase distribution (1.331) is formed on the line $x = 0$ situated directly behind the lens. This phase consists of the transmittance phase of the lens $\varphi(l,h)$ and the phase shift acquired by SPP over the distance $L-l$ to the line $x = 0$. The exponential factor in (1.333) determines the SPP decay upon propagation overthe distance $L-l$. According to Fig. 1.28, there is a set of points (l, h), providing the predetermined phase shift with respect to the modulus 2π.

Figure 1.44 shows the calculated functions of the length of the relief, the relief height and the absolute value of the transmission coefficient of the lens obtained from the condition of maximizing transmittance (1.333) under the condition (1.334). Maximization was carried out by a brute force search of the values (l,h) in domain D, marked by the dashed rectangle in Fig. 1.28a. The optimal values (l,h) are on the curve shown by the dashed line in Fig. 1.28a. This curve passes through the local maxima of the transmittance modulus. This modulus in Fig. 1.44a exceeds 0.83, which is significantly greater than the lenses in Figs. 1.42 and 1.43. The generated distribution of $|H_y(x,y)|$ is shown in Fig. 1.44b. The energy efficiency of the lenses

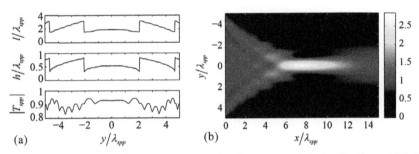

Fig. 1.44. Functions of the lens length (top), the lens height (in the middle), normalized by the wavelength of the SPP and the absolute value of the transmission coefficient function (below) within the lens aperture (a) forming the distribution of $|H_y(x,y)|$ (b).

(1.332) is 80.6%. This is more than 12% greater than that of the lenses based on the change of only one parameter (length or height). Optimization of (l,h) in the wider domain can increase the efficiency of the lens by another 3–4%, but the functions of the length and height of the relief become significantly irregular. Thus, the method of modulation of SPP by simultaneously changing the length and height of the step is the most effective in terms of achieving the high energy efficiency of the diffractive structures.

1.4.4.3. The calculation of diffractive lenses for surface plasmon polaritons with parasitic scattering suppression

Consider now the design of lenses for SPP using the method of suppressing scattering described above. As the dielectric constants of the materials we selected the values $\varepsilon_m = -24.2 + 1.442i$, (Au), $\varepsilon_d = 1$, $\varepsilon_1 = 2.1$, $\varepsilon_2 = 2.89$, coinciding with the material parameters of one of the above examples of suppression of scattering. The free-space wavelength, as before, was 800 nm.

Consider first the lens with $h_1 = 0$ (Fig. 1.30). Fig. 1.45a shows the phase function (top), the length of the microrelief of the lens (middle) and the absolute value of the transmission coefficient (bottom) calculated for the focal length $f = 8\lambda_{spp}$, and the aperture size of the lens $2a = 10\lambda_{spp}$. The height of the lens was constant, 1.5 μm. Fig. 1.45b shows the distribution of $|H_y(x,y)|^2$ generated by the lens and calculated according to the formulas (1.329) and (1.330). Fig. 1.45b shows the focus on the axis on the line $x = f = 8\lambda_{spp}$.

Consider now a two-layer lens (Fig. 1.46), calculated at $h_1 = 65$ nm (the optimum value for the above example of the scattering suppression). Fig. 1.47a shows the function of the height of the microrelief of the lens and transmission coefficient. The parameters of

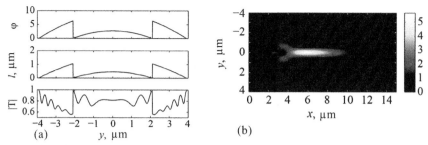

Fig. 1.45. Phase function (top), length of the microrelief of the lens (middle) and the absolute value of the transmission coefficient (bottom) (a), the generated distribution of $|H_y(x,y)|^2$ at $h_1 = 0$ (b).

Fig. 1.46. Double-layer dielectric diffractive structure for the SPP of varying length and a fixed height.

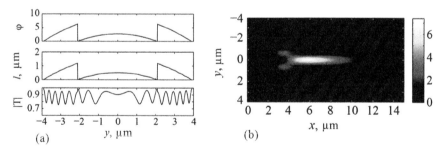

Fig. 1.47. Phase function (top), length of the microrelief of the lens (middle) and the absolute value of the transmission coefficient (bottom) (a), the generated distribution of $|H_y(x,y)|^2$ at $h_1 = 65$ nm (b).

the lens coincide with the previous example. The generated distribution of $|H_y(x,y)|^2$ is shown in Fig. 1.47b and the focusing to the point is also shown. The graphs in Figs. 1.45b and 1.47b are similar in structure.

The energy efficiency (1.332) is 54.5% for the lens shown in Fig. 1.45 and 74% for the lens shown in Fig. 1.47. Thus, the proposed method of suppressing the scattering allows one to increase the efficiency of the lens by 20%. Note that the main reason for the lower efficiency of the lens shown in Fig. 1.45, compared with the lens shown in Fig. 1.42, is the large absorption loss in the metal (Au).

   One of the promising applications of the lenses for focusing SPP are plasmonic circuits in which SPP are used to transmit information between the elements of the optical integrated circuit [104]. Such applications require lenses for the SPP with multiple foci, so as an example we calculated the bifocal lenses. The modulation function was the phase function of the two-order lens [12]:

$$\varphi(y) = \mathrm{mod}_{2\pi}\left\{\left[\varphi_1(y) + \varphi_2(y)\right]/2 + \Phi\left[\mathrm{mod}_{2\pi}\left(\left[\varphi_1(y) - \varphi_2(y)\right]/2\right)\right]\right\}$$

(1.335)

where

$$\varphi_1(y) = -\mathrm{Re}\left(k_{spp}\right)\sqrt{(y-y_1)^2 + f_1^2},$$
$$\varphi_2(y) = -\mathrm{Re}\left(k_{spp}\right)\sqrt{(y-y_2)^2 + f_2^2}$$

(1.336)

and the phase functions of the lenses with foci at the points $(y_1, f_1), (y_2, f_2)$

$$\Phi(\xi) = \begin{cases} 0, & \xi \in [0, \pi), \\ \pi, & \xi \in [\pi, 2\pi). \end{cases}$$

(1.337)

   The calculated distributions of $|H_y(x,y)|^2$ generated by the lenses with two foci at $h_1 = 0$ and $h_1 = 65$ nm are shown in Figs. 1.48a and 1.48b, respectively. The Figures show that the transition to a two-layer structure leads to a 1.4-fold increase in the focal intensity. The parameters of the lenses: $f_1 = f_2 = 10\lambda_{spp}$, $y_1 = -5\lambda_{spp}$, $y_2 = 5\lambda_{spp}$, $2a = 20\lambda_{spp}$.

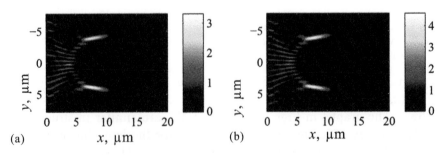

Fig. 1.48. Generated distributions $|H_y(x,y)|^2$ for bifocal lenses at (a) $h_1 = 0$ and (b) $h_1 = 65$ nm.

1.4.5. Calculation of reflective Bragg gratings for surface plasmon–polaritons

Consider the calculation of reflective Bragg gratings (Fig. 1.49) for the case of normal incidence of SPP with a free-space wavelength of 800 nm. In the simulations, the following values of the material parameters of the structure were used: ε_m = −24.2 + 1.442i (corresponds to the dielectric constant of gold), ε_d = 1, ε_{gr} = 2.22. The aspect ratio of the gratings $h/\min\{l_1,l_2\}$ was limited to the value of 3, easily attainable by means of modern technologies of production of nanostructures. The values of l_1 and l_2 were calculated from the equation

$$\mathrm{Re}\left(k_{SPP}^1\right)l_1 = \mathrm{Re}\left(k_{SPP}^2\right)l_2 = \frac{2n+1}{2}\pi \qquad (1.338)$$

where k_{SPP}^1 and k_{SPP}^2 are the propagation constants of the plasmonic modes in their respective grating fields, n = 0,1,... is the order of Bragg reflection. The values of l_1 and l_2 calculated from (1.338) ensure fulfilling the condition of constructive interference in the reflected wave or, what is the same, the condition of the appearance of the band gap:

$$\mathrm{Re}\left(k_{SPP}^1\right)l_1 + \mathrm{Re}\left(k_{SPP}^2\right)l_2 = (2n+1)\pi. \qquad (1.339)$$

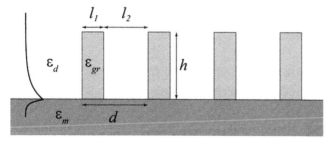

Fig. 1.49. Dielectric Bragg grating for SPP.

In this example, the Fourier modal method was used to calculated and simulate two Bragg gratings for the cases n = 0 and n =1. The geometric parameters of the gratings were: d^0 = 384 nm, l_1^0 = 129 nm, l_2^0 = 196 nm, h^0 = 387 nm, d^1 = 972 nm, l_1^1 = 384 nm, l_2^1 = 588 nm, h^1 = 1152 nm (the superscript corresponds to n). Fig. 1.50 shows the dependences of the reflection coefficients of the SPP (R, solid lines), transmittance of SPP (T, dashed lines) and the scattering loss (S, dotted lines) on the length of the structure n = 0 (thin lines) and n = 1 (thick lines). The crosses show the lengths

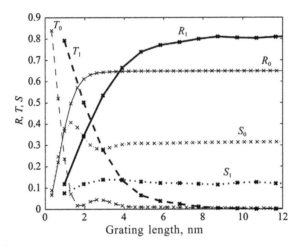

Fig. 1.50. Dependences of the reflection coefficient of SPP (*R*, solid lines), SPP transmittance (*T*, dashed lines) and the scattering losses (*S*, dotted lines) on the length of the structure at *n* = 0 (thin lines) and *n* = 1 (thick lines).

corresponding to gratings consisting of an integer number of the periods. According to Fig. 1.50, an increase in the grating period at a constant aspect ratio increases the reflectance by more than 15% and reduces the scattering by more than 3 times.

Thus, the main reason for reducing the efficiency of the dielectric Bragg gratings for the SPP is parasitic scattering. Using Bragg reflecting gratings with a long period can significantly reduce parasitic scattering and increase the reflectance.

1.5. Generation of interference patterns of evanescent electromagnetic waves by means of diffraction gratings

This section studies the formation of high-frequency interference patterns of evanescent electromagnetic waves (EEW), corresponding to the higher evanescent diffraction orders, in the near-field of diffraction gratings with one-dimensional and two-dimensional periodicity. It is shown that the generation of interference patterns of higher evanescent diffractive orders occurs when quasi-guided modes are excited in the gratings. Methods for controlling the type and period of the produced patterns by changing the parameters of the incident radiation (wavelength, polarization, angle of incidence) have been proposed.

Subsection 1.5.1 describes theoretically the interference patterns of damping diffraction orders of the gratings with one-dimensional and two-dimensional periodicity, assuming that plasmonic modes or modes with the structure of the field below the grating similar to the plane-parallel waveguide modes are excited in the gratings. Subsection 1.5.2 shows the results of investigating the possibility of formation of one-dimensional and two-dimensional interference patterns of the plasmonic modes excited by higher evanescent diffraction orders with a metal layer and methods for controlling the pictures formed by changing the parameters of the incident wave. Subsection 1.5.3 shows the results of investigating the possibility of formation of interference patterns of higher evanescent diffraction orders in the near-field of the dielectric diffraction gratings with one-dimensional and two-dimensional periodicity.

1.5.1. Theoretical description of the interference patterns of evanescent diffraction orders of gratings with one-dimensional and two-dimensional periodicity

This subsection shows the theoretical description of the interference patterns evanescent diffraction orders of diffraction gratings with one-dimensional and two-dimensional periodicity. Approximate analytical expressions for the distributions of the electric field of the generated patterns are derived and their contrast is evaluated. The possibility of controlling configuration of the generated two-dimensional interference patterns by changing the polarization of the incident wave is predicted theoretically.

1.5.1.1. One-dimensional interference patterns

Consider a one-dimensional interference pattern formed by two transmitted evanescent diffraction with numbers $\pm n$ in the case of normal incidence on the grating with a one-dimensional periodicity wave with TE- or TM-polarization. At the same time, we assume that the quasi-guided modes are excited in the grating (in particular, plasmonic modes) and these modes have in the region below the grating the field distribution close to the field of the considered orders, so for ease of analysis, we neglect the contribution of the other transmitted diffraction orders. In accordance with (1.21) and (1.22) the time and coordinate dependences of the components of the electric and magnetic fields of the order with number n have the form $\exp\left[-i\left(\omega t - \mathbf{k}_{n,T}\mathbf{r}'\right)\right]$, where $\mathbf{k}_{n,T} = \left(k_{x,n}, 0, i\kappa_n\right)$, $\kappa_n = \sqrt{k_{x,n}^2 - k_0^2 n_2^2}$. Considering the last expression, we obtain:

$$\mathbf{E}_n, \mathbf{H}_n \sim T_n \exp(-i\omega t)\exp(ik_{x,n}x)\exp(-\kappa_n z'). \qquad (1.340)$$

Note that in the case of normal incidence

$$k_{x,n} = -k_{x,-n} = \frac{2\pi}{d}n. \qquad (1.341)$$

In the case of a symmetrical structure $|T_n| = |T_{-n}|$ and the amplitudes of the $\pm n$-th orders can be represented as

$$T_n = T_{\pm n} \exp\left(i\frac{\varphi}{2}\right),$$
$$T_{-n} = T_{\pm n} \exp\left(-i\frac{\varphi}{2}\right), \qquad (1.342)$$

where values $T_{\pm n}$ and φ are determined by the expressions

$$T_{\pm n} = (T_n T_{-n})^{1/2},$$
$$\varphi = \arg(T_n) - \arg(T_{-n}). \qquad (1.343)$$

At the same time, according to (1.343), $|T_{\pm n}| = |T_n| = |T_{-n}|$.

Consider the case in which the incident wave is TE-polarized $\mathbf{E}_n = (0, E_{y,n}, 0)$. Then:

$$E_{y,n} = T_n \exp(-i\omega t)\exp(ik_{x,n}x)\exp(-\kappa_n z'). \qquad (1.344)$$

Taking into account (1.342) and (1.344), we write the expression for the electric field intensity of the interference pattern:

$$I(x,z') = |E_{y,n} + E_{y,-n}|^2 =$$

$$= |T_{\pm n}|^2 \left|\exp\left[i\left(k_{x,n}x + \frac{\varphi}{2}\right)\right] + \exp\left[-i\left(k_{x,n}x + \frac{\varphi}{2}\right)\right]\right|\exp(-2\kappa_n z') = \qquad (1.345)$$

$$= 2|T_{\pm n}|^2 \cos^2\left(k_{x,n}x + \frac{\varphi}{2}\right)\exp(-2\kappa_n z').$$

Applying to (1.345) the double-angle formulae, we finally obtain:

$$I(x,z') = |T_{\pm n}|^2 \left[\cos(2k_{x,n}x + \varphi) + 1\right]\exp(-2\kappa_n z'). \qquad (1.346)$$

From (1.346) it follows that the period of the formed interference pattern is $d_{ip} = \pi/k_{x,r}$. Taking into account (1.341) in the expression for the period, we have:

$$d_{ip} = \frac{d}{2n}. \qquad (1.347)$$

Thus, the period of the generated interference pattern is $2n$ times smaller than the period of the grating used. From (1.346) it also follows that the contrast of the interference pattern, defined by the expression

$$K = \frac{\max_x I(x,0) - \min_x I(x,0)}{\max_x I(x,0) + \min_x I(x,0)},$$ (1.348)

in the case of TE-polarization is

$$K_{TE} = 1.$$ (1.349)

It should be noted that the phase difference φ between the amplitudes of the diffraction orders T_n, T_{-n} affects only the position of the maxima and minima of the interference pattern on the period, but not its form and contrast. Hereinafter, we assume that the intensity of the wave incident on the structure is equal to 1, and the expression (1.346), and similar expressions to be obtained below, will describe the intensity of the interference pattern, normalized by the intensity of the incident wave.

Now consider the case where the incident wave on the grating is TM-polarized. At the same time, $\mathbf{E}_n = (E_{x,n}, 0, E_{z,n})$, $\mathbf{H}_n = (0, H_{y,n}, 0)$ and with the component $H_{y,n}$ has the form

$$H_{y,n} = T_n \exp(-i\omega t) \exp(ik_{x,n} x) \exp(-\kappa_n z').$$ (1.350)

Substituting (1.350) into the Maxwell's equations (1.1), we obtain expressions for the components of the electric field $E_{x,n}$ and $E_{z,n}$:

$$\mathbf{E}_n = \left(\frac{i\kappa_n}{k_0 n_2^2}, 0, -\frac{k_{x,n}}{k_0 n_2^2} \right) T_n \exp(-i\omega t) \exp(ik_{x,n} x) \exp(-\kappa_n z').$$ (1.351)

According to (1.351),

$$E_{x,\pm n} = E_{x,n} + E_{x,-n} =$$

$$= \frac{i\kappa_n}{k_0 n_2^2} T_{\pm n} \left\{ \exp\left[i\left(k_{x,n} x + \frac{\varphi}{2} \right) \right] + \exp\left[-i\left(k_{x,n} x + \frac{\varphi}{2} \right) \right] \right\} \exp(-\kappa_n z') =$$

$$= \frac{2i\kappa_n}{k_0 n_2^2} T_{\pm n} \cos\left(k_{x,n} x + \frac{\varphi}{2} \right) \exp(-\kappa_n z'),$$

$$E_{z,\pm n} = E_{z,n} + E_{z,-n} =$$

$$= -\frac{k_{x,n}}{k_0 n_2^2} T_{\pm n} \left\{ \exp\left[i\left(k_{x,n}x + \frac{\varphi}{2} \right)\right] - \exp\left[-i\left(k_{x,n}x + \frac{\varphi}{2} \right)\right] \right\} \exp(-\kappa_n z') =$$

$$= -\frac{2ik_{x,n}}{k_0 n_2^2} T_{\pm n} \sin\left(k_{x,n}x + \frac{\varphi}{2} \right) \exp(-\kappa_n z'). \qquad (1.352)$$

We now write the expression for the intensity of the generated interference pattern:

$$I(x,z') = \left| E_{x,\pm n} \right|^2 + \left| E_{z,\pm n} \right|^2 =$$

$$= \frac{2|T_{\pm n}|^2}{k_0^2 n_2^4} \left\{ \kappa_n^2 \left[\cos(2k_{x,n}x + \varphi) + 1 \right] + k_{x,n}^2 \left[1 - \cos(2k_{x,n}x + \varphi) \right] \right\} \cdot$$

$$\cdot \exp(-2\kappa_n z') = \frac{2|T_{\pm n}|^2}{k_0^2 n_2^4} \left\{ \kappa_n^2 + k_{x,n}^2 + \cos(2k_{x,n}x + \varphi)\left(\kappa_n^2 - k_{x,n}^2 \right) \right\} \exp(-2\kappa_n z').$$

$$(1.353)$$

Given the expression for κ_n, we rewrite (1.353) as

$$I(x,z') = \frac{2|T_{\pm n}|^2}{k_0^2 n_2^4} \left\{ 2k_{x,n}^2 - k_0^2 n_2^2 - k_0^2 n_2^2 \cos(2k_{x,n}x + \varphi) \right\} \exp(-2\kappa_n z').$$

$$(1.354)$$

According to (1.354), the period of the interference pattern in this case is also determined by the expression (1.347). To find the contrast of the interference pattern (1.348), we write down the expressions for the maximum and minimum intensities of the interference pattern:

$$\max_x I = \frac{4|T_{\pm n}|^2 k_{x,n}^2}{k_0^2 n_2^4} \exp(-2\kappa_n z'),$$

$$(1.355)$$

$$\min_x I = \frac{4|T_{\pm n}|^2 \left(k_{x,n}^2 - k_0^2 n_2^2 \right)}{k_0^2 n_2^4} \exp(-2\kappa_n z').$$

From (1.355) we obtain the expression for the contrast of the pattern formed in the case of TM-polarization of the incident wave:

$$K_{TM} = \frac{n_2^2}{2\left(k_{x,n}/k_0 \right)^2 - n_2^2} = \frac{n_2^2}{2\left(\lambda n/d \right)^2 - n_2^2}. \qquad (1.356)$$

Once again we note that the expression for the interference patterns of the type (1.346), (1.354) and contrast (1.349), (1.356) correspond to the 'ideal' interference patterns in the absence of diffraction orders with numbers other than ±n, therefore the study of interference patterns, formed by particular diffraction gratings, should be carried out on the basis of simulation within rigorous electromagnetic diffraction theory. The simulation results for several diffraction gratings by the Fourier modal method, as described in section 1.1, are given below in the sections 1.5.2 and 1.5.3.

1.5.1.2. Two-dimensional interference patterns

Let us now study the configuration of the two-dimensional interference patterns of the transmitted evanescent diffraction orders generated in the near field of diffraction gratings with two-dimensional periodicity at normal incidence. In this case, the time and coordinate dependences of the components of the electromagnetic field of the transmitted diffraction orders are given by $\exp\left[-i\left(\omega t - \mathbf{k}_{n,m,T}\mathbf{r}'\right)\right]$, $\mathbf{k}_{n,m,T} = \left(k_{x,n}, k_{y,m}, i\kappa_{n,m}\right)$ where

$$k_{x,n} = \frac{2\pi}{d_x}n,$$

$$k_{y,m} = \frac{2\pi}{d_y}m, \tag{1.357}$$

$$\kappa_{n,m} = \sqrt{k_{x,n}^2 + k_{y,n}^2 - k_0^2 n_2^2},$$

where d_x, d_y are the grating periods along the axes Ox and Oy, respectively. Here

$$\mathbf{E}_{n,m}, \mathbf{H}_{n,m} \sim T_{n,m}\exp\left(-i\omega t\right)\exp\left[i\left(k_{x,n}x + k_{y,n}y\right)\right]\exp\left(-\kappa_{n,m}z'\right). \tag{1.358}$$

We assume that the gratings are symmetrical about the planes parallel to the coordinate planes xOz and yOz and assume that the periods along the axes are equal: $d_x = d_y = d$, in this case $k_{x,n} = k_{y,n} = k_n$. Similarly to the one-dimensional case, in incidence on such gratings the waves with circular or mixed linear polarization we have $|T_{n,m}| = |T_{-n,-m}| = |T_{n,-m}| = |T_{-n,m}|$ [105,106]. At mixed linear polarization the angle between the vector of the electric field vector and the Ox axis is 45°.

Let us consider the superposition of four diffraction orders with numbers $(\pm n,0)$, $(0, \pm n)$ and, similar to the one-dimensional case, we assume that the normal incidence on the grating of a wave with circular or mixed linear polarization results in the excitation in the structure of quasi-guided modes, with the field distribution below the grating close to the field of the considered diffraction orders. Consider first the case when the excited modes have the polarization close to TE-polarization. Thus, we consider only the y-component of the electric field for the modes propagating along the Ox axis and the x-component for the modes propagating along the Oy axis. Then the electric field corresponding to the transmitted evanescent diffraction orders with numbers $(\pm n, 0)$, $(0, \pm n)$ will have the form

$$E_{y,n,0} = T_{n,0} \exp(-i\omega t)\exp(ik_n x)\exp(-\kappa_n z'),$$
$$E_{x,0,n} = T_{0,n} \exp(-i\omega t)\exp(ik_n y)\exp(-\kappa_n z'),$$

(1.359)

where $\kappa_n = \kappa_{n,0} = \kappa_{0,n}$.

We assume that in the considered structure quasi-guided modes propagating along the Ox and Oy axes are excited by only TE- and TM-components of the incident wave, respectively. At the same time, the phase difference between the excited modes is equal to the phase difference between the components of the incident wave φ and for the amplitudes of diffraction orders with numbers $(\pm n, 0)$, $(0, \pm n)$ the following expressions will be valid:

$$T_{n,0} = \exp(i\phi)T_{0,n} = T_{\pm n} \exp\left(i\frac{\varphi}{2}\right),$$
$$T_{-n,0} = \exp(i\phi)T_{0,-n} = T_{\pm n} \exp\left(-i\frac{\varphi}{2}\right),$$

(1.360)

where the values $T_{\pm n}$ and φ are defined similarly to the one-dimensional case (equation (1.343)). Taking into account (1.360) in (1.359), we write the expression for the components of the electric field of superposition of diffraction orders:

$$E_{y,\pm n,0} = E_{y,n,0} + E_{y,-n,0} = 2T_{\pm n} \exp(-i\omega t)\cos\left(k_n x + \frac{\varphi}{2}\right)\exp(-\kappa_n z'),$$

$$E_{x,0,\pm n} = E_{x,0,n} + E_{x,0,-n} = 2\exp(-i\phi)T_{\pm n} \exp(-i\omega t)\cos\left(k_n y + \frac{\varphi}{2}\right)\exp(-\kappa_n z').$$

(1.361)

We now write down the expression for the electric field intensity of the interference pattern:

$$I(x,y,z') = |E_{y,\pm n,0}|^2 + |E_{x,0,\pm n}|^2 = 4|T_{\pm n}|^2 \left[\cos^2\left(k_n x + \frac{\varphi}{2}\right) + \cos^2\left(k_n y + \frac{\varphi}{2}\right)\right]\exp(-2\kappa_n z') =$$

$$= |T_{\pm n}|^2 \left[2 + \cos(2k_n x + \varphi) + \cos(2k_n y + \varphi)\right]\exp(-2\kappa_n z').$$

$$(1.362)$$

From (1.362) it follows that similar to the one-dimensional case, the periods of the interference pattern formed along the Ox and Oy axes are equal to $d_{ip,x} = d_{ip,y} = d_{ip} = \pi/k_n$, and the contrast of the interference pattern at the interface between the grating and the substrate in the two-dimensional case defined by the expression

$$K = \frac{\max\limits_{x,y} I(x,y,0) - \min\limits_{x,y} I(x,y,0)}{\max\limits_{x,y} I(x,y,0) + \min\limits_{x,y} I(x,y,0)},$$

$$(1.363)$$

is equal to 1.

We find the position of the maxima and minima of the interference pattern, described by (1.362). For this we use a sufficient condition for the extremum of the function of two variables [107]: for the existence of an extremum of the function F of two variables at the point (x_0, y_0) it is sufficient to satisfy the following conditions:

$$F'_x(x_0, y_0) = 0,$$
$$F'_y(x_0, y_0) = 0,$$

$$(1.364)$$

$$F''_{xx}(x_0, y_0) \cdot F''_{yy}(x_0, y_0) - \left(F''_{xy}(x_0, y_0)\right)^2 > 0,$$

$$(1.365)$$

where F'_x, F'_y, and F''_{xx}, F''_{yy}, F''_{xy} are the first and second partial derivatives of the function F, respectively. Thus if $F''_{xx}(x_0, y_0) > 0$ at the point (x_0, y_0) the studied function has a minimum, if $F''_{xx}(x_0, y_0) < 0$ a maximum. Without loss of generality we set $\varphi = 0$ and examine the function

$$F = \cos(2k_n x) + \cos(2k_n y),$$

$$(1.366)$$

and the positions of the extrema of this function coincide with the positions of the extrema of the intensity of the electric field of the formed interference pattern.

It is easy to show that the condition of equality of the first derivatives to zero (1.364) is performed at the points

$$\begin{cases} x = \dfrac{dl_x}{4n}, l_x \in \mathbb{Z}, \\[2mm] y = \dfrac{dl_y}{4n}, l_y \in \mathbb{Z}. \end{cases} \qquad (1.367)$$

At these points we need to check the fulfillment of the condition (1.365). For the considered function (1.366), it takes the form

$$\cos(2k_n x)\cos(2k_n y) > 0. \qquad (1.368)$$

Substituting the set of points (1.367) into (1.368), we obtain the inequality

$$\cos(\pi l_x)\cos(\pi l_y) > 0. \qquad (1.369)$$

The inequality (1.369) is satisfied when l_x and l_y have the same parity. In the case where l_x and l_y are even the function F (and therefore the electric field intensity of the interference pattern) reaches its maximum and when l_x and l_y are odd a minimum is reached.

The positions of the extrema of the function I at $n = 3$ are shown in Fig. 1.51. The maxima are indicated by the symbols '+', the minima by dots. Fig. 1.52 shows the intensity distribution (1.362) within the period on the grating at $n = 3$ and $\varphi = 0$ normalized by the maximum value.

Note that the expression (1.362) shows that when the modes excited in the grating are TE-polarized, the form of the pattern

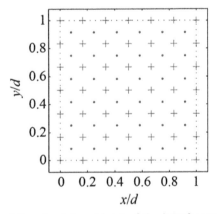

Fig. 1.51. Maxima (+) and minima (dots) of the interference pattern in the case of excitation of quasi-guided modes of the TE-type at $n = 3$ in the grating. The dashed rectangle shows the boundaries of the grating period.

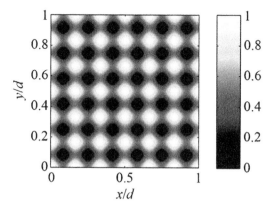

Fig. 1.52. A theoretical estimate of the form of the electric field intensity distribution *I* in the substrate at the grating/substrate interface within one grating period upon excitation of quasi-waveguide TE-modes at $n = 3$.

does not depend on the phase difference between the TE- and TM-components of the incident wave.

Consider now the case where the quasi-guided modes with the polarization close to the TM-polarization are excited in the grating. Thus, we consider only the y-component of the magnetic field for the modes propagating along the axis Ox and the x-component for the modes propagating along the axis Oy. Then the magnetic field components corresponding to the transmitted evanescent diffraction orders with numbers $(\pm n, 0)$, $(0, \pm n)$ will have the form

$$H_{y,n,0} = T_{n,0} \exp(-i\omega t) \exp(ik_n x) \exp(-\kappa_n z'),$$
$$H_{x,0,n} = T_{0,n} \exp(-i\omega t) \exp(ik_n y) \exp(-\kappa_n z'). \tag{1.370}$$

Substituting (1.370) into (1.1), we write the expression for the components of the electric field of the considered diffraction orders:

$$\mathbf{E}_{n,0} = \left(\frac{i\kappa_n}{k_0 n_2^2}, 0, \frac{-k_n}{k_0 n_2^2} \right) T_{n,0} \exp(-i\omega t) \exp(ik_n x) \exp(-\kappa_n z'),$$
$$\mathbf{E}_{0,n} = \left(0, \frac{-i\kappa_n}{k_0 n_2^2}, \frac{k_n}{k_0 n_2^2} \right) T_{0,n} \exp(-i\omega t) \exp(ik_n y) \exp(-\kappa_n z'), \tag{1.371}$$

Assuming, as in the case of excitation of the TE-type modes, that the equalities (1.360) are fulfilled, we write the expressions for the components of the electric field in superposition of diffraction orders

$$E_{\pm n,0} = E_{n,0} + E_{-n,0} = \left(\kappa_n \cos\left(k_n x + \frac{\varphi}{2}\right), 0, -k_n \sin\left(k_n x + \frac{\varphi}{2}\right)\right) \frac{2iT_{\pm n}}{k_0 n_2^2} \exp(-i\omega t)\exp(-\kappa_n z'),$$

$$E_{0,\pm n} = E_{0,n} + E_{0,-n} = \left(0, -\kappa_n \cos\left(k_n y + \frac{\varphi}{2}\right), k_n \sin\left(k_n y + \frac{\varphi}{2}\right)\right) \frac{2iT_{\pm n}}{k_0 n_2^2} \exp(-i\varphi)\exp(-i\omega t) \cdot$$

$$\cdot \exp(-\kappa_n z'). \tag{1.372}$$

In view of (1.372), we obtain representations for the intensity of the electric field generated by the interference pattern:

$$I(x,y,z') = \frac{4|T_{\pm n}|^2}{\left(k_0 n_2^2\right)^2}\left\{\kappa_n^2\left[\cos^2\left(k_n x + \frac{\varphi}{2}\right) + \cos^2\left(k_n y + \frac{\varphi}{2}\right)\right] + \right.$$

$$\tag{1.373}$$

$$\left. + k_n^2\left|\exp(-i\varphi)\sin\left(k_n y + \frac{\varphi}{2}\right) - \sin\left(k_n x + \frac{\varphi}{2}\right)\right|^2\right\}\exp(-2\kappa_n z').$$

Consider the case where the incident wave has mixed linear polarization at $\varphi = 0$. In this case, the expression (1.373) at $\varphi = 0$ takes the form

$$I(x,y,z') = \frac{4|T_{\pm n}|^2}{\left(k_0 n_2^2\right)^2}\left\{2k_n^2 - k_0^2 n_2^2 - \left[k_0^2 n_2^2 \frac{\cos(2k_n x) + \cos(2k_n y)}{2} + 2k_n^2 \sin(k_n x)\sin(k_n y)\right]\right\} \cdot$$

$$\cdot \exp(-2\kappa_n z'). \tag{1.374}$$

Similarly to the case of excitation of the TE-type modes, we find the positions of the extrema of the interference pattern. Condition (1.364) will have the form

$$\begin{cases} \cos(k_n x)\left[k_n^2 \sin(k_n y) - k_0^2 n_2^2 \sin(k_n x)\right] = 0, \\ \cos(k_n y)\left[k_n^2 \sin(k_n x) - k_0^2 n_2^2 \sin(k_n y)\right] = 0. \end{cases} \tag{1.375}$$

Solutions of of the system (1.375) are the two sets of points:

$$\begin{cases} x = \dfrac{d}{4n} + \dfrac{dl_x}{2n}, l_x \in \mathbb{Z}, \\ \\ y = \dfrac{d}{4n} + \dfrac{dl_y}{2n}, l_y \in \mathbb{Z} \end{cases} \tag{1.376}$$

and

$$\begin{cases} x = \dfrac{dl_x}{2n}, l_x \in \mathbb{Z}, \\ y = \dfrac{dl_y}{2n}, l_y \in \mathbb{Z}. \end{cases} \quad (1.377)$$

Condition (1.365) in this case takes the form

$$4k_n^4 \left[k_n^2 \sin(k_n x) \sin(k_n y) + k_0^2 n_2^2 \cos(2k_n x) \right]$$
$$\left[k_n^2 \sin(k_n x) \sin(k_n y) + k_0^2 n_2^2 \cos(2k_n y) \right] - \quad (1.378)$$
$$-4k_n^8 \cos^2(k_n x) \cos^2(k_n y) > 0.$$

Direct substitution shows that the points (1.376) satisfy the inequality (1.378), points (1.377) – do not satisfy. Thus, the positions of the extrema of the interference pattern are determined by the expression (1.376). By analyzing the sign of the derivative F''_{xx}, it is easy to show that in the case where the values l_x and l_y have different parity intensity maxima are reached, and otherwise – its minima. The positions of the extrema of the function I at $n = 3$ are shown in Fig. 1.53. As in the previous case, the symbols '+' show the maxima, the points show the minima.

Figure 1.54 shows the intensity distribution (1.374) within one grating period $n = 3$, normalized by the maximum intensity.

In this case, the period of the interference pattern is the distance between adjacent peaks located on a straight line at an angle of 45°

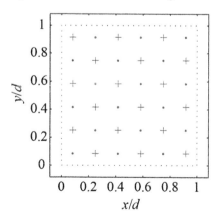

Fig. 1.53. Maxima (+) and minima (dots) of the intensity of the interference pattern in the case of excitation in the grating of quasi-guided modes of the TM-type with a mixed linear polarization of the incident wave at $n = 3$. The dashed rectangle shows the boundaries of the period.

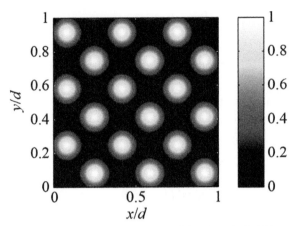

Fig. 1.54. A theoretical estimate of the form of the electric field intensity distribution *I* in the substrate at the grating/substrate interface within one period of the grating upon excitation in the structure of quasi-guided modes of the TM-type ay *n* = 3. The incident wave has a mixed linear polarization.

to the coordinate axes (see Fig. 1.53). In this case, the period of the pattern equals to

$$d_{ip} = \frac{d}{\sqrt{2}n}. \tag{1.379}$$

Substituting the coordinates of the extrema (1.376) in the expression for the intensity (1.374), we find that the intensity at the minima of the interference pattern is equal to 0, so the contrast of the pattern, defined by (1.363), in the case of mixed linear polarization of the incident wave is equal to 1.

Now consider the case where the wave incident on the grating wave has circular polarization. In this case, $\varphi = \pi/2$ the expression (1.373) $\varphi = 0$ will look at

$$I(x,y,z') = \frac{4|T_{\pm n}|^2}{\left(k_0 n_2^2\right)^2}\left\{2k_n^2 - k_0^2 n_2^2\left[\cos^2\left(k_n x\right) + \cos^2\left(k_n y\right)\right]\right\}\exp\left(-2\kappa_n z'\right). \tag{1.380}$$

The maxima of the interference pattern intensity in this case are located at the points (1.376), while the minima correspond to the points (1.377), and there are no further restrictions on the values of l_x and l_y. The positions of the extrema of the function *I* at *n* = 3 are shown in Fig. 1.55. The symbols '+' show the maxima, the dots – the minima.

Figure 1.56 shows the normalized to the maximum intensity distribution (1.380) on the same period in the lattice *n* = 3.

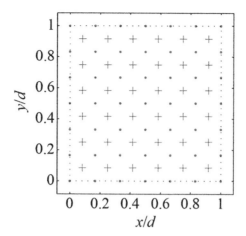

Fig. 1.55. Maxima (+) and minima (dots) of the interference pattern in the case of excitation in the grating of quasi-guided modes of the TM-type at circular polarization of the incident wave at $n = 3$. The dashed rectangle shows the boundaries of the period.

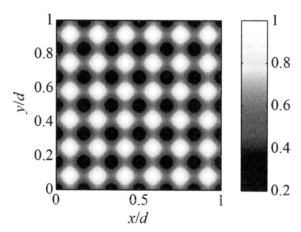

Fig. 1.56. A theoretical estimate of the form of the electric field intensity distribution I in the substrate at the grating/substrate interface within one grating period in excitation in the structure of quasi-guided modes of the TM-type at $n = 3$. The incident wave has circular polarization. $\lambda/d \approx 0.596$, $n_2 = 1.6$.

Substituting the coordinates of the extremes (1.376) and (1.377) into (1.380), we find the maximum and minimum values of the intensity of the interference pattern:

$$\max_{x,y} I(x,y,0) = \frac{8k_n^2 |T_{\pm n}|^2}{\left(k_0 n_2^2\right)^2},$$

$$\min_{x,y} I(x,y,0) = \frac{8\left(k_n^2 - k_0^2 n_2^2\right)|T_{\pm n}|^2}{\left(k_0 n_2^2\right)^2}.$$

(1.381)

From (1.381) it follows that in this case, the contrast of the interference pattern coincides with the expression for the one-dimensional interference pattern (1.356).

From a comparison of the expressions (1.362) and (1.380) and Figs. 1.52 and 1.56 it follows that the form of the interference pattern formed in the excitation of the quasi-waveguide modes of the TM-type with circular polarization, is similar to the form of the patterns, formed in the excitation of the TE-type modes. The period of the patterns is $d/2n$. Furthermore, from a comparison of the expressions (1.374) and (1.380) and Figs. 1.54 and 1.56 it follows that the form of the interference patterns generated during excitation of the TM-type modes depends on the polarization of the incident wave. This property may allow to create different interference patterns using a single diffraction grating.

Once again, we note that the expressions for the intensity and contrast of the interference patterns, derived in this subsection, are estimates that do not take into account the contribution of diffraction orders with numbers other than $\pm n$ (in the case of one-dimensional interference patterns), or $(\pm n, 0)$, $(0, \pm n)$ (in the case of two-dimensional interference patterns) and assume that the quasi-waveguide modes, excited in the diffraction grating, are similar in field configuration and polarization to the plane-parallel waveguide or plasmonic modes. To test them, we must use a numerical simulation based on the rigorous electromagnetic theory of diffraction to calculate the intensity distribution of the electric field generated by the particular diffraction gratings with excitation quasi-waveguide modes in them.

1.5.2. Formation of interference patterns of evanescent electromagnetic waves in the metal–dielectric diffraction gratings

1.5.2.1. Generation of one-dimensional interference patterns
The generation of one-dimensional interference patterns of evanescent

electromagnetic waves by means of a diffraction structure consisting of a dielectric diffraction grating with a one-dimensional periodicity and a uniform metal layer disposed on a dielectric substrate (Fig. 1.57) [108,109]. The period of the grating is d, the lattice has one ridge per period with the width w. The dielectric constant of the grating is ε_{gr}, step height is h_{gr}, the thickness of a homogeneous dielectric layer underneath the grating is h_1. The thickness of the homogeneous metal film located beneath the grating and having a permittivity ε_m is h_m.

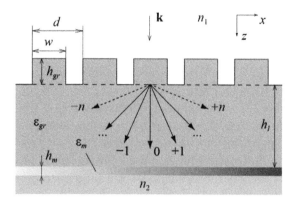

Fig. 1.57. Diffractive structure for generation of one-dimensional interference patterns of plasmonic modes.

The theoretical description of the generation of interference patterns given in subsection 1.5.1 was obtained under the assumption that at normal incidence of a plane wave on the structure a quasi-guided mode is excited in it. Plasmonic modes with the field structure similar to SPP propagating at the interface between the metal layer and the substrate can be excited in the structure under consideration. This approximate condition of the excitation of modes by evanescent diffraction orders with numbers $\pm n$ has the form

$$k_{x,n} = \mathrm{Re}\{k_{spp}\}, \qquad (1.382)$$

where the propagation constant of the diffraction order with number n is defined by the expression (1.341), and the propagation constant of SPP – by the expression (1.284), wherein value n_2^2 has to be used as the dielectric permittivity. Given a specific value of n, from (1.382) we can obtain an expression for the period of the diffraction grating:

$$d = \lambda n \operatorname{Re}\left\{\sqrt{\frac{\varepsilon_m + n_2^2}{\varepsilon_m n_2^2}}\right\}. \qquad (1.383)$$

Once again, we note that the expression for the grating period (1.383) corresponds to the approximate condition of excitation of plasmonic modes in the structure, however, as will be shown below, using this expression to calculate the grating period provides a structure in which the interference patterns of high intensity and contrast form.

Consider the example of the generation of one-dimensional interference patterns of the plasmonic modes using the structure shown in Fig. 1.57. In the simulations, the wavelength of the incident light was equal to 550 nm, and the following values of the material parameters of the structure were selected: $n_1 = 1$, $n_2 = 1.6$, $\varepsilon_m = -12.922 + 0.44727i$. The value n_2 corresponds to a photoresist, and ε_m is the permittivity of silver for the selected wavelength [110]. The dielectric permititivity of the material lattice ε_{gr} was also chosen equal to 2.56. Note that the selected wavelength is far from the resonant value which for the pair of the studied materials is equal to 368 nm. The propagation length of the SPP, given by (1.291), is in this case 5739 nm. At the selected value $n = 5$ the period of the diffraction grating, calculated by the formula (1.383), is equal to 1539 nm. According to (1.347), the period of the interference pattern of the SPP is 154 nm, that is an order of magnitude less than the period of the diffraction grating generating the pattern. Thus, assuming that the size of pattern details is equal to half of its period, we obtain a value of 77 nm.

A theoretical estimate of the contrast of the pattern, calculated using the formula (1.356), is equal to 0.67.

It should be noted that in [111] plasmonic modes were excited by higher propagating orders formed by the diffraction grating in a thick dielectric layer above the metal film. In this case, the dielectric constant of the material of the grating ε_{gr} should be greater than the dielectric constant of the material n_2^2 beneath the film. This subsection deals with the case where the diffraction orders used to excite the plasmonic modes are evanescent. In this case ε_{gr} may be equal to or smaller than n_2^2.

Values of other geometrical parameters of the structure were determined from the condition of maximizing the quality of the interference pattern. The following function was selected as the merit function:

$$F\left(h_{gr},h_{l},h_{m},w\right)=\frac{\int\limits_{0}^{d}\left|I\left(x,0\right)-I_{RCWA}\left(x,0\right)\right|dx}{\max\limits_{x}\left\{I\left(x,0\right)\right\}}\cdot\frac{1}{\max\limits_{x}\left\{I\left(x,0\right)\right\}}\to\min\limits_{h_{gr},h_{l},h_{m},w}. \qquad (1.384)$$

The first factor in (1.384) is a measure of the closeness of the calculated interference pattern $I_{RCWA}(x,0)$ to the 'perfect' interference pattern $I(x,0)$ generated upon interference of two SPPs excited by two transmitted grating orders and given by (1.354). The second factor allows to find the structures that generate interference patterns with high field intensity at the maxima. The merit function was minimized using numerical optimization procedures of the MATLAB software. Note that in the optimization of the geometrical parameters of the structure we considered the limitations on the minimum thickness of the metal film, related to the penetration depth of the SPP into the metal defined by expression (1.297) and equal to 22 mm for the selected wavelength and the pair of materials.

As a result of optimization we found a structure having the following geometric parameters: h_{gr} = 435 nm, h_{1} = 0, h_{m} = 65 nm, w = 0.5d. Fig. 1.58 shows a graph of the field intensity normalized to the intensity of the incident wave. This field forms directly under the metal layer at the above parameters. The intensity distribution formed by only taking into account the diffraction orders with numbers $\pm n$ is also shown.

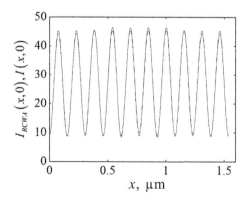

Fig. 1.58. The intensity of the electric field at the metal layer/substrate interface (solid line – the full field, dashed line – diffraction orders with numbers ± 5).

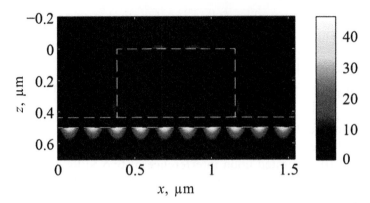

Fig. 1.59. Distribution of the electric field intensity in the structure (the structure is indicated by dashed lines).

Note that the period of the interference pattern d_{ip} is not only 10 times smaller than the period of the diffractive grating generating the pattern, but is also 3.57 times smaller than the wavelength of the incident light. The ratio of the intensity of the interference maxima to the intensity of the incident wave exceeds 45. The contrast of the resulting interference pattern is equal to 0.69, which is close to the theoretical estimate of 0.67 obtained by the formula (1.356). Thus, the proposed structure allows one to create an interference pattern of high quality. At the same time, the period of the structure can be determined used expression (1.383), following from approximate conditions of excitation of the plasmonic modes (1.382).

Figure 1.59 shows the distribution of the normalized intensity of the electric field in the structure under consideration. The penetration depth of the SPP in the medium with a refractive index n_2 (photoresist) is equal to 111 nm, which agrees with the theoretical value for the selected wavelength, calculated by the formula (1.296) and equal to 110 nm.

Figure 1.60 shows the dependence of the electric field intensity at the maxima of the interference pattern and the contrast on the distance from the interface between the metal layer and the substrate. Note that the total intensity of the transmitted diffraction orders is 0.019, so high contrast is maintained throughout the decay distance of the interference pattern.

Consider another example. Let the value of the wavelength be 436 nm, which corresponds to the InGaN/GaN semiconductor laser [112], and the materials of the grating, the metal film and the substrate are the same as in the previous case. The propagation

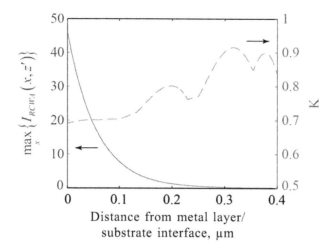

Fig. 1.60. Dependence of the intensity at the maxima of the interference pattern (solid line) and contrast (dashed line) on the distance from the interface between the metal layer and the substrate.

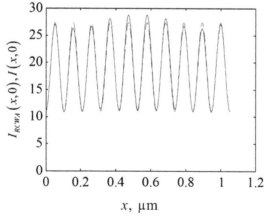

Fig. 1.61. The intensity of the electric field at the interface between the metal layer and the substrate (solid line – the full field, dashed line – diffraction orders with numbers ±5) at λ = 435 nm.

length of SPP is 1521 nm. The grating period, equal to 1051 nm, was calculated by the formula (1.383) at n = 5. The geometrical parameters of the structure h_{gr} = 835 nm, h_l = 0, h_m = 73 nm, w = 0.425d were found using the optimization procedure as in the previous case. The calculated interference pattern is shown in Fig. 1.61.

The period of the interference pattern is 105 nm, which is 4.15 times smaller than the wavelength of incident light. The electric field intensity, normalized by the incident wave intensity, at the maxima of the interference pattern was greater than 25, the contrast was 0.45, which is close to the theoretical estimate of 0.43.

Note that when the wavelength of the incident radiation to the resonant value of 368 nm the propagation length of the SPP was reduced, which leads to the deterioration of the generated interference patterns and decrease of the intensity of the electric field at their maxima. For example, at a wavelength of 420 nm the propagation length of the SPP (1.291) becames smaller than the diffraction grating period (1.383), calculated at $n = 5$. This limits the range of wavelengths that can be used in practice.

1.5.2.2. Controlling the period of one-dimensional interference patterns by changing the parameters of the incident wave

In this subsection we suggest two ways to control the frequency of the interference patterns by changing the parameters of the incident wave. This approach enables to generate interference patterns with different periods using the same diffractive structure. The generation of interference patterns with different periods is achieved by the excitation of plasmonic modes in the structure by various diffraction orders. As will be shown below, this is possible by changing the angle of incidence or wavelength of the incident light.

The geometrical parameters of the structures discussed as examples were found by optimization procedures similar to the previous case. The merit function in this case has the form similar to (1.384):

$$F\left(h_{gr}, h_l, h_m, w\right) = \frac{\int_0^d \left|I(x,0) - I_{RCWA}(x,0)\right| dx}{\max_x \left\{I(x,0)\right\}} \cdot \frac{1}{\max_x \left\{I(x,0)\right\}} \to \min_{h_{gr}, h_l, h_m, w}.$$

(1.385)

where $\mathbb{N}$ is the set of numbers of diffraction orders, exciting the plasmonic modes in the structure at various parameters of the incident wave, and the upper index (n) indicates the number of the order.

Frequency control of the interference pattern by changing the wavelength of the incident wave

Consider the use of incident light with different wavelengths to generate interference patterns of different periods. Let the diffraction grating period d, defined by the formula (1.383), be calculated from the excitation conditions of SPP by orders with the numbers $\pm n$ at some wavelength. Since the propagation constant of SPP k_{spp} (1.284) depends on the wavelength, it is possible to excite plasmonic modes at a different wavelength using the transmitted diffraction orders with numbers $\pm m$, $n \neq m$. Thus, in accordance with the expressions (1.341) and (1.382), the wavelengths that will excite plasmonic modes by orders with numbers $\pm m$ can be found from the equation

$$\frac{2\pi m}{d} = \mathrm{Re}\{k_{spp}(\lambda)\}. \qquad (1.386)$$

In particular, for the above case of the excitation of the plasmonic modes at $\lambda = 550$ nm, $n = 5$, SPP will also be excited by ± 4-th transmitted orders at $\lambda = 659$ nm and orders with numbers ± 3 at $\lambda = 852$ nm. The corresponding interference patterns will have periods of 192 nm and 257 nm (6 and 8 times less than the diffraction grating period). Theoretical estimates of the contrast of the resultant interference patterns are equal to 0.77 and 0.87 for the wavelengths of 659 nm and 852 nm, respectively.

The geometrical parameters of the structure were determined by minimizing the criterion (1.385). The following values were found: $w = 0.37$ nm, $h_{gr} = 1000$ nm, $h_l = 0$, $h_m = 70$ nm. The calculated graphs of the normalized intensity of the interference patterns in the dielectric directly under the metal film and the intensity distribution in the structure at wavelengths of 550 nm, 659 nm and 852 nm are shown in Figs. 1.62–1.64.

The values of the contrast and the normalized intensity at the maxima of the pattern are (0, 69, 19), (0, 78, 31) and (0, 87, 73), respectively. Note that the contrast values of the generated interference patterns are close to the theoretically calculated values. The penetration depth of the SPP in the medium with refractive index n_2 is equal to 172 nm and 305 nm for the wavelengths of incident light 659 nm and 852 nm, respectively.

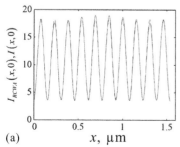
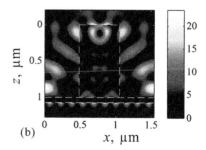

(a)

(b)

Fig. 1.62. The intensity of the electric field at the metal layer/substrate interface (solid line – the total field, the dashed line – diffraction orders with numbers ± 5) (a) and the intensity distribution of the electric field in the structure (b) at $\lambda = 550$ nm.

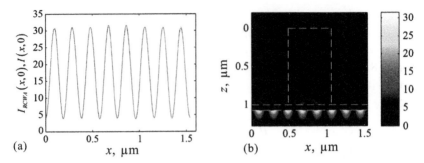

Fig. 1.63. The intensity of the electric field at the metal layer/substrate interface (solid line – the total field, the dashed line – diffraction orders with numbers ±4) (a) and the intensity distribution of the electric field in the structure (b) at $\lambda = 659$ nm.

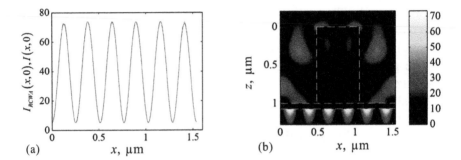

Fig. 1.64. The intensity of the electric field at the metal layer/substrate interface (solid line – the total field, the dashed line – diffraction orders with numbers ±3) (a) and the intensity distribution of the electric field in the structure (b) at $\lambda = 852$ nm.

Controlling the interference pattern by changing the angle of incidence at conical incidence

Note that the above method of controlling the frequency of the interference pattern is based on using light sources with different wavelengths, which might be an inconvenience in its practical implementation. Therefore, we consider the second way based on changing the angle of incidence at a fixed wavelength. It is the case of conical diffraction, when the projection of the wave vector of the incident wave is parallel to the grooves of the grating. The incidence geometry is shown in Fig. 1.65. In this case, the time and coordinate dependences of the field of transmitted diffraction orders have the form

$$\mathbf{E}_n, \mathbf{H}_n \sim \exp(-i\omega t)\exp\left[i\left(k_{x,n}x + k_y y\right)\right]\exp(-\kappa_n z'), \quad (1.387)$$

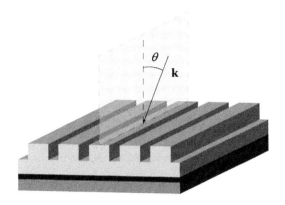

Fig. 1.65. The geometry of conical incidence.

where $k_{x,n}$ is determined as before by (1.341), and the value k_y is found from the expression [2]

$$k_y = k_0 n_1 \sin\theta, \qquad (1.388)$$

where θ is the angle of incidence.

In this case the approximate excitation condition of the plasmonic modes (1.382) takes the form

$$\sqrt{k_{x,n}^2 + k_y^2} = \mathrm{Re}\{k_{spp}\}. \qquad (1.389)$$

Consider an example and show that (1.389) can be fulfilled for various combinations of the order number n and the angle of incidence θ. Figure 1.66 shows the graphs of the dependence of the modulus of the projection of the wave vector $k_{\|,n}(\theta) = \sqrt{k_{x,n}^2 + k_y^2(\theta)}$ on

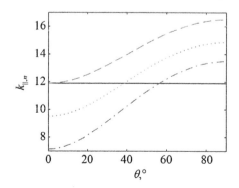

Fig. 1.66. The dependence $k_{\|,n}(\theta)$ of the angle of incidence ($n = 3$ – dot-dash line, $n = 4$ – dotted line, $n = 5$ – the dashed line, the value $\mathrm{Re}\{k_{spp}\}$ – solid line).

the incidence angle at $n = 3$ (dot-dash line), $n = 4$ (dotted line) and $n = 5$ (dashed line). The solid line corresponds to the propagation constant of the SPP. The graphs were plotted at $n_2 = 1$, $d = 2641$ nm, and the values λ, ε_m, n_1 and ε_{gr} coincide with the previous example. The conditions (1.389) are satisfied at the points 56.4° ($n = 3$), 38.7° ($n = 4$) and 0° ($n = 5$). This means that at these angles the plasmonic modes are excited in the structure by ±3-rd, ±4-th and ±5-th diffraction orders. The periods of interference patterns were equal to 440 nm, 330 nm and 264 nm, respectively.

Similar to the previous example, the geometric parameters of the structure $w = 0.554d$, $h_{gr} = 668$ nm, $h_l = 100$ nm, $h_m = 40$ nm were found in the optimization process from the condition of maximizing the quality of the interference patterns (1.385). The calculated normalized intensity plots at the metal layer/substrate interface are shown in Figs. 1.67–1.69. The values of the contrast and normalized maximum intensity are (0,89,58), (0,98,26), (0,97,32), respectively.

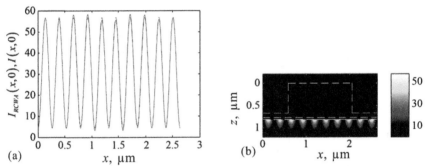

Fig. 1.67. The intensity of the electric field at the metal layer / substrate interface (solid line – the total field, the dotted line – diffraction orders with numbers ±5) (a) and the intensity distribution of the electric field in the structure (b) when $\theta = 0°$.

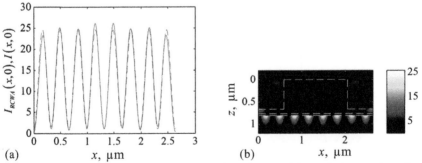

Fig. 1.68. The intensity of the electric field at the metal layer /substrate interface (solid line – the total field, the dashed line – diffraction orders with numbers ±4) (a) and the intensity distribution of the electric field in the structure (b) at $\theta = 38.7°$.

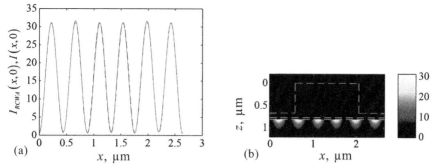

Fig. 1.69. The intensity of the electric field at the metal layer / substrate interface (solid line – the total field, the dashed line – diffraction orders numbers ±3) (a) and the intensity distribution of the electric field in the structure (b) at $\theta = 56.4°$.

Thus, both the proposed method allow to generate interference patterns with varying period with high contrast and intensity.

1.5.2.3. Generation of two-dimensional interference patterns and control of their shape by changing the parameters of the incident wave

Let us consider the structure containing a diffraction grating with a two-dimensional periodicity for generating two-dimensional interference patterns of evanescent diffraction orders (see Fig. 1.70) [113,114]. As in subsection 1.5.1.2, we assume that the grating periods along the Ox and Oy axes are d and the grating has a single square hole on the period with a side length of w. Consider the example of a grating, calculated for the wavelength and material parameters that coincide with the parameters of the first examples of subsection 1.5.2.1. The grating period 923 nm was calculated according to the expression (1.383) at $n = 3$ 923 nm. The rest of the geometrical parameters of the structure $h_{gr} = 260$ nm, $h_l = 0$, $h_m = 70$ nm, $w = 0.26d$ were found by the optimization procedure with the merit function similar to (1.384).

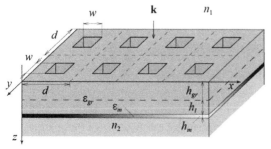

Fig. 1.70 Diffractive structure for the generation of two-dimensional interference patterns of plasmonic modes.

Figures 1.71 and 1.72 show the intensity distributions of the electric field at the interface between the metal layer and the substrate in the case of normal incidence of a wave with TM- and TE-polarization, respectively. The Figures show that in this case the interference patterns are one-dimensional and their period is $d_{ip} = d/6 = 154$ nm. The values of the contrast and normalized peak intensity are thus 0.73 and 10.7. Generation of the one-dimensional interference patterns in the incidence of the waves with TM- and TE-polarization confirms the assumption made in the subsection 1.5.1.2 that the quasi-guided modes of the structure, propagating along the Ox and Oy axes, are excited only by TE- and only by TM-components of the incident wave, respectively.

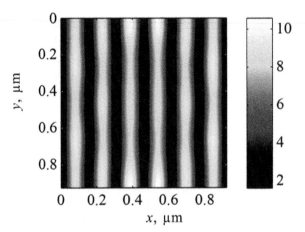

Fig. 1.71. The intensity of the electric field at the metal layer/substrate interface at the TM-polarization of the incident wave.

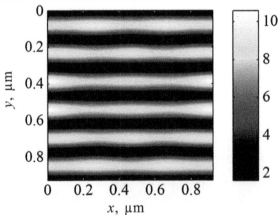

Fig. 1.72. The intensity of the electric field at the metal layer/substrate interface at the TE-polarization of the incident wave.

Now consider the case when the amplitudes of the TM- and TE-components of the incident wave are equal. Figures 1.73 and 1.74 show the calculated interference patterns formed directly under the metal film for the case of linear and circular polarization of the incident wave. The period of the interference pattern for the case of linear and circular polarizations of the incident wave is equal to 218 nm and 154 nm, respectively. The values of the contrast and the normalized electric field intensity at the maxima of the interference pattern are equal to (0,99,19,5) and (0,73,10,8), respectively. The form of the patterns coincides with the configuration of the theoretically obtained patterns (Figs. 1.54 and 1.56), the values of the contrast are close to the theoretical estimates.

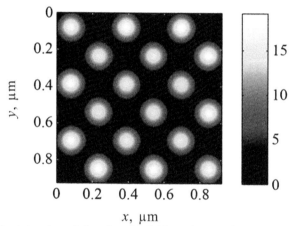

Fig. 1.73. The intensity of the electric field at the metal layer/substrate interface in the case of mixed linear polarization of the incident wave.

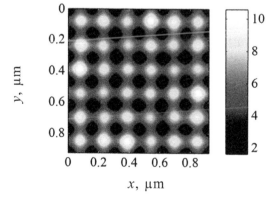

Fig. 1.74. The intensity of the electric field at the metal layer/substrate interface in the case of circular polarization of the incident wave.

Note that in the case of incidence on the structure of a circularly polarized wave the magneto-optical inverse Faraday effect is also enhanced [115].

Let's look at two more examples of intensity distributions of the electric field under the structure. Figure 1.75 shows the intensity distribution for the case where the incident wave is linearly polarized and the angle between the electric field and the Ox axis is 20°. The contrast of the generated interference pattern in this case is 0.86 and normalized maximum intensity is 16.2.

Figure 1.76 shows the calculated interference pattern for an elliptically polarized incident wave. The amplitudes of the TE- and TM-components in the present case are equal and the phase difference between them is 50°. The contrast of the interference pattern is 0.83, the intensity at the maxima of the interference pattern is 16.7 times higher than the intensity of the incident wave.

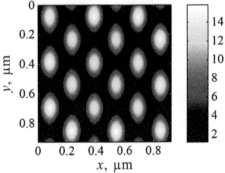

Fig. 1.75. The intensity of the electric field at the metal layer/substrate interface in the linear polarization of the incident wave (the angle between the vector of the electric field strength and the Ox axis is 20°).

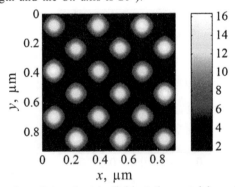

Fig. 1.76. The intensity of the electric field at the metal layer/substrate interface with the elliptical polarization of the incident wave (phase difference between TM- and TE-components is 50°).

Thus, the proposed structure allows one to create two-dimensional interference patterns of the evanescent electromagnetic waves with high contrast and intensity. The form of the pattern can be controlled by changing the polarization of the incident wave. The ratio of the electric field intensity at the maxima of the patterns to the intensity of the incident wave exceeds 10.

Note that in the two-dimensional case, as in the case of forming one-dimensional interference patterns, the period of the interference patterns can be controlled by changing the wavelength of incident light. In general, the plasmonic modes can be excited by orders with the numbers $(\pm m, \pm l)$, $(\pm l, \pm m)$, $m \neq n$, $l \neq n$. The approximate condition for excitation of the modes takes the form

$$\frac{2\pi}{d}\sqrt{m^2 + l^2} = \mathrm{Re}\{k_{spp}(\lambda)\}. \qquad (1.390)$$

According to (1.390), for the structure under consideration the plasmonic modes are excited by the diffraction orders with the numbers $(\pm 2, 0)$, $(0, \pm 2)$ at the wavelength of 774 nm and by the orders with the numbers $(\pm 3, \pm 1)$, $(\pm 1, \pm 3)$ at the wavelength of 532 nm. In the latter case, the interference pattern will be formed by eight SPPs.

Distributions of the electric field intensity immediately below the structure at the wavelength of incident light of 774 nm are presented in Figs. 1.77 and 1.78. At linear polarization of the incident wave the period of the pattern is $d_{ip} = d/2\sqrt{2} = 326\,\mathrm{nm}$ and the normalized intensity at the maxima of the pattern is 47 times higher than the intensity of the incident wave, the contrast is close to 1. In the case of circular polarization the

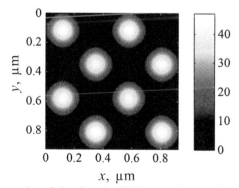

Fig. 1.77. The intensity of the electric field at the the metal layer/substrate interface in the case of mixed linear polarization of the incident wave at $\lambda = 774$ nm.

period of the pattern is 231 nm, the contrast of the pattern and the normalized intensity at the maxima are equal to 0.85 and 24 respectively. The configuration of the patterns, as in the case of the wavelength of 550 nm, coincides with the theoretically predicted configuration. Note that in contrast to the structure with one-dimensional periodicity, in this case the additional optimization of the geometrical parameters of the structure with the merit function similar (1.385) was not carried out.

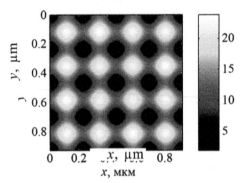

Fig. 1.78. The intensity of the electric field at the the metal layer/substrate interface in the case of circular polarization of the incident wave at $\lambda = 774$ nm.

Fig. 1.79 shows the distribution of the electric field at the metal layer/substrate interface for the wavelength of 532 nm and circular polarization. In this case, the interference pattern has a complicated form.

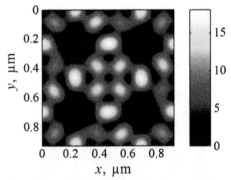

Fig. 1.79. The intensity of the electric field at the the metal layer/substrate interface in the case of circular polarization of the incident wave at $\lambda = 532$ nm.

Finally, consider the case where the excitation condition of the plasmonic modes in the structure (1.390) is not satisfied for

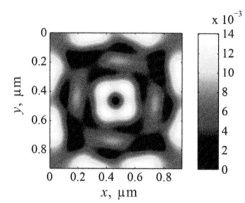

Fig. 1.80. The intensity of the electric field at the metal layer/ substrate interface in the case of circular polarization of the incident wave at $\lambda = 660$ nm.

any values of the numbers of diffraction orders m and l. Fig. 1.80 shows the intensity of the electric field at the metal layer / substrate interface in the case of an incident wave with circular polarization and the wavelength of 600 nm. In this case the SPP excitation does not occur and the maximum intensity under the metal layer is 70 times less than the intensity of the incident wave. Note that the order of values of the intensity does not depend on the polarization of the incident wave, so in the case of linear polarization the maximum normalized intensity is 0.03.

Thus, the formation of high-frequency interference patterns of evanescent higher diffraction orders in the considered structures, consisting of dielectric gratings with one-dimensional or two-dimensional periodicity and the metal layer is possible when excited plasmonic modes similar to the SPP propagating at the interface between the metal layer and the substrate. This is confirmed by the agreement between the calculated interference pattern and the theoretical estimates obtained by analyzing the interference of the evanescent transmitted diffraction orders corresponding to the excited modes. The minimum feature size of the interference patterns generated in the above examples is 50–55 nm, which is more than 8 times smaller than the wavelength of the incident light and 10 times smaller than the size of parts of the used diffraction grating.

1.5.3. Generation of interference patterns of evanescent electromagnetic waves in dielectric diffraction gratings

This subsection discusses the formation of interference patterns of evanescent electromagnetic waves in dielectric guided-mode resonant gratings [116,117]. The guided-mode resonant gratings have been intensively studied over the last decade as narrow-band spectral filters with high reflectivity close to 100% in the vicinity of certain wavelengths of incident light [118–120], but thus far relatively little attention has been paid to the study of the distribution of the electromagnetic field in such gratings [121,122]. The high reflection effect is associated with the resonant excitation of quasi-guided eigenmodes in the structure [120]. Below we show the high efficiency of these structures in for the generation of formation of interference patterns of evanescent electromagnetic waves (EEW).

The considered guided-mode resonant gratings are shown in Figs. 1.81–1.83. The motivation for using these structures to generate interference patterns of the EEW is the following. Under certain conditions, provided by the selection of the geometrical and physical parameters of the structure, quasi-guided modes, similar in the field structure to the slab waveguide modes, are excited in the structure [120–122]. The structure in Fig. 1.81, which is a binary dielectric diffraction grating (hereinafter – the structure A), is assumed to support the modes, which in the near field of the grating have the form, close to the superposition of two evanescent diffraction orders with the numbers $\pm n$. Indeed, for relatively small slits or small optical contrast (i.e., a small difference in dielectric constants of the materials of the grating) the structure A is close to the slab waveguide. At the subwavelength period the permittivity of the waveguide can be assessed using the theory of the effective medium. For the structure in Fig. 1.82 composed of a binary dielectric grating with a waveguide layer (hereinafter – structure B) we can assume the existence of modes, which are located in the waveguide layer and in the substrate are close to the superposition of two evanescent diffraction orders. The structure shown in Fig. 1.83 is similar to the structure B and is intended to generate two-dimensional interference patterns of the EEW.

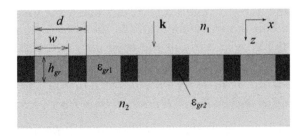

Fig. 1.81. The dielectric diffraction grating to generate one-dimensional interference patterns of the EEW (structure A).

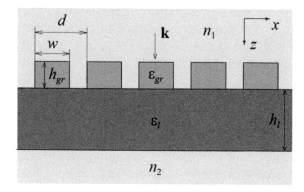

Fig. 1.82. Diffractive structure for generating one-dimensional interference patterns of EEW, which consists of a dielectric diffraction grating with a waveguide layer (structure B).

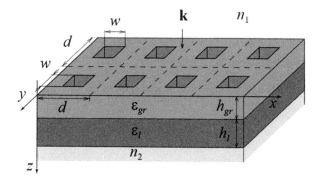

Fig. 1.83. Diffractive structure for the generation of two-dimensional interference patterns of the EEW.

Upon excitation of the modes the electric field intensity inside the structure can greatly increase. In particular, according to the calculation results of the electromagnetic field inside the waveguide grating shown in [121,122], the electric field amplitude at the maxima of the interference patterns of modes within the grating is 20–45 times greater than the amplitude of the wave incident on the structure. Thus, upon excitation of modes we should also expect a significant increase in the amplitude of evanescent waves (diffraction orders) in the near field of the structure.

The possibility of obtaining a large amplitude of the evanescent orders upon excitation of the eigenmodes in the diffraction gratings follows from the description of their optical properties on the basis of the scattering matrix method [60,123]. The frequencies of the eigenmodes of the periodic structure correspond to the poles of the scattering matrix. The complex amplitude of the n-th diffraction order (for structure with one-dimensional periodicity) in the vicinity of the resonance has the form [70, 124, 125]:

$$A_n(\omega) = a_n + \frac{b_n}{\omega - \omega_0}, \tag{1.391}$$

where ω_0 is the complex frequency of the eigenmode of the grating, a_n and b_n are slowly varying functions of the real frequency w of the wave. The maximum values of the squared absolute values of the amplitudes of the propagating orders are bounded by the law of conservation of energy. The sum of the intensities of the propagating orders in the gratings made of non-absorbent material is 1. Large values of the amplitudes of evanescent orders do not contradict the law of conservation of energy and, according to (1.391), at $Im(\omega_0) \ll 1$ (i.e., in the case of high-Q modes [124]) in the near field they can be much greater than the amplitude of the incident wave.

In general, the field in the substrate is a superposition of an infinite number of the transmitted diffraction orders (propagating and evanescent). However, due to the effect of field enhancement at the excitation of the modes [121,122], the near-field distribution with the structure close to the structure of the superposition of evanescent orders corresponding to the excited quasi-guided modes can be expected and, depending on the polarization of the incident wave (and the polarization of the excited mode), this distribution is described by (1.346), (1.354) in the case of generation of one-dimensional interference patterns, and expressions (1.362), (1.374) and (1.380) in the case of the generation of two-dimensional interference patterns. It should be noted that the diffraction structures in Figs. 1.81–1.83 may support modes of complex configurations that are substantially different from the distribution of the field of the slab waveguide modes. The interference pattern of the EEW for such modes will also have complex configuration. These interference patterns are of less interest for practical applications in interference lithography, so here we investigate the formation of interference patterns of the form (1.346), (1.354), (1.362), (1.374) and (1.380).

1.5.3.1. Generation of one-dimensional interference patterns
The possibility of generating one-dimensional interference patterns of the EEWs (1.346), (1.354) was studied for a normal incident

plane wave with wavelength λ = 543 nm (corresponding to the InGaN/GaN quantum heterostructure laser) for the structure A and λ = 441.6 nm (corresponds to a helium–cadmium laser) for the structure B. It should be noted that these values are in the range of wavelengths used in the near-field interference lithography [126-129]. The dielectric constants of materials were [130]: n_1^2 = 2.25 (SiO$_2$), ε_{gr1} = 4.41 (ZnO), ε_{gr2} = n_2^2 = 2.56 for structure A, and n_1 = 1, ε_{gr} = ε_l = 4.41, n_2 = 1.6 for structure B. Note that the value n_2 = 1.6 corresponds to the standard photoresist.

As in previous examples, the Fourier modal method was used for simulating the diffraction of light on the structures and the calculation of the intensity of the interference patterns and the distribution of the electric field in the structures. The geometrical parameters of the structures were determined using the optimization procedure with the merit function (1.384).

Initially, calculations were carried out to determine the diffraction structures in geometry B to generate the interference patterns of the EEWs corresponding to $\pm$1st diffraction orders. The grating period was equal to 240 nm, in this case, the period of the generated interference patterns was d_{ip} = d/2 = 120 nm.

Figure 1.84b shows the electric field distribution in the structure optimized for TE-polarization of the incident wave. The following values were found in optimization: h_{gr} = 123 nm, h_l = 545 nm, w = 98 nm. The interference pattern at the interface between the waveguide layer and the substrate (photoresist) is shown in Fig. 1.84 (a). The shape of the interference pattern corresponds to the expression (1.346) with high accuracy. As in the previous examples,

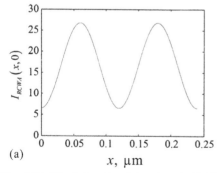
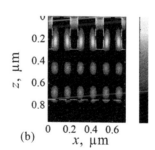

(a)

(b)

Fig. 1.84. The intensity of the electric field at the waveguide layer/substrate interface (one period) (a) and the intensity distribution of the electric field in the structure B (three periods), (b), calculated for the case of TE-polarized incident wave at n = 1.

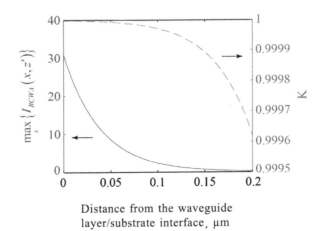

Distance from the waveguide
layer/substrate interface, μm

Fig. 1.85. The intensity of the maxima of the interference pattern (solid line) and contrast (the dashed line) vs. the distance from the interface between the waveguide layer and the substrate (structure B for TE-polarization of the incident wave at $n = 1$).

the intensity values in Fig. 1.84 are normalized by the intensity of the incident wave. The maximum intensity at the interface is more than 30 times the intensity of the incident wave, i.e. it is comparable to the intensity of the interference patterns of surface plasmon–polaritons [109,114] (see also Figs. 1.58 and 1.61). Fig. 1.85 shows the dependence of the contrast and intensity at the maxima of the interference pattern on the distance from the interface between the waveguide layer and the substrate. Fig. 1.85 shows that in accordance with the theoretical estimate in subsection 1.5.1.1 the contrast of the pattern is close to unity. Note that the field distribution corresponding to the structure, shown in Fig. 1.84b, is close to the field distribution in the interference of two modes in a planar waveguide [131,132].

Figures 1.86a,b show the interference pattern and the distribution of the electric field generated by the structure optimized for the case of TM-polarization of the incident wave. The structure parameters: h_{gr} = 290 nm, h_l = 456 nm, w = 166 nm. The interference pattern corresponds to the formula (1.354), and the intensity of the interference peaks is 27 times higher than the intensity of the incident wave. The value of the contrast of the interference pattern coincides with the theoretical estimate (1.356) and is 0.608 (Fig. 1.87).

Note that for both structures the intensity of the 0-th transmitted diffraction order (the only propagating) is less than 0.0001 and therefore the contrast of the interference patterns remains almost unchanged with increasing distance from the interface.

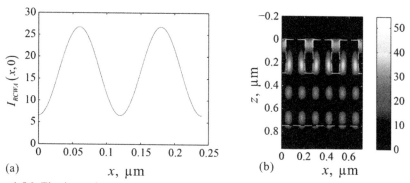

Fig. 1.86. The intensity of the electric field at the interface between the waveguide layer and the substrate (one period) (a) and the intensity distribution of the electric field in the structure B (three periods), (b), calculated for the case of TM-polarization of the incident wave at $n = 1$.

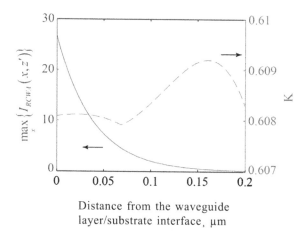

Distance from the waveguide
layer/substrate interface, μm

Fig. 1.87. Dependence of he intensity of the maxima of the interference pattern (solid line) and contrast (the dashed line) on the distance from the interface between the waveguide layer and the substrate (structure B for the TM-polarization of the incident wave at $n = 1$).

Of particular interest is the formation of interference patterns of the EEW when the eigenmodes are excited by the diffraction orders with the numbers $\pm n$, $n > 1$. In this case, the period of the interference pattern of the EEW (1.347) will be $2n$ times smaller than the grating period. Thus, the use of higher orders reduces the requirements on the technological implementation of the structure and enables to generate high-frequency interference patterns of the EEW using a low-frequency diffraction relief.

Calculation of the interference patterns of the evanescent waves generated by the structure of A was carried out at the period of the grating d = 750 nm and n = 3 and the TM-polarized incident wave. The period of the generated interference pattern is d_{ip} = $d/6$ = 125 nm which corresponds to the feature size of about 60 nm (over 7 times smaller than the wavelength of the incident light). The geometrical parameters of the structure h_{gr} = 141 nm, w = 675 nm were found as a result of the optimization criterion (1.384). Fig. 1.88b shows the distribution of the electric field in the calculated structure. The interference pattern at different distances from the interface between the grating and the substrate (photoresist) is shown in Fig. 1.88a. The intensity of the interference maxima at the interface is more than 100 times greater than the intensity of the incident wave. Such high values of the intensity are characteristic of the plasmon interference lithography [109,114]. The contrast at the interface is 0.68 and is close to the theoretical estimate of 0.64 obtained by the formula (1.356). With increasing distance from the interface the contrast decreases and at 100 nm is approximately 0.54. The decrease of the contrast is due to the presence of the propagating diffraction orders (total intensity of the propagating orders is 0.76). At a distance of 145 nm the contrast decreases to 0.2 (Fig. 1.89). The value of contrast 0.2 is the minimum required for registration of the interference pattern using standard photoresists [133].

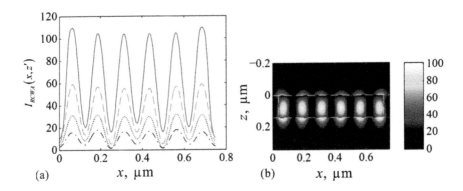

Fig. 1.88. The intensity of the electric field at distances of 0 nm (solid line) and 30 nm (dashed line), 60 nm (dotted line), 90 nm (dash-dot line) from the grating/substrate interface (a) and the intensity distribution of the electric field in the structure A (b), calculated for the case of TM-polarization of the incident wave at n = 3.

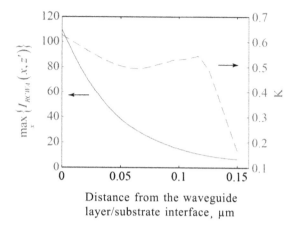

Distance from the waveguide
layer/substrate interface, μm

Fig. 1.89. The intensity of the maxima of the interference pattern (solid line) and contrast (the dashed line) of the distance from the interface between the grating and the substrate (structure A for TM-polarized incident wave at $n = 3$).

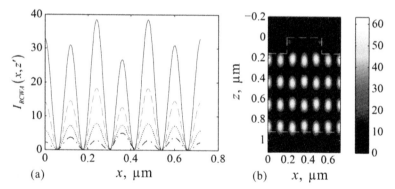

(a) x, μm   (b) x, μm

Fig. 1.90. Intensity of the electric field at distances of 0 nm (solid line) and 30 nm (dashed line), 60 nm (dotted line), 90 nm (dash-dot line) from the interface between the waveguide layer/substrate (a) and the electric field intensity distribution in the structure B (b), calculated for the case of TE-polarized incident wave at $n = 3$.

Figure 1.90b shows the electric field distribution in the structure B with period $d = 720$ nm, formed upon incidence of the TE-polarized wave. Similarly to the previous example, the geometrical parameters of the structure were found as a result of the optimization criterion (1.384) at: $n = 3$, $h_{gr} = 155$ nm, $h_1 = 763$ nm, $w = 349$ nm. The intensity of the interference pattern generated underneath the structure at different distances from the interface between the waveguide layer and the photoresist is shown in Fig. 1.90a. The period of the generated pattern is 120 nm (6 times smaller than the grating period) that corresponds to the feature size of 60 nm. The intensity of the interference peaks at the interface is more than 25 times the intensity of the incident wave, and the contrast was 0.99, which also agrees with the theoretical estimate. With increasing distance from the interface the contrast is reduced and at a distance of 127 nm is 0.2 (Fig. 1.91).

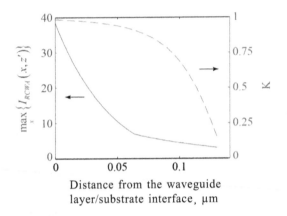

Fig. 1.91. Dependence of the intensity at the maxima of the interference pattern (solid line) and contrast (the dashed line) on the distance from the interface between the waveguide layer and the substrate (structure B for the TE-polarization of the incident wave at $n = 3$).

Calculation of the eigenmodes was carried out to further confirm the formation of interference patterns due to the excitation in the structure of quasi-guided eigenmodes. The modes were calculated using the method described above in subsection 1.2.2. As a result of the calculations it was found that in the above structure there exist quasi-guided modes with the propagation constants close to the propagation constants of the corresponding diffraction orders at normal incidence of the wave (± 1-st or ± 3-rd). Fig. 1.92 shows the distribution of the field in the structures shown in Figs. 1.86b and 1.90b, corresponding to the interference of two quasi-guided eigenmodes propagating in the positive and negative directions of the Ox axis. In Fig. 1.92 the electric field intensity is normalized by the maximum value in the structure. The agreement of the field distributions in Figs. 1.86b, 1.92a and 1.90b 1.92b confirms the assumption of the excitation of the modes.

Note that the diffraction structures were also calculated for the cases of excitation of the modes by the orders with numbers ± 5 at the above parameters. Thus, the total intensity of the propagating transmitted orders increases, which leads to a more rapid decrease in the contrast in comparison with Figs. 1.89, 1.91. The contrast

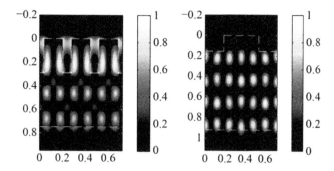

Fig. 1.92. The intensity of the electric field upon the interference of the two eigenmodes in structures designed to form interference patterns ±1-st of the evanescent diffraction orders in the case of TM-polarization of the incident wave (a) and ±3-rd orders in the TE-polarization of the incident wave.

value at $n = 5$ becomes less than 0.2 already at a distance of 40–50 nm from the interface.

1.5.3.2. Generation of two-dimensional interference patterns

The proposed approach can be generalized to the case of the generation of two-dimensional interference patterns of the EEWs. In this case, we use a three-dimensional diffraction grating (Fig. 1.83), the grating parameters are calculated from the condition for the generation below the structure of the interference pattern of diffraction orders with numbers $(\pm n,0)$, $(0,\pm n)$, based on the optimization criterion identical to the criterion (1.384). As an example, we calculated the diffractive structure with the period $d = 660$ nm and material parameters $\varepsilon_{super} = n_1^2 = 1.69$, $\varepsilon_{sub} = n_2^2 = 2.56$, $\varepsilon_{gr} = \varepsilon_l = 4.41$ at the wavelength $\lambda = 441.6$ nm for generating a two-dimensional interference pattern at $n = 3$. The geometrical parameters of the structure, calculated in the optimization process, were $h_{gr} = 287$ nm, $h_l = 305$ nm, $w = 200$ nm. Note that in this structure, as in the generation of the interference patterns of the plasmonic modes, the configuration of the interference patterns of the EEWs depends on the polarization of the incident wave, which means that the quasi-guided modes with the polarization close to TM-polarization are excited in the structure. Fig. 1.93 shows the interference pattern at the interface between the waveguide layer and the photoresist generated in the case of mixed linear polarization of the incident wave. In this case, in accordance with the theoretical prediction (expression (1.374), Fig. 1.54), the interference pattern is rotated

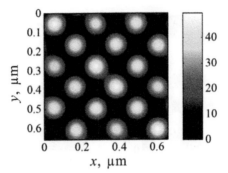

Fig. 1.93. The intensity of the electric field at the interface between the waveguide layer and the substrate at mixed linear polarization of the incident wave in the case of excitation of quasi-guided modes of the TM-type.

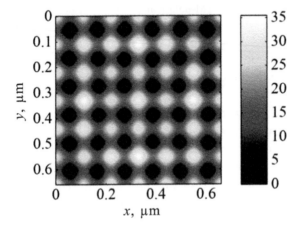

Fig. 1.94. The intensity of the electric field at the interface between the waveguide layer and the substrate at mixed linear polarization of the incident wave in the case of excitation of the quasi-guided modes of the TE-type.

relatively to the coordinate axes by an angle of 45°, and its period is equal to $d_{ip} = d/3\sqrt{2} = 156$ nm. The intensity of the interference peaks in Fig. 1.93 is more than 45 times the intensity of the incident wave, and the contrast of the generated pattern is close to 1.

Consider another example: the structure with geometrical parameters $h_{gr} = 152$ nm, $h_l = 962$ nm, $w = 273$ nm. Other parameters coincide with those of the previous example. The interference pattern formed at the interface between the waveguide layer and the photoresist in mixed linear polarization of the incident wave is shown in Fig. 1.94. The form of the pattern is close to the theoretical estimate, corresponding to the case of excitation of quasi-guided modes with polarization close to the TE-polarization (the expression

(1.362), Fig. 1.52). Calculations show that, in accordance with the theoretical description, the configuration of the pattern in this case does not vary with the polarization of the incident wave. Note that similarly to the metal-dielectric structure discussed above, in the case where the incident wave is circularly polarized, the inverse Faraday effect is also enhanced in the near field [134].

Thus, this section shows the possibility of the generation of interference patterns of the EEWs using resonant diffraction gratings. The configuration the patterns corresponds to the theoretical estimates obtained in subsection 1.5.1. For the considered wavelengths of incident light the period of the interference patterns of the EEWs is 120-160 nm, which corresponds to the feature size of about 60 nm (~λ/7). The contrast of the interference patterns, generated in the case of TE-polarization of the incident wave is close to unity, and the intensity of the electric field at the interference maxima in the near field is more than 35 times higher than the intensity of the incident wave, which is comparable with the intensity at the maxima of the interference patterns of the plasmonic modes (subsection 1.5.2). The possibility of the generation of one-dimensional and two-dimensional interference patterns corresponding to higher evanescent diffraction orders was demonstrated. The use of higher orders reduces the requirements for the fabrication technology of the gratings and allows to generate high-frequency interference patterns of the evanescent waves with the help of a low-frequency grating.

References

1. M. G. Moharam, T. K. Gaylord. Rigorous coupled-wave analysis of planar-grating diffraction. J. Opt. Soc. Am., 71(7), 811–818, 1981.

2. M. G. Moharam and Eric B. Grann and Drew A. Pommet and T. K. Gaylord. Formulation for stable and efficient implementation of the rigorous coupled-wave analysis of binary gratings. J. Opt. Soc. Am. A, 12(5),1068–1076, 1995.

3. M. G. Moharam, Drew A. Pommet and Eric B. Grann and T. K. Gaylord. Stable implementation of the rigorous coupled-wave analysis for surface-relief gratings: enhanced transmittance matrix approach. J. Opt. Soc. Am. A, 12(5), 1077–1086, 1995.

4. S. Peng, G.M. Morris. Efficient implementation of rigorous coupled-wave analysis for surface-relief gratings. J. Opt. Soc. Am. A, 12(5), 1087–1096, 1995.

5. Lifeng Li. Use of Fourier series in the analysis of discontinuous periodic structures. J. Opt. Soc. Am. A, 13(9), 1870–1876, 1996.

6. Lifeng Li. New formulation of the Fourier modal method for crossed surface-relief gratings. J. Opt. Soc. Am. A, 14(10), 2758–2767, 1997.

7. E. Popov, M. Nevière. Grating theory: new equations in Fourier space leading to

fast converging results for TM polarization. J. Opt. Soc. Am. A, 17(10), 1773–1784, 2000.

8. Lifeng Li. Fourier modal method for crossed anisotropic gratings with arbitrary permittivity and permeability tensors. J. Opt. A: Pure Appl., 5(4), 345–355, 2003.

9. E. Popov, M. Nevière. Maxwell equations in Fourier space: fast-converging formulation for diffraction by arbitrary shaped, periodic, anisotropic media. J. Opt. Soc. Am. A, 18(11), 2886–2894, 2001.

10. K. Watanabe, R. Petit, M. Nevière Differential theory of gratings made of anisotropic materials J. Opt. Soc. Am. A, 19(2), 325–334, 2002.

11. C. Zhou, L. Li Formulation of the Fourier modal method for symmetric crossed gratings in symmetric mountings. J. Opt. A: Pure Appl. Opt., 6, 43–50, 2004.

12. Diffractive computer optics. Ed. V.A. Soifer, Moscow, Fizmatlit, 2007.

13. Methods for Computer Design of Diffractive Optical Elements. Edited by Victor A. Soifer. A Wiley Interscience Publication. John Wiley & Sons, Inc., New York, 2002, pp.159–266.

14. Gans, M. J. A General Proof of Floquet's Theorem. IEEE Transactions on Microwave Theory and Techniques, 13(3):384–385, 1965.

15. Electromagnetic Theory of Gratings: Topics in current physics (22, Ed. by R.Petit, N.Y.: Springer-Verlag, 1980)

16. S. Višovský, K. Postava, T. Yamaguchi, R. Lopusník. Magneto-Optic Ellipsometry in Exchange-Coupled Films. Appl. Opt., 41(19), 3950–3960, 2002.

17. A. K. Zvezdin and V. A. Kotov, Modern Magneto-Optics and Magneto-Optical Materials, IOP, Bristol, 1997.

18. L. Li Formulation and comparison of two recursive matrix algorithms for modeling layered diffraction gratings J. Opt. Soc. Am. A, 13(5), 1024–1035,1996

19. T. Vallius, J. Tervo, P. Vahimaa, J. Turunen Electromagnetic field computation in semiconductor laser resonators J. Opt. Soc. Am. A, 23(4), 906–911

20. M. Born, E. Wolf. Principles of optics. Fourth Edition, Pergamon Press, 1968.

21. Silberstein E. et al. Use of grating theories in integrated optics, Journal of the Optical Society of America A. 2001. Vol. 18, No. 11. P. 2865–2875.

22. Berenger J.-P. A perfectly matched layer for the absorption of electromagnetic waves, Journal of Computational Physics. 1994. Vol. 114, No. 2. P. 185–200.

23. J. P. Hugonin and P. Lalanne, "Perfectly matched layers as nonlinear coordinate transforms: a generalized formalization," J. Opt. Soc. Am. A 22, 1844–1849 (2005).

24. S. Fan, P. R. Villeneuve, J. D. Joannopoulos, and H. A. Haus, "Channel drop tunneling through localized states," Phys. Rev. Lett., vol. 80, pp. 960–963, Feb. 1998.

25. W. Suh and S. Fan, "Mechanically switchable photonic crystal filter with either all-pass transmission or flat-top reflection characteristics," Opt. Lett., vol. 28, no. 19, pp. 1763–1765, Oct. 2003.

26. N. A. Gippius and S. G. Tikhodeev, "The scattering matrix and optical properties ofmetamaterials," Phys. Usp., vol. 52, no. 9, pp. 1027–1030, 2009.

27. N. A. Gippius, T. Weiss, S. G. Tikhodeev, and H. Giessen, "Resonant mode coupling of optical resonances in stacked nanostructures," Opt. Exp., vol. 18, no. 7, pp. 7569–7574, 2010.

28. S. G. Tikhodeev, A. L. Yablonskii, E. A. Muljarov, N. A. Gippius, and T. Ishihara, "Quasiguided modes and optical properties of photonic crystal slabs," Phys. Rev. B, vol. 66, no. 4, p. 045102, Jul. 2002.

29. D. A. Bykov and L. L. Doskolovich, "Magneto-optical resonances in periodic dielectric structures magnetized in plane," J. Mod. Opt., vol. 57, no. 17, pp. 1611–1618, 2010.

30. E. Centeno and D. Felbacq, "Optical bistability in finite-size nonlinear bidimensional photonic crystals doped by a microcavity," Phys. Rev. B, vol. 62, pp. R7683–R7686, Sep. 2000.

31. E. Centeno and D. Felbacq, "Rabi oscillations in bidimensional photonic crystals," Phys. Rev. B, vol. 62, pp. 10 101–10 108, Oct. 2000.

32. M. Nevière, E. Popov, and R. Reinisch, "Electromagnetic resonances in linear and nonlinear optics: Phenomenological study of grating behavior through the poles and zeros of the scattering operator," J. Opt. Soc. Amer. A, vol. 12, no. 3, pp. 513–523, 1995.

33. M. Liscidini, D. Gerace, L. C. Andreani, and J. E. Sipe, "Scattering matrix analysis of periodically patterned multilayers with asymmetric unit cells and birefringentmedia," Phys. Rev. B, vol. 77, Jan. 2008, Art. ID 035324.

34. D. Felbacq, "Numerical computation of resonance poles in scattering theory," Phys. Rev. E, vol. 64, Sep. 2001, Art. ID 047702.

35. T. Weiss, N. A. Gippius, S. G. Tikhodeev, G. Granet, and H. Giessen, "Derivation of plasmonic resonances in the Fourier modal method with adaptive spatial resolution and matched coordinates," J. Opt. Soc. Amer. A, vol. 28, no. 2, pp. 238–244, Feb. 2011.

36. L. Li, "Field singularities at lossless metal-dielectric arbitrary-angle edges and their ramifications to the numerical modeling of gratings," J. Opt. Soc. Amer. A, vol. 29, no. 4, pp. 593–604, Apr. 2012.

37. T. Weiss, "Advanced numerical and semi-analytical scattering matrix calculations for modern nano-optics," Ph.D. dissertation, Fakultät Mathematik und Physik, Physikalisches Institut der Universität, Stuttgart, Germany, 2011.

38. E. Anemogiannis, E. Glytsis, and T. Gaylord, "Efficient solution of eigenvalue equations of optical waveguiding structures," J. Lightw. Technol., vol. 12, no. 12, pp. 2080–2084, Dec. 1994.

39. C. Chen, P. Berini, D. Feng, S. Tanev, and V. Tzolov, "Efficient and accurate numerical analysis of multilayer planar optical waveguides in lossy anisotropic media," Opt. Exp., vol. 7, no. 8, pp. 260–272, Oct. 2000.

40. F. Zolla, G. Renversez, A. Nicolet, B. Kuhlmey, S. Guenneau, and D. Felbacq, Foundations of Photonic Crystal Fibres. London, U.K.: Imperial College, 2005.

41. D. Felbacq, "Finding resonance poles bymeans of cauchy integrals," in Proc. 13th Int. Conf. Transparent Optical Networks, Jun. 2011, paper We.B4.2.

42. V. Lomakin and E. Michielssen, "Transmission of transient plane waves through perfect electrically conducting plates perforated by periodic arrays of subwavelength holes," IEEE Trans. Antennas Propag., vol. 54, no. 3, pp. 970–984, Mar. 2006.

43. A. Akimov, N. Gippius, and S. Tikhodeev, "Optical Fano resonances in photonic crystal slabs near diffraction threshold anomalies," JETP Lett., vol. 93, pp. 427–430, 2011.

44. F. R. Gantmacher, The Theory of Matrices, Nauka, Moscow, 1988, p. 548.

45. A. S. Householder, The Numerical Treatment of a Single Nonlinear Equation, ser. Int. Series in Pure Appl. Math.. New York: McGraw-Hill, 1970.

46. Bykov D. A., Doskolovich L. L. Numerical methods for calculating poles of the scattering matrix with applications in grating theory, J. Lightw. Technol. 2013.—March. Vol. 31, no. 5. Pp. 793–801.

47. Y. Hua and T. K. Sarkar, "Matrix pencil and system poles," Signal Process., vol. 21, no. 2, pp. 195–198, 1990.

48. Y. Hua and T. K. Sarkar, "Matrix pencil method for estimating parameters of ex-

ponentially damped/undamped sinusoids in noise," IEEE Trans. Acoust., Speech, Signal Process., vol. 38, no. 5, pp. 814–824, May 1990.

49.   Y. Hua and T. K. Sarkar, "On SVD for estimating generalized eigenvalues of singular matrix pencil in noise," IEEE Trans. Signal Process., vol. 39, no. 4, pp. 892–900, Apr. 1991.

50.   A.D.Rakic,A.B.Djurišic, J. M. Elazar, andM. L. Majewski, "Optical properties of metallic films for vertical-cavity optoelectronic devices," Appl. Opt., vol. 37, no. 22, pp. 5271–5283, Aug. 1998.

51.   L. Yau and A. Ben-Israel, "The Newton and Halley methods for complex roots," Am. Math. Mon., vol. 105, no. 9, pp. 806–818, 1998.

52.   Q. Cao, P. Lalanne, and J.-P. Hugonin, "Stable and efficient Bloch-mode computational method for one-dimensional grating waveguides," J. Opt. Soc. Am. A 19, 335–338 (2002).

53.   E. A. Bezus and L. L. Doskolovich, "Stable algorithm for the computation of the electromagnetic field distribution of eigenmodes of periodic diffraction structures," J. Opt. Soc. Am. A 29, 2307–2313 (2012).

54.   T. W. Ebbesen, H. J. Lezec, H. F. Ghaemi, T. Thio, P. A. Wolff, "Extraordinary optical transmission through subwavelength hole arrays," Nature, 391, 667-669 (1998)

55.   H. F. Ghaemi, T. Thio, and D. E. Grupp, "Surface plasmons enhance optical transmission through subwavelength holes," Phys. Rev. B 58, 6779-6782 (1998)

56.   J. Le Perchec, P. Qu´emerais, A. Barbara, and T. Lopez-Rios, "Why metallic surfaceswith grooves a few nanometers deep and wide may strongly absorb visible light," Phys.Rev. Lett. 100, 066408 (2008).

57.   Diffractive nanophotonics. Edited by V.A. Soifer, Moscow, Fizmatlit, 2011.

58.   V. I. Belotelov, D. A. Bykov, L. L. Doskolovich, A. N. Kalish, and A. K. Zvezdin, "Giant transversal Kerr effect in magneto-plasmonic heterostructures: The scattering-matrix method," J. Exp. Theor. Phys. 110, 816–824 (2010).

59.   V.I. Belotelov, D.A. Bykov, L.L. Doskolovich, A.N. Kalish, V.A. Kotov, A.K. Zvezdin, Giant magneto-optical orientational effect in plasmonic heterostructures, Optics Letters, Vol. 34, Issue 4, pp. 398-400 (2009)

60.   M. Sarrazin, J.-P. Vigneron, and J.-M. Vigoureux, "Role of Wood anomalies in optical properties of thin metallic films with a bidimensional array of subwavelength holes," Phys. Rev. B 67, 085415 (2003).

61.   T. Vallius, P. Vahimaa, J. Turunen, "Pulse deformations at guided-mode resonance filters," Opt. Express 10, 840-843 (2002)

62.   P.P. Vabishchevich, V.O. Bessonov, F.Yu. Sychev, M.R. Shcherbakov, T. V. Dolgova, A.A. Fedyanin "Femtosecond Relaxation Dynamics of Surface Plasmon–Polaritons in the Vicinity of Fano-Type Resonance", JETP Letters 92, 575–579 (2010).

63.   J. Azaña, Proposal of a uniform fiber Bragg grating as an ultrafast all-optical integrator, Optics Letters, Vol. 33, No. 1, pp. 4-6 (2008).

64.   M.A. Preciado, M.A. Muriel. Ultrafast all-optical integrator based on a fiber Bragg grating: proposal and design, Optics Letters/ Vol. 33, No. 12, pp. 1348-1350 (2008).

65.   M. Li, D. Janner, J. Yao, V. Pruneri. Arbitrary-order all-fiber temporal differentiator based on a fiber Bragg grating: design and experimental demonstration, Opt. Express 17(22), 19798-807 (2009).

66.   N.K. Berger, B. Levit, B. Fischer, M. Kulishov, D.V. Plant, J. Azaña. Temporal differentiation of optical signals using a phase-shifted fiber Bragg grating, Opt. Express 15(2), 371-381 (2007).

67.   Y. Park, M. Kulishov, R. Slavík, J. Azaña. Picosecond and sub-picosecond flat-top pulse generation using uniform long-period fiber gratings, Opt. Express 14(26),

12670-78 (2006).

68. M. Kulishov, J. Azaña, Design of high-order all-optical temporal differentiators based on multiple-phase-shifted fiber Bragg gratings, Opt. Express 15(10), 6152-6166 (2007).

69. A. Papoulis, The Fourier integral and its applications, New York: McGraw-Hill, 1962.

70. N.A. Gippius, S.G. Tikhodeev, and T. Ishihara, "Optical properties of photonic crystal slabs with an asymmetrical unit cell," Phys. Rev. B 72, 045138 (2005).

71. D. A. Bykov, L. L. Doskolovich, and V. A. Soifer, "Temporal differentiation of optical signals using resonant gratings," Opt. Lett. 36, 3509-3511 (2011).

72. D.A. Bykov, et al., Capacity of resonant diffractive gratings to differentiate the pulsed optical signa, Zh. Eksp. Teor. Fiz., 2012. V. 141, No. 5. 832–839.

73. M. Abramowitz and I. A. Stegun, Handbook of Mathematical Functions (Dover Publications, New York, 1964).

74. D.A. Bykov, et al., Single-resonance diffraction gratings for time-domain pulse transformations: integration of optical signals, J. Opt. Soc. Am. A. 2012.—August. Vol. 29, no. 8. Pp. 1734–1740.

75. D.A. Bykov, et al., Pis'ma Zh. Eksp. Teor. Fiz., 2012. V. 95, No. 1. . 8–12.

76. Soifer V., Kotlyar V., Doskolovich L. Iterative Methods for Diffractive Optical Elements Computation. Taylor & Francis, 1997. 244 p.

77. Goodman J.W. Introduction to Fourier Optics. New York: McGraw-Hill, 2005. 491 p.

78. Зверев В.А. Радиооптика. Москва: Советское радио, 1975. 304 с.

79. Raether H. Surface plasmons on smooth and rough surfaces and on gratings. Springer-Verlag, 1988. 136 p.

80. Jackson J.D. Classical Electrodynamics. Third Edition. New York: Wiley, 1998. 808 p.

81. Johnson P.B., Christy R.W. Optical Constants of the Noble Metals, Physical Review B. 1972. Vol. 6, No. 12. P. 4370–4379.

82. Barnes W.L. Surface plasmon–polariton length scales: a route to sub-wavelength optics, Journal of Optics A: Pure and Applied Optics. 2006. Vol. 8, No. 4. P. S87–S93.

83. Berini P. Long-range surface plasmon polaritons, Advances in Optics and Photonics. 2009. Vol. 1. P. 484–588.

84. Burke J., Stegeman G., Tamir T. Surface-polariton-like waves guided by thin, lossy metal films, Physical Review B. 1986. Vol. 33, No. 8. P. 5186–5201.

85. Zia R. et al. Geometries and materials for subwavelength surface plasmon modes, Journal of the Optical Society of America A. 2004. Vol. 21, No. 12. P. 2442–2446.

86. Avrutsky I. et al. Highly confined optical modes in nanoscale metal-dielectric multilayers, Physical Review B. 2007. Vol. 75. P. 241402(R) (4pp).

87. Pollock C.R. Fundamentals of Optoelectronics. Chicago: Irwin, 1995. 592 p.

88. Kekatpure R.D. et al. Solving dielectric and plasmonic waveguide dispersion relations on a pocket calculator, Optics Express. 2009. Vol. 17, No. 26. P. 24112–24129.

89. Berini P. Plasmon–polariton modes guided by a metal film of finite width, Optics Letters. 1999. Vol. 24, No. 15. P. 1011–1013.

90. Holmgaard T., Bozhevolnyi S. Theoretical analysis of dielectric-loaded surface plasmon-polariton waveguides, Physical Review B. 2007. Vol. 75, No. 24. P. 245405 (12 pp).

91. Kurihara K., Suzuki K. Theoretical Understanding of an Absorption-Based Surface Plasmon Resonance Sensor Based on Kretchmann's Theory, Analytical Chemistry. American Chemical Society, 2002. Vol. 74, No. 3. P. 696–701.

92. Bezus E.A., Doskolovich L.L., Kazanskiy N.L. Scattering suppression in plasmonic optics using a simple two-layer dielectric structure, Applied Physics Letters. 2011. Vol. 98, No. 22. P. 221108 (3pp).

93. Bezus E.A., et al., Pis'ma Zh. Tekh. Fiz., 2011. V. 37, No. 23. 10–18.

94. Bezus E.A., et al., Izv. RAN. Ser. Fiz., 2011. V. 75, No. 12. C. 1674–1677.

95. Liu Y. et al. Transformational Plasmon Optics, Nano Letters. 2010. Vol. 10, No. 6. P. 1991–1997.

96. Sannikov D.G., et al., Pis'ma Zh. Teor. Fiz., 2003. V. 29, No. 9. 1–8.

97. Avrutsky I., Soref R., Buchwald W. Sub-wavelength plasmonic modes in a conductor-gap-dielectric system with a nanoscale gap, Optics Express. 2010. Vol. 18, No. 1. P. 348–363.

98. Sámson Z.L. et al., Femtosecond surface plasmon pulse propagation, Optics Letters. 2011. Vol. 36, No. 2. P. 250–252.

99. Zia R., Brongersma M.L. Surface plasmon polariton analogue to Young's double-slit experiment, Nature Nanotechnology. 2007. Vol. 2. P. 426–429.

100. Kim H., Hahn J., Lee B. Focusing properties of surface plasmon polariton floating dielectric lenses, Optics Express. 2008. Vol. 16, No. 5. P. 3049–3057.

101. Feng L. et al. Fourier plasmonics: Diffractive focusing of in-plane surface plasmon polariton waves, Applied Physics Letters. 2007. Vol. 91, No. 8. P. 081101 (3 pp).

102. Bezus E.A. et al. Design of diffractive lenses for focusing surface plasmons, Journal of Optics. 2010. Vol. 12, No. 1. P. 015001 (7pp).

103. Bezus E.A. et al., Komp. optika, 2009. V. 33, No. 2. 185–192.

104. Gramotnev D.K., Bozhevolnyi S.I. Plasmonics beyond the diffraction limit, Nature Photonics. Nature Publishing Group, 2010. Vol. 4, No. 2. P. 83–91.

105. Bai B., Li L. Reduction of computation time for crossed-grating problems: a group-theoretic approach, Journal of the Optical Society of America A. 2004. Vol. 21, No. 10. P. 1886–1894.

106. Bai B., Li L. Group-theoretic approach to enhancing the Fourier modal method for crossed gratings with square symmetry, Journal of the Optical Society of America A. 2006. Vol. 23, No. 3. P. 572–580.

107. Fikhtengol'ts G.M., Lectures in differential and integral calculus. Vol 1, Moscow, Fizmatlit, 2001.

108. Bezus E.A. et al., Komp. Optika, 2008. V. 32, No. 3. 234–237.

109. Bezus E.A. et al. Diffraction gratings for generating varying-period interference patterns of surface plasmons, Journal of Optics A: Pure and Applied Optics. 2008. Vol. 10, No. 9. P. 095204 (5 pp).

110. Palik E.D. Handbook of Optical Constants. Academic press, 1985. Vol. 1. 294 p.

111. Doskolovich L.L., Kadomina E.A., Kadomin I.I. Nanoscale photolithography by means of surface plasmon interference, Journal of Optics A: Pure and Applied Optics. 2007. Vol. 9. P. 854–857.

112. Weber M.J. Handbook of Laser Wavelengths. CRC Press, 1998. 784 p.

113. Bezus E.A., Doskolovich L.L., Komputernaya optika. 2009. Vol. 33, No. 1. P. 10–16.

114. Bezus E.A., Doskolovich L.L. Grating-assisted generation of 2D surface plasmon interference patterns for nanoscale photolithography, Optics Communications. 2010. Vol. 283, No. 10. P. 2020–2025.

115. Belotelov V.I. et al. Inverse Faraday effect in plasmonic heterostructures, Journal of Physics: Conference Series. 2010. Vol. 200, No. 9. P. 092003 (4 pp).

116. Bezus E.A., Doskolovich L.L., Kazanskiy N.L. Evanescent-wave interferometric nanoscale photolithography using guided-mode resonant gratings, Microelectronic

Engineering. 2011. Vol. 88, No. 2. P. 170–174.

117. Bezus E.A., Doskolovich L.L., Kazanskiy N.L. Interference pattern generation in evanescent electromagnetic waves for nanoscale lithography using waveguide diffraction gratings, Quantum Electronics. 2011. Vol. 41, No. 8. P. 759–764.

118. Brundrett D.L. et al. Effects of modulation strength in guided-mode resonant subwavelength gratings at normal incidence, Journal of the Optical Society of America A. OSA, 2000. Vol. 17, No. 7. P. 1221–1230.

119. Magnusson R., Shin D., Liu Z.S. Guided-mode resonance Brewster filter, Optics Letters. 1998. Vol. 23, No. 8. P. 612–614.

120. Tamir T., Zhang S. Resonant scattering by multilayered dielectric gratings, Journal of the Optical Society of America A. 1997. Vol. 14, No. 7. P. 1607–1616.

121. Wei C. et al. Electric field enhancement in guided-mode resonance filters, Optics Letters. 2006. Vol. 31, No. 9. P. 1223–1225.

122. Sun T. et al. Electric field distribution in resonant reflection filters under normal incidence, Journal of Optics A: Pure and Applied Optics. 2008. Vol. 10. P. 125003 (5pp).

123. Gippius N.A., Tikhodeev S.G., Usp. Fiz. Nauk, 2009. Vol. 179. P. 1027–1030.

124. Belotelov V.I., et al. Zh. Eksper. Teor. Fiz. 2010. Vol. 137, No. 5. P. 932–942.

125. Bykov D.A., et al., Zh. Eksper. Teoret. Fiz., 2010. V. 138, No. 6. 1093–1102.

126. Blaikie R.J., McNab S.J. Evanescent interferometric lithography, Applied Optics. 2001. Vol. 40, No. 10. P. 1692–1698.

127. Luo X., Ishihara T. Surface plasmon resonant interference nanolithography technique, Applied Physics Letters. 2004. Vol. 84, No. 23. P. 4780–4782.

128. Guo X. et al. Large-area surface-plasmon polariton interference lithography, Optics Letters. 2006. Vol. 31, No. 17. P. 2613–2615.

129. Jiao X. et al. Numerical simulation of nanolithography with the subwavelength metallic grating waveguide structure, Optics Express. 2006. Vol. 14, No. 11. P. 4850–4860.

130. Handbook of Optics / ed. Bass M. New York: McGraw-Hill, 1995. Vol. 2. 1568 p.

131. Lifante G. Integrated photonics: fundamentals. Wiley, 2003. 184 p.

132. Belotelov V.I. et al. Magnetooptical effects in the metal-dielectric gratings, Optics Communications. 2007. Vol. 278, No. 1. P. 104–109.

133. Madou M.J. Fundamentals of microfabrication. CRC Press, 1997. 589 p.

134. Bezus E.A. et al. Komp. optika, 2011. V. 35, No. 4. 432–437.

Components of photonic crystals

2.1 One-dimensional and two-dimensional photonic crystals

Photonic crystals (PC) [1,2] are structures with the nanoresolution and periodic modulation of the refractive index having a photonic bandgap. Bandgaps determine the range of frequencies of electromagnetic radiation that can not exist in the structure. The size of the bandgap of the optical PCs with a wavelength of 1.3 µm is tens of nanometers. Accordingly, in the incidence of electromagnetic radiation on PC whose frequency lies in the bandgap complete reflection takes place. This property determines the prospects of using photonic crystal structures as waveguides, antireflection coatings, frequency filters, metamaterials, photonic–crystal lens working at a given light frequency.

2.1.1 Photonic bandgaps

Based on the general theory of light propagation in the PC [3], we consider the solution of Maxwell's equations for a dielectric medium without free charges and currents, simulating the photonic crystal. The system of Maxwell's equations in this case has the form (in the SI system):

$$
\begin{cases}
\nabla \mathbf{D} = 0, \\
\nabla \mathbf{B} = 0, \\
\nabla \times \mathbf{E} = -\dfrac{\partial \mathbf{B}}{\partial t}, \\
\nabla \times \mathbf{H} = \dfrac{\partial \mathbf{D}}{\partial t},
\end{cases}
\tag{2.1}
$$

where $\mathbf{D} = \varepsilon\varepsilon_0\mathbf{E}$, $\mathbf{B} = \mu\mu_0\mathbf{H}$, $\sqrt{(\varepsilon_0\mu_0)}^{-1} = c$

From (2.1) we obtain the following relationship:

$$\varepsilon^{-1}(r)(\nabla \times \mathbf{H}) = \varepsilon_0 \frac{\partial \mathbf{E}}{\partial t}. \tag{2.2}$$

Applying the 'rot' operation to the expression (2.2) with (2.1) we get:

$$\nabla \times (\varepsilon^{-1}(r)(\nabla \times \mathbf{H})) = -\frac{\mu}{c^2} \frac{\partial^2}{\partial t^2} \mathbf{H}. \tag{2.3}$$

Hence, for monochromatic waves we have:

$$\frac{\partial^2}{\partial t^2} \mathbf{H}(r,t) = -w^2 \mathbf{H}(r,t). \tag{2.4}$$

where w is the frequency of the monochromatic wave. Taking into account (2.4), instead of (2.3) for $\mu = 1$ can be written:

$$\nabla \times (\varepsilon^{-1}(r)(\nabla \times \mathbf{H})) = (\frac{w}{c})^2 \mathbf{H}. \tag{2.5}$$

Since the value of $\varepsilon(r)$ in this case is real, then the equation (2.5) is the problem of finding the eigenvalues of ω^2/c^2 of the Hermitian operator $L = \nabla \times (\varepsilon^{-1}(r)\nabla \times)$ in the equation

$$LH = \left(\frac{w}{c}\right)^2 \mathbf{H}. \tag{2.6}$$

Consider a one-dimensional PC with the structure period $d = a + b$, where a and b is the size sections having dielectric constants ε_1 and ε_2, respectively. It is known [1,3] that the eigenfunctions of the equation (2.6) in a periodic medium have the Bloch form:

$$\phi = e^{ikx}u(x), \tag{2.7}$$

where x is the coordinate, $k = 2\pi/\lambda$ is the wave number. The eigenfunctions of the operator L are determined on the basis of their form (2.7) and the boundary conditions defined by the function

$$\varepsilon(x) = \begin{cases} \varepsilon_1, nd \leq x < a + nd, \\ \varepsilon_2, a + nd \leq x < (n+1)d, \end{cases} \tag{2.8}$$

where n is an integer. The eigenfunctions in regions with dielectric constants ε_1 and ε_2 will be, respectively,

$$\phi_1(x) = Ae^{ik_1 x} + Be^{-ik_1 x},$$

$$\phi_2(x) = Ce^{ik_2 x} + De^{-ik_2 x}, \quad (2.9)$$

wherein A, B, C, D are the unknown coefficients.

Since the boundaries of the bands with different permittivities must have continuous both the eigenfunctions and their derivatives, it is possible to create a system of equations:

$$\begin{cases} A + B = e^{-iQd}(Ce^{ik_2 d} + De^{-ik_2 d}), \\ k_2(A - B) = k_2 e^{-iQd}(Ce^{ik_2 d} - De^{-ik_2 d}), \\ Ae^{ik_1 a} + Be^{-ik_1 a} = Ae^{ik_2 a} + Be^{-ik_2 a}, \\ k_1(Ae^{ik_1 a} - Be^{-ik_1 a}) = k_2(Ae^{ik_2 a} - Be^{-ik_2 a}). \end{cases} \quad (2.10)$$

In matrix form the system of equations for A, B, C and D can be written as:

$$M(k_1, k_2, Q)V = 0, \quad (2.11)$$

where

$$M(k_1, k_2, Q) = \begin{pmatrix} 1 & 1 & -e^{id(k_2 - Q)} & -e^{-id(k_2 + Q)} \\ k_1 & -k_1 & -k_2 e^{id(k_2 - Q)} & k_2 e^{-id(k_2 + Q)} \\ e^{ik_1 a} & e^{-ik_1 a} & -e^{ik_2 a} & -e^{-ik_2 a} \\ k_1 e^{ik_1 a} & -k_1 e^{-ik_1 a} & -k_2 e^{ik_2 a} & k_2 e^{-ik_1 a} \end{pmatrix} \quad (2.12)$$

$$V = \begin{pmatrix} A \\ B \\ C \\ D \end{pmatrix}.$$

This system of equations has a non-trivial solution if det $M = 0$. Expanding the determinant we obtain the dispersion law in implicit form $\omega(Q)$:

$$\cos(k_1 a)\cos(k_2 b) - \frac{1}{2}\frac{\varepsilon_1 + \varepsilon_2}{\sqrt{\varepsilon_1 \varepsilon_2}}\sin(k_1 a)\sin(k_2 b) = \cos(Qd), \quad (2.13)$$

where $k_i = \sqrt{\varepsilon_i}\, w/c$, $i = 1,2$, Q is Bloch's wave number. Since $|\cos(\Omega d) \leq 1|$, bandgaps form in the spectrum, i.e. the values k_i for which

$$\left| \cos(k_1 a)\cos(k_2 b) - \frac{1}{2}\frac{\varepsilon_1 + \varepsilon_2}{\sqrt{\varepsilon_1 \varepsilon_2}} \sin(k_1 a)\sin(k_2 b) \right| > 1. \qquad (2.14)$$

In these bands, the propagation of radiation in the crystal becomes impossible. Or vice versa, if a one-dimensional photonic crystal (a, b, ε_1, ε_2) is considered, from the inequality (2.14) we find $w = 2\pi v$ – the cyclic frequency of light that can pass through a photonic crystal.

Considering the simplest one-dimensional theory, we can now consider some examples of the model. Let us consider the diffraction of a plane wave on photonic crystals in the two-dimensional case. In all the examples in this section we considered the wave of unit intensity ($E_0 = 1$ V/m). The simulation was performed using the program FullWAVE 6.0 and the FDTD method included in this program.

2.1.2. Diffraction of a plane wave on a photonic crystal with no defects

Consider the fall of a plane electromagnetic wave with TE polarization on a photonic crystal (PC). PC parameters are taken from [4] to obtain comparable results: $n_1 = 3.25$ the refractive index of the medium, $n_2 = 1$ the refractive index of the holes, $r = 0.25$ μm the radius of the holes, $T_z = 0.6$ μm and $T_x = 1$ μm is the distance between the centres of the holes on the optical axis z and x, respectively. The fill factor on the axis z: $\Lambda_z = 0.83$, axis x: $\Lambda_x = 0.5$, was also taken from the article [4]. Figure 2.1 shows a photon bandgap for this crystal for TE polarization. The length of the electromagnetic wave $\lambda = 1.55$ μm. Figure 2.1 shows that the wavelength lies in the photon bandgap, the reflection coefficient $R \approx 0.89$. Figure 2.2 shows that the light having a given wavelength does not penetrate into the crystal further than the first 3 layers of the holes. Now consider the same photonic crystal, but the wavelength is changed to $\lambda = 2.2$ μm. From Fig. 2.3 it is clear that at this wavelength the PC passes electromagnetic radiation, the reflectance is low, $R \approx 0.2$.

That is, the bandgap at Fig. 2.1 is not an accurate function-rectangle (rect) and the reflectance of light does not reach the maximum value 1 due to the fact that in modelling we consider in the finite dimensions of the PC in the coordinates x and z.

Reflectance, R

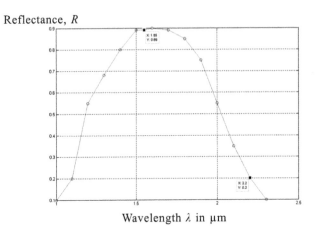

Wavelength λ in μm

Fig. 2.1. Photonic bandgap.

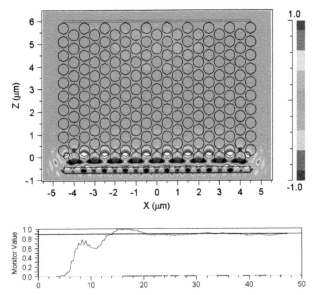

Fig. 2.2. Propagation of light in a photonic crystal in the bandgap. Light travels from the bottom up.

2.1.3. The propagation of light in a photonic crystal waveguide

Consider the same photonic crystal as in the preceding paragraph and remove from it three central rows of holes. The wavelength was selected based on Fig. 2.1 so that the transmittance was minimal.

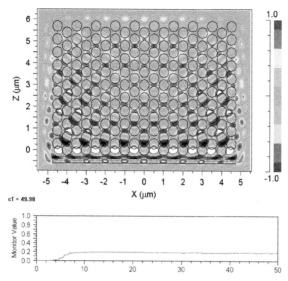

Fig. 2.3. The propagation of light in a photonic crystal outside the bandgap. Light travels from the bottom up.

This condition is satisfied when $\lambda = 1.55$ µm. Figure 2.4 shows the distribution of light in a photonic crystal waveguide. From Fig. 2.4 it can be seen that light travels only in the way that we have created in a photonic crystal by removing the three adjacent rows of holes.

Now reduce the size of the radiation source to the dimensions of the waveguide to evaluate power losses when light passes through the waveguide (Fig. 2.5).

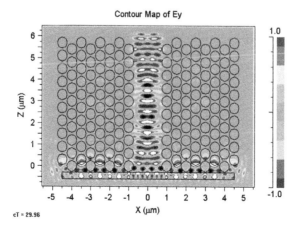

Figure. 2.4. Light propagation in a photonic crystal waveguide.

The energy losses at a distance of 5 μm (the length of the waveguide) were approximately 0.51%. That is, if the light is focused on the input of the photonic crystal waveguide about 1 μm wide, the light almost through the entire crystal to the exit.

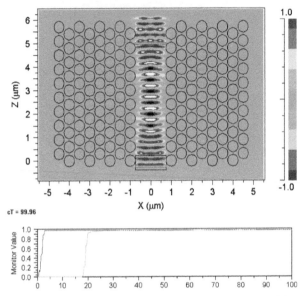

Fig. 2.5. Propagation of light in a photonic crystal waveguide in illumination with a narrow source. Light travels from the bottom up.

2.1.4 Photonic crystal collimators

Recently optimization techniques have been developed [5] for the structure of photonic crystal fibres in order to reduce the divergence of radiation at the exit of the optical waveguide. For conventional light fibres this is achieved by structuring the output end of the fibre.

Figure 2.6a shows a schematic of a two-dimensional photonic crystal waveguide, the shell of which consists of periodically arranged (period 228 nm) dielectric nanorods (ε = 3.38, silica) with a diameter of 114 nm. To create a waveguide one row of the nanorods is removed. The size of this 'defect' in the periodic structure of the nanorods has a value of 342 nm or one and a half period. The wavelength of light is 633 nm. Figure 2.6b shows the instantaneous pattern of diffraction of light on the structure, calculated by the FDTD using the FullWAVE program. It is seen that almost no light enters the shell and propagates inside the waveguide portion having a refractive index of 1. At the output of the waveguide the light wave rapidly diverges propagating in an angle of 140 deg.

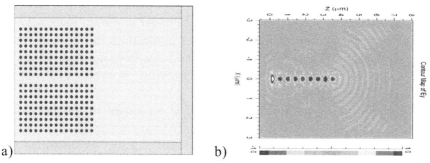

Fig. 2.6. Photonic crystal waveguide (a) and diffraction of light inside the waveguide, and when leaving it (b). Light travels from left to right.

Note that the width of the waveguide (342 nm) is slightly more than half the length of light (633 nm). Some modernization of the waveguide structure near its output significantly reduces the radiation divergence. Thus, Fig. 2.7 a shows a PC waveguide from which two rods were removed in the last row near the waveguide portion. Consequently, the radiation after the waveguide diverges at an angle of only 30 degrees (Fig. 2.7b). Note that according to the scalar diffraction theory, the full divergence angle can be estimated as $2\lambda/(\pi r) = 2.35$ or 130 degrees, r is the radius of the waveguide.

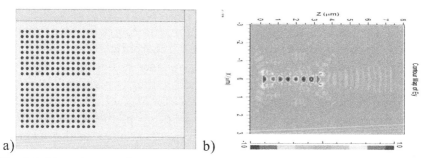

Fig. 2.7. Photonic crystal waveguide with a collimator (a) and diffraction of light inside the waveguide and an exit from it (b). Light travels from left to right.

2.2. The two-dimensional photonic crystal gradient Mikaelian lens

Modern technologies allow to create optical micro- and nano-objects with the dimensions comparable to the wavelength of light. Therefore, the question arises of computer simulation of light diffraction on such objects. To solve this problem, it is necessary solve directly the system of Maxwell equations. One of the most common methods for the numerical solution of these equations is the finite-difference time-domain method (FDTD) [6]. This method has been quite successful thanks to its versatility in solving diffraction problems [7,8]. It is used in this section.

The photonic crystals that we briefly discussed in the previous section are structures with a periodically varying refractive index (this period should be less than the wavelength), which allow to manipulate the light on the nanometer scale [9]. Recently, they have attracted increasing attention due to a number of interesting properties. One of the fundamental properties of these materials is that they do not transmit light in a specific wavelength range. This spectral region is called the photonic band gap. Currently, the most interest is attracted by photonic crystals, the bandgap of which lies in the visible or near infrared [10–13].

With the development of the production technology of photonic crystals it is possible to create photonic crystal lenses. The PC lens is a photonic crystal in which the radius of the holes varies according to a specific law to ensure focusing of the light. The period of the crystal lattice remains constant. Such lenses, for example, solve the problem of focusing the light on the input of the photonic crystal waveguide, as a more compact alternative to the microlenses and tapered waveguides.

It is known that the gradient medium with the radial dependence of the refractive index in the form of a hyperbolic secant, proposed in [14], is used for the self-focusing of laser radiation. A gradient lens with such a refractive index which gathers rays parallel to the axis into focus at the surface is called the Mikaelian lens. In [15] it is suggested to look for the modal solution of the wave equation for a 2D gradient medium with the refractive index in the form of a hyperbolic secant similar to finding soliton solutions of the non-linear Schrödinger equation.

This section discusses the paraxial and non-paraxial solutions in the form of a hyperbolic secant for a gradient two-dimensional waveguide whose refractive index depends on the transverse coordinate in the form of a hyperbolic secant. An identical photon crystal lens is formed for a cylindrical gradient lens which can be manufactured by photo- or electronic lithography. The programming language C++ of the FDTD method is used for a comparative modelling of passage of a plane wave through both microlenses.

2.2.1. The modal solution for a gradient secant medium

In 1951, A.L. Mikaelian showed [14] that in the gradient medium with cylindrical symmetry and the dependence of the refractive index of the radial coordinate as a function of the hyperbolic secant all the rays emanating from a single axial point again gather in the axial focus at a certain distance. This phenomenon is called the self-focusing of light in the gradient medium.

It can be shown that a light field which retains its structure propagates in the two-dimensional gradient medium whose refractive index depends on the transverse coordinate as a function of the hyperbolic secant. This light field shows modal (solitons) properties and its complex amplitude is proportional to the same function of the hyperbolic secant.

Indeed, let the refractive index distribution in a 2D model gradient medium depends only on the transverse coordinate as a function of the hyperbolic secant:

$$n(y) = n_0 ch^{-1}\left(\frac{kn_0 y}{\sqrt{2}}\right) \tag{2.15}$$

where n_0 is the maximum refractive index in the medium on the optical axis, k is the wave number of light in vacuum. In the case of TE-polarization the only non-zero projection of the electric field of a monochromatic electromagnetic wave $E_x(y,z)$ satisfies the Helmholtz equation:

$$\left[\frac{\partial^2}{\partial z^2} + \frac{\partial^2}{\partial y^2} + \frac{k^2 n_0^2}{ch^2\left(kn_0 y/\sqrt{2}\right)}\right] E_x(y,z) = 0 \tag{2.16}$$

where z is the direction along the optical axis. Then the modal solution of the equation (2.16) in the form of 'a soliton' will look like this:

$$E_x(y,z) = E_0 ch^{-1}\left(\frac{kn_0 y}{\sqrt{2}}\right) \exp\left(\frac{ikn_0 z}{\sqrt{2}}\right) \tag{2.17}$$

where E_0 is a constant. The word 'soliton' is in quotes, as in this case there is no non-linearity and the solution (2.17) is only similar to the soliton solution and is the mode of the given gradient medium. Interestingly, the solution (2.17) holds in the paraxial case. If instead of the gradient medium (2.15) we select a slightly different refractive index on the transverse coordinates:

$$n_1(y) = n_0 \sqrt{1 + ch^{-2}\left(\frac{kn_0 y}{\sqrt{2}}\right)} \qquad (2.18)$$

where $n_1(0) = \sqrt{2}n_0$ is the maximum refractive index of the medium, and $n_1(\infty) = n_0$ is the minimum refractive index, then the solution of the paraxial equation

$$\left[2ik\frac{\partial}{\partial z} + \frac{\partial^2}{\partial y^2} + \frac{k^2 n_0^2}{ch^2\left(kn_0 y / \sqrt{2}\right)}\right] E_{1x}(y,z) = 0 \qquad (2.19)$$

will be similar to the complex amplitude (2.17):

$$E_{1x}(y,z) = E_0 ch^{-1}\left(\frac{kn_0 y}{\sqrt{2}}\right)\exp\left(\frac{ikn_0^2 z}{4}\right). \qquad (2.20)$$

Note that the solutions (2.17) and (2.20) have a finite energy:

$$W = \int_{-\infty}^{\infty}\left|E_x(y,z)\right|^2 dy = \left|E_0\right|^2 \int_{-\infty}^{\infty} ch^{-2}\left(kn_0 y / \sqrt{2}\right) dy = 2\left|E_0\right|^2. \qquad (2.21)$$

The modal solutions similar to (2.17) and (2.20) can be found for the 3D gradient waveguide with a refractive index:

$$n(x,y) = n_0 ch^{-1}\left(bx + y\sqrt{\left(kn_0\right)^2 / 2 - b^2}\right) \qquad (2.22)$$

where b is an arbitrary parameter. A method for producing such a ch^{-1} solutions can be found in recent works of I.V. Alimenkov [16,17] who found 3D-soliton solutions to the non-linear Schrödinger equation with the Kerr non-linearity of the third order, when the refractive index of the non-linear medium is described by the expression:

$$n^2(x,y,z) = n_0^2 + \alpha I(x,y,z) \qquad (2.23)$$

where α is a constant, $I(x,y,z) = \left|E(x,y,z)\right|^2$ is the intensity of one of the components of the vector of the electric field strength of the light wave. The analogy between the soliton ch^{-1}-solution for the non-linear medium (2.23) and for the linear gradient medium with a refractive index (2.15) or (2.18) was reported for the first time by A.W. Snayder [15].

The next section shows how to replace the cylindrical gradient lens (GL) by a 2D photonic crystal lens.

2.2.2. Photonic crystal gradient lenses

The two-dimensional photonic crystal gradient lenses (PCGL) consists of a photonic crystal in which the radius of the holes varies according to a certain law. Like a conventional lens, the PCGL allows to focus the parallel light beam to a point. However, the PCGL can be more compact and can be more easily manufactured. Figure 2.8 schematically illustrates the PCGL.

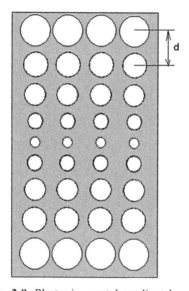

Fig. 2.8. Photonic crystal gradient lenses.

The cylindrical secant GL [14] is a gradient lens whose refractive index varies from the centre to the edge of the lens according to the law:

$$n(y) = \frac{n_0}{ch\left(\frac{\pi |y|}{2L}\right)},$$

(2.24)

wherein L is the width of the lens along the axis z, n_0 is refractive index at its centre.

We choose an equivalent PCGL made of a material with refractive index n and the thickness along the optical axis a so that it could replace the GL. To do this, the optical path length at discrete points of this lens should equal to the optical path length in the GL. The optical path length in the GL will be equal to:

$$\Delta_1 = \frac{Ln_0}{ch\left(\dfrac{\pi|y|}{2L}\right)}.$$

(2.25)

The optical path length in PCGL is:

$$\Delta_2 = N\left[2r(y) + (d - 2r(y))n\right]$$

(2.26)

where N is the number of holes in a row, d is a constant of the crystal or the distance between the centres of holes, $r(y)$ is the radius of holes which varies from row to row. Equating the optical lengths (2.25) and (2.26), we obtain the following relation for the radius

$$r(y) = \frac{d}{2(n-1)}\left(n - n_0 \frac{1}{ch\left(\dfrac{\pi|y|}{2L}\right)}\frac{L}{a}\right).$$

(2.27)

Suppose that in each column of the lens there are M holes. Then, the obtained dependence should be fulfilled at the points $y=\pm dm$, m varies from 0 to $M/2$. Thus, the radius of the holes must also be subject to certain conditions. Firstly, the radius must be non-negative. From (2.27) it is clear that the minimum value of the radius is reached at $y = 0$. Imposing the condition of non-negativity on the radius, we obtain the following relation for the parameters of the GL and the corresponding PCGL:

$$na \geq n_0 L.$$

(2.28)

Secondly, the diameter of the hole must obviously be less than the constant of the crystal. The maximum value of the radius is attained at $y = b/2$ where b is the aperture of the lens. This condition imposes the following restriction on the aperture of the lens:

$$ch\frac{\pi b}{4L} < n_0\frac{L}{a}.$$

(2.29)

Third, the period of the lattice, as mentioned above, should be subject to the condition $d < \lambda$. In addition, in the numerical modelling of the photonic crystal lens the step of the grid must be selected so small that the radius changes from row to row. The point is that it may happen to the extent that the change in the radius from row to row may be smaller than the sampling interval. In this case, the radius does not change and the desired effect will not be achieved.

Modelling of the diffraction of light on 2D microlenses was carried out using the finite difference solution of Maxwell's equations by the FDTD method. The C++ language in MS Visual Studio 6.0 was used to apply the Yee algorithm [18] in the two-dimensional case for TE-polarization. The radiation was applied in the computational domain using the 'total field–scattered field' condition [19]. The boundary conditions were represented by perfectly absorbing Berenger layers (J.P. Berenger) [20].

GL has the property of focusing the light to a point on the surface. In our numerical experiment we used the GL with the following parameters: wavelength $\lambda = 1.5$ µm, $L = 3$ µm, $n_0 = 1.5$ µm, $b = 4$ µm. The distribution of the square of the modulus of the complex amplitude of the electric field when light passes through such a lens is shown in Fig. 2.9, and the intensity of the cross-section – in Fig. 2.10.

The graph shows that the focus of this lens coincides exactly with its front surface.

Now we model the transmission of light through a PCGL with the parameters $a = l = 3$ µm, $n = n_0 = 1.5$, $d = 0.25$ µm. Figure 2.11 shows the dependence of the radius of the holes (Fig. 2.8) in the PCGL on the row number of the holes.

However, even with a large sample $\lambda/h = 100$ we obtain quite inaccurate approximation of the radius. This is shown in Fig. 2.12.

The period of the nanostructure holes is 250 nm, the minimum hole diameter 10 nm, the maximum diameter 40 nm. Figures 2.13 and 2.14 show the distribution of the square of the modulus of the complex amplitude of light passing through the lens.

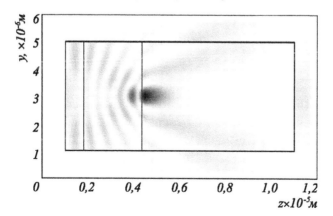

Fig. 2.9. The distribution of the square of the modulus of the electric field $|Ex|^2$ (negative) of the GL, the location of the lens is specified by the gray rectangle.

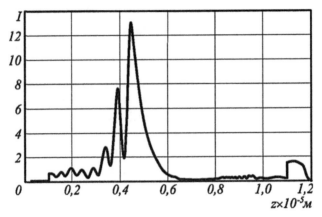

Fig. 2.10. The distribution of the square of the modulus of the electric field $|E_x|^2$ in the cross section along the main axis of GL.

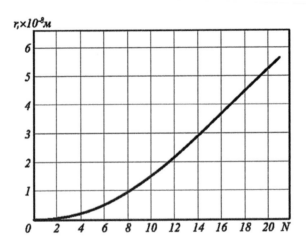

Fig. 2.11. The dependence of the radius of the holes on the row number.

It is seen that the focus of this lens is located $f = 3.3$ μm from the start of the lens, i.e. it accurately corresponds to the GL with $L = 3$ μm. The intensity at the focus is $I_f = 7.5$ a.u. which is less than in the GL (PCGL efficiency is 70% of the efficiency of GL), and the focal length of the PCGL in the z axis is twice that in the GL. This is due to small sampling and rough approximation (Fig. 2.12) of the curve in Fig. 2.11. Figure 2.14 shows the 'noise' caused by the coupling conditions of the 'full field–scattered field' procedure (it is not physical noise). The distribution of the square of the modulus of the electric field in the focal plane is shown in Fig. 2.15.

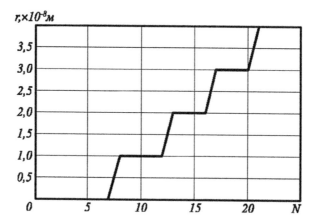

Fig. 2.12. The dependence of the radius of the holes on the row number for 100 samples per wavelength.

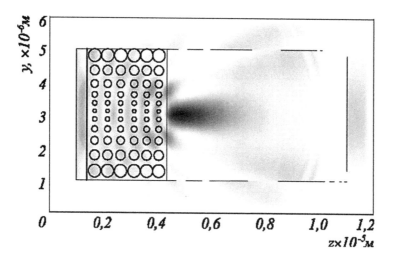

Fig. 2.13. The distribution of the square of the modulus of the electric field $|E_x|^2$ (negative) in PCGL.

From Fig. 2.15 it can be seen that the diameter of the focal spot of the PCGL at half intensity is FWHM = 0.42λ (FWHM – full width half maximum). Note that in the scalar case, the focal spot whose intensity is described by the sinc-function, it is known that the width of the focal spot at half is $0.44\lambda/NA$, where NA is the numerical aperture of the lens. In the PCGL NA = 0.67, so the half width of the spot is equal to FWHM = $0.29\lambda/NA$. This is 1.5 times less than in the scalar case.

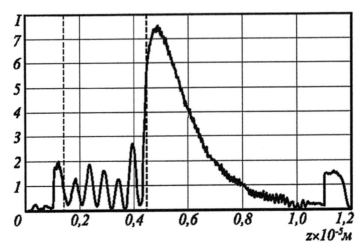

Fig. 2.14. The distribution of the square of the modulus of the electric field $|E_x|^2$ along the main optical axis of the PCGL.

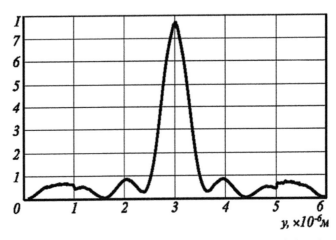

Fig. 2.15. The distribution of the square of the modulus of the electric field $|E_x|^2$ in the focal plane of the PCGL.

2.2.3. Photonic crystal lens for coupling two waveguides

In recent years, extensive research has been carried out into various micro- and nanophotonics devices for coupling two waveguides of different types, such as a conventional single-mode fibre with a wire or a planar waveguide or a planar waveguide with a photonic crystal (PC) waveguide. The following nanophotonic devices are available for coupling the two waveguide structures: adiabatic taper ridge waveguide for coupling with the PC waveguides [21–27]; in this case, the waveguide structure can not only be joined with each by output to input but can also be superimposed parallel to each other [28]; Bragg diffractive gratings in the waveguide [29–32] for the emission of radiation from the fibre wherein the fibre with the Bragg grating can lie on the surface of the planar waveguide [33]; a parabolic micromirror at an angle for entry into the planar waveguide [34]; conventional refractive lenses or microlenses [35–38]; Veselago superlens with negative refraction: flat [39–46] or with a concave surface [47–49]; interface devices in the millimeter range of the spectrum: superlens [50,51] and PC-lens [52]. Work has also been carried out to pair two different PC waveguides [53].

The tapered waveguide may have a high coupling efficiency, if the width of the modes in the ridge waveguides and PC waveguides is comparable. In this case, the coupling efficiency (i.e. the ratio of the energy at the output to the input energy) can reach 80% [21], 90% [24], 95% [22] and even almost 100% [26]. If the width of the ridge waveguide (1.6 μm) is several times greater than the width of the PC waveguide (200 nm), the coupling efficiency is reduced to 60% [23]. When the difference of the widths of the coupled waveguides is even larger, the size of the adiabatically tapering portion of the waveguide is relatively large: in the compression of the mode of the single-fibre with the core diameter of 4.9 μm to the size of the mode of the planar waveguide with a width of 120 nm, the length of the taper is 40 μm [25] and in the waveguide with a cross section of 0.3 × 0.5 μm decreases to 75 nm at a distance of 150 μm [27].

Coupling devices that introduce radiation from a single-mode fibre to planar waveguides or PC waveguides with diffraction gratings on a waveguide, also have tapered zones. For example, the narrowing of the Gaussian beam with a waist diameter of 14 μm to the size of the waveguide with a width of 1 μm is ensured by a taper only 14 μm long [29,30]. In this case the experimental coupling efficiency is 35% [10], with no mirror layer on the reverse side of the waveguide, and 57% [29] with a mirror. The Gaussian beam with a wavelength of 1.3 μm was supplied to the waveguide using a diffraction grating on the waveguide [30]. A similar device for coupling with a grating on a silicon waveguide with a period of 630 nm and a 20–40 μm taper, but for the wavelength of 1.55 μm, had the experimental efficiency of 33% (with the mirror 54%) [31]. The device with higher quality of manufacture for supplying radiation from a single-mode fibre with grating on a silicon with a period of 610 nm and a width of 10 μm to a wire waveguide with a width of 10 μm had an experimental coupling efficiency of 69% [32]. The calculated coupling efficiency of more than 90% has a J-coupler, which connects a waveguide (10 μm) with the PC waveguide (420 nm) using a parabolic mirror with the size of 15×20 μm for a wavelength of 1.3 μm [34]. Thus, both waveguides and the parabolic mirror are made of silicon film (refractive index $n = 3.47$).

Conventional refractive lenses and microlenses are also successfully used in the coupling problems. For example, the waveguide in silicon ($n = 3.092$) with a width of 1–2 μm has a lens at the end which allows, with an estimated efficiency of 90%, to couple this waveguide with the PC waveguide also in silicon ($n = 3.342$) [35]. Simulations have shown [37] that the single-mode fibres with a diameter of 10.3 μm (wavelength $\lambda = 1.55$ μm) using a collimating lens made of glass VK7 (numerical aperture NA = 0.1) with a radius $R = 1.77$ mm, and using a focusing silicon microlens with a radius of aperture of 123 μm can be coupled with the PC waveguide with a cross section of the mode of 0.19×0.27 μm with an efficiency of 80%. In this case, the microlens creates within the PC waveguide a focal spot with a diameter at half intensity FWHM = 0.24λ (numerical aperture of the waveguide NA = 2.2).

A special place among the coupling devices are devices based on 2D superlens (or Veselago lenses), which are based on the phenomenon of negative refraction. The superlens with the effective refractive index close to −1 can be produced using the photonic crystals. The superlens is used to image a point source. The first image is formed inside the lens, and the second image – behind the lens at distance $2B$-A, wherein B is the thickness of the plane-parallel lenses, A is the distance from the lens to the source [39,43]. In [41] it is shown that if the 2D point source of light is described by the Hankel function $H_0(kr)$, k is the wave number, r is the distance from the source to the observation point, the image will be proportional to the Bessel function $J_0(kr)$. That is, the image spot, formed by the superlens, has a diameter FWHM = 0.35λ. In [44] simulation of the 2D photonic-crystal superlens has shown that if this lens consists of only two layers of dielectric rods (dielectric permittivity $\varepsilon = 12.96$) for the wavelength $\lambda = 1.55$ μm with the radius $r = 0.45a$, where a is the period of the grating rods, then at the cyclic frequency $w = 0.293a/\lambda$ the refractive index is $n = -1$, and a point source located at a distance of $A = 0.26\lambda$ from the lens is imaged at roughly the same distance on the other side of the lens, and the width of the image spot is FWHM $= 0.36\lambda$. Some studies examined the Veselago lens not as a plane-parallel PC-layer and one of its surfaces is concave. Thus, in [47] it is shown that the PC-lens made of a rectangular grating of rods with $\varepsilon = 10$ and the magnetic permeability $\mu = 1.5$ with a period $a = 0.48$ cm, the radius of the columns $r = 0.4a$, has an effective refractive index $n = -0.634$. If this 2D lens in plane-concave with the radius of curvature $R = 3.31$ cm, the focus of this superlens will be located at distance $f = R/(1-n)$, for TE-polarization $f = 1.69$ cm, and for TM polarization $f = 2.38$ cm. The radiation frequency is $w = 0.48a/\lambda$. Study [48] presents the results of simulation of the input of radiation to the PC waveguide via a concave superlens. The PC-lens had a thickness $8.6a$ and aperture $38a$, and the PC consisted of 2D grating holes with a period $a = 465$ nm in GaAs ($\varepsilon = 12.96$) and a diameter $2r = 372$ nm. In the focus of the lens at a distance of 7.56λ ($\lambda = 1.55$ μm) there was a focal spot with a radius of 0.5λ, if the lens is illuminated with a Gaussian beam with a waist radius 3λ. Then, the radiation fell behind the lens in a 3W PC waveguide (3W – it means that the width of the waveguide is equal to three periods of the PC grating) with a width of $3a$ (about λ). Unfortunately, the efficiency of input of such a structure is not given in [48]. Study [49] also discusses the results of the simulation of input of radiation from a single-mode fibre in a PC waveguide via PC-superlens (plane-concave, $n = -1$). The thickness of the lens $16a = 4.8\lambda$, aperture $25a$, and it consists of a triangular grating of holes with the period $a = 0.305\lambda$ and a radius $r = 0.4a$ in GaAs. The radius of curvature of the concave surface of the lens $R = 2.1\lambda$, the focal length $f = 1.05\lambda$.

The calculated efficiency of input in the PC waveguide with $\varepsilon = 12.96$, $r' = 0.2a$, $a' = 0.312\lambda$ was equal to 95%. The width of the waveguide was equal to one period of the PC grating a, and the angular frequency $w = 0.315a/\lambda$. Unfortunately, the size of the focal spot of such a lens is omitted.

In [54–56] the authors considered another type of PC lens. The grating of holes of such 2D PC-lens has a constant period, but the size of the holes is changed in accordance with a certain function. In the gradient Mikaelian lens [14] in which all the rays parallel to the optical axis and incident perpendicular to its flat surface are collected a point on the optical axis on the opposite flat surface. This axisymmetric gradient lens has the refractive index of the radial coordinate (distance from the optical axis) in the form (2.24). Study [5)3] simulated a 2D Mikaelian lens with an aperture of 12 µm, consisting of 7 columns of holes with a period of 0.81 µm for a wavelength λ = 1.55 µm. The effectiveness of input for a wide waveguide (12 µm) to the PC waveguide 1.5 µm wide and with an effective refractive index n = 1.73 was 55%. The PC waveguide consisted of a grating of holes with a period of 0.63 µm and a diameter of 0.4 µm. Again, the focal spot characteristics of the lens were not given. In [55, 56] similar PCGLs but with different parameters were simulated. The thickness of the lens was 3 µm, 12 columns of holes, the aperture of the lens 4 µm, the refractive index 1.5, the wavelength 1.5 µm. The focal spot diameter FWHM = 0.42λ, and from zero to zero intensity the focal spot diameter was equal to 0.8λ.

This section discussed ultra-compact nanophotonics devices, allowing effective coupling of the 2D waveguides of different widths using the PCGL. The device was made by the technology 'silicon-on-silica' technology, the width of the input waveguide was 4.5 µm, the width of the output waveguide 1 µm, the size of the PCGL 3×4 µm. The lens is composed of a matrix of 12×17 holes with a grating period of 250 nm, and the hole diameter varied from the center towards the periphery from 160 to 200 nm. The device operates in the wavelength range of 1.5–1.6 µm. The calculated coupling efficiency varied from 40% to 80%, depending on the width of the output waveguide. The PCGL focuses the light into a small focal spot in the air immediately behind the lens that at half intensity is equal to FWHM = 0.36λ, which is 1.4 times less than the scalar diffraction resolution limit in the 2D case, which is determined by the width of the *sinc*-function and is equal to FWHM = 0.44λ.

Modelling PC-lens in the waveguide

The photonic crystal gradient lens which is simulated in the work consisted of a matrix of 12×17 holes in silicon (the effective refractive index for the TE-wave n = 2.83), the lattice constant of the holes 250 nm, the minimum diameter of the holes on the optical axis 186 nm, the maximum diameter of the holes at the edge of the lens 250 nm. The thickness of the lens along the optical axis 3 µm, the width of the lens (aperture) 5 µm. The wavelength λ = 1.55 µm.

The simulation was performed using the difference method of solution of Maxwell's equations FDTD, implemented in the programming language C++. Figure 2.16a shows 2D PC-lens in silicon described above, and Fig. 2.16b the two-dimensional halftone

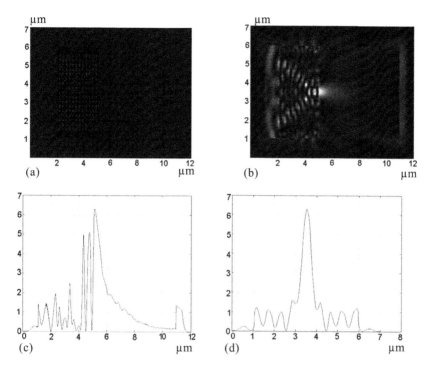

Fig. 2.16. 2D PCGL 12×17 holes in silicon, size of 3–4 μm (a), the diffraction field of light (plane TE-wave) or 2D intensity distribution $|E_x|^2$, y – vertical axis, z – horizontal axis (b), the intensity distribution along the optical axis (a) and in the focal plane (z).

diffraction pattern (time-averaged) of the plane wave with TE-polarization with amplitude E_x (the x-axis is perpendicular to the plane of Fig. 2.16). Figures 2.16c and d show the distribution of intensity $|E_x(y,z)|^2$ along the optical z-axis and along the line y perpendicular to the optical axis, with the focus located on the line y. Figure 2.16d shows that the size of the focal spot at half intensity is FWHM = 0.36λ, and the longitudinal dimension of the focus is FWHM = 0.52λ.

The PCGL with the parameters from the previous example (Fig. 2.16a) was modelled, but the PCGL was located at the exit of the waveguide in silicon with a width of 5 μm and a length of 5 μm (plus the length of the lens 3 μm, with a total length of the waveguide with the lens along the optical axis of 8 μm) (Fig. 2.17a).

The diffraction field (intensity $|E_x(y,z)|^2$) calculated by the FDTD method and averaged over time is shown in Fig. 2.17b (wavelength 1.45 μm). Figure 2.17c shows the intensity distribution along the optical axis. From the comparison of Fig. 2.16c and 2.17c it can be seen that the intensity at the focus increased, and the amplitude of modulation of the intensity inside the lens decreased. This is due to the fact that the difference in the refractive indices between the lens and the waveguide (Fig. 2.16c) is much less than the difference between the lens and the air (Fig. 2.17c) and, therefore, the amplitude of the wave reflected from the media interface is smaller. Figure 2.17d shows the distribution of intensity in the focus lens along a line parallel to the axis y. Figure 2.17d also shows that the diameter of the focal spot at half intensity is FWHM = 0.31λ. From the comparison of Figs. 2.17d and 2.16d we can see that besides reducing the diameter of the focal spot in the case of the PCGL in the waveguide the side lobes of the diffraction patterns in the focus also decreased in size.

Note that the scalar theory in the 2D case describes the diffraction-limited focus by the *sinc*-function: $E_x(y,z) = sinc\ (2\pi y/(\lambda/\mathrm{NA}))$, which

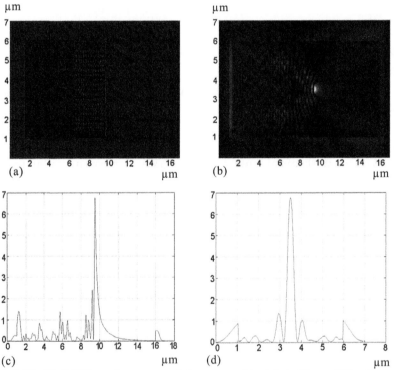

Fig. 2.17. 2D PCGL at the output of the waveguide (a), halftone diffraction of a plane TE-wave with an amplitude E_x, incident to the input of the waveguide with a length of 5 μm, the output of which has a lens 3 μm long (b), the intensity distribution $|E_x|^2$ along the optical axis (a) and at the focus of the lens (d). The intensity is given in arbitrary units.

gives at the maximum numerical aperture NA = 1 the diffraction limit of the focal spot with a diameter at half intensity FWHM = 0.44λ. The limiting value of the focusing spot for the superlens [41] is described by the Bessel function $J_0(kr)$ and gives the value of the diameter at half intensity FWHM = 0.35λ. Thus, the lens in Fig. 2.17a focuses the light into a spot smaller than the diffraction limit.

Simulations have shown that in the wavelength range of 1.3–1.6 μm the intensity at the focus has two maximal values for the wavelengths of 1450 nm and 1600 nm (both maxima have a width of about 20 nm). At other wavelengths of this range the intensity of the focus is 2–3 times lower. With increasing wavelength the focus shifts to the surface of the lens, and at λ = 1.6 μm the focus is already inside the lens.

Modelling of coupling of two waveguides using the PCGL

Figure 2.18a shows the device for coupling of 2D waveguides using PCGL. The width of the input waveguide 5 μm, output – 0.5 μm. The PCGL in silicon (n = 2.83) has a matrix of 12×19 holes with a grating period of 0.25 μm. The diameters of the holes are the same as in the previous examples. The wavelength of 1.55 μm. The length of both waveguides 6 μm.

The simulation was performed using the FDTD method, implemented in the program FullWAVE 6.0 (company RSoft). Figure 2.18b shows the instantaneous diffraction pattern of the TE wave. Coupling efficiency was 45%. Part of the radiation (about 20%) is reflected from the lens back into the input waveguide and part of the radiation passes the lens, but misses the narrow waveguide. Figure 2.18c shows an enlarged fragment of the diffraction pattern in Fig. 2.18b at the output from the narrow output waveguide. Unfortunately, in this program the transverse axis is not y, as in Figs. 2.16 and 2.17, and is has the x axis. The intensity distribution $\left|E_y(x,z)\right|^2$ along the transverse axis x at the outlet of a narrow waveguide shown in Fig. 2.18d. Figure 2.18d shows that the diameter of the laser spot at the output at half intensity is FWHM = 0.32λ. Note that the focus within the output waveguide with a width of 1 μm (ceteris paribus) has a smaller diameter FWHM = 0.21λ, where λ is the wavelength in vacuum. This is lower than previously reported in [37] (FWHM = 0.24λ).

Modelling of the effect of the gap between the waveguides

Figure 2.19 shows a 2D diagram of coupling of two coaxial waveguides with a gap between them. The width of the input waveguide with the PCGL is W_1 = 4.6 µm, the output W_2 = 1 µm, the gap between the waveguide is Δz = 1 µm. Other parameters: λ = 1.55 µm, n = 1.46. The PC-lens consisted of a 12×17 matrix of holes with a period a = 0.25 µm and the diameter of the holes of 186 to 250 nm. The white colour in Fig. 2.19a shows the waveguide material (n = 1.46), and black and gray – air (n = 1). Figure 2.19b shows the instantaneous

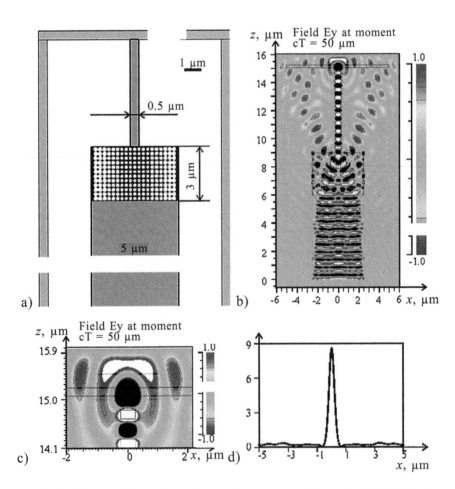

Fig. 2.18. Scheme of coupling of two planar waveguides using PCGL (a), the instantaneous diffraction pattern of TE-wave calculated by using the FDTD method with FullWAVE 6.0 program (b) and the enlarged portion of this pattern at the output of the waveguide with a width of 0.5 µm (c), the intensity distribution at the output of the fibre (d).

pattern of the amplitude $E_y(x,z)$ for the TE-wave calculated using FullWAVE 6.0 for the circuit in Fig. 2.19a. Figure 2.19c shows the dependence of the effectiveness of coupling (the ratio of the light intensity at the output of the narrow waveguide I to the intensity at the input of the wide waveguide I_0) on the distance between waveguides Δz. Figure 2.19c shows that the maximum coupling efficiency of 73% is achieved when the gap between the waveguides is equal to 0.6 µm. Note that in the gap between the waveguides is the waveguide material ($n = 1.46$) and not air.

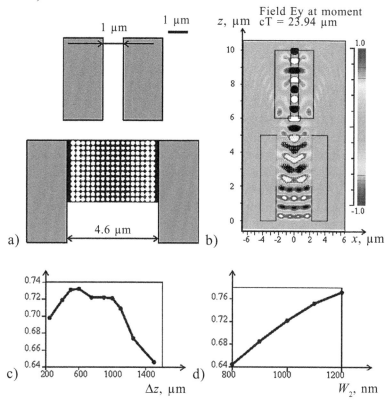

Fig. 2.19. 2D coupling of two waveguides by means of PCGL with a gap Δz = 1 µm between the waveguides (white – material, black – air) (a); instantaneous amplitude distribution $E_y(x,z)$ of the TE-wave calculated using FullWAVE (b); dependence of coupling efficiency of the size of the gap between the waveguides Δz (c) and on the width of the output waveguide W_2 (g).

Figure 2.19d shows the calculated dependence of the efficiency of coupling for the case of Fig. 2.19a on the width of the output waveguide W_2 at the size of the gap $\Delta z = 1$ µm. Figure 2.19d shows that with increasing width of the output waveguide W_2 the coupling efficiency increases almost linearly. Study [57] modelled a similar nanophotonic device not in 2D and in the 3D variant.

Making of two 2D waveguides connected by PCGL

Planar waveguides according to the scheme in Fig. 2.19a were recorded on a PMMA resist by direct electron beam writing at a voltage of 30 kV using a ZEP520A lithographer (University of St. Andrews, Scotland) [58]. Processing of the resist in order to eliminate sections 'illuminated' by an electron beam was carried using xylene. Thereafter, reactive ion etching (RIE) in a mixture of CHF_3 and SF_6 gases was carried out in further plasma etching of the materials. That is, the 2D pattern of the waveguides with the PCGL (Fig. 2.19a) was transferred to a silicon film (SOI technology: silicon-on-insulator): the Si film 220 nm thick on a layer of fused silica with a thickness of 2 μm. The etching depth was about 300 nm. The diameter of the holes in the PCGL ranged from 160 nm to 200 nm. The length of the entire sample (the length of the two waveguides) was 5 μm. On one substrate we simultaneously fabricated several similar structures differing in the spacing between the waveguides $\Delta z = 0$ μm, 1 μm, 3 μm and several structures differing in the offset between the axes of two waveguides $\Delta x = 0$ μm, ±0.5 mm, ±1 μm. Figure 2.20 shows an enlarged (×7000) photograph (top view) of the two waveguides with a gap $\Delta z = 1$ μm and with the PCGL obtained with a scanning electron microscope.

The parameters of the sample in Fig. 2.20 were as follows. The design width of the waveguides $W_1 = 4.5$ μm and $W_2 = 1$ μm, the PCGL consisted of a 12×17 matrix of holes with a period of 250 nm.

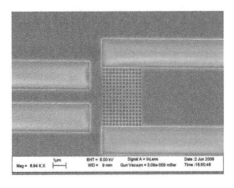

Fig. 2.20. Photograph of two planar waveguides in a Si film coupled with a PCGL; scanning electron microscope, ×7000.

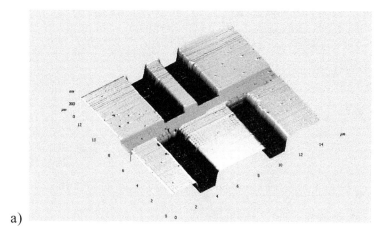

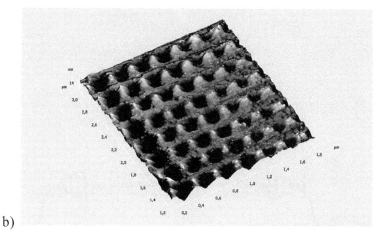

Fig. 2.21. Profile of the relief of 2D waveguides with the PCGL obtained with a scanning probe microscope (atomic power microscope) SolverPro (Zelinograd): profile of the waveguides (a); 6×6 section of the matrix of the holes in PCGL (b). The horizontal axes give microns, the vertical axis – nanometers.

Figure 2.21 shows a section of the profile relief of two waveguides fabricated in a silicon film on fused quartz (a) and a portion of a the 6×6 matrix of holes in the PCGL (b) obtained with a scanning probe microscope.

Figure 2.22 shows the cross sections of the output (a) and input (b) waveguides. It can be seen that the depth of etching of both waveguides is about the same and is equal to 300 nm, and the width of the output waveguide at the top is 1 μm, and at the bottom – 2 m

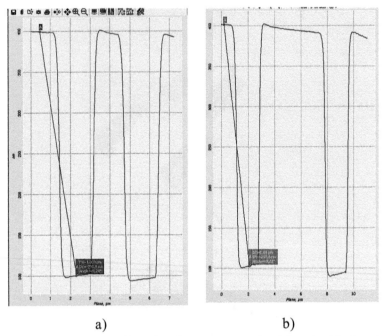

a)   b)

Fig. 2.22. Profile cross sections for the narrow output (a) and wide input (b) waveguides.

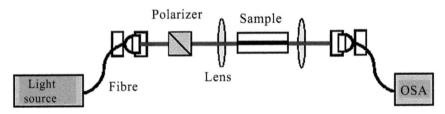

Fig. 2.23. The optical setup for research of nanophotonic devices, comprising two waveguides and the PCGL.

(Fig. 2.22a). Similarly, the width of the input waveguide at the apex was 4.5 μm, and at the base of the trapezium 5 μm (Fig. 2.22b).

Characterization of two waveguides with the PCGL

Figure 2.23 shows the optical circuit for producing the transmission spectrum for the two planar waveguides coupled with the PCGL. A wide-band light source (1450–1700 nm), working on the basis of amplified spontaneous emission, was coupled with the optical fibre. The light at the output from the fibre was collimated and was incident on the polarizer which transmits only the TE-polarization.

Next, using a microlens, the radiation was focused onto the surface of the input waveguide. A small part of the energy of light entered the waveguide and passed through the sample.

At the output from the narrow waveguide there was a second microlens collecting the light and focusing on the input end of the multimode optical fibre coupled with the optical spectrum analyzer (OSA). Figure 2.24 shows the spectrum of the radiation source the maximum of which is at the wavelength of 1.55 μm. The emission intensity is given in arbitrary units.

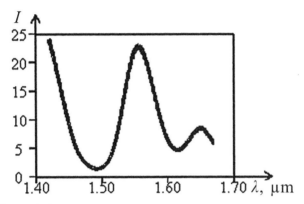

Fig. 2.24. The emission spectrum of the light source useα ın the optical circuit shown in Fig. 2.23.

Figure 2.25 shows the transmission spectra of the samples in the range of 1.5–1.6 μm at the following gaps Δz between the waveguides on the optical axis (a): 0 μm (curve 1) and 1 μm (curve 2) and 3 μm (curve 3), and also at the following offset values Δx from the optical axis of the output waveguide (b): 0 μm (curve 3), −0.5 μm (curve 2), +0.5 μm (curve 4), −1 μm (curve 5) and +1 μm (curve 1). Figure 2.25a (curve 1) shows that the spectrum of the transmission has four local maxima at wavelengths of approximately 1535 nm, 1550 nm, 1565 nm and 1590 nm. Moreover, two of these maxima (at wavelengths of 1550 nm and 1565 nm) have intensities three times greater than the other two. Most likely this is due to the fact that the emission of the source (Fig. 2.24) is several times weaker at these wavelengths.

By increasing the axial distance $\Delta z = 1$ μm between the waveguides (Fig. 2.25a, curve 2) the transmission spectrum retains on average its structure, but the local maxima decrease in size and are shifted into the 'red' region of the spectrum. With further increase

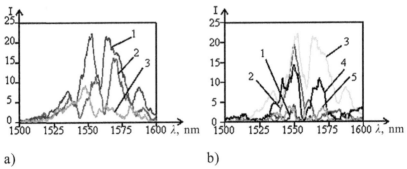

a)   b)

Fig. 2.25. The transmission spectra measured by the optical circuit in Fig. 2.23 for the test samples, similar to those shown in Fig. 2.20, with the following gaps between the waveguides (a): $\Delta z = 0$ µm (curve 3), $\Delta z = 1$ µm (curve 2) and $\Delta z = 3$ µm (curve 4), and also at the following offset from the optical axis of the output waveguide (b): $\Delta x = 0$ µm (curve 1), $\Delta x = -0.5$ µm (curve 2), $\Delta x = +0.5$ µm (curve 3), $\Delta x = -1$ µm (curve 5) and $\Delta x = +1$ µm (curve 1).

of the gap $\Delta z = 3$ µm between the waveguides (Fig. 2.25a, curve 3) the local maxima not only further decrease, but also show the 'blue' shift. The 'red' shift is about 10 nm, and the 'blue' shift is also −10 nm (for a maximum near the central wavelength – 1.55 µm). Figure 2.25b shows that when the output waveguide is shifted by 1 µm from the optical axis (curves 4 and 1) the intensity of the output decreases by 8 times (wavelength 1.55 µm). This means that the diameter of the focal spot, formed by the PCGL in silicon, is less than 1 µm.

2.3. Simulation of three-dimensional nanophotonic devices for input of radiation in a planar waveguide

This section presents the results of three-dimensional modelling of the complete coupler including also the grating for inputting the focused Gaussian beam into a wide planar waveguide, and a Mikaelian PC-lens to match the modes of wide and narrow waveguides. Using the FullWAVE program it was shown that the effectiveness of such a nanophotonic device is 32%, and thus there is a 'compression' (125 times) of the Gaussian beam cross-sectional area (3 × 4.6 µm) in the area of the narrow waveguide mode (0.22 × 0.5 mm). If the diffractive grating is illuminated by a portion of a plane wave, the efficiency rises to 52%. Moreover, the efficiency of input by the grating of the focused Gaussian beam into a wide planar waveguide is equal to 62.5%, and the efficiency of coupling of wide (4.6 µm) and narrow (0.5 µm) waveguides using a Mikaelian PC-lens is 46%. For a plane wave the last two digits increase by 78% and 67%, respectively.

2.3.1. The two-dimensional simulation of the input of light in a planar waveguide using a grating

Figure 2.26 shows a 2D-setup for the input light into a thin planar waveguide with a binary grating (the width of the steps is equal to the width of the grooves) on a thin substrate. Along the Y-axis of the waveguide the grating steps and the substrate have unlimited dimensions. The substrate has a multilayer structure (3 periods of the Bragg mirror) to increase the light input efficiency into the waveguide. The structure parameters in Fig. 2.26 were selected by the quasi-optimal selection procedure. The materials were selected according to the widely used silicon-on-insulator technology.

The waveguide was made of silicon (Si) with a refractive index $n_1 = 3.479$ for the wavelength of light of $\lambda = 1.57$ µm, thickness $h = 220$ µm on quartz (SiO$_2$) having a refractive index $n_2 = 1.44$ and width $W_1 = 1.03$ µm. The substrate again contained a silicon layer with a width of $W_2 = 0.55$ µm. These two layers in the substrate (silicon + silica) with the thickness $W_1 + W_2$ were repeated two more times. When adding the fourth period of the Bragg mirror the input efficiency of light to the waveguide did not increase. The silicon film had a subwavelength surface binary grating with a period $T = 750$ nm, the depth of the grooves of the lattice was $D = 90$ nm. The working length of the grating was 7–8 periods. The light in the form of a cylindrical Gaussian beam with TM-polarization

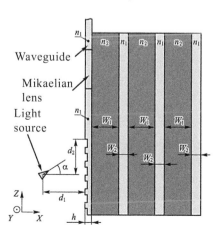

Fig. 2.26. Two-dimensional diagram of the input of light with TM-polarization ($E_y \neq 0$) to a silicon waveguide using a binary surface subwavelength grating.

(electric field vector is directed along the lines of the grating and along the Y axis, while TE-polarized light is hardly introduced into the waveguide). The radius of the Gaussian beam along the Z axis is $b = 3$ μm. The centre of the waist of the Gaussian beam is located at a distance from the grating (along X axis) $d_1 = 2.78$ μm and at a distance from the edge of the grating (Z axis) $d_2 = 4.15$ μm. The angle of incidence of the laser beam on the grating $\alpha = 34°$.

Simulation was carried out using the FullWAVE 6.0 program (company RSoft, USA) with the FDTD-difference method for solving Maxwell's equations. In this case at the above input parameters, the efficiency of input of light into a waveguide using the grating was $\eta = 66.3\%$. Efficiency is defined as the ratio of power (or total intensity) of the waveguide mode to the power of the Gaussian beam, multiplied by 100%.

Simulation in this work was carried out with the following discretization steps: steps in X and Y axes equal to $\lambda/87$, the step along the longitudinal Z-axis equal to $\lambda/78$, and the step in the timeline set equal to $cT/196$, where c is the speed of light in vacuum, T is the period of oscillation of the light wave. The thickness of the absorbing layer on all sides of the calculation domain was 0.5 μm. Table 2.1 shows the efficiency of input on the sampling steps. It is evident that at the chosen values the efficiency is resistant to changes in the grating (the penultimate line of Table 2.1).

Table 2.1. Dependence of efficiency of input of light into the waveguide on the selected sampling steps

Δx	Δz	Δt	η, %
$\lambda / 20$	$\lambda / 20$	$cT / 64$	22
$\lambda / 30$	$\lambda / 30$	$cT / 80$	52
$\lambda / 40$	$\lambda / 40$	$cT / 80$	63
$\lambda / 50$	$\lambda / 50$	$cT / 100$	65
$\lambda / 87$	$\lambda / 78$	$cT / 196$	66,3
$\lambda / 100$	$\lambda / 100$	$cT / 200$	67

Figure 2.27 shows the dependence of efficiency η on the quartz layer thickness in the substrate W_1 – Bragg mirror in Fig. 2.26. Figure 2.27 shows that there are resonance (narrow) reflection peaks (as in the Bragg mirror or a 1D photonic crystal) when $W_1 = 0{,}47$ μm, 1.03 μm and 1.6 μm. Maximum efficiency is achieved at $W_1 = 1.03$ μm.

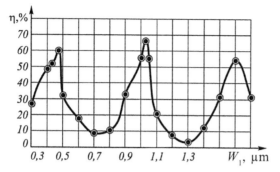

Fig. 2.27. The dependence of the efficiency η of input in the 2D waveguide on the thickness of the silicon layer on the substrate W_1.

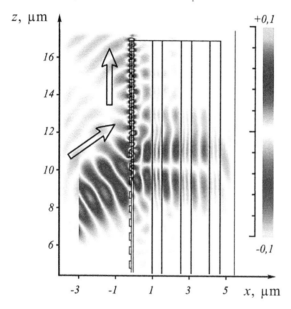

Fig. 2.28. The instantaneous amplitude distribution $E_y(x, z)$ (the arrows indicate the direction of light propagation).

Figure 2.28 shows the result of simulation of input of light into the 2D waveguide: the picture shows the instantaneous amplitude of the electric field $E_y(x, z)$. From Fig. 2.28 it is clear that the Gaussian beam is attenuated when passing into the substrate and hardly penetrates in the three periods of the Bragg mirror. The Bragg mirror operates like the antireflection coating and, therefore, the Gaussian beam is not reflected from the lattice and is effectively introduced into the silicon film.

Figure 2.29 shows the dependence of the efficiency of input of a focused Gaussian beam into the waveguide in the 3D case when increasing the thickness h_2 of the waveguide (Fig. 2.26) along the Y axis. Figure 2.29 shows that when the thickness of the waveguide grating is h_2 = 4.6 μm the efficiency is η = 60%, and at a thickness of h_2 = 8 μm the input efficiency is η = 65% is slightly lower than in the 2D case η = 66.3%. A further increase in the thickness h_2 of the device (Figure 2.26) was not simulated because of the limited RAM of the computer. Thus, 3D simulation using the FullWAVE program showed that 60% of the light energy of the Gaussian beam focused on the grating (waist area 3 × 4.6 μm) can be introduced into the planar silicon waveguide (Fig. 2.26) 4.6 μm wide.

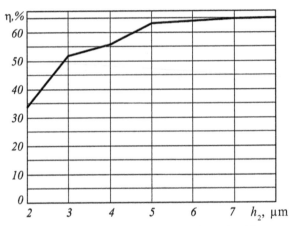

Fig. 2.29. Dependence of the efficiency η of input into the 3D waveguide on the thickness h_2 of the waveguide in the third measurement along the axis Y.

2.3.2. Three-dimensional modelling Mikaelian photonic crystal lens for coupling two waveguides

In [58] the authors simulated a 2D Mikaelian PC-lens connecting two planar waveguides in silicon with a thickness of 5 μm and 0.5 μm. Coupling efficiency was η = 45%. This section presents the results of three-dimensional modelling of a similar nanophotonic device.

Figure 2.30 shows a diagram of the planar waveguides, connected via PC lens. The width of the input waveguide – 4.6 μm, the output waveguide – 0.5 μm, the thickness of the silicon film – 220 nm, the refractive index n = 3.47 for the wavelength λ = 1.55 μm, the number of holes – 12 × 17, diameter of holes 186 nm at the lens axis to 240 nm at the periphery, the grating period of holes – 250 nm. A quartz substrate with a refractive index of 1.44 had a a thickness of

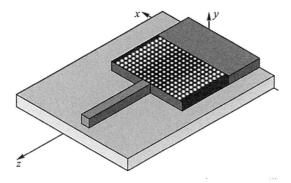

Fig. 2.30. 3D diagram of two planar silicon waveguides (dark gray) coupled with a PC-lens (black) on a substrate made of quartz (light gray).

1 μm. The input of a wide waveguide received TE-polarized light (the electric vector E lies in the plane of the PC-lens) of the elliptical Gaussian laser beam with radii in X and Y of 4.6 μm and 0.22 μm. The total 3D area of the calculation by program FullWAVE was equal $X \times Y \times Z = 6 \times 3.1 \times 9$ μm³. The calculation of the components of the electromagnetic field for the nanophotonic device, shown in Fig. 2.30, took about 18 min in an Intel®Celeron® processor, 3.06 GHz, 512 MB RAM.

The coupling efficiency the two waveguides (Fig. 2.30) was $\eta = 46\%$. This is almost the same value as in the two-dimensional case [58]. In both cases, in this work and in [58], Fresnel reflection is not taken into account. The coincidence of the results suggests that the three-layer substrate is chosen so that the light is almost never leaves waveguides and lenses in the direction of 'up' and 'down' along the axis Y and is only reflected from the boundaries between the media and goes from the lens past the waveguide in the XZ plane, as in the two-dimensional case.

If we replace the Gaussian beam by a portion of the plane wave with the area 4.6×0.22 μm², which illuminates the end of the input waveguide, the coupling efficiency (ignoring Fresnel reflection from the waveguide end face) increases to 67%. Such an increase in the coupling efficiency is due to the fact that a plane wave is focused by the Mikaelian lens to a spot with a smaller diameter than the Gaussian beam of the same size. A focal spot with the smaller diameter changes more efficiently to the narrow waveguide mode.

Figure 2.31 shows the result of simulation of the nanophotonic device shown in Fig. 2.30 in illuminating a wide Gaussian beam waveguide with a waist of 4.6×0.22 μm².

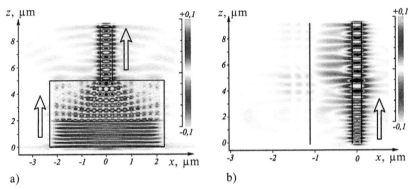

Fig. 2.31. Instantaneous amplitude distribution $E_x(x,y,z)$ in the plane ZX (a) and the plane ZY (b), the vertical line (b) shows the boundary of the substrate, and the arrows show the direction of light propagation.

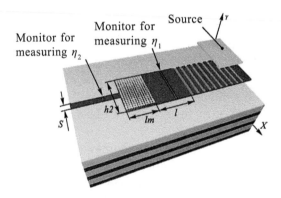

Fig. 2.32. Diagram of the entire nanophotonic device, including the grating, the waveguide, the PC-lens, the narrow waveguide and the substrate – the Bragg mirror.

Figure 2.31 which shows the instantaneous amplitude distribution $E_x(x,y,z)$ indicates that the light propagates inside the waveguide and only a small part of light energy propagates outside the waveguide (in the substrate). Most of the lost energy is reflected from the boundary of the PC-lens and goes back into the waveguide. We also see that the energy loss in the narrow waveguide (withdrawal of light into the substrate) is larger than in the wide waveguide.

2.3.3. Three-dimensional modeling of the entire device nanophotonics

Figure 2.32 shows a 3D diagram of the entire nanophotonic device which includes the grating, the Mikaelian PC-lens and two coupled planar waveguides. On the substrate (three periods of the Bragg mirror) made of silicon and quartz there is a thin film (220 nm thick) of silicon, width h_2 = 4.6 µm in surface gratings, the lenses and a wide waveguide and the width s = 0.5 µm in the output of the narrow waveguide. The length of the entire device is 17 µm. For PC-lens l_m = 3 µm, the gap between the waveguide grating and the lens l = 4 µm (the size of this area does not affect the efficiency, and its length can be changed). An elliptical Gaussian beam with linear polarization (electric vector has only one projection E_x) with an area of the waist 3 × 4.6 µm² (in this case the maximum efficiency) falls on the device at an angle of 34° in the area of the grating.

Figure 2.33 shows the instantaneous distribution of the amplitude of the electrical field $E_x(x,y,z)$ in the ZY plane (a) and the ZX plane (b).

Figure 2.33 also shows that the light travels in the substrate beneath the grating and below the waveguide and the lens it hardly diffuses into the substrate.

The efficiency of input of light into the waveguide by the grating (calculated in the broad waveguide after the grating) is η_1 = 62.5% of the total optical power of the Gaussian beam (in this time, the efficiency is given with the Fresnel reflection from the grating taken into account), and the efficiency of the entire device (calculated at the output of the narrow waveguide) is η_2 = 32%. If the same device (Fig. 2.32) is illuminates by a plane linearly polarized wave with the cross sectional area of 3 × 4.6 µm², the efficiency increases: η_1 = 78% and η_2 = 52%. Note for comparison that if we focus the same Gaussian beam on the end of a wide waveguide of the device shown in Fig. 2.30 (without the grating), the total efficiency will be only η = 8%.

2.4. Photonic crystal fibres

Photonic crystal fibres (PCF) are a relatively new class of optical fibres, using the properties of photonic crystals [59].

In the cross section the PCF are quartz or glassy microstructures with a perodic or aperiodic system of microinclusions, especially cylindrical microperforations oriented along the fibre axis. The defect of the microstructure corresponding to the absence of one or more elements in its centre, is the core of the fibre, providing the waveguide mode of propagation of electromagnetic radiation.

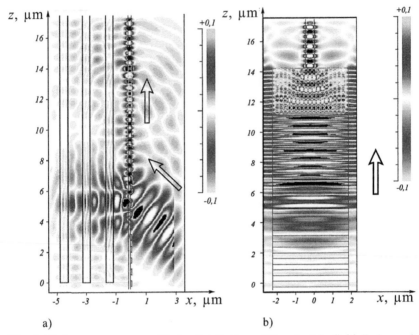

Fig. 2.33. Instantaneous amplitude distribution of the electric field E_x (x, y, z) for the entire device: in the ZY plane (a) and the ZX plane (b), arrows indicate the direction of light propagation.

PCF is divided into two basic types according to the mechanism of retention and guiding of light. There are optical fibres with a solid core or a core with a higher average refractive index with respect to the microstructured (perforated) shell that use total internal reflection, as well as the conventional optical fibres, but unlike the latter, they may have better ability to retain the light in the core as it propagates as a result of the larger local difference of the refractive indices in the core and shell. Another type of PCF for confine light uses the effect of photonic bandgaps (Bragg reflection) produced by the microstructured shell so that the light may be retained and propagate even in the core with a lower relative refractive index in comparison with that of the shell, including a hollow core. The photonic bandgap arising in the transmission spectrum (the dependence of the transmittance on the wavelength) of the two-dimensional periodic shell of this type of the fibre provides high reflection coefficients for the radiation propagating along the hollow core, allowing to significantly reduce the losses inherent in the modes in normal hollow optical waveguides with a solid shell and quickly growing [60] with decreasing diameter of the hollow core. There are several groups of the fibres of the second type.

Fibres with a hollow core – the central hole, as a rule, with a larger radius, surrounded by 'rings' of the microholes.

The Bragg fibre [61–65] has a hollow core surrounded by a cladding layer formed by alternating rings of low and high refractive indices. In [66] the authors describe Bragg fibres with a filled (solid) core whose working where can be shifted into the visible range, by varying the thickness of the cladding of the Bragg reflective rings.

Solid fibres [67,68] in which the cladding contains rods of a high refractive index, are arranged in the optical fibre base material with a low refractive index.

Moreover, in both types of fibres we can vary a number of parameters (Fig. 2.34) such as the form the microperforations (round, square, elliptical) [69], the circuit for localization of the microperforations (regular hexagonal and rectangular gratings, irregular grating) [70.71], the number of cores [72,73], the number of defects that form the core [74,75], the material filling the microperforations [76,77]. For example, in [78] the cavities in the cladding of the PCF were filled with liquid crystals, which gave rise to a hybrid mechanism of propagation of light: the effect of total internal reflection and the effect of photonic bandgaps act respectively on the orthogonally polarized modes. The application of the PCFs filled with liquid crystals, in optical switches is discussed in [79–81].

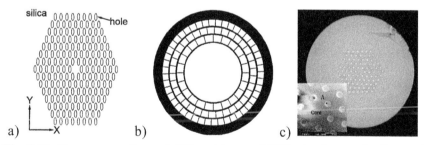

Fig. 2.34. The cross sections of different types of PCF: a) scheme of a flattened hexagonal lattice with elliptical holes [70], b) scheme of the Bragg fibre with three concentric cylindrical quartz layers [64], c) optical micrograph of the section of a solid-state optical fibre, the light and dark areas in the image correspond to regions of high and low refractive indices [68].

The development of the PCFs resulted in new degrees of freedom to control the characteristics of the fibre with effective control of the dispersion, making the fibre with very high or very low non-linearity, large or small effective mode area, etc.

Theoretical interest in the PCF is caused largely by the fact that their structure has a considerably larger contrast of the refractive index than conventional weakly guiding fibres and as a consequence, it is necessary to develop special performance methods for their study. It should be noted that the ability of the PCF to confine light depends on many factors such as the number of microinclusions, their location, the ratio of the diameter and distance between centres. This gives scope for the design of optical fibres with predetermined properties.

The PCF modes are electromagnetic waves that can be excited and distributed in the fibre. Any radiation beam directed into the fibre will initiate a set of spatial modes in the fibre.

There are several methods for calculating the PCF modes, which can be divided into three groups: the approximate analytical methods or expansion methods, integral methods and finite difference methods.

Expansion methods. The basic idea exploited in this group of methods is the possibility of representing the mode field of the fibre in the form of an expansion in some basis. As a result, the search for modes is reduced to the problem of the eigenvalues and eigenvectors of a matrix.

The expansion in plane waves [82] with periodic boundary conditions gives a solution for the infinite fibre, periodically repetitive in the transverse plane, which makes it impossible to obtain the data in principle by the method of the imaginary part of the propagation constant, corresponding to the losses in propagation of the leaking or non-eigen mode.

The method of expansion of Gauss–Hermite modes [83,84] is more suitable for describing the complex structure of the cross section of the PCF rather than the method of expansion in plane waves, but this method is limited only to the use for the PCF with the holes in the cladding, located at the nodes of a regular hexagonal grating, so the distance between the centres of any two adjacent holes shall be fixed, and this value is included in the expression for the basic functions.

A generalization of the multipole method, commonly used for optical fibres with multiple cores, for the case of PCF was considered in [85]. A key aspect of the multipole method proposed by the authors is that it uses to advantage the fact of rounded inclusions and symmetry properties inherent in many of the PCFs in connection with regularly spaced microholes [86]. The strength of the method also consists in the fact that it allows us to calculate both the real and imaginary parts of the propagation constant. Unlike the multipole approach for optical fibres with multiple cores, demonstrated in [87] and using the technique of pointwise joining of the field at the boundaries of the inclusions, the method [85] treats the boundary conditions by expanding the field components in an orthonormal basis. Application of the method can be extended also to the fibres with an arbitrary form of the inclusions in accordance with the strategy proposed in [88].

A characteristic feature of the method of the matched sinusoidal modes [89] is the technique of splitting the inhomogeneous section of the waveguide structure into rectangular regions with a constant value of the refractive index of the medium. In each of these areas the field of the mode is approximated by a superposition of the factored harmonic functions. The propagation constants of the modes are found by minimizing the discrepancy of the representations of the field on the borders of neighbouring areas using an integrated approach. The method of the matched sinusoidal modes uses the search procedure to find the roots of the equations and is therefore inferior in the speed to the methods based solely on finding the eigenvalues of matrices.

Integral methods. Integral methods are grid methods, that is, in contrast to the methods of the previous group, in this case the solution of the problem of finding the mode field is the grid function, not specified analytically.

Among this group we should mention the finite element method [90,91]. It is a powerful tool for vector analysis, capable to take into account all the peculiarities of the geometry of the microholes and their location in the structure of the section. Fast enough and flexible, it is often used to model the properties of the PCF. The shortcomings of the finite element method include memory demands because description of the structure of the cross section of the PCF requires detailed sampling and a large number of variables, as well as the need for human intervention in the operation of the algorithm to better determine the boundary conditions (perfectly matched layer) and the sampling grid.

The boundary element method [92], where the cross section is divided into homogeneous areas and the eigenvalue problem is the result of the application of Green's theorem, is less demanding on the memory. However, a significant drawback is the possibility of false solutions.

In the method of Green's functions [93] the problem of finding propagation constants of the modes can also be reduced to the problem of eigenvalues of the matrix with a special fast algorithm developed to solve this problem. This method is efficient in the case of complex geometries of the microholes, although at a slower rate of convergence than in the case of circular holes.

Finite difference methods. Finite difference methods,like the integral methods, provide grid solutions.

The finite difference method is widely used for all sorts of solutions of equations. Due to the simplicity of implementation, this method has become a convenient tool for the calculation of the optical fibre modes, especially those for which there is no analytical solution, such as the PCF.

The presence of large refractive index contrasts in the structure of the cross section of the PCF requires the use of a fully vector approach in the calculation of the modes, instead of the scalar approach often used for weakly guiding fibres. However, as shown in [94], the scalar finite-difference method can be used to produce at least a qualitative evaluation of the mode distribution of the PCF, including those in which the light propagation mechanism is based on the effect of photonic bandgaps.

Vector finite-difference schemes have been proposed for a more accurate analysis [95]. Discretization was carried out for differential operators and the functions included in the Helmholtz equation and the wave equation. In [96] the authors presented by the FDTD-method of calculating the modes of the PCF with shifted grids (Yee cells). An enhanced FDTD method, which takes into account the dispersion of the material, is described in [97].

The result of the application of special finite difference schemes for time-dependent wave equations and Maxwell's equations is a family of methods of propagation of the beam [98–100]. The method is based on modelling the propagation of a coherent light beam along the fibre, resulting in the modes of the structure, like a posteriori. The method is convenient to investigate the energy loss during the passage of light through the optical fibre, although this may be difficult due to the problem of convergence of the method.

This section describes the approximate analytical method for the calculation of modes of optical fibres: the method of matched sinusoidal modes. The basic idea of the method of matched sinusoidal modes (MSM method), also known as the cross-resonance technique [101,102], was first formulated in [103]. The subsequent development of the method was in [104], where it was used to calculate the radiation loss due to leaky modes in step fibres. Then, in [99,105] the authors introduced the descriptive term 'matching of sinusoidal modes' and presented an exact mathematical formulation. The MSM method has been modified [106] using an iterative Krylov method [107] at the most computationally difficult stage of solving a non-linear problem on the eigenvalues of a large matrix, to which the problem of finding propagation constants of modes was reduced. The MSM method can be used to calculate both scalar and vector modes [108] of the conventional round fibres [109] and the PCFs [110].

2.4.1. Calculation of modes of photonic-crystal fibres by the method of matched sinusoidal modes (MSM method)

The method of matched sinusoidal modes [89,105] differs favourably from many other approaches of numerical investigation of homogeneous longitudinal fibres by the possibility of analytical representation of the field obtained as a result.

The MSM method is based on the representation of solutions for the spatial mode as a superposition of sinusoidal local modes that are eigenmodes of homogeneous, with a constant refractive index, rectangular parts of the fibre with the non-uniform cross-section.

MSM method in the scalar case
We formulate the problem of finding eigenmodes of a dielectric waveguide uniform in the longitudinal direction and non-uniform in cross-section surrounded by a perfect conductor, so-called 'electric walls' or the ideal magnetic material – 'magnetic walls'.

The fibres homogeneous in the longitudinal direction have a constant distribution of the refractive index of the material along the length in the cross section. The cylindrical symmetry of the fibre allows the separation of the variables and to present the mode field in the form of

$$\mathbf{E}_j(x,y,z,t) = \mathbf{E}_j(x,y)\exp(-ik_{zj}z)\exp(i\omega t),$$
$$\mathbf{H}_j(x,y,z,t) = \mathbf{H}_j(x,y)\exp(-ik_{zj}z)\exp(i\omega t),$$

$$(2.30)$$

where $\mathbf{E}_j(x,y,z,t)$ – the electric field strength; $\mathbf{H}_j(x,y,z,t)$ – the magnetic field strength; k_{zj} – propagation constant, the eigenvalue of j-th mode, or the projection of the wave vector $\mathbf{k}_j$ on the longitudinal axis z, ω – radiation frequency.

The propagation constants of different modes differ.

The spatial component of the field of the j-th mode can be decomposed into transverse and longitudinal components, which are denote by the indices t and z respectively. As a result, we have:

$$\mathbf{E}_j(x,y,z) = \mathbf{E}_j(x,y)\exp(ik_{zj}z) = [\mathbf{e}_{tj}(x,y) + e_{zj}(x,y)\mathbf{z}]\exp(ik_{zj}z),$$

(2.31a)

$$\mathbf{H}_j(x,y,z) = \mathbf{H}_j(x,y)\exp(ik_{zj}z) = [\mathbf{h}_{tj}(x,y) + h_{zj}(x,y)\mathbf{z}]\exp(ik_{zj}z),$$

(2.31b)

where $\mathbf{z}$ is the unit vector parallel to the fibre axis.

The modes whose longitudinal component of the magnetic vector is zero, called the transverse–magnetic modes (TM), and those modes for which the longitudinal component of the electric vector is zero, – transverse–electric (TE). In general, the TE and TM modes are not the optical waveguide modes. The modes of the fibres are generally hybrid and contain longitudinal components of both the electric and magnetic vectors. They are called HE or EH modes.

We begin with a scalar approach, according to which the distribution of light is described by the scalar Helmholtz equation:

$$\nabla^2 E(x,y,z) + k_0^2 \varepsilon E(x,y,z) = 0 \qquad (2.32)$$

where $E(x,y,z)$ is the complex amplitude of the light, which can be associated with any component of the electric and magnetic field vectors; $k_0 = \dfrac{2\pi}{\lambda_0}$; where λ_0 is the wavelength in a vacuum; ε the dielectric constant of the medium.

A solution of equation (2.32) for the rectangular region with a constant value of the refractive index $n = \sqrt{\varepsilon}$ is a function of the form

$$E = u(x)\varphi(y)e^{-ik_z z} . \qquad (2.33)$$

Substituting (2.33) into (2.32)

$$\nabla^2 (u(x)\varphi(y)e^{-ik_z z}) + k_0^2 \varepsilon u(x)\varphi(y)e^{-ik_z z} = 0 .$$

$$\nabla^2 (u(x)\varphi(y))e^{-ik_z z} - k_z^2 u(x)\varphi(y)e^{-ik_z z} + k_0^2 \varepsilon u(x)\varphi(y)e^{-ik_z z} = 0 .$$

$$\nabla^2 (u(x)\varphi(y)) - k_z^2 u(x)\varphi(y) + k_0^2 \varepsilon u(x)\varphi(y) = 0 .$$

$$\ddot{u}(x)\varphi(y) + \ddot{\varphi}(y)u(x) + (k_0^2 \varepsilon - k_z^2)u(x)\varphi(y) = 0 .$$

Let us omit the arguments x and y at the functions u and φ to shorten the records. So, we got

$$\ddot{u}\varphi + \ddot{\varphi}u + (k_0^2\varepsilon - k_z^2)u\varphi = 0 \qquad (2.34)$$

hereinafter points over the designation of the function denote the derivative.

Separation of variables divide (2.34) to $u\varphi$

$$\frac{\ddot{\varphi}}{\varphi} + \frac{\ddot{u}}{u} + (k_0^2\varepsilon - k_z^2) = 0 \qquad (2.35)$$

or

$$\frac{\ddot{\varphi}}{\varphi} + (k_0^2\varepsilon - k_z^2) = -\frac{\ddot{u}}{u}. \qquad (2.36)$$

Let

$$\frac{\ddot{u}}{u} = -k_x^2. \qquad (2.37)$$

Equation (2.36) can be divided into two equations:

$$\begin{cases} \ddot{\varphi} + (k_0^2\varepsilon - k_z^2 - k_x^2)\varphi = 0, \\ \ddot{u} + k_x^2 u = 0. \end{cases} \qquad (2.38)$$

Let

$$k_z^2 + k_x^2 = k_k^2 \qquad (2.39)$$

then

$$\begin{cases} \ddot{\varphi} + (k_0^2\varepsilon - k_k^2)\varphi = 0, \\ \ddot{u} + (k_k^2 - k_z^2)u = 0. \end{cases} \qquad (2.40)$$

We introduce another designation

$$k_0^2\varepsilon - k_k^2 = k_y^2 \qquad (2.41)$$

so now

$$k_0^2\varepsilon = k_x^2 + k_y^2 + k_z^2. \qquad (2.42)$$

That is, k_x, k_y, k_z are the projections of the wave vector $\mathbf{k}$ on the relevant axes; the modulus of this wave vector is $|\mathbf{k}| = \sqrt{k_0^2\varepsilon}$.

If the cross section of the fibre can be divided into N rows and M columns so that none of the rectangular cells of this partition contains irregularities (2.35), then the solution in each of these cells can be represented in a fairly simple manner.

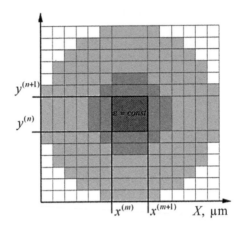

$y^{(n+1)}$

$y^{(n)}$

$\varepsilon = const$

$x^{(m)}$ $x^{(m+1)}$ X, μm

Figure 2.35. Schematic of the cross section of the fibre, dark gray area shows the value of the refractive index $n_{co} = 1.47$, light gray $n_{cl} = 1.463$, white $n_v = 1$.

Each cell located at the intersection of the *n*-th row and *m*-th column can be assigned a certain value of the dielectric constant $\varepsilon^{(m,n)}$ constant for a given cell. Let the coordinate axes are located as shown in Fig. 2.35. Then the thickness of the *n*-th line is

$$d_y^{(n)} = y^{(n+1)} - y^{(n)} \tag{2.43}$$

where $y^{(n)}$– the coordinate plane, $n - 1$ and n dividing lines. Similarly, the thickness $m-$ of the column

$$d_x^{(m)} = x^{(m+1)} - x^{(m)} \tag{2.44}$$

where $x^{(m)}$ is the coordinate plane dividing $m-1$ and m columns.

For a uniform rectangular portion (cell) of the section (Fig. 2.35), where the value of the dielectric constant is constant, $\varepsilon(x, y) = \varepsilon^{(m,n)} = const$ the first equation (2.40) takes the form

$$\ddot{\varphi}(y) + (k_0^2 \varepsilon^{(m,n)} - k_k^2)\varphi(y) = 0 \tag{2.45}$$

and its solution can be written as

$$\varphi(y) = \varphi_s^{(n,l)} \cos[k_y^{(n)}(y - y^{(n)})] + \frac{\varphi_a^{(n,l)}}{k_y^{(n)}} \sin[k_y^{(n)}(y - y^{(n)})] \tag{2.46}$$

where

$$k_y^{(n)} = \sqrt{\varepsilon^{(m,n)}k_0^2 - k_k^2} ; \tag{2.47}$$

$$\varphi_s^{(n,l)} = \varphi(y^{(n)} + 0) \tag{2.48a}$$

– bottom or left value of the function $\varphi(y)$ in the rectangular patch;

$$\varphi_a^{(n,l)} = \dot{\varphi}(y^{(n)} + 0) \qquad (2.48b)$$

– bottom or left value of the derivative $\dot{\varphi}(y)$ in the same cell section.

Similarly, (2.48a) and (2.48b) can determine the right values $\varphi(y)$ and its derivative:

$$\varphi_s^{(r)} = \varphi(y^{(n+1)} - 0) \qquad (2.48a)$$

$$\varphi_a^{(r)} = \dot{\varphi}(y^{(n+1)} - 0). \qquad (2.48b)$$

Using these values, it is possible to offer an alternative form of the solution (2.46) and its derivative, which is computationally more convenient when the quantity is $k_y^{(n)}$ purely imaginary:

$$\varphi(y) = (\varphi_s^{(n,l)} \sin[k_y^{(n)}(y^{(n+1)} - y)] + \varphi_s^{(n,r)} \sin[k_y^{(n)}(y - y^{(n)})]) / \sin(k_y^{(n)}d_y^{(n)}) \qquad (2.49a)$$

$$\dot{\varphi}(y) = (\varphi_a^{(n,l)} \sin[k_y^{(n)}(y^{(n+1)} - y)] + \varphi_a^{(n,r)} \sin[k_y^{(n)}(y - y^{(n)})]) / \sin(k_y^{(n)}d_y^{(n)}) . \qquad (2.49b)$$

Illustration of the location of $\varphi_s^{(n,l)}$, $\varphi_a^{(n,l)}$, $\varphi_s^{(n,r)}$, $\varphi_a^{(n,r)}$ is given in Fig. 2.36.

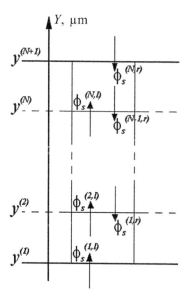

Fig. 2.36. Designation of the fields on the borders of partition of the section of the fibre.

The second equation of (2.40) in a homogeneous region takes the form

$$\ddot{u}_k^{(m)}(x) + (k_k^{(m)2} - k_z^2)u_k^{(m)}(x) = 0 \tag{2.50}$$

and its solution similarly to (2.46):

$$u_k^{(m)}(x) = u_{sk}^{(m,l)}\cos[k_{xk}^{(m)}(x - x^{(m)})] + \frac{u_{ak}^{(m,l)}}{k_{xk}^{(m)}}\sin[k_{xk}^{(m)}(x - x^{(m)})] \tag{2.51}$$

where

$$k_{xk}^{(m)} = \sqrt{k_k^{(m)2} - k_z^2} \tag{2.52}$$

$$u_{sk}^{(m,l)} = u_k^{(m)}(x^{(m)} + 0) \tag{2.53a}$$

$$u_{ak}^{(m,l)} = \dot{u}_k^{(m)}(x^{(m)} + 0). \tag{2.53b}$$

Similarly, we introduce notations:

$$u_{sk}^{(m,r)} = u_k^{(m)}(x^{(m+1)} - 0) \tag{2.53c}$$

$$u_{ak}^{(m,r)} = \dot{u}_k^{(m)}(x^{(m+1)} - 0). \tag{2.53d}$$

The presence of the index k at the function $u(x)$ in the expressions (2.50)–(2.53) will be explained hereinafter.

Thus, the Helmholtz equation in a homogeneous rectangular cell is satisfied by the product of two harmonic functions.

In accordance with the scalar approximation the modes of the given fibre structure are the solutions of the Helmholtz equation (2.32) in its cross section. Let us assume that the assumptions made earlier about the possibility of splitting the optical fibre section to a finite number of rectangular areas with a constant value of the refractive index of the medium are valid, that is, we consider the fibre with piecewise constant filling of the cross-section. At the edges of the cross section we assume the presence of electric or magnetic walls, providing the vanishing of the field function or its derivative with $x = 0$, $y = 0$, $x = x^{(M+1)}$, $y = y^{(N+1)}$.

In the scalar case the field in m-th column can be expressed as

$$\xi^{(m)}(x,y) = \sum_{k=1}^{\infty} u_k^{(m)}(x)\varphi_k^{(m)}(y) \tag{2.54}$$

here we omitted factor $\exp(-ik_z z)$. Each of the plurality of functions $u_k^{(m)}(x)$ satisfies the second equation and each of $\varphi_k^{(m)}(y)$ – the first equation (2.40) in the column m, that is, for both, $x^{(m)} \le x < x^{(m+1)}$ and $y^{(1)} \le y < y^{(N+1)}$, and in this area all of these functions are continuous together with their first derivatives. $u_k^{(m)}(x)$ and $\varphi_k^{(m)}(x)$ are the local sinusoidal modes, the index k determines the number of the local mode and is directly connected with the value k_k introduced by (2.39); their relationship will be shown later.

We take into account the proposed form of the solution of (2.45) in the form of a sinusoidal mode (2.46). Thus, to solve (2.45) it is necessary to determine k_x and $2N-2$ of the constants $\varphi_s^{(n,l)}$ and $\varphi_a^{(n,l)}$ so as the function $\varphi(y)$ satisfies (2.45) and the boundary conditions

$$\begin{cases} \varphi(y^{(1)}) = 0, \\ \varphi(y^{(N+1)}) = 0, \end{cases} \text{or} \quad \begin{cases} \dot{\varphi}(y^{(1)}) = 0, \\ \dot{\varphi}(y^{(N+1)}) = 0, \end{cases}, \text{respectively, the for electrical or}$$

magnetic walls.

The requirement of continuity of the solution (2.46) and its derivative leads to the following relations

$$\varphi_s^{(n,r)} = \varphi_s^{(n+1,l)} \tag{2.55a}$$

$$\varphi_a^{(n,r)} = \varphi_a^{(n+1,l)} \tag{2.55b}$$

In addition, for $\varphi_s^{(n,l)}$, $\varphi_a^{(n,l)}$, $\varphi_s^{(n,r)}$, $\varphi_a^{(n,r)}$ equalities:

$$\varphi_s^{(n,r)} = \varphi_s^{(n,l)} \cos[k_y^{(n)} d_y^{(n)}] + \frac{\varphi_a^{(n,l)}}{k_y^{(n)}} \sin[k_y^{(n)} d_y^{(n)}] \tag{2.56a}$$

$$\varphi_a^{(n,r)} = \varphi_a^{(n,l)} \cos[k_y^{(n)} d_y^{(n)}] - \varphi_s^{(n,l)} k_y^{(n)} \sin[k_y^{(n)} d_y^{(n)}] \tag{2.56b}$$

$$\varphi_s^{(n,l)} = \varphi_s^{(n,r)} \cos[k_y^{(n)} d^{(n)}] - \frac{\varphi_a^{(n,r)}}{k_y^{(n)}} \sin[k_y^{(n)} d^{(n)}] \tag{2.57a}$$

$$\varphi_a^{(n,l)} = \varphi_a^{(n,r)} \cos[k_y^{(n)} d^{(n)}] + \varphi_s^{(n,l)} k_y^{(n)} \sin[k_y^{(n)} d^{(n)}]. \tag{2.57b}$$

We consider the matrix the determinant of which is equal to one

$$\mathbf{P}^{(i)} = \begin{bmatrix} \cos[k_y^{(i)} d^{(i)}] & \sin[k_y^{(i)} d^{(i)}]/k_y^{(i)} \\ -\sin[k_y^{(i)} d^{(i)}] k_y^{(i)} & \cos[k_y^{(i)} d^{(i)}] \end{bmatrix}, \tag{2.58}$$

$$\mathbf{Q}^{(i)} = \begin{bmatrix} \cos[k_y^{(i)} d^{(i)}] & -\sin[k_y^{(i)} d^{(i)}]/k_y^{(i)} \\ \sin[k_y^{(i)} d^{(i)}] k_y^{(i)} & \cos[k_y^{(i)} d^{(i)}] \end{bmatrix}. \tag{2.59}$$

The validity of the equalities $\begin{bmatrix} \varphi_s^{(n,r)} \\ \varphi_a^{(n,r)} \end{bmatrix} = \mathbf{P}^{(n)} \begin{bmatrix} \varphi_s^{(n,l)} \\ \varphi_a^{(n,l)} \end{bmatrix}$ and

$\begin{bmatrix} \varphi_s^{(n,l)} \\ \varphi_a^{(n,l)} \end{bmatrix} = \mathbf{Q}^{(n)} \begin{bmatrix} \varphi_s^{(n,r)} \\ \varphi_a^{(n,r)} \end{bmatrix}$ is obvious; they are equivalent to (2.56) and

(2.57). Using the matrices $\mathbf{P}^{(i)}$ and $\mathbf{Q}^{(i)}$, we can easily express through

the values $\begin{bmatrix} \varphi_s^{(j,r)} \\ \varphi_a^{(j,r)} \end{bmatrix}$ through $\begin{bmatrix} \varphi_s^{(i,l)} \\ \varphi_a^{(i,l)} \end{bmatrix}$, $i < j$, and the values $\begin{bmatrix} \varphi_s^{(j,l)} \\ \varphi_a^{(j,l)} \end{bmatrix}$

through $\begin{bmatrix} \varphi_s^{(i,r)} \\ \varphi_a^{(i,r)} \end{bmatrix}$, $i \geq j$

$$\begin{bmatrix} \varphi_s^{(n',r)} \\ \varphi_a^{(n',r)} \end{bmatrix} = \mathbf{P}^{(n')}\mathbf{P}^{(n'-1)} \cdots \mathbf{P}^1 \begin{bmatrix} \varphi_s^{(1,l)} \\ \varphi_a^{(1,l)} \end{bmatrix} \tag{2.60}$$

$$\begin{bmatrix} \varphi_s^{(n'+1,l)} \\ \varphi_a^{(n'+1,l)} \end{bmatrix} = \mathbf{Q}^{(n'+1)}\mathbf{Q}^{(n'+2)} \cdots \mathbf{Q}^{(N)} \begin{bmatrix} \varphi_s^{(N,r)} \\ \varphi_a^{(N,r)} \end{bmatrix}. \tag{2.61}$$

For numerical calculation of k_k or, more accurately, the square of this value since only the square appears in the above formulas, we use the following method. We define the function on the basis of the conditions (2.55) in the border of the areas:

$$\Delta^{(n)}(k_k^2) = \varphi_s^{(n',r)}\varphi_a^{(n'+1,l)} - \varphi_a^{(n',r)}\varphi_s^{(n'+1,l)} \tag{2.62}$$

wherein $n' \in (1,N)$, the values $\varphi_s^{(n',r)}$, $\varphi_a^{(n',r)}$ and $\varphi_s^{(n'+1,l)}$, $\varphi_a^{(n'+1,l)}$ are calculated using (2.60) and (2.61) and therefore depend on k_k^2. The zeros of the function $\Delta^{(n)}(k_k^2)$ corresponding to the equality of he left and right values of the local mode at the boundary of $n'+1$ and n' partition lines, are the desired values. The search of zero of (2.62) is equivalent to the solution of the characteristic equation

$$\varphi_s^{(n',r)}\varphi_a^{(n'+1,l)} - \varphi_a^{(n',r)}\varphi_s^{(n'+1,l)} = 0 \tag{2.63}$$

and n' it may be an arbitrary number in the range $(1,N)$.

It is not difficult to make sure that the dependence of the solution

(2.63) of the non-zero values, $\begin{bmatrix} \varphi_s^{(1,l)} \\ \varphi_a^{(1,l)} \end{bmatrix}$ and $\begin{bmatrix} \varphi_s^{(N,r)} \\ \varphi_a^{(N,r)} \end{bmatrix}$ does not exist,

so solving (2.63), in the case of electric walls we can assume

$$\begin{bmatrix} \varphi_s^{(1,l)} \\ \varphi_a^{(1,l)} \end{bmatrix} = \begin{bmatrix} 0 \\ 1 \end{bmatrix}, \quad \begin{bmatrix} \varphi_s^{(N,r)} \\ \varphi_a^{(N,r)} \end{bmatrix} = \begin{bmatrix} 0 \\ 1 \end{bmatrix},$$ and in the case of magnetic walls

$$\begin{bmatrix} \varphi_s^{(1,l)} \\ \varphi_a^{(1,l)} \end{bmatrix} = \begin{bmatrix} 1 \\ 0 \end{bmatrix}, \quad \begin{bmatrix} \varphi_s^{(N,r)} \\ \varphi_a^{(N,r)} \end{bmatrix} = \begin{bmatrix} 1 \\ 0 \end{bmatrix}.$$

Different values k_k^2 satisfying (2.63) are infinitely many, but they are all on the interval $(-\infty, k_{max}^2]$, where they can be sorted in descending order. Each of these values k_k^2 defines a local mode $\varphi_k^{(m)}(y)$ in the m-th column of the section of the fibre. The accuracy of constructing the field by the formula (2.64) depends on the number of local modes. But no matter how many of these modes we might consider (how many roots of (2.63) are found), it will always be a finite number K.

Search for the roots (2.63) is complicated by the fact that their distribution within the interval $(-\infty, k_{max}^2]$ is irregular. For some structures of the columns (especially with multiple alternation of fragments with different refractive indices) two close roots of the characteristic equation (2.63) can differ by the order 10^{-11}. Skipping the roots and, consequently, the relevant local modes in the expansion (2.64) can lead to a significant distortion of the results.

Avoid missing the roots and reducing the number of calculations allows to use the following simple algorithm for adaptive selection of the step in the localization of the roots of the characteristic equation. The initial (random) step value L is defined. Then, on the real axis, starting at k_{max}^2, we examine sequentially the segments of length L in the direction of decreasing. At each of these points P_1 we the calculate values of the left side of equation (2.63). For the set of the resultat values we then determined the statistical properties such as variance and expectation. In the same interval we then determine the values of the left part of the equation at the points P_2 and for the points $P_1 + P_2$ we determine the same statistical characteristics. If the relative change of the expectation and variance does not exceed a predetermined threshold, we localize the roots on the basis of the determined values $P_1 + P_2$, otherwise the original segment is divided into parts and all the operations are repeated for each part. This recursive procedure improves the accuracy where it is needed.

When all K values of k_k^2 are found, we are interested in constants $\varphi_s^{(n,l)}$, $\varphi_a^{(n,l)}$, $\varphi_s^{(n,r)}$, $\varphi_a^{(n,r)}$ which can be found from the normalization of function $\varphi(y)$ with respect to unity:

$$\frac{1}{y^{(N+1)} - y^{(1)}} \int_{y^{(1)}}^{y^{(N+1)}} \varphi^2(y)dy = 1 .$$ (2.64)

Here

$$I = \int_{y^{(1)}}^{y^{(N+1)}} \phi^2(y)dy = \sum_{n=1}^{N} \int_{y^{(n)}}^{y^{(n+1)}} \phi^2(y)dy =$$

$$= \sum_{n=1}^{N} [\frac{1}{2(k_y^{(n)})^2} (\phi_s^{(n,l)} \phi_a^{(n,l)} - \phi_s^{(n,r)} \phi_a^{(n,r)}) + \frac{d^{(n)}}{2} ((\phi_s^{(n,l)})^2 + (\frac{\phi_a^{(n,l)}}{k_y^{(n)}})^2)] .$$

When we know k_k^2 the value $k_y^{(n)}$ for each row of the partition can be calculated using the formula (2.47). Taking into account the relations

$$\begin{bmatrix} \varphi_s^{(n',r)} \\ \varphi_a^{(n',r)} \end{bmatrix} = \mathbf{P}^{(n')} \mathbf{P}^{(n'-1)} \cdots \mathbf{P}^1 \begin{bmatrix} \varphi_s^{(1,l)} \\ \varphi_a^{(1,l)} \end{bmatrix}, \quad n' = \overline{1, N}$$ (2.65)

$$\begin{bmatrix} \varphi_s^{(n,l)} \\ \varphi_a^{(n,l)} \end{bmatrix} = \begin{bmatrix} \varphi_s^{(n-1,r)} \\ \varphi_a^{(n-1,r)} \end{bmatrix}, \quad n = \overline{2, N}$$ (2.66)

the value of the integral I can be considered a function of $\varphi_s^{(1,l)}$ or $\varphi_a^{(1,l)}$ depending on the type of boundary. Then (2.64) can be solved by standard numerical methods, such as Newton's method, and has exactly two real roots, differing only in the sign, and any of them can be selected as the value $\varphi_s^{(1,l)}$ (or $\varphi_a^{(1,l)}$), and the other constants are calculated using (2.65), (2.66) and the resultant value.

Acting on the above algorithm, for each column of the fibre cross-section it is possible to find K values of k_k^2 and for each of them to construct a function $\varphi(y)$ that satisfies (2.45) for this k_k^2.

When the local y-modes are found, the number of x-modes of type (2.51) is not sufficient for constructing the field (2.54). The function $u_k^{(m)}(x)$ is the k-th mode the column m, and corresponds to $k_k^{(m)}$ as $\varphi_k^{(m)}(y)$.

The difference between equations (2.45) and (2.50) is that the solution of (2.50) is not a continuous function of the form (2.51) and it is a set M of defined and continuous functions at appropriate intervals. Formula (2.51) combines them into one but a discontinuous function, therefore the relationship similar to (2.55) are not fulfilled for $u_{sk}^{(m,l)}$, $u_{sk}^{(m,r)}$ and $u_{ak}^{(m,l)}$ $u_{ak}^{(m,r)}$, however, due to the requirement of the continuity of the field (2.54) at the boundary of the partition column, the following equalities hold

$$u_{ak}^{(m,l)} = -\frac{k_{xk}^{(m)} u_{sk}^{(m,l)}}{tg(k_{kx}^{(m)} d_x^{(m)})} + \frac{k_{xk}^{(m)} u_{sk}^{(m,r)}}{\sin(k_{kx}^{(m)} d_x^{(m)})} \qquad (2.67a)$$

$$u_{ak}^{(m,r)} = -\frac{k_{xk}^{(m)} u_{sk}^{(m,l)}}{\sin(k_{kx}^{(m)} d_x^{(m)})} + \frac{k_{xk}^{(m)} u_{sk}^{(m,r)}}{tg(k_{kx}^{(m)} d_x^{(m)})} \qquad (2.67b)$$

Because the relationships (2.67) are satisfied for all $k = \overline{1,K}$, it is convenient to use the matrix notation. We introduce the diagonal matrices $\mathbf{T}^{(m)}$ and $\mathbf{S}^{(m)}$ with the dimension $K \times K$ with the diagonal elements

$$T_{kk}^{(m)} = k_{xk}^{(m)} / tg(k_{kx}^{(m)} d_x^{(m)}) \qquad (2.68a)$$

$$S_{kk}^{(m)} = k_{xk}^{(m)} / \sin(k_{kx}^{(m)} d_x^{(m)}). \qquad (2.68b)$$

Now (2.67) takes the form

$$\mathbf{U}_a^{(m,l)} = -\mathbf{T}^{(m)} \mathbf{U}_s^{(m,l)} + \mathbf{S}^{(m)} \mathbf{U}_s^{(m,r)} \qquad (2.69a)$$

$$\mathbf{U}_a^{(m,r)} = -\mathbf{S}^{(m)} \mathbf{U}_s^{(m,l)} + \mathbf{T}^{(m)} \mathbf{U}_s^{(m,r)} \qquad (2.69b)$$

Here

$$\mathbf{U}_a^{(m,l)} = [u_{a1}^{(m,l)} \quad u_{a2}^{(m,l)} \quad \cdots \quad u_{aK}^{(m,l)}]^{\mathrm{T}},$$

$$\mathbf{U}_a^{(m,r)} = [u_{a1}^{(m,r)} \quad u_{a2}^{(m,r)} \quad \cdots \quad u_{aK}^{(m,r)}]^{\mathrm{T}},$$

$$\mathbf{U}_s^{(m,l)} = [u_{s1}^{(m,l)} \quad u_{s2}^{(m,l)} \quad \cdots \quad u_{sK}^{(m,l)}]^{\mathrm{T}},$$

$$\mathbf{U}_s^{(m,r)} = [u_{s1}^{(m,r)} \quad u_{s2}^{(m,r)} \quad \cdots \quad u_{sK}^{(m,r)}]^{\mathrm{T}}.$$

To build relations connecting the left and right values of the x-modes from the adjacent the columns of the section, we consider the overlap integral

$$< f_1 | f_2 > = \frac{1}{y^{(N+1)} - y^{(1)}} \int_{y^{(1)}}^{y^{(N+1)}} f_1(y) f_2(y) dy \qquad (2.70)$$

and form the matrix from the elements of the type (2.70), where the f_1 and f_2 ar represented by the previously determined modes $\varphi_k^{(m)}(y)$:

$$O_{pq}^{(m,m')} = < \varphi_q^m | \varphi_p^{m'} > =$$

$$\sum_{n=1}^{N}(\varphi_{sq}^{(n,r)(m)}\varphi_{ap}^{(n,r)(m')} - \varphi_{aq}^{(n,r)(m)}\varphi_{sp}^{(n,r)(m')} - \varphi_{sq}^{(n,l)(m)}\varphi_{ap}^{(n,l)(m')} + \varphi_{aq}^{(n,l)(m)}\varphi_{sp}^{(n,l)(m')}) / (k_{kq}^{(m)2} - k_{kp}^{(m')2}) =$$

$$\frac{\sum_{n=1}^{N}(\varphi_{sq}^{(n,r)(m)}\varphi_{ap}^{(n,r)(m')} - \varphi_{aq}^{(n,r)(m)}\varphi_{sp}^{(n,r)(m')} - \varphi_{sq}^{(n,l)(m)}\varphi_{ap}^{(n,l)(m')} + \varphi_{aq}^{(n,l)(m)}\varphi_{sp}^{(n,l)(m')})}{(k_{kq}^{(m)} + k_{kp}^{(m')})(k_{kq}^{(m)} - k_{kp}^{(m')})}. \tag{2.71}$$

We obtain the square matrix $O^{(m,m')}$ with the size $K \times K$ and the property

$$(\mathbf{O}^{(m,m')})^{\mathrm{T}} = \mathbf{O}^{(m',m)}, \tag{2.72}$$

where the symbol 'T' denotes transposition.

Furthermore, since orthogonal functions $\varphi_k^{(m)}(y)$ are normalized, each of the matrices $O^{(m,m')}$ is unitary:

$$\mathbf{O}^{(m,m')}\mathbf{O}^{(m',m)} = \begin{bmatrix} 1 & & & & \\ & 1 & & 0 & \\ & & \ddots & & \\ & 0 & & 1 & \\ & & & & 1 \end{bmatrix}. \tag{2.73}$$

In the case of finite values of the dimension $K \times K$ the ratio (2.73) is satisfied with sufficient accuracy only for the adjacent columns, that is if $m' = m+1$ or $m' = m-1$. From the orthonormality of the functions $\varphi_k^{(m)}(y)$ and requirements of the continuity of the field $\xi(x,y)$, from (2.54), together with the derivatives normal to the interface of the columns the following conditions are fulfilled:

$$u_{sk}^{(m,r)} = \sum_{p=1}^{K} O_{kp}^{(m,m+1)} u_{sp}^{(m+1,l)} \tag{2.74a}$$

$$u_{ak}^{(m,r)} = \sum_{p=1}^{K} O_{kp}^{(m,m+1)} u_{ap}^{(m+1,l)}. \tag{2.74b}$$

Or in the matrix notation

$$\mathbf{U}_s^{(m,r)} = \mathbf{O}^{(m,m+1)}\mathbf{U}_s^{(m+1,l)} \tag{2.75a}$$

$$\mathbf{U}_a^{(m,r)} = \mathbf{O}^{(m,m+1)}\mathbf{U}_a^{(m+1,l)} \tag{2.75b}$$

Combining (2.69) and (2.75) and excluding $\mathbf{U}_a^{(m,l)}$, $\mathbf{U}_a^{(m,r)}$, $\mathbf{U}_s^{(m,r)}$ we obtain

$$(\mathbf{O}^{(m,m-1)}\mathbf{T}^{(m-1)}\mathbf{O}^{(m-1,m)} + \mathbf{T}^{(m)})\mathbf{U}_s^{(m,l)} =$$
$$\mathbf{O}^{(m,m-1)}\mathbf{S}^{(m-1)}\mathbf{U}_s^{(m-1,l)} + \mathbf{S}^{(m)}\mathbf{O}^{(m,m+1)}\mathbf{U}_s^{(m+1,l)} \qquad (2.76a)$$

for $2 < m < M$;

$$(\mathbf{O}^{(2,1)}\mathbf{T}^{(1)}\mathbf{O}^{(1,2)} + \mathbf{T}^{(2)})\mathbf{U}_s^{(2,l)} = \mathbf{S}^{(2)}\mathbf{O}^{(2,3)}\mathbf{U}_s^{(3,l)} \qquad (2.76b)$$

for $m = 2$ when $M > 2$;

$$(\mathbf{O}^{(M,M-1)}\mathbf{T}^{(M-1)}\mathbf{O}^{(M-1,M)} + \mathbf{T}^{(M)})\mathbf{U}_s^{(M,l)} =$$
$$\mathbf{O}^{(M,M-1)}\mathbf{S}^{(M-1)}\mathbf{U}_s^{(M-1,l)} \qquad (2.76c)$$

for $m = M$.

In the simplest case, when $M = 2$, we obtain

$$(\mathbf{O}^{(1,2)}\mathbf{T}^{(2)} + \mathbf{T}^{(1)}\mathbf{O}^{(1,2)})\mathbf{U}_s^{(2,l)} = 0. \qquad (2.76d)$$

Equations (2.76a)–(2.76d) are fulfilled in the case of electrical walls when the boundary conditions are:

$$\begin{cases} \mathbf{U}_s^{(1,l)} = [0 \quad 0 \quad \cdots \quad 0]^{\mathrm{T}}, \\ \mathbf{U}_s^{(M,r)} = [0 \quad 0 \quad \cdots \quad 0]^{\mathrm{T}}. \end{cases}$$

If the wall are magnetic, i.e. the boundary conditions:

$$\begin{cases} \mathbf{U}_a^{(1,l)} = [0 \quad 0 \quad \cdots \quad 0]^{\mathrm{T}}, \\ \mathbf{U}_a^{(M,r)} = [0 \quad 0 \quad \cdots \quad 0]^{\mathrm{T}}, \end{cases}$$

then for $m = 2$ when $M = 2$:

$$(\mathbf{O}^{(2,1)}(\mathbf{T}^{(1)} - \mathbf{S}^{(1)}(\mathbf{T}^{(1)})^{-1}\mathbf{S}^{(1)})\mathbf{O}^{(1,2)} + \mathbf{T}^{(2)})\mathbf{U}_s^{(2,l)} = \mathbf{S}^{(2)}\mathbf{O}^{(2,3)}\mathbf{U}_s^{(3,l)} \qquad (2.76d)$$

and for $m = M$:

$$(\mathbf{O}^{(M,M-1)}\mathbf{T}^{(M-1)}\mathbf{O}^{(M-1,M)} + \mathbf{T}^{(M)} - \mathbf{S}^{(M)}(\mathbf{T}^{(M)})^{-1}\mathbf{S}^{(M)})\mathbf{U}_s^{(M,l)} =$$
$$\mathbf{O}^{(M,M-1)}\mathbf{S}^{(M-1)}U_s^{(M-1,l)}. \qquad (2.76e)$$

In the simplest case, however, for $M = 2$ for the magnetic boundary conditions, we have:

$$\mathbf{O}^{(2,1)}[\mathbf{T}^{(1)} - \mathbf{S}^{(1)}(\mathbf{T}^{(1)})^{-1}\mathbf{S}^{(1)}]\mathbf{O}^{(1,2)} - \mathbf{S}^{(2)}(\mathbf{T}^{(2)})^{-1}\mathbf{S}^{(2)})\mathbf{U}_s^{(2,l)} = 0. \qquad (2.76f)$$

The problem, described by the relation (2.76), in the general form can be written as

$$\Lambda(k_z)U = 0 \tag{2.77}$$

wherein matrix $\Lambda(k_z)$ consists of $(M-1)\times(M-1)$ blocks each with the dimension $K\times K$. The structure of the matrix is of the form:

$$
\Lambda(k_z) =
\begin{bmatrix}
\mathbf{A}^{(2)} & \mathbf{C}^{(2)} & \mathbf{O} & \mathbf{O} & \mathbf{O} & \mathbf{O} & \mathbf{O} & \mathbf{O} \\
\mathbf{B}^{(3)} & \mathbf{A}^{(3)} & \mathbf{C}^{(3)} & \mathbf{O} & \mathbf{O} & \ddot{\mathbf{Y}} & \mathbf{O} & \mathbf{O} \\
\ddot{\mathbf{Y}} & \mathbf{B}^{(4)} & \mathbf{A}^{(4)} & \mathbf{C}^{(4)} & \mathbf{O} & \mathbf{O} & \mathbf{O} & \mathbf{O} \\
 & & \mathbf{O} & & & & & \\
\mathbf{O} & \mathbf{O} & \mathbf{O} & \mathbf{O} & \mathbf{O} & \mathbf{B}^{(M-1)} & \mathbf{A}^{(M-1)} & \mathbf{C}^{(M-1)} \\
\mathbf{O} & \mathbf{O} & \mathbf{O} & \mathbf{O} & \mathbf{O} & \mathbf{O} & \mathbf{B}^{(M)} & \mathbf{A}^{(M)}
\end{bmatrix}. \tag{2.78}
$$

Here

$$\mathbf{A}^{(m)} = \mathbf{O}^{(m,m-1)}\mathbf{T}^{(m-1)}\mathbf{O}^{(m-1,m)} + \mathbf{T}^{(m)} \tag{2.79a}$$

in the case of magnetic walls the expressions for $\mathbf{A}^{(2)}$ and $\mathbf{A}^{(M)}$ do not fit the general scheme, so they should be written separately:

$$\mathbf{A}^{(2)} = \mathbf{O}^{(2,1)}(\mathbf{T}^{(1)} - \mathbf{S}^{(1)}(\mathbf{T}^{(1)})^{-1}\mathbf{S}^{(1)})\mathbf{O}^{(1,2)} + \mathbf{T}^{(2)} \tag{2.79b}$$

$$\mathbf{A}^{(M)} = \mathbf{O}^{(M,M-1)}\mathbf{T}^{(M-1)}\mathbf{O}^{(M-1,M)} + \mathbf{T}^{(M)} - \mathbf{S}^{(M)}(\mathbf{T}^{(M)})^{-1}\mathbf{S}^{(M)}. \tag{2.79c}$$

$$\mathbf{B}^{(m)} = -\mathbf{O}^{(m,m-1)}\mathbf{S}^{(m-1)} \tag{2.80}$$

$$\mathbf{C}^{(m)} = -\mathbf{S}^{(m)}\mathbf{O}^{(m,m-1)} \tag{2.81}$$

$\mathbf{O}$ is the zero matrix of dimension $K\times K$.

In (2.77), the vector

$$
\mathbf{U} =
\begin{bmatrix}
U_s^{(2,l)} \\
U_s^{(3,l)} \\
\vdots \\
U_s^{(M,l)}
\end{bmatrix}. \tag{2.82}
$$

We have a problem (2.77), which has only the trivial solution $\mathbf{U} = 0$, $\det(\Lambda(k_z)) \neq 0$. The value of the parameter k_z for which there is no trivial solution is called the eigenvalue of the matrix $\Lambda(k_z)$. To find these values, we can use the iterative Krylov method, which will be discussed later.

For each of the obtained eigenvalues k_z the matrix $\Lambda(k_z)$ becomes numeric, and determining eigenvector $\mathbf{U}$, we get some of the values of the constants required to construct K functions of the type (2.51). The rest can be found from (2.69) and (2.75), in the following order.

1. For $m=M$

$U_s^{(M,l)}$ is known as the part U.

1.1. If the walls are electrical then

$$U_s^{(M,r)} = [0 \quad 0 \quad \cdots \quad 0]^T \text{ then}$$

$$U_a^{(M,l)} = -T^{(M)}U_s^{(M,l)} \text{ and } U_a^{(M,r)} = -S^{(M)}U_s^{(M,l)}.$$

12. If the walls are magnetic, then

$$U_a^{(M,r)} = [0 \quad 0 \quad \cdots \quad 0]^T \text{ then}$$

$$U_a^{(M,l)} = (-T^{(M)} + S^{(M)}S^{(M)}(T^{(M)})^{-1}U_s^{(M,l)} \text{ and } U_s^{(M,r)} = (T^{(M)})^{-1}S^{(M)}U_s^{(M,l)}$$

2. $1 \le m < M$.

2.1. $U_s^{(m,r)} = O^{(m,m+1)}U_s^{(m+1,l)}$.

2.2. $U_a^{(m,r)} = O^{(m,m+1)}U_a^{(m+1,l)}$.

2.3. $U_s^{(m,l)} = (S^{(m)})^{-1}(-U_a^{(m,r)} + T^{(m)}U_s^{(m,r)})$ – this item is fulfilled only for $m = 1$ because for other values m the vector $U_s^{(m,l)}$ is known as part of eigenvector U.

2.4. $U_a^{(m,l)} = -T^{(m)}U_s^{(m,l)} + S^{(m)}U_s^{(m,r)}$.

Thus, for each value k_z we can construct K modes of the type (2.51) and combining them with K modes of the form (2.46) (they are the same for different k_z) from the formula (2.54), we obtain the field extending in the axial direction z, with the projection of the wave vector k_z on this axis.

2.4.2. Method of matched sinusoidal modes in the vector case

The principal difference between the vector case and the scalar case is that we must consider the local modes of two different polarizations – TE and TM, as they both contribute to the formation of the hybrid fibre mode. Accordingly, expression (6.54) is transformed as follows:

$$F^{(m)}(x,y) = \sum_{p=e,h} \sum_{k=1}^{\infty} [u_{pk}^{(m)}(x)F_{spk}^{(m)}(y) + \dot{u}_{pk}^{(m)}(x)F_{apk}^{(m)}(y)] \qquad (2.83)$$

where F refers to any of the electric or magnetic components of the mode field and the outer sum corresponds to summation over the polarizations: TE $-p = h$, TM $-p = e$. Let us consider the local modes in the expression (2.83). The local x-mode is now as follows:

$$u_{pk}^{(m)}(x) = \left(\frac{k_0}{k_{pk}^{(m)}}\right)^2 u_{spk}^{(m,l)} \cos[k_{xpk}^{(m)}(x-x^{(m)})] +$$

$$\frac{u_{apk}^{(m,l)}}{k_{xpk}^{(m)}} \sin[k_{xpk}^{(m)}(x-x^{(m)})] \qquad (2.84)$$

Here we take into account polarization and introduced factor $\left(\dfrac{k_0}{k_{pk}^{(m)}}\right)^2$, the need for which will be explained below. The expressions for $F_{sp}^{(m)}$ and $F_{ap}^{(m)}$ are defined in Table 2.2, in terms of local y-modes.

Table 2.2 and subsequent relations use the following symbols: $B_x = \mu_0 H_x$, $B_y = \mu_0 H_y$, $B_z = \mu_0 H_z$, where μ_0 is the magnetic permeability of free space.

Thus, the expression (2.83) takes the well-defined form for each of the components of the vector $\bar{E}(x,y)$ and $c\bar{B}(x,y) = c\mu_0\bar{H}(x,y)$:

$$E_x^{(m)}(x,y) = \sum_{k=1}^{\infty} u_{hk}^{(m)}(x)\frac{k_z}{k_0}\phi_k^{(m)}(y) - \sum_{k=1}^{\infty} \dot{u}_{ek}^{(m)}(x)\frac{\dot{\psi}_k^{(m)}(y)}{k_0^2\varepsilon^{(m)}(y)} \qquad (2.85a)$$

$$E_y^{(m)}(x,y) = -\sum_{k=1}^{\infty} u_{ek}^{(m)}(x)\left(\frac{k_{ek}^{(m)}}{k_0}\right)^2 \frac{\psi_k^{(m)}(y)}{\varepsilon^{(m)}(y)} \qquad (2.85b)$$

$$E_z^{(m)}(x,y) = -\sum_{k=1}^{\infty} \dot{u}_{hk}^{(m)}(x)\frac{i\phi_k^{(m)}(y)}{k_0} + \sum_{k=1}^{\infty} u_{ek}^{(m)}(x)\frac{ik_z}{k_0^2}\frac{\dot{\psi}_k^{(m)}}{\varepsilon^{(m)}(y)} \qquad (2.85c)$$

$$cB_x^{(m)}(x,y) = \sum_{k=1}^{\infty} \dot{u}_{hk}^{(m)}(x)\frac{\dot{\phi}_k^{(m)}(y)}{k_0^2} + \sum_{k=1}^{\infty} u_{ek}^{(m)}(x)\frac{k_z}{k_0}\psi_k^{(m)}(y) \qquad (2.85d)$$

$$cB_y^{(m)}(x,y) = \sum_{k=1}^{\infty} u_{hk}^{(m)}(x)\left(\frac{k_{hk}^{(m)}}{k_0}\right)^2 \phi_k^{(m)}(y) \qquad (2.85e)$$

$$cB_z^{(m)}(x,y) = -\sum_{k=1}^{\infty} u_{hk}^{(m)}(x)\frac{ik_z}{k_0^2}\dot{\phi}_k^{(m)}(y) - \sum_{k=1}^{\infty} \dot{u}_{ek}^{(m)}(x)\frac{i\psi_k^{(m)}(y)}{k_0} . \qquad (2.85f)$$

Table 2.2. Symmetric and antisymmetric y-components of the field, expressed through local modes

F	TE		TM	
	$F_{sh}^{(m)}$	$F_{ah}^{(m)}$	$F_{se}^{(m)}$	$F_{ae}^{(m)}$
E_x	$\left(\dfrac{k_z}{k_0}\right)\phi^{(m)}(y)$	0	0	$-\dfrac{\psi^{(m)}(y)}{k_0^2\mu^{(m)}(y)}$
E_y	0	0	$-\left(\dfrac{k_{ck}^{(m)}}{k_0}\right)\dfrac{\dot\psi^{(m)}(y)}{\mu^{(m)}(y)}$	0
E_z	0	$-\dfrac{i\phi^{(m)}(y)}{k_0}$	$\left(\dfrac{ik_z}{k_0^2}\right)\dfrac{\dot\psi^{(m)}(y)}{\mu^{(m)}(y)}$	0
cB_x	0	$-\dfrac{\dot\phi^{(m)}(y)}{k_0^2}$	$\left(\dfrac{k_z}{k_0}\right)\psi^{(m)}(y)$	0
cB_y	$\left(\dfrac{k_{hk}^{(m)}}{k_0}\right)^2\phi^{(m)}(y)$	0	0	0
cB_z	$\left(\dfrac{-ik_z}{k_0^2}\right)\dot\phi^{(m)}(y)$	0	0	$-\dfrac{i\psi^{(m)}(y)}{k_0}$

These are the components of the hybrid mode.

Table 2.2 and expressions (2.85) $\varphi(y)$– considered in detail in paragraph 2.4.1 local fashion $-y$, in this case, the relevant TE polarization. Mode $\psi(y)$, representing the TM polarization are the same (2.46) in the form of a column m:

$$\psi(y)=\psi_s^{(n,l)}\cos[k_y^{(n)}(y-y^{(n)})]+\frac{\psi_a^{(n,l)}}{k_y^{(n)}}\sin[k_y^{(n)}(y-y^{(n)})] \quad (2.86)$$

but the conditions (2.55) at the interface between the homogeneous regions for them take a different form:

$$\psi_s^{(n,r)}=\psi_s^{(n+1,l)} \quad (2.87a)$$

$$\frac{\psi_a^{(n,r)}}{\varepsilon^{(n)}}=\frac{\psi_a^{(n+1,l)}}{\varepsilon^{(n+1)}}. \quad (2.87b)$$

This involves changes in the matrix relations (2.60) and (2.61) for the modes $\varphi(y)$, they look like this:

$$\begin{bmatrix} \psi_s^{(n',r)} \\ \psi_a^{(n',r)} \end{bmatrix} = \mathbf{P}^{(n')}\mathbf{W}^{(n')}\mathbf{P}^{(n'-1)}\mathbf{W}^{(n'-1)}\cdots\mathbf{W}^{(2)}P^1\begin{bmatrix} \psi_s^{(1,l)} \\ \psi_a^{(1,l)} \end{bmatrix} \qquad (2.88)$$

$$\begin{bmatrix} \psi_s^{(n'+1,l)} \\ \psi_a^{(n'+1,l)} \end{bmatrix} = \mathbf{Q}^{(n'+1)}\mathbf{V}^{(n'+1)}\mathbf{Q}^{(n'+2)}\mathbf{V}^{(n'+2)}\cdots\mathbf{V}^{(N-1)}\mathbf{Q}^{(N)}\begin{bmatrix} \psi_s^{(N,r)} \\ \psi_a^{(N,r)} \end{bmatrix}. \qquad (2.89)$$

Here $\mathbf{P}^{(i)}$ and $\mathbf{Q}^{(i)}$ are as previously matrices of the form (2.58) and (2.59) respectively;

$$\mathbf{W}^{(i+1)} = \begin{bmatrix} 1 & 0 \\ 0 & \dfrac{\varepsilon^{(i+1)}}{\varepsilon^{(i)}} \end{bmatrix}, i = \overline{1, N-1} ; \qquad (2.90)$$

$$\mathbf{V}^{(i)} = \begin{bmatrix} 1 & 0 \\ 0 & \dfrac{\varepsilon^{(i)}}{\varepsilon^{(i+1)}} \end{bmatrix}, i = \overline{1, N-1} . \qquad (2.91)$$

It is obvious that the matrices $\mathbf{W}^{(i+1)}$ and $\mathbf{V}^{(i)}$ are reciprocal and depend solely on the dielectric structure of the cross section of the waveguide. The relations (2.88) and (2.89), as well as their analogues for TE polarization, are used to calculate the quantities in the characteristic equation for k_k^2:

$$\varepsilon^{(n')}\psi_s^{(n',r)}\psi_a^{(n'+1,l)} - \varepsilon^{(n'+1)}\psi_a^{(n',r)}\psi_s^{(n'+1,l)} = 0 . \qquad (2.92)$$

There is no doubt the fact that, in general, the roots of (2.91) and (2.92) are different, and because they characterize the local y-modes of different polarizations, we must enter the appropriate designations to avoid confusion. Let k_{hk}^2 be the set of solutions of (2.63) corresponding to the TE case, and k_{ek}^2 the solution of (2.92), describing the local TM modes.

Changes are also mode in the formula for calculating the normalization integral:

$$\frac{1}{y^{(N+1)} - y^{(1)}} \int_{y^{(1)}}^{y^{(N+1)}} \frac{\psi^2(y)}{\varepsilon(y)} dy = 1 . \qquad (2.93)$$

Thus, when calculating the vector fields the problem is complicated exactly twice. Now the algorithm for finding local y-modes is as follows:

- For each column of the section of the fibre is necessary to find K roots k_{hk}^2 of (2.63), defining the modes $\varphi(y)$ with the given boundary conditions (electric or magnetic walls);
- Then calculate the constants $\varphi_s^{(n,l)}$, $\varphi_a^{(n,l)}$, $\varphi_s^{(n,r)}$, $\varphi_a^{(n,r)}$ based on the normalization condition (2.64) to finally form K y-modes of the type (2.46), corresponding to the case of TE-polarization;
- For each column section of the fibre we need to find exactly the same number K of the roots k_{ek}^2 of the equation (2.92), defining the modes $\psi(y)$ with the boundary conditions different from those in the TE-case, that is, if the condition of equality to zero of the function at the interface is fulfilled for the $\varphi(y)$ modes, then for $\psi(y)$ we need to use the condition of zero derivative on the at the interface, and vice versa;
- Continue to calculate the constants $\psi_s^{(n,l)}$, $\psi_a^{(n,l)}$, $\psi_s^{(n,r)}$, $\psi_a^{(n,r)}$ based on the normalization condition (2.93), to finally form K y-modes of the form (2.86), corresponding to the case of TM-polarization.

Substantial modifications are also required in the algorithm for finding the propagation constants and the actual 'cross-linking' of local modes in the locally continuous functions describing the components of the vector field. The local x-modes of both polarizations are descibed by the expressions

$$\mathbf{U}_{ap}^{(m,l)} = -\mathbf{T}_p^{(m)}\mathbf{U}_{sp}^{(m,l)} + \mathbf{S}_p^{(m)}\mathbf{U}_{sp}^{(m,r)} \tag{2.94a}$$

$$\mathbf{U}_{ap}^{(m,r)} = -\mathbf{S}_p^{(m)}\mathbf{U}_{sp}^{(m,l)} + \mathbf{T}_p^{(m)}\mathbf{U}_{sp}^{(m,r)} . \tag{2.94b}$$

Here

$$T_{pkk}^{(m)} = (k_0 / k_{pk}^{(m)})^2 k_{xpk}^{(m)} / tg(k_{xpk}^{(m)} d_x^{(m)}) \tag{2.95a}$$

$$S_{pkk}^{(m)} = (k_0 / k_{pk}^{(m)})^2 k_{xpk}^{(m)} / \sin(k_{xpk}^{(m)} d_x^{(m)}) . \tag{2.95b}$$

The equalities, similar to (2.46), linking the constants of the local x-modes of both polarizations of adjacent columns, are shown below:

$$\mathbf{U}_{sh}^{(m,r)} = \mathbf{O}_{hh}^{(m,m+1)}\mathbf{U}_{sh}^{(m+1,l)} , \tag{2.96a}$$

$$\mathbf{U}_{ah}^{(m,r)} = \mathbf{O}_{hh}^{(m,m+1)}\mathbf{U}_{ah}^{(m+1,l)} - k_z\mathbf{O}_{he}^{(m,m+1)}\mathbf{U}_{se}^{(m+1,l)} , \tag{2.96b}$$

$$\mathbf{U}_{se}^{(m,r)} = \mathbf{O}_{ee}^{(m,m+1)} \mathbf{U}_{se}^{(m+1,l)},$$
(2.96c)

$$\mathbf{U}_{ae}^{(m,r)} = \mathbf{O}_{ee}^{(m+1,m)} \mathbf{U}_{ae}^{(m+1,l)} + k_z \mathbf{O}_{he}^{(m+1,m)} \mathbf{U}_{sh}^{(m+1,l)},$$
(2.96d)

wherein the matrices $\mathbf{O}_{hh}^{(m,m')}, \mathbf{O}_{ee}^{(m,m')}$ and $\mathbf{O}_{he}^{(m,m')}$ have the following elements:

$$O_{hhkp}^{(m,m')} = < \varphi_k^{(m)} \mid \varphi_p^{(m')} >$$
(2.97a)

$$O_{eekp}^{(m,m')} = < \psi_k^{(m)} \mid \psi_p^{(m')} / \varepsilon^{(m')}(y) >$$
(2.97b)

$$O_{hekp}^{(m,m')} = < \dot{\phi}_k^{(m)} \mid \psi_p^{(m')} / \varepsilon^{(m')}(y) > / k_{hk}^{(m)2} + < \phi_k^{(m)} \mid \ddot{\psi}_p^{(m')} / \varepsilon^{(m')}(y) > / k_{ep}^{(m')2}.$$
(2.97c)

The expression $< \varphi_k \mid \varphi_p >$ for the values of the functions and derivatives at the boundaries of the lines (constants $\varphi_s^{(n,l)}$, $\varphi_a^{(n,l)}$, $\varphi_s^{(n,r)}$, $\varphi_a^{(n,r)}$) has already been previously determined by formula (2.81). Next, we present formulas convenient for calculating the remaining matrix elements of the overlap integrals:

$$< \psi_k \mid \psi_p / \varepsilon(y) > = \sum_{n=1}^{N} \frac{(\psi_{sk}^{(n,r)} \psi_{ap}^{(n,r)} - \psi_{ak}^{(n,r)} \psi_{sp}^{(n,r)} - \psi_{sk}^{(n,l)} \psi_{ap}^{(n,l)} + \psi_{ak}^{(n,l)} \psi_{sp}^{(n,l)})}{\varepsilon^{(n)}(k_{kk}^2 - k_{kp}^2)}$$
(2.98)

$$< \dot{\phi}_k \mid \psi_p / \varepsilon(y) > = \sum_{n=1}^{N} \frac{(-\phi_{ak}^{(n,l)} \psi_{ap}^{(n,l)} - k_{kk}^2 \phi_{sk}^{(n,l)} \psi_{sp}^{(n,l)} + \phi_{ak}^{(n,r)} \psi_{ap}^{(n,r)} + k_{kk}^2 \phi_{sk}^{(n,r)} \psi_{sp}^{(n,r)})}{\varepsilon^{(n)}(k_{kk}^2 - k_{kp}^2)},$$
(2.99)

$$< \varphi_k \mid \dot{\psi}_p / \varepsilon(y) > = \sum_{n=1}^{N} \frac{(-\varphi_{ak}^{(n,l)} \psi_{ap}^{(n,l)} - k_{kp}^2 \varphi_{sk}^{(n,l)} \psi_{sp}^{(n,l)} + \varphi_{ak}^{(n,r)} \psi_{ap}^{(n,r)} + k_{kp}^2 \varphi_{sk}^{(n,r)} \psi_{sp}^{(n,r)})}{\varepsilon^{(n)}(k_{kk}^2 - k_{kp}^2)}.$$
(2.100)

It should be noted that the introduction of a factor $\left(\dfrac{k_0}{k_{pk}^{(m)}} \right)^2$ by which distinguishes the expression for the x-modes (2.84) differs from the scalar analogue (2.51) allows one to retain the simple expressions for the normalization (2.64) and (2.93) of the local y-modes of the TE- and TM-polarizations, respectively, to obtain expressions for the matrix elements $O_{hhkp}^{(m,m')}$ and $O_{eekp}^{(m,m')}$ which do not depend on the roots of the characteristic equations (2.63) and (2.92).

Similar to the problem (2.77) for finding propagation constants for the scalar case, we can formulate the problem of solving a homogeneous algebraic system of equations

$$\Xi(k_z)U = 0, \tag{2.101}$$

where the matrix $\Xi(k_z)$ as well as $\Lambda(k_z)$ has a band structure and depends on the parameter k_z:

$$\Xi(k_z) = \begin{bmatrix} \mathbf{A}_h^{(2)} & \mathbf{C}_h^{(2)} & \mathbf{B}_h^2 & \mathbf{O} & \mathbf{O} & \mathbf{O} & \cdots & \cdots & \mathbf{O} & \mathbf{O} & \mathbf{O} & \mathbf{O} \\ \mathbf{C}_e^{(2)} & \mathbf{A}_e^{(2)} & \mathbf{O} & \mathbf{B}_e^{(2)} & \mathbf{O} & \mathbf{O} & \cdots & \cdots & \mathbf{O} & \mathbf{O} & \mathbf{O} & \mathbf{O} \\ \mathbf{D}_h^{(3)} & \mathbf{O} & \mathbf{A}_h^{(3)} & \mathbf{C}_h^{(3)} & \mathbf{B}_h^{(3)} & \mathbf{O} & \cdots & \cdots & \mathbf{O} & \mathbf{O} & \mathbf{O} & \mathbf{O} \\ \mathbf{O} & \mathbf{D}_e^{(3)} & \mathbf{C}_e^{(3)} & \mathbf{A}_e^{(3)} & \mathbf{O} & \mathbf{B}_e^{(3)} & \cdots & \cdots & \mathbf{O} & \mathbf{O} & \mathbf{O} & \mathbf{O} \\ & & & \mathbf{O} & & & & & \mathbf{O} & \mathbf{O} & \mathbf{O} & \mathbf{O} \\ \mathbf{O} & \mathbf{O} & \mathbf{O} & \mathbf{O} & \mathbf{O} & \mathbf{O} & \cdots & \cdots & \mathbf{D}_h^{(M)} & \mathbf{O} & \mathbf{A}_h^{(M)} & \mathbf{C}_h^{(M)} \\ \mathbf{O} & \mathbf{O} & \mathbf{O} & \mathbf{O} & \mathbf{O} & \mathbf{O} & \cdots & \cdots & \mathbf{O} & \mathbf{D}_e^{(M)} & \mathbf{C}_e^{(M)} & \mathbf{A}_e^{(M)} \end{bmatrix}. \tag{2.102}$$

The components of the block matrix $\Xi(k_z)$ are expressed in terms of the already defined matrices of the overlap integrals and diagonal matrices $\mathbf{T}_p^{(m)}$ and $\mathbf{S}_p^{(m)}$.

$$\mathbf{A}_h^{(m)} = \mathbf{O}_{hh}^{(m,m-1)}\mathbf{T}_h^{(m-1)}\mathbf{O}_{hh}^{(m-1,m)} + \mathbf{T}_h^{(m)} \tag{2.103a}$$

$$\mathbf{A}_e^{(m)} = \mathbf{O}_{ee}^{(m-1,m)T}\mathbf{T}_e^{(m-1)}\mathbf{O}_{ee}^{(m-1,m)} + \mathbf{T}_e^{(m)} \tag{2.103b}$$

Likewise, as noted in paragraph 2.4.1, in the case of magnetic walls the expressions $\mathbf{A}_p^{(2)}$ and $\mathbf{A}_p^{(M)}$ differ from the general scheme:

$$\mathbf{A}_h^{(2)} = \mathbf{O}_{hh}^{(2,1)}(\mathbf{T}_h^{(1)} - \mathbf{S}_h^{(1)}(\mathbf{T}_h^{(1)})^{-1}\mathbf{S}_h^{(1)})\mathbf{O}_{hh}^{(1,2)} + \mathbf{T}_h^{(2)} \tag{2.103c}$$

$$\mathbf{A}_e^{(2)} = \mathbf{O}_{ee}^{(1,2)T}(\mathbf{T}_e^{(1)} - \mathbf{S}_e^{(1)}(\mathbf{T}_e^{(1)})^{-1}\mathbf{S}_e^{(1)})\mathbf{O}_{ee}^{(1,2)} + \mathbf{T}_e^{(2)} \tag{2.103d}$$

$$\mathbf{A}_h^{(M)} = \mathbf{O}_{hh}^{(M,M-1)}\mathbf{T}_h^{(M-1)}\mathbf{O}_{hh}^{(M-1,M)} + \mathbf{T}_h^{(M)} - \mathbf{S}_h^{(M)}(\mathbf{T}_h^{(M)})^{-1}\mathbf{S}_h^{(M)} \tag{2.103e}$$

$$\mathbf{A}_e^{(M)} = \mathbf{O}_{ee}^{(M-1,M)T}\mathbf{T}_e^{(M-1)}\mathbf{O}_{ee}^{(M-1,M)} + \mathbf{T}_e^{(M)} - \mathbf{S}_e^{(M)}(\mathbf{T}_e^{(M)})^{-1}\mathbf{S}_e^{(M)} \tag{2.103f}$$

$$\mathbf{B}_h^{(m)} = -\mathbf{S}_h^{(m)}\mathbf{O}_{hh}^{(m,m+1)} \tag{2.103g}$$

$$\mathbf{B}_e^{(m)} = -\mathbf{S}_e^{(m)}\mathbf{O}_{ee}^{(m,m+1)} \tag{2.103h}$$

$$\mathbf{C}_h^{(m)} = k_z\mathbf{O}_{hh}^{(m,m-1)}\mathbf{O}_{he}^{(m-1,m)} \tag{2.103i}$$

$$\mathbf{C}_e^{(m)} = -k_z\mathbf{O}_{ee}^{(m-1,m)T}\mathbf{O}_{he}^{(m,m-1)T} \tag{2.103j}$$

$$\mathbf{D}_h^{(m)} = -\mathbf{O}_{hh}^{(m,m-1)}\mathbf{S}_h^{(m-1)} \tag{2.103k}$$

$$\mathbf{D}_e^{(m)} = -\mathbf{O}_{ee}^{(m-1,m)\mathrm{T}}\mathbf{S}_e^{(m-1)}. \tag{2.103l}$$

When constructing the matrix $\Xi(k_z)$ of problem (2.101) we must also take into account the rule of defining the boundary conditions stated in Table 2.3.

Table 2.3. Two possible options for specifying the boundary conditions

	$\phi(y)$	$\psi(y)$	$u_h(x)$	$u_e(x)$
'Electric walls'	$\begin{cases} \phi_s^{(1,l)} = 0, \\ \phi_s^{(N,r)} = 0. \end{cases}$	$\begin{cases} \psi_a^{(1,l)} = 0, \\ \psi_a^{(N,r)} = 0. \end{cases}$	$\begin{cases} u_{ha}^{(1,l)} = 0, \\ u_{ha}^{(N,r)} = 0. \end{cases}$	$\begin{cases} u_{es}^{(1,l)} = 0, \\ u_{es}^{(N,r)} = 0. \end{cases}$
'Magnetic walls'	$\begin{cases} \phi_a^{(1,l)} = 0, \\ \phi_a^{(N,r)} = 0. \end{cases}$	$\begin{cases} \psi_s^{(1,l)} = 0, \\ \psi_s^{(N,r)} = 0. \end{cases}$	$\begin{cases} u_{hs}^{(1,l)} = 0, \\ u_{hs}^{(N,r)} = 0. \end{cases}$	$\begin{cases} u_{ea}^{(1,l)} = 0, \\ u_{ea}^{(N,r)} = 0. \end{cases}$

In turn, the vector $\mathbf{U}$ (2.101) contains the constants as the local modes of both TE- and TM-polarization.

$$\mathbf{U} = \begin{bmatrix} U_{sh}^{(2,l)} \\ U_{se}^{(2,l)} \\ U_{sh}^{(3,l)} \\ U_{se}^{(3,l)} \\ \vdots \\ U_{sh}^{(M,l)} \\ U_{se}^{(M,l)} \end{bmatrix}. \tag{2.104}$$

The problem (2.101) is solved in the same manner as in the scalar case (2.77). As a result, after determining the propagation constant and also local x-modes, we calculate the components of the vector field E_x, E_y, E_z, cB_x, cB_y, and cB_z from the resulting set of functions using formulas (2.85).

2.4.3. Krylov method for solving non-linear eigenvalue problems

At the stage of solving of the non-linear eigenvalue problem (2.77), (2.101) it is convenient to use the iterative Krylov method [107] which allows precise calculations (avoiding gaps) of the eigenvalues of large dimension.

Interpolating a non-linear matrix operator $\Lambda(k_z)$ between two arbitrary values σ and μ as follows:

$$\Lambda(k_z) \approx \tilde{\Lambda}(k_z) = \frac{k_z - \sigma}{\mu - \sigma}\Lambda(\mu) + \frac{\mu - k_z}{\mu - \sigma}\Lambda(\sigma) \qquad (2.105)$$

we obtain a linear eigenvalue problem:

$$\tilde{\Lambda}(k_z)\mathbf{U} = 0 \qquad (2.106)$$

which we will solve iteratively as:

$$\left[\frac{\mu_{k+1} - \sigma}{\mu_k - \sigma}\Lambda(\mu_k) + \frac{\mu_k - \mu_{k+1}}{\mu_k - \sigma}\Lambda(\sigma)\right]\mathbf{U}_k = 0 \qquad (2.107)$$

where μ_{k+1} is the approximate value of the sought k_z obtained in the $(k + 1)$-th iteration. We introduce the notation:

$$\theta = \frac{\mu_{k+1} - \mu_k}{\mu_{k+1} - \sigma}. \qquad (2.108)$$

Equation (2.107) can be written as:

$$[\Lambda(\sigma)^{-1}\Lambda(\mu_k) - \theta\mathbf{I}]\mathbf{U}_k = 0. \qquad (2.109)$$

Let μ_1 is some initial approximation for the eigenvalue k_z of the matrix $\Lambda(k_z)$, let σ be the fixed value close to μ_1, then the iterative procedure of clarifying the eigenvalue k_z is as. In the k-th step we solve with respect to θ:

$$\Lambda(\sigma)^{-1}\Lambda(\mu_k)\mathbf{U} = \theta\mathbf{U} \qquad (2.110)$$

and calculate a new estimate for the eigenvalues k_z of the problem (2.87) or (2.101):

$$\mu_{k+1} = \mu_k + \frac{\theta}{1 - \theta}(\mu_k - \sigma). \qquad (2.111)$$

The iterations are repeated as long as the sequence of estimates $\{\mu_k\}$ converges. In [111] it is shown that the iterative procedure in the Krylov method converges to the desired eigenvalue.

In [90] the authors examine another method for solving non-linear eigenvalue problems. The method will be called the zero function method. The method is as follows. We choose an arbitrary vector $\mathbf{V}$

with non-zero components, such as a unit vector. We will solve the inhomogeneous equation

$$\Lambda(k_z)\mathbf{U}' = \mathbf{V} \qquad (2.112)$$

with respect to $\mathbf{U}'$. We obtain different solutions of equation (2.112) for different values of parameter k_z. We define the function

$$f(k_z) = 1/U'_p \qquad (2.113)$$

where U'_p is the p-th component of the vector $\mathbf{U}'$. In the vicinity of the desired values k_z the function $f(k_z)$ is a continuous function of a scalar argument, and its zeros are the sought k_z.

Search for zeros $f(k_z)$ is carried out by standard methods.

The discontinuity of the function $f(k_z)$ makes it difficult to find the propagation constants in this way. Therefore, even at a sufficiently small sampling step, much smaller than that needed for the separation of adjacent zeros of function $f(k_z)$, there is the possibility of missing the roots because of breaks located close to zero. This problematic situation is shown in Fig. 2.37, where in the area of the root $k_z = 6.9951$ μm^{-1} the function has a break.

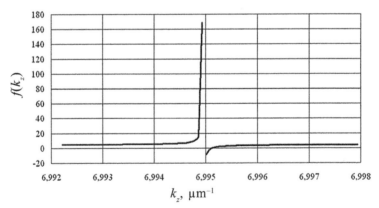

Fig. 2.37. An example of discontinuity $f(k_z)$ near zero.

Numerous gaps also do not allow the use of statistical estimates of the behaviour of the function in the interval, and a decrease in the sampling step increases the time spent on calculations.

Combined use of (2.112)–(2.113) with Krylov's method (2.105)–(2.111) presumably could outperform both of these methods.

Let us use the Krylov method to indicate the presence of a root on some interval where further clarification of the values of the roots is conducted by means of the function (2.113). For a given accuracy of the separation of roots we choose a discretization step in the Krylov method so as to minimize the computation time

$$t(h_k) = Mt_{ki} + Int_{si} \rightarrow \min .$$ (2.114)

In (2.114) $M = \frac{L}{h_k}$ is the number of intervals analyzed by the Krylov method, where L is the length of the interval in which we search eigenvalues k_z, t_{ki} is the time required to perform one iteration of the Krylov method, I is the estimated number of roots on the length interval L, $n = \frac{h_k}{h}$ is the number of intervals in the range of length h_k considered in refining roots with the step h, t_{si} is the average time to complete one iteration of detection of the zero of function $f(k_z)$ on the interval with length h.

The best value is the value h_k of the sampling step:

$$h_k^{opt} = \sqrt{\frac{Lht_{ki}}{It_{si}}} .$$ (2.115)

A practical example of the calculation of multiple eigenmodes of a photonic fibre (Fig. 2.38) with the use of these three methods for solving non-linear eigenvalue problems in the method of matched sinusoidal modes is discussed later.

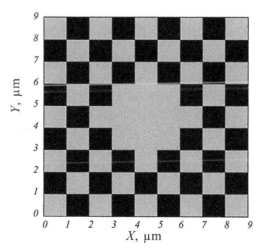

Fig. 2.38. Model of the section of the photonic fibre, light areas show areas with the refractive index $n_1 = 1.47$, dark – with $n_2 = 1$.

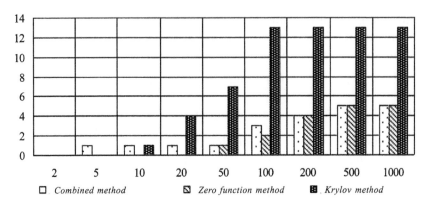

Fig. 2.39. Dependence of the number of determined propagation constants for several first modes of the PCF (Fig. 2.38) on $1/h$, inversely proportional to the accuracy of separation of the roots.

Approximation using formula (2.54) was carried out using thirty local modes that, according to a study conducted in [110], is enough to obtain estimates of the propagation constants of the modes with an error of not more than 10^{-4}, that is, up to three decimal places.

In accordance with the MSM method we considered the problem (2.77), where the elements of the matrix $\Lambda(k_z)$ depend in a non-linear manner on the parameter k_z which converts $\Lambda(k_z)$ to the degenerate numerical matrix whose determinant is zero.

For a given cross-section structure of the dimension of the reduced number of employed local modes of the matrix $\Lambda(k_z)$ composition 480×480.

In the interval of unit length the search for the propagation constants is carried out by three methods: by determining the zeros of function (2.113), the Krylov method (2.105)–(2.111) and the combined method with the optimal discretization step (2.115).

A diagram showing the number of roots detected with an accuracy of 10^{-4} for each method with different numbers of points of the partition – the value inversely proportional to the accuracy of separation of roots h – is shown in Fig. 2.39.

As can be seen from the diagram, the Krylov method is substantially better than the other two methods in detecting the desired values of parameter k_z, but, as can be seen from the graph in Fig. 2.40, is the most time-consuming method for a given value h. For instance, the calculation of the thirteen root by the Krylov method with $h = 10^{-3} \, \mu\text{m}^{-1}$ in a PC takes about three hours. It should be noted (Fig. 2.39) that the Krylov method also find 13 roots but the

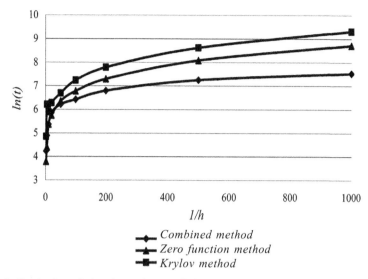

Combined method
Zero function method
Krylov method

Fig. 2.40. A plot of the dependence of the natural logarithm of time in seconds spent on the calculation of zeros on the value $1/h$, inversely proportional to the accuracy of separation of roots.

time is eight times less. At the same time, the zero function method and the combined method have approximately the same frequency of detection of roots and the slight advantage of the combined method in the field of low precision of separation of the roots is due to accidental better positioning of the length interval h, in which the secondary iterations of the zero function method are performed, relative to the root. In this case, the combined method is more time-efficient than the zero function method, as follows from Fig. 2.40.

Thus, the application in the MSM of the iterative procedure of finding the propagation constants by the Krylov method solves the problem of missing roots similar in value or location near the discontinuity of function (2.113) which arises when using the zeros function method, by increasing the calculation time.

2.4.4. Calculation of stepped fibre modes

Consider the use of the MSM method for calculating the modes of weakly stepped fibres of fused silica SiO_2. The radiation source is infrared light having a wavelength in vacuum of $\lambda_0 = 1.3$ μm, corresponding to a minimum dispersion of the material.

The optical fibres whose cross sections are shown in Fig. 2.41 (model 1) and Fig. 2.42 (model 2), are approximate models of the

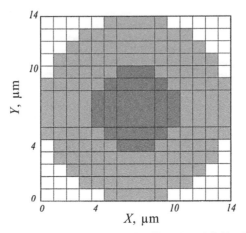

Fig. 2.41. Schematic cross-section of the fibre (model 1), dark-gray regions are values of the refractive index n_{co} = 1.47, light gray – n_{cl} = 1.463, white – n_v = 1.

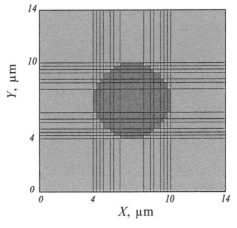

Fig. 2.42. Schematic of the cross-section of the fibre (model 2), dark-gray region are the values of the refractive index n_{co} = 1.47, light gray – n_{cl} = 1.463.

stepped waveguide having a circular section, with a refractive index of the material in the core and the cladding of respectively, n_{co} = 1.47 and n_{cl} = 1.463. Since the refractive indices of the core and the cladding differ slightly

$$n_{co} \cong n_{cl} \qquad (2.116)$$

this is a weakly guiding waveguide.

In addition, the fibre is also not a multi-mode fibre [112], since the condition

$$V = \frac{2\pi\rho\sqrt{n_{co}^2 + n_{cl}^2}}{\lambda_0} \gg 1 \qquad (2.117)$$

is not satisfied. In this case ρ is the characteristic dimension of the core (the radius) is equal to 3 μm, and the fibre parameter V [112]:

$$V = \frac{2\pi\sqrt{1.47^2 + 1.463^2} \cdot 3\ \mu\text{m}}{1.3\ \mu\text{m}} \approx 2.078 . \qquad (2.118)$$

Therefore, in the calculations we pay attention only to a few lower-order modes. Figure 2.43 shows the intensity distribution of the fundamental modes for both models, resulting from the application of the scalar version of the MSM method, that is function $|\xi(x,y)|^2$ of (2.54). Figure 2.43 shows the modes in halftones: black colour corresponds to the maximum intensity of the mode, white – the minimum (zero).

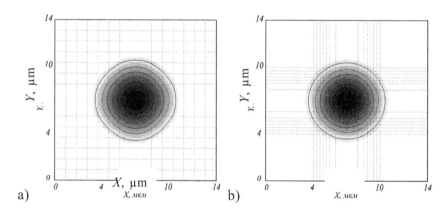

a)   b)

Fig. 2.43. The distribution of the field intensity of the fundamental mode: a) model 1, b) model 2.

Since there are no principal differences between these two results, further calculations were carried out using a more convenient model 2 with the cross section shown in Fig. 2.42. With the same number of rows and columns in the models 1 and 2, the accuracy of the approximation of the form of the core for model 2 is higher and, at the same time, the radius of the cladding can be considered much greater than the radius of the core, because that is usually the case in practice. The intensity distribution of the first ten modes of the model is shown in Fig. 2.44, of which only the fundamental mode is guided, as only its propagation constant satisfies the cutoff condition [112] $n_{cl}k_0 < k_z$ (Table 2.4).

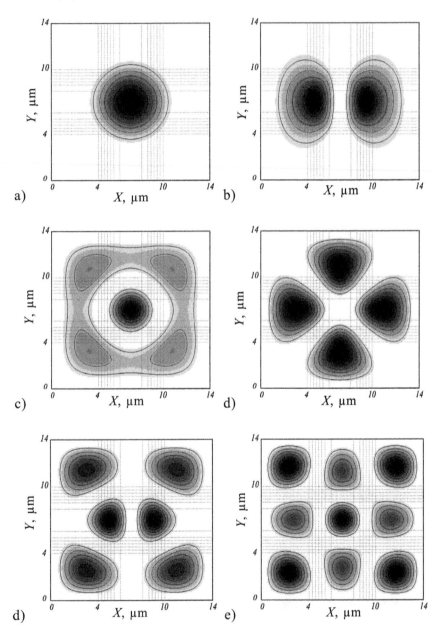

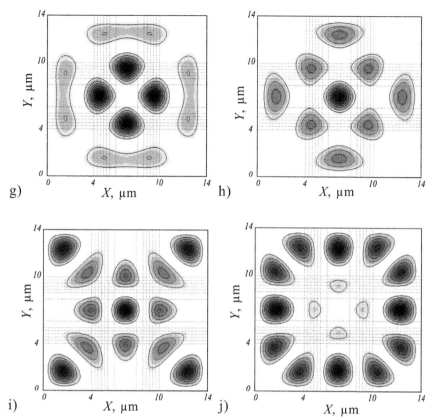

Fig. 2.44. Field intensity distribution of the first ten modes for model 2.

Table 2.4. The values of propagation constants of the first ten modes of model 2

Number of mode	k_z, μm^{-1}	$n_{eff} = k_z/k_0$
1	7.0879	1.4665
2	7.0617	1.4611
3	7.0447	1.4576
4	7.0389	1.4564
5	7.0301	1.4545
6	7.0102	1.4504
7	6.9890	1.4460
8	6.9846	1.4451
9	6.9551	1.4390
10	6.9512	1.4382

The determined spatial modes are a system of mutually orthogonal functions in the section. The matrix of the values of integrals of the form

$$\int_{x^{(1)}}^{x^{(M+1)}} \int_{y^{(1)}}^{y^{(N+1)}} \xi_{k1}(x,y)\xi_{k2}(x,y)dxdy \quad \text{where } k1, k2 = \overline{1,10} \quad (2.119)$$

is shown in Table 2.5. From Table 2.5 it is seen that the calculated modes are orthogonal with the accuracy up to the fourth decimal place.

Table 2.5. Matrix of the values of overlap integrals for the amplitudes of the first ten modes of model 2

Number of modes	1	2	3	4	5	6	7	8	9	10
1	8.4803	0	0	0	0	0	0	0	0	0
2	0	7.7895	0	0	0	0	0	0	0	0
3	0	0	14.7646	0	0	0	0	0	0	0
4	0	0	0	13.3078	0	0	0	0	0	0
5	0	0	0	0	10.5809	0	0	0	0	0
6	0	0	0	0	0	40.0445	0	0	0	0
7	0	0	0	0	0	0	9.9666	0	0	0
8	0	0	0	0	0	0	0	10.4592	0	0
9	0	0	0	0	0	0	0	0	19.4402	0
10	0	0	0	0	0	0	0	0	0	21.7969

After the normalization of each mode we get a system of orthonormal functions.

2.4.5. Calculation of modes of photonic crystal fibres

Application of the vector calculation methods of the eigenmodes is especially important for the study of optical fibres the cross section of which contain large (more than 20%) 'jumps' in the refractive index, such as the PCF. Further, the vector and scalar modes of the PCF, calculated by the MSM method, are compared. The sections of these modes are shown in Fig. 2.38.

The intensity distribution of the component cB_y of the fundamental mode with an effective refractive index $n_{eff} = 1.4473$ and the intensity of the corresponding scalar field eith $n_{eff} = 1.4491$, calculated for radiation with a wavelength $\lambda_0 = 1.3$ μm are shown in Figure 2.45.

There is a fairly strong resemblance of the configuration of the scalar field and vector component cB_y. The standard deviation between the two solutions, normalized to unity, in the region $W_x \times W_y = 9 \ \mu m \times 9 \ \mu m$ is 0.0000014 or 0.00014%. For the other three modes the distribution of the square of the amplitude of one of the vector components and the scalar field intensity (Figs. 2.46–2.48) are also in agreement.

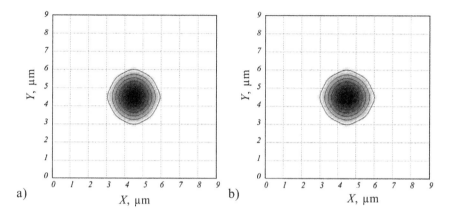

Fig. 2.45. Distribution of: a) square of the amplitude of the component of the fundamental vector mode cB_y, b) the intensity of the corresponding scalar mode.

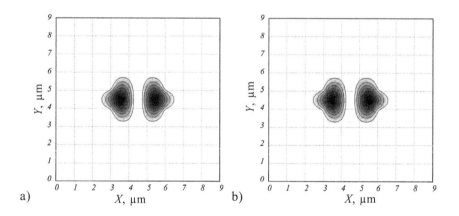

Fig. 2.46. Distribution of: a) square of the amplitude of the component cB_y of the second vector mode, b) the intensity of the corresponding scalar mode.

The similarity of the configuration of individual components of the vector and scalar modes is not accidental. The corresponding vector component is the most powerful of the six vector components of the field, i.e. it has the highest value of the integral in the cross section of the region (Table 2.6).

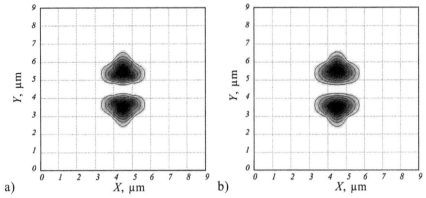

Fig. 2.47. Distribution of: a) square of the amplitude of the component cB_x of the third vector mode, b) intensity of the corresponding scalar mode.

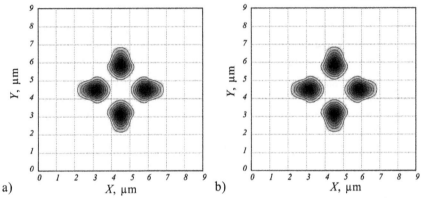

Fig. 2.48. Distribution a) square of the amplitude component cB_y of the fourth vector mode, b) corresponding to the intensity of the scalar mode

Table 2.6. The values of the integrals of the square of the amplitude of the components of the first four modes

F	$I(|F_1|^2)$	$I(|F_2|^2)$	$I(|F_3|^2)$	$I(|F_4|^2)$
E_x	4.7671	4.7263	0.2488	5.9050
E_y	0.0662	0.0421	4.8538	0.2406
E_z	0.1364	0.2805	0.2899	0.9118
cB_x	0.2799	0.1442	10.6741	1.7101
cB_y	10.1416	10.3738	0.4344	11.1618
cB_z	0.2481	0.2208	0.3118	1.8900

Characteristically, the integrals of the longitudinal components constitute a small value relative to the dominant component of the number of the transverse magnetic components.

Thus, the calculations show that the scalar MSM method provides a sufficiently good approximation of the most intensive component of the vector mode, even in the case of fibres with a rapidly changing refractive index profile. Nevertheless, the presence of several non-zero vector components makes it application of the vector approach feasible for PCFs.

2.4.6. Calculation of modes using Fimmwave

The commercial software for the simulation of the propagation of light FIMMWAVE v.4.6 http:,www.photond.com/products/fimmwave. htm can calculate the modes of dielectric waveguides with arbitrary cross sections uniform in the longitudinal direction, including the PCF, by implementing the vector method of matched sinusoidal modes (FMM Solver (real)) and the effective index method (Eff. Idx. Solver (real)).

For the model of the PCF the cross section of which is shown in Fig. 2.38 and the wavelength $\lambda = 1.3$ μm in FIMMWAVE using the FMM Solver (real), with the number of the local y-mode equal to sixty, we obtained a fundamental mode with the effective index of 1.4477.

The results obtained in [109] by implementing MSM- and Raman techniques in Matlab for this model are compared with the results of FIMMWAVE in Table 2.7.

Table 2.7. Absolute and relative deviations of the effective index values of the fundamental mode of the PCF (Fig. 2.38), calculated by different methods, from the result obtained using FIMMWAVE

Method	Efficient index n_{eff}	Absolute deviation, Δ	The relative deviation, δ
Raman method [113–115] $n_x \times n_y = 52 \times 52$	1.4480	0.0003	0.02%
MSM method (vector)	1.4473	0.0004	0.03%
MSM method (scalar)	1.4491	0.0014	0.10%
FMM Solver (real)	1.4477	0	0

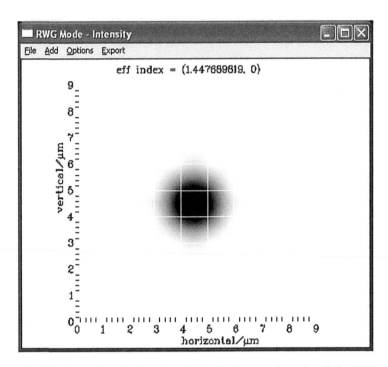

Fig. 2.49. The intensity distribution of the fundamental mode of the PCF model (Fig. 2.38) obtained in the FIMMWAVE software.

As shown in Table 2.7, the calculated values differ from the FMM Solver (real) results by no more than one-tenth of a percent. The intensity distribution of the modes (Fig. 2.49), obtained in FIMMWAVE, also agrees well with the intensity distribution of the scalar mode and the main vector components shown in Fig. 2.45.

In the accompanying FIMMWAVE documentation, the recommended minimum number of local y-modes when calculating using FMM Solver (real) is equal to thirty, which is consistent with studies of convergence of the method conducted in [110].

Let us compare the two implementations of the MSM method: the one proposed in [108] and the commercial program FIMMWAVE – for the convergence with respect to the number of local modes. Changes of the relative error of calculation one of the output parameters – the propagation constant, in relation to the number of local modes in the range of from 10 to 60 for the PCF (Fig. 2.38) is shown in Fig. 2.50.

From a comparison of the graphs of the dependence of the relative error of calculating the propagation constants on the number of local modes for the two implementations of the MSM method it follows that the Matlab proposed in [108] provides a more stable and monotonic convergence, as well as a significantly smaller error for a small number of local modes than the commercial program FIMMWAVE.

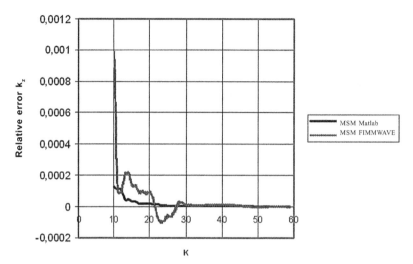

Fig. 2.50. The graphs of the dependence of the relative error of the propagation constant of the fundamental mode of the PCF (Fig. 2.38) on the number of local y-modes have two implementations of the MSM method.

References

1.   J.D. Joannopoulos, S.G. Johonson, J.N. Winn, Photonic Crystals: Molding the Flow of Light, sec. ed., Princeton Univ. Press, 304p, 2008.
2.   P.N. Prasad, Nanophotonics, Wiley, 432p, 2004.
3.   A.Yariv, P. Yeh , Optical wave in crystals: propagation and control of laser radiation, Wiley-Interscience, 589p. (2002).
4.   Y.Fukaya, D.Ohsaki and T.Baba. Two-dimensional photonic cryctal waveguide with 600 bends in a thin slab structure, J. Jap. Soc. Appl. Phys., v.39, no.5A, p.2619-2623 (2000).
5.   W.R. Frei, D.A. Tortorelli, H.T. Johnson, Geometry projection method for optimizing photonic nanostructures, Opt. Lett., v. 32, no.1, p.77-79 (2007).
6.   Taflove A. Computational Electrodynamics: the finite-difference time-domain method. – M.: Artech House, Inc. 1995.
7.   Pernice W.H., Payne F.P., Gallagher D.F. Numerical investigation of field enhancement by metal nano-particles using a hybrid FDTD-PSTD algorithm, Optics Express. - 2007.-Vol.15.- P.11433-11443.
8.   Kim J.H., Chrostowski L., Bisaillon E., Plant D.V. DBR, sub-wavelength grating, and photonic crystal slab Fabry-Perot cavity design using phase analysis by FDTD,Optics Express. – 2007.- Vol.15.-P.10330-10339.
9.   Yablonovitch E. Inhibited spontaneous emission in solid-state physics and electron-

ics, Phys. Rev. – 1987. – Vol.58. – P. 2059-2062.

10. Hugonin J.P., Lalanne P., White T.P., Krauss T.F. Coupling into slow-mode photonic crystal waveguides , Opt. Lett.-2007.-Vol.32.-P.2639-2640.

11. Kwan K.C., Tao X.M., Peng G.D. Transition of lasing modes in disordered active pho-tonic crystals ,Opt. Lett.-2007.-Vol.32.-P.2720-2722.

12. Zabelin V., Dunbar L.A.,Thomas N.L., Houndre R., Kotlyar M.V., O'Faolain L., Krauss T.F. Self-collimating photonic crystal polarization beam splitter, Opt. Lett.-2007.-Vol.32.-P.530-532.

13. Li Y., Jin J. Fast full-wave analysis of large-scale three-dimensional photonic crystal device , J. Opt. Soc. Am. B.-2007.-Vol.24.-P.2406-2415.

14. Mikaelyan A.L., Dokl. AN SSSR, 1951, No. 81, 569-571.

15. Snyder A.W., Mitchell D.J. Spatial solitons of the power-law nonlinearity ,Opt. Lett. – 1993.-Vol.18.-P.101-103.

16. Alimenkov I.V., Komp. optika, - 2005. - No. 28. 45-54.

17. Alimenkov I.V., Komp. optika, - 2005. -No. 28. 55-59.

18. Yee K. S. Numerical solution of initial boundary value problems involving Maxwell's equations in isotropic media , IEEE Trans. Antennas and Propagation. – 1966. – AP-14. – P.302-307.

19. Moore G. Absorbing boundary conditions for the finite-difference approximation of the time-domain electromagnetic field equations , IEEE Trans. Electromagnetic Compatibility. – 1981. – Vol.23. – P. 377-382.

20. Berenger J. P. A perfectly matched layer for the absorption of electromagnetic waves, Computational Physics. – 1994. – Vol.114. – P. 185-200.

21. Y. Xu, R.K. Lee, A. Yariv, Adiabatic coupling between conventional dielectric wave-guides with discrete translational symmetry, Opt. Lett., v.25, no.10, p.755-757 (2000).

22. A Mekis, J.D. Joannopoulos, Tapered couplers for efficient interfacing between dielec-tric and photonic crystal waveguides, J. Light Techn., v.19, no.6, p.861-865 (2001).

23. T.D. Happ, M. Kamp, A. Forchel, Photonic crystal tapers for ultracompact mode con-version, Opt. Lett., v.26, p.14, p.1102-1104 (2001).

24. A. Talneau, P. Lalanne, M. Agio, C.M. Soukoulis, Low-reflection photonic crystal taper for efficient coupling between guide sections of arbitrary widths, Opt. Lett., v.27, no.17, p.1522-1524 (2002).

25. V.R. Almeida, R.R. Panepucci, M. Lipson, Nanotaper for compact mode conversion, Opt. Lett., v.28, no.15, p.1302-1304 (2003).

26. P. Bienstman, S.Assefa, S.G. Johson, J.D. Joannopoulos, G.S. Petrich, L.A. Koloziejski, Taper structures for coupling into photonic crystal slab waveguides, J. Opt. Soc. Am. B, v.20, no.9, p.1817-1821 (2003).

27. S.J. MacNab, N. Moll, Y.A. Vlasov, Ultra-low loss photonic integrated circit with membrane-type photonic crystal waveguide, Opt. Express, v.11, no.22, p.2927-2939 (2003).

28. P.E. Barclay, K. Srinivasan, o. Painter, Design of photonic crystal waveguide for evanescent coupling to optical fiber tapers and integration with high-Q cavities, J. Opt. Soc. Am. B, v.20, no. 11, p.2274-2284 (2003).

29. R. Orobtchouk, A. Layadi, H. Gualous, D. Pascal, A. Koster, S. Laval, High-efficiency light coupling in a submicrometric silicon-on-insulator waveguide, Appl. Opt., v.39, no.31, p.57-73-5777 (2000).

30. S. Lardenois, D. Pascal, L. Vivien, E.Cassan, S. Laval, R.Orobtchouk, M. Heitzmann, N. Bonzaida, L. Mollard, Low-loss submicrometer silicon-on-insulator rib wave-

guides and corner mirrors, Opt. Lett., v.28, no.13, p.1150-1153 (2003).

31.   D. Taillaert, F. Vanlaere, M. Ayre, W. Bogaerts, D. VanThourhout, P. Bienstman, R. Baets, Grating couplers for couping between optical fiber and nanophotonic waveguides, Jap. J. Appl. Phys., v.45, no.8, p.6071-6077 (2006).

32.   F. Van Laere G. Roelkens, M. Ayre, D.Taillaert, D. Van Thourhout, T.F. Krauss, R. Baets, Compact and high efficient grating couplers between optical fiber and nanophotonic waveguides, J. Light. Techn., v. 25, no.1, p.151-156 (2007).

33.   B.L. Bachim, O.O. Ogunsola, T.K. Gaylord, Optical fiber-to-waveguide coupling using carbon-dioxide-laser-induced long-period fiber gratings, Opt. Lett., v.30, no.16, p.2080-2082 (2005).

34.   D.W. Prather, J. Murakowski, S.Shi, S. Venkataraman, A. Sharkawy, C Chen, D. Pustai, High-efficiency coupling structure for a single-line-defect photonic crystal waveguide, Opt. Lett., v.27, no.18, p.1601-1603 (2002).

35.   H. Kim, S. Lee, B.O, S. Park, E. Lee, High efficiency coupling technique for photonic crystal waveguides using a waveguide lens, OSA Techn. Digest: Frontiers in optics, 2003, MT68.

36.   J.C.W Corbett, J.R. Allington-Smith, Coupling starlight into single-mode photonic crystal fiber using a field lens, Opt. Express, v.13, no.17, p.6527-6540 (2005).

37.   D. Michaelis, C. Wachter, S. Burger, L. Zschiedrich, A. Brauer, Micro-optical assisted high-index waveguide coupling, Appl. Opt., v.45, no.8, p.1831-1838 (2006).

38.   G. Kong, J. Kim, H. Choi, J.E. Im, B. Park, V. Paek, B. H. Lee, Lensed photonic crystal fiber obtained by use of an arc discharge, Opt. lett., v.31, no.7, p.894-896 (2006).

39.   A.L. Pokro0vsky, A.L. Efros, Lens based on the use of left-handed materials, Appl. Opt., v.42, no.28, p.5701-5705 (2003).

40.   N. Fabre, S. Fasquel, C. Legrand, X. Melique, M. Muller, M. Francois, O. Vanbesien, D. Lippens, Toward focusing using photonic crystal flat lens, Opto-electronics Review, v.14, no.3, p.225-232 (2006).

41.   C. Li, M. Holt, A.L. Efros, Far-field imagimg by the Veselago lens made of a photonic crystal, J. Opt. Soc. Am. B, v.23, no.3, p.490-497 (2006).

42.   T. Matsumoto, K. Eom, T. Baba, Focusing of light by negative refraction in a photonic crystal slab superlens on silicon-on-insulator substrate, Opt. Lett.,v.31, no.18, p.2786-2788 (2006).

43.   C.Y. Li, J.M. Holt, A.L. Efros,Imaging by the Veselago lens based upon a two-dimensional photonic crystal with a triangular lattice, J. Opt. Soc. Am. B, v.23, no.5, p.963 -968 (2006).

44.   T. Geng, T. Lin, S. Zhuang, All angle negative refraction with the effective phase index of -1, Chinese Opt. lett., v.5, no.6, p.361-363 (2007).

45.   T. Asatsume, T. Baba, Abberation reduction and unique light focusing in a photonic crystal negative refractive lens, Opt. Express, v.16, no.12, p.8711-8718 (2008).

46.   N. Fabre, L.Lalonat, B. Cluzel, X. melique, D. Lippens, F. deFornel,O. Vanbesien « Measurement of a flat lens focusing in a 2D photonic crystal at optical wavelength», OSA Digest, CLEO/QELS, 2008, CTuDD6, CA.

47.   S. Yang, C. Hong, H.Yang, Focusing concave lens photonic crystals with magnetic ma-terials, J. Opt. Soc. Am. A,v.23, no.4, p.956-959 (2006).

48.   P. Luan, K.Chang, Photonic crystal lens coupler using negative refraction, Prog. In Electr. Res., v.3, no.1, p.91-95 (2007).

49.   S. Haxha, F. AbdelMalek, A novel design of photonic crystal lens based on negative re-fractive index, Prog. In Electr. Res., v.4, no.2, p.296-300 (2008).

50.   Z. Lu, S. Shi, C.A. Schuetz, J.A. Murakowski, D. Prather, Three-dimensional pho-

tonic crystal flat lens by full 3D negative refraction, Opt. Express, v.13, no.15, p.5592-5599 (2005).

51. Z. Lu, S. Shi, C.A. Schuetz, D.W. Prather, Experimental demonstration of negative re-fraction imaging in both amplitude and phase, Opt. Express, v.13, no.6, p.2007-2012 (2005).

52. I.V. Minin, O.V. Minin, Y.R. Triandafilov, V.V. Kotlyar, Subwavelength diffractive photonic crystal lens, Prog. In Electr. Res. B, v.7, p.257-264 (2008).

53. E.Pshenay-Severin, C.C. Chen, T. Pertsch, M. Augustin, A. Chipoline, A. Tunnermann, Photonic crystal lens for photonic crystal waveguide coupling, OSA Techn. Di-gest:CLEO, 2006, CThK3.

54. J.P. Hugonin, P. Lalanne, T.P. White, T.F. Krauss, Coupling into clow-mode photonic crystal waveguide, Opt. Lett., v.32, no.18, p.2638-2640 (2007).

55. Ya. R. Triandafilov, et al., Komp. optika, v.31, No. 3, 27-31 (2007).

56. Y.R. Triandafilov, V.V. Kotlyar, Photonic crystal Mikaelian lens, Opt. Mem. Neur. Net., v.17, no.1, p.1-7 (2008).

57. A.G. Nalimov, A.A. Kovalev, V.V. Kotlyar, V.A. Soifer, Three-dimensional simulation of a nanophotonics device with use of fullwave software, Opt. Mem. Neur. Netw. (Inform.Opt.), v.18, no.2, pp.85-92 (2009).

58. M.I. Kotlyar, Y.R. Traindaphilov, A.A. Kovalev, V.A. Soifer, M.V. Kotlyar, L. O'Faolain, Photonic crystal lens for coupling two waveguides, Appl. Opt., v.48, no.19, p.3722-3730 (2009).

59. Knight J.C., Birks T.A., Russel P.S.J., Atkin D.M. All-silica single mode optical fiber photonic crystal cladding , Opt. Lett. 1996. V. 21, No. 19. P. 1547-1549.

60. Adams M.J. An Introduction to Optical Waveguides. – New York: Wiley, 1981.

61. Yeh P., Yariv A., Marom E. Theory of Bragg fiber, J. Opt. Soc. Am. 1978. V. 68. P. 1196-1201.

62. Ibanescu M., Fink M., Fan S., Thomas E.L., Joannopoulos J.D. All-dielectric coaxial waveguide , Science. 2000. No. 289. P. 415-419.

63. Cojocaru E. Dispersion analysis of hollow-core modes in ultralarge-bandwith all-silica Bragg fibers, with nanosupports, Appl. Opt. 2006. V. 45. No. 9. P. 2039 – 2045.

64. Foroni M. et al., Confinement loss spectral behavior in hollow-core Bragg fiber , Opt. Lett. 2007. V. 32. No. 21. P. 3164 – 3166.

65. Zhelticov A.M., Ray-optic analysis of the (bio)sensing ability of ring-cladding hollow waveguides , Appl. Opt. 2008. V. 47. No. 3. P. 474-479.

66. Dupuis A. et al., Guiding in the visible with, colorful solid-core Bragg fiber , Opt. Lett. 2007. V. 32. No. 19. P. 2882-2884.

67. Fang Q. et al. Dispersion design of all-solid photonic bandgap fiber, J. Opt. Soc. Am. A. 2007. V. 24. No. 11. P. 2899-2905.

68. Ren G. et al. Low-loss all-solid photonic bangap fiber, Opt. Lett. 2007. V. 32. No. 9. P. 1023-1025.

69. Yang R. et al. Research of the effects of air hole shape on the properties of microstructured optical fibers, Opt. Eng. 2004. V. 43. No. 11. P. 2701-2706.

70. Yue Y. et al. Highly birefringent elliptical-hole photonic crystal fiber with squeezed hexagonal lattice, Opt. Lett. 2007. V. 32. No. 5. P. 469-471.

71. Choi H.-G. et al. Discpersion and birefringence of irregularly microstructured fiber with elliptical core, Appl. Opt. 2007. V. 46. No. 35. P. 8493-8498.

72. Mafi A., Moloney J.V. Shaping Modes in Multicore Photonic Crystal Fiber, IEEE Photonics Tech. Lett. 2005. V. 17. P. 348-350.

73. Michaille L. et al., Characteristics of a Q-switched multicore photonic crystal fiber

laser with a very large mode field area, Opt. Lett. 2008. V. 33. No. 1. P. 71-73.

74. Szpulak M. et al. Experimental and theoretical investigations of birefringent holey fibers with a triple defect, Appl. Opt. 2005. V. 44. P. 2652-2658.

75. Eguchi M., Tsuji Y. Geometrical birefringence in square-lattice holey fibers having a core consisting of multiple defect, J. Opt. Soc Am. B. 2007. V. 24. No. 4. P.750-755.

76. Zhang Ch. et al. Design of tunable bandgap guidance in high-index filled micro-structure fibers, J. Opt. Soc. Am. A. 2006. V. 23. No. 4. P. 782-786.

77. Sun J. et al. Refractive index measurement using photonic crystal fiber, Opt. Eng. 2007. V. 46. No. 1. P.014402.

78. Sun J., Chan C.C., Hybrid guiding in liquid-crystal photonic crystal fibers, J. Opt. Soc. Am. A. 2007. V. 24. No. 10. P. 2640-2646.

79. Larsen T. et al. Optical devices based on liquid crystal photonic bandgap fibres, Opt. Express. 2003. V. 11. No. 20. P. 2589-2596.

80. Haakestad M. W. et al. Electrically tunable photonic bandgap guidance in a liquid-crystal-filled photonic crystal fiber, IEEE Photon. Technol. Lett. 2005. V. 17. No. 4. P. 819-821.

81. Domachuk P., Nguyen H.C., Eggleton B.J. Transverse probed microfluidic switch-able photonic crystal fiber devices, Photon. Technol. Lett. 2004. V. 16. No. 8. P. 1900-1902.

82. Ferrando A. et al. Nearly zero ultraflattened dispersion in photonic crystal fibers, Opt. Lett. 2000. V. 25. P. 790-792.

83. Broderick AN.G.R. Modeling large air fraction holey optical fiber, J. Opt. Tech. 2000. V. 18. P. 50-56.

84. Broderick AN.G.R. et al. Nonlinearity in holey optical fibers: measurement and fu-ture opportunities, Opt. Lett. 1999. V. 24. P. 1395.

85. White T.P. Multipole method for microstructured optical fibers, J. Opt. Soc. Am. A. 2002. V. 19. No. 10. P. 2322-2330.

86. Steel M.J. et al. Symmetry and degeneracy in microstructured optical fibers, Opt. Lett. 2001. V. 26. P. 488-490.

87. Yamashita E., Ozeki S., Atsuki K. Modal analysis method for optical fibers with sym-metrically distributed multiple cores, J. Lighhtwave Techn. 1985. V. 3. P. 341-346.

88. Tayed, G. Scattering by a random set of parallel cylinders,G. Tayed et al., J. Opt. Soc. Am. A, 1994. – Vol. 11. – PP. 2526-2538.

89. Sudbo, A.S. Film mode matching: A versatile method for mode field calculations in die-lectric waveguides, A.S. Sudbo , Pure Appl. Opt. (J. Europ. Opt. Soc. A), 1993. – Vol. 2. – PP. 211-233.

90. Cucinotta, A. Holey fiber analysis through the finite element method, A. Cucinotta [and other] , IEEE Photon. Technol. Lett., 2002. – Vol. 14. – PP. 1530-1532.

91. Brechet, F. Complete analysis of characteristics of propagation into photonic crystal fi-bers by the finite element method, F. Brechet [and other] , Opt. Fiber Technol., 2000. – Vol. 6(2). –PP. 181-191.

92. Guan, N. Boundary Element Method for Analysis of Holey Optical Fibers, N. Guan [and other], J. Lightwave Technol., 2003. – Vol. 21(8).

93. Cheng, H. Fast, accurate integral equation methods for the analysis of photonic crys-tal fibers, H. Cheng [and other] , Opt. Express, 2004. – Vol. 12(16) – PP. 3791-3805.

94. Riishede, J. A Poor Man's Approach to Modeling Micro-Structured Optical Fibers, J. Riishede, N.S. Mortensen and J. Legsgaard, J. Opt. A: Pure Appl. Opt., 2003. – Vol. 5. – PP. 534-538.

95. Hardley, G.R. Full-vector waveguide modeling using an iterative finite-difference

method with transparent boundary conditions, G.R. Hardley and R.E. Smith , J. Lightwave Technol., 1994. – Vol. 13. – PP. 465-469.

96. Zhu, Z. Full-vectorial finite-difference analysis of microstructured optical fibers, Z. Zhu and T.G. Brown, Opt. Express, 2002. – Vol. 10(17). – PP. 853-864.

97. Jiang, W. An Extended FDTD Method With Inclusion of Material Dispersion for the Full-Vectorial Analysis of Photonic Crystal Fibers, W. Jiang [and other] , J. Lightwave Technol., 2006. – Vol. 24(11). – PP. 4417-4423.

98. Xu, C.L. Full-vectorial mode calculations by finite difference method, C.L. Xu [and other] , Inst Elec. Eng., Proc.-J., 1994. – Vol. 141. – PP. 281-286.

99. Huang, W.P. The finite-difference vector beam propagation method. Analysis and As-sessment, W.P. Huang [and other], J. Lightwave Technol., 1992. – Vol. 10. – PP. 295-305.

100. Xu, C.L. Efficient and accurate vector mode calculations by beam propagation method, C.L. Xu , J. Lightwave Technol., 1993. – Vol. 11(9). – PP. 1209-1215.

101. Itoh, T. Numerical techniques for microwave and millimeter-wave passive structures, T. Itoh – New York, Wiley, 1988.

102. Sorrentino, R. Transverse resonance technique, R. Sorrentino, Ch. 11 in Itoh's book [101].

103. Schlosser, W. Partially filled waveguides and surface waveguides of rectangular cross section, W. Schlosser and H.G. Unger – New York, Advances in Microwaves – Academic Press, 1966.

104. Peng, S.T. Guidance and leakage properties of a class of open dielectric waveguides: Part I – Mathematical formulations, S.T. Peng and A.A. Oliner, IEEE Trans. Microwave Theory Techn., 1981. – Vol. MTT-29. – PP. 843-855.

105. Sudbo, A.S. Improved formulation of the film mode matching method for mode field calculations in dielectric waveguides, A.S. Sudbo, Pure Appl. Opt. (J. Europ. Opt. Soc. A), 1994. – Vol. 3. – PP. 381-388.

106. Kotlyar V.V., Computer Optics, ISOI RAN, 2007. – V. 31. – 27-30.

107. Ruhe, A. A Rational Krylov Algorithm For Nonlinear Matrix Eigenvalue Problems:/ A. Ruhe, Zapiski Nauchnih Seminarov, Steklov Mathematical Institute, 2000 – Vol. 268. - PP.176-180.

108. Kotlyar V.V., Computer Optics, ISOI RAN, 2005. V. 27. – 89-94.

109. Kotlyar V.V., Computer Optics, ISOI RAN, 2005. T. 27. – 84-88.

110. Kotlyar V.V., Computer Optics, ISOI RAN, 2003. T. 25. – 41-48.

111. Jarlebring, E., Rational Krylov for nonlinear eigenproblems, an iterative projection method, E. Jarlebring, H. Voss, Applications of Mathematics, 2005. - Vol.50, PP. 543-554.

112. Snaider A., Theory of optical waveguids, Moscow, Radio i svyaz', 1987.

113. Kotlyar, V.V. Calculating spatial modes in photonic crystal fibers based on applying finite difference method to wave equations, V.V. Kotlyar, Y.O. Shuyupova , Proceedings of ICO Topical Meeting on Optoinformatics/Information Photonics 2006, September 4-7, 2006, St. Petersburg, Russia, PP. 483-485.

114. Kotlyar V.V., Computer Optics, ISOI RAN, 2005. – V. 28. – 41-44.

115. Kotlyar V.V., Opticheskii zhurnal, 2007. – V. 74, No. 9. – 600-608.

3

Sharp focusing of light with microoptics components

3.1. Numerical and experimental methods for studying sharp focusing of light

3.1.1. BOR-FDTD method

Since the focusing is often carried out near the interface of two media in order to overcome the diffraction limit where it is necessary to take into account the vector nature of the electromagnetic field, the numerical investigations are carried using facilities unit based on the exact solution of Maxwell's equations. To calculate the sharp focusing of laser radiation, we mostly used a modification of the finite-difference method for solving Maxwell's equations (FDTD-method) for the cylindrical coordinate system. Modification of the method for the cylindrical coordinate system was proposed in 1994 and named BOR-FDTD [1]. BOR means Body Of Revolution. Using BOR-FDTD compared to the original three-dimensional FDTD-method allows by eliminating the derivative with respect to the azimuthal angle to reduce the three-dimensional problem to a two-dimensional one thus reducing the computational time and costs. Among the shortcomings of the method one can mention, firstly, that the formulas used in the main computed field can not be applied on the axis. This problem is solved by using in calculations of the components of the field at the axis the finite-difference equations other than the equations of the calculated area. Second, when calculating the propagation of electromagnetic waves by BOR-FDTD the wave velocity depends on the direction of propagation (or numerical dispersion). This problem occurs because the cell size (3D cell in a cylindrical coordinate system) near the axis is less than the size of a cell at a considerable distance from the axis. The influence of the numerical dispersion becomes smaller with the decrease of the size of the calculated area along the radial coordinate.

The key stages in the development BOR-FDTD method are the creation of modifications for the oblique incidence of the input radiation [2], modified to calculate the distribution of the components of the field at a distance from the calculated field [3], and modification of perfectly matched layers (PML), which was later applied for the cylindrical coordinate system. [4]

As can be seen from the above references, the BOR-FDTD method has been mostly developed for the microwave or radio-frequency range and not for the optical range. For the optical range it has been applied, for example, in [5], in which, first, describes a detailed derivation of the finite-difference equations, secondly, it is shown that the solution is consistent with the known analytic solution on the sphere, and thirdly, the method is used to analyze the passage of light through the diffractive lens. The same group of researchers later applied the method to calculate the oblique incidence of radiation on the DOE [6]. As an analytical solution with which a comparison was made was the oblique incidence of a plane wave at the interface of two media. The solution in this case can be found by the known Fresnel coefficients. Furthermore, it was shown that, firstly, the increase of the angle of incidence increases the simulation time (discretization step in time depends on the angle of incidence): increase of the angle between zero and 10 degrees resulting increased time up to 600 times. Secondly, despite this the BOR-FDTD method allows one to calculate the passage of light through a diffractive lens with much lower computational cost than the three-dimensional FDTD method: calculation by the BOR-FDTD method would require 90 times less RAM and would be 4.5 times faster .

In [7] the BOR-FDTD method (modification) is used to analyze non-linear Bragg gratings and Bragg cavity. In [8] the method is used to calculate the optimum settings for the apertureless probe for scanning near-field optical microscopy. Surface plasmons are studied in [9]. Calculations by the BOR-FDTD method of the light intensity distribution after passing through the diffractive lens are presented in [10]. The dispersion characteristics of diffractive lenses were studied in [11]. In [12] the method was used to test single-mode fibres. This work deserves special attention, because it also compares the intensity distributions obtained by the BOR-FDTD with the known analytical solution. Simulation of the passage of light having a wavelength $\lambda = 1.55$ μm through a single mode fibre (refractive index of the core $n_c = 1.460$, refractive index of the cladding = 1.455, the core diameter 10 μm) showed that the correlation coefficient is VAF = 0.99 (i.e. curves coincide up to 99%). Such a good result was achieved despite a relatively large size of the Yee cell in space: $\Delta r = \Delta z = \lambda/10$.

Finite difference schemes for Maxwell's equations in cylindrical coordinates

Maxwell's equations in differential form in the SI system in the absence of external currents are as follows:

$$\begin{cases} rot\mathbf{H} = \dfrac{\partial \mathbf{D}}{\partial t} + \mathbf{J}; \\[2mm] rot\mathbf{E} = -\dfrac{\partial \mathbf{B}}{\partial t}; \\[2mm] div\mathbf{D} = \rho; \\[2mm] div\mathbf{B} = 0, \end{cases} \qquad (3.1)$$

where $\mathbf{E}$ is the electric field vector, $\mathbf{H}$ is the magnetic field vector, $\mathbf{D}$ is the electric induction vector , $\mathbf{B}$ is the magnetic induction vector, $\mathbf{J}$ is the conduction current density vector, ρ is density of the free electric charge.

For further discussion, we need only the first two equations of the system (3.1):

$$\begin{cases} rot\mathbf{H} = \dfrac{\partial \mathbf{D}}{\partial t} + \mathbf{J}; \\[2mm] rot\mathbf{E} = -\dfrac{\partial \mathbf{B}}{\partial t}. \end{cases} \qquad (3.2)$$

To these equations we should added three constitutive equations describing the substance with which the electromagnetic wave interacts:

$$\begin{cases} \mathbf{D} = \varepsilon\varepsilon_0 \mathbf{E}; \\ \mathbf{J} = \sigma\mathbf{E}; \\ \mathbf{B} = \mu\mu_0 \mathbf{H}. \end{cases} \tag{3.3}$$

where ε and μ are the permittivity and permeability of the medium, $\varepsilon_0 = 8.85418782 \cdot 10^{-12}$ and $\mu_0 = 1.25663706 \cdot 10^{-6}$ are the dielectric and magnetic constants, σ is electrical conductivity.

We will assume that we are considering an inertialess medium, that is, the permittivity and permeability are independent of time, and isotropic, that is, the values of permittivity and permeability are independent of the choice of the coordinates. Then, taking into account the material equations (3.3), the system (3.2) takes the form:

$$\begin{cases} rot\mathbf{H} = \varepsilon\varepsilon_0 \dfrac{\partial \mathbf{E}}{\partial t} + \sigma\mathbf{E}; \\ rot\mathbf{E} = -\mu\mu_0 \dfrac{\partial \mathbf{H}}{\partial t}. \end{cases} \tag{3.4}$$

In a cylindrical coordinate system, the operator of the rotor has the form:

$$rot(\mathbf{A}) = \left(\frac{1}{r}\frac{\partial A_z}{\partial \varphi} - \frac{\partial A_\varphi}{\partial z} \right)\mathbf{e}_r + \left(\frac{\partial A_r}{\partial z} - \frac{\partial A_z}{\partial r} \right)\mathbf{e}_\varphi + \frac{1}{r}\left(\frac{\partial (rA_\varphi)}{\partial r} - \frac{\partial A_r}{\partial \varphi} \right)\mathbf{e}_z. \tag{3.5}$$

The system of equations (3.4) can be written as:

$$\frac{1}{r}\frac{\partial H_z}{\partial \varphi} - \frac{\partial H_\varphi}{\partial z} = \varepsilon\varepsilon_0 \frac{\partial E_r}{\partial t} + \sigma E_r \tag{3.6}$$

$$\frac{\partial H_r}{\partial z} - \frac{\partial H_z}{\partial r} = \varepsilon\varepsilon_0 \frac{\partial E_\varphi}{\partial t} + \sigma E_\varphi \tag{3.7}$$

$$\frac{1}{r}\frac{\partial (rH_\varphi)}{\partial r} - \frac{1}{r}\frac{\partial H_r}{\partial \varphi} = \varepsilon\varepsilon_0 \frac{\partial E_z}{\partial t} + \sigma E_z \tag{3.8}$$

$$\frac{1}{r}\frac{\partial E_z}{\partial \varphi} - \frac{\partial E_\varphi}{\partial z} = -\mu\mu_0 \frac{\partial H_r}{\partial t} \tag{3.9}$$

$$\frac{\partial E_r}{\partial z} - \frac{\partial E_z}{\partial r} = -\mu\mu_0 \frac{\partial H_\varphi}{\partial t} \tag{3.10}$$

$$\frac{1}{r}\frac{\partial (rE_\varphi)}{\partial r} - \frac{1}{r}\frac{\partial E_r}{\partial \varphi} = -\mu\mu_0 \frac{\partial H_z}{\partial t} \tag{3.11}$$

We expand all components of the field in a Fourier series in the azimuthal angle φ:

$$E_\gamma(r,z,\varphi,t) = \frac{E_{\gamma 0}(r,z,t)}{2} + \sum_{k=1}^{\infty} E_{\gamma,k}^{(1)}(r,z,t)\cos(k\varphi) + E_{\gamma,k}^{(2)}(r,z,t)\sin(k\varphi)$$

$$\tag{3.12}$$

$$H_\gamma(r,z,\varphi,t) = \frac{H_{\gamma 0}(r,z,t)}{2} + \sum_{k=1}^{\infty} H_{\gamma,k}^{(1)}(r,z,t)\cos(k\varphi) + H_{\gamma,k}^{(2)}(r,z,t)\sin(k\varphi)$$

$$\tag{3.13}$$

where
$$E_{\gamma,k}^{(1)}(r,z,t) = \frac{1}{\pi}\int_0^{2\pi} E_{\gamma,k}(r,z,\varphi,t)\cos(k\varphi),$$

$$E_{\gamma,k}^{(2)}(r,z,t) = \frac{1}{\pi}\int_0^{2\pi} E_{\gamma,k}(r,z,\varphi,t)\sin(k\varphi),$$

for the magnetic components.

Substituting (3.12) and (3.13) in the equation (3.6)–(3.11) would find the derivative of the azimuthal angle φ, and the coefficients of the respective multipliers sin $(k\varphi)$ and cos $(k\varphi)$ form four independent systems of equations:

$$\begin{cases} -\dfrac{\partial H_{\varphi,0}}{\partial z} = \varepsilon\varepsilon_0 \dfrac{\partial E_{r,0}}{\partial t} + \sigma E_{r,0} \\[2ex] \dfrac{1}{r}\dfrac{\partial (rH_{\varphi,0})}{\partial r} = \varepsilon\varepsilon_0 \dfrac{\partial E_{z,0}}{\partial t} + \sigma E_{z,0} \\[2ex] \dfrac{\partial E_{r,0}}{\partial z} - \dfrac{\partial E_{z,0}}{\partial r} = -\mu\mu_0 \dfrac{\partial H_{\varphi,0}}{\partial t} \end{cases} \tag{3.14}$$

$$
\begin{cases}
\dfrac{\partial H_{r,0}}{\partial z} - \dfrac{\partial H_{z,0}}{\partial r} = \varepsilon\varepsilon_0 \dfrac{\partial E_{\varphi,0}}{\partial t} + \sigma E_{\varphi,0} \\[3mm]
-\dfrac{\partial E_{\varphi,0}}{\partial z} = -\mu\mu_0 \dfrac{\partial H_{r,0}}{\partial t} \\[3mm]
\dfrac{1}{r}\dfrac{\partial\left(rE_{\varphi,0}\right)}{\partial r} = -\mu\mu_0 \dfrac{\partial H_{z,0}}{\partial t}
\end{cases}
\tag{3.15}
$$

$$
\begin{cases}
-\dfrac{1}{r}kH_{z,k}^{(1)} - \dfrac{\partial H_{\varphi,k}^{(2)}}{\partial z} = \varepsilon\varepsilon_0 \dfrac{\partial E_{r,k}^{(2)}}{\partial t} + \sigma E_{r,k}^{(2)} \\[3mm]
\dfrac{\partial E_{r,k}^{(2)}}{\partial z} - \dfrac{\partial E_{z,k}^{(2)}}{\partial r} = -\mu\mu_0 \dfrac{\partial H_{\varphi,k}^{(2)}}{\partial t} \\[3mm]
\dfrac{1}{r}\dfrac{\partial\left(rH_{\varphi,k}^{(2)}\right)}{\partial r} + \dfrac{1}{r}kH_{r,k}^{(1)} = \varepsilon\varepsilon_0 \dfrac{\partial E_{z,k}^{(2)}}{\partial t} + \sigma E_{z,k}^{(2)} \\[3mm]
\dfrac{\partial H_{r,k}^{(1)}}{\partial z} - \dfrac{\partial H_{z,k}^{(1)}}{\partial r} = \varepsilon\varepsilon_0 \dfrac{\partial E_{\varphi,k}^{(1)}}{\partial t} + \sigma E_{\varphi,k}^{(1)} \\[3mm]
-\dfrac{1}{r}kE_{z,k}^{(2)} - \dfrac{\partial E_{\varphi,k}^{(1)}}{\partial z} = -\mu\mu_0 \dfrac{\partial H_{r,k}^{(1)}}{\partial t} \\[3mm]
\dfrac{1}{r}\dfrac{\partial\left(rE_{\varphi,k}^{(1)}\right)}{\partial r} - \dfrac{1}{r}kE_{r,k}^{(2)} = -\mu\mu_0 \dfrac{\partial H_{z,k}^{(1)}}{\partial t}
\end{cases}
\tag{3.16}
$$

$$
\begin{cases}
\dfrac{\partial H_{r,k}^{(2)}}{\partial z} - \dfrac{\partial H_{z,k}^{(2)}}{\partial r} = \varepsilon\varepsilon_0 \dfrac{\partial E_{\varphi,k}^{(2)}}{\partial t} + \sigma E_{\varphi,k}^{(2)} \\[3mm]
-\dfrac{1}{r}kE_{z,k}^{(1)} - \dfrac{\partial E_{\varphi,k}^{(2)}}{\partial z} = -\mu\mu_0 \dfrac{\partial H_{r,k}^{(2)}}{\partial t} \\[3mm]
\dfrac{1}{r}\dfrac{\partial\left(rE_{\varphi,k}^{(2)}\right)}{\partial r} - \dfrac{1}{r}kE_{r,k}^{(1)} = -\mu\mu_0 \dfrac{\partial H_{z,k}^{(2)}}{\partial t} \\[3mm]
-\dfrac{1}{r}kH_{z,k}^{(2)} - \dfrac{\partial H_{\varphi,k}^{(1)}}{\partial z} = \varepsilon\varepsilon_0 \dfrac{\partial E_{r,k}^{(1)}}{\partial t} + \sigma E_{r,k}^{(1)} \\[3mm]
\dfrac{1}{r}\dfrac{\partial\left(rH_{\varphi,k}^{(1)}\right)}{\partial r} - \dfrac{1}{r}kH_{r,k}^{(2)} = \varepsilon\varepsilon_0 \dfrac{\partial E_{z,k}^{(1)}}{\partial t} + \sigma E_{z,k}^{(1)} \\[3mm]
\dfrac{\partial E_{r,k}^{(1)}}{\partial z} - \dfrac{\partial E_{z,k}^{(1)}}{\partial r} = -\mu\mu_0 \dfrac{\partial H_{\varphi,k}^{(1)}}{\partial t}
\end{cases}
\tag{3.17}
$$

For further calculations we do not need all of the equations (3.14)–(3.17). The choice of the equation will depend on the polarization of light whose propagation is to be studied. We will consider linear and radial polarizations.

Difference schemes for linear polarization

Consider normal incidence of an electromagnetic wave with linear polarization in the initial plane $z = 0$. The electric vector of this wave will have only one projection $E^{inc} = E_y = E_0(r) \cos(\omega t)$, where ω is the angular frequency of a monochromatic wave, $E_0(r)$ is the amplitude of the wave in the plane $z = 0$. We express the incident field E^{inc} through cylindrical components:

$$\begin{cases} E_r^{inc} = E^{inc} \sin \varphi, \\ E_\varphi^{inc} = E^{inc} \cos \varphi. \end{cases} \tag{3.18}$$

Expand the components of the incident field in a Fourier series:

$$E_{\varphi,0}^{inc} = \frac{1}{\pi} \int_{-\pi}^{\pi} E^{inc} \cos \varphi \, d\varphi = 0 \tag{3.19}$$

$$E_{r,0}^{inc} = \frac{1}{\pi} \int_{-\pi}^{\pi} E^{inc} \sin \varphi \, d\varphi = 0 \tag{3.20}$$

$$E_{\varphi,k}^{inc(1)} = \frac{1}{\pi} \int_{-\pi}^{\pi} E^{inc} \cos \varphi \cos k\varphi \, d\varphi = \begin{cases} E^{inc}, k = 1 \\ 0, k \neq 1 \end{cases} \tag{3.21}$$

$$E_{\varphi,k}^{inc(2)} = \frac{1}{\pi} \int_{-\pi}^{\pi} E^{inc} \cos \varphi \sin k\varphi \, d\varphi = 0, \tag{3.22}$$

$$E_{r,k}^{inc(1)} = \frac{1}{\pi} \int_{-\pi}^{\pi} E^{inc} \sin \varphi \cos k\varphi \, d\varphi = 0, \tag{3.23}$$

$$E_{r,k}^{inc(2)} = \frac{1}{\pi} \int_{-\pi}^{\pi} E^{inc} \sin \varphi \sin k\varphi \, d\varphi = \begin{cases} E^{inc}, k = 1 \\ 0, k \neq 1 \end{cases}. \tag{3.24}$$

Thus, the expansion of the components of the incident field is represented as:

$$E_{\varphi,1}^{inc(1)} = E_0 \cos \omega t, \tag{3.25}$$

$$E_{r,1}^{inc(2)} = E_0 \cos \omega t . \tag{3.26}$$

Therefore, from the equations (3.14)–(3.17) for further calculations we need only the equation for calculating the components: $E_{\varphi,1}^{(1)}; E_{r,1}^{(2)}; E_{z,1}^{(2)}; H_{r,1}^{(1)}; H_{z,1}^{(1)}; H_{\varphi,1}^{(2)}$, i.e. the system of equations (3.16) for the case $k = 1$. Its finite-difference approximation with the Yee grid for the cylindrical coordinate system (Fig. 3.1) will have the following form:

$$\varepsilon(i+\frac{1}{2},j)\varepsilon_0 \frac{E_{r,1}^{(2)n+1}(i+\frac{1}{2},j)-E_{r,1}^{(2)n}(i+\frac{1}{2},j)}{\Delta t} =$$

$$= -\frac{1}{r(i+\frac{1}{2})} H_{z,1}^{(1)n+\frac{1}{2}}(i+\frac{1}{2},j)- \tag{3.27}$$

$$-\frac{H_{\varphi,1}^{(2)n+\frac{1}{2}}(i+\frac{1}{2},j+\frac{1}{2})-H_{\varphi,1}^{(2)n+\frac{1}{2}}(i+\frac{1}{2},j-\frac{1}{2})}{\Delta z}$$

$$\varepsilon(i,j+\frac{1}{2})\varepsilon_0 \frac{E_{z,1}^{(2)n+1}(i,j+\frac{1}{2})-E_{z,1}^{(2)n}(i,j+\frac{1}{2})}{\Delta t} =$$

$$= \frac{1}{r(i)} \frac{r(i+\frac{1}{2})H_{\varphi,1}^{(2)n+\frac{1}{2}}(i+\frac{1}{2},j+\frac{1}{2})-r(i-\frac{1}{2})H_{\varphi,1}^{(2)n+\frac{1}{2}}(i-\frac{1}{2},j-\frac{1}{2})}{\Delta r} +$$

$$+\frac{1}{r(i)} H_{r,1}^{(1)n+\frac{1}{2}}(i,j+\frac{1}{2})$$

$$\tag{3.28}$$

$$\varepsilon(i,j)\varepsilon_0 \frac{E_{\varphi,1}^{(1)n+1}(i,j)-E_{\varphi,1}^{(1)n}(i,j)}{\Delta t} =$$

$$= \frac{H_{r,1}^{(1)n+\frac{1}{2}}(i,j+\frac{1}{2})-H_{r,1}^{(1)n+\frac{1}{2}}(i,j-\frac{1}{2})}{\Delta z} - \tag{3.29}$$

$$-\frac{H_{z,1}^{(1)n+\frac{1}{2}}(i+\frac{1}{2},j)-H_{z,1}^{(1)n+\frac{1}{2}}(i-\frac{1}{2},j)}{\Delta r}$$

$$-\mu_0 \frac{H_{\varphi,1}^{(2)n+\frac{1}{2}}(i+\frac{1}{2},j+\frac{1}{2}) - H_{\varphi,1}^{(2)n-\frac{1}{2}}(i+\frac{1}{2},j+\frac{1}{2})}{\Delta t} =$$

$$= \frac{E_{r,1}^{(2)n}(i+\frac{1}{2},j+1) - E_{r,1}^{(2)n}(i+\frac{1}{2},j)}{\Delta z} - \qquad (3.30)$$

$$- \frac{E_{z,1}^{(2)n}(i+1,j+\frac{1}{2}) - E_{z,1}^{(2)n}(i,j+\frac{1}{2})}{\Delta r}$$

$$-\mu_0 \frac{H_{r,1}^{(1)n+\frac{1}{2}}(i,j+\frac{1}{2}) - H_{r,1}^{(1)n-\frac{1}{2}}(i,j+\frac{1}{2})}{\Delta t} =$$

$$= \frac{1}{r(i)} E_{z,1}^{(2)n}(i,j+\frac{1}{2}) - \frac{E_{\varphi,1}^{(1)n}(i,j+1) - E_{\varphi,1}^{(1)n}(i,j)}{\Delta z} \qquad (3.31)$$

$$-\mu_0 \frac{H_{z,1}^{(1)n+\frac{1}{2}}(i+\frac{1}{2},j) - H_{z,1}^{(1)n-\frac{1}{2}}(i+\frac{1}{2},j)}{\Delta t} =$$

$$= \frac{1}{r(i+\frac{1}{2})} \frac{r(i+1)E_{\varphi,1}^{(1)n}(i+1,j) - r(i)E_{\varphi,1}^{(1)n}(i,j)}{\Delta r} \qquad (3.32)$$

$$- \frac{1}{r(i+\frac{1}{2})} E_{r,1}^{(2)n}(i+\frac{1}{2},j)$$

where Δt, Δz, Δr are the discrete steps in the corresponding coordinates: $z = i\Delta z$, $r = j\Delta r$, $t = n\Delta t$. The samples of the electrical component are calculated in integer times $t = n\Delta t$, and the samples of the magnetic vectors are computed in half-integer times $t = (n + 1/2)\,\Delta t$. Equations (3.27)–(3.32) are an example of the quasi-stable difference scheme, which is solved by the sweep method with the boundary conditions taken into account. For the stable convergence of the solutions of (3.27)–(3.32) the discretization steps should be selected to satisfy the inequality [5]:

$$c\Delta t \le \frac{\min(\Delta r, \Delta z)}{2} \qquad (3.33)$$

where c is the velocity of light in vacuum.

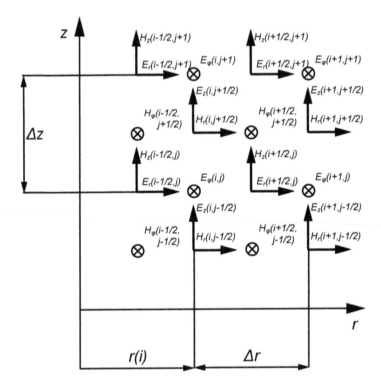

Fig. 3.1. The grid with Yee half step in a cylindrical coordinate system.

Difference schemes for radial polarization

If an optical element whose axis of symmetry coincides with the optical axis is impacted by normally incident monochromatic electromagnetic wave with radial polarization, the electric vector of the incident wave retains only one radial component:

$$E^{inc} = E_r = E_0(r)\cos\omega t .\qquad(3.34)$$

For the radial polarization the electric vector of the incident wave also retains only one Fourier component: $E^{inc} = E_{r,0}$, and from the equations (3.14)–(3.17) only the system of equations (3.32) will remain. Its finite-difference approximation has the form:

$$\varepsilon\left(i+\frac{1}{2},j\right)\varepsilon_0 \frac{E_{r,0}^n\left(i+\frac{1}{2},j\right)-E_{r,0}^{n-1}\left(i+\frac{1}{2},j\right)}{\Delta t}=$$

$$=\frac{H_{\varphi,0}^{n-\frac{1}{2}}\left(i+\frac{1}{2},j+\frac{1}{2}\right)-H_{\varphi,0}^{n-\frac{1}{2}}\left(i+\frac{1}{2},j-\frac{1}{2}\right)}{\Delta z}$$

(3.35)

$$\varepsilon\left(i,j+\frac{1}{2}\right)\varepsilon_0 \frac{E_{z,0}^n\left(i,j+\frac{1}{2}\right)-E_{z,0}^{n-1}\left(i,j+\frac{1}{2}\right)}{\Delta t}=$$

$$=\frac{1}{r(i)}\frac{r\left(i+\frac{1}{2}\right)H_{\varphi,0}^{n-\frac{1}{2}}\left(i+\frac{1}{2},j+\frac{1}{2}\right)-r\left(i-\frac{1}{2}\right)H_{\varphi,0}^{n-\frac{1}{2}}\left(i-\frac{1}{2},j-\frac{1}{2}\right)}{\Delta r}$$

(3.36)

$$-\mu_0 \frac{H_{\varphi,0}^{n+\frac{1}{2}}\left(i+\frac{1}{2},j+\frac{1}{2}\right)-H_{\varphi,0}^{n-\frac{1}{2}}\left(i+\frac{1}{2},j+\frac{1}{2}\right)}{\Delta t}=$$

$$=\frac{E_{r,0}^n\left(i+\frac{1}{2},j+1\right)-E_{r,0}^n\left(i+\frac{1}{2},j\right)}{\Delta z}-\frac{E_{z,0}^n\left(i+1,j+\frac{1}{2}\right)-E_{z,0}^n\left(i,j+\frac{1}{2}\right)}{\Delta r}$$

(3.37)

The Yee grid will take the form shown in Fig. 3.2.

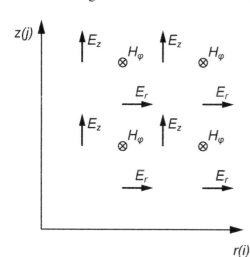

Figure 3.2. Grid with the Yee half step in cylindrical coordinates for radial polarization.

Features of calculating the field on the optical axis

From the equations (3.27)–(3.32) it follows that it is impossible to use them when $r = 0$, that is on the axis of the radially symmetric axis of the optical element.

The first Maxwell equation (3.1) has the form:

$$\nabla \times \mathbf{H} = \frac{\partial \mathbf{D}}{\partial t} + \mathbf{J} \tag{3.38}$$

Consider the contour in the form of a circle with a radius equal to half the Yee grid cell and located perpendicular to the axis (Fig. 3.3).

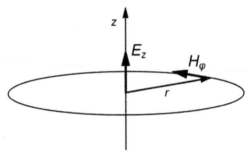

Fig. 3.3. Layout of the optical axis.

Integrating the equation (3.38) on the area bounded by the radius $r = \Delta r/2$:

$$\int_s (\nabla \times \mathbf{H}) ds = \int_s \left(\frac{\partial \mathbf{D}}{\partial t} + \mathbf{J} \right) ds \tag{3.39}$$

According to the Stokes law we can reduce the integral over the area to an integral over the contour:

$$\int_l (\nabla \times \mathbf{H}) d\mathbf{l} = \int_s \left(\frac{\partial \mathbf{D}}{\partial t} + \mathbf{J} \right) dS \tag{3.40}$$

From Fig. 3.3 it is clear that

$$H_\varphi \left(2\pi \frac{\Delta r}{2} \right) = \left(\frac{\partial D_z}{\partial t} + J_z \right) \pi \left(\frac{\Delta r}{2} \right)^2 \tag{3.41}$$

or

$$H_\varphi \frac{4}{\Delta r} = \left(\frac{\partial D_z}{\partial t} + J_z \right) \tag{3.42}$$

In the absence of current in the calculated field:

$$H_\varphi \frac{4}{\Delta r} = \frac{\partial D_z}{\partial t} \qquad (3.43)$$

The finite-difference analogue of this equation would be:

$$\frac{D_z^{n+1}\left(j+\frac{1}{2},i\right) - D_z^{n}\left(j+\frac{1}{2},i\right)}{\Delta t} = \frac{4}{\Delta r} H_\varphi \qquad (3.44)$$

This equation gives the value of the longitudinal component of the electric displacement on the axis:

$$D_z^{n+1}\left(j+\frac{1}{2},i\right) = D_z^{n}\left(j+\frac{1}{2},i\right) + \frac{4\Delta t}{\Delta r} H_\varphi \qquad (3.45)$$

The boundary conditions in the form of absorbing layers
The absorbing layers are arranged on the borders of the calculated area (Fig. 3.4) and are designed to simulate the propagation of electromagnetic waves in perpetuity. The absorbing layers for BOR-FDTD are calculated using equations other than normal FDTD formulas. In this study we will a modification for PML calculations proposed in [5].

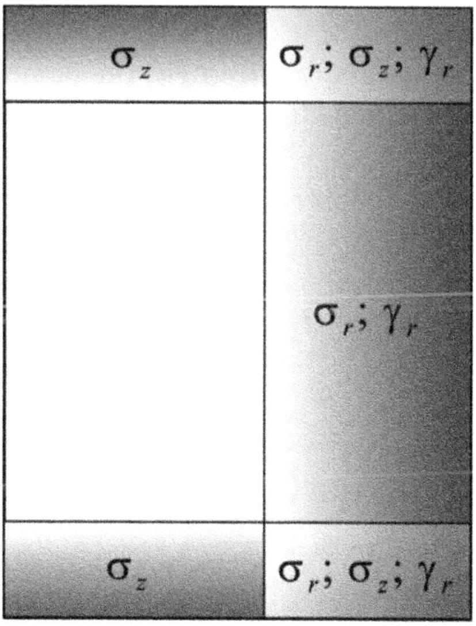

Fig. 3.4. Location of the PML layers for the radial-symmetric case.

The calculation is carried out in two stages: the first stage calculates the induction values of the electric and magnetic fields using the voltage value at the previous time step, and in the second stage we already calculated new field strength values.

Maxwell's equations for a monochromatic field are rewritten as

$$\begin{cases} \nabla \times \tilde{\mathbf{H}} = i\omega\varepsilon_0\tilde{\varepsilon}\tilde{\mathbf{E}} \\ \nabla \times \tilde{\mathbf{E}} = i\omega\mu_0\tilde{\mu}\tilde{\mathbf{H}} \end{cases} \tag{3.46}$$

where

$$\tilde{\mu} = \tilde{\varepsilon} = \begin{pmatrix} \dfrac{s_\varphi s_z}{s_r} & 0 & 0 \\ 0 & \dfrac{s_z s_r}{s_\varphi} & 0 \\ 0 & 0 & \dfrac{s_r s_\varphi}{s_z} \end{pmatrix}, \quad s_r = 1 + \dfrac{\sigma_r}{i\omega\varepsilon_0},$$

$$s_\varphi = 1 + \dfrac{\gamma_r}{i\omega\varepsilon_0} \quad \text{и} \quad s_z = 1 + \dfrac{\sigma_z}{i\omega\varepsilon_0}.$$

Expanding the operator of the rotor and getting rid of the derivatives in the azimuthal angle φ (in the same way as it was in free space), we rewrite the Maxwell equations in the form:

$$i\omega\frac{s_\varphi s_z}{s_r}\mu_0 H_{r,k} = \frac{k}{r}E_{z,k} + \frac{\partial E_{\phi,k}}{\partial z} \tag{3.47}$$

$$i\omega\frac{s_r s_z}{s_\varphi}\mu_0 H_{\varphi,k} = -\frac{\partial E_{r,k}}{\partial z} + \frac{\partial E_{z,k}}{\partial r} \tag{3.48}$$

$$i\omega\frac{s_r s_\varphi}{s_z}\mu_0 H_{z,k} = -\frac{1}{r}\frac{\partial\left(rE_{\varphi,k}\right)}{\partial r} - \frac{k}{r}E_{r,k} \tag{3.49}$$

$$i\omega\frac{s_\varphi s_z}{s_r}\varepsilon_0 E_{r,k} = -\frac{k}{r}H_{z,k} - \frac{\partial H_{\varphi,k}}{\partial z} \tag{3.50}$$

$$i\omega\frac{s_r s_z}{s_\varphi}\varepsilon_0 E_{\varphi,k} = \frac{\partial H_{r,k}}{\partial z} - \frac{\partial H_{z,k}}{\partial r} \tag{3.51}$$

$$i\omega \frac{s_r s_\phi}{s_z} \varepsilon_0 E_{z,k} = \frac{1}{r} \frac{\partial \left(r H_{\phi,k} \right)}{\partial r} - \frac{k}{r} H_{r,k} \qquad (3.52)$$

A two-step method is based on the calculation of the intermediate values of electric and magnetic induction:

$$B_{r,k} = \mu_0 \frac{s_\phi}{s_r} H_{r,k} \qquad (3.53)$$

$$B_{\phi,k} = \mu_0 \frac{s_z}{s_\phi} H_{\phi,k} \qquad (3.54)$$

$$B_{z,k} = \mu_0 \frac{s_r}{s_\phi} H_{z,k} \qquad (3.55)$$

$$D_{r,k} = \varepsilon_0 \frac{s_\phi}{s_r} E_{r,k} \qquad (3.56)$$

$$D_{\phi,k} = \varepsilon_0 \frac{s_z}{s_\phi} E_{\phi,k} \qquad (3.57)$$

$$D_{z,k} = \varepsilon_0 \frac{s_r}{s_\phi} E_{z,k} \qquad (3.58)$$

Substituting them in the equation above:

$$i\omega s_z \mu_0 B_{r,k} = \frac{k}{r} E_{z,k} + \frac{\partial E_{\phi,k}}{\partial z} \qquad (3.59)$$

$$i\omega s_r \mu_0 B_{\phi,k} = -\frac{\partial E_{r,k}}{\partial z} + \frac{\partial E_{z,k}}{\partial r} \qquad (3.60)$$

$$i\omega s_\phi \mu_0 B_{z,k} = -\frac{1}{r} \frac{\partial \left(r E_{\phi,k} \right)}{\partial r} - \frac{k}{r} E_{r,k} \qquad (3.61)$$

$$i\omega s_z \varepsilon_0 D_{r,k} = -\frac{k}{r} H_{z,k} - \frac{\partial H_{\phi,k}}{\partial z} \qquad (3.62)$$

$$i\omega s_r \varepsilon_0 D_{\phi,k} = \frac{\partial H_{r,k}}{\partial z} - \frac{\partial H_{z,k}}{\partial r} \qquad (3.63)$$

$$i\omega s_{\phi}\varepsilon_0 D_{z,k} = \frac{1}{r}\frac{\partial\left(rH_{\phi,k}\right)}{\partial r} - \frac{k}{r}H_{r,k} \tag{3.64}$$

To update the electric induction components, the finite-difference formulas for the first stage will take the form:

$$D_r^{n+1}\left(j,i+\frac{1}{2}\right) = \frac{2\varepsilon\varepsilon_0 - \Delta t\sigma_z}{2\varepsilon\varepsilon_0 + \Delta t\sigma_z}D_r^n\left(j,i+\frac{1}{2}\right) -$$

$$-\frac{2\varepsilon\varepsilon_0\Delta t}{r\left(i+\frac{1}{2}\right)\left(2\varepsilon\varepsilon_0 + \Delta t\sigma_z\right)}H_z^{n+\frac{1}{2}}\left(j,i+\frac{1}{2}\right) -$$

$$-\frac{2\varepsilon\varepsilon_0\Delta t}{\Delta z\left(2\varepsilon\varepsilon_0 + \Delta t\sigma_z\right)}\left(H_\phi^{n+\frac{1}{2}}\left(j+\frac{1}{2},i+\frac{1}{2}\right) - H_\phi^{n+\frac{1}{2}}\left(j-\frac{1}{2},i+\frac{1}{2}\right)\right) \tag{3.65}$$

$$D_z^{n+1}\left(j+\frac{1}{2},i\right) = \frac{2\varepsilon\varepsilon_0 - \Delta t\gamma_r}{2\varepsilon\varepsilon_0 + \Delta t\gamma_r}D_z^n\left(j+\frac{1}{2},i\right) -$$

$$-\frac{2\varepsilon\varepsilon_0\Delta t}{r(i)\left(2\varepsilon\varepsilon_0 + \Delta t\gamma_r\right)}H_r^{n+\frac{1}{2}}\left(j+\frac{1}{2},i\right) +$$

$$+\frac{2\varepsilon\varepsilon_0\Delta t}{r(i)\Delta r\left(2\varepsilon\varepsilon_0 + \Delta t\gamma_r\right)}\left(r\left(i+\frac{1}{2}\right)H_\phi^{n+\frac{1}{2}}\left(j+\frac{1}{2},i+\frac{1}{2}\right) - \tag{3.66}$$

$$-r\left(i-\frac{1}{2}\right)H_\phi^{n+\frac{1}{2}}\left(j+\frac{1}{2},i-\frac{1}{2}\right)\right)$$

$$D_\phi^{n+1}(j,i) = \frac{2\varepsilon\varepsilon_0 - \Delta t\sigma_r}{2\varepsilon\varepsilon_0 + \Delta t\sigma_r}D_\phi^n(j,i) +$$

$$+\frac{2\varepsilon\varepsilon_0\Delta t}{\Delta z\left(2\varepsilon\varepsilon_0 + \Delta t\sigma_r\right)}\left(H_r^{n+\frac{1}{2}}\left(j+\frac{1}{2},i\right) - H_r^{n+\frac{1}{2}}\left(j-\frac{1}{2},i\right)\right) - \tag{3.67}$$

$$-\frac{2\varepsilon\varepsilon_0\Delta t}{\Delta r\left(2\varepsilon\varepsilon_0 + \Delta t\sigma_r\right)}\left(H_z^{n+\frac{1}{2}}\left(j,i+\frac{1}{2}\right) - H_z^{n+\frac{1}{2}}\left(j,i+\frac{1}{2}\right)\right)$$

To update magnetic induction components:

$$B_r^{n+\frac{1}{2}}\left(j+\frac{1}{2},i\right) = \frac{2\varepsilon\varepsilon_0 - \Delta t \sigma_z}{2\varepsilon\varepsilon_0 + \Delta t \sigma_z} B_r^{n-\frac{1}{2}}\left(j+\frac{1}{2},i\right) +$$

$$+ \frac{2\varepsilon\varepsilon_0 \Delta t}{r(i)\left(2\varepsilon\varepsilon_0 + \Delta t \sigma_z\right)} E_z^n\left(j+\frac{1}{2},i\right) + \tag{3.68}$$

$$+ \frac{2\varepsilon\varepsilon_0 \Delta t}{\Delta z\left(2\varepsilon\varepsilon_0 + \Delta t \sigma_z\right)}\left(E_\varphi^n\left(j+1,i\right) - E_\varphi^n\left(j,i\right)\right)$$

$$B_z^{n+\frac{1}{2}}\left(j,i+\frac{1}{2}\right) = \frac{2\varepsilon\varepsilon_0 - \Delta t \gamma_r}{2\varepsilon\varepsilon_0 + \Delta t \gamma_r} B_z^{n-\frac{1}{2}}\left(j,i+\frac{1}{2}\right) +$$

$$+ \frac{2\varepsilon\varepsilon_0 \Delta t}{r\left(i+\frac{1}{2}\right)\left(2\varepsilon\varepsilon_0 + \Delta t \gamma_r\right)} E_r^n\left(j,i+\frac{1}{2}\right) -$$

$$- \frac{2\varepsilon\varepsilon_0 \Delta t}{r\left(i+\frac{1}{2}\right)\Delta r\left(2\varepsilon\varepsilon_0 + \Delta t \gamma_r\right)}\left(r\left(i+1\right)E_\varphi^n\left(j,i+1\right) - r\left(i\right)E_\varphi^n\left(j,i\right)\right)$$

$$\tag{3.69}$$

$$B_\varphi^{n+\frac{1}{2}}\left(j+\frac{1}{2},i+\frac{1}{2}\right) = \frac{2\varepsilon\varepsilon_0 - \Delta t \sigma_r}{2\varepsilon\varepsilon_0 + \Delta t \sigma_r} B_\varphi^{n-\frac{1}{2}}\left(j+\frac{1}{2},i+\frac{1}{2}\right) -$$

$$- \frac{2\varepsilon\varepsilon_0 \Delta t}{\Delta z\left(2\varepsilon\varepsilon_0 + \Delta t \sigma_r\right)}\left(E_r^n\left(j+1,i+\frac{1}{2}\right) - E_r^n\left(j,i+\frac{1}{2}\right)\right) - \tag{3.70}$$

$$+ \frac{2\varepsilon\varepsilon_0 \Delta t}{\Delta r\left(2\varepsilon\varepsilon_0 + \Delta t \sigma_r\right)}\left(E_z^n\left(j+\frac{1}{2},i+1\right) - E_z^n\left(j+\frac{1}{2},i\right)\right)$$

The second step is the calculation of the stress components by the formulas (3.53)–(3.58). Then the finite-difference formula will look like:

$$E_r^{n+1} = \frac{2\varepsilon\varepsilon_0 - \Delta t \gamma_r}{2\varepsilon\varepsilon_0 + \Delta t \gamma_r} E_r^n + \frac{1}{\varepsilon\varepsilon_0}\left(\frac{2\varepsilon\varepsilon_0 + \Delta t \sigma_r}{2\varepsilon\varepsilon_0 + \Delta t \gamma_r} D_r^{n+1} - \frac{2\varepsilon\varepsilon_0 - \Delta t \sigma_r}{2\varepsilon\varepsilon_0 + \Delta t \gamma_r} D_r^n\right)$$

$$\tag{3.71}$$

$$E_z^{n+1} = \frac{2\varepsilon\varepsilon_0 - \Delta t \sigma_r}{2\varepsilon\varepsilon_0 + \Delta t \sigma_r} E_z^n + \frac{1}{\varepsilon\varepsilon_0}\left(\frac{2\varepsilon\varepsilon_0 + \Delta t \sigma_z}{2\varepsilon\varepsilon_0 + \Delta t \sigma_r} D_z^{n+1} - \frac{2\varepsilon\varepsilon_0 - \Delta t \sigma_z}{2\varepsilon\varepsilon_0 + \Delta t \sigma_r} D_z^n\right)$$

$$\tag{3.72}$$

$$E_\varphi^{n+1} = \frac{2\varepsilon\varepsilon_0 - \Delta t\sigma_z}{2\varepsilon\varepsilon_0 + \Delta t\sigma_z} E_\varphi^n + \frac{1}{\varepsilon\varepsilon_0}\left(\frac{2\varepsilon\varepsilon_0 + \Delta t\gamma_r}{2\varepsilon\varepsilon_0 + \Delta t\sigma_z} D_\varphi^{n+1} - \frac{2\varepsilon\varepsilon_0 - \Delta t\gamma_r}{2\varepsilon\varepsilon_0 + \Delta t\sigma_z} D_\varphi^n\right)$$

$$\tag{3.73}$$

$$H_r^{n+\frac{1}{2}} = \frac{2\varepsilon\varepsilon_0 - \Delta t\gamma_r}{2\varepsilon\varepsilon_0 + \Delta t\gamma_r} H_r^{n-\frac{1}{2}} + \frac{1}{\mu\mu_0}\left(\frac{2\varepsilon\varepsilon_0 + \Delta t\sigma_r}{2\varepsilon\varepsilon_0 + \Delta t\gamma_r} B_r^{n+\frac{1}{2}} - \frac{2\varepsilon\varepsilon_0 - \Delta t\sigma_r}{2\varepsilon\varepsilon_0 + \Delta t\gamma_r} B_r^{n-\frac{1}{2}}\right)$$

$$\tag{3.74}$$

$$H_z^{n+\frac{1}{2}} = \frac{2\varepsilon\varepsilon_0 - \Delta t\sigma_r}{2\varepsilon\varepsilon_0 + \Delta t\sigma_r} H_z^{n-\frac{1}{2}} + \frac{1}{\mu\mu_0}\left(\frac{2\varepsilon\varepsilon_0 + \Delta t\sigma_z}{2\varepsilon\varepsilon_0 + \Delta t\sigma_r} B_z^{n+\frac{1}{2}} - \frac{2\varepsilon\varepsilon_0 - \Delta t\sigma_z}{2\varepsilon\varepsilon_0 + \Delta t\sigma_r} B_z^{n-\frac{1}{2}}\right)$$

$$\tag{3.75}$$

$$H_\varphi^{n+\frac{1}{2}} = \frac{2\varepsilon\varepsilon_0 - \Delta t\sigma_z}{2\varepsilon\varepsilon_0 + \Delta t\sigma_z} H_\varphi^{n-\frac{1}{2}} + \frac{1}{\mu\mu_0}\left(\frac{2\varepsilon\varepsilon_0 + \Delta t\gamma_r}{2\varepsilon\varepsilon_0 + \Delta t\sigma_z} B_\varphi^{n+\frac{1}{2}} - \frac{2\varepsilon\varepsilon_0 - \Delta t\gamma_r}{2\varepsilon\varepsilon_0 + \Delta t\sigma_z} B_\varphi^{n-\frac{1}{2}}\right)$$

$$\tag{3.76}$$

3.1.2. Richards–Wolf vector formula

Most studies of the modelling of sharp focusing of the laser beam used the Debye vector theory or the Richards–Wolf similar to it. In these theories the electromagnetic field in the image of a point source, located at infinity, is expressed by an aplanatic optical system in integral form as the expansion in plane waves. Some studies used the Rayleigh–Sommerfeld diffraction formula. Thus, in [13] on the basis of the Debye formulas that are true if the focal length is much larger than the wavelength, it is shown that using a parabolic mirror or a flat diffractive lens with a numerical aperture $NA = 0.98$, the radially polarized hollow Gaussian beam with the amplitude $r \exp(-r^2/w^2)$, where r is the radial coordinate, w is the Gaussian beam waist radius, can be focused with an aplanatic lens in the focal spot area HMA $= 0.210\lambda^2$ and HMA $= 0.157\lambda^2$ respectively. It was also calculated [13] that for a parabolic mirror with a numerical aperture $NA = 1$ the area of the focal spot is less than HMA $= 0.154\lambda^2$. And if the Gaussian beam is restricted with a narrow annular aperture, the area of the focal spot will be even less HMA $= 0.101\lambda^2$.

In [14] attention was given to the non-paraxial propagation of helically polarized Laguerre–Gaussian (LG) beams. It is shown that such beams are also candidates for sharp focusing. Radially polarized laser beams can produced using a conventional interferometer, in which the shoulders have two phase steps, giving a half wavelength delay, and rotated relative to each other by 90° around the optical axis [15]. In [16] the Richards–Wolf (RW)formulas are used to model focusing a linearly polarized plane wave using an aplanatic lens with a high numerical aperture together with a ring (2 or 3 rings) step phase mask. The parameters of the mask resulting in a super-resolution of 20% along the optical axis are calculated. Using the RW formulas in [17] it is shown that when focusing the radially polarized laser mode TEM_{11}, which has in its cross-section two light rings, using an aplanatic lens with $NA = 1.2$ in water ($n = 1.33$) a dark area appears in the focal region and is surrounded on all sides by a light region (optical bottle). Moreover, the longitudinal dimension of the area 2λ, and transverse λ. In [18] the RW formulas are used to study the propagation of an optical vortex with circular polarization. It is shown that at the topological charge $n = 1$ and selecting a sign at which the spiral rotation of the phase of the optical vortex compensates rotation of polarization rotation in the opposite direction, the focal plane ($NA = 0.9$) shows the formation of a circular focal spot with a diameter smaller than the wavelength.

With the help of the Rayleigh–Sommerfeld integral (RS-integral) the distribution of the LG mode with radial polarization, but without the spiral phase component was studied in [19]. It is shown that when we select the non-paraxiality paramater $f = (kw)^{-1}$, where k is the wave number of light, w is the Gaussian beam waist, equal to 0.5, the light spot diameter at the Fresnel distance from the waist is about 0.4λ for $p = 3$, where p is the order of the Laguerre polynomial. In [20] Maxwell's equations in cylindrical coordinates were solved by a series expansion in the non-paraxiality parameter $f = \theta/2$, where θ is the angle of diffraction, up to θ^5. The diffraction of the beam of the axicon–Gauss was used as an example. It is shown that at the diffraction angle $\theta = 0.75$ the waist radius is 0.424λ. The RS-integral was used in [21] to obtain analytical expressions describing the non-paraxial propagation of elegant LG modes. The cross-section of these modes always shows an annular intensity distribution. In [22] the authors reported a new form of a resist for lithography PMMA-DR1, which has polarization-filtering properties, and only responds to the longitudinal component of the electric vector of the electromagnetic wave. The paper shows by experiments that the radially polarized beam of an argon laser $\lambda = 514$ nm, passing an axicon with $NA = 0.67$, forms a focal spot with diameter FWHM $= 0.89\lambda$, but after writing to the resist a spot with a size of 0.62λ forms.

Using the RW formulas it is shown in [23] that the radially polarized higher modes of laser radiation R-TEM$_{pl}$ can reduce the diameter of the focal spot. For example, when $NA = 1$ and focusing using an aplanatic lens for mode numbers $p = 0,1,2,3$ focal spots with a diameter FWHM = 0.582λ, 0.432λ, 0.403λ, 0.378λ were produced. In [24] the RW formulas were used to study vector diffraction and focusing by an aplanatic lens of a linearly polarized beam with the elliptical radial symmetry and eccentricity 0.87. The numerical aperture was $NA = 0.9$. In this case, an elliptical focal spot with the area HMA = $0.56\lambda^2$ was produced. In [25] attention was given to the non-paraxial distribution (amendments of the 5th order) of radially polarized beams LG R-TEM$_{pl}$. It is shown that at the diffraction angle $\theta = 2(kw)^{-1}$ greater than 0.5 the non-paraxial amendments of the 5th order are not enough to describe the mode R-TEM$_{21}$. In [26] using the RW formula it is shown that when the exit pupil of a spherical lens is illuminated with a plane, Gaussian or Bessel–Gaussian beam of radially polarized light the focal spot diameter is equal to FWHM = 0.6λ, 1.2λ, 1.4λ respectively, when $NA = 1.4$, $\lambda = 632.8$ nm, $n = 1.5$. In [27] also with the help of the RW formulas it is shown that for the incident radially polarized Bessel-Gaussian beam $J_1(2r)\exp(-r^2)$ and a binary phase Fresnel plate the focal spot has a diameter FWHM = $0.425\lambda/NA$. After adding a three-zone optimized plate to the Fresnel zone plate the diameter of the focal spot was even smaller FWHM = $0.378\lambda/NA$.

In [28] using a parabolic mirror with a diameter of 19 mm and $NA = 0.999$, and with the help of a radially polarized laser beam with a wavelength of 632.8 nm, the authors experimentally obtained the smallest focal spot today with the area HMA = $0.134\lambda^2$. Radial polarization was produced from the linear polarization of the laser beam with four half-wave plates arranged in the four quadrants of the beam aperture and rotated by 45 degrees (along the bisector in each quadrant). The incident beam has the Bessel–Gauss amplitude. Modelling was performed using the Debye formulas. Intensity distribution in the focal plane was measured using a fluorescence sphere with a diameter of 40 nm. For comparison, the radius of the Airy disk in the scalar approximation is 0.61λ when $NA = 1$, and in [28], a focal spot with a radius from the maximum to the first minimum equal to 0.45λ was produced. Recall that the best experimental result for the aplanatic lens is HMA = $0.16\lambda^2$ [29].

An interesting result was obtained in [30], where it is shown with the help of the Debye formula that the radially polarized LG modes of even orders LG^0_p at the numerical aperture $NA = 0.85$ after passage of a special ring amplitude mask are focuses on a small spot with almost no side lobes, with an area of half intensity $HMA = 0.276\lambda^2$. Using the amplitude mask does not reduce the size of the focal spot but reduces the level of the sidelobes in the focal diffraction pattern and also reduces 5 times the depth of focus. In [31], based on the scalar version of the RW-formula, the authors analytically investigated the optimal functions of the exit pupil for obtaining high resolution. In the two-dimensional case, the radiation propagating in a planar waveguide, and focusing of this radiation at the output waveguide can be studied using the photonic crystal lens.

According to the theory of Debye vector electric field vector electromagnetic waves in the focus area in cylindrical coordinates (r, ψ, z) is expressed in terms of the amplitude $l(\theta)$ converging spherical wave in the coordinates of the exit pupil aplanatic optical system in the form (linear polarization vector directed along the y axis):

$$E_x(r,\psi,z) = \frac{-iA}{2\pi} \int_0^\alpha d\theta \int_0^{2\pi} d\varphi \sin\theta \sqrt{\cos\theta} \sin 2\varphi$$
$$(1-\cos\theta)l(\theta)\exp\left[ikz\cos\theta - ikr\sin\theta\cos(\psi-\varphi)\right], \tag{3.77}$$

$$E_y(r,\psi,z) = \frac{iA}{2\pi} \int_0^\alpha d\theta \int_0^{2\pi} d\varphi \sin\theta \sqrt{\cos\theta}$$
$$\times\left[(1+\cos\theta) + (1-\cos\theta)\cos 2\varphi\right]$$
$$\times l(\theta)\exp\left[ikz\cos\theta - ikr\sin\theta\cos(\psi-\varphi)\right], \tag{3.78}$$

$$E_z(r,\psi,z) = \frac{iA}{\pi} \int_0^\alpha d\theta \int_0^{2\pi} d\varphi \sin^2\theta \sqrt{\cos\theta} \cos\varphi$$
$$l(\theta)\exp\left[ikz\cos\theta - ikr\sin\theta\cos(\psi-\varphi)\right], \tag{3.79}$$

where A is a constant, $\alpha = \arcsin(NA)$, k is the wave number of the light. For example, the Gaussian function in the pupil plane will be:

$$l(\theta) = \exp\left(\frac{-\rho^2}{w^2}\right) = \exp\left[-\left(\frac{\beta\sin\theta}{\sin\alpha}\right)^2\right], \tag{3.80}$$

where β is a constant.

On the basis of the Debye formulas (3.30)–(3.32) Richards and Wolf derived simpler formulas by doing the integration over the azimuthal angle φ for the radially polarized light [80]:

$$E_r(r,z) = A \int_0^\alpha \sin 2\theta \sqrt{\cos \theta}\, l(\theta) \exp\left[ikz \cos \theta\right] J_1(kr \sin \theta) d\theta, \quad (3.81)$$

$$E_z(r,z) = 2iA \int_0^\alpha \sin^2 \theta \sqrt{\cos \theta}\, l(\theta) \exp\left[ikz \cos \theta\right] J_0(kr \sin \theta) d\theta, \quad (3.82)$$

where $J_0(x)$ and $J_1(x)$ are the Bessel functions. From (3.81) and (3.82) it follows that the radially polarized wave does not depend on the angle ψ, has only two electrical components E_r and E_z, and also that $E_r(r = 0) = 0$ for all z and any function $l(\theta)$. The focus is at the origin $(r, \psi, z) = (0, \psi, 0)$. When replacing the aplanatic lens by the Fresnel zone plate, instead of the factor $(\cos\theta)^{1/2}$ in (3.81), (3.82) we should use the other factor $(\cos \theta)^{-3/2}$ [13].

In the Cartesian coordinates for linearly polarized light (polarization vector is directed along the axis y) the RW equations will take the form:

$$E_x(r,\psi,z) = -iA \sin 2\psi \int_0^\alpha \sin \theta \sqrt{\cos \theta}$$
$$\times l(\theta)(1 - \cos \theta) \exp\left[ikz \cos \theta\right] J_2(kr \sin \theta) d\theta, \quad (3.83)$$

$$E_y(r,\psi,z) = -iA \cos 2\psi \int_0^\alpha \sin \theta \sqrt{\cos \theta}\, l(\theta)(1 - \cos \theta)$$
$$\exp\left[ikz \cos \theta\right] J_2(kr \sin \theta) d\theta -$$
$$iA \int_0^\alpha \sin \theta \sqrt{\cos \theta}\, l(\theta)(1 + \cos \theta)$$
$$\exp\left[ikz \cos \theta\right] J_0(kr \sin \theta) d\theta, \quad (3.84)$$

$$E_z(r,\psi,z) = -2A \cos \psi \int_0^\alpha \sin^2 \theta \sqrt{\cos \theta}\, l(\theta) \exp\left[ikz \cos \theta\right] J_1(kr \sin \theta) d\theta \quad (3.85)$$

In the case of azimuthal polarization only the azimuthal component of the electric field will differ from zero:

$$E_\psi(r,z) = 2A \int_0^\alpha \sin \theta \sqrt{\cos \theta}\, l(\theta) \exp\left[ikz \cos \theta\right] J_1(kr \sin \theta) d\theta. \quad (3.86)$$

From (3.86) it is clear that the azimuthally polarized wave does not depend on angle ψ. These formulas (3.81)–(3.86) were used for simulation in [13,16–18, 21,23,24,26–30] studying sharply focused laser light.

The minimum focal spot: analytical evaluation
In [31] the intensity distribution of light in the focus of a radially symmetric optical system with a high numerical aperture was analyzed by the scalar form of the RW formula:

$$U(r,z) = -ikf \int_0^\alpha P(\theta) \exp[ikz \cos \theta] J_0(kr \sin \theta) d\theta, \quad (3.87)$$

where $U(r,z)$ is the complex amplitude of the light near the focus, $P(\theta)$ is the pupil function of the optical system, f is the focal length. If we compare (3.87) with (3.82) for the longitudinal component of the radially polarized light, it is possible to conclude that the scalar amplitude, describing non-paraxial light focusing (3.87), is proportional to the longitudinal component of the converging spherical wave with radial polarization (3.82). Using the reference integrals in [32], we can estimate the smallest possible diameter of the focal spot at the focus of the non-paraxial optical system. Using the reference integral

$$\int_0^\pi \begin{Bmatrix} \sin(bx) \\ \cos(bx) \end{Bmatrix} J_v(c \sin x) dx = \pi \begin{Bmatrix} \sin(b\pi/2) \\ \cos(b\pi/2) \end{Bmatrix} J_{(v-b)/2}(c/2) J_{(v+b)/2}(c/2),$$

$$(3.88)$$

and putting in (3.87) $P(\theta) = \sin \theta$, $a = \pi$, and in (3.88) $v = 0$, $b = 1$, $c = kr$, we obtain following the complex amplitude from equation (3.87) for the uniform pupil in the focal plane $z = 0$:

$$U_1(r, z = 0) = -2ikf \sin(kr) / (kr) \qquad (3.89)$$

From equation (3.89) it follows that the minimum diameter of the focal spot (twice the distance from the maximum to the first minimum) equals

$$D_1 = \lambda \qquad (3.90)$$

the diameter of such a focal spot is at half intensity FWHM = 0.44λ, and the area of the spot at half of intensity HMA = $0.152\lambda^2$. The latter figure is consistent with the calculation [13].

The same order of magnitude result can be obtained if we select a uniform pupil function $P(\theta) = 1$. Then, instead of (3.87) with (3.88) taken into account, we obtain ($v = 0$, $b = 0$):

$$U_2(r, z = 0) = -ikf\,\pi J_0^2(kr\,/\,2) \qquad (3.91)$$

From (3.91) it follows that in this particular case, the diameter of the focal spot (twice the distance from the maximum to the first intensity minimum) is equal to

$$D_2 = 1.53\lambda \qquad (3.92)$$

If the focusing lens is illuminated with a narrow annular field with the pupil function $P(\theta) = \delta(\theta-\alpha)$, then from (3.87) we obtain for the amplitude at the focus:

$$U_3(r, z = 0) = -2ikf J_0(krNA)\,. \qquad (3.93)$$

From (3.93), it follows that the diameter of the focal spot, identical to (3.90) and (3.92), is equal to ($NA = 1$)

$$D_3 = 0.76\lambda \qquad (3.94)$$

and the diameter of the focal spot is FWHM = 0.36λ, and the area of the spot is HMA = $0.101\lambda^2$. The latter figure is consistent with the calculation [13].

The formulas (3.90), (3.92) and (3.94) give only an estimate of the minimum diameter of the focal spot by the scalar formula (3.87), but in sharp focusing it is required to take into account the vector nature of the field, when all three components of the electric fields make comparable contributions to the formation of the focal pattern. The value (3.94) can be regarded as the accurate minimum size of the focal spot, which can be formed by the focusing optical system illuminated by radially polarized light. This follows from the fact that the scalar equation (3.87) is identical with the expression for the longitudinal component of the field (3.82), and the radial component (3.81) of the radially polarized light on the optical axis is zero. But the Debye and Richards–Wolf formulas are approximate (they were obtained under the condition that the focal length of the optical system is much larger than the wavelength), so we will continue to consider the rigorous solution of the diffraction problem based on the numerical solution of Maxwell's equations. Only when we have the exact solution of the diffraction problem we can, in the case of the focal length comparable to the wavelength, hope to obtain the area of the focal spot smaller than HMA = $0.101\lambda^2$.

3.1.3. Scanning near-field optical microscopy

As part of the study of the objects we carried out focusing in the near field where there are evanescent waves, and for the experimental verification of the results we used scanning near-field optical microscopy (SNOM), enabling the detection of evanescent waves. The SNOM concept is to use probes which, being in close proximity to the sample surface, are able to transform the evanescent wave to a propagating one. The first scanning near-field microscope was designed in 1984 by Dieter Pohl and his colleagues [33]. Thus, the first SNOM was created only two years later after the construction of the scanning tunneling microscope – the first probe microscope [34].

In our studies used a Ntegra Spectra (NT-MDT) microscope, its image and the optical scheme of the experiments are shown in Fig. 3.5.

Measurements of the intensity distribution at the focus, formed by the object under investigation, were conducted as follows. Linearly polarized light from a solid state laser with a wavelength of 532 nm or a helium–neon laser with a wavelength of 633 nm was focused by the lens $L1$ on the surface of a glass substrate on which the microoptics object to be analyzed was placed. Behind it was the cantilever C (Fig. 3.6), through which the scanning parallel to the substrate surface at different distances from it was carries out. The light that passed through the hole in the cantilever was then collected

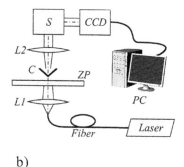

a)   b)

Fig. 3.5. (a) General view of the microscope Ntegra Spectra, (b) the optical scheme of the experiment: L – laser, $L1$, $L2$ – lens, $A1$ – the object under study, ZP – cantilever with a hole, S – spectrometer, CCD – CCD camera, PC – the computer.

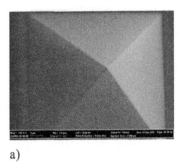 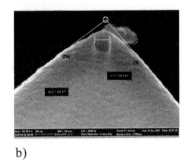

a)   b)

Fig. 3.6. A tetrahedral metal cantilever with a 100 nm hole used in the near-field microscope: top view in an electron microscope (a) and an enlarged side view (b). The hole is shown by the horizontal segment.

by the lens $L2$, passed through the spectrometer S (to filter extraneous radiation) and recorded in the CCD-camera.

3.2. Axicon

To focus light using evanescent waves it is required to form a focal spot near the media interface. The simplest refractive optical element that can focus light near the surface is a conical axicon. The axicon for use in optical applications was first proposed in 1954 [35] (diffraction analogue – in 1958 [36]). Note also that the analytical expressions for the axial intensity at diffraction of a plane wave and a Gaussian beam on a conventional conical axicon were first obtained in [37]. However, despite a considerable age, the interest in the study of axicons and similar structures is not diminishing. For example, in [38] the authors developed an approximate theory, which describes well the diffraction axicon with the period of rings $T <$ 5λ. In this case, the binary axicon can be regarded as a diffraction grating, if the central portion of axicon is covered with an opaque disk. It is shown that the axicon with period $T = 5\lambda$, radius 40λ at a distance from the surface of the axicon focal spot diameter is equal to FWHM $= 0.88\lambda$. In [39] the authors studied experimentally a binary axicon (diameter 30 mm) with a period $T = 33$ μm (it corresponds to the conical axicon of glass with an apex angle of 88°), formed on the resist ZEP520A (refractive index $n = 1.46$). It has been shown that the diameter of the laser beam does not depend on the wavelength, that at a distance of 0 to $z = 50$ mm the radius of the Bessel beam increases from 1.2 μm ($\lambda = 532$ nm) to 12.5 μm, and further from $z = 50$ mm to $z = 100$ mm retains this radius.

Furthermore, the axicon is used to generate a diffraction-free Bessel beam at a certain segment of the optical axis which is also of considerable interest. For example, in [40] a 30 μm diameter fibre with a circular cross section of fused silica (ring thickness 3 μm) at the end of which there was a lens with a radius of curvature of 70 μm, was used to form a Bessel beam with a diameter of 20 μm and a length of 500 μm (wavelength $\lambda = 1.55$ μm). In [41] the FDTD method was used to simulate a 2D photonic crystal of a rectangular grid of dielectric rods in the form of the axicon: base $20a$, the height of the axicon $10a$, the refractive index of the rods $n = 3.13$, the radius of the rods $0.22a$, wavelength $\lambda = a/0.36$, a is the period of the grating of rods. It is shown that at a distance $z < 30a$ a divergent Bessel beam with a diameter at half intensity FWHM = 1.5λ formed. In [42] the experiments were carried out using a radially polarized laser beam ($\lambda = 532$ nm) a conical axicon and an immersion microlens with a numerical aperture $NA = 1.25$ in a silver film with thickness of 50 nm (with a dielectric constant $\varepsilon = -10.1786 - i0.8238$) to produce a surface plasmon wave in the form of concentric rings, described by the Bessel function of the first order. The radiud of the central ring 278 nm, thickness 250 nm $\approx 0.5\ \lambda$. The pattern of the surface plasmon was observed in the near-field microscope Veeco Aurora 3 with a resolution of 50–100 nm. Similarly, in [43] using the He–Ne laser ($\lambda = 632.8$ nm) emitting radially polarized light, the axicon and an immersion lens with $NA = 1.4$ a surface plasmon with a focal point in the centre with diameter FWHM = 0.22 μm = 0.35 λ was produced in a gold film with a thickness 44 nm ($\varepsilon = 0.3 + i3.089$). The plasmon was observed using a latex sphere with a diameter of 175 nm.

3.2.1. Sharp focusing of light with radial polarization using a microaxicon

The possibility to overcome the diffraction limit by changing the state of polarization was shown for the first time in [29] where a radially polarized beam was used to produce a focal spot with the subwavelength diameter. The resultant the spot area was 35% smaller than that of spots obtained by linearly polarized light. The linearly polarized light was converted to radially polarized using a four-sector plate with each sector containing a half-wave plate rotating the plane of polarization by 90 degrees. The method of focusing with radially polarized light has been named by this group of researchers 'sharp focusing' [29]) or 'tight focusing' [44]. In this paper we will used the first term – sharp focusing, however, it should be noted that the second name is used more often in the foreign literature.

Subsequent studies of sharp focusing have mainly been aimed at reducing the size of the focus by adjusting the polarization, the amplitude and phase of the illuminating beam, and also the parameters of the focusing element. Often the purpose of work was, along with a decrease in the width of the focal spot, to increase the depth of focus.

In early studies the focusing element was a wide-aperture aplanatic lens [29]. Later, in [1] in focusing a radially polarized Hermite–Gaussian beam (the amplitude of the incident beam is proportional to $\rho \cdot \exp\{-\rho^2/\omega^2\}$, where ρ is the radial coordinate and ω the beam radius) it was shown numerically (simulation was conducted using the Richards–Wolf formulas (RW formulas)) that the parabolic mirror and the flat diffractive lens can focus light into a smaller spot than the aplanatic lens. So when focusing with the flat diffractive lens with a numerical aperture $NA = 0.98$ the area of the focal spots was HMA = $0.14\lambda^2$ (this is 2.38 times smaller than the area formed by the aplanatic lens with the same numerical aperture). In [28] it was experimentally confirmed that a parabolic mirror with a diameter of 19 mm and a numerical aperture $NA = 0,999$ allows one to focus a radially polarized laser beam with a wavelength $\lambda = 632.8$ nm into a spot with an area of HMA = $0.134\lambda^2$ (FWHM = 0.41λ). The radial polarization was obtained from the linear polarization of the laser beam by means of four half-wave plates arranged in the four quadrants of the beam aperture and rotated by 45 degrees (along the bisector in each quadrant). The incident beam has the Bessel–Gauss amplitude. Modelling was performed using the Debye formulas. The intensity distribution in the focal plane was measured using a fluorescence sphere with a diameter of 40 nm. In [45], the authors numerically compared the focusing by the phase zone plate (with eight zones and a numerical aperture $NA = 0.996$) of a plane linearly polarized wave with a radially polarized Bessel-Gaussian beam. It has been shown that such a phase zone plate can focus the Bessel–Gaussian beam into a focal spot with a diameter at half intensity FWHM = 0.39λ, and a linearly polarized plane wave into an elliptical spot with diameters FWHM$_x$ = 0.87λ and FWHM$_y$ = 0.39λ. In [46] the focusing element was a lens axicon (lens + axicon), and the width of the resulting focal spot at half intensity was equal to FWHM = 0.43λ, and the depth of focus, too, at half intensity DOF = 1.8λ (DOF = Depth Of Focus).

Furthermore, it was shown that the addition to the binary phase plates to the focusing elements also helps reduce the size of the focal spot. Such phase plates act as a polarizing filter to select the longitudinal component of the focused beam. In [47] a radially polarized Bessel–Gaussian beam was focused by a lens with a numerical aperture of $NA = 0.95$, in front of which there was a binary phase plate consisting of 5 annular zones (even zones provide the phase shift by π compared with odd ones). In [47] the calculated focus had a diameter FWHM = 0.43λ and depth DOF = 4λ. In another study [48] when focusing the Bessel–Gaussian beam with the same parameters but using a binary phase plate with four zones (instead of five as in [47]), it was possible to calculate the focus with a diameter FWHM = 0.44λ and depth DOF = 3λ. In [27] the focusing of the Bessel–Gaussian beam with the phase zone plate and aplanatic lens was also simulated. The size of the focus decreased when adding to the focusing element a phase three-zone plate: using the aplanatic lens the diameter of the spot could be reduced from FWHM = 0.584λ,NA to FWHM = $0.413\lambda/NA$, and when using the zone plate – from FWHM = $0.425\lambda/NA$ to FWHM = $0.378\lambda/NA$. A similar approach is proposed in [50]. In [50] the authors used the concept of 'outward-input-outward polarization', to show that after the passage of the binary phase element with phase shift π, radial polarization reverses direction in each jump of the relief. As a result, the focal spot can be obtained with the radius $\Delta r = 0.29\lambda$ (0.46λ – without a phase element), but the ratio of the energy in the first sidelobe to the main lobe energy is 0.54 (compared to 0.06 without the use of the phase element). An interesting result was obtained in [50], where with the help of the Debye formula it was shown that the radially polarized Laguerre–Gaussian mode of even orders LG_p^0 at the numerical aperture $NA = 0.85$ after the passage of a special annular amplitude mask are focused on a small spot with almost no side lobes with area at half intensity HMA = $0.276\lambda^2$. Using the amplitude mask does not reduce the size of the focal spot but reduces the size of the sidelobes in the focal diffraction pattern and 5 times decreases the depth of focus.

The illuminating beam can be the radially polarized Bessel-Gaussian beam [51], radially polarized modes R-TEM$_{p1}$ (Laguerre-Gaussian) [23] (increasing the order of the modes with $p = 1$ to $p = 5$, it is possible to reduce the focal spot from FWHM = 0.582λ

to FWHM = 0.378λ). In [26] using the RW formula it is shown that when illuminating the exit pupil of a spherical lens with flat, Gaussian or Bessel–Gaussian beams of radially-polarized light, the diameter of the focal spot is equal to FWHM = 0.6λ, 1.2λ, 1.4λ respectively, at NA = 1.4, λ = 632.8 nm, n = 1.5. In [52] it is proposed to carry out subwavelength focusing using polarization-inhomogeneous Bessel–Gaussian beams. In [24] it is proposed for a similar purpose to use beams with 'elliptical symmetry of linear polarization': the RW formula was used to simulate the diffraction pattern and focus with the aplanatic lens a linearly polarized beam with an elliptical radial symmetry with an eccentricity of 0.87. The numerical aperture was NA = 0.9. In this case, an elliptical shaped focal spot with the area HMA = 0.56λ² was produced.

It is also important to mention briefly how to obtain a radially polarized light. There are two methods light: intracavity and extracavity. In the first case, the conversion of light occurs within the laser cavity, in the second – outside the laser. Radially polarized laser beams can be produced using a conventional interferometer, in which the shoulders contain two phase steps, giving a delay at half wavelength and they are rotated relative to each other by 90 degrees around the optical axis [15]. Study [53] proposed and experimentally tested a way to convert linear polarization to radial or azimuthal using a photonic crystal fibre with a length of 24 mm. In [54] a beam of a Nd:YAG laser passed through a conical prism with the Brewster angle at the base: the azimuthally polarized component underwent total internal reflection and the radially polarized component passed through a prism. The result was a radially polarized mode R-TEM$_{01}$.

It was shown experimentally in [54] that the central section of the radially polarized mode coincides with high accuracy with the Hermite–Gauss mode (0,1). Therefore, we can write analytically the distribution of such mode and then simulate it with the help of microoptics. The Hermite–Gauss mode (HG) has the form:

$$E_{mn}(x,y,z) = \left[\frac{\omega_0}{\omega(z)}\right] \exp\left\{i(m+n+1)\eta(z) - \frac{ik(x^2+y^2)}{2R(z)} - \frac{(x^2+y^2)}{\omega^2(z)}\right\}$$

$$H_m\left(\frac{\sqrt{2}x}{\omega(z)}\right) H_n\left(\frac{\sqrt{2}y}{\omega(z)}\right), \tag{3.95}$$

where $\eta(z)$ = arctg (z/z_0), $R(z)$ = $z(1 + z_0^2/z^2)$, $\omega(z)$ = $\omega_0(1+z^2/z_0^2)^{1/2}$, $z_0 = k(\omega_0)^2/2$, $H_n(x)$ – Hermite polynomial, $H_0(x)$ = 1,

$H_1(x) = 2x$, ω_0 – the Gaussian beam waist radius. The laser mode with radial polarization is the sum of two linearly polarized modes $E_{1,0}$ and $E_{0,1}$, one of which is polarized in x and the other in y:

$$\mathbf{E}_r(x,y,z) = \left[\frac{2\sqrt{2}\omega_0}{\omega^2(z)}\right] \exp\left\{i2\eta(z) - \frac{ik(x^2+y^2)}{2R(z)} - \frac{x^2+y^2}{\omega^2(z)}\right\}(x\mathbf{e}_x + y\mathbf{e}_y),$$

(3.96)

where in the last parentheses there are two unit vectors of the Cartesian axes. Introducing the notation of the unit vector directed along the radius of the polar coordinate system $\mathbf{e}_r$, we obtain the final expression for the electric field vector of radially polarized laser mode R-TEM$_{01}$

$$\mathbf{E}_r(x,y,z) = \left[\frac{2\sqrt{2}\omega_0}{\omega^2(z)}\right] \exp\left\{i2\eta(z) - \frac{ikr^2}{2R(z)} - \frac{r^2}{\omega^2(z)}\right\} r\mathbf{e}_r. \quad (3.97)$$

At $z = 0$ (the waist plane where the wavefront is flat), the expression (3.97) can be written as:

$$\mathbf{E}_r = Ar\exp\left\{-\frac{r^2}{\omega_0^2}\right\}\mathbf{e}_r. \quad (3.98)$$

Below we numerically investigate the focusing of the mode (3.98) by microoptics objects such as a axicon (refraction, diffraction, and logarithmic) and the Fresnel zone plate.

The refractive axicon
Below we consider focusing a radially polarized light by a refraction microaxicon with the parameters close to those under which total internal reflection takes place. With these parameters of the axicon it is possible to form a sharp focus close to its peak.

We use the BOR-FDTD method to model the propagation of light through a refractive glass conical ($n = 1.5$) axicon with a height $h = 6$ μm and a base radius $R = 7$ μm. Figure 3.7 shows the profile of the axicon and its location in the calculated area. The numerical aperture of the conical axicon is calculated as follows:

$$NA = \frac{\dfrac{nh}{R} - \sqrt{\left(\dfrac{h}{R}\right)^2 - n^2 + 1}}{\left(\dfrac{h}{R}\right)^2 + 1}. \quad (3.99)$$

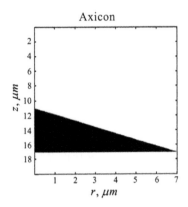

Fig. 3.7. Radial–axial cross-section of the refractive conical axicon and its location in the calculated area.

In this case, NA = 0.6.

The illuminating beam was the mode R-TEM$_{01}$ with a wavelength $\lambda = 1$ μm and the radius $\omega = 3$ μm. In this simulation the dimensions of the region along the z axis constituted 20 μm (or 20λ), along the axis r 9 μm (or 9λ), at the edges of the region there were ideal absorbing layers 1 μm thick (or 1λ). The discreteness of the partition of the region in space was 50 samples per wavelength. The simulation results of passage of the mode R-TEM$_{01}$ through the axicon are shown in Figs. 3.8–3.11. Figure 3.8 shows the distribution of the intensity along the axis of the axicon, and Fig. 3.9 – the intensity distribution at the focus.

Figure 3.8 (b) shows that the subwavelength focusing of the light occurs at a distance from the top of the axicon of 0.02 μm and the axial (longitudinal) width of the focus at half intensity is DOF = 0.12λ. Figure 3.10 shows the instantaneous amplitude distribution E_z and E_r in the calculated area, and Fig. 3.11 – the intensity distribution in the calculated area.

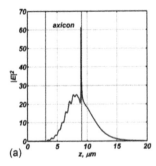
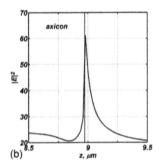

Figure 3.8. The intensity distribution along the axis of the axicon: (a) along the calculated area, (b) enlarged portion near the tip of the axicon. The dotted line shows the boundaries of the axicon.

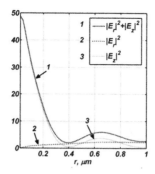

Fig. 3.9. The distribution of the intensity $|E|^2$ at the focus (curve 1), $|E_r|^2$ (curve 2), $|E_z|^2$ (curve 3).

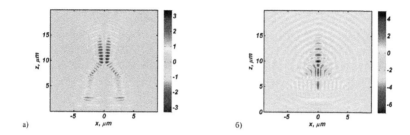

Fig. 3.10. The instantaneous amplitude distribution: (a) E_r and (b) E_z in the calculated area.

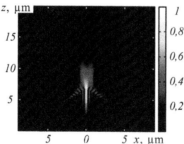

Fig. 3.11. The intensity distribution in the calculated area.

The diameter of the resulting focal spot (Fig. 3.9) at half intensity is FWHM = 0.30λ, and the area of the focal spot at half intensity is HMA = 0.071λ². The radius of the focal spot from the maximum to the first minimum is Δr = 0.40λ. For comparison it should be mentioned that we obtained the diameter of the focal spot at half intensity 1.7 times smaller than the diameter of the Airy disk (FWHM = 0.51λ), and the area was 2.87 times lower than that of the Airy disk (HMA = 0.204λ²).

Figure 3.12 shows that the chosen parameters of the axicon are optimal: changing the height of the axicon h (at constant radius of the base) increases the size of the focal spot and decreases the intensity at the focus.

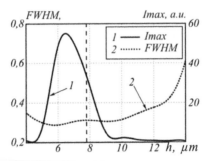

Fig. 3.12. Dependence of intensity change in focus I_{max} (curve 1, right axis) and the width of the focal spot FWHM (curve 2, left axis) on the height h of the glass axicon with the radius of the base 7 μm. The dotted line shows the height of the axicon at which the angle of total internal reflection of the incident beam from the conical surface of the axicon is reached.

The binary axicon

The binary microaxicon is easier to manufacture than the conical axicon discussed earlier. For example, the binary axicon may be manufactured by electronic lithography or photolithography using a binary amplitude mask (or direct electron beam writing) in the form of light and dark concentric rings of the same thickness. Figure 3.13 shows a radial section of a binary microaxicon.

This binary microaxicon (Fig. 3.13) has a height of steps $H = \lambda/2 \ (n-1) \approx 633$ nm for a wavelength $\lambda = 633$ nm and the refractive index $n = 1.5$, width of steps $d = 0.74$ μm is equal to the width of the 'pits' $D-d = 0.74$ μm, and the period is equal to $D = 2d = 1.48$

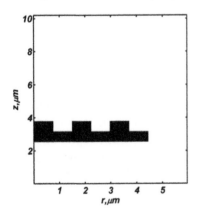

Fig. 3.13. Profile of the binary microaxicon and its location in a calculated field.

Table 3.1. The dependence of the diameter of the focal spot at half intensity (FWHM) and focal length (L) on wavelength (λ)

λ, μm	FWHM, λ	L, μm
0.600	0.69	0.64
0.630	0.56	0.53
0.700	0.45	0.19
0.750	0.46	0.19
0.850	0.39	0.08

μm. The total height of the axicon with the substrate (along the z axis) is $2H = 1.266$ μm. The radius of the axicon is three periods $R = 3D = 4.44$ μm. This section presents the results of modelling sharp focusing of the R-TEM$_{01}$ laser mode using a binary axicon (Fig. 3.13). The microaxicon was illuminated with radially polarized mode R-TEM$_{01}$ with the waist radius $\omega = 1.9$ μm.

Table 3.1 shows the data for the dependence of the distance from the surface of the axicon to the maximum intensity along the axis of the axicon lens (focal length) (L) and the diameter of the focal spot at half intensity (FWHM) on the wavelength of the radiation illuminating the axicon (λ). The table shows that the chromatic dispersion of the binary microaxicon is the same (in the sign) as that of the conventional diffractive grating: the longer wavelengths are diffracted at a large angle to the optical axis. Therefore, Table 3.1 shows that with increasing wavelength λ ranging from 0.600 μm to 0.850 μm focal spot is formed closer to the top of the axicon and its diameter (in wavelengths) decreases. The minimum focal spot diameter is equal to FWHM = 0.39λ and forms in the vicinity of the vertices of the axicon ($L = 0.08$ μm). Figure 3.14 shows the distribution of the intensity $|E|^2$ along the optical axis (a) and radial intensity distribution $|E_z|^2$ (curve 3), $|E_r|^2$ (curve 2) and $|E|^2 = |E_r|^2 + |E_z|^2$ (curve 1) in the plane of the focus (b) for a wavelength of 850 nm.

From Fig. 3.14 it is seen that near the central circular step of the binary axicon (Fig. 3.13) there is a sharp focus whose diameter at half intensity is FWHM = 0.39λ, and the area at half intensity is HMA = $0.119\lambda^2$.

Figure 3.15 shows the instantaneous distribution of amplitude E_r (a) and amplitude E_z (b), and Fig. 3.16 – the intensity distribution in the diffraction of the R-TEM$_{01}$ mode with a wavelength of 850 nm in the binary axicon in the calculated area. Figure 3.15 shows that

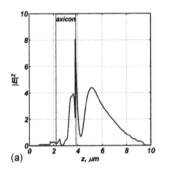
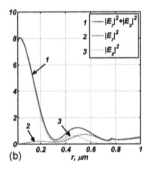

(a)   (b)

Fig. 3.14. The distribution of the intensity $|E|^2$ along the optical axis (a) and radial intensity distribution $|E_z|^2$ (curve 3), $|E_r|^2$ (curve 2) and $|E| = |E_z|^2 + |E_r|^2$ (curve 1) in the focal plane (b) for the wavelength of 850 nm.

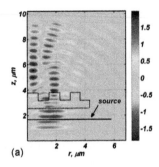
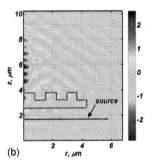

(a)   (b)

Fig. 3.15. The instantaneous distribution of amplitude E_r (a) and amplitude E_z (b) in the diffraction of the R-TEM$_{01}$ mode with a wavelength of 850 nm on a binary glass microaxicon with a period of 1.48 μm in the calculated area.

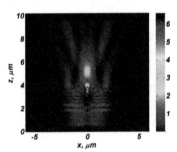

Fig. 3.16. The average intensity distribution for a binary microaxicon with a period of 1.48 μm in the calculated area.

each step of the profile of the axicon produces mainly a lobe in the diffraction pattern: the central near-axial step forms the first main lobe of the focal pattern, the second annular step of the axicon forms a second annular lobe and the third step – the third lobe. Thus, the binary microaxicon is easier to manufacture, but is less effective in sharp focusing than the conical axicon, although the former (binary axicon) forms the focal spot (FWHM = 0.39λ) smaller than the diffraction limit (FWHM = 0.51λ).

The numerical aperture of the binary axicon is calculated from the equation of the diffraction grating and for the first diffraction order is $NA = \sin\theta = \lambda/D$. Consequently, to ensure that the numerical aperture of a binary axicon with the period $D = 1.48$ μm is the same as that of the conical axicon described earlier (Fig. 3.7, $NA = 0.6$), the binary axicon should be illuminated with the light of wavelength $\lambda = 0.888$ μm. This estimate is in good agreement with the optimum wavelength from Table 3.1 $\lambda = 0.850$ μm. For the wavelength $\lambda = 0.850$ μm used in this case the numerical aperture of the binary microaxicon is equal to $NA = 0.57$.

3.2.2. Diffractive logarithmic microaxicon

This section considers diffraction on a logarithmic axicon (LA). Previously, LAs were considered, for example, in [55], where a diamond cutter was used to produce a plastic logarithmic axicon (LA) with a radius of 6.5 mm having the form of the phase function $S(r) = \gamma \ln (a + br^2)$, wherein γ, a, b are constants, r is the transverse radial coordinate in a cylindrical coordinate system. This axicon focused light from a He–Ne laser in an axial section 10 cm long and with a diameter of 10 μm. In [56] an LA with a quadratic dependence on the radial coordinate $S(r) = \gamma \ln (a + br^2)$ was investigated for the first time with the help of a digital hologram and focusing of light in the axial segment. In [57] the authors simulated LA and showed that the axial intensity along the section is on average more constant, in contrast to the linear conical axicon in which the average intensity increases along the axial section.

We will consider the scalar paraxial diffraction of the Gaussian beam on a logarithmic axicon which is described by the phase function, as in [55–57], but at $a = 0$. This axicon focuses the light into an axial segment that starts just behind the axicon. The phase function of the axicon has the form $S(r) = \gamma \ln(r/\sigma)$ and has a singularity at the origin when $r = 0$. At this point, the phase $S(r)$ approaches the plus (or minus) infinity. However, this feature is observed only in the initial plane at $z = 0$. In any other plane ($z > 0$), with illumination of such LA by the Gaussian beam, the light field has finite energy and has no singularities. An explicit analytic expression for the complex amplitude of such field in the Fresnel diffraction zone was presented. An estimate for the diameter of the transverse intensity, which is inversely proportional to the parameter of the LA $|\gamma|^{-1/2}$ was obtained. That is, for sufficiently large $|\gamma| \gg 1$ we can obtain the subwavelength laser beam diameter near the LA. Numerical examples confirm this. Numerical simulation of hypergeometric beams has already carried been out in [58] which also showed the possibility to overcome the diffraction limit by using the logarithmic axicon, but the formulas for the axial intensity and width of the focal spot have not been derived.

The general expression for amplitude
In the scalar paraxial approximation we consider the diffraction of the Gaussian beam on a spiral logarithmic axicon (SLA) and simply on the logarithmic axicon. The transmission function in the approximation of a transparant for SLA in polar coordinates (r, φ) is as follows:

$$T(r,\phi) = \exp\left[i\gamma \ln\left(\frac{r}{\sigma}\right) + in\phi \right],\qquad(3.100)$$

where n is the integer (the topological charge of the optical vortex), γ is the real number (parameter of the 'force' of the axicon), σ is scaling parameter of the axicon. Then immediately behind the SLA the complex amplitude of the monochromatic light field would be:

$$E_0(r,\phi) = \exp\left[-\left(\frac{r}{w}\right)^2 + i\gamma \ln\left(\frac{r}{\sigma}\right) + in\phi \right],\qquad(3.101)$$

where w is the Gaussian beam waist radius. The Fresnel transform of the function (3.101) has the form:

$$E(\rho,\theta,z) = \frac{(-i)^{n+1}}{n!}\left(\frac{z_0}{z}\right)\left(\frac{w}{\sigma}\right)^{i\gamma}\left(\frac{kw\rho}{2z}\right)^n \times$$

$$\times \Gamma\left(\frac{n+2+i\gamma}{2}\right)\left(1-\frac{iz_0}{z}\right)^{-\frac{n+2+i\gamma}{2}} \exp\left(in\theta + \frac{ik\rho^2}{2z}\right) \times\qquad(3.102)$$

$$\times {}_1F_1\left[\frac{n+2+i\gamma}{2}; n+1; -\left(\frac{kw\rho}{2z}\right)^2\left(1-\frac{iz_0}{z}\right)^{-1}\right],$$

where (ρ, θ) are the transverse polar coordinates in the observation plane, z is the coordinate along the optical axis, $k = 2\pi/\lambda$ is the wave number of the light with a wavelength λ, $z_0 = kw^2/2$ is the Rayleigh distance, $\Gamma(x)$ is the gamma function, $_1F_1(a, c; x)$ is the confluent hypergeometric function [59]. Note that the expression (3.102) is the exact solution of the paraxial wave equation (such as the Schrödinger equation), and it is a special case of the previously obtained solution for a family of hypergeometric laser beams [60, 61]. The intensity of the light field $I(\rho, z) = |E(\rho, z)|^2$, derived from (3.102), is as follows:

$$I(\rho,z) = \frac{1}{(n!)^2}\left(\frac{z_0}{z}\right)^2\left(\frac{kw\rho}{2z}\right)^{2n}\left|\Gamma\left(\frac{n+2+i\gamma}{2}\right)\right|^2 \times$$

$$\times\left(1+\frac{z_0^2}{z^2}\right)^{-\frac{n+2}{2}}\exp\left[-\gamma\operatorname{arctg}\left(\frac{z_0}{z}\right)\right]\times \qquad (3.103)$$

$$\times\left|_1F_1\left[\frac{n+2+i\gamma}{2}; n+1; -\left(\frac{kw\rho}{2z}\right)^2\left(1-\frac{iz_0}{z}\right)^{-1}\right]\right|^2.$$

Axial intensity

Note that the intensity (3.103) does not depend on the scaling parameter of the axicon σ in (3.100). Intensity (3.103) everywhere on the optical axis (except for initial plane $z = 0$) is equal to zero when $n \neq 0$. To view the focusing of light with the help of LA we will set $n = 0$. Then, instead of (3.103), the intensity of the light field will be:

$$I_0(\rho,z) = \frac{z_0^2}{z^2+z_0^2}\left|\Gamma\left(1+\frac{i\gamma}{2}\right)\right|^2\exp\left[-\gamma\operatorname{arctg}\left(\frac{z_0}{z}\right)\right]\times$$

$$\times\left|_1F_1\left[1+\frac{i\gamma}{2}; 1; -\left(\frac{kw\rho}{2z}\right)^2\left(1-\frac{iz_0}{z}\right)^{-1}\right]\right|^2. \qquad (3.104)$$

Putting $\rho = 0$ (3.104), we obtain an expression for the axial intensity:

$$I_0(z) = \left(\frac{\pi\gamma}{2}\right)\operatorname{sh}^{-1}\left(\frac{\pi\gamma}{2}\right)\frac{z_0^2}{z_0^2+z^2}\exp\left[-\frac{\pi\gamma}{2}+\gamma\operatorname{arctg}\left(\frac{z}{z_0}\right)\right]. \qquad (3.105)$$

To derive (3.105) it was considered that $_1F_1(a, c; 0) = 1$, $|\Gamma(1 + ix)|^2 = (\pi x)\operatorname{sh}^{-1}(\pi x)$ and $\operatorname{arctg}(x) = \pi/2 - \operatorname{arctg}(1/x)$.

In Fig. 3.17 the axial intensity values at the origin ($\rho = 0$, $z = 0$) in (3.105) are:

$$I_0(0) = \pi\gamma\left[\exp(\pi\gamma)-1\right]^{-1} \geq 1.$$

(3.106)

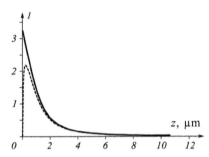

Fig. 3.17. The axial intensity ($\rho = 0$) for the collecting LA ($\gamma < 0$) ($\gamma = -1$, $w = \lambda$, $\lambda = 532$ nm), calculated by the formula (3.105) (solid line) and on the basis of the Rayleigh–Sommerfeld integral(dashed curve).

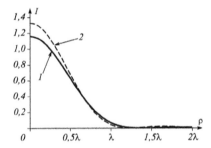

Fig. 3.18. The intensity in the transverse plane $z = 2\lambda$ ($\gamma = -1$, $w = \lambda$, $\lambda = 532$ nm), obtained by the Rayleigh–Sommerfeld (curve 1) and Fresnel (curve 2) integrals.

The axial intensity in Fig. 3.17 (dashed curve) is calculated using the Rayleigh–Sommerfeld non-paraxial integral to show that, firstly, the paraxial and non-paraxial axial intensities differ considerably from each other at $z < 2\lambda$ and, secondly, the paraxial curve (solid curve) gives a wrong result at zero $I_0(0) \geq 1$ ($z = \rho = 0$).

Figure 3.18 shows the intensity distribution in the transverse plane $z = 2\lambda$, obtained by the Rayleigh–Sommerfeld (curve 1) and Fresnel (curve 2) integrals. Other simulation parameters are the same as in Fig. 3.17.

Figure 3.18 also shows that the LA generates a beam of light with almost no side lobes, unlike the conventional axicon which forms a beam with an amplitude proportional to the Bessel function $J_0(\alpha\rho)$, the side lobes of which are 0.4 from the axial amplitude.

Phase radial singularity in the centre

The value of $I_0(0)$ in (3.106) is greater than one (or equal to 1 at $\gamma = 0$), and is therefore contrary to the expression (3.101), which implies that the intensity at zero $I_0(0)$ must be equal to zero, since the phase at the origin of the coordinates is not defined and in any small neighborhood of zero ($\rho = z = 0$) there are all the values of the phase of the section $[0, 2\pi]$. The fact is that the transition from (3.104) to (3.105) is possible for any z and $\rho = 0$, except for $z = 0$. With both z and ρ tending to zero, the result of the conversion of (3.104) will depend on the relative velocity at which both variables z and ρ tend to zero. For example, if z tends to zero at fixed $\rho \neq 0$, then instead of (3.105) from (3.104), we obtain:

$$I_0(\rho, z) = \frac{z_0^2}{z_0^2 + z^2} \exp\left[-\frac{2\rho^2}{w^2} + 2\gamma \operatorname{arctg}\left(\frac{z}{z_0} \right) \right]. \tag{3.107}$$

Equation (3.107) was derived using the asymptotic expression for the hypergeometric function at $x \to \infty$ [59]:

$$_1F_1(a;c;x) = \frac{\Gamma(c)\exp(x)}{\Gamma(a)x^{c-a}}. \tag{3.108}$$

At $z = 0$ from (3.107) follows the correct expression for the initial field intensity (3.101) $I_0(\rho, z = 0) = \exp[-2(\rho/w)^2]$. Equation (3.107) is true at $z \to 0$ and $\rho \neq 0$. Figure 3.19 shows the function (3.107) with ρ close to zero.

At first glance, Figs. 3.17 and 3.19 contradict each other, but it is not so. Since for $z = 0$ and $\rho = 0$ the phase radial singularity appears (feature), that the phase is (3.101) and its derivative at $n = 0$ tend to infinity at $r \to 0$, this means that when $z = 0$ the intensity in the centre ($\rho = 0$) is zero $I_0(0) = 0$, and in the adjacent

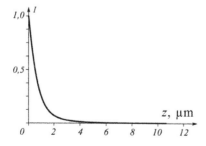

Fig. 3.19. The dependence of the intensity of z near $z = 0$ and $\rho = 0.1\lambda$ ($\gamma = -1$, $w = \lambda$, $\lambda = 532$ nm).

points ($\rho \neq 0$) $I_0(\rho \approx 0) \approx 1$. Therefore, the continuous passage to the limit along the optical axis toward the point $z = 0$ can not be used to obtain the initial intensity $I_0(0) = 1$. But if the limiting transition takes place at $\rho \neq 0$ (next to the optical axis), the value of intensity at $z = 0$ turns out to be correct $I_0(0) \sim 1$ (Fig. 3.19).

In the case where ρ is strictly zero and z is close to zero, the calculation of the Rayleigh–Sommerfeld integral gives zero axial intensity (Fig. 3.17), which can be explained by the uncertainty of the phase in the centre of the initial field at $\rho = z = 0$. However, if the phase singularity at the centre in the field (3.101) is eliminated, replacing it with a plane wave with a constant phase inside the circle $r < r_0$, i.e., field (3.101) is replaced by

$$E_0(r,\phi) = \begin{cases} \exp\left[-\left(\dfrac{r}{w}\right)^2 + i\gamma \ln\left(\dfrac{r}{\sigma}\right) + in\phi \right], r \geq r_0, \\ 1, r < r_0, \end{cases} \tag{3.109}$$

the calculation of the Rayleigh–Sommerfeld integral provides a single axial intensity $I_0(0) = 1$, which decreases rapidly as a function of the circle radius r_0, reaching a local minimum. Figure 3.20 shows the axial intensity for several values of r_0: $r_0 = 0.011\ \lambda$ (curve 1), $r_0 = 0.031$ (curve 2), $r_0 = 0.051$ (curve 3).

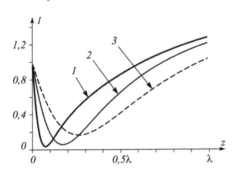

Fig. 3.20. Axial intensity for the field (3.109).

The radius of the transverse intensity distribution

We estimate the width of the transverse intensity distribution near the optical axis. It is known that the first approximation of coordinate x_1 of the first zero of the hypergeometric function $_1F_1\ (a,\ c;\ x)$ is given by [59]:

$$x_1 = \frac{\gamma_{c-1,1}^2}{2(c-2a)}, \tag{3.110}$$

where $\gamma_{c-1,\,1}$ is the first root of the Bessel function of the $(c-1)$-th order: $J_{c-1}\,(\rho_{c-1,\,1}) = 0$. In view of (3.104), into (3.110) we must substitute the values $a = 1 + i\gamma/2$, $c = 1$, $\gamma_{0,1} = 2.4$. Since the values of a and x in this case are complex, the formula (3.110) should be regarded as the modulus of complex numbers:

$$|x_1| = \frac{\gamma_{c-1,1}^2}{2|c-2a|}. \qquad (3.111)$$

Then the real coordinate of the first integrated zero (local minimum) ρ_1 of the intensity (3.104) can be estimated by the expression:

$$\rho_1 = 2,4w\left[\frac{1+\dfrac{z^2}{z_0^2}}{2(1+\gamma^2)}\right]^{1/4}. \qquad (3.112)$$

From (3.112) it follows that at parameters gamma $|\gamma| \gg 1$, the dependence of the effective radius of the transverse distribution of intensity is proportional to $(z \ll z_0)$:

$$\rho_1 \approx 2w|\gamma|^{-1/2}. \qquad (3.113)$$

From (3.112) it also follows that by the choice of a sufficiently large value of $|\gamma|$ ($\gamma < 0$), we can use LA to get near the plane $z = 0$ a light spot with any small subwavelength diameter. For example, if $w = \lambda$, $z_0 = kw^2/2 = \pi\lambda$, $z = \lambda$ and $\gamma = -400$ we can get a light spot with a diameter at half intensity FWHM $= \rho_1 = \lambda/10$. This follows from the fact that the phase gradient (3.100) for $n = 0$ is $\gamma\sigma/r$ and tends to infinity at $r \to 0$. That is the LA not only collects the propagating waves on the optical axis, but also excites evanescent surface waves with a high value of the projection of the wave vector on the transverse axis $k_r \gg k$. The presence of these surface waves

Table 3.2. The radii of the transverse intensity distribution

m	1	2	3	4	5
$-\gamma_m$	1	3	5	7	9
ρ_m, λ	2.80	1.93	1.53	1.29	1.13
ρ_m/ρ_{m+1}	1.45	1.26	1.19	1.14	–
$(\gamma_{m+1}/\gamma_m)^{1/2}$	1.52	1.29	1.18	1.13	–

near $z = 0$ provides subwavelength dimensions of the focal spot. We now check the correctness of dependence (3.113) with the help of simulation.

In Table 3.2: m – the number of the order; γ_m – LA parameter; ρ_m – the range in wavelengths of the first zero (or the first local minimum) intensity at $z = 10\lambda$ (other parameters $w = 2\lambda$, $z_0 = 4\pi\lambda$) is numerically calculated using Fresnel transform; ρ_m/ρ_{m+1} – the ratio of the two adjacent radii; $(\gamma_{m+1}/\gamma_m)^{1/2}$ – the square root of the ratio of two neighbouring parameters of the LA. In accordance with (3.113) the values in the third and fourth rows of Table 3.2 in each column must be the same as $\rho_m/\rho_{m+1} = (\gamma_{m+1}/\gamma_m)^{1/2}$. From the comparison of the third and fourth rows of Table 3.2 we can see that these lines converge with an increase in the value of the modulus of the parameter γ.

Simulation by FDTD-method

Consider passage of a radially polarized mode R-TEM$_{01}$ through a diffraction logarithmic microaxicon. The simulation was performed using the BOR-FDTD method. Simulation parameters: dimensions of the calculation domain $20\lambda \times 20\lambda$, discrete samples in space $\lambda/20$, the discrete samples of time $\lambda/40c$, where c is the speed of light in vacuum. Axicon parameters: maximum height $h_{max} = \lambda/(n-1)$, where n is the refraction index of the substance ($n = 1.5$), the current height of the profile is calculated by the formula where $\lambda \bmod \left[\gamma \ln(r/\sigma)/(2\pi)\right]/\left[2\pi(n-1)\right]$, $\gamma = -20$ and $\sigma = 20$ μm, and mod (.) represents the remainder of the division. The parameters of the mode: the radius $w = 6\lambda$, the wavelength $\lambda = 532$ nm. Figure 3.21 shows the radial section of the LA. Figure 3.22 shows the radial section of the amplitude of the beam illuminating the axicon. Figure 3.23 shows the axial intensity distribution, and Fig. 3.24 the radial cross section of intensity in the focal plane (in the plane with maximum intensity).

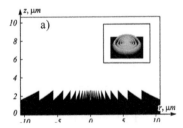 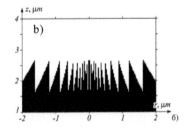

Fig. 3.21. Logarithmic microaxicon (a) in the calculated area and its enlarged detail (b) close to the axis. Inset – a three-dimensional view of the logarithmic axicon.

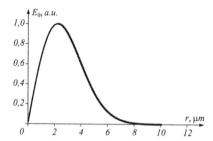

Fig. 3.22. The radially polarized laser mode R-TEM$_{01}$.

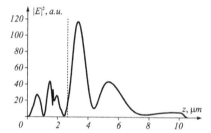

Fig. 3.23. The intensity distribution along the axis z (vertical line – the line where the vertices of the LA relief are located).

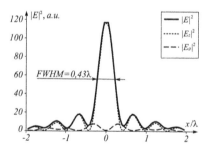

Fig. 3.24. The intensity distribution in the focal spot.

The simulation showed that the depth of focus at half intensity (Fig. 3.23) is equal to DOF = 2λ, and the diameter of the focal spot at a distance of a wavelength from the 'serrated' axicon surface at half intensity is FWHM = 0.43λ. It is smaller than the diffraction limit FWHM = 0.51λ.

We give another example. In this case, the LA is two-dimensional and simulation is performed using FullWAVE (RSoft).

Figure 3.25 shows a two-dimensional axison relief. The border of the modeling domain: $[-4\lambda, +4\lambda] \times [0, 3\lambda]$, modelling time: $cT = 30l$, sampling by x (horizontal axis) and z (vertical) $-\lambda/50$, in time

$-T/100$. The relief height: $\lambda/(n-1) = 2\lambda$, refractive index: $n = 1.5$, wavelength of the illumination light 532 nm, the parameters of the axicon $\gamma = -20$ and $\sigma = 4\lambda$. LA is illuminated by a Gaussian beam of light with TE-polarization and waist radius $w = 3\lambda$.

The width at half central maximum of intensity in Fig. 3.26 is FWHM = 108 nm = 0.20 λ. This is less than the diffraction limit in a 2D environment FWHM = $0.44\lambda/n$ = 0.2931 (for glass $n = 1.5$).

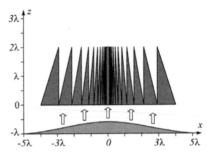

Fig. 3.25. Relief of two-dimensional LA with the parameters: radius 4λ, the height of the relief $\lambda/(n-1) = 2\lambda$, setting $\gamma = -20$.

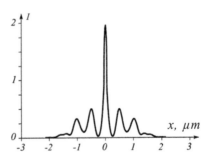

Fig. 3.26. Averaged intensity distribution immediately after the axicon (at $z = 2\lambda$).

3.2.3. Binary axicons with period 4, 6 and 8 µm

Non-paraxial scalar diffraction binary axicon
The analysis of the field near the optical axis and near the axicon can not be carried out using the electromagnetic theory [38], as in the central part the axicon can not be regarded as a diffractive grating.

Let the axicon receives a plane wave with linear polarization, then in the initial plane $z = 0$, which coincides with the output surface of the axicon in the approximation of the transparant the component of the electric field for the binary axicon will have the form:

$$E_{x0}(r) = \left(1 - e^{i\varphi}\right) \sum_{n=0}^{N-1} (-1)^n \, circl\left(\frac{r}{r_n}\right),$$ (3.114)

where $r_n = (n + 1)r_0$, n is an integer, r_n are the radii of jumps of the binary axicon relief along the radial coordinate r, φ is the phase delay caused by the projections of the axicon relief with respect to the relief depressions, N is the number of jumps of the relief of the axicon. The amplitude of the spectrum of plane waves for the initial field (3.114) is found from the expression:

$$A(\rho) = \frac{k^2}{2\pi}\left(1 - e^{i\varphi}\right) \sum_{n=0}^{N-1} (-1)^n \int_0^\infty cicrl\left(\frac{r}{r_n}\right) J_0(kr\rho)r\,dr =$$

$$= \left(1 - e^{i\varphi}\right) k^2 r_0^2 \sum_{n=0}^{N-1} (-1)^n (n+1)^2 \frac{J_1\left[k\rho(n+1)r_0\right]}{\left[k\rho(n+1)r_0\right]},$$ (3.115)

where ρ is the dimensionless variable.

Then the amplitude of the x-th component of the electric field on any plane z is found from the expression:

$$E_x(r,z) = \left(1 - e^{i\varphi}\right) k^2 r_0^2 \sum_{n=0}^{N-1} (-1)^n (n+1)^2 \int_0^\infty \frac{J_1\left[kr_0(n+1)\rho\right]}{\left[kr_0(n+1)\rho\right]} J_0(k\rho r)e^{ikz\sqrt{1-\rho^2}}\rho\,d\rho$$ (3.116)

The problem of finding the amplitude of the light field in the vicinity of the binary axicon (3.116) is reduced to the calculation of the integral:

$$I = \int_0^\infty J_1(\alpha x)J_0(\beta x)e^{i\gamma\sqrt{1-x^2}}\,dx,$$ (3.117)

where α, β, γ are constants. The integral (3.117) could not be found in the reference literature, so we will calculate it in extreme cases. Let $\gamma = kz \ll 1$ – we want to find a field near the axicon. Then, expanding the exponent into a Taylor series, instead of (3.117), we obtain, retaining only two members of the series:

$$I \approx \int_0^\infty J_1(\alpha x)J_0(\beta x)\,dx + i\gamma \int_0^\infty \sqrt{1-x^2}\, J_1(\alpha x)J_0(\beta x)\,dx.$$ (3.118)

The first integral can be found in the handbook [62]:

$$\int_0^\infty J_1(\alpha x)J_0(\beta x)dx = \begin{cases} \dfrac{1}{\alpha}, & \alpha > \beta, \\ 0, & \alpha < \beta. \end{cases} \qquad (3.119)$$

In our case: $\alpha = kr_0(n + 1)$, $\beta = kr$. The second integral in (3.118) we find near the optical axis, assuming that the radial coordinate r is much smaller than the radius of the first jump of the relief of the axicon r_0: $r \ll r_0$, then the zero-order Bessel function can be replaced by its square dependence: $J_0(x) \approx 1-(x/2)^2$ at $x \ll 1$. Then, to calculate the second integral in (3.118) we can use the reference integral [62]:

$$\int_0^1 x^{a-1}(1-x^2)^{b-1}J_n(cx)dx =$$

$$= \frac{c^n}{2^{n+1}} \frac{\Gamma(b)\Gamma\left(\dfrac{n+a}{2}\right)}{\Gamma\left(b+\dfrac{n+a}{2}\right)\Gamma(n+1)} \,_1F_2\left[\frac{n+a}{2}, \frac{n+a}{2}+b, n+1, -\frac{c^2}{4}\right],$$

$$(3.120)$$

where $_1F_2 (a, b, c, x)$ is the hypergeometric function.

Then, for the second integral in (3.118) we can write:

$$\int_0^1 \sqrt{1-x^2}J_1(\alpha x)dx - \left(\frac{kr}{2}\right)^2 \int_0^1 x^2\sqrt{1-x^2}J_1(\alpha x)dx =$$

$$= \left(\frac{kr_0}{6}\right)^2 \left\{ \,_1F_2\left(1, \frac{5}{2}, 1, -y^2\right) - \frac{(kr)^2}{10} \,_1F_2\left(2, \frac{7}{2}, 2, -y^2\right) \right\},$$

$$(3.121)$$

where $y = kr_0/2$.

Replacing the infinite limits of integration in (3.118) by the finite limits of integration in (3.121) is the usual procedure in such cases, the essence of which is that we neglect the contribution to the amplitude of the decaying inhomogeneous waves.

Taking (3.109) and (3.121) into account, for the field near the optical axis and near the axicon ($kz \ll 1$, $r \ll r_0$) we obtain:

$$E_x(r,z) \approx \frac{1}{kr_0} + ikz\frac{(kr_0)}{6}\left[\,_1F_2\left(1, \frac{5}{2}, 2, -y^2\right) - \right.$$

$$\left. -\frac{(kr)^2}{10} \,_1F_2\left(2, \frac{7}{2}, 2, -y^2\right) \right].$$

$$(3.122)$$

From (3.122) we can obtain an expression for the estimation of the diameter of the central maximum of the light field near the optical axis:

$$2r = k^{-1} \left[\frac{10\,_1F_2\left(1,\frac{5}{2},2,-y^2\right)}{\,_1F_2\left(2,\frac{7}{2},2,-y^2\right)} \right]^{\frac{1}{2}}, \qquad (3.123)$$

from which you can obtain a numerical estimate of the diameter of focus:

$$2r \approx 0.6\lambda. \qquad (3.124)$$

From (3.124) it follows that the diameter of the central maximum of the light field near the axicon does not depend on its period $T = 2r_0$ and is equal to the diffraction limit (FWHM = $0.51\lambda/NA$).

To test the expression (3.124), we estimate the diameter of the central maximum of the field intensity by other considerations.

Paraxial estimate of the diameter of the axial beam

In [39], based on the scalar paraxial theory, it in shown that as the axicon generates a Bessel beam, the diameter of the Bessel beam can be estimated from the expression:

$$J_0^2(k\sin\theta\cdot r) = 0 \qquad (3.125)$$

Then we obtain

$$2r = \frac{2.4\lambda}{\pi\sin\theta}, \qquad (3.126)$$

where θ is the half of the apex angle of the conical wave formed by the axicon.

For a binary axicon, treating it as a diffraction grating [38], we can assume that the angle θ of the conical wave at the same time is the angle of the diffraction grating with a period T:

$$\sin\theta_m = \frac{\lambda m}{T}, \qquad (3.127)$$

where m is the number of diffraction order.

In view of (3.126) and (3.127), we obtain the final expression for estimating the diameter of the light field on the axicon optical axis:

$$2r = \frac{2.4}{\pi}\frac{T}{m} \approx 0.774\frac{T}{m} \qquad (3.128)$$

From (3.128) we see that for the binary axicon the Bessel beam diameter does not depend on the wavelength [39], and is determined only by the period of the axicon and the number of the diffraction order. From (3.128) it follows that the point on the optical axis near the axicon receives different diffraction orders from different points of the axicon. Therefore, near the axicon at $z < z_0$, where $z_0 = \dfrac{RT}{2\lambda}$ is the distance at which the light field produces only one diffraction order of the axicon, R is the radius of the axicon, there will form a light field whose diameter on the optical axis is changed in a complex manner at $0 < z < z_0$.

From (3.128) it follows that for $z \geq z_0$ the diameter of the axial beam is equal to $2r = 6.2\lambda$ for an axicon with the period $T = 4$ μm $= 8\lambda$.

Evaluation of axial beam diameter as the waveguide mode
Near the axicon $r < r_0 = T/2$ and at $z < T$ the diameter of the light field on the optical axis can be estimated from the waveguide theory, since the depth of the relief axicon $H = \dfrac{\lambda}{2(n-1)} = \lambda$ where $n = 1.5$ is the index of refraction, and the central part of the axicon it can be considered as a portion of the round fiber with the core radius equal to $r_0 = T/2$.

The number of modes of round fibers having a stepped refractive index is calculated based on the dispersion equation [63]:

$$\frac{uJ_0(u)}{J_1(u)} = -\frac{wI_0(w)}{I_1(w)}, \qquad (3.129)$$

where J_0, J_1, I_0, I_1 are the conventional and modified Bessel functions of zero and first order, $u^2 + w^2 = V^2$, $V = kr_0\sqrt{n_1^2 + n_2^2}$, where n_1 and n_2 are the refractive indices of core and the cladding of the fibre, r_0 is its radius.

The maximum root of the equation (3.129) for the mode is smaller than the cutoff number $u_{max} < V$, so the waveguide mode with the minimum diameter has the amplitude proportional to Bessel function:

$J_0\left(\dfrac{Vr}{r_0}\right)$, $r < r_0$. Then the diameter of the mode with the maximum number equals ($n = 1.5$):

$$2r = 2\frac{2.4r_0}{V} = \frac{2.4\lambda}{\pi\sqrt{n^2 - 1}} \approx 0.7\lambda \qquad (3.130)$$

Interestingly, the estimate of the diameter (3.130) is also independent of the axicon period T, as the estimate (3.124).

Non-paraxial expressions for axial intensity

The dependence of the diameter of the central maximum of the light field on the distance from the axicon should be correlated with the dependence of the axial intensity on the distance from the axicon. Indeed, if the light ring is situated on the axis, its diameter is larger as compared with the diameter at the focus and the axial intensity, on the contrary, in the case of the focus reaches a local maximum, and in the case of a ring – the minimum. Therefore, we obtain an expression for the axial intensity of light for a binary axicon. The expression for the axial amplitude of the scalar non-paraxial field from a circular hole of radius R as a function of the longitudinal coordinate [64]:

$$E_x(z) = e^{ikz} - \frac{ze^{ik\sqrt{R^2+z^2}}}{\sqrt{R^2+z^2}}. \qquad (3.131)$$

For a binary axicon with transmittance (3.114) we can similarly find an expression for the amplitude of the electric vector on the optical axis:

$$E_x(z) = e^{i\varphi}\left[e^{ikz} - \frac{ze^{ik\sqrt{r_{2N+1}^2+z^2}}}{\sqrt{r_{2N+1}^2+z^2}}\right] + \left(e^{i\varphi} - 1\right)z\sum_{m=1}^{2N}\frac{e^{ik\sqrt{r_m^2+z^2}}}{\sqrt{r_m^2+z^2}}(-1)^m,$$

$$(3.132)$$

where $r_m = mr_0$ is the radius of the jump of the relief of the binary axicon, $r_0 = T/2$ is the half of the period of the axicon.

From (3.132) it follows that at $\varphi = \pi$ any point on the z axis will receive a contribution from spherical waves emanating from the secondary sources from each point of the jumps of the relief (that is, half a period of the axicon). Because the number of these points will be $2N + 1$, the result of such multibeam interference would be difficult to predict. The only thing that can be said is that when $z \ll R$, equation (3.132) reduces to equation (3.131) which shows that the minimum period of oscillations of the axial intensity close to the surface of the axicon is equal to λ.

Production of binary axicons

The first object of the study were binary axicons with periods 4, 6 and 8 μm and 4 mm in diameter. To produce them, a quartz substrate (1 mm thick, 30 mm diameter) with a refractive index $n = 1.46$ was deposited with a chromium layer with a thickness of 100 nm in UVM2M1 equipment. This plate with chromium was deposited with a photomask in a circular laser recording system CLWS-200 (Novosibirsk) with a minimum diameter of the laser spot of 0.8 μm and a positioning accuracy of 20 nm. Recording of the photomask was in fact the creation of a protective oxide film on the surface of chromium by illuminating it with a focused beam of an argon laser with a wavelength of 500 nm. Thereafter the photomask was subjected to plasmochemical etching. The etching process was carried out at the USP PDE-125-009 facility, which allows to obtain the details of the relief with the horizontal size of not less than 100 nm. The etching parameters of quartz: the power of high-frequency current 800 W, preliminary vacuum of $2.1 \cdot 10^{-2}$ Pa, the electromagnet current confining the plasma 0.8 A. Etching was carried out in an atmosphere of Freon-12 for 21 min. The etching rate of the quartz substrate was equal to 20–25 nm per minute. Thus, three axicons with a diameter of 4 mm each, a period 4 μm, 6 μm and 8 μm were fabricated.

Figure 3.27 shows a top view of a binary axicon on quartz with a period of 4 μm, produced by a SUPRA-25-30-85 scanning electron microscope with a magnification of 1000 times. Figure 3.28 also shows the top view but of a binary axicon with a period of 8 μm, obtained using an atomic force microscope Solver Pro (Zelenograd).

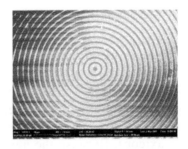

Fig.. 3.27. Top view of the binary axicon with a period of 4 μm in a scanning electron microscope Supra-25-30-85 with a magnification of 1000.

Fig. 3.28. Top view of the binary axicon with a period of 8 μm obtained using a Solver Pro atomic force microscope.

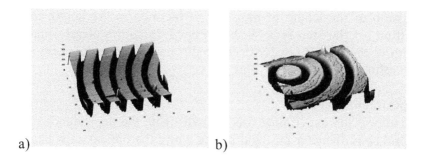

a)   b)

Fig. 3.29. View at an angle of the peripheral portion of the binary axicon with a period of 6 μm (a) and the central part of the axicon with a period of 8 μm (b) obtained with a Solver Pro microscope.

Figure 3.29 shows the binary axicon relief picture taken with a Solver Pro microscope at an angle to the peripheral portion of the axicon with a period of 6 μm (a) and the central part of the axicon with a period of 8 μm (b).

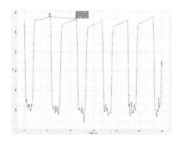

Fig. 3.30. The cross section (profile) of the relief of the binary axicon with a period of 6 μm shown in Figure 3.29a, obtained using a Solver Pro microscope.

Figure 3.310 shows the radial section of the peripheral part of the relief of the binary axicon with a period of 6 μm (Fig. 3.29*a*) from which it is seen that the relief depth is 450–500 nm. All three axicons were calculated for a wavelength of 532 nm, so the estimated depth of the relief is to be

$$h = \frac{\lambda}{2(n-1)} \approx 578 \text{ nm} \qquad (3.133)$$

Therefore, the error in the height of etching of the relief (incomplete etching) is about 20%. Figure 3.30 also shows that the upper faces of the relief have a slope of about 1/150 of a radian, and at the bottom of the relief there is visible roughness with an amplitude of 30 nm. Also, there the binary relief is trapezoidal: the ratio of the length of the 'top' of the step to the length of the 'bottom' of the step is 3: 4. Also Fig. 3.30 shows clearly that the width of the step is greater than the distance between adjacent steps (the ratio is about 2:1).

Experimental results

The aim of the experiment was to investigate the dependence of the central spot diameter on the distance along the optical axis. Custom made binary axicons were alternately placed in the optical setup in Fig. 3.31 and a CCD camera was used to measures the diffraction pattern in the near field at different distances with the axicons illuminated by a collimated laser light with a wavelength of 532 nm.

A solid-state laser beam λ = 532 nm with a diameter of 1.4 mm was focused with a microlens $L1$ to a pinhole with a diameter of 15 μm. A uniform light spot formed after passing through this aperture and its central part was collimated by lens $L2$. The formed beam was an almost plane wave with a limited aperture. The central part of this beam was introduced into the optical system of a Biolam-M microscope where it was focused with microlens $L3$, so that the diameter of the laser beam was equal to the diameter of the axicon D_3 (alignment of the light beam and the axicon was performed in order to reduce the energy loss). The resulting diffraction pattern was recorded by the CCD camera through the microscope objective $L4$. At the same time, shifting the axicon D_3 it was possible to produce diffraction patterns at different distances from the element. The reference point was the plane of the bottom of the microrelief. This plane was determined by obtaining a sharp image of the microrelief bottom in the white light. Axicon D_3 was shifted with a micrometric screw with 1 µm steps. The resolution of the CCD Camera was 2048 × 1536 pixels, with a pixel size of 6.9 µm.

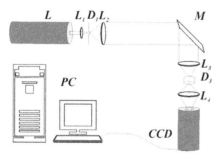

Fig. 3.31. The optical scheme for measuring the diffraction pattern in the near field for binary axicons: L – laser, L_1 – microscope objective (×20, $NA = 0.4$), D_1 – pinhole (diameter 15 µm), L_2 – collimating lens ($f = 100$ mm), M – swiveling mirror, L_3 – focusing microlens (×8, $NA = 0.2$), L_4 – imaging microscope objective (×20, $NA = 0.4$), D_3 – binary axicon.

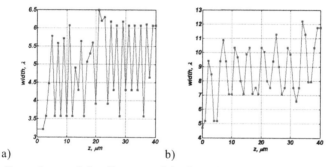

a)   b)

Fig. 3.32. Dependence of the diameter of the intensity spot on the optical axis (in wavelengths) on the distance to the binary axicon with a period: 4 µm (a) and 8 µm (b).

Figure 3.32 shows the distribution of the diameter of the central maximum (in one of the transverse Cartesian coordinates) on the distance from the surface of the axicons with a period of 4 µm (a) and 8 µm (b).

Figure 3.32 shows that at a distance of 0 to 40 µm the diameter of the central maximum of intensity varies quasi-periodically with the exemplary oscillation period of 2 µm (Fig. 3.33 (a)) and 4 µm (Fig. 3.32b). Moreover, the diameters larger than 5λ (Fig. 3.33a) and 9λ (Fig. 3.33b) correspond to the diameter of the circular intensity distribution on the axis (i.e. in this case, the intensity on the axis has a local minimum), and the diameters less than 4.5λ (Fig. 3.32a) and

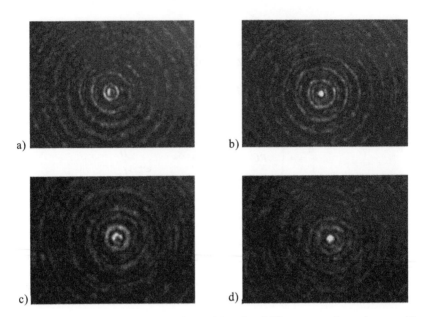

Fig. 3.33. Diffraction patterns registered by the CCD camera for axicons with a period of 4 μm (a, b) and 8 μm (c, d) at different distances: 5 μm (a), 8 μm (b), 16 μm (c) and 18 μm (d).

8λ (Fig. 3.32b) corresponds to the diameters of local maxima (focal points). Analysis of Fig. 3.22 suggests that the diameter of the axial focal spots in the near zone (at a distance of up to 40 μm) changes from 3.5λ to 4.5λ (Fig. 3.32a) with an error of 0.4λ and from 5λ to 8λ (Fig. 3.32b) with an error of 0.5λ, respectively, for the axicon period of 4 μm and 8 μm.

Figure 3.33 shows the registered diffraction pattern of laser light on binary axicons wth a period of 4 μm (a, b) and 8 μm (c, d) at distances of 5 μm (a), 8 μm (b), 16 μm (c) and 18 μm (d). Figure 3.33 shows that the annular intensity distribution on the optical axis (a, c) is replaced by the formation of the focal spot (the central peak) of 2–3 μm at distances of less than 40 μm. The scale of the diffraction patterns in Fig. 3.33 is the same, so we can see that the diameters of the ring and the focus for the axicon with a period of 8 μm is 1.5 times greater than for the axicon with a period of 4 μm.

Figure 3.34 shows the diffraction pattern for the axicon with a period of 4 μm, recorded at a distance of 2 μm from its surface (a) and its cross section along the axes x (b) and y (c). From Fig. 3.34 it can be seen that at small distances $z < 5$ μm (in Fig. 3.34 the ellipticity is no longer found) the diffraction pattern shows an elliptic focal spot (eccentricity $\varepsilon = 5/7$), associated with the linear

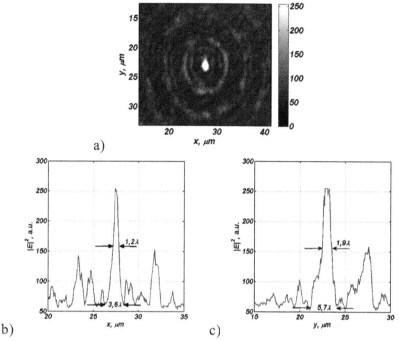

Fig. 3.34. The intensity distribution registered at a distance of 2 μm from the binary axicon with a period of 4 μm (a) and its cross section along the axes x (b) and y (b).

polarization of the incident laser light on the axicon (the electric vector of the incident beam is directed along the y axis in Fig. 3.35).

Simulation results

Modelling of diffraction of a linearly polarized plane wave on binary diffractive axicons was carried out by two similar, but different methods: by BOR-FDTD, implementing in Matlab the algorithm of the difference solution of Maxwell's equations in the cylindrical coordinates, and the 3D-FDTD method, implemented in the program FullWAVE (company RSoft, USA).

Figure 3.35 shows the distribution of the square of the modulus of the electric vector in the XZ plane of the linearly polarized plane wave transmitted through a binary axicon with a period of 4 μm, calculated by two different methods. From Fig. 3.35 it is clear that qualitatively both diffraction patterns are consistent, since both patterns show that each projection of the relief of the axicon focuses light in the local foci, at a distance of 2 μm (see Fig. 3.35a) and 1.7 μm (Fig. 3.35b) from the surface of the axicon .

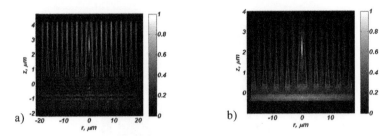

Fig. 3.35. The calculated intensity distribution in the *XZ* plane in the diffraction of a linearly polarized plane wave (the electric vector is directed along the *Y* axis) on the binary axicon with a period of 4 μm: Matlab2008a (a) and FullWAVE (RSoft) (b).

Figure 3.36 shows the diffraction patterns in the *XY* plane at a distance $z = 1.7$ μm (in local focal plane) (a) and its radial cross section (b); for comparison the figure also shows the intensity distribution in the *XY* plane at a distance $z = 0.3$ μm (c) where a light ring forms on the optical axis.

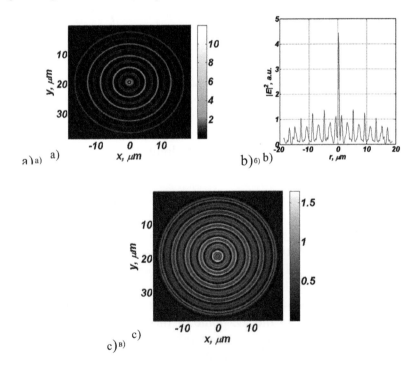

Fig. 3.36. Intensity distribution in the transverse *XY* plane calculated using FullWAVE (RSoft) at a distance of 1.7 μm (a) and its cross section (b), and at a distance of 0.3 μm (c) for a binary axicon with a period of 4 μm.

Figures 3.36*a* and *b* can be used to estimate the local maximum diameter which is equal to 1.25λ ($\lambda = 532$ nm). If we compare this value with the experimentally registered local diameter of the focal spot in Fig. 3.33 (the diameter along the x axis is 3.3λ), then it would be 2.5 times lower.

Figure 3.34 shows the axial intensity distribution for the axicon with a period of 4 μm calculated using the BOR-FDTD method.

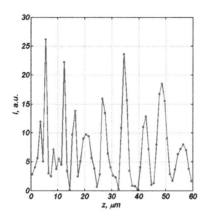

Fig. 3.37. The axial intensity distribution for an axicon with a period of 4 μm illuminated by a Gaussian beam with linear polarization and calculated by the BOR-FDTD method.

Figure 3.37 shows that the local extrema along the optical axis follow each other in a quasi-periodic manner. The distance between the adjacent maxima increases from 2 μm to 8 μm at a distance $z < 60$ μm.

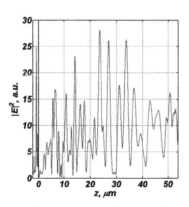

Fig. 3.38. The intensity distribution along the optical axis, calculated using the R-FDTD method for the axicon with a period of 4 μm, axicon radius 28 μm, illuminated with a divergent Gaussian beam with linear polarization.

Figure 3.38 shows the distribution of light intensity (arbitrary units) along the optical axis for the binary axicon with a period of 4 μm and 28 μm radius. In contrast to Fig. 3.37, in this case (Fig. 3.38) the axicon was illuminated with a divergent Gaussian beam with linear polarization with the amplitude of the electric vector in the form of:

$$E_x(r, z = 0) = \exp\left(-\frac{r^2}{w^2} + i\frac{kr^2}{2f}\right),$$
(3.134)

where $w = 2$ mm is the Gaussian beam waist radius, $f = 2.5$ w is the focal length of the parabolic lens. The divergent beam (as opposed to the flat beam, which was investigated so far) has been used to better align calculations with the experimental results obtained by the circuit of Fig. 3.32. In the setup in Fig. 3.32 the binary axicon D_3 4 mm in diameter was illuminated with divergent laser light after the microlens L_3. The divergent light beam was used to match the diameter of the illuminating light spot with the diameter of the axicon D_3. In Fig. 3.38 the distance between the neighbouring maxima of intensity increases from 2 μm to 4 μm for $z < 50$ μm. Reducing the distance between the maxima is associated with the use of the divergent Gaussian beam instead of the Gaussian beam without divergence.

Figure 3.39 shows the calculated (by the BOR-FDTD method) diameters of the central spots of the diffraction pattern of the diverging wave (3.134) on the axicon with a period of $T = 4$ μm, depending on the distance from the axicon.

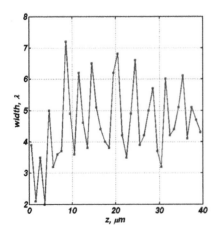

Fig. 3.39. The distribution of the full width (diameter along the x axis in the wavelengths) of the central maximum of intensity in the diffraction pattern of a divergent laser beam in a binary axicon with a period of 4 μm, calculated using the BOR-FDTD method.

In Fig. 3.39 local maxima correspond to the annular intensity distribution in the central spot of the diffraction pattern, and local minima correspond to the focal spots on the optical axis.

Figure 3.39 also shows that the diameter of the focal spot on the optical axis for the binary axicon with a period of 4 μm in the near field ($z < 40$ μm) varies from 4.3λ to 2λ with the longitudinal period of about 3 μm (at 40 μm there are 12 local minima in Fig. 3.38).

From the comparison of the experimental curve of the diameters of the central spots of the diffraction patterns (Fig. 3.32a) with the calculated curve of the diameters of the central spot (Fig. 3.39) it follows that there is agreement of the diameters of the focal spot at a distance to 40 μm: from 3.5λ to 4.5λ (Fig. 3.32a) and from 2λ to 4.3λ (Fig. 3.39). Although the longitudinal period of changes in the diameter of the focal spots in Fig. 3.39 was 3 μm, and in the experiment was less than 2 μm.

From the comparison of Figs. 3.38 and 3.39 it is clear that there is agreement of the number of local foci (minima in Fig. 3.38) and the local maxima of intensity (Fig. 3.39): in the interval between 30 μm $< z <$ 40 μm in Fig. 3.40 there are 4 minima and in Fig. 3.39 4 maxima; between 20 μm $< z <$ 30 μm in Fig. 3.39 there are only 2 minima, and in Figure 3.38 two large maxima.

From (3.128) it follows that the diameter of the focal spot for the axicon with a period of 4 μm will be equal to: $2r = 0.774T = 3.1$ μm $= 6\lambda$, with $m = 1$ and 3λ when $m = 2$. Comparing these numbers with the experiment (Fig. 3.32a) and calculation (Fig. 3.39), we can conclude that in the near diffraction field the contribution to the formation of foci along the optical axis comes from the first and second diffraction order of the binary axicon.

3.2.4. Binary axicon with a period of 800 nm
Manufacture

A glass substrate was coated with thin ZEP520A resist layer, which was then warmed for 10 min at a temperature of 180°C to dry the solvent. The height of the resist was adjusted so that it was sufficient for the required phase delay of the axicon. The pattern of the concentric rings was 'drawn' on a resist with an electron beam in a Zeiss Gemini electron microscope with the lithographic attachment RAITH ELPHY PLUS at a voltage of 30 kV. The pixel size was 10 nm, and the exposure energy 45 mAs/cm². After exposure, the sample was etched in xylene at a temperature of 23°C and then washed in isopropanol to fix the development process. Washing also removes the overexposed areas of the resist in which the original long polymer molecules were broken during exposure in the electron beam exposure. Thereafter, the remaining resist formed an axicon with a period of 800 nm (the refractive index of the resist $n = 1.5$). Figure 3.40 shows an image of the tested axicon. The height of the profile of this axicon is equal to 465 nm.

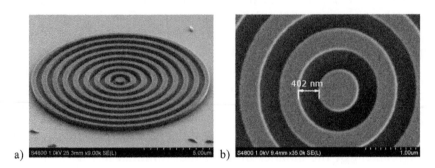

Fig. 3.40. Image of the tested binary axicon with a period of 800 nm obtained in an electron microscope: in an oblique view (a) and a top view, enlarged (b).

Experiment and numerical simulation

Next, using the near-field microscope NT-MDT the transmission of the linearly polarized Gaussian beam with a wavelength of 0.532 μm via a binary axicon with a period of 800 nm was studied.

Figure 3.41a shows the intensity distribution measured at a distance of 1 μm from the axicon. Figure 3.41b shows the horizontal and vertical sections of the two-dimensional intensity pattern (Fig. 3.41a). From a comparison of the cross sections in Fig. 3.41 it is clear that the focal spot has an elliptical cross section due to linear polarization (vertical polarization plane). The smallest spot diameter at half intensity is equal to FWHM = 0.58λ. The measurement error is 10%. Intensity is given in Fig. 3.41 in relative units. The intensity at the focus was 5 times greater than the maximum intensity of the Gaussian beam illuminating the microaxicon, although the effectiveness of focusing was low, 6%. This means that the formation of the focal spot (Fig. 3.41a) is determined only by the first circle and ring of the microaxicon relief (Fig. 3.40).

Figure 3.42 shows a comparison of the experimental intensity distribution at the focus (a distance of 1 μm from the axicon surface) with the distribution obtained by simulation- by the BOR-FDTD method. The standard deviation of the curves in Figure 3.42 is 6%.

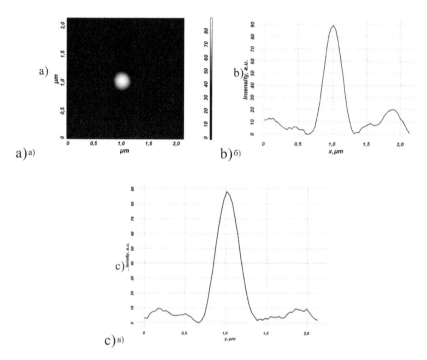

Fig. 3.41. The registered intensity pattern with a focal spot at a distance of 100 nm from the axicon (3.41): 2D image (a) and the cross section along the *x* axis (b), and along the *y* axis (c).

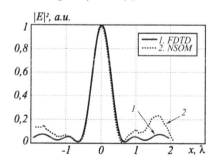

Fig. 3.42. The normalized transverse intensity distribution at a distance of 1 μm from the surface of the microaxicon.

The main difference between the curves in Fig. 3.42 is observed in the side lobes of the diffraction patterns. The main maximum intensity at $|x| \leq 1$ μm differs from the calculated maximum by only 1%. In the simulation, it was assumed that the axicon is illuminated by a Gaussian beam with a wavelength of 532 nm with

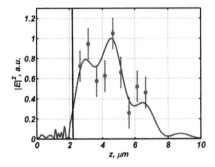

Fig. 3.43. The intensity distribution (normalized) along the optical axis of the axicon, obtained by simulation in Matlab (solid line) and experiments (separate boxes with vertical bars that show the range of measurement errors). The vertical lines show the border of the axicon.

linear polarization and a radius of 2.5 μm (this value was chosen to match the simulation results with the experiment).

Figure 3.43 shows the calculated intensity distribution along the optical *z* axis. The vertical line in Fig. 3.43 shows the surface of the axicon. The squares with vertical bars in Fig. 3.43 mark the experimental intensity values. The vertical segments indicate the amount of allowable measurement error rate (error ±0.1). The longitudinal depth of focus of the microaxicon at half intensity is DOF = 3 μm = 5.6λ.

We estimate the diameter of the focal spot of our binary microaxicon with a period $T = 800$ nm, according to (3.128). Since $NA = \lambda/T = 0.665$ for $m = 1$: FWHM = 0.36λ/NA = 0.54λ. Thus, the experimentally obtained value of the focal spot diameter (FWHM = 0.58λ) differs from the theoretical estimate (FWHM = 0.54λ) by only 8%.

3.3. Fresnel zone plate

The focusing element in this section will be used the zone plate (ZP). Earlier in [13] the ZP was used to obtain a focal spot with a diameter of 0.42λ, and in [27] it is shown that the addition to the ZP of a phase plate consisting of three annular zones reduces the spot size to 0.37λ. In the above works focusing was simulated using the Richards–Wolf (RW) formula [65], modified for the radially polarized light [51]. It should be noted that the vector Debye theory [66] on which the RW formulas are based is valid only if the focal distances are much larger than the wavelength. At distances comparable to the wavelength one should use, for example, the FDTD method (= finite difference time domain), which is a numerical solution of Maxwell's equations and is now widely used due to its universatility. With reference to the sharp focusing of light by the zone plates this method is used, in particular, for the calculation of focusing light by the plasmonic lenses. Thus, in [67] it is shown theoretically and by the FDTD and experimentally using near-field optical microscopy how to overcome the diffraction limit by the plasmon lens (diameter of the spot at half intensity is theoretically FWHM = 0.41λ, by experiments 0.48λ). In [68] the FDTD method is used for the plasmon lens to obtain focal spot 0.33λ wide, and in [69] the authors show numerically the possibility of focusing in the near field using a simple phase ZP (the diameter of the spot at the same time is 0.52λ). In [70] the focusing properties of the amplitude ZP are analyzed. In [71] the FDTD-method was used to simulate focusing a zone plate with the radius of the rings $r_n^2 = 2nf\lambda + n^2\lambda^2$, $f = 1$ μm, $\lambda = 633$ nm, formed in thin films of silver (50 nm) and gold (50 nm) deposited on quartz. The diameter of the annular structure was 13 μm. It is shown that at a distance $z = 1.5$ μm from the plate there is a focal spot with the diameter at half intensity FWHM = 0.3λ (full width of the spot 0.7λ). In [72] experiments were conducted with similar ring structures (diameter 8 μm) in a gold film (100 nm). A focal spot with a diameter at half intensity FWHM = 1.7λ (total diameter 5λ), $\lambda = 633$ nm, was studied in a near-field microscope NTEGRA (NT-MDT) with a resolution of 100 nm at a distance $z = 1.6$ μm Although the theory predicts the size of the focal spot FWHM = 0.5λ. In [73] an amorphous silicon film with a thickness of 120 nm was used to produce a Fresnel lens with a focal length $f = 5$ μm and a diameter of 50 μm for a wavelength $\lambda = 575$ nm (26% transmission). The numerical aperture of the lens in immersion was $NA = 1.55$, and the lens focused light in a focal spot with a diameter FWHM = 0.9λ. Interestingly, this spot was measured by a fluorescent sphere with a diameter of 0.5 μm.

3.3.1. Comparison with the Richards–Wolf formulas

As previously mentioned, using a binary zone plate (ZP) the radially polarized light is focused to a spot with the size smaller than when focusing using an aplanatic lens with the same numerical aperture [13]. Focusing of light by the zone plate was modelled out by two different methods: the BOR-FDTD method and the Richards–Wolf formulas. The results were compared by the example of the glass zone plate with a refractive index $n = 1.5$, radius $R = 20\lambda$ and the height of the relief $h = \lambda/2(n-1) = 0.532$ μm, which provides the focusing of the radially polarized mode R-TEM$_{01}$ with a wavelength $\lambda = 0.532$ μm, radius $\omega = 10\lambda$. The radii of the rings of the zone plate were calculated by the known formula (see Fig. 3.44):

$$r_m^2 = m\lambda f + m^2\lambda^2/4,$$ (3.135)

where f is the focal length of the zone plate, m is an integer. The dependence of the changes in the diameter of the focal spot at half intensity on the focal length of the ZP is shown in Fig. 3.45a (curve 1 – BOR-FDTD-method, curve 2 – RW formula).

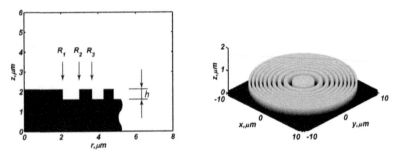

Fig. 3.44. The radial profile (a) and three-dimensional image (b) of a binary zone pla   with a focal length $f = 15\lambda$.   b)

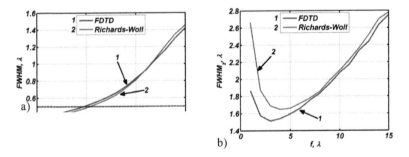

Fig. 3.45. Dependence of the change (a) of the diameter of the focal spot FWHM and (b) the depth of the focal spot DOF at half intensity on the focal length of the zone plate with a radius $R = 20\lambda$ in illumination with a radially polarized mode R-TEM$_{01}$ with a radius $\omega = 10\lambda$. Simulation using the BOR-FDTD method (curve 1) and RW-formulas (curve 2). The dotted curve in Fig. 3.46 a shows the diffraction limit.

Figure 3.45b shows the dependence of the change of the longitudinal width of the focal spot (depth of field focus) at half intensity on the focal length.

The simulation parameters by the FDTD-method: discreteness of partition in space $\lambda/50$, the discreteness of partition in time $\lambda/100c$, where c is the speed of light in vacuum, perfectly matched layers with a thickness λ were placed on the boundary.

The RMS relative deviation δ of the diameter of the focal spot, calculated by the BOR-FDTD method and the RW formula, does not exceed 6% (Fig. 3.46, curve 1), and the relative deviation of the depth of focus FWHM increases with decreasing focal length of up to 30% at $f = \lambda$ (Fig. 3.46, curve 2). At the focal length greater than $f \geq 4\lambda$ ($NA \leq 0.98$) the two errors do not exceed 6%. Thus, it can be argued that the RW-formula can be applied for modeling sharp focus of light, if the distance of the focus of the ZP from the surface is more than 4λ.

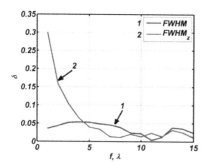

Fig. 3.46. The dependence of the mean-square relative error of calculating the diameter of the focal spot (curve 1) and the depth of the focal spot (line 2) on the focal length of the zone plate.

As seen in Fig. 3.45a, both simulations show overcoming of the diffraction limit. When calculating by the BOR-FDTD method overcoming of the diffraction limit is observed for the focal length $f = 4.7\lambda$ ($NA = 0.97$), and when using the RW formulas – at $f = 5.4\lambda$ ($NA = 0.96$). The minimum values of the diameter of the focal spot were FWHM = 0.37λ when calculating by the RW formulas and FWHM = 0.39λ when calculating by the FDTD method (in agreement with [45]) for the numerical aperture $NA = 0.999$ ($f = \lambda$). Figure 3.48 shows the intensity distribution in the focal plane of the zone plate with a focal length $f = \lambda$. The figure shows that overcoming the diffraction limit is due to the increase in the energy of the side lobes (or increasing the depth of focus), and as a result the effectiveness of focusing decreases. Figure 3.45b shows that the depth of focus calculated by the FDTD-method (curve 1) is always less than that calculated by the RW formulas (curve 2). This is because the ZP limits by its surface the light beam in the longitudinal direction. This is evident from Fig. 3.47b.

Figure 3.47*b* shows that there is a minimum value of the longitudinal width of the focus – in simulation by the FDTD method it was DOF = 1.51λ for the focal length $f = 3λ$ ($NA = 0.99$), in the calculation by the RW formulas DOF = 1.65λ for the focal length $f = 4λ$ ($NA = 0.98$). Figure 3.47*b* (curve 1) shows that with decreasing focal length of the ZP to $f = 3λ$ the depth of focus is also reduced, but with decreasing focal length at $f < 3λ$ the depth of focus increases. This confirms the general idea that reducing the diameter of the focal spot to less than the diffraction limit the depth of focus begins to grow in a way that preserves the limiting diffraction volume of the focus.

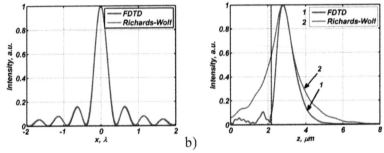

a)   b)

Fig. 3.47. The intensity distribution (a) in the focal plane along the x axis and (b) along the symmetry axis of the ZP with the radius $R = 20λ$ and the focal length $f = λ$ in illumination with the radially-polarized R-TEM$_{01}$ mode with $\omega = 10λ$. Simulation by the FDTD method (curve 1) and RW formulas (curve 2). The dashed vertical line in Fig. 3.48 b marks the edge of the ZP.

The effectiveness of focusing can be estimated as $\eta = W_1/W_0$, where $W_0 = \int_0^\infty |E_r|^2 r dr$ is the energy of the input beam, E_r is the radial component of the amplitude of the R-TEM$_{01}$ mode, $W_1 = \int_0^{r_0} I_{FDTD} r dr$ is the energy in focus, I_{FDTD} is the intensity distribution at the focus (Fig. 3.47a), $r_0 = 0.4λ$ is the point of the first minimum of function I_{FDTD}. When focusing the mode with the radius $\omega = 10λ$ by the zone plate with a focal length $f = λ$ and radius $R = 20λ$ the efficiency of focusing was $\eta = 42\%$.

Our results are in agreement with the results of [45], where the RW formulas and the FDTD method were used to produce a spot with a diameter FWHM = 0.39λ in illumination of the zone plate with a radially polarized Bessel–Gaussian beam. In [13] the RW formulas were also used when focusing ZP with a numerical aperture $NA = 0.98$ to produce a focal spot with a width FWHM = 0.42λ. In our case, for the numerical aperture $NA = 0.98$ (it is achieved when the radius of the ZP is $R = 20\lambda$ and the focal length $f = 4\lambda$) the calculation by the RW formula gives the same result, if the radius of the mode is $\omega = 15\lambda$. Simulation by the FDTD method for such parameters of the zone plate and the input beam showed that the diameter of the spot is FWHM $= 0.44\lambda$. Thus, the relative error in calculating FWHM equal to 4.55%. If the mode radius is reduced to $\omega = 10\lambda$, simulation by the FDTD method shows an increase in the diameter of the spot to FWHM $= 0.47\lambda$.

Figures 3.47 and 3.48 show the effect of ZP radius and width of the illuminating beam on the characteristics of the focal spot (at half the diameter and depth at half intensity).

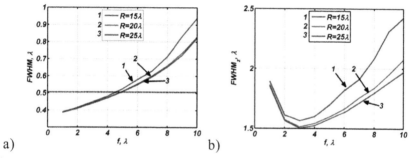

Fig. 3.48. The dependence of the focal spot diameter (a) at half intensity and the depth of the focal spot (b) at half intensity on the focal length of ZP with the radius $R = 15\lambda$ (curve 1), $R = 20\lambda$ (curve 2) and $R = 25\lambda$ (curve 3) in illumination with the radially polarized mode R-TEM$_{01}$ with $\omega = 10\lambda$.

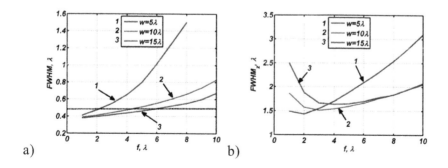

Fig. 3.49. The dependence of the diameter of the focal spot (a) at half the intensity and depth of the focal spot (b) at half the intensity on the focal length of the ZP with a radius $R = 20\lambda$ in illumination with the radially polarized mode R-TEM$_{01}$ with $\omega = 5\lambda$ (curve 1), $\omega = 10\lambda$ (curve 2) and $\omega = 15\lambda$ (curve 3).

As seen in Figure 3.48, the increase in the radius of the zone plate at constant parameters of the illuminating beam reduces the diameter of the focal spot. However, when $R > 20\lambda$ the diameter of the focal spot is almost unchanged (curves 2 and 3 in Fig. 3.48a). This means that the value of the radius $R = 20\lambda$ of the ZP and the beam radius $\omega = 10\lambda$ were optimum. Figure 3.49 shows that the increase in the radius of the illuminating beam ω reduces the diameter of the focal spot although it happens in a non-linear manner. Thus, an increase in ω by a factor of 1.5 (curves 2 and 3 in Figure 3.49a) reduces the diameter of the focal spot by 20%–25% when $f > 6\lambda$ and there is almost no change in the diameter of the focal spot at $f < 2\lambda$. The depth of focus at $f > 6\lambda$ remains almost unchanged (curves 2 and 3 in Figure 3.49b). Figure 3.49 also shows that as soon as the diameter of the focal spot becomes smaller than the diffraction limit (curves 1, 2, 3 are below the dotted line in Figure 3.49a), with the further reduction of f the depth of focus starts to rise (curves 1, 2, 3 'pass' through a minimum in Fig. 3.49b).

3.3.2. Symmetry of the intensity and power flow of the subwavelength focal spot

The following question may arise in the study of sharp focusing of light by scanning near-field optical microscopy: what does the near-field microscope register – energy (power) density, or the energy (power) flux? To answer this question, in this chapter we carry out experiments using a near-field microscope for focusing a linearly polarized Gaussian beam by binary glass zone plate with a focal length equal to the wavelength. Comparison of the experimental data with the simulation results by the FDTD-method enabled to unambiguously conclude that the near-field microscope measures the transverse intensity (power density) rather than the power flux or full intensity. The fact that the metal cantilever with a small aperture measures the transverse intensity follows from the Bethe–Boukamp theory.

In this section we will also answer another important question: why, for linearly polarized light, the intensity (power density) of the focal spot has the form of an ellipse, and the projection on the optical axis of the Umov–Poynting vector (power flux) in the focal spot has the form of a circle? The answer to this question will be searched with the help of the expansion of the light field with linearly polarized plane waves. It will be shown that the elliptical form of the cross section of intensity is determined by the longitudinal component of the electric vector, and as the projection of the Umov–Poynting vector on the optical axis (power flux) is independent of the longitudinal component of the electric vector, the section of the power flux is circular.

The intensity and projection of the Umov–Poynting vector for linearly-polarized light

We consider the propagation of light along the optical axis z from a transverse plane P_1 (we call it the initial plane) to the other transverse plane P_2, located parallel to the initial plane and located from it at distance z. We introduce in these planes the Cartesian coordinates (x, y) and (u, v) and polar coordinates (r, φ) and (r, θ). Let the linearly polarized electromagnetic field with radial symmetry forms in the initial plane:

$$\begin{cases} E_x(r,\varphi,0) \equiv E_x(r), \\ E_y(r,\varphi,0) \equiv 0, \\ E_z(r,\varphi,0) \equiv 0. \end{cases} \tag{3.136}$$

In plane P_2 we will measure the intensity

$$I = |\mathbf{E}|^2 = |E_x|^2 + |E_y|^2 + |E_z|^2 \tag{3.137}$$

and power flux (the component of the Umov–Poynting vector parallel to the optical axis z)

$$S_z = \frac{1}{2}\mathrm{Re}\left\{(\mathbf{E}\times\mathbf{H}^*)_z\right\} = \frac{1}{2}\mathrm{Re}\left\{E_x H_y^* - E_y H_x^*\right\}. \tag{3.138}$$

According to the Rayleigh–Sommerfeld diffraction integrals, component E_y in the plane P_2 will also be zero. Therefore, equtions (3.137) and (3.138) can be rewritten as:

$$I = |E_x|^2 + |E_z|^2, \tag{3.139}$$

$$S_z = \frac{1}{2}\mathrm{Re}\left\{E_x H_y^*\right\}. \tag{3.140}$$

Using Maxwell's equation for monochromatic light with frequency ω

$$\text{rot } \mathbf{E} = -i\omega\mu_0\mu\mathbf{H}, \tag{3.141}$$

where μ is the permeability of the medium, μ_0 is the magnetic constant, we obtain from (3.140):

$$S_z = \text{Re}\left\{\frac{-i}{2\omega\mu_0\mu}E_x\left(\frac{\partial E_x^*}{\partial z} - \frac{\partial E_z^*}{\partial u}\right)\right\}. \tag{3.142}$$

We expand the component E_x in the angular spectrum of plane waves. We introduce for this the Cartesian (α, β) and polar (ζ, ϕ) coordinates in the spectral plane. Then

$$E_x(u,v,z) = \iint_{\mathbb{R}^2} A(\alpha,\beta)\exp\left\{ik\left[\alpha u + \beta v + z\sqrt{1-\alpha^2-\beta^2}\right]\right\}d\alpha d\beta. \tag{3.143}$$

From the third Maxwell equation

$$\frac{\partial E_x}{\partial u} + \frac{\partial E_y}{\partial v} + \frac{\partial E_z}{\partial z} = 0 \tag{3.144}$$

given the fact that $E_y \equiv 0$, we obtain an expression for the component E_z:

$$E_z(u,v,z) = -\iint_{\mathbb{R}^2}\frac{\alpha}{\sqrt{1-\alpha^2-\beta^2}}A(\alpha,\beta)\times \tag{3.145}$$

$$\times\exp\left\{ik\left[\alpha u + \beta v + z\sqrt{1-\alpha^2-\beta^2}\right]\right\}d\alpha d\beta + C(u,v).$$

Constant $C(u, v)$, which arose in the integration in z, denotes a constant field of infinite extent along the z axis. From physical considerations we equate it to zero. We also transform the expression in brackets in (3.142):

$$\frac{\partial E_x^*}{\partial z} - \frac{\partial E_z^*}{\partial u} = -ik\iint_{\mathbb{R}^2}\frac{1-\beta^2}{\sqrt{1-\alpha^2-\beta^2}}A^*(\alpha,\beta)\times \tag{3.146}$$

$$\times\exp\left\{-ik\left[\alpha u + \beta v + z\sqrt{1-\alpha^2-\beta^2}\right]\right\}d\alpha d\beta.$$

Since component E_x is radially symmetric in the initial plane, its angular spectrum is also radially symmetric, i.e. $A(\zeta, \phi) \equiv A(\zeta)$. With this in mind, we can rewrite (3.143), (3.145) and (3.146) in polar coordinates. In this case all the integrals over ϕ are expressed in terms of the Bessel function:

$$E_x(\rho,\theta,z) = 2\pi \int_0^\infty A(\zeta)\exp\left(ikz\sqrt{1-\zeta^2}\right)J_0(k\rho\zeta)\zeta d\zeta, \quad (3.147)$$

$$E_z(\rho,\theta,z) = -2\pi i \cos\theta \int_0^\infty A(\zeta)\exp\left(ikz\sqrt{1-\zeta^2}\right)J_1(k\rho\zeta)\frac{\zeta^2 d\zeta}{\sqrt{1-\zeta^2}}.$$

$$(3.148)$$

$$\frac{\partial E_x^*}{\partial z} - \frac{\partial E_z^*}{\partial u} =$$

$$= -2\pi i k \int_0^\infty \left[\left(1-\frac{\zeta^2}{2}\right)J_0(k\rho\zeta) - \frac{\zeta^2}{2}J_2(k\rho\zeta)\cos(2\theta)\right]\frac{\zeta d\zeta}{\sqrt{1-\zeta^2}}$$

$$(3.149)$$

Substituting (3.147)–(3.149) into (3.139) and (3.141):

$$I = 4\pi^2 \left|\int_0^\infty A(\zeta)\exp\left(ikz\sqrt{1-\zeta^2}\right)J_0(k\rho\zeta)\zeta d\zeta\right|^2$$

$$+4\pi^2 \cos^2\theta \left|\int_0^\infty A(\zeta)\exp\left(ikz\sqrt{1-\zeta^2}\right)J_1(k\rho\zeta)\frac{\zeta^2 d\zeta}{\sqrt{1-\zeta^2}}\right|^2,$$

$$(3.150)$$

$$S_z = -\frac{2\pi^2 k}{\omega\mu_0\mu}\mathrm{Re}\left(\left\{\int_0^\infty A(\zeta)\exp\left(ikz\sqrt{1-\zeta^2}\right)J_0(k\rho\zeta)\zeta d\zeta\right\}\times\right.$$

$$\left.\times\left\{\int_0^\infty A^*(\zeta)\exp\left(-ikz\sqrt{1-\zeta^2}\right)\left[\left(1-\frac{\zeta^2}{2}\right)J_0(k\rho\zeta)-\frac{\zeta^2}{2}J_2(k\rho\zeta)\cos(2\theta)\right]\frac{\zeta d\zeta}{\sqrt{1-\zeta^2}}\right\}\right).$$

$$(3.151)$$

From (3.150) and (3.151) it follows that both the intensity and power flux are symmetrical with respect to axis x, i.e. for any pair of points with the polar coordinates (ρ, θ) and $(\rho, -\theta)$ they are equal. From (3.150) is can also be seen that for a fixed ρ the intensity at the points $(\rho, 0)$ and (ρ, π) is maximal, and at points $(\rho, \pi/2)$ and $(\rho, 3\pi/2)$ is minimal. This explains the formation of the focal spot in the shape of an ellipse elongated along the x axis (the polarization plane xz). Note that the radial symmetry violation in the expression for the intensity is determined by the second term, and in terms of the power flux by the second factor. In the case of the small numerical aperture factor $\zeta^2/(1-\zeta^2)^{-1/2}$ in the second integral in (3.150) is close to zero and prevents the second term to give a significant contribution to the intensity. Therefore, the focal spot is circular. If the focusing is sharp, the wave with ζ, close to unity, occupy a significant part in the angular spectrum. In this case, the contribution of the second term increases and may exceed several time the contribution of the first term, asymmetry appears, and the spot takes an elliptical shape, or even the form of 'bone' or 'dumbbell'.

For the power flux the dependence is different. The factor $\zeta/(1-\zeta^2)^{-1/2}$ equally affects both symmetrical and asymmetrical part to the expression in brackets in (3.151). In the case of a small numerical aperture ($\zeta \ll 1$) the power flux is defined by the radially symmetrical term $(1- \zeta^2/2) J_0 (k\rho\zeta)$. Therefore, the focal spot measured not in terms of intensity but the power flux, has a circular shape. In the case of a large numerical aperture, when ζ is close to unity, the contribution of both terms in the square brackets is approximately the same, but the contribution of the second term can not substantially exceed the contribution of the first term, and, besides, near the focus ($\rho = 0$), the zero-order Bessel function has a greater effect compared to the second-order Bessel function. Therefore, the spot is a more circular shape than in the case of measuring the intensity.

For example, consider the Bessel beam. Its angular spectrum is annular

$$A(\zeta) = \delta(\zeta - \alpha),\tag{3.152}$$

where $\delta(x)$ is the Dirac delta function, α is the parameter of the Bessel beam. In this case, the integrals in (3.150)–(3.151) vanish, and the intensity with the power flux can be easily calculated:

$$I = \left[2\pi\alpha J_0\left(k\alpha\rho\right)\right]^2 + \left[\frac{2\pi\alpha^2}{\sqrt{1-\alpha^2}}J_1\left(k\alpha\rho\right)\cos\theta\right]^2,\tag{3.153}$$

$$S_z = -\frac{2\pi^2 k}{\omega\mu_0\mu} J_0(k\alpha\rho) \left[\left(1-\frac{\alpha^2}{2}\right)J_0(k\alpha\rho) - \frac{\alpha^2}{2}J_2(k\alpha\rho)\cos(2\theta)\right]\frac{\alpha^2}{\sqrt{1-\alpha^2}}.$$

(3.154)

To simulate the case of a small numerical aperture we choose $\alpha = 0.3$ (Fig. 3.50). To simulate sharp focusing we choose $\alpha = 0.8$ (Fig. 3.51) and $\alpha = 0.9$ (Fig. 3.52). The intensity and power flux for different values of α are shown in Figs. 3.50–3.52: the horizontal axis is the x axis, the vertical axis – the y axis. Other simulation parameters: the wavelength $\lambda = 532$ nm, the distance $z = 10\lambda$, the simulation domain $5\lambda \le x \le + 5\lambda$, $-5\lambda \le y \le + 5\lambda$.

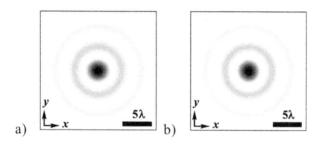

Fig. 3.50. The intensity (a) and power flux (b) for the Bessel beam with $\alpha = 0.3$.

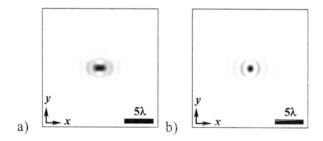

Fig. 3.51. The intensity (a) and power flux (b) for the Bessel beam with $\alpha = 0.8$.

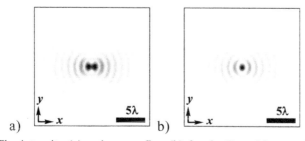

Fig. 3.52. The intensity (a) and power flux (b) for the Bessel beam with $\alpha = 0.9$.

Figures 3.50–3.52 confirm earlier assumptions. When increasing the numerical aperture of the focal spot, measured by intensity, first had a round shape, then an elliptical shape and then the form of 'bones'. The focal spot, measured in terms of the power flux, all the while remained almost round (in Fig. 3.52b the focal spot has the form of a weak ellipse stretched along the y axis; it is easy to show this with the help of equation (3.154)).

From Figs. 3.51 and 3.52 it can be seen that the size of the focal spots for the intensity and power flux along the vertical axis are similar.

Modeling sharp focusing of linearly polarized light by the zone plate

We consider focusing a linearly polarized Gaussian beam with a wavelength $\lambda = 532$ nm and the radius $\omega = 7\lambda$ by a binary zone plate (ZP) with a focal length equal to one wavelength of the focused light $f = \lambda$, radius of 10.64 μm (20λ), the refractive index of the material 1.52. Figure 3.54 shows a ZP template in the calculation domain. The numerical aperture of the ZP is $NA = 0.997$. The calculated efficiency of focus is 42%.

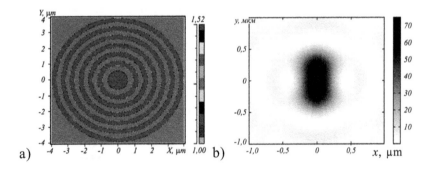

a)   b)

Fig. 3.53. ZP template with a focal length equal to the wavelength $f = \lambda$ (a), and the intensity distribution in the focal plane (b). The y axis is the axis of polarization.

The radii of the ZP (Fig. 3.54 a) were calculated by the known formula $r_m = (m\lambda f + m^2\lambda^2/4)^{1/2}$, where $f = \lambda = 532$ nm is the focal length, m is the number of of the zone radius. The simulation was performed by the BOR-FDTD method and the results obtained in different discrete partitions of the computed field in space were compared. Figure 3.54b shows in pseudocolours the intensity distribution at the focus.

Figure 3.54 shows the cross-section of the Cartesian axes: the x axis (φ = 0) and the y axis (φ = π/2) of the intensity (a) and power flux (projection on the z axis of the Umov–Poynting vector) (b). From Figs. 3.53b and Fig. 3.54a it can be seen that the focal spot of intensity is elliptical. Figure 3.54b also shows that the focal spot of the power flux is circular. In Table 3.3 the first two lines show the values of the focal spot size at half intensity (first line) and the power flux (second line).

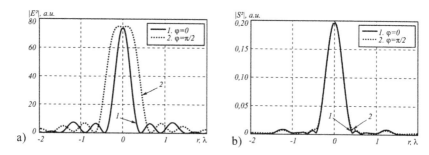

Fig. 3.54. Distributions of (a) intensity and (b) the modulus of the projection on the z axis of the Umov–Poynting vector in the focus at discreteness of the partition Δr = λ/50. The section along the x axis (φ = 0) and the y axis (φ = π/2)

Table 3.3. The diameters of the focal spot

	FWHM$_x$ (φ=0), λ	FWHM$_y$ (φ=π/2), λ	DOF, λ
Intensity	0.42±0.01	0.84±0.01	0.86±0.01
Modulus of projection on the z axis of the Umov–Poynting vector	0.45±0.01	0.45±0.01	–
Experiment with NSOM	0.44±0.02	0.52±0.02	0.75±0.02

Figure 3.55 shows the values calculated using the BOR-FDTD method in the Matlab package (curve 1) and the FDTD method, implemented in the FullWAVE program (curve 2).

The depth of focus for the ZP is shown in the last column of Table 5.3.

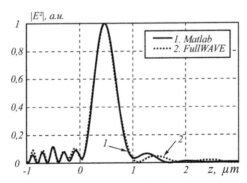

Fig. 3.55. Comparison of the intensity distribution along the optical axis for the ZP in modeling in BOR-FDTD Matlab (curve 1) and FDTD FullWAVE (curve 2).

Experimental research of focusing of linearly polarized light by a zone plate with focal length $f = \lambda$

The high-quality zone plate was made by lithography from ZEP resist (the refractive index of the resist $n = 1.52$). Figure 3.56 shows an image of the ZP in the electron microscope: the depth of the relief 510 nm, diameter 14 μm, the outer zone $0.5\lambda = 266$ nm. ZP has 12 rings and a central disk. Figure 3.57 shows a side view (a) and the view from above (b) of the ZP in an atomic force scanning microscope.

In experiments, a scanning near-field optical microscope (SNOM) Ntegra Spectra was used to study the passage of a linearly polarized Gaussian beam with a wavelength $\lambda = 532$ nm through a zone plate with a focal length $f = \lambda$.

Figure 3.58*a* shows experimental intensity distribution along the optical axis for the ZP (curve, left axis) and the diameter of the focal spot (squares, right axis). Figure 3.58*b* shows an example of the intensity distribution at the focus (the image obtained directly from the microscope).

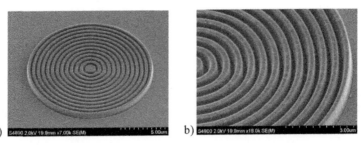

Fig. 3.56. Image of ZP in an electron microscope with a magnification of 7000 (a) and 18000 (b).

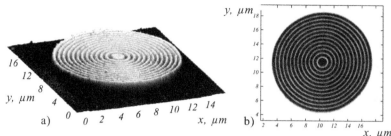

Fig. 3.57. Side view (a) and top view (b) of the ZP, atomic force scanning microscope.

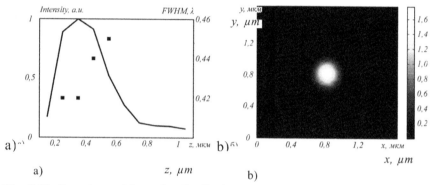

Fig. 3.58. Experimental intensity distribution along the optical axis (a) for the ZP (curve, left axis) and the values of smaller diameters of focal spots (squares, right axis); a cross section of the focal spot at the focal length $f = \lambda = 532$ nm (vertical axis coincides with the plane of polarization) (b).

Figure 3.58*a* shows the experimental intensity distribution along the optical axis for the ZP (curve, left axis) and the diameter of the focal spot (squares, right axis). Figure 3.58*b* shows an example of the intensity distribution at the focus (the image obtained directly from the microscope).

The averaged values of the diameter of the elliptic focal spot of the ZP are given in the third row of values in Table 3.3. Figure 3.58*a* shows that the maximum intensity on the axis is displaced from the plane of the geometrical focus $f = \lambda = 532$ nm closer to the ZP ($z = 400$ nm). Note that at this distance ($z = 400$ nm), the smaller calculated and experimental diameters of the focal spot match FWHM = 0.42λ.

For comparison, Fig. 3.59 shows the cross section along the x axis, perpendicular to the polarization plane, intensity (curve 1), the power flux (curve 3) and the experimental curve measured using NSOM (curve 2). The curves are almost identical (see the second column of Table 3.3): their difference is less than the measurement

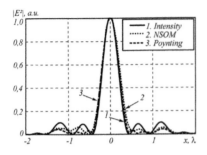

Fig. 3.59. Comparison of experimental and calculated distributions at the focus along the *x* axis: the calculated intensity distribution (curve 1), the experimental intensity distribution (curve 2) and the calculated distribution of the modulus of projection of the Umov–Poynting vector (line 3) on the *z* axis.

error (±0.02λ). The differences are noticeable only in the side lobes. But these side lobes cannot be used to say what is measured in the experiment – the intensity or power flux, since the sidelobes of the calculated power flux are slightly smaller than those of the experimental curve and the side lobes of the calculated intensity curve are slightly larger than those of the experimental curve.

Figure 3.60 shows the cross-section curves in focus along the *y* axis (parallel to the plane of polarization): the calculated distribution of the modulus of projection of the Umov–Poynting vector on the *z* axis (line 3), the experimental intensity distribution (curve 2) and the calculated intensity distribution (curve 1), calculated as the sum of all the components (a) and the sum of only the transverse components (b).

From Fig. 3.60*a* it can be seen that in the experiments we do not measure the longitudinal component of intensity (see the third column of Table 3.3), as the curve of the total intensity (FWHM = 0.84λ) is wider than the experimental curve (FWHM = 0.52λ) by an amount greater than the measurement error (±0.021).

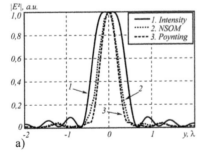
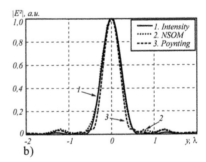

a)   b)

Fig. 3.60. Comparison of experimental and calculated distributions at the focus along the axis *y* (parallel to the plane of polarization): experimental intensity distribution (curve 2), the calculated distribution of the modulus of projection of the Umov–Poynting vector on the *z* axis (line 3) and the calculated intensity distribution (curve 1), calculated as a the sum of all the components (a) and the sum of only the transverse components (b).

But, in turn, the experimental curve is wider than the calculated power flux curve (FWHM = 0.45λ) by an amount greater than the experimental error. But what is measured in the experiment? Figure 3.60*b* shows the comparison of the experimental curve (curve 2) with transverse intensity $|E_x|^2 + |E_y|^2$ (curve 1). They have the same width (FWHM = 0.52λ). Therefore, on the basis of Fig. 3.60 it can be clearly concluded that the near-field microscope NSOM with a cantilever in the form of a metal pyramid with an aperture of 100 nm (Fig. 3.60*b*) measures the transverse intensity $|E_x|^2 + |E_y|^2$ (power density) rather than the power flux and full intensity $|E_x|^2 + |E_y|^2 +|E_z|^2$. Therefore, the ellipse of the focal spot in Fig. 3.58*b* is less pronounced than the calculated ellipse in Fig. 3.53*b*.

The penetration of the electromagnetic field in a small hole in a metal screen is described in the framework of the Bethe–Bowkamp theory [74–76]. According to this theory, a linearly polarized plane wave incident at an angle on a metal screen with a small round hole with the diameter $a \ll \lambda$, induces an electric dipole oriented perpendicular to the hole, and a magnetic dipole located in the plane of the hole. Therefore, the field in the far zone for a small hole, illuminated with an inclined plane wave **E** is given by the radiation of electric **P** and magnetic **M** dipoles with moments

$$\mathbf{P}=-\frac{4}{3}\varepsilon_0 a^3\left(\mathbf{En}_z\right)\mathbf{n}_z,$$

$$\mathbf{M}=-\frac{8}{3}a^3\left[\mathbf{n}_z\times\left[\mathbf{E}\times\mathbf{n}_z\right]\right],\tag{3.155}$$

where $\mathbf{n}_z$ is a unit vector along the optical axis (perpendicular to the plane of the hole). From (3.155) we see that the electric dipoleis generated only by the longitudinal component of the electric field **E**. However, the dipole oriented along the optical axis, emits in the transverse direction and does not emit along the optical axis itself. The magnetic dipole in (3.155) is formed, on the contrary, only by the transverse components of the electric field as the inner vector product on the right in formula (3.155) is equal to zero for the longitudinal component of the electric field. That is, the longitudinal component of the electric field will not be registered by a photodetector disposed at a distance from the small hole in the metal along the optical axis.

3.4. Focusing light with gradient lenses

3.4.1. Mechanism of superresolution in a planar hyperbolic secant lens

Using a DOE in conjunction with the far-field refractive focusing optics we can achieved superresolution, for example, obtain a focal spot with a diameter at half intensity equal to FWHM = 0.44λ [48], where λ is the wavelength of light in vacuum, rather than the diffraction limit (FWHM = 0.51λ). But at the same time the intensity side lobes of the diffraction pattern at the focus will exceed the value of 0.1 of the intensity at the focus. It is possible to achieve even smaller diameters of the focus in the far field, but this will increase the proportion of light energy in the side lobes so that their intensity can equal to or exceed the intensity at the focus [77]. Note that there are several diffraction resolution limits in optics: it is the Rayleigh resolution limit $0.61\lambda/NA$ [78], the Houston limit $0.5\lambda/NA$ and the Sparrow limit $0.475\lambda/NA$ [79], where NA is the numerical aperture of the focusing system. We will use our own resolution limits that are defined for 3D fields at half a square of the Airy function ($0.51\lambda/NA$), and for the 2D field at half $sinc$-function ($0.44\lambda/NA$).

To overcome the resolution limit without increasing the side lobes, the optical element should be closer to the light source. This area of optics is explored by near-field microscopy [74]. If we exclude from consideration metallic surfaces and surface plasmons [80], which allows to achieve a resolution of λ/50 and remain in the domain of refractive and gradient optics, the high resolution can be achieved by using the near-field lens: SIL (solid immersion lens) [81], NAIL (numerical aperture immersion lens) [82–84] and nSIL (nano solid immersion lens) [78,85].

Using experimentally SILs for a wavelength of λ = 633 nm and a glass LASFN9 hemisphere with a radius of 5 mm and a refractive index n = 1.845, the resolution FWHM = 190 nm = 0.298λ was obtained in [81]. In experiments with the help of a silicon hemisphere for a NAIL with a radius of 1.6 mm the resultant resolution was FWHM = 250 nm = 0.23λ (the theoretical limit of silicon FWHM = 0.147λ when n = 3.4 and λ = 1 μm) [82]. In a later paper [83] using a ring aperture and a silicon NAIL the resolution FWHM = 145 nm = 0.11λ (λ = 1.3 μm) was achieved. If the lens is illuminated by an annular beam, a Bessel beam is generated in the focus with the resolution limit FWHM = 0.36λ/NA. Using near-field optics (nSIL) on model examples it was shown that for a hemisphere with a radius of 1–2 μm [78] made of glass (n = 1.6) it is possible to produce a focal spot with the diameter FWHM = 126 nm = 0.235λ (λ = 532 nm) [85]. Note that the refractive near-field optics increases the numerical aperture of the already convergent light beam, and in the propagation of light from the source this type of optics is able to collect only light but additional optics is required to focus the light.

Tunneling evanescent waves of the source in a medium
In the two-dimensional (2D) case, the strength of the electric field of a monochromatic TE wave at a distance z from the initial plane is:

$$E_1(x,z) = \frac{k}{2\pi} \int\limits_{-\infty}^{+\infty} \int\limits_{-\infty}^{\infty} E_0(x',z=0)\exp\left[-ik\xi(x'-x)+ikz\sqrt{1-\xi^2}\right]dx'd\xi .$$

(3.156)

If the source in the initial plane is a point source:

$$E_0(x,z=0) = E_0\delta\left(\frac{kx}{2\pi}\right),$$

(3.157)

where $\delta(x)$ is the Dirac delta function, the amplitude of the field at a distance z is equal to the sum of plane waves and evanescent waves:

$$E_1(x,z) = E_0 \int\limits_{-\infty}^{\infty} \exp\left[-ik\xi x + ikz\sqrt{1-\xi^2}\right]d\xi .$$

(3.158)

Note that since the Hankel function of zero order and the first kind is given by [86]

$$H_0^1(k\sqrt{x^2+z^2}) = \frac{1}{\pi} \int\limits_{-\infty}^{\infty} \frac{\exp\left[-ik\xi x + ikz\sqrt{1-\xi^2}\right]}{\sqrt{1-\xi^2}}d\xi,$$

(3.159)

then (3.158) can be expressed through the derivative of Hankel's function:

$$E_1(x,z) = -\frac{i\pi}{k} \frac{\partial}{\partial z} H_0^1(k\sqrt{x^2+z^2})$$

(3.160)

Suppose that on the path of the radiation the boundary between two media appears at distance z from the source. That is, the radiation from the source passes into a medium with refractive index n. Then, the amplitude of the electric vector in a medium at a distance z from the source is equal to

$$E_2(x,z) = E_0 \int_{-\infty}^{\infty} T_1(\xi) \exp\left[-ik\xi x + ikz\sqrt{n^2 - \xi^2}\right] d\xi, \quad (3.161)$$

where

$$T_1(\xi) = \begin{cases} \dfrac{2\sqrt{1-\xi^2}}{\sqrt{1-\xi^2} + \sqrt{n^2 - \xi^2}}, & 0 < |\xi| < 1, \\[4mm] \dfrac{2i\sqrt{\xi^2 - 1}}{i\sqrt{\xi^2 - 1} + \sqrt{n^2 - \xi^2}}, & 1 < |\xi| < n, \\[4mm] \dfrac{2\sqrt{\xi^2 - 1}}{\sqrt{\xi^2 - 1} + \sqrt{\xi^2 - n^2}}, & |\xi| > n. \end{cases} \quad (3.162)$$

The values of $T_1(\xi)$ are coefficients obtained from the Fresnel formulas for three different cases: the transformation of a propagating plane wave in a propagating plane wave; transformation of a evanescent wave in a propagating plane wave; transformation of a decaying plane wave in a decaying plane wave in the medium.

Indeed, from (3.162) it can be seen that the propagating waves from a point source in a medium with $n = 1$ and $0 < |\xi| < 1$, where $\xi = k_x/k$, enter the medium at $0 < \theta < \theta_1$, where $\theta_1 = \arcsin(1/n)$, and will continue to propagate in the medium with $n > 1$. Surface waves of the first type from the source with the projection of the wave number in the range $1 < |\xi| < n$ enter the medium under the angles $\theta_2(\xi) = \arcsin(\xi/n)$, within the range of $\theta_1 < \theta_2(\xi) < \pi/2$ as the maximum angle $\theta_2(\xi)$ is $\pi/2$ at $\xi = n$. These waves are converted from evanescent to propagating waves, will continue to propagate in a medium with $n > 1$. The remaining surface waves (the second type) from source with $|\xi| > n$ are surface waves of the medium and will propagate along the interface.

The numerical aperture of the hyperbolic secant lens
Consider a two-dimensional hyperbolic secant (HS) lens whose refractive index is as follows:

$$n(x) = \frac{n}{\mathrm{ch}\left(\dfrac{\pi x}{2L}\right)}, \quad (3.163)$$

where n is the refraction index on the lens axis, L is the length of the lens, and x is the coordinate in the transverse plane.

The numerical aperture of the HS-lens can be found from the beam equation in a gradient medium: $n(x) \cos \theta (x) = \text{const}$, where θ is the angle between the tangent to the beam and the optical axis z. Suppose that the beam is incident on the lens parallel to the optical axis at a distance $x = R$ from it, where R is the radius of the HS lens, which can be found from the condition $n(R) = 1$: $R = \dfrac{2L}{\pi} ch^{-1}(n)$. Then the beam equation becomes: where $n \cos \theta_0 = n(R) \cos \theta (R) = 1$, θ_0 is the angle between the tangent to the beam and the optical z-axis at the point of intersection of the beam axis. It follows that $\cos \theta_0 = 1/n$, i.e. the numerical aperture of the HS lens is equal to $NA = n \sin \theta_0 = (n^2-1)^{1/2}$.

Since the numerical aperture of such lens is $NA = (n^2-1)^{1/2} = n \sin \theta_0$, wherein $\theta_0 = \arcsin [(n^2-1)^{1/2}/n]$, the plane waves with propagation angles θ_1 and θ_2 smaller than θ_0 will contribute to the focal spot at the output of the HS lens. Let us find the maximum ξ_{max} for light waves which are involved in the formation of the focus of the HS lens. We presume equality $\theta_2(\xi_{max}) = \theta_0$ and we obtain $\arcsin(\xi/n) = \arcsin\left[\left(n^2 -1\right)^{1/2}\big/n \right]$. The last equality implies an expression $\xi_{max} = (n^2-1)^{1/2}$. For silicon and the wavelength of $\lambda = 1550$ nm $\xi_{max} = 3.32$, since $n = 3.47$. Then, the minimum diameter at half the intensity of the focal spot at the output of the planar HS-lens is ($n = 3.47$):

$$\text{FWHM} = 0.44\frac{\lambda}{NA} = 0.44\frac{\lambda}{\sqrt{n^2 -1}} = 0.44\frac{\lambda}{\xi_{max}} = 0.133\lambda . \quad (3.164)$$

The theoretical resolution limit (in the planar case), which can be obtained using a solid-state immersion near-field optics: SIL (solid immersion lens) [81], NAIL (numerical aperture increasing lens) [82] is equal to ($n = 3.47$):

$$\text{FWHM} = 0.44\frac{\lambda}{n} = 0.127\lambda . \quad (3.165)$$

Equation (3.165) follows from the fact that the numerical aperture of the SIL and NAIL $NA_{SIL} = n\sin \theta \le NA_{NAIL} = \left(n^2 - \cos^2 \theta\right)^{1/2}$ to the limit $(\theta \rightarrow \pi/2)$ tends to $NA_{max} = n$. The limiting focal spot (3.165) is only 5% smaller than the focal spot of the HS-lenses (3.164). We estimate the maximum angle at which the beams propagate in the HS-lens. The propagating plane waves from a source with respect to the projection of the wave vectors in the interval $0 < |\xi| < 1$ are also converted in a homogeneous medium ($n = 3.47$) to propagating waves travelling at angles in the range $0 < \theta < \theta_1 = \arcsin(1/n) \cong 17°$ and the maximum angle at which the beams can be propagate in the HS-lens is $\theta_0 = \arcsin\left[(n^2-1)^{1/2}\big/n\right] \cong 74°$.

Note that if the SIL or NAIL is illuminated with an annular light beam, a diffraction pattern forms in the focus which is not described by the Airy function (not by *sinc*-function in the 2D case) and is described by the zero-order Bessel function. Therefore, the theoretical resolution limit in the medium will be equal to:

$$\text{FWHM} = 0.36\frac{\lambda}{n} = 0.104\lambda. \qquad (3.166)$$

Three types of waves propagating in the hyperbolic secant lens
Not all the waves that enter the HS-lens will come out of it. If the medium which receives the waves from the source has finite dimensions along the optical axis z, then let it be a plane-parallel plate with thickness d. Then the waves that have passed through the plate are described at the ouput by the expression:

$$E_3(x,z) = E_0 \int_{-\infty}^{\infty} T_2(\xi)\exp\left[-ik\xi x + ikz\sqrt{1-\xi^2}\right]d\xi, \qquad (3.167)$$

where

$$T_2(\xi) = \begin{cases} \dfrac{2\sqrt{(1-\xi^2)(n^2-\xi^2)}}{2\sqrt{(1-\xi^2)(n^2-\xi^2)}\cos A - i(n^2+1-2\xi^2)\sin A}, & 0 < |\xi| < 1, \\[4mm] \dfrac{2\sqrt{(\xi^2-1)(n^2-\xi^2)}}{2\sqrt{(\xi^2-1)(n^2-\xi^2)}\cos A - (n^2+1-2\xi^2)\sin A}, & 1 < |\xi| < n, \\[4mm] \dfrac{2\sqrt{(\xi^2-1)(\xi^2-n^2)}}{2\sqrt{(\xi^2-1)(\xi^2-n^2)}\,\text{ch}\,B - (n^2+1-2\xi^2)\,\text{sh}\,B}, & |\xi| > n, \end{cases}$$

$$(3.168)$$

$A = iB = kd(n^2-\xi^2)^{1/2}$. The values $T_2(\xi)$, as well as T_1 in the formula (3.162) are coefficients obtained from the Fresnel formulas for three different cases. From (3.168) it follows that the plane waves propagating from the source ($0 < |\xi| < 1$) pass through the plate, and again will propagate behind the plate at the same angles (the medium in front and behind the plate is air with $n = 1$). The surface waves of the first type ($1 < |\xi| < n$) are converted to modes inside a parallel plate and to surface waves on the opposite side (in relation to the source) of the plate. That is, these waves will not propagate in the space behind the plate. The surface waves of the second type ($|\xi| > n$) are converted to surface waves at the nearest (to the source) side of the plate, and only the exponentially decaying 'tails' will reach to the opposite side of the plate. Therefore, in the HS-lens the central rays from a source propagating through the HS-lens at an angle to the optical axis smaller than $\theta_1 = \arcsin(1/n) \cong 17°$ for $n = 3.47$, will pass through the lens and further extend behind the lens. The surface waves of the first type will propagate from the source to the HS-lens as in the annular cavity [87] and will not be released therefrom. The surface waves of the second type will propagate from the source as the surface wave in the HS-lens scattered partly on its sharp corners, since the lens is not a plane-parallel plate and is restricted in the transverse coordinates.

Therefore, if we assume that the focal spot in the HS-lens is created only by propagating waves with a maximum slope equal to θ_1, the width of the focus should be equal to ($n = 3.47$; $\theta_1 = 17°$) FWHM $= 0.44\lambda/n \sin\theta_1 = 0.43\lambda$. And when we consider that the surface wave of the first type, which propagate in the HS-lens with a maximum inclination to the optical axis equal to $\theta_0 = 74°$, contribute significantly to the formation of the focus, the diameter of the focal spot should be equal to FWHM $= 0.44\lambda/(n \sin\theta_0) = 0.132\lambda$. This number is consistent with the formula (3.164), which actually leads to the same result, only in a different way.

Figure 3.61 shows a typical dependence of the width of the focal spot on the width of the spectrum of plane waves that contribute to this focus, according to (3.164).

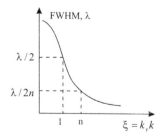

Fig. 3.61. The dependence of the diameter of the image of a point source (3.164) on the width of the spectrum of plane waves, including surface evanescent waves reaching the image plane.

Thus, if the simulation shows that the spot diameter in the HS-lens is smaller than the diffraction limit in the medium (3.165), it means that the surface waves of the second type take part in the formation of this focal spot.

The numerical aperture of near-field optics

Compare the numerical apertures of the near-field lens (SIL and NAIL) and HS-lens. Suppose that a convergent light beam with a numerical aperture $NA = \sin\theta$ propagates in vacuum, where θ is the maximum angle which the rays of this beam form with the optical axis. If such a beam enters the medium with a refractive index n and a plane interface between the media, its numerical aperture does not change $NA = n \sin\beta = \sin\theta$, where β is the maximum angle of the rays of the beam in the medium. SIL is used to increase the numerical aperture of the initial beam. This hemisphere made of a material with refractive index n is located with its spherical surface to the incident rays so that the rays are incident on the surface in the normal direction (Fig. 3.62a). Then the rays of the converging beam will gather in the focus at the centre of the hemisphere and the numerical aperture of the light beam inside the hemisphere will be equal to $NA_{SIL} = n \sin\theta$. Thus using the near-field optics (near-field because the focus is formed on the flat surface of the hemisphere) the numerical aperture of the initial light beam can be increased n time, or the diameter of the focal spot can be reduced n times.

Another type of near-field lens – NAIL – allows to further increase the numerical aperture of the beam. Unlike the SIL, a segment of a sphere (smaller than the hemisphere) is situated in the path of the converging rays of the light beam. Thus, the rays should fall on the spherical surface in such a way that their angle with the optical axis is smaller than the angle formed with this axis normal to the the spherical surface passing through the point of intersection

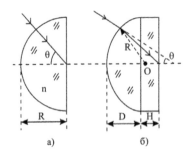

a)   б)

Fig. 3.62. Incidence and refraction of rays in the SIL (a) and NAIL (b).

of the ray with the sphere. Thus, to the segment of the sphere we must add a cylinder with the radius equal to that of the sphere and the height H, made of the same material as that of the sphere. The cylinder is adjusted to ensure that the rays are focused exactly on the intersection of the back plane of the cylinder with the optical axis. Let the sphere (ball) has radius R, the segment height $D < R$, then if the height of the cylinder is determined from the equation $D + H = R (1 + 1/n)$, the focus is formed on the output (for the rays) plane of the cylinder (Fig. 3.63 b). Figure 3.62b shows that the ray is refracted in the lens and crosses the optical axis at an angle greater than the angle θ in the ray incident on the lens. Moreover, if the numerical aperture of the light beam incident on NAIL is equal to $NA = \sin\theta$, the numerical aperture of the beam of rays in the cylinder converging at the output flat surface is $NA_{NAIL} = (n^2 - \cos^2\theta)^{1/2}$. We can show that $NA_{NAIL} \geq NA_{SIL}$. If we compare these numerical apertures with the numerical aperture of the HS-lens $NA = (n^2 - 1)^{1/2} = n \sin \theta_0$, the difference is not great: the maximum numerical aperture of the refractive optical near-field $NA_{NAIL} = NA_{SIL} = n = 3.47$ differs from the numerical aperture of the gradient lens $NA = (n^2 - 1)^{1/2}$ by only 5%. However, note that the SIL and NAIL lenses only collect light from the source and convert the first type of surface waves to propagating waves, but do not focus the collected light. Additional refractive optics is required for subsequent focusing. In contrast, the HS-lens also collects the light from the source and focuses it on its output surface.

Decreasing the focal spot using refractive index modulation
Recent studies [88,89] demonstrated numerically that the subwavelength gratings can be used to transform the surface waves from the source to propagating waves and achieve superresolution $\lambda/20$. In [88] for this purpose, several diffraction gratings with unlimited apertures and different subwavelength periods were stacked on each other in [88], and in [89] the authors used a subwavelength metallic grating with a very high dielectric constant ($\varepsilon = -100$). However, optical elements were not used in these studies to focus or image.

Simulation using FullWAVE with the FDTD-method showed that the addition to the gradient HS-lens of a subwavelength diffraction grating or producing the HS-lenses in binary form with subwave inhomogeneities reduces the width of the focal spot by 10% and 20%.

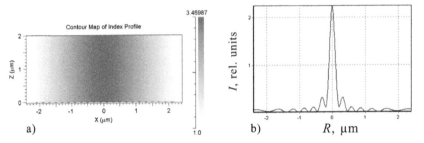

Fig. 3.63. The distribution of the refractive index in the gradient HS-lens (a) (on the axis the refractive index of 3.47), the horizontal size of 4.8 μm, vertical 2 μm. Light travels vertically. Transverse intensity distribution $|E|^2$ at the output from the lens (10 nm from it) (b).

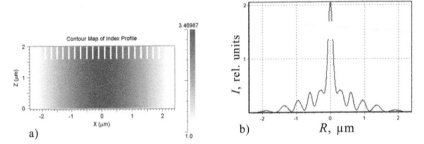

Fig. 3.64. The distribution of the refractive index of the gradient HS-lens (a) (on the axis refractive index of 3.47), the horizontal size 4.8 μm, vertical 2 μm. The upper part of the HS-lenses has a diffraction grating with a depth of 0.4 μm with a period of 0.2 μm and 0.05 μm width of the grooves (air in within the grooves). Light travels from the bottom up. Transverse intensity distribution $|E|^2$ at the output from the lens (at 10 nm from it) (b).

Figure 3.63 is shown in gray tones the refractive index profile of the HS-lens (a) and the intensity distribution at the exit of the lens (b). The width of the focal spot in Fig. 3.63*b* is FWHM = 191 nm = 0.123λ, λ = 1.55 μm. This value is slightly smaller than the diffraction limit of silicon (3.165), which proves that the surface waves of the second type take part in the formation of the focal spot in the HS-lens.

Figure 3.64 shows the same HS-lens as in Fig. 3.63, but in the upper part (outlet) there is a subwavelength diffractive grating (a), and Fig. 3.64b shows the intensity distribution at the exit from this lens. The width of the focal spot in Fig. 3.64*b* is FWHM = 177 nm = 0.114λ. It is 8% less than the width of the focus in Fig. 3.64b, and 10% less than the diffraction limit (3.165).

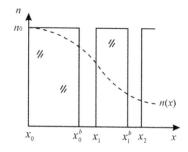

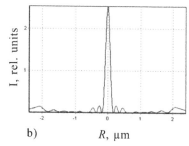

Fig. 3.65. Refractive index profiles of the gradient (dashed line) and binary (solid line) HS-lens. The shaded area is silicon.

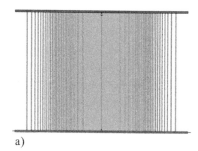

a)   b) R, μm

Fig. 3.66. Binary HS-lens (a) (refractive index 3.47), the horizontal size 4.8 μm, vertical size 2 μm, the minimum groove 20 nm (within the grooves is the air). Light travels vertically. Transverse intensity distribution $|E|^2$ at the output from the lens (at 10 nm from it) (b).

Consider a two-dimensional zone lens in which the zone sizes are selected so that the effective refractive index is described by the formula (3.162). This zone lens will be called the binary HS-lens.

Figure 3.65 shows the profile of binary and gradient HS-lenses. The radius of the lens is split into segments $[x_m, x_{m+1}]$, $m = 0,1,2,..$ M, and in each segment we select a point $x_m < x^b_m < x_{m+1}$, such that in the interval $[x_m, x^b_m]$ the lens material is silicon, and in the interval $[x^b_m, x_{m+1}]$ it is air. Point x^b_m is chosen to approximate the refractive index (3.163)

$$\int_{x_m}^{x_{m+1}} n(x) = n_0 \left(x^b_m - x_m \right) + 1 \cdot \left(x_{m+1} - x^b_m \right). \qquad (3.169)$$

Fig. 3.66a shows a binary GS-lens obtained by the formula (3.169) for a lens in Fig. 3.63a and Fig. 3.66b intensity distribution at the focus at the exit from this lens. The width of the focal spot in Fig. 3.66b is FWHM = 159 nm = 0,102λ. This is 17% less than the width of the focus in Fig. 3.63b, and 20% less than the diffraction limit (3.165). That is, according to Fig. 3.61, the surface waves of the second type ($k_x > nk$) provide a 20% contribution in the formation of the focal spot (Fig. 3.66b) at the output of the binary HS-lens.

3.4.2. Gradient elements of micro-optics for achieving superresolution

Focusing coherent laser light into a spot smaller than the diffraction limit or image the point light source with superresolution has become possible in recent years with the development of micro-optics and nanophotonics. The diffraction limit of $0.51\lambda/n$, where λ is the the wavelength of light in vacuum, n is the refractive index of the material in the area of the focus, can be overcome using superlens [90]. The superlens is a 2D flat plate of a metamaterial which consists of alternating subwavelength layers of a metal and a dielectric. The dielectric constants of these layers are selected so that the effective refractive index of the composite material is equal to $n = -1$. Experiments with superresolution using such lenses were made in [91, 92]. A superresolution equal to 0.4λ was achieved in the experiments [91]. A similar experiment was carried out in [93] using a subwavelength silver layer which also acts as a superlens. In this work two lines were resolved; the lines were separated by a distance of 145 nm when illuminated with ultraviolet light with a wavelength of 365 nm. Therefore, the superresolution equal to 0.4λ was also achieved in [93].

A hyperlens that depicts with superresolution in the far field was made in [94]. Thus, two lines with a width of 35 nm each, separated by a distance of 150 nm, were resolved using light with a wavelength of 365 nm. That is, the superresolution equal to 0.4λ was again achieved. However, in modeling a hyperlens in the form of a grating a superresolution equal to 0.05λ was achieved at a distance of 1.5λ from the surface in [95]. Apparently, the absorption and scattering of light by the metamaterial which occurs in actual experiments was not taken into account in this work. This is indirectly confirmed by the results of [72]. In [72] the authors studied the focusing of the laser light by a zone plate of a gold film with a thickness of 100 nm. The calculation showed that the diameter of the focal spot at half intensity should be equal 0.35λ, and in the experiment it was found that the diameter of the focal spot at half intensity is λ.

In [96–98] the multilayered and anisotropic nanostructures that allow subwave resolution to be achieved were numerically analysed. Thus, in [96] the authors selected the parameters of the eight-layer

1D layered structure based on Ag/SiO$_2$ which had a thickness of 400 nm and showed a source with a width of 0.4λ in the light spot of the same width. In [97] an anisotropic 2D nanostructure was proposed in which the components of the permittivity tensor were ε_x = 0.01–i0.01 and ε_z = –100, and the thickness of the structure along the z-axis was 400 nm (the wavelength λ = 700 nm). The simulation showed that this structure resolves two lines with a thickness of 3 nm each, separated by a distance of 23 nm. Therefore, the superresolution of 0.03λ was achieved. In [98], the same authors proposed another 2D nanostructure consisting of two different anisotropic layers, which resolved two narrow slits, separated by a distance of 50 nm when illuminated with light having a wavelength of 1550 nm.

Another candidate for achieving the superresolution are photonic crystals. In [99] using simulation it is shown that the 2D photonic crystal layer of a dielectric material with a permittivity ε = 12 and a triangular grating of circular holes with a radius r = 0.4a, where a is the period of the grating with the holes, has an effective refractive index n = –1 and is an imaging a lens for wavelength λ = a/0.3. The point source is represented as a spot with the diameter at half intensity equal to 0.3λ, and two point sources are resolved under the criterion of 20%, if the distance between them is 0.5λ.

In recent experiments using a 2D photonic crystal layer as a superlens the image of a point source with a width 0.4λ was obtained [100]. There are also works that propose to improve the superlens with a special nano-sized shell [101] or using a gradient boundary for the layer of a material with negative refraction [102]. In [103] it is demonstrated experimentally that the layer of the anisotropic substance is able to enhance and transform the surface evanescent waves into propagating light waves.

This section shows numerically that the Mikaelian gradient lens and Maxwell's 'fish-eye' can also be considered as candidates for producing images with superresolution. For a 2D Mikaelian microlens it is shown that a point light source near the surface of the lens is represented in the form of a light spot with a width at half intensity FWHM = 0.12λ. This is less than the diffraction limit of silicon (n = 3.47): 0.5λ,n = 0.144λ and less than that reported in [72,96,99,100]. It is also shown that the Mikaelian microlens resolves at half intensity two close point sources separated by a distance of 0.3λ, which is less than that reported in [91,93,94,99].

In [104,105] the authors obtained analytical expressions for calculating the modes in a gradient planar waveguide. In [104] the approximation of a continuous function of the refractive index of the gradient waveguide by a piecewise constant function and by applying the method of transfer matrices recurrent equations are written for finding the undetermined constants of local modes in each layer with a constant refractive index. In [105] the modified Airy functions and the WKB method were used to derive equations for the eigenvalues for calculating the parameters of the mode functions of the gradient medium. A common limitation in these studies is the assumption of the existence of one [104] or two [105] turns points of the function of the distribution of the refractive index of the planar waveguide.

This work presents general analytical expressions for modal solutions in a planar waveguide with no restrictions on the number of turn points in the function of the refractive index.

The solution of the Helmholtz equation for 2D gradient waveguides

Figure 3.67 is the diagram of the problem. Consider a two-dimensional gradient medium with a refractive index $n = n(x)$, in which the electromagnetic wave with TE-polarization propagates and the vector of the electric field strength is directed along the y axis.

The amplitude of the electric vector $\underline{E}_y(x,y)$, satisfies the Helmholtz equation [106]:

$$\left[\frac{\partial^2}{\partial x^2} + \frac{\partial^2}{\partial z^2} + k^2 n^2(x) \right] E_y(x,z) = 0 , \qquad (3.170)$$

where k is the wave number. We expand the amplitude of the electromagnetic wave in the transverse modes of the gradient medium:

$$E_y(x,z) = \sum_{n=0}^{\infty} C_n(x) \exp(i\beta_n z) , \qquad (3.171)$$

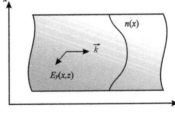

Fig. 3.67. Diagram of the problem of propagation of the TE-polarized waves in a 2D gradient waveguide.

where $\beta_n = k_{zn}$ is the propagation constant of the n-th mode. From (3.171) it follows that the light field has a longitudinal period T, so that $\beta_n = 2\pi n/T$. For example, the modes for the gradient medium with a quadratic dependence of the refractive index are Hermite-Gaussian functions that make up a countable basis [107].

Substituting (3.171) into (3.170), we obtain the equation for the amplitude of the modes of the gradient medium:

$$\frac{d^2 C_n(x)}{dx^2} + p_n(x)C_n(x) = 0,$$ (3.172)

where

$$p_n(x) = k^2 n^2(x) - \beta_n^2.$$ (3.173)

The change of variables

$$C_n(x) = C_n(0)\exp\left[\int_0^x f_n(\xi)d\xi\right]$$ (3.174)

transforms equation (3.172) reduces to a Whittaker non-linear differential equation [108] for the function $f_n(x)$:

$$\frac{df_n(x)}{dx} + f_n^2(x) + p_n(x) = 0.$$ (3.175)

Equation (3.175) can be solved by using the expansion of functions $f_n(x)$ and $p_n(x)$ to a Taylor series:

$$f_n(x) = \sum_{m=0}^{\infty} C_m^{(n)} x^m,$$ (3.176)

$$p_n(x) = \sum_{m=0}^{\infty} p_m^{(n)} x^m,$$ (3.177)

where $C_m^{(n)}$ and $p_m^{(n)}$ are unknown and known expansion coefficients of the appropriate functions. Substituting (3.176) and (3.177) into (3.175) we obtain the recurrence relations ($m > 0$) for the unknown coefficients of the series (3.176):

$$C_m^{(n)} = -m^{-1}\left(p_{m-1}^{(n)} + \sum_{s=0}^{m-1} C_s^{(n)} C_{m-1-s}^{(n)}\right),$$ (3.178)

where n is a positive integer, $C_0^{(n)}$ are undetermined constants. Then the mode amplitudes of the gradient waveguide medium can be expressed by the explicit expression:

$$C_n(x) = C_n(0) \exp\left(\sum_{m=0}^{\infty} C_m^{(n)} \frac{x^{m+1}}{m+1} \right), \tag{3.179}$$

in which $C_m^{(n)}$ is determined from the recurrent relations (3.178). The coefficients in (3.177) and (3.178) are given by:

$$p_m^{(n)} = \frac{k^2}{m!} \frac{d^m n^2(x)}{dx^m}\bigg|_{x=0} \tag{3.180}$$

for $m > 0$ and for $m = 0$ from the relations

$$p_0^{(n)} = k^2 n^2(0) - \beta_n^2. \tag{3.181}$$

The final expression for the amplitude of the light field of the TE-polarized wave in a 2D gradient medium is as follows:

$$E_y(x,z) = \sum_{n=0}^{\infty} C_n(0) \exp\left[i\beta_n z + \sum_{m=0}^{\infty} C_m^{(n)} \frac{x^{m+1}}{m+1} \right]. \tag{3.182}$$

In (3.182) there are two undetermined constants $C_n(0)$ and $C_0^{(n)}$ for each mode. The modes (3.182) are not orthogonal and normalized, so the expansion of the field (3.182) for these modes (that is, finding the coefficients $C_n(0)$) required truncation of the two series in (3.182) to the final sums and the solution of linear algebraic equations.

Constant $C_0^{(n)}$ for each mode must be chosen in a special way. For example, we consider two particular cases.

1) Let $n(x) = n_0$ and the modes are distributed in a homogeneous space. From (3.180) and (3.182) we obtain that $p_0^{(n)} = k^2 n^2(0) - \beta_n^2$ at $m = 0$ and $p_m^{(n)} = 0$ at $m > 0$. From (3.178) it follows that

$$C_1^{(n)} = -\left(p_0^{(n)} + (C_0^{(n)})^2 \right). \tag{3.183}$$

An arbitrary choice, put:

$$C_0^{(n)} = i\sqrt{p_0^{(n)}} = i\sqrt{k^2 n_0^2 - \beta_n^2}. \tag{3.184}$$

Then from (3.183) it follows that $C_1^{(n)} = 0$, as from (3.178) that all other factors are equal to zero: $C_m^{(n)} = 0$ when $m > 0$. Thus, for a homogeneous space $n(x) = n_0$ instead of (3.182), we obtain:

$$E_y(x,z) = \sum_{n=0}^{\infty} C_n(0) \exp\left[i\beta_n z + ix\sqrt{k^2 n_0^2 - \beta_n^2} \right]. \quad (3.185)$$

The solution (3.185) is a well-known solution of the Helmholtz equation (3.170) as a superposition of plane waves.

2) Consider another special case $n^2(x) = n_0^2 - \alpha^2 x^2$ is a quadratic medium. In this case $p_n(x) = p_0^{(n)} + p_1^{(n)} x + p_2^{(n)} x^2$, where $p_0^{(n)} = k^2 n_0^2 - \beta_n^2$, $p_1^{(n)} = 0$, $p_2^{(n)} = -k^2 \alpha^2$, and $p_m^{(n)} = 0$ when $m > 2$. Suppose $C_0^{(n)} = 0$ then $C_2^{(n)} = -\left(C_0^{(n)} C_1^{(n)} \right) = 0$, and $C_1^{(n)} = -p_0^{(n)} = \beta_n^2 - k^2 n_0^2$. To ensure that all other factors were equal to zero $C_m^{(n)} = 0$ for m> 2, it is enough to put to zero the third factor $C_3^{(n)} = -p_2^{(n)}/3 - (C_1^{(n)})^2/3 = 0$. Then for the medium parameter α we obtain the condition: $k^2 \alpha^2 = (k^2 n_0^2 - \beta_n^2)^2$. Suppose $\beta_n = kn_0/\sqrt{2}$, then we get that $\alpha = kn_0^2/2$. So, it turns out that the mode of the quadratic waveguide a refractive index $n^2(x) = n_0^2 \left(1 - k^2 n_0^2 x^2/4 \right)$ is a Gaussian exponent:

$$E_y(x,z) = C(0) \exp\left(\frac{ikn_0 z}{\sqrt{2}} - \frac{k^2 n_0^2 x^2}{4} \right). \quad (3.186)$$

To conclude this section we note that since the derivative at zero of the mode function (3.179) is proportional to the zero factor in the series (3.176):

$$\frac{dC_n(x)}{dx} \bigg|_{x=0} = C_n(0) C_0^{(n)}, \quad (3.187)$$

selecting $C_0^{(n)} = 0$ (as was done in the last example) we ensure the presence of the extremum at the mode on the optical axis.

General solution for a secant gradient waveguide

For a gradient waveguide with a secant dependence of the refractive index on the transverse coordinate

$$n(x) = n_0 \, \mathrm{ch}^{-1}\left(\frac{kn_0 x}{\sqrt{2}} \right) \quad (3.188)$$

we know a particular solution of equation (3.170) of the following form [109]:

$$E_{1y}(x,z) = \exp\left(\frac{ikn_0 z}{\sqrt{2}} \right) \mathrm{ch}^{-1}\left(\frac{kn_0 x}{\sqrt{2}} \right). \quad (3.189)$$

The Helmholtz equation (3.170) in this case is:

$$\left[\frac{\partial^2}{\partial x^2}+\frac{\partial^2}{\partial z^2}+k^2 n_0^2 \operatorname{ch}^{-2}\left(\frac{kn_0 x}{\sqrt{2}}\right)\right]E_y(x,z)=0. \qquad (3.190)$$

We seek a common solution (3.190) in the form

$$E_{1y}(x,z)=A(x)\exp(i\gamma z). \qquad (3.191)$$

Substituting (3.191) into (3.190), we obtain the equation

$$\frac{d^2 A(x)}{dx^2}+g(x)A(x)=0, \qquad (3.192)$$

where

$$g(x)=k^2 n_0^2 \operatorname{ch}^{-2}\left(\frac{kn_0 x}{\sqrt{2}}\right)-\gamma^2. \qquad (3.193)$$

It is well known [108] that the equation of the form (3.192) has the general solution

$$A(x)=A_1(x)\left[C_1+C_2\int_0^x A_1^{-2}(\xi)d\xi\right], \qquad (3.194)$$

where $A_1(x)$ is a partial solution of (3.192), C_1, C_2 are undetermined constants. In this case, as the partial solution we can choose the solution (3.189), i.e.,

$$A_1(x)=\operatorname{ch}^{-1}\left(\frac{kn_0 x}{\sqrt{2}}\right), \quad \gamma=\frac{kn_0}{\sqrt{2}}. \qquad (3.195)$$

Then the general solution for the mode of the secant gradient waveguide has the form:

$$E_{1y}(x,z)=\exp\left(\frac{ikn_0 z}{\sqrt{2}}\right)\operatorname{ch}^{-1}\left(\frac{kn_0 x}{\sqrt{2}}\right)\times$$

$$\times\left\{C_1+\frac{C_2}{kn_0\sqrt{2}}\left[\frac{1}{2}\operatorname{sh}(kn_0 x\sqrt{2})+\frac{kn_0 x}{\sqrt{2}}\right]\right\}, \qquad (3.196)$$

where

$$C_1=\left|E_{1y}(x=0,z)\right|, \quad C_2=\left|\frac{dE_{1y}(x=0,z)}{dx}\right|. \qquad (3.197)$$

From (3.189) we can obtain an expression for the width of the mode at half intensity in the secant-gradient waveguide:

$$\text{FWHM} = \frac{\ln(3+2\sqrt{2})}{\pi n_0 \sqrt{2}} \approx \frac{0.4\lambda}{n_0} \tag{3.198}$$

where λ is the wavelength of light in vacuum, n_0 – as before, the refractive index at the axis of the waveguide.

A particular solution for the quadratic waveguide
The first part of this section shows that the mode for a quadratic medium with certain parameters is the Gaussian exponent (3.186). In this section, we show that this also holds true for the quadratic medium with arbitrary parameters:

$$n^2(x) = n_0^2 \left(1 - w^2 x^2\right), \tag{3.199}$$

where w is an arbitrary constant. Then the equation (3.170) will be:

$$\left[\frac{\partial^2}{\partial x^2} + \frac{\partial^2}{\partial z^2} + k^2 n_0^2 (1 - w^2 x^2) \right] E_y(x,z) = 0, \tag{3.200}$$

and its solution will be sought in the form of:

$$E_{2y}(x,z) = E_0 \exp\left(ipz - q^2 x^2\right). \tag{3.201}$$

Substituting (3.201) into (3.200), we obtain the equation:

$$\left(-2q^2 - p^2 + k^2 n_0^2 + 4q^4 x^2 - w^2 k^2 n_0^2 x^2\right) E_0 \exp\left(ipz - q^2 x^2\right) = 0. \tag{3.202}$$

From (3.202) determine the unknown parameters:

$$q^2 = \frac{wkn_0}{2}, \quad p = kn_0 \sqrt{1 - \frac{w}{kn_0}}.$$

Thus, a particular modal solution of the equation (3.200) has the form:

$$E_{2y}(x,z) = E_0 \exp\left(ikn_0 z \sqrt{1 - \frac{w}{kn_0}} - \frac{wkn_0}{2} x^2 \right). \tag{3.203}$$

Note that when $w = kn_0/2$ the solution (3.203) coincides with the solution (3.186). When $w = kn_0/2$ from (3.203) it follows that the width (diameter) of the Gaussian mode at half intensity is:

$$\text{FWHM} = \frac{\sqrt{\ln 4}\,\lambda}{\pi n_0} \approx \frac{0.38\lambda}{n_0}.$$

(3.204)

Comparing (3.204) to (3.198) we see that the width of the two modes (secant and Gaussian) is almost identical. The effective width of the quadratic waveguide is found from the condition $n(x_0) = 1$ and is equal to

$$2x_0 = \frac{2\sqrt{n_0^2 - 1}}{\pi n_0}\lambda.$$

(3.205)

When $n_0 = 1.5$ from (3.205) we obtain: $2x_0 \approx 0.48\lambda$. That is, the effective width of the glass planar waveguide with the quadratic dependence of the refractive index in which only the Gaussian mode (3.203) can propagate, is almost equal to half the wavelength of light in vacuum. Note that such 'half-wave' waveguides are being actively used in applications [110, 111].

Simulation of passage of light through gradient elements of micro-optics: superresolution

A few years ago, experiments were conducted with superresolution in the optical range by superlens [92, 94]. In [92] the superlens was a thin layer of silver 50 nm thick. The experiment consisted of obtaining images with superresolution in the resist layer of an amplitude grating with a period of 145 nm in the vicinity of the thin silver film. After irradiation of a three-layer structure (amplitude grating + silver film + resist) with ultraviolet light with a wavelength of 365 nm, and after development, a grating with a period of 170 nm was 'recorded' in the resist layer and restored using Fourier transformation. In [94] the authors used a hyperlens in the form of a half-cylinder with a circular cut on a quartz substrate, consisting of alternating layers of silver (35 nm thick) and an Al_2O_3 insulator (thickness 35 nm). Such a multilayer structure has an anisotropic dielectric constant. The object for resolution was in the form of two amplitude lines produced lithographically and having a width of 35 nm, and separated by a distance of 150 nm. This object was illuminated with ultraviolet light of TE polarization and a wavelength of 365 nm. A conventional lens was used to obtain an enlarged image of these two lines separated now by a distance of 350 nm. Thus, the superresolution obtained in these experiments [92,94] was 0.4λ.

Theoretically (without considering absorption of the material), using the superlens we can obtain any resolution. Thus, simulation in [97] showed that the hyperbolic lens, forming in the near field the enlarged subwavelength image, can be realized in the form of a plane-parallel layer. It was shown that the layer of anisotropic material with dielectric constants $\varepsilon_x = 0.01 - i0.01$ and $\varepsilon_z = -100$ and a thickness of 400 nm allows to resolve two slits with a width of 3 nm (elongated along the y axis) in a metal screen with a dielectric constant $\varepsilon = 1 - i10^4$, separated by a distance of 23 nm, if they are illuminated with the light with TM polarization and a wavelength of 700 nm. The value of superresolution can be estimated as 0.05λ.

In the following sections we will show numerically (by the well-known FDTD method) that with the help of gradient 2D micro-optics we can also achieve superresolution. The simulation was performed using FullWAVE software by RSoft (USA).

Superresolution using Mikaelian microlenses
Because of the diffraction of light in a homogeneous space two close point sources of light are no longer resolved at a distance much smaller than the wavelength. Thus, Fig. 3.69 a shows the five initial (in the plane $z = 5$ nm) Gaussian coherent light sources with the width $\lambda/200$ each, which are separated by a distance $\lambda/50$. Figure 3.69 b shows the intensity of light from these sources at a distance $z = 30$ nm for a wavelength $\lambda = 1550$ nm. From Fig. 3.68 it is clear that the close point sources are no longer resolved at a distance approximately equal to the gap between them ($z = \lambda/50$).

Figure 3.69 shows the result of simulation of the imaging of two point light sources using the Mikaelian lens (the width of each of them is 35 nm) and separated by a distance of 150 nm [112,113]. The refractive index of such a secant-gradient 2D microlens is:

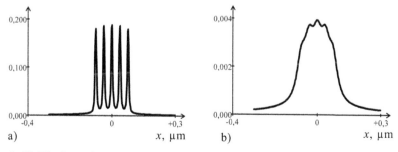

Fig. 3.68. The intensity of the light field near the five Gaussian light sources with width $\lambda/200$, separated by a distance of $\lambda/50$ at different distances: $z = 5$ nm (a), $z = 30$ nm (b), $\lambda = 1550$ nm (on the ordinate axis there are arbitrary units).

$$n(x) = n_0 \, \text{ch}^{-1}\left(\frac{\pi x}{2L}\right),$$

(3.206)

where $2L$ is the length of the lens. Lens width $2R$ is determined from the condition that $n(R) = 1$. For the lens in Fig. 3.70 the refractive index on the axis is $n_0 = 2.1$, lens width $2R = 1$ mm, length of the lens $2L = 1.144$ μm, wavelength $\lambda = 365$ nm. The applied parameters coincide with the parameters of the experiment in [94].

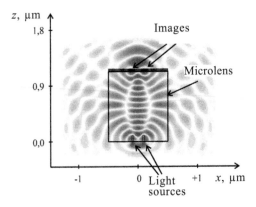

Fig. 3.69. Arrangement of the 2D gradient Mikaelian microlenes in the window of the FullWAVE program for simulating images of two close point light sources. The instantaneous amplitude $E_y(x, z)$ at the time when the light has passed a distance of 18.2398 μm is shown.

Figure 3.69 shows an instantaneous image of the amplitude of the electric component of an electromagnetic wave with TE polarization (light travels in Fig. 3.70 from bottom to top). It is seen that the image of the light sources forms on the opposite side of the lens with respect to the light sources. Figure 3.70a shows the distribution of the time-averaged intensity of the electric field $I(x, z = z_0) = \left|E_y(x, z = z_0)\right|^2$ directly on the 'back' side of the lens, i.e. at a distance $z_0 = 2L$ from the 'front' side of the lens. Two light source are in front of the 'front' side of the lens at a distance of 20 nm.

Figure 3.70a shows that the two point sources separated by a distance of 150 nm (between the centres of the sources 180 nm) are confidently resolved. If the intensity recording plane is moved from the 'rear' plane of the lens to a distance $z = 10$ nm (Fig. 3.70 b), the sources continue to be resolved, although the magnitude of 'the dip' in the distribution of the intensity is slightly reduced. The

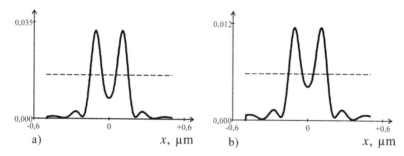

Fig. 3.70. Averaged intensity distributions of the electric component of the TE-polarized light wave calculated at a distance $z = 0$ (a) and $z = 10$ nm (b) from the 'rear' surface of the Mikaelian lens (Fig. 3.70), at the input of the lens there are two closely space point light sources (the horizontal line divides the intensity at half, the ordinate axis gives the arbitrary units).

resolution achieved in Fig. 3.70 is 0.41λ. From Fig. 3.70a we can also determine the width of the image by the Mikaelian lens of the point source at half intensity FWHM = 100 nm = 0.27λ. The resulting value of resolution 0.41λ is almost equal to the resolution obtained in [91,92,94] (0.4λ). It should be noted that the width of the image 0.27λ of the point source is consistent with the minimum width of the mode that can propagate in the secant-gradient waveguide $0.4\lambda/n_0 = 0.27\lambda$ when $n_0 = 1.5$ (see Eq. (3.199)).

In order to increase the resolution of the Mikaelian lens, as indicated by equations (3.199) and (3.205), the refractive index at the axis was increased $n_0 = 3.47$ (silicon, Si). Other simulation parameters also changed (Fig. 3.71a): wavelength $\lambda = 1$ μm, the width of the lens $2R = 6$ μm, length of the lens $2L = 4.92$ μm. The step of simulation on the spatial axes in all the examples was $\lambda/100$. The width of the Gaussian point light source in the initial plane was $\lambda/20$. Figure 3.71a shows the instantaneous amplitude pattern of the light wave electric vector in the Mikaelian lens, calculated at the time point when the light wave passed 200 μm from the source. Figure 3.71b shows the distribution of average light intensity of the TE polarized wave at the output from the lens (Fig. 3.72 a) on the 'rear' plane of the Mikaelian lens. The calculation showed that for the intensity shown in Fig. 3.71b the width of the central maximum at half of intensity on the image of the point source is FWHM = 0.12λ. However, the intensity (or the power density of light) is not the value that indicates what fraction of the radiation power of the radiation source propagates in the space along the z axis. Therefore, Fig. 3.72 shows just such a value: the projection of the optical axis of the Umov–Poynting vector calculated at the output of the Mikaelian lens (Fig. 3.71*a*) at the entrance to which there a point light source.

The width of the central maximum of the power flux along the *z*-axis in Fig. 3.72 is the same as in Fig. 3.71*b* and is FWHM = 0.12λ.

The value of the diffraction limit, which can be achieved by focusing the light in a homogeneous medium, is known and equal 0.5λ/*n*, where *n* is the refraction index of the homogeneous medium. In our case n_0 = 3.47, so the diffraction limit is FWHM = 0.144λ. Thus, Figs. 3.71*b* and 3.72 show that using the Mikaelian microlenses (Fig. 3.71*a*) the diffraction limit can be overcome not only in free space but also in the medium. The reason for this, in our opinion, is that by focusing the light at the interface of two media the formation of the focus is also contribute to by surface evanescent waves that can generate interference and diffraction patterns with a period much smaller than the wavelength. The fact that the surface waves also take

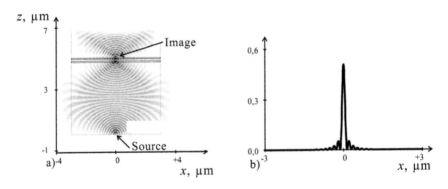

Fig. 3.71. Instantaneous pattern of the electric field amplitude of the TE-polarized light wave in a Mikaelian lens with a point source at the front plane of the lens (a) and the average intensity distribution in the rear plane of the lens (b) (on the ordinate axis there are arbitrary units).

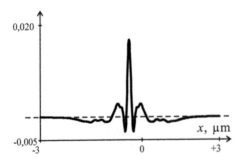

Fig. 3.72. The distribution along the *x* axis of the projection on the *z* axis of the Umov–Poynting vector (in relative units), calculated at the output of the Mikaelian lens (Fig. 3.71*a*) with the input point source.

part is the formation of the image of a point source is seen from Fig. 3.73. The projection of the Umov–Poynting vector on the z-axis in some region on the x-axis is negative, that is, near the exit surface of the Mikaelian lens light not only comes out of the lens but also comes into it, i.e. a surface wave propagates along the surface of the microlens. In the propagation of light in free space the increase of the distance from the surface of the lens results in a rapid increase of the width of the image of the point source, and at a distance from the surface, approximately equal to the wavelength, the width of the image reaches the diffraction limit of 0.5λ.

Note that the change of the signs of the projection of the Umov–Poynting vector, similar to Fig. 3.73, was previously observed in [96] and was named the optical vortex and interpreted as the interference of the propagating wave and enhanced surface wave. However, in [96], a 1D multilayer structure (1D photonic crystal) was studied.

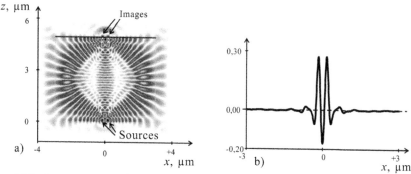

Fig. 3.73. Instantaneous picture of the amplitude of the electric vector of light waves with TE-polarization in the Mikaelian len (Fig. 5a???), when in front of its input (the lower horizontal line in the figure) surface at a distance of 10 nm there are two point sources (with a width of 50 nm each), separated by a distance of 300 nm (a); the time-averaged distribution of the projection of the Umov–Poynting vector on the optical axis, calculated at a distance of 10 nm from the rear (the upper horizontal line in the figure) plane of the lens (on the axis of ordinates there are arbitrary units) (b).

Figure 3.74 a shows the instantaneous amplitude pattern of the electric vector of the light wave in the Mikaelian lens (parameters are the same as for Fig. 3.71a), when in front of its input (bottom in Fig. 3.73a) surface at a distance of 10 nm there are two point sources (width of 50 nm), separated by a distance of 300 nm. Figure 3.73b shows the average distribution of the projection on the optical axis of the Umov–Poynting vector calculated behind the output surface of the microlens at a distance of 10 nm. From Fig. 3.73b it can be seen that the two sources are resolved, with the value of superresolution being 0.3λ, which is less than in [91,92,93].

Superresolution via the 'fish eye' microlens

Apart from the Mikaelian microlens considered in the previous section, we can also use other gradient imaging optical elements in which the dependence of the refractive index on the coordinates is obtained in an explicit analytic form.

One such optical element is the Maxwell 'fish eye' [96,114,115]. The function of the refractive index of this element in the two-dimensional case in the polar coordinates has the form:

$$n(r) = n_0 \left[1 + \left(\frac{r}{R} \right)^2 \right]^{-1},\tag{3.207}$$

where n_0 is the refractive index in the centre of the circle, R is the radius of an element. Equation (3.207) shows that at $r = R$ the refractive index is halved. The drawback of this element is the limited 'differential' of the refractive index at the centre and the edge of the optical element: it can not change by more than 2 times. In the Mikaelian lens the differential refractive index is determined only by the material. For example, for silicon the index varies from 3.47 to 1. However, the 'fish-eye' has a circular symmetry, and therefore any point source located on the surface is imaged 'ideally' in the diametrically opposed point on its surface.

Figure 3.74a shows the instantaneous amplitude distribution of the electric field of the TE-polarized light wave in a 2D 'fish-eye' microlens on the surface of which there are two close point light sources. The simulation parameters: the refractive index at the centre of the lens $n_0 = 3.47$, lens radius $R = 2,5$ mm, the wavelength of light $\lambda = 1$ μm, the width of point sources 0.05λ, and they are separated by a distance of 440 nm or 0.44λ. Figure 3.74b shows the distribution of the time-averaged intensity in the image plane (the upper horizontal line in Fig. 3.74a). It can be seen that the two sources are resolved (the resolution is 0.44λ by the Rayleigh criterion of 20%). Thus, the microlens slightly (0.44λ) overcomes the diffraction limit in resolution (0.5λ).

When an individual image of a point source using this microlens (Fig. 3.74a) is obtained the intensity distribution shown in Fig. 3.75. The width of the central maximum at half intensity is FWHM = 0.3λ. From the comparison of Fig. 3.76 and Fig. 3.72 b it can be concluded

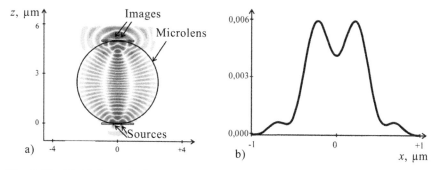

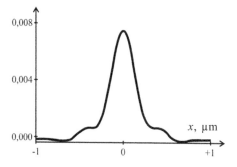

Fig, 3.74. 2D 'fish-eye' microlens in the window of FullWAVE software, and the instantaneous picture of the electric field amplitude of the TE polarized wave within the lens in which there are two point sources at the input (light propagates upwards) (a), and the average intensity distribution in the image plane of the lens (on the ordinate axis there are arbitrary units) (b).

Fig. 3.75. The time-averaged electric field intensity distribution in the image plane of the 'fish-eye' microlens (Fig. 5.74 a) in which the surface has only one point source of light (on the ordinate axis are arbitrary units).

that the 'fish-eye' microlens forms a wider image (approximately 2 times wider) of the point source than the Mikaelian lens under comparable parameters. These magnitudes of resolution 0.44λ (Fig. 3.74 b) and 0.3λ (Fig. 3.75) were comparable to the superresolution values obtained in [91,92,93,96,99,100].

3.4.3. Construction of an enlarged image with superresolution using planar Mikaelian lenses

This section describes the construction of an enlarged image by using Mikaelian lenses, consisting entirely of dielectrics. Using the Fermat principle, we predicted the position of the image of an off-axis source power generated by two silicon Mikaelian lenses, located close to each other. Silicon at a wavelength of $\lambda > 1$ µm hardly absorbs light. It is shown that in the optical system of two Mikaelian lenses, one of which is twice as large as the other, the image formed by an off-axis point source located at a distance y from the optical axis is not situated at a distance twice that of the optical axis (i.e., $2y$) and represents a light spot, the basic energy of which is contained within the range $(0.37y; 2y)$. Imaging with two lenses and with a single linearly expanded Mikaelian lens was investigated numerically using the finite-difference FDTD-method. For a system of two Mikaelian lenses, one of which is 2.5 times larger than the other, we obtained the two image of two point light sources with a linear magnification of 2.1. For a linearly expanding Mikaelian lens the linear magnification was 3.14. In both cases, point sources separated by a distance 0.28λ were resolved.

The equation of the beam path in a gradient waveguide

Given a planar gradient waveguide with an optical axis x and the dependence of the refractive index of the transverse coordinate $n(y)$. Suppose the light beam connects the points A and B with the coordinates (x_1, y_1) and (x_2, y_2) (Fig. 3.76).

According to the Fermat's principle, the light beam passes through the trajectory $y(x)$, providing a minimum optical path, i.e. the minimum of the following functionality:

$$L[y(x)] = \int_{x_1}^{x_2} n(y)\sqrt{1+\left(\frac{dy}{dx}\right)^2}\,dx. \qquad (3.208)$$

It is known that such problems can be solved using the Euler-Lagrange equation [116], which for the functional (3.208) is as follows:

$$\frac{1}{n(y)}\frac{dn}{dy} = \frac{y''}{1+(y')^2}. \qquad (3.209)$$

Multiplying both sides by $2y'$, and integrating with respect to variable x, we obtain an expression for the derivative of the beam path $y(x)$:

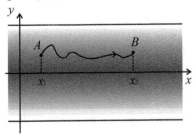

Fig. 3.76. Homogeneous gradient waveguide and a light beam connecting arbitrary two points A and B.

$$\frac{dy}{dx} = \pm\sqrt{Cn^2(y)-1}, \tag{3.210}$$

where C is a constant.

This differential equation can be solved by radicals, if we look for the beam path through the inverse function $x(y)$:

$$x(y) = \pm\int_0^y \frac{dt}{\sqrt{Cn^2(t)-1}} + D, \tag{3.211}$$

where D is a constant.

This equation is the equation of the beam path for an arbitrary planar gradient waveguide. Next we get the equation for the gradient secant waveguide.

The equation of the beam path in a gradient secant waveguide
In [117, 118] the authors devied the beam equation in a gradient secant waveguide. We briefly repeat it with the equation (3.211).

Let the distribution of the refractive index is set in the form of a hyperbolic secant:

$$n(y) = \frac{n_0}{\operatorname{ch}\left(\dfrac{y}{a}\right)}, \tag{3.212}$$

where n_0 is the refractive index on the axis of the waveguide (i.e. for $y = 0$), a is a parameter that determines the width of the waveguide and defines the rate of decrease in the refractive index from the waveguide axis toward its edges.

Substituting this expression into the equation for the beam path (3.211), we obtain:

$$x(y) = D \pm a\arcsin\left[\frac{\operatorname{sh}\left(\dfrac{y}{a}\right)}{\sqrt{n_0^2 C-1}}\right]. \tag{3.213}$$

From this equation we express explicitly $\operatorname{sh}(y/a)$ through x:

$$\operatorname{sh}\left(\frac{y}{a}\right) = \sqrt{n_0^2 C-1}\,\sin\left(\frac{x-D}{a}\right) \tag{3.214}$$

Sign '±' in (3.214) is omitted, since the change from '+' to '−' is achieved by adding πa to constant D. We denote $(n_0^2 C-1)^{1/2} \cos(D/a)$ as C, and $(n_0^2 C-1)^{1/2} \sin(D/a)$ as D. Then the equation of the beam path in a gradient secant waveguide takes the form:

$$\text{sh}\left(\frac{y}{a}\right) = C\sin\left(\frac{x}{a}\right) + D\cos\left(\frac{x}{a}\right). \tag{3.215}$$

Let the beam in the $x = x_1$ passes through the point with the coordinates (x_1, y_1) and the tangent of the slope of the ray to the optical axis is a (Fig. 3.77).

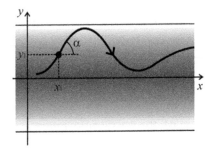

Fig. 3.77. Setting the beam propagating in the gradient hyperbolic secant waveguide, the distance from the optical axis y_1 and the angle of inclination to α in some reference plane $x = x_1$.

We obtain the equation for this beam, differentiating (3.215) for x:

$$\text{sh}\left(\frac{y}{a}\right) = \alpha\,\text{ch}\left(\frac{y_1}{a}\right)\sin\left(\frac{x-x_1}{a}\right) + \text{sh}\left(\frac{y_1}{a}\right)\cos\left(\frac{x-x_1}{a}\right). \tag{3.216}$$

By analogy with the matrix methods of calculating optical systems, the beam will be described by two coordinates that specify the point of intersection of the beam with a reference plane and also the inclination of the beam to the optical axis. The first coordinate is sh(y/a), and the second ch(y/a) (dy/dx).

From (3.216) it is easy to get that in the matrix form the beam propagation is described by the following equation:

$$\begin{bmatrix} \text{sh}\left(\dfrac{y}{a}\right) \\ \text{ch}\left(\dfrac{y}{a}\right)\dfrac{dy(x)}{dx} \end{bmatrix} = T\left(\frac{x-x_1}{a}\right)\begin{bmatrix} \text{sh}\left(\dfrac{y_1}{a}\right) \\ \text{ch}\left(\dfrac{y_1}{a}\right)\dfrac{dy(x_1)}{dx} \end{bmatrix}, \tag{3.217}$$

where

$$T(\varphi) = \begin{bmatrix} \cos\varphi & \sin\varphi \\ -\sin\varphi & \cos\varphi \end{bmatrix}. \tag{3.218}$$

From this equation it follows that when $x = x_1 + \pi m a$ (m is an integer) $y = (-1)^m y_1$, regardless of the angle of the beam α. This means that even for an off-axis point light source the homocentricity beam is retained and a point image is formed.

The numerical aperture for imaging in a gradient hyperbolic secant waveguide

Obviously, the actual width of the waveguide is limited by that which for the waveguide is determined (3.212), for example, by the decline of the refractive index to one. In this case not every ray which passes through a point remains in the waveguide, part of the rays coming out through the edge of the waveguide. To determine the most extreme beam, we find the point of the beam path in which the beam is farthest away from the optical axis. For this we differentiate both sides (3.217) with respect to x and equate in the left side the derivative $y'(x)$ to zero. Then we find that the beam moves away to the maximum distance from the optical axis at the points (the turning point of the beam) with coordinates x_{extr}, satisfying the equation

$$\text{th}\left(\frac{y_1}{a}\right)\text{tg}\left(\frac{x_{extr} - x_1}{a}\right) = \alpha \tag{3.219}$$

Next, we consider a gradient lens, which is a truncated waveguide with length πa. Equation (3.219) shows that if $y1$ and π have the same sign, then $x_{extr} < x_1 + \pi a/2$ (beam 1 in Fig. 3.78), and if y_1 and α have different signs, then $x_1 + \pi a/2 < x_{extr} < x_1 + \pi a$ (beam 2 in Fig. 3.78).

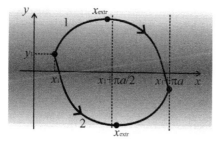

Fig. 3.78. The planar hyperbolic secant waveguide and light rays connecting point of the object (x_1, y_1) and the point of the image $(x_1 + \pi a, -y_1)$.

According (3.216) and (3.219), the maximum distance of the ray from the axis is described by the following equation:

$$\text{sh}\left(\frac{y_{extr}}{a}\right) = \pm\sqrt{\text{sh}^2\left(\frac{y_1}{a}\right) + \alpha^2\,\text{ch}^2\left(\frac{y_1}{a}\right)}.$$ (3.220)

At the edge of the waveguide $n(y_{extr}) = 1$, then $\text{sh}(y_{extr}/a) = (n_0^2 - 1)^{1/2}$. From this condition we can find the maximum angle of inclination α_{max}, wherein the beam does not come out of the waveguide, and will propagate therein, forming at point $(x_1 + \pi a, -y_1)$ the image of a point source located at the point (x_1, y_1):

$$\alpha_{max} = \pm\sqrt{\frac{n_0^2}{\text{ch}^2\left(\frac{y_1}{a}\right)} - 1}.$$ (3.221)

This means that when the point source is far away from the optical axis the numerical aperture of the rays, forming the image of such a source, will decrease. At the point of the image the sine of the maximum angle of the beam with the optical axis is equal to:

$$\sin\theta = \frac{|\alpha_{max}|}{\sqrt{1 + \alpha_{max}^2}} = \sqrt{1 - \frac{\text{ch}^2\left(\frac{y_1}{a}\right)}{n_0^2}},$$ (3.222)

and the numerical aperture is equal to

$$NA = n(y)\sin\theta = \sqrt{\frac{n_0^2}{\text{ch}^2\left(\frac{y_1}{a}\right)} - 1}.$$ (3.223)

For axial light source $(y_1 = 0)$ is equal to the numerical aperture $NA = (n_0^2 - 1)^{1/2}$, which coincides with [117]. For a source at the edge of the waveguide (i.e. $\text{ch}(y_1/a) = n_0$) the numerical aperture tends to zero and the produced light spot of the image spreads endlessly.

Construction of an enlarged image by using two Mikaelian lenses
Consider a gradient lens, which is a truncated hyperbolic secant waveguide with length $\pi a/2$. Such a lens is known as the Mikaelian lens [118]. Suppose we have an optical system consisting of two such lenses with parameters $a_1 = 2L_1/\pi$ and $a_2 = 2L_2/\pi$ $(a_1 < a_2,$

L_1, L_2 are the lengths of the lens), located close to each other (Fig. 3.79). Suppose that the front plane of the first lens ($x = x_1$) contains a point source located near the optical axis at a distance y_1 ($y_1 \ll a_1$).

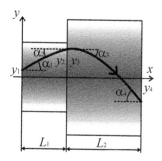

Fig. 3.79. The optical system of two Mikaelian lenses.

According to (3.218) for both lens the matrix T has the form:

$$T(\varphi) = \begin{bmatrix} 0 & 1 \\ -1 & 0 \end{bmatrix}. \tag{3.224}$$

From (3.217) it follows that the light rays in the back plane of the first lens will be described by the coordinates:

$$\begin{cases} \mathrm{sh}\left(\dfrac{y_2}{a_1}\right) = \alpha_1 \, \mathrm{ch}\left(\dfrac{y_1}{a_1}\right), \\[2mm] \alpha_2 \, \mathrm{ch}\left(\dfrac{y_2}{a_1}\right) = -\mathrm{sh}\left(\dfrac{y_1}{a_1}\right). \end{cases} \tag{3.225}$$

where α_1 and α_2 are the tangents of the angles of inclination of the beam to the optical axis in the planes $x_2 = x_1$ and $x_2 = x_1 + \pi a_1/2$, respectively.

From the second equation (3.225) and the proximity of the source to the optical axis ($y_1 \ll a_1$) it follows (since $\mathrm{ch}(y_2/a_1) \le 1$) that $\alpha_2 \ll 1$. Therefore, when recording the law of refraction at the boundary of the first and second lens we can write tangents instead of sines:

$$\frac{n_0}{\mathrm{ch}\left(\dfrac{y_2}{a_1}\right)} \alpha_2 = \frac{n_0}{\mathrm{ch}\left(\dfrac{y_3}{a_2}\right)} \alpha_3, \tag{3.226}$$

where y_3 and α_3 are the distance to the optical axis of the beam and the tangent of the inclination of the beam to the optical axis in the plane of the two lenses. Obviously, $y_3 = y_2$, i.e. notation y_3 is introduced for the uniformity of the indices. Again, we apply the transformation of the beams (3.217), but now for the second lens. Let y_4 and α_4 be the distance to the optical axis of the beam and the slope of the beam to the optical axis in the back plane of the second lens (i.e. in the plane $x_2 = x_1 + \pi a_1/2 + \pi a_2/2$):

$$
\begin{cases}
\mathrm{sh}\left(\dfrac{y_4}{a_2}\right) = \alpha_3 \,\mathrm{ch}\left(\dfrac{y_3}{a_2}\right), \\[2mm]
\alpha_4 \,\mathrm{ch}\left(\dfrac{y_4}{a_2}\right) = -\mathrm{sh}\left(\dfrac{y_3}{a_2}\right).
\end{cases}
\tag{3.227}
$$

Expressing explicitly from (3.226) α_3 through α_2, expressing explicitly α_2 from (3.225) and substituting the resultant expression for α_3 in the first equation (3.227), we obtain:

$$
\mathrm{sh}\left(\frac{y_4}{a_2}\right) = -\frac{\mathrm{ch}^2\left(\dfrac{y_2}{a_2}\right)}{\mathrm{ch}^2\left(\dfrac{y_2}{a_1}\right)}\,\mathrm{sh}\left(\frac{y_1}{a_1}\right).
\tag{3.228}
$$

This equation makes it possible to determine the position of the point of intersection of the beam at the back plane of the second lens, knowing the position of the source in the front plane of the first lens. It is easy to see that if the light source is axial ($y_1 = 0$), then all the beams from it meet in a single point and form an axial image ($y_4 = 0$). It is also seen that at $a_1 = a_2$ beams again gather at a single point and form an image of the source symmetrical about the optical axis, regardless of whether the source is axial or not. However, at $a_1 \neq a_2$ and $y_1 \neq 0$ the image does not form because the the coordinate y_4 depends on what angle is between the beam and the optical axis in the initial plane $x = x_1$.

The ratio of the squares of the hyperbolic cosines in (3.228) can not exceed unity, as $a_1 < a_2$. The light source is located near the axis, i.e. $y_1 \ll a_1$. This implies that the hyperbolic sines of both sides of (3.228) is approximately equal to their arguments, and the point of intersection of the beam back plane of the second lens can be approximately defined as follows:

$$y_4 = -\frac{a_2}{a_1} \frac{\text{ch}^2\left(\dfrac{y_2}{a_2}\right)}{\text{ch}^2\left(\dfrac{y_2}{a_1}\right)} y_1 . \tag{3.229}$$

For the beam emitted from the source parallel to the optical axis ($a_1 = 0$), from (3.225) it follows that $y_2 = 0$ and, according to (3.229)

$$y_4 = -\frac{a_2}{a_1} y_1 . \tag{3.230}$$

For the beam passing through the very edge of the first lens, the relation $\text{ch}(y_2/a_1) = n_0$ holds and therefore

$$\text{ch}\left(\frac{y_2}{a_2}\right) = \frac{1}{2}\left[\left(n_0 + \sqrt{n_0^2 - 1}\right)^{\frac{a_1}{a_2}} + \left(n_0 - \sqrt{n_0^2 - 1}\right)^{\frac{a_1}{a_2}}\right]. \tag{3.231}$$

In particular, when $a_2 = 2a_1$ obtain

$$\text{ch}\left(\frac{y_2}{a_2}\right) = \sqrt{\frac{n_0 + 1}{2}} , \tag{3.232}$$

i.e.

$$y_4 = -\frac{n_0 + 1}{n_0^2} y_1 . \tag{3.233}$$

This means that when $a_2 = 2a_1$ the image of a point source located at a distance y_1 from the axis of the double Mikaelian lens made of silicon ($n_0 = 3.47$) will not be at a distance from the axis of $2y_1$, and will spread, with most of the energy concentrated at distances from the optical axis of between $0.37y_1$ and $2y_1$.

Numerical modeling was performed by the FDTD-method. Simulation parameters were as follows (Fig. 3.79): the material of the two lenses was silicon (refractive index on the axis $n_0 = 3.47$), the light wavelength 1.55 µm, the length of the first lens 1 µm (width 2.44 µm), the second lens 2 µm (width 4.88 µm), the calculated area $-3\lambda \le x \le 3\lambda$, $0.21\lambda \le x \le 2.5\lambda$, the simulation time $60\lambda/c$, where c is the speed of light in vacuum, sampling step for both coordinates $\lambda/50$, in time $\lambda/(100c)$. The light source was a plane wave $\lambda/50$ wide, the distance from the source to the front plane of the first lens $\lambda/50$. The distance from the centre of the light source to the optical axis was varied and equalled 0.1λ, 0.2λ, 0.3λ, 0.4λ, 0.5λ, 0.6λ. The measuring screen was placed at a distance λ/50 from the back plane of the second lens. Figure 3.81 shows the light intensity distribution in the plane of the measuring screen for all the six positions of the source.

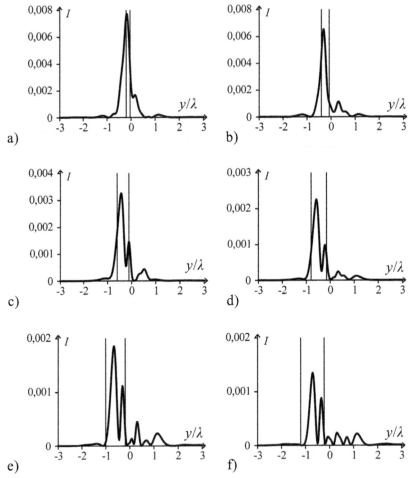

Fig. 3.80. The intensity distribution of light in the plane of the measuring screen for the six positions of the source, defined by the distance from the optical axis: 0.1λ (a), 0.2λ (b), 0.3λ (c), 0.4λ (d), 0.5λ (e), 0.6λ (f).

The vertical lines indicate the boundaries of the spot predicted by the equations (3.231) and (3.234). From Fig. 3.80 we see that most of the energy is concentrated precisely in these boundaries, except for Fig. 3.80 a. In Fig. 3.80 a the boundaries are located only 0.163λ from each other, which is close to the two-dimensional diffraction limit in silicon. In this case, the geometrical optics approach is not applicable and the light spot is not located in the specified ranges. Furthermore, when removing the source from the axis two instead of one intensity peak form. This can be explained by the presence of coma-type aberrations.

Numerical simulation of constructing the enlarged image with superresolution with two Mikaelian lenses

In the previous section we showed that when using two Mikaelian lenses of different sizes the formed image has a linear magnification less than the aspect ratio of the lenses.

This section presents the results of modelling the construction of an enlarged image with superresolution using two gradient Mikaelian microlenses, one of which is 2.5 times larger than the other. The simulation schema is shown in Fig. 3.81. The light from two point sources S_1 and S_2 (Fig. 3.81) illuminates the Mikaelian lens ML_1 (sources could be, for example, two slits 140 nm in diameter, separated by a distance of 140 nm, as in [119]). In the front plane of the lens the light intensity is recorded by the screen M_1. Then the light propagated through the lens ML1 with length L_1 and illuminates the Mikaelian lens ML_2 whose length $L2$ is about 2.5 times larger than L_1. In the back plane of the lens ML_2 there is the measuring screen M_2. The dotted line shows the boundaries of the field of the simulation domain.

The refractive indices of the two lenses are distributed according to the law of the hyperbolic secant, i.e. the distribution of the refractive index of the entire element is:

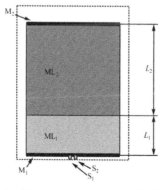

Fig. 3.81. The optical circuit of the imaging system with the magnification, consisting of two Mikaelian microlens.

$$n(x,z) = \begin{cases} \dfrac{n_0}{\cosh\left(\dfrac{\pi x}{2L_1}\right)}, & 0 \le z \le L_1, \\[4mm] \dfrac{n_0}{\cosh\left(\dfrac{\pi x}{2L_2}\right)}, & L_1 < z \le L_2. \end{cases} \qquad (3.234)$$

The simulation parameters for TE-polarization were as follows: wavelength of light in vacuum $\lambda = 1$ μm, the lens material – silicon, the refractive index on the optical axis $n_0 = 3.47$, the width of the two lenses $W = 4.8$ μm, the calculated length of the first lens $L_1 = 2$ μm, the calculated length of the second lens $L_2 = 4.45$ μm. The sampling step on both coordinates $\lambda/50 = 20$ nm, the distance between the sources 0.28λ, the width of the sources 0.14λ. The field intensity ($|E_y|^2$) at the input of the lens (i.e. in the plane of the screen M1 in Fig. 3.81) is shown in Fig. 3.82.

Figure 3.83 shows the instantaneous amplitude distribution E_y at time 30.6λ/c, where c is the speed of light in vacuum.

From Fig. 3.83 it can be seen that two light spots form in the image plane (i.e., in the back plane of the second lens). The time-averaged intensity $|E_y|^2$ in this plane is shown in Fig. 3.84 and this confirms that the sources are resolved on the level of 1/4 of the maximum.

From Fig. 3.84 it is clear that the distance between the source images is about 0.59λ, i.e. the linear magnification of approximately 2.1. These two images can already be resolved by conventional high-aperture optics with a numerical aperture of 0.86.

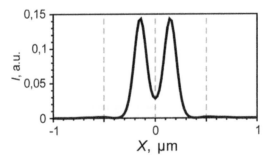

Fig. 3.82. The intensity of the light field in the front plane of the first lens.

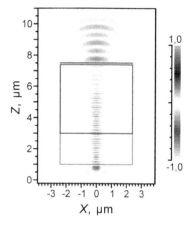

Fig. 3.83. Instantaneous distribution of the amplitude E_y at time $30.6\lambda/c$.

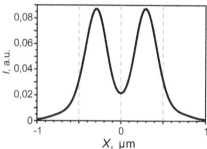

Fig. 3.84. Intensity of the light field in the back plane of the second lens.

Construction of an enlarged image with superresolution using expanding Mikaelian lens

Fresnel losses occur at the junction of the two lenses discussed in the previous section. To avoid this, we examine a Mikaelian lens, expanding linearly along the optical axis, i.e. add the dependence on the coordinate z to the refractive index:

$$n = \frac{n_0}{\cosh\left(\dfrac{\pi x}{2L(z)}\right)} \tag{3.235}$$

where $L(z) = 2 + 0.5755z$. The distribution of the refractive index of this lens is shown in Fig. 3.85 (in halftones).

The length of the lens in Fig. 3.85 is 15.2 µm, the width in the front plane 4.8 µm, in the back plane 48 µm. Figure 3.86 shows the intensity distribution in the back plane of the lens on the images of the same two sources with a width of 0.14 µm with the distance between their centres of 0.28λ.

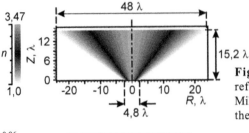

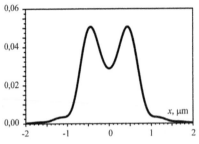

Fig. 3.85. The distribution of the refractive index (in halftones) of the Mikaelian lens linearly expanding along the optical axis.

Fig. 3.86. The intensity distribution in the back plane of the lens (3.86) for the images of two sources with a width of 0.14λ with the distance between their centres of 0.28λ.

The distance between the centres of the images is 0.88λ, i.e. magnification of such a lens is approximately 3.14. These two images can already be resolved by a conventional microlens with a numerical aperture of 0.58.

It has been previously shown that with increasing distance from the optical axis, even if the lens does not magnify, when the light beam which preserves its homocentricity and converges in the plane of the image, the formed image of the point source is large size because of the diffraction blur. This follows from the fact that the diffraction limit is inversely proportional to the numerical aperture, which has the form of (3.223) for a system of two identical Mikaelian lenses. If we connect two lenses of different sizes, the rays from the off-axis source no longer intersect in the image plane, i.e. a spot on the image will be expanded not only because of diffraction, but also because of a violation of the homocentricity of the beam of rays. Figure 3.88 shows the dependence of the width (at half intensity of FWHM) of the light spot formed by the off-axis light source using linearly expanding lenses (3.235). The calculation was performed by the FDTD-method. The calculation parameters are the same as for Fig. 3.84.

The position of the source changed from 0 to 2.4 μm, i.e. along the entire front plane of the lens (Fig. 3.87). Figure 3.87 shows that the width of the light spot on the image in the rear plane of the lens ranged from 0.51λ to 0.58λ, i.e. it remained approximately

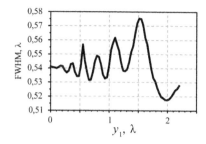

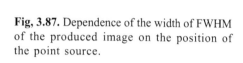

Fig, **3.87.** Dependence of the width of FWHM of the produced image on the position of the point source.

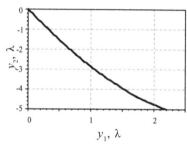

Fig. **3.88.** The dependence of the centre of the image spot on the position of the centre of the light source.

constant, which distinguishes the lens (3.235) from the system of the two lenses (3.234).

Figure 3.88 shows the calculated dependence of the position of the centre of the light spot on the position of the centre of the light source in the formation of an image using the lens (3.86).

From Fig. 3.88 it can be seen that the dependence is close to linear. This leads us to expect that in the presence of multiple sources the ratios on the image will be saved.

3.4.4. Hyperbolic secant lens with a slit for subwavelength focusing of light

This section summarizes the useful qualities of the gap of a few tens of nanometers for the localization of light and gradient lenses for sharp focusing of light, for example, Mikaelian lenses [120, 121]. We considered sharp focusing of light with a planar gradient lens using a slit for the localization of light in a narrow spot with the FWHM width close to the width of the slit. For example, for the planar binary microlens in silicon with a slit 50 nm wide with an energy efficiency of 44% a wide focal spot with the width FWHM = $\lambda/23$ forms in the vicinity of the lens surface. Focusing is carried out at the border of the lens, so the sharp focal light spot is available for various applications of nanophotonics.

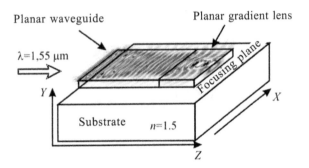

Fig. 3.89. Schematic of the problem.

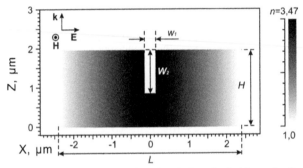

Fig. 3.90. The distribution of the refractive index (in halftones) of a gradient lens with a slit, the direction of the vectors of electric and magnetic fields are shown for the case of TM-polarization.

The schematic of the problem is shown in Fig. 3.89.

The light emitted from the planar waveguide made of silicon ($n_0 = 3.47$) on a substrate of fused silica is focused with the gradient lens. The maximum refractive index of the gradient lens on the optical axis is equal to the refractive index in the waveguide n_0.

Figure 3.90 shows a diagram of a planar gradient lens with a slit. As the gradient lens we consider a planar hyperbolic secant (HS) lens whose refractive index depends on the transverse coordinate x in the following way [120, 121]:

$$n(x) = n_0 \frac{1}{ch\left(\dfrac{\pi x}{2H}\right)}$$

(3.236)

where H is the length of the lens.

A slit in the planar HS lens with the width W_1 is located on the optical axis of the lens and reaches the output focal plane of the lens. The slit may extend through the entire lens ($W_2 = H$), or be

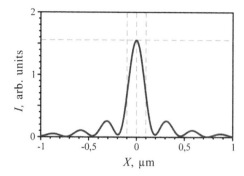

Fig. 3.91. The intensity distribution of radiation in the focus of the lens $|E_y|^2$ in the absence of the slit, the incident wave is TE-polarized.

in the last part of the lens ($W_2 < H$). The slit width W_1 may be less than the diffraction limit of light focusing in the material of the lens along the optical axis.

The propagation of light through the lens was modelled by the FDTD method, implemented in commercial software FullWave (company RSoft). Figure 3.91 shows a graph of the intensity distribution in the focal plane of the lens without the slit at a distance of 10 nm behind the output plane of the lens. Simulation parameters: $H = 1.95$ µm, $L = 4.8$ µm, refractive index on the optical axis $n = 3.47$ (silicon), wavelength $\lambda = 1.55$ µm, the incident plane wave, TE-polarization. The length of the lens H is chosen for the optimum focal spot. Note that in the case of TM-polarized light the quality of focusing light by a gradient lens without a slit will be much worse (focal spot will be four times wider, and the side lobes near the focus will form approximately half of the focal spot).

In Fig. 3.92 the width of the focal spot at half intensity FWHM = 0.181 µm = 0.117λ. This value is smaller than the diffraction limit in the focus which has the following value for the given refractive index FWHM = 0.44λ/3.47 = 0.127λ. We show that by using the slit on the optical axis of the lens with the refractive index in the gap $n = 1$ and the initial light field with TM-polarization, it is possible to achieve sharper focusing of light. Figure 3.92a shows the dependence of the width of the focal spot FWHM on the width of the slit W_1. The length of the gap was taken equal to the length of the lens $W_2 = H = 2.2$ µm, the other simulation parameters are the same as for Fig. 3.90.

Figure 3.92a shows that the width of the focal spot at half intensity is linearly dependent on the width of the gap in the lens

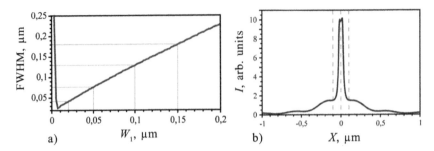

Fig. 3.92. The dependence of the width of the focal spot at half-intensity on the width of the slit W_1 (a), the intensity distribution $I = |E_x|^2 + |E_z|^2$ in the focal plane (10 nm from the lens) at $W_1 = 50$ nm (b).

W_1, the focal spot is slightly wider than the width of the slit. Figure 3.93 b is a plot of the intensity in the focus of the lens at $W_1 = 50$ nm. It is seen that the light intensity in the focus of the lens with a slit is approximately 6.5 times greater than in the case of lens without slits (Fig. 3.91), all other parameters being equal.

The diffraction efficiency of focusing light n_D by the planar gradient lens with a slit in a region close to the width of the slit also depends on the width of the slit W_1. The maximum efficiency of light focusing is observed at $W_1 \approx 40$ nm, and with increasing width of the slit there is a decline in efficiency n_D (Fig. 3.93 a). This effect is due, firstly, to the decrease of the light intensity at the focus by increasing the width of the slit W_1 (Fig. 3.93 b) and secondly the change of side lobes, and as a result, the change of n_D. Efficiency n_D is calculated as the ratio of the energy contained in the central lobe of the diffraction pattern at the focus (approximately -75 nm $< X <75$ nm in Fig. 3.92b) to the total energy supplied to the exit plane with width L.

Figure 3.92a also shows that when the width of the slit is $W_1 < 5$ nm the width of the focus begins to grow. The minimum width of the focal spot is achieved with $W_1 = 5$ nm and is equal to FWHM $= 13$ nm $= \lambda/119$. For comparison, in Fig. 3.91 the focal spot has a width FWHM $= \lambda/8$ and efficiency $n_D = 60\%$.

Figure 3.93 shows that for the considered parameters the maximum efficiency is $n_D = 39.9\%$, and the width of the focal spot is then equal to FWHM $= 55$ nm $= \lambda/28$ (see Fig. 3.92a).

Figure 3.94 shows the dependence of the efficiency of focusing light n_D (a) and the light intensity in the focus of the lens (b) on the length of the slit W_2 for the fixed slot width $W_1 = 50$ nm. For the calculation of the efficiency n_D here, as in Fig. 3.93, we take into

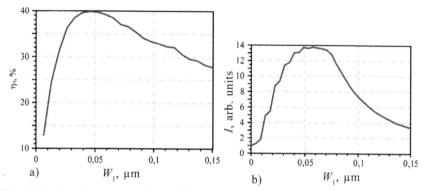

Fig. 3.93. The dependence of the efficiency of focusing light n_D (a) and the intensity in arbitrary units at the focus of the lens (b) on the slit width, W_1, $W_2 = H$.

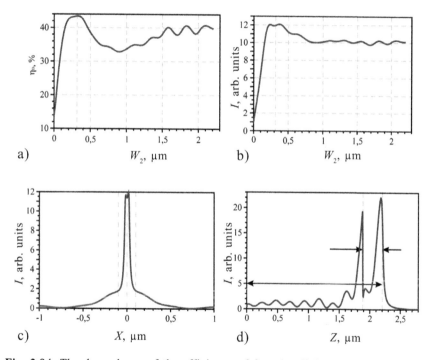

Fig. 3.94. The dependence of the efficiency of focusing light n_D (a) and the light intensity at the focus of the lens (b) on slit length W_2 at $W_1 = 50$ nm; light intensity distribution 10 nm behind the lens at $W_1 = 50$ nm, $W_2 = 0.31$ μm (in) and on the optical axis (d).

account the sharpest area of the focal spot (to the nearest side lobes), and use -75 nm $< X < 75$ nm. Figure 3.94 shows that in the absence of the slit $W_1 = W_2 = 0$ the diffraction efficiency of light focusing is about 10%.

It can be seen that the maximum efficiency of focusing n_D and the intensity of the light field form approximately at the length of the slit providing a delay of light in it at $\lambda/2$ which at the considered wavelength and the refractive index corresponds to $W_2 = \lambda/[2\,(n_0-1)]$ = 0.314 μm. The efficiency of focus the light in this case reaches n_D = 43.4%. The intensity of light in the focus of the lens is greater by about 20% (Fig. 3.94 c)), and the side lobes are less pronounced than in the case of $W_2 = H$ (Fig. 3.92 b). Figure 3.94d shows that the highest values of the intensity of the light field are obtained at a distance of up to about 30 nm from the edge of the lens inside the slit.

Since a lens having a gradient refractive index is difficult to produce with modern lithography tools, sharp focusing of TM-polarized waves can be carried out using a photonic crystal lens with a slit on the optical axis similar to the gradient lens at the average gradient distribution of the refractive index. Figure 3.95a shows the distribution of the refractive index in the XZ plane of the photonic crystal lens similar to the gradient lenses (Fig. 3.90).

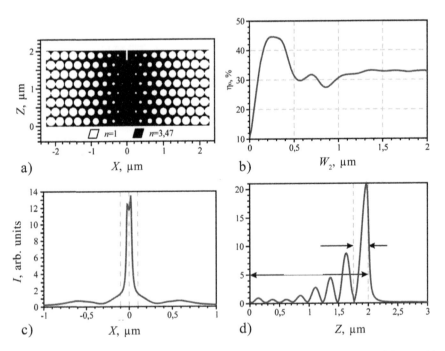

Fig. 3.95. The distribution of the refractive index in the photonic crystal lens with a slit (a), dependence of diffraction efficiency n_D on the length of the slit W_2 at a slit width W_1 = 50 nm (b), the intensity distribution in the focus in the transverse plane 10 nm behind the lens (c) and on the optical axis (d).

The parameters of the lens in Fig. 3.95 a: 8 rows of holes along the Z axis, 20 staggered rows of holes along the X axis, the minimum hole diameter 30 nm, the maximum – 250 nm, the length of the lens 2 μm, width 4.8 μm, the refractive index of the lens material $n = 3.47$, $W_1 = 50$ nm, $W_2 = 0.25$ μm. For this lens the dependence of the efficiency of focusing light on the length η_D of the slit at a fixed slit width $W_1 = 50$ nm looks similar to the gradient lens option (Fig. 3.95 b). The maximum efficiency of focusing the light in the focal spot with a width of the slit $W_1 = 50$ nm is $\eta_D = 44.3\%$ at $W_2 = 0.25$ μm. The width of the focal spot FWHM = 0.044λ.

3.4.5. Sharp focusing of radially polarized light with a 3D hyperbolic secant lens

We consider the focusing of a radially polarized wave incident normally on a flat surface of a secant hyperbolic lens with cylindrical symmetry.

Figure 3.96 shows the radial distribution of the refractive index in the gradient microlens. The parameters of the microlens (Fig. 3.96): $n_0 = 1.5$, $L = 10$ μm, the radius of the lens aperture $R = 6$ μm. Lens radius R is determined in the ideal case by equation (3.25) with the proviso that $n(R) = 1$.

The microlens (Fig. 3.96) was covered by a radially-polarized circular Gaussian beam of the type

$$E_r = \exp\left\{-\frac{(r-r_0)^2}{\omega^2}\right\}, \qquad (3.237)$$

where $r_0 = 5.5$ μm – the centre, $\omega = 1.5$ μm – waist radius. The wavelength was chosen to be $\lambda = 1$ μm. Figure 3.97 shows the profile of the beam.

Figure 3.98 shows an instantaneous picture of the distribution of the amplitudes for this case. Figure 3.99 shows the radial distribution

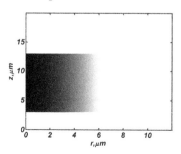

Fig. 3.96. The radial cross-section of the cylindrical gradient secant lens and its location in the calculated area (the dependence of the refractive index on the radial coordinate is shown in halftones).

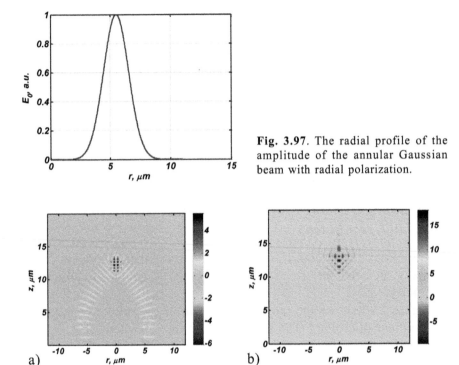

Fig. 3.97. The radial profile of the amplitude of the annular Gaussian beam with radial polarization.

Fig. 3.98. The instantaneous amplitude distribution E_r (a) and E_z (b) in the diffraction of the annular Gaussian beam with radial polarization on the cylindrical gradient secant lens.

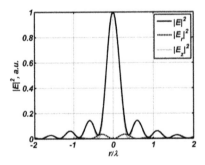

Fig. 3.99. The radial intensity distribution in the focal plane (just behind the lens at $z = 13$ μm) of the gradient secant lens in incidence of the annular Gaussian beam with radial polarization.

of the intensity in the focal plane, and Figure 3.100 shows the intensity distribution along the axis of the gradient secant lens. The diameter of the focal spot at half intensity is FWHM = 0.40λ, and the area of the focal spot at half intensity is HMA = 0.126λ².

It should be noted that in practice it is not possible to implement the annular Gaussian beam of form (3.237). In practice, for example,

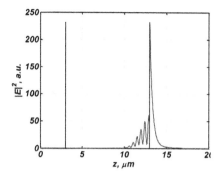

Fig. 3.100. The intensity distribution along the axis of the gradient secant lens when the annular Gaussian beam with radial polarization passes through it.

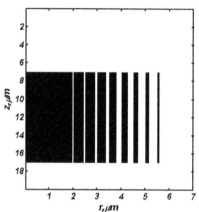

Fig. 3.101. Binary analogue HS-lens.

a substrate on which the microlens is fabricated, can be sprayed with a thin metal film with an annular diaphragm.

The binary analogue of the gradient HS-lens is a fragment of the optical Bragg fibre (Fig. 3.101). The thickness of the step of the binary analogue of the HS-lens is calculated as follows:

$$x(r) = \Delta r \left[\frac{n(r)-1}{n_0 -1} \right],$$

(3.238)

where n_0 and $n(r)$ are taken out of the equation (3.236), $\Delta r = 0.5$ μm is a discrete step.

The results of focusing with the binary analogue of the HS-lens are shown in Fig. 3.96 – the intensity distribution along the axis of Fig. 3.102 a and the intensity distribution in the focus in Fig. 3.102 b. Figure 3.102 shows that the substitution of the gradient lens by the binary analogue reduces the intensity at the focus (by approximately 20%), the spot size remains substantially unchanged.

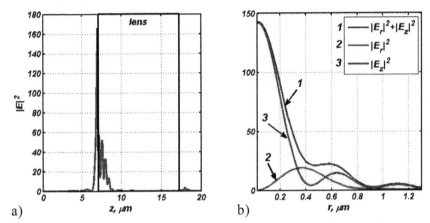

Fig. 3.102. (a) The intensity distribution along the z axis, and (b) in the focal spot.

3.4.6. Optimizing the parameters of the planar binary lens for the visible radiation range

This section describes 3D modelling of the focusing of linearly polarized light by a planar binary microlens or photonic crystal (PC) lens, similar to the Mikaelian gradient lens for the wavelength $\lambda = 0.532$ μm, on its border. Photonic crystals were used to focus the light in the past, for instance in [122] the authors simulated the focusing of light pulses by a multilayer photonic crystal, but the resulting focal spot was of the order of the light wavelength of 1.55 μm. In addition, the photonic crystal in [122] was not calculated as an approximation of the gradient secant lenses.

It is assumed that the lens designed in this section is made in the electron resist film having a refractive index of $n = 1.56$, deposited on a fused quartz substrate ($n_0 = 1.46$). Such a simulation option was selected because this lens can be manufactured by electron beam lithography. The simulation was performed by the FDTD-method implemented in the FullWave program (RSoft).

Modelling of the two-dimensional case
First, consider the two-dimensional version. Figure 3.103 shows a diagram of the problem in the 2D case. Light with a wavelength $\lambda = 532$ nm propagates in a planar waveguide with the width W and is focused by a gradient secant lens. A plane wave with TE polarization falls on the lens. The electric field vector is directed along the Y axis. The waveguide is an electron resist film deposited

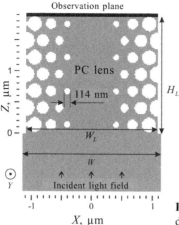

Fig. 3.103. Diagram of the problem in the two dimensions (gray n = 1.56, white n = 1).

on fused silica. The refractive index of the resist is n_1= 1.56. For ease of manufacturing the lens is in the form of a photonic crystal (PC). The gradient refractive index of the lens is formed by etching the holes in the electron resist to the substrate. The lens is calculated assuming the possibility of manufacturing by means of electronic lithography with a resolution of 100 nm.

The minimum diameter of the holes in the gradient photonic crystal forming the lens is 114 nm. The refractive index of the substrate n_0 = 1.46. For sharp light focusing at the boundary of the PC lens its width was chosen as W_L = 2.55 μm, length H_L = 1.83 μm. With these parameters in the two-dimensional case the width of the focal spot at half intensity was FWHM = 0.361λ, the effectiveness of focusing at half intensity was 65%. The location and diameter of the holes in the PC were calculated for the width and height of the lens and this was followed by optimizing the value W_L and H_L to reduce the focal spot. To speed up the calculations, optimization of the size of the PC lens was carried out in two-dimensional case for TE-polarization.

When changing the width W_L and height H_L of the lens the coordinates of the centres of the holes were proportionally shifted but their diameters remained unchanged. Figure 3.104 shows plots of the intensity distribution of radiation in the plane of observation 10 nm behind the lens before and after optimization.

At the initial design parameters the observation plane contains only the focal spot with almost complete absence of the side lobes (Fig. 3.104 a). However, if some side lobes appear (their height is about 30% of the main peak intensity) it is possible to reduce

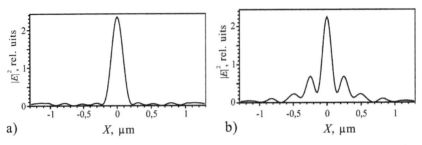

a)   b)

Fig. 3.104. The intensity distribution 10 nm behind the lens in the two dimensional case prior to optimization (a) and after optimization (b) of the size of the lens. The incident wave on the waveguide is planar, the partitioning interval of the grid samples in all three coordinates is $\lambda/30$.

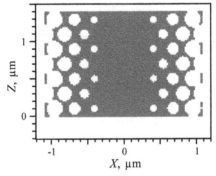

Fig. 3.105. The distribution of the refractive index in the lens in the XZ plane after optimizing the width and height (gray $n = 1.56$, white $n = 1$).

the width of the focal spot. The minimum width of the focal spot (Fig. 3.104 b) is FWHM = 0.28λ and is achieved at the width and height of the lens W_L = 2.652 μm, H_L = 1.39 μm (Fig. 3.105). The effectiveness of the focusing thus fell to 39.8%.

The optimized refractive index profile of the lens is shown in Fig. 3.105.

The highest value for sharp focusing is achieved for the part of the light incident on the edge of the lens. If the lens is illuminated with light having the Gaussian intensity distribution with radius σ along the X axis, equal to half the width of the lens, the sharpness of the focus falls significantly. Figure 3.106 shows the distribution of the field intensity 10 nm behind the lens for a Gaussian light source with the radius along the X axis σ = 1.275 μm in the two-dimensional case. In Fig. 3.106 the width of the focal spot is FWHM = 0.35λ, which is about 30% more than in the case of a plane incident wave. However, in a waveguide with a constant refractive index the fundamental mode with the almost Gaussian intensity

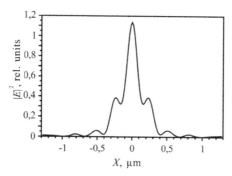

Fig. 3.106. The intensity distribution of the lens illuminated by a Gaussian beam with a radius along the X axis $\sigma = 1.275$ µm in two dimensions, $W = W_L$.

distribution is set. To increase the intensity of incident light on the edge of the lens the waveguide width W was selected larger than the width of the lens and the radius of the Gaussian beam along the X axis was accepted as $\sigma = 2,75$. Thus, the lens receives the central part of the Gaussian beam, the amplitude of which at the edges of the lens falls approximately 0.8 from the maximum at the centre of the lens. This enhances the sharpness of the focus, although it reduces its effectiveness.

Simulation of three-dimensional case

The three-dimensional optical layout and the external appearance of the lens with the waveguide are shown in Fig. 3.107. The Z axis passes through the optical axis of the system – in the middle of the planar waveguide (although in Fig. 3.107 it is drawn on the side).

The main waveguide with width W extends along the sides of the lens and there are slits with width x_1 between the waveguide and PC lens. Thus some of the light which does not fall on the lens extends further beyond the lens and can be removed in two waveguides outside the focal region. The simulation showed that in the three dimensional case the width of the focal spot of the PC lens is affected by both the thickness of the waveguide film and the width of the slit x_1 on the sides of the lens.

Simulation of the three-dimensional optical setup was conducted with the following parameters: the radius of the Gaussian beam propagating in the waveguide $\sigma = 2.75$ µm, the length of the waveguide between the source and the lens $L = 6$ µm, the width of the waveguide $W = 5.5$ µm, the step of the grid of samples in all three coordinates is $\lambda/20$. The dimensions of the

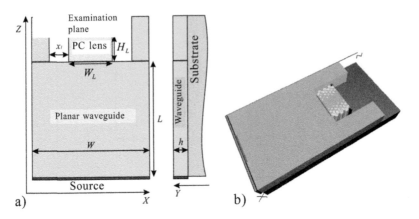

Fig. 3.107. Three-dimensional optical setup of the planar PC lens on a fused quartz substrate (a) and 3D view (b).

lens W_L and H_L are the same as in Fig. 3.106. Figure 3.108 shows the instantaneous distribution of the field amplitude E_y at the time $cT = 12$ μm, where c is the speed of light in vacuum. The waveguide is illuminated with a wave with the electric field strength E_y and the intensity $I = |E^2| = |E_y|^2 + |E_z|^2$ was calculated at the focus of the lens, where E_z is the longitudinal component of the electric vector.

From the instantaneous field distributions E_y it can be seen that after reaching the end of the lens part of the light enters the substrate and begins to spread inside, the other part goes into the surrounding space above the substrate. It is also clear that due to the excess width of the waveguide W in relation to the width of the lens W_L, the lens receives the central part of the Gaussian beam in the waveguide. This results in a more uniform intensity of the incident field both in the centre and on the edges of the PC lens. This improves the sharpness of focus.

Figure 3.109 a shows the dependence of the width of the focal spot at half intensity on the waveguide thickness measured at the centre of the waveguide. The gap at the edges of the lens in this case was assumed to be $x_1 = 0.11$ μm.

Figure 3.109 a shows that the smaller the waveguide thickness h, the wider the focal spot along the X axis. On the other hand, the smaller the waveguide thickness h, the smaller the fraction of the light propagating in the waveguide film as a mode. The dependence of the maximum time-averaged intensity in the focus of the lens on the thickness of the waveguide is shown in Fig. 3.109 b. It is evident that with the thickness h of the waveguide decreasing to less than

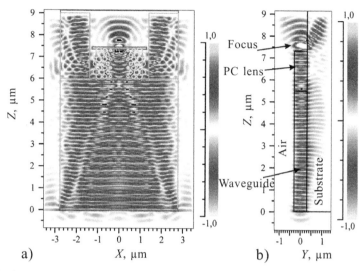

Fig. 3.108. Instantaneous amplitude distribution of the field E_y at time $cT = 12$ μm in the plane XZ, $Y = 0$ (a) and in the plane YZ, $X = 0$ (b), TE-polarization.

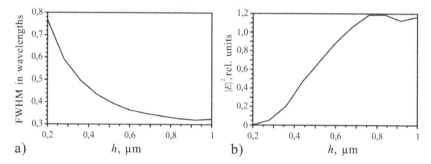

Fig. 3.109. The dependence of the width of the focal spot of the lens on the thickness of the waveguide film (a); the dependence of peak intensity $|E|^2$ at the focus on the thickness of the waveguide (b); $x_1 = 0.11$ μm for both graphs.

0.75 μm the light intensity at the focus of the lens falls, indicating the loss of light passing into the substrate. However, increasing the thickness of the waveguide increases the ratio of the etching depth to the diameter of the holes forming the photonic crystal (the aspect ratio), which complicates the manufacturing of the PC lens by electron lithography. The optimum selected height was $h = 0.6$ μm.

Figure 3.110 shows the dependence of the width of the focal spot along the X axis through the centre of the waveguide at half intensity on the distance x_1.

It can be seen that the optimum distance is $x_1 = 0.11$ μm. This results in the minimum focal spot FWHM = 0.365λ and maximum

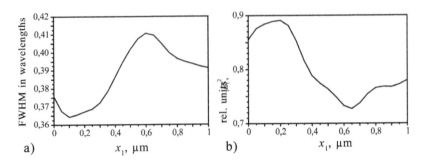

Fig. 3.110. The dependence of the focal width FWHM along the axis X (a) and the intensity $|E|^2$ in the central part of the focal spot (b) of the distance between the lens x_1 and the lens units along the waveguide, the waveguide thickness $h = 0.)6$ mm.

intensity in the focus. The efficiency of focusing at half intensity is then equal to 41%. At a different waveguide thickness h only the intensity in the focal spot changes, the optimal distance x_1 is kept constant.

Figure 3.111 shows the shape of the focal spot behind the lens at the optimum settings: $x_1 = 0.11$ µm, $h = 0.6$ µm.

It is seen that in the three-dimensional case the side lobes appear weaker (15%) than in the two-dimensional case (30%), both in the case of a plane wave incident on the lens (Fig. 3.104 b) and in the case of a Gaussian distribution (Fig. 3.105). The width of the focal spot along the X axis is FWHM $= 0.365\lambda$. Since the lens is planar, there is no focus on the Y axis and the width at half intensity along the Y axis is FWHM $= 0.74\lambda$. If the photonic crystal is removed (the holes are filled with the same material), the focus in the observation plane is converted into an interference pattern consisting of three intensity peaks of equal values, wherein the central intensity peak becomes larger, in which case its width is FWHM $= 0.46\lambda$. This is shown in Fig. 3.112. Other modelling parameters are the same as in Fig. 3.111.

From Fig.3.112 it is also seen that in the absence of the PC lens the asymmetry of the optical circuit along the Y axis (the presence of the substrate with a similar refractive index) affects the position of intensity maxima and they are displaced toward the substrate by about 60 nm. In the presence of the photonic crystal, forming the lens, this effect is not observed.

3.5. Formation of a photon nanojet by a sphere

The fact that a sphere with radius R focuses a plane wave near

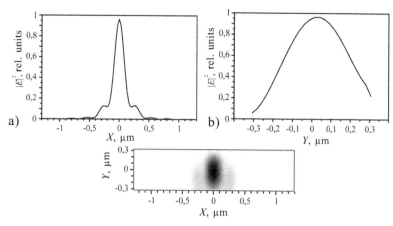

Fig. 3.111. The distribution of intensity $|E|^2$ behind the lens in a line along the X axis (a) through the maximum intensity (in the middle of the waveguide) with the PC lens, along the Y axis (b) through the centre of the waveguide; two-dimensional distribution of the intensity $|E|^2$ in the observation plane, negative (c).

its surface follows from the formula for the focal length $f = R/(n-1) \approx 2R$, where $n \approx 1.5$ is the refractive index. In recent years, there has been increased interest in focusing light by microspheres. In [123] the authors studied the transmission of a plane TM-polarized wave through a two-dimensional dielectric cylinder and first introduced the concept of the photon nanojet. The generalization of the two-dimensional case to the three-dimensional one is given in the paper [124]. In [124] it was shown that by focusing the linearly polarized plane wave with a wavelength $\lambda = 400$ nm by a sphere with a refractive index $n = 1.59$ and the diameter $2R = 1$ μm the diameter of the waist of the photonic jet is FWHM $= 0.325\lambda$ of the wavelength. The first direct experimental observation of the photon nanojet was carried out in [125] by confocal microscopy. In [125] a latex ($n = 1.6$) sphere with $2R = 3$ μm was used to obtain a photon jet with a waist diameter FWHM $= 0.52\lambda$. Also of interest is the experimental work [126], in which a high-precision interferometer was used to study the effect of the radiation input parameters on the photon jet. The formation of a photonic jet at focusing of a plane wave, a spherical wave, a Bessel beam, and an azimuthally polarized beam was studied. In [127] experiments were carried out using an optical microscope with an ×100 lens ($NA = 0.7$) to investigate the photonic jet formed using a silicon nitride disc ($n = 2.1$) 400 nm thick with $2R = 9.5$ μm. This allows to produce with the smallest focal spot diameter at half intensity equal to FWHM $= 0.86\lambda$. Another

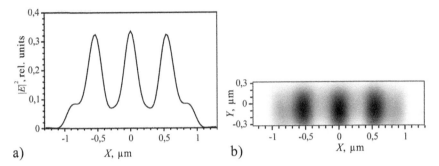

Fig. 3.112. The intensity distribution of the lens ($|E|^2$) in the direction along the X-axis through a maximum intensity (at a distance of 0.24 m from the substrate) without a photonic crystal (no holes); two-dimensional distribution $|E|^2$ in the observation plane, negative (b).

important characteristic of the photon nanojet is its length (depth of focus, DOF) which was studied in [128]. It was shown that the use of a gradient microsphere in which the refractive index varies linearly from 1.43 to 1.59 can increase the length of the photon nanojet to DOF = 11.8λ. It should be noted that the length was determined as the distance from the sphere to the point where the intensity fell by half compared to the beam illuminating the sphere. In [129], the authors set themselves the opposite problem – to achieve reduction of the size of the photonic nanojet both in the transverse and spatial coordinates. The microsphere (radius 2.5λ) was illuminated with a Gaussian beam focused by a wide-aperture lens ($NA ≈ 1$). The length of the photon nanojet in this case was DOF = 0.88λ.

The photonic nanojets may find use in Raman spectroscopy [130], for optical memory systems [131] as well as in nanolithography. In [132] the microspheres were used construct a lithographic apparatus in which movement of the microspheres was performed by an optical tweezer (λ = 532 nm, the accuracy of positioning the spheres 40 nm). The microsphere, moved to the desired point and illuminated by a laser pulse (λ = 355 nm), formed behind itself a focal spot used for burning a groove in the substrate (a pattern with the dimensions of parts of 100 nm formed).

3.5.1. Numerical simulation of passage of continuous radiation through the microsphere

The radial method of the difference solution of Maxwell's equations in cylindrical coordinates (BOR-FDTD) was used for numerical simulated of the passage of linearly polarized (polarization plane *YZ*) Gaussian beam with a wavelength λ = 633 nm, radius ω = 7λ, through a polystyrene microsphere (n = 1.59) with a diameter $2R$ = 5 μm using the BOR-FDTD method with the following parameters: grid sampling in space $\lambda/50$, in time of $\lambda/100c$, where c is the speed of light in vacuum, with absorbent Berenguer layers with thickness λ placed at the edges. The geometry of the problem is shown in Fig. 3.113.

The simulation results: the size of the focal spot along the x axis FWHM$_x$ = (0.49 $\pm$ 0.02)λ, and along the y axis FWHM$_y$ = (0.78 $\pm$ 0.02) λ. The maximum intensity was located at a distance of 0.134 μm from the sphere. This point is called in different sources the waist area of the photon nanojet [124], or the focus area [133], we will adhere to the second name. Then, under the focal length we mean the distance from the surface of the microsphere to the focus.

Figure 3.114 shows the time-averaged intensity distribution in the computed field in the *XZ* plane, perpendicular to the plane of polarization of the input light (Fig. 3.114 a) and in the *YZ* plane, parallel to the plane of polarization of the input light (Figure 3.114 b). Figure 3.115 shows the cross-section of intensity in the focus along the x axis (Fig. 3.115 a) and the y axis (Fig. 3.115 b) and Fig 3.116 – intensity distribution along the z axis. Figure 3.116 can be used to assess the depth of focus or the length of the photon nanojet at half intensity along the z axis: DOF = (1.06 $\pm$ 0.02)λ.

In order to verify the modelling results and also the effect of the width of the input beam on the focus spot size we compared the results obtained using the BOR-FDTD method and also the FullWave commercial software for focusing a plane wave. In both cases, it counts only the intensity distribution in the plane perpendicular to the direction of polarization of the input light (*XZ* plane) was calculated.

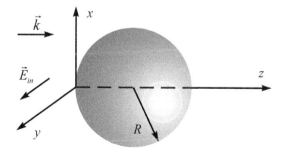

Fig. 3.113. The geometry of the problem.

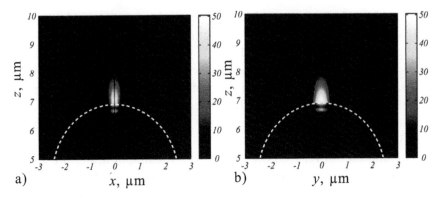

Fig.. 3.114. The intensity distribution in the calculated field in (a) the plane perpendicular to and (b) parallel to the plane of polarization of inpur radiation.

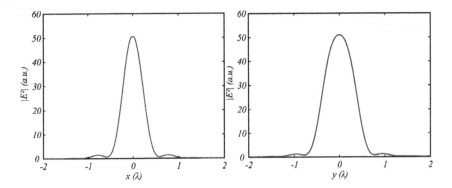

Fig. 3.115. The intensity distribution in the focus.

Modelling by the BOR-FDTD was carried out with the same parameters of the calculated area as in the previous section. The figures overleaf show the simulation results: Figure 3.117 – intensity distribution in the plane perpendicular to the direction of polarization of the input light (where the minimum diameter of the focal spot is found), Fig. 3.118 – intensity distribution along the z axis, Fig. 3.119 – intensity distribution in the XZ plane of the focus. In this case, the focal length was 0.108 μm (from the surface of the sphere), and the smallest diameter of the focal spot is equal to FWHM = $(0.46 \pm 0.02)\lambda$; depth of focus DOF = (0.99 ± 0.02) λ.

The following parameters were chosen in the simulation by the FDTD-method using FullWave: the step of grip sampling in space 0.05 μm (or $\approx\lambda/13$), in time 0.025 μm/c, where c is the speed of light in vacuum. The intensity distribution in the plane perpendicular to the direction of polarization of the input light is shown in Fig. 3.121.

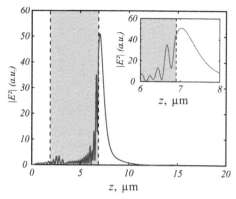

Fig. 3.116. The intensity distribution along the z axis (the inset shows the enlarged fragment near the surface of the sphere). Grey color shows the area inside the microspheres.

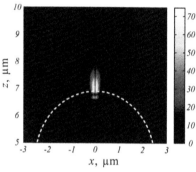

Fig. 3.117. The intensity distribution in the calculated field in the XZ plane.

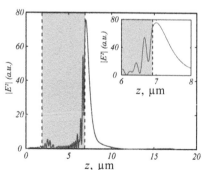

Fig. 3.118. The intensity distribution along the z axis, (the inset shows the enlarged portion near the surface of the sphere). The grey color shows the area inside the microsphere.

Figure 3.122 shows the cross-sectional of intensity, and Fig. 3.123 – intensity distribution along the z axis. The smallest diameter of the focal spot at half intensity was FWHM = $(0.55 \pm 0.08)\lambda$, the depth of focus DOF = $(0.68 \pm 0.08)\lambda$, and the distance from the sphere to the maximum intensity 0.05 μm.

The step of the grid sampling in the FullWave program was chosen almost 4 bigger than in the BOR-FDTD program because of

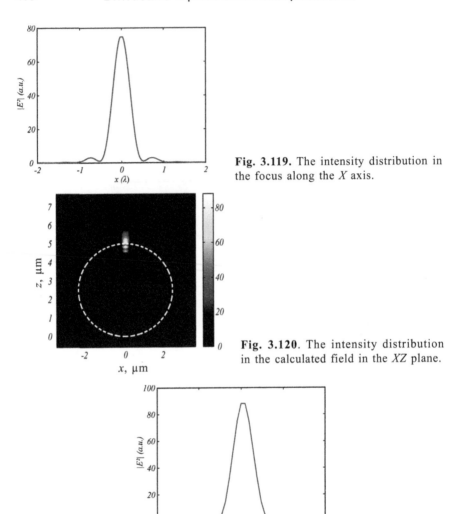

Fig. 3.119. The intensity distribution in the focus along the X axis.

Fig. 3.120. The intensity distribution in the calculated field in the XZ plane.

Fig. 3.121. The intensity distribution in the focus along the X axis.

the fact that the computing time of the first program is tens of times longer than that of the second software.

The MEEP program (Massachussets Institute of Technology) was used to study the focusing of a Gaussian beam with the width $\omega = 12\lambda$ by spheres with a refractive index $n = 1.59$; the radius of the spheres ranged from 0.5λ to 4λ. The size of the calculation area was $10 \times 10 \times 10$ wavelengths and the mesh pitch in space $\lambda/20$. In calculating the intensity the averaging was carried out over 20 periods. Figure 3.123 shows the dependence of the distance from

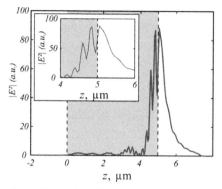

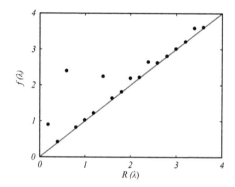

Fig. 3.122. The intensity distribution along the z axis, (the inset shows the enlarged portion near the surface of the sphere). The grey color shows the area inside the microsphere.

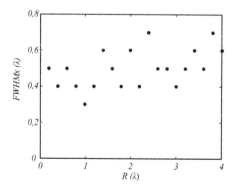

Fig. 3.123. Dependence of the distance from the centre of the sphere to the maximum intensity point outside the sphere.

Fig. 3.124. Dependences of the diameters of the focal spot FWHM along the X axis on the radius of the sphere.

the centre of the sphere to the point of maximum intensity outside the sphere. Plots of the dependence of the diameters of the focal spot of the radius of the sphere are shown in Fig. 3.124.

Figure 3.123 shows that if the radius of the sphere is equal to the wavelength of focused light $R = \lambda$, the diameter of the focal spot is at its lowest value FWHM = 0.3λ. The results obtained for a sphere of radius $R = 2.5$ μm = 3.95λ, which was seen in the previous sections, were: FWHM = (0.50 ± 0.05) λ.

3.5.2. Numerical simulation of passage of pulsed radiation through a microsphere

The focusing of pulses of a linearly polarized plane wave with a length of 1.25 fs at half intensity ($\lambda = 633$ nm) by a polystyrene sphere ($n = 1.59$) with a diameter $2R = 5$ μm was simulated. Numerical simulation of the focusing of pulsed laser light by the sphere was conducted in the FullWave program with the same parameters as continuous radiation: mesh pitch in space 0.05 μm (or $\approx\lambda/13$), in time 0.025 μm/c (or $\approx\lambda/26$), where c is the speed of light in vacuum.

The considered pulse of 1.25 fs has the shape shown in the Fig. 3.125. At the moment of focusing by the microsphere $t = 30$ fs the instantaneous intensity distribution in the plane perpendicular to the plane of polarization of the input light will be as shown in Figure 3.126. The cross section of the instantaneous intensity distribution along the Z axis at the moment of focusing is shown in Fig. 3.127, and Fig. 3.128 shows the intensity distribution in the focus along the X axis. The maximum intensity in the focus in arbitrary units is 134.22 (which is almost 100 times greater than the maximum intensity of the incident pulse), and the focal length is 0.05 μm. The smallest diameter of the focal spot at half intensity is FWHM = $(0.55 \pm 0.08)\lambda$.

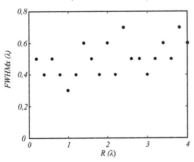

Fig. 3.125. Dependence of the diameter of the focusing spot FWHM along the X axis on the sphere radius.

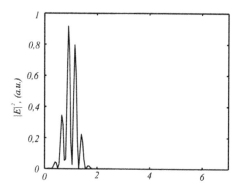

Fig. 3.126. The instantaneous intensity distribution of the incoming pulse of 1.25 fs along the Z axis.

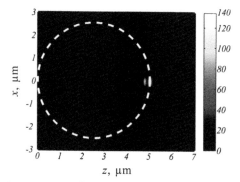

Fig. 3.127. The instantaneous intensity distribution in the XZ plane ($y = 0$ μm) for a pulse with a duration of 1.25 fs at focusing $t = 30$ fs.

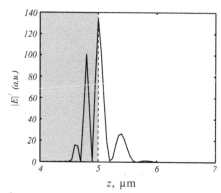

Fig. 3.128. The instantaneous intensity distribution along the z-axis for a pulse with the duration of 1.25 fs at the focus $t = 30$ fs. Dashed line marks the boundary of the sphere.

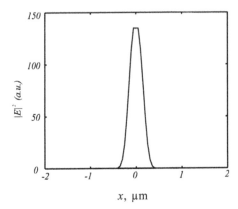

Fig. 3.129. The instantaneous intensity distribution in the focus along the y axis for the 1.25-fs pulse duration at the moment of focusing $t = 30$ fs.

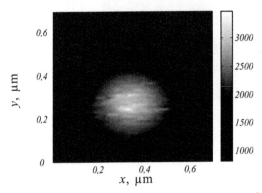

Fig. 3.130. The intensity distribution at the focus measured experimentally using a near-field scanning optical microscope.

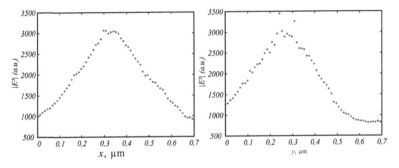

Fig. 3.131. Experimental cross sections of the intensity at the focus (Fig. 3.129) along the axis (a) x and (b) y.

3.5.3. Experiments with focusing of light by a microsphere

The experiment was conducted using a scanning near-field optical microscope Integra Spectra (company NT-MDT), the optical circuit is shown in Fig. 3.5. In this scheme, the linearly polarized light from the LGN-215 laser with a wavelength $\lambda = 633$ nm passed through a polystyrene sphere, forming a photon nanojet. The transverse intensity distribution in the nanojet was measured by a metal pyramidal cantilever with a hole with the size of 100 nm at different distances from the surface of the sphere. The focus was located at a distance of 100 nm from the surface of the sphere.

Figure 3.130 shows the distribution of intensity obtained in the focal plane. The focal spot was elliptical, with the dimensions along the x axis FWHM$_x$ = (0.73 ± 0.08) λ, and the y axis FWHM$_y$ = (0.60 ± 0.08) λ. Figure 3.131 shows the cross sections of the intensity of the along the x and y axes in Fig. 3.130.

The intensity distribution in Figs. 3.130 and 3.131 was obtained with a small number of samples with a resolution equal to the size of the holes of the cantilever – 100 nm. The error in measuring the diameter of the focal spot is 0.08λ.

Table 3.4 shows for comparison the data obtained in the section.

In this section, it was shown that for some values of the radius of the sphere is possible to overcome the diffraction limit. The minimum possible value of the diameter of the focus is achieved with a radius $R = \lambda$ and is equal FWHM = 0.3λ. Modelling using the FullWave program showed that when focusing continuous laser (a plane wave with a wavelength of 633 nm) the smaller focus diameter

Table 3.4. The smaller diameter of the focal spot formed in the vicinity of the sphere surface with the radius 3.95λ ($n = 1.59$) obtained by different methods

Method for obtaining the diameter	FWHM, λ
BOR-FDTD, Gaussian beam	0.49 ± 0.02
BOR-FDTD, the plane wave	0.46 ± 0.02
FullWave, plane wave	0.55 ± 0.08
FullWave, plane wave, pulse	0.54 ± 0.08
MEEP, wide Gaussian beam of	0.50 ± 0.05
The experiment	0.60 ± 0.08

for a sphere with radius $R = 2.5$ μm $= 3.95\lambda$ is FWHM= $(0.55\pm0.08)\lambda$, and in focusing the radiation pulse with a duration of 1.25 fs FWHM = (0.54 ± 0.08) λ. When calculating using the MEEP program it was found that the smaller focal spot size for a sphere with the same radius is FWHM = (0.50 ± 0.05) λ. All these values are slightly smaller than the diameter of the focal spot, as measured by an optical near-field scanning microscope FWHM = (0.60 ± 0.08) λ.

References

1.   Davidson, D.B. Body-of-revolution finite-difference time-domain modeling of space-time focusing by a three-dimensional lens, D.B. Davidson, R.W. Ziolkowski, J. Opt. Soc. Am. A. – 1994. – Vol. 11(4). – P. 1471-1490.

2.   Yu, W. On the solution of a class of large body problems with full or partial circular symmetry by using the finite-difference time-domain (FDTD) method, W. Yu, D. Arakaki, R. Mittra, IEEE Trans. on An. And Prop. – 2000. – Vol. 48(12). – P. 1810-1817.

3.   Farahat, N. A fast near-to-far-field transformation in body of revolution finite-difference time-domain method, N. Farahat, W. Yu, R. Mittra, IEEE Trans. on An. And Prop. – 2003. – Vol. 51(9). – P.2534-2540.

4.   Gedney, S.D. An anisotropic perfectly matched layer-absorbing medium for the trun-cation of FDTD lattices, S.D. Gedney, IEEE Trans. on An. And Prop. – 1996. – Vol. 44(12). – P. 1630-1639.

5.   Prather, D.W. Formulation and application of the finite-difference time-domain meth-od for the analysis of axially symmetric diffractive optical elements, D.W. Prather, S. Shi, J. Opt. Soc. Am. A. – 1999. – Vol. 16(5). – P. 1131-1142

6.   Shi, S. Electromagnetic analysis of axially symmetric diffractive optical elements illu-minated by oblique incident plane waves, S. Shi, D.W. Prather, J. Opt. Soc. Am. A. – 2001. – Vol. 18(11). – P. 2901-2907.

7.   EL_Mashade, M.B. BOR-FDTD analysis of nonlinear Fiber Bragg grating and dis-tributed Bragg resonator, M.B. EL_Mashade, M. Nady, Optics & Laser Technology. – 2011. – Vol. 43(7). – P. 1065-1072.

8.   Antosiewicz, T.J. Dielectric-metal-dielectric nanotip for SNOM, T.J. Antosiewicz, P. Wróbel, T. Szoplik, Proc. of SPIE. – 2009. – Vol. 7353. – P. 73530I.

9.   Olkkonen, J.T. On surface plasmon enhanced near-field transducers, J.T. Olkkonen, K.J. Kataja, J. Aikio, D.G. Howe, Proc. of SPIE. – 2004. – Vol.5380. – P. 360-367.

10.   Liu, Y. Analysis of a diffractive microlens using the finite-difference time-domain method, Y. Liu, H. Liu, J. Micro/Nanolith. MEMS MOEMS. – 2010. – Vol. 9(3). – P. 033004.

11.   Liu, Y. Broadband dispersion characteristics of diffractive microlenses based on the finite difference time-domain method, Y. Liu, H. Liu, L. He, H. Zhou, C. Sui, Optics & Laser Technology. – 2010. – Vol. 42(8). – P. 1286–1293

12.   Pérez-Ocón, F. Fast Single-Mode Characterization of Optical Fiber by Finite-Dif-ference Time-Domain Method, F. Pérez-Ocón, A.M. Pozo, J.R. Jiménez, E. Hita, J. Lightwave Technol. – 2006. – Vol. 24(8). – P. 3129.

13.   Davidson, N. High-numerical-aperture focusing of radially polarized doughnut beams with a parabolic mirror and a flat diffractive lens, N. Davidson, N. Bokor,

Opt. Lett. 2004.– Vol. 29(12). – P.1318-1320.

14. R. Borghi, M.Santarsiero, Nonparaxial propagation of spirally polarized optical beams, J. Opt. Soc. Am. A , v.21, no.10, p.2029-2037 (2004).

15. N. Passilly, R.S. Denis, K. Ait-Ameur, Simple interferometric technique for generation of a radially polarized light beam, J. Opt. Soc. Am. A, v.22, no.5, p.984-991 (2005).

16. T.G. Jabbour, S.M. Kuebler, Vector diffraction analysis of high numerical aperture focused beams modified by two- and three-zone annular multi-phase plates, Opt. Ex-press, v.14, no.3, p.1033-1043 (2006) .

17. Y. Kozawa, S. Sato, Focusing property of a double-ring-shaped radially polarized beam, Opt. Lett., v.31, no.7, p.867-869 (2006).

18. Q. Zhan, Properties of circularly polarized vortex beams, Opt. Lett., v.31, no.7, p.867-869 (2006).

19. D. Deng, Nonparaxial propagation of radially polarized light beams, J. Opt. Soc. Am. B, v.23, no.6, p.1228-1234 (2006).

20. Y.I. Salamin, Fields of a radially polarized Gaussian laser beam beyond the paraxial approximation, Opt. Lett., v.31, no.17, p.2619-2621 (2006).

21. D. Deng, Q. Guo, L.Wu, X. Yang, Propagation of radially polarized elegant light beams, J. Opt. Soc. Am. B, v.24, no.3, p.636-643 (2007).

22. T. Grosjean, D. Courjon, C. Banier, Smallest lithographic marks generated by optical focusing systems, Opt. Lett., v.32, no.8, p.976-978 (2007).

23. Y. Kozawa, S. Sato, Sharper focal spot formed by higher-order radially polarized laser beams, J. Opt. Soc. Am. A, v.24, no.6, p.1793-1798 (2007) .

24. G.M. Lerman, U. Levy, Tight focusing of spatial variant vector optical fields with elliptical symmetry of linear polarization, Opt. Lett., v.32, no.15, p.2194-2196 (2007).

25. S. Yan, B. Yao, Description of a radially polarized Laguerre-Gauss beam beyond the paraxial approximation, Opt. Lett., v.32, no.22, pp.3367-3369 (2007).

26. E.Y.S. Yew. C.J.R. Sheppard, Tight focusing radially polarized Gaussian and Bessel-Gauss beams, Opt. Lett., v.32, no.23, p.3417-3419 (2007) .

27. V. P. Kalosha, I. Golub, Toward the subdiffraction focusing limit of optical super-resolution, v.32, no.24, p.3540-3542 (2007).

28. J. Stadler, C. Stanciu, C. Stupperich, A.J. Meixner „Tighter focusing with a parabolic mirror", Opt. Lett., v.33, no.7, p.681-683 (2008).

29. R. Dorn, S. Quabis, G. Leuchs, Sharper focus for a radially polarized light beam, Phys. Rev. Lett., v.91, p.233901 (2003).

30. Y. Ohtaka, T. Ando, T. Inone, N. Matsumoto, H. Toyoda, Sidelobe reduction of tightly focused radially higher-order Laguerre-Gaussian beams using annular masks, Opt. Lett., v.33, no.6, p.617-619 (2008).

31. C.J.R. Sheppard, M.A. Alonso, N.J. Moore, Lacalization measures for high-aperture wave fields based on pupil moments, J. Opt. A: Pure Appl. Opt.,v.10, p.033001 (2008).

32. A.P. Prudnikov, et al., Integrals and series: special functions. Nauka, Moscow, 1983.

33. Pohl, D.W. Optical spectroscopy: image recording with resolution $\lambda/20$, D.W.Pohl, W.Denk, M.Lanz, Appl. Phys. Lett. – 1984. – Vol. 44. - P. 651 – 653.

34. Binnig, G. Tunneling through a controllable vacuum gap, G.Binnig, H.Rohrer, Ch.Gerber, E.Weibel, Appl. Phys. Lett. – 1982. – Vol. 40(2). – P. 178.

35. McLeod, J.H. The Axicon: A New Type of Optical Element, J.H. McLeod, J. Opt. Soc. Am. – 1954. – Vol. 44(8). – P. 592-597.

36. Dyson, J. Circular and Spiral Diffraction Gratings, J. Dyson, Proc. R. Soc. Lond. A. – 1958. – Vol. 248. – P. 93-106.

37. Lit, J.W. Focal depth of a transmitting axicon, J.W. Lit, R. Tremblay, J. Opt. Soc. Am. – 1973. – Vol. 63, No. 4. – P. 445-449.

38. Vahimaa, P. Electromagnetic analysis of nonparaxia 1 Bessel beams generated by diffractive axicon, P.Vahimaa, V. Kettunen, M.Knittinm, J.Turunen, J.Opt.Soc.Am. A. – 1997. – V.4, -No. 8. – P. 1817-1824.

39. Kizuka, T. Characteristics of a laser beam spot focused by a binary diffractive axicon, T. Kizuka, M. Yamanchi, Y.Matsuoka, Opt.Eng. – 2008. – V.45. -No 5. – P.053401.

40. Kim, J.K. Compact all-fiber Bessel beam generator based on hollow optical fiber combined with a hybrid polymer fiber lens, J.K. Kim, J.Kim, Y.Jung, W.Ha, Y.S.Jeong, S.Lee, A.Tunnermann, K.Oh, Opt.Lett. – 2009. – V.34, - No. 19. – P.2973-2975.

41. Kurt, H. Limited-diffraction light propagation with axicon-shape photonic crystal, H. Kurt, J. Opt. Soc. Am. B. – 2009. – Vol. 26. -No 5. – P. 981-986

42. Chen, W. Realization of an evanescent Bessel beam via surface Plasmon interference exited by a radially polarized beam, W. Chen, Q.Zhan, Opt. Lett. – 2009. – V. 34. -No. 6. – P.722-724.

43. Watanabe, K. Localized surface plasmon microscope with an illumination system employing a radially polarized zeros-order Bessel beam, K.Watanabe, G. Terakedo, H.Kano, Opt.Lett. – 2009. – vol.34, no.8. – p.1180-1182.

44. Quabis, S. Focusing light to a tighter spot, S. Quabis, R. Dorn, M. Eberler, O. Glöckl, G. Leuchs, Opt. Commun. – 2000. – Vol.179. – P.1-7.

45. Mote, R.G. Subwavelength focusing behavior of high numerical-aperture phase Fresnel zone plates under various polarization states, R.G. Mote, S.F. Yu, W. Zhou, X.F. Li, Appl. Phys. Let. – 2009. – Vol. 95. – P. 191113.

46. Rajesh, K.B. Generation of sub-wavelength and super-resolution longitudinally polarized non-diffraction beam using lens axicon, K.B. Rajesh, P.M. Anbarasan, Chin. Opt. Lett. – 2008. – Vol. 6(10). – P. 785-787.

47. Wang, H. Creation of a needle of longitudinally polarized light in vacuum using binary optics, H. Wang, L. Shi, B. Lukyanchuk, C. Sheppard, C.T. Chong, Nature photonics. – 2008. – Vol.2. – P. 501-505.

48. Huang, K. Design of DOE for generating a needle of a strong longitudinally polarized field, K. Huang, P. Shi, X.-L. Kang, X. Zhang, Y.-P. Li, Opt. Lett. – 2010. – Vol. 35(7). – P. 965-967.

49. Sun, C.-C. Ultrasmall focusing spot with a long depth of focus based on polarization and phase modulation, C.-C. Sun and C.-K. Liu, Opt. Lett. – 2003. – Vol. 28(2). – P. 99-101.

50. Ohtake, Y. Sidelobe reduction of tightly focused radially higher-order Laguerre-Gaussian beams using annular masks, Y. Ohtake, T. Ando, T. Inoue, N. Matsumoto, H. Toyoda, Opt. Lett. – 2008. – Vol. 33(6). – P.617-619.

51. Youngworth, K.S. Focusing of high numerical aperture cylindrical vector beams, K.S. Youngworth, T.G. Brown, Opt. Expr. – 2000. – Vol. 7. – P. 77-87.

52. Huang, K. Vector-vortex Bessel–Gauss beams and their tightly focusing properties, K. Huang, P. Shi, G.W. Cao, K. Li, X.B. Zhang, Y.P. Li, Opt. Lett. – 2011. – Vol. 36(6). – P. 888-890.

53. Witkowska, A. All-fiber LP11 mode convertors, A. Witkowska, S. G. Leon-Saval, A. Pham, T. A. Birks, Opt. Lett. – 2008. – Vol. 33(4). – P.306-308.

54. Kozawa, Y. Generation of a radially polarized laser beam by use of a conical Brewster prizm, Y. Kozawa, S. Sato, Opt. Lett. – 2005. – Vol. 30(22). – P. 3063-3065.

55. Golub, I. Characterization of a refractive logarithmic axicon, I. Golub, B. Chebi, D.

Shaw, D. Nowacki, Opt. Lett. – 2010. – Vol. 35, No. 16. – P. 2828-2830.

56. Golub, M.A. Focusing light into a specified volume by computer-synthesized holo-gram, M.A. Golub, S.V. Karpeev, A.M. Prokhorov, I.N. Sisakyan, V.A. Soifer, Sov. Tech. Phys. Lett. – 1981. – Vol. 7. – P. 264-266.

57. Staronski, L.R. Lateral distribution and flow of energy in uniform-intensity axicon, L.R. Staronski, J. Sochacki, Z. Jroszewicz, A. Kolodziejcziwz, J. Opt. Soc. Am. A. – 1992. – Vol. 9, No. 11. – P. 2091-2094.

58. Khonina S.N., et al., Komp. Optika, 2009. – V. 33, No 4. – P. 427-435.

59. Handbook of Mathematical Functions, edited by M. Ab¬ramowitz, I.A. Stegun – Na-tional Bureau of Standards, Washington, DC, 1964.– 1044 p.

60. Kovalev A.A., et al., Komp. Optika, 2007. – V. 31, No 4. – 29-32.

61. Kotlyar, V.V. Family of hypergeometric laser beams, V.V. Kotlyar, A.A. Kovalev, J. Opt. Soc. Am. A. – 2008. – Vol. 25. – P. 262-270.

62. Prudnikov, A.P. Integrals and Series. Special functions, A.P. Prudnikov, Y.A. Brych-kov, O.I. Marichev. -M.: Science, 1983. – (in Russian)

63. Snyder, A. Optical waveguide theory, A. Snyder, J. Love. - M.: Radio and Commu-ni-cations, 1987. – (in Russian)

64. Osterberg H. Closed solutions of Rayleigh's diffraction integral for axial points, H. Osterberg, L.Smith, J.Opt.Soc.Am. – 1961. – V.51. – P.1050-1054.

65. Richards, B. Electromagnetic diffraction in optical systems II. Structure of the im-age field in an aplanatic system, B. Richards, E. Wolf, Proc. R. Soc. London A. – 1959. –Vol. 253. – P. 358-379.

66. Debay, P. Das Verhalten von Lichtwellen in der Nähe eines Brennpunktes oder einer Brennlinie, P. Debay, Ann. d. Phys. – 1909. – Vol. 335, N 14. – P. 755-776.

67. Fu, Y. Experimental investigation of superfocusing of plasmonic lens with chirped circular nanoslits, Y. Fu, Yu Liu, X. Zhou, Z. Xu, F. Fang, Opt. Exp. – 2010. – Vol. 18. – P. 3438-3443.

68. Fu, Y. Plasmonic microzone plate: Superfocusing at visible regime, Y. Fu, W. Zhou, L.E.N. Lim, C.L. Du, X.G. Luo, Appl. Phys. Let. – 2007. – Vol. 91. – P. 061124.

69. Mote, R.G. Near-field properties of zone plates in visible regime – New insights, R.G. Mote, S.F. Yu, B.K. Ng, W. Zhou, S.P. Lau, Opt. Express. – 2008. – Vol. 16. – P. 9554-9564.

70. Lopez, L.C. Vectorial diffraction analysis of near-field focusing of perfect black Fresnel zone plates under various polarization states, L.C. Lopez, M.P. Molina, P.A. Gon-zalez, S.B. Escarre, A.F. Gil, R.F. Madrigal, A.M. Cases, J. Light Technol. – 2011. – Vol. 29. –P. 822-829.

71. Fu, Y. Hybrid Au-Ag subwavelength metallic structures with variant periods for superfocusing, Y. Fu, W. Zhou,J. Nanophotonics. – 2009. – Vol. 3. – P. 033504.

72. Fu, Y. Experimental study of plasmonic structures with variant periods for sub-wavelength focusing: analyses of characterization errors, Y. Fu, R.G. Mote, Q. Wang, W. Zhou, J. Mod. Opt. – 2009. – Vol. 56(14). – P. 1550-1556.

73. Schonbrun, E. Scanning microscopy using a short-focallength Fresnel zone plate, E. Schonbrun, W.N. Ye, K.B. Crozier, Opt. Lett. – 2009. – Vol. 34(14). – P. 2228-2230.

74. Novotny, L. Principles of nano-optics, L. Novotny, B. Hecht. – Cambridge Univer-sity Press, 2006. – 539 p.

75. Michalski, K.A. Complex image method analysis of a plane wave-exited subwave-length circular aperture in a planar screen, K.A. Michalski, Prog. Electromag. Res. B. – 2011. – Vol. 27. – P. 253-272.

76. Wu, J.H. Modeling of near-field optical diffraction from a subwavelength aperture in a thin conducting film, J.H. Wu, Opt. Lett. – 2011. – Vol. 36, N 17. – P. 3440-3442.

77. Khonina, S.N. Controlling of the components of vectors of electrical and magnetic fields in a focus of lens with a high aperture with aim of binary phase structures, S.N. Khonina, S.G. Volotovsky, Computer optics, 2010. – V. 34. – No. 1. – PP. 58-68.

78. Lee, J.Y. Near-field focusing and magnification through self-assembled nanoscale spherical lenses, J.Y. Lee, B.H. Hong, W.Y. Kim, S.K. Min [and oth.], Nature, 2009. – V. 460,08173. – PP. 498-501.

79. Goldstein, D.J. Resolution in light microscopy studied by computer simulations, D.J. Godstein, J. Microsc., 1992. - V. 166. – PP. 185-197.

80. Bouhelier, A. Plasmon-coupled tip-enhanced near-field optical microscopy, A. Bouhelier, J. Renger, M.R. Beversluis, L. Novotny, J. Microsc., 2003. – V. 210. – PP. 220-224.

81. Karrai, K. Enhanced reflectivity contrast in confocal solid immersion lens microscopy, K. Karrai, X. Lorenz, Appl. Phys. Lett., 2000. – V. 77. – No. 21. – PP. 3459-3461.

82. Ippolito, S.B. High spatial resolution subsurface microscopy, S.B. Ippolito, B.B. Goldberg, M.S. Unlu, Appl. Phys. Lett., 2001. – V. 78. – No. 26. – PP. 4071-4073.

83. Koklu, F.H. Subsurface microscopy of integrated circuits with angular spectrum and polarization control, F.H. Koklu, S.B. Ippolito, B.B. Goldberg, M.S. Unlu, Opt. Lett., 2009. – V. 34. – No. 8. – PP. 1261-1263.

84. Karabacak, D.M. Diffraction of evanescent waves and nanomechanical displacement detection, D.M. Karabacak, K.L. Ekinci, C.H. Gan, G.J. Gbur [and oth.], Opt. Lett., 2007. – V. 32. – No. 13. – PP. 1881-1883.

85. Mason, D.R. Enchanced resolution beyond the Abbe diffraction limit with wavelength-scale solid immersion lenses, D.R. Mason, M.V. Jouravlev, K.S. Kim, Opt. Lett., 2010. – V. 35. – No. 12. – PP. 2007-2009.

86. Zverev, V.A. Radio optics. M.: Sov. Radio, 1975. – 304 p.

87. Kotlyar, V.V. Subwave light localization in waveguide structures, V.V. Kotlyar, A.A. Kovalev, Y.O. Shujupova, A.G. Nalimov [and oth.], Computer optics, 2010. – V. 34. – No. 2. – PP. 169-186.

88. Handmer, C.J. Blazing evanescent gtrating orders: a spectral approach to beating the Rayleigh limit, C.J. Handmer, C. Martijn de Sterke, R.C. McPhedran, L.S. Botten [and oth.], Opt. Lett., 2010. – V. 35. – No. 17. – PP. 2846-2848.

89. Thongrattanasiri, S. Analytical technique for subwavelength far field imaging, S. Thongrattanasiri, N.A. Kuhta, M.D. Escarra, A.J. Hoffman [and oth.], App. Phys. Lett., 2010. – V. 97. – PP. 101103.

90. Pendry, J.B. Negative refraction makes a perfect lens, J.B. Pendry, Phys. Rev. Lett. 2000. – Vol. 85(18). – P. 3966-3969.

91. Blaikie, R.G. Imaging through planar silver lenses in the optical near field, R.G. Blaikie, D.O.S. Melville, J. Opt. A: Pure Appl. Opt. 2005. – Vol. 7(2). – P. S176-S183.

92. Melvile, D.O.S. Super-resolution imaging through a planar silver layer, D.O.S. Melvile, R.J. Blaikie, Opt. Express 2005. – Vol. 13(6). – P. 2127-2134.

93. Fang, N. Sub-diffraction-limited optical imaging with a silver superlens, N. Fang, H. Lee, C. Sun, X. Zhang, Science 2005. – Vol. 308(5721). – P. 534-537.

94. Liu, Z. Far-field optical hyperlens magnifying sub-diffraction-limited object, Z. Liu, H. Lee, Y. Xiong, C. Sun, X. Zhang, Science 2007. – Vol. 315(5819). – P. 1686.

95. Thongrattanasiri, S. Hypergratings: nanophotonics in planar anisotropic metamaterials, S. Thongrattanasiri, V.A. Podolskiy, Opt. Lett. 2009. – Vol. 34(7). – P. 890-892.

96. Webb, K.J. Subwavelength imaging with a multilayer silver film structure, K.J. Webb, M. Yang, Opt. Lett. 2006. – Vol. 31(14). – P. 2130-2132.

97. Liu, H. Webb. Submevelength imaging opportunities with planar uniaxial anisotropic lenses, H. Liu, Shivananad, K.J. Webb, Opt. Lett. 2008. – Vol. 33(21). – P. 2568-2570.

98. Liu, H. Subwavelength imaging with nonmagnetic anisotropic bilayers, H. Liu, Shivanand, K.J. Webb, Opt. Lett. 2009. – Vol. 34(14). – P. 2243-2245.

99. Wang, X. Unrestricted superlensing in a triangular two-dimensional photonic crystal, X.Wang, Z.F. Ren, K. Kempa, Opt. Express 2004. – Vol. 12(13). – P. 2919-2924.

100. Casse, B.D.F. Imaging with subwavelength resolution by a generalized superlens at infrared wavelengths, B.D.F. Casse, W.T. Lu, R.K. Banyal, Y.J. Huang, S. Selvarasah, M.R. Dokmeci, C.H. Perry, S. Sridhar, Opt. Lett. 2009. – Vol. 34(13). – P. 1994-1996.

101. Tsukerman, I. Superfocusing by nanoshells, I. Tsukerman, Opt. Lett. 2009. – Vol. 34(7). – P. 1057-1059.

102. Ingrey, P.C. Perfect lens with not so perfect boundaries, P.C. Ingrey, K.I. Hopcraft, O. French, E. Jakeman, Opt. Lett. 2009. – Vol. 34(7). – P. 1015-1017.

103. Ray, E.A. Simple demonstration of visible evanescent-wave enhancement with far-field detection, E.A. Ray,M. J. Hampton, R. Lopez, Opt. Lett. 2009. – Vol. 34(13). – P. 2048-2050.

104. Cao, Z. Exact analytical method for planar optical waveguides with arbitrary index profile, Z. Cao, Y. Jiang, Q. Shen, X. Dou, Y. Chen, J. Opt. Soc. Am. A 1999. – Vol. 16(9). – P. 2209-2212.

105. Chung, M. General eigenvalue equations for optical planar waveguides with arbitrarily graded-index profiles, M. Chung, C. Kim, J. Lightwave Techn. 2000. – Vol. 18(6). – P. 878-885.

106. Born, M. Principles of optics, M. Born, E. Wolf – Moscow.: Nauka, 1973. – 719 p. – (in Russian).

107. Miller, W. Symmetry and Separation of Variables, W. Miller – Moscow.: Mir, 1981. – 342 p. – (in Russian).

108. Korn, G. Mathematical handbook, G. Korn, T. Korn – Moscow.: Nauka, 1968. – 720 p. – (in Russian).

109. Triandafilov, Y.R. Photonic-crystal Mikaelian lens, Y.R. Triandafilov, V.V. Kotlyar, Computer Optics. – 2007. – V. 31, N 3. – P. 27-31. – ISSN 0134-2452. – (in Russian).

110. He, J. Wavelength switchable semiconductor laser using half-wave V-coupled cavities, J. He, D. Liu, Opt. Express 2008. – Vol. 16(6). – P. 3896-3911.

111. Lin, X. Design and analysis of 2x2 half-wave waveguide couplers, X. Lin, D. Liu, J. He, Appl. Opt. 2009. – Vol. 48(25). – P. F18-F23.

112. Mikaelian, A.L. Application of stratified medium for waves focusing, A.L. Mikaelian, Doklady Akademii Nauk SSSR 1951. – Vol. 81. – P. 569–571.

113. Kotlyar, V.V. Abel's transform in tasks of synthesis of gradient-index optical elements, V.V. Kotlyar, A.S. Melekhin, Computer Optics. – 2001. – N 22. – P. 29-36. – ISSN 0134-2452. – (in Russian).

114. Kotlyar, V.V. Abel's transform for calculation of gradient-index optical elements with spherically-symmetric index distribution, V.V. Kotlyar, A.S. Melekhin, Computer Optics. – 2002. – N 24. – P. 48-52. – ISSN 0134-2452. – (in Russian).

115. Kotlyar, V.V. Calculation of Maxwell's, Fish eye and Iton-Lipman's generalized lenses, V.V. Kotlyar, A.S. Melekhin, Computer Optics. – 2002. – N 24. – P. 53-57. – ISSN 0134-2452. – (in Russian).

116. Alekseev, V.M. Optimal control, V.M. Alekseev, V.M. Tikhomirov, S.V. Fomin — Moscow.: Nauka, 1979 (in Russian).

117. Beliakov, G. Analysis of inhomogeneous optical systems by the use of ray tracing.I. Planar systems, G. Beliakov, D. Chan, Appl. Opt. 1997. – Vol. 36. – P. 5303-5309.

118. Mikaelian, A.L. Self-focusing media with variable index of refraction, A.L. Mikaelian, Prog. Opt. 1980. – Vol. 17. – P. 279–345.

119. Rho, J. Spherical hyperlens for two-dimensional sub-diffractional imaging at visible frequencies, J. Rho, Z. Ye, Y. Xiong, X. Yin, Z. Liu, H. Choi, G. Bartal, X. Zhang, Nature Communications 2010. – Vol. 1. – P. 443.

120. Kotlyar, V.V. et al., Komp. Optika 2012. - Т. 36, N. 3. - С. 327-332.

121. Kotlyar, V.V. High resolution through graded-index microoptics, V.V. Kotlyar, A.A. Kovalev, A.G. Nalimov, S.S. Stafeev, Advances in Optical Technologies, 2012. - V. 2012. - P. 1-9.

122. Chien, H.T. Focusing of electromagnetic waves by periodic arrays of air holes with gradually varying radii, H.T. Chien and C.C. Chen, Opt. Exp. - 2006. - V. 14. – P. 10759.

123. Chen, Z. Photonic nanojet enhancement of backscattering of light by nanoparticles: a potential novel visible-light ultramicroscopy technique, Z. Chen, A. Taflove, V. Backman, Opt. Exp. – 2004. – v.12. – p.1214.

124. Li, X. Optical analysis of nanoparticles via enhanced backscattering facilitated by 3-D photonic nanojets, X. Li, Z. Chen, A. Taflove, V. Backman, Opt. Exp. – 2005. – v.13. – p.526.

125. Ferrand, P. Direct imaging of photonic nanojets, P. Ferrand, J.Wenger, A. Devilez, M. Pianta,B. Stout, N. Bonod, E. Popov, H. Rigneault, Opt. Exp. – 2008. – v.16. – p.6930.

126. Kim, M.-S. Engineering photonic nanojets, M.-S. Kim, T. Scharf, S. Mühlig, C. Rockstuhl, H.P. Herzig, Optics Express. – 2011. – v.19. – p.10206.

127. McCloskey, D. Low divergence photonic nanojets from Si3N4 microdisks, D. Mc-Closkey, J.J. Wang, J.F. Donegan, Optics Express. – 2012. – Vol. 20. – p.128.

128. Kong, S.-C. Quasi one-dimensional light beam generated by a graded-index microsphere, S.-C. Kong, A. Taflove, V. Backman, Optics Express. – 2009. – v.17, p.3722.

129. Devilez, A. Three-dimensional subwavelength confinement of light with dielectric microspheres, A. Devilez, N. Bonod, J. Wenger, D. Gérard, B.S., H. Rigneault, E. Popov, Optics Express. – 2009. – v.17. – p.2089.

130. Yi, K.J. Enhanced Raman scattering by self-assembled silica spherical microparticles, K.J. Yi, H. Wang, Y.F. Lu, Z.Y. Yang, J. Appl. Phys. – 2007. – v.101. – p.063528.

131. Kong, S.-C. Photonic nanojet-enabled optical data storage, S.-C. Kong, A. Sahakian, A. Taflove, V. Backman, Opt. Exp. – 2008. – V.16. – p.13713.

132. McLeod, E. Subwavelength direct-write nanopatterning using optically trapped microspheres, E. McLeod, C.B. Arnold, Nature Nano. – 2008. – v.3. – p.413.

133. Wang, T. Subwavelength focusing by a microsphere array, T. Wang, C. Kuang, X. Hao, X. Liu, J. Opt. – 2011. – v.13. – p.035702.

4

Focusing vortex beams and overcoming the diffraction limit

Introduction

Objects with a vortex structure exist in a variety of areas of the material world: in the macrocosm (the spiral shape of galaxies and nebulae), in the microcosm (elementary particles, light fields) and in our daily lives (cyclones and anticyclones, tornadoes and typhoons). Their structure and behaviour have not as yet been exhaustively studied and represent a vast field for research. The section of optics dealing with the study of light beams with screw phase singularities (i.e. vortex laser beams) is called singular optics.

At the point of singularity the intensity of the light field is zero, and the phase is not defined. In the neighbourhood of this point there are abrupt phase changes.

Singular peculiarities in light fields can appear as they pass through randomly inhomogeneous and non-linear media. It is also possible to excite eddy fields in laser resonators and multimode optical fibres. The most simple and manageable method of forming the vortex fields is to use spiral diffractive optical elements (DOE), and also dynamic liquid crystal transparants (energy efficiency of the latter is still quite low). The simplest DOE is a spiral phase plate (SPP) and a helical axicon. Vortex laser beams have been the subject of numerous studies and publications by both Russian scientists and their foreign counterparts. There are now also studies of the properties of such beams based on Bessel, Laguerre–Gauss, hypergeometric and other modes.

The application of optical vortices is constantly expanding. In particular, in the tasks of nanophotonics it is proposed to use them to manipulate dielectric micro- and nano-objects. Also, more attention is paid to the study of the possibilities of using plasmonic effects as nano-tweezers. The use of optical vortices in photolithography can achieve resolutions of $\lambda/10$ (λ is the wavelength of light). It is possible to effectively use spiral optical structures even with a small number of quantization levels.

Other applications of optical vortices include, for example, interferometry: using a SPP placed in the plane of the spatial spectrum of the optical $4f$-system (f is the focal length of the spherical lens), a method of producing spiral interferograms was proposed. They can be used to readily distinguish convex and concave portions of the wavefront. Spiral filters are used for contrasting and relief images of phase objects of the nanometer size.

Using optical vortex interferometers which are based on the generation of light fields representing regular gratings or meshes of optical vortices (i.e., the measurements are taken on the node positions with the luminous intensity, not maximum) it is possible to determine the rotation angles within 0.03 arcseconds and measure the angles of inclination of the wavefront with an accuracy of 0.2 arcseconds.

SPP is also used in a star coronagraph, in which the light from a bright star is converted into a ring and stopped down, and the faint light from the planets of the star passes through the diaphragm and is recorded. It is known that the vortex waves in a coherent system have a well-defined phase which, however, is poorly defined in a partially coherent system. In the limit, for the completely incoherent light neither the helical phase nor zero intensity is observed. This allows the use of optical vortices for exclusion from the observation domain of the coherent radiation to enhance the incoherent signal. It is this effect that is used in the coronagraph.

Plates with a spiral phase relief have been successfully used for the optical performance of the radial Hilbert transform. Hilbert-optics, as well as shadow optics, close by transformation in the frequency domain, have been used successfully for image pre-processing and phase analysis. Hilbert spectroscopy allows to achieve nano-resolution in spectral analysis. Using the radial Hilbert transform, including the fractional one, the SPPs open new possibilities for solving the problems mentioned above.

The phase dislocations, determining the zero intensity, are a promising tool in metrology. Since the accuracy of determining the position of the dislocation is not limited by the classical diffraction limit (the gradient of the phase change in this case increases without bounds) and is limited only by the signal/noise ratio, the geometry of the object subject to the availability of *a priori* information about the object can be determined with very high accuracy. This approach is based on the method of optical vortex metrology successfully applied to optical vortex interferometers to track the displacement of objects with nanometer precision. The sensitivity of singular beams to changes of the wavefront and various defects can be used for testing of surfaces and analysis of optical systems.

In the non-linear optical media the optical vortices can be used to form waveguide structures and 'labyrinths', as well as for the study of various physical phenomena.

Of particular interest is the use of singular optics elements in sharp focusing when the combination of the polarization and phase characteristics leads to a variety of effects, including the diffraction limit overcoming in the far diffraction field.

4.1. Formation of vortex laser beams by using singular optics elements

The simplest elements of singular optics are the spiral phase plate and the spiral axicon (see Fig. 4.1.1). The theoretical discussion of the possibility of applying the SPP was presented in 1984 in the Doklady Akademii Nauk (Proceedings of the Academy of Sciences) [1]. The SPP was made for the first time and experimentally investigated in 1992 [2]. Until now, the manufacture of multilevel SPPs is associated with technological complexities, but often they

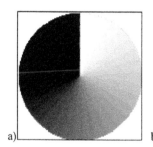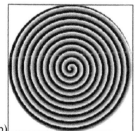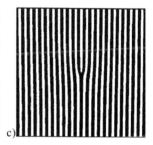

Fig. 4.1.1. Elements of singular optics: a spiral phase plate (a) and a helical axicon (b), the binary model with a 'fork' (c).

cannot be replaced by dynamic liquid crystal modulators due to the low efficiency and spatial resolution of the latter. Therefore, the encoding of the multilevel phase in the binary distribution is often carried out (see Fig. 4.1.1 c).

The properties of vortex beams are often considered on the basis of Bessel, Laguerre–Gauss, hypergeometric and other laser modes. And if the generation of higher modes and their superposition in the laser resonator require special skills and fine-tuning, the means of diffractive optics allow us to solve this problem by the most simple and flexible method [3].

This section examines the theoretical and numerical formation of the main types of paraxial vortex laser beams by using diffractive elements of singular optics.

4.1.1. Theoretical analysis of the diffraction of plane and Gaussian waves on the spiral phase plate

Complex transmission function of the spiral phase plate has the form:

$$f_n(r,\varphi) = f_n(\varphi) = \exp(in\varphi) \tag{4.1.1}$$

Fresnel transform that describes the propagation of light in free space, of a light field (4.1.1) is:

$$F_n(\rho,\theta,z) = \frac{(-i)^{n+1}k}{z} \exp\left(in\theta + \frac{ik}{2z}\rho^2\right) \int_0^\infty \exp\left(\frac{ik}{2z}r^2\right) J_n\left(\frac{k}{z}r\rho\right) r\,dr, \tag{4.1.2}$$

where $k = 2\pi/\lambda$ is the wave number of the light; z is the coordinate along the optical axis; (ρ, θ) are the transverse polar coordinates in the transverse plane at a distance z; $J_n(x)$ is the Bessel function of the first kind of n-th order.

The integral in expression (4.1.2) can be calculated [4], and instead of (4.1.2) we get:

$$F_n(\rho,\theta,z) = (-i)^{n+1}\sqrt{\frac{\pi}{2}}\sqrt{\frac{k\rho^2}{4z}} \exp\left(in\theta + \frac{i\pi n}{4} + \frac{ik}{4z}\rho^2\right)\left\{iJ_{\frac{n-1}{2}}\left(\frac{k\rho^2}{4z}\right) + J_{\frac{n+1}{2}}\left(\frac{k\rho^2}{4z}\right)\right\}. \tag{4.1.3}$$

Equation (4.1.3) shows that at $\rho = 0$ the field at any z (except $z = 0$) is zero. The field is the result of interference between two divergent light beams such as Bessel modes. But, in contrast to the Bessel mode beams, the order of the Bessel functions and the order of the angular harmonic (phase singularity) in equation (4.1.3) are not matched:

$$J_{\frac{n\pm 1}{2}}(x)\exp(in\theta)$$

For this reason, the resulting interference of the two beams of the Bessel type must change with the growth z not only in the scale, but also the structure. However, this does not happen due to the fact that the function $F_n(\rho,\theta,z)$ in equation (4.1.3) is homogeneous, since it depends only on the combination of variables ρ^2/z. That is, the light field (4.1.3) is a paraxial mode of the free space, as while spreading, it retains its structure and changes only the scale, similar to the usual Gaussian modes. Also the field (4.1.3) is a solution of the paraxial wave equation satisfying the boundary conditions (4.1.1) with $z = 0$.

Equation (4.1.3) leads to the expression for the intensity distribution:

$$\hat{I}_n(\rho,z)=\left|F_n(\rho,\theta,z)\right|^2 =\frac{\pi}{2}x\left\{J_{\frac{n-1}{2}}^2(x)+J_{\frac{n+1}{2}}^2(x)\right\} \qquad (4.1.4)$$

where $x=k\rho^2/(4z)$

Let us find the equation for determining the local maxima of intensity (4.1.4). The derivative function (4.1.4) with respect to the variable ρ has the form:

$$\frac{d\hat{I}_n(\rho,z)}{d\rho}=\frac{\pi k\rho n}{4z}\left\{J_{\frac{n-1}{2}}^2\left(\frac{k\rho^2}{4z}\right)-J_{\frac{n+1}{2}}^2\left(\frac{k\rho^2}{4z}\right)\right\}. \qquad (4.1.5)$$

Equating the right-hand side of equation (4.1.5) to zero, we obtain the equation for finding the local maxima:

$$\left|J_{\frac{n-1}{2}}(x)\right|=\left|J_{\frac{n+1}{2}}(x)\right|. \qquad (4.1.6)$$

Informative is the position of only the first maximum that determines the radius of the light 'funnel', where the intensity drops from a maximum to zero at $\rho = 0$. Equation (4.1.6) cannot be used to determined analytically exactly the coordinate of the first maximum, but can be roughly estimated. For example at $z \to 0$ the arguments of functions in the equation (4.1.6) with a constant ρ will tend to infinity. Therefore, we can use the asymptotic behaviour of the Bessel functions [5]:

$$J_\nu(x) \approx \sqrt{\frac{2}{\pi x}} \cos\left(x - \frac{\nu\pi}{2} - \frac{\pi}{4}\right)$$ (4.1.7)

when $x \to \infty$.

Equation (4.1.6) for the first maximum can be rewritten as:

$$\text{tg}\left[x - (n-1)\frac{\pi}{4} - \frac{\pi}{4}\right] = 1,$$ (4.1.8)

from which the estimate for the radius of the light 'funnel' ($n > 0$):

$$\rho_n \approx \sqrt{\frac{(n+1)\lambda z}{2}}$$ (4.1.9)

Equation (4.1.9) shows that the size of the 'funnel' can be smaller than the wavelength at

$$z < \frac{2\lambda}{n+1}$$ (4.1.10)

i.e. at a distance from the plane $z = 0$ smaller than the wavelength.

For specific orders of the singularity $n = 1, 2$ it is possible to obtain estimates of the radius of the 'funnel' for any z. When $n = 1$ the equation (4.1.6) takes the form:

$$J_0(x) = J_1(x),$$ (4.1.11)

from which the estimate for the radius of the first light ring is:

$$\rho_1 \approx 0.94\sqrt{\lambda z}$$ (4.1.12)

For $n = 2$ instead of the equation (4.1.6), we obtain

$$\text{ctg}(x) = \frac{1-x}{x}$$ (4.1.13)

from which the estimate for the first light ring:

$$\rho_2 \approx 1.13\sqrt{\lambda z}$$ (4.1.14)

From (4.1.9), correct only if $z \to 0$, at $n = 1$ and $n = 2$ we get

$$\rho_1 \approx \sqrt{\lambda z}$$ (4.1.15)

$$\rho_2 \approx 1.22\sqrt{\lambda z}$$ (4.1.16)

Figure 4.1.2 shows the distribution of the intensity of a plane wave with a singularity of the first (a) and second (b) orders at a distance of 10 mm from the plane $z = 0$ and wavelength 0.633 μm.

The numerical plotting of the graph using formula (4.1.4) shows that the radius of the first light ring or radius of the 'funnel' is $\rho_1 \approx 0.076$ mm, and the formula (4.1.12) gives 0.074 mm, and the formula (4.1.15) gives 0.0795 mm. For $n = 2$ graphically $\rho_2 \approx 0.091$ mm, the value obtained by the formula (4.1.14) is 0.0899 mm, and by the formula (4.1.16) 0.097 mm.

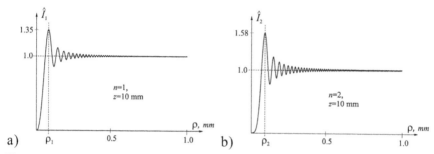

a)   b)

Fig. 4.1.2. The intensity distribution of a plane wave with a singularity of the first (a) and second (b) orders at a distance of 10 mm from the plane $z = 0$.

We obtain expressions for the intensity (4.1.4) at $\rho \to \infty$ and $\rho \to 0$. For this we use the asymptotic behaviour of Bessel functions (4.1.7) with $\rho \to \infty$, and

$$J_v(x) \approx \left(\frac{x}{2}\right)^v \Gamma^{-1}(v+1) \qquad (4.1.17)$$

in which $\rho \to 0$, $\Gamma(x)$ is the the gamma function.

From (4.1.7) it follows that the intensity of a plane wave with a singularity at $\rho \to \infty$ tends to unity:

$$\hat{I}_n(\rho,z) \approx \frac{\pi}{2} x \left\{ \frac{2}{\pi x} \left[\cos^2\left(x - \frac{(n-1)\pi}{4} - \frac{\pi}{4}\right) + \cos^2\left(x - \frac{(n+1)\pi}{4} - \frac{\pi}{4}\right) \right] \right\} = 1.$$

$$(4.1.18)$$

Figure 4.1.2 shows this clearly.

From (4.1.17) it follows that at $\rho \to 0$ intensity (4.1.4) can be approximated by the expression:

$$\hat{I}_n(\rho,z) \approx \frac{\pi}{\Gamma\left(\dfrac{n+1}{2}\right)} \left(\frac{k\rho^2}{8z}\right)^n \qquad (4.1.19)$$

Equation (4.1.19) shows that with an increase in the order of singularity the 'funnel walls' become more vertical, i.e., tend to a 'step' for the same z (Fig. 4.1.3).

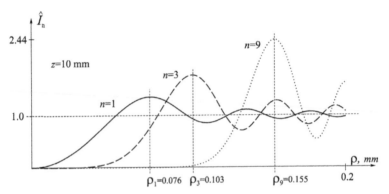

Fig. 4.1.3. The intensity distribution of a plane wave with the singularity $n = 1, 3, 9$ orders at a distance of 10 mm.

Note that the equation (4.1.4) does not describe the behaviour of the intensity in the far-field diffraction zone, because at $z \to \infty$ equation (4.1.4) gives the approximation (4.1.19). That is, it turns out that when $z \to \infty$ the radius of the 'funnel' (area with low intensity) tends to infinity, and the intensity at any fixed ρ tends to zero.

In order to evaluate the behaviour of a plane wave with a singularity in the far field, it is necessary to limit the plane wave at $z = 0$ by a certain circle of radius R. Then, instead of equation (4.1.1) we should consider the light field with the amplitude

$$F(r,\varphi) = \mathrm{circl}\left(\frac{r}{R}\right)\exp(in\varphi) = \begin{cases} \exp(in\varphi), & r \le R, \\ 0, & r > R. \end{cases} \qquad (4.1.20)$$

Analytical study of the Fraunhofer diffraction of the light field (4.1.20) is described later, but the Fresnel diffraction cannot be studied analytically. But if instead of a function circl(x) we use a Gaussian function, that is carry out a 'soft' restriction of the plane wave

$$F_n^0(r,\varphi) = \exp\left[-\left(\frac{r}{w}\right)^2 + in\varphi\right]_0, \qquad (4.1.21)$$

we can get some analytical expressions.

Consider the Fresnel diffraction of the Gaussian beam at the SPP. At the distance z the complex amplitude of the light field in the paraxial approximation has the form [6]:

$$E_n(\rho,\theta,z) = \frac{(-i)k}{2\pi z} \int_0^\infty \int_0^{2\pi} E_n^0(r,\varphi) \exp\left\{\frac{ik}{2z}\left[r^2 + \rho^2 - 2r\rho\cos(\varphi-\theta)\right]\right\} r\,dr\,d\varphi =$$

$$= \frac{(-i)^{n+1}\sqrt{\pi}}{2}\left(\frac{z_0}{z}\right)^2\left(\frac{\rho}{w}\right)\left[1+\left(\frac{z_0}{z}\right)^2\right]^{-\frac{3}{4}} \exp\left[i\frac{3}{2}\mathrm{arctg}\left(\frac{z_0}{z}\right) + i\frac{k\rho^2}{2R(z)} - \frac{\rho^2}{w^2(z)} + in\theta\right] \times$$

$$\times \left\{I_{\frac{n-1}{2}}\left[\rho^2\left(\frac{1}{w^2(z)} + \frac{ik}{2R_0(z)}\right)\right] - I_{\frac{n+1}{2}}\left[\rho^2\left(\frac{1}{w^2(z)} + \frac{ik}{2R_0(z)}\right)\right]\right\},$$

$$(4.1.22)$$

where

$$w^2(z) = 2w^2\left[1 + \left(\frac{z}{z_0}\right)^2\right]$$

$$R(z) = 2z\left[1 + \left(\frac{z_0}{z}\right)^2\right] \cdot \left[2 + \left(\frac{z_0}{z}\right)^2\right]^{-1}$$

$$R_0(z) = 2z\left[1 + \left(\frac{z}{z_0}\right)^2\right]$$

$$z_0 = \frac{kw^2}{2}$$

$$E_n^0(r,\varphi) = \exp\left[-\left(\frac{r}{w}\right)^2 + in\varphi\right]_0$$

$I_v(x)$ is the Bessel function of the second kind of vth order.

Transferring from the diffraction in the Fresnel zone to the far field we obtain an expression for the Fraunhofer diffraction of Gaussian beam on SPP [7].

At $z \gg z_0$

$$w^2(z) \approx 2w^2 \frac{z^2}{z_0^2}$$

$$R(z) \approx z$$

$$R_0(z) \approx \frac{2z^3}{z_0^2}$$

$$E_n(\rho,\theta,z \to \infty) = \frac{(-i)^{n+1}\sqrt{\pi}}{2}\left(\frac{z_0}{z}\right)^2\left(\frac{\rho}{w}\right) \times$$

$$\times \exp(in\theta)\exp\left[-\frac{\rho^2}{w^2(z)}\right]\left\{I_{\frac{n-1}{2}}\left(\frac{\rho^2}{w^2(z)}\right) - I_{\frac{n+1}{2}}\left(\frac{\rho^2}{w^2(z)}\right)\right\}. \quad (4.1.23)$$

From (4.1.23) we obtain an expression for the intensity of the Gaussian beam with a phase singularity in the far-field diffraction zone:

$$\hat{I}_n(\rho,z \to \infty) \approx \frac{\pi}{4}\left(\frac{z_0}{z}\right)^4\left(\frac{\rho}{w}\right)^2\exp\left[-\frac{2\rho^2}{w^2(z)}\right]\left\{I_{\frac{n-1}{2}}\left(\frac{\rho^2}{w^2(z)}\right) - I_{\frac{n+1}{2}}\left(\frac{\rho^2}{w^2(z)}\right)\right\}^2 =$$

$$= \frac{\pi}{2}\left(\frac{z_0}{z}\right)^2 x\exp(-2x)\left\{I_{\frac{n-1}{2}}(x) - I_{\frac{n+1}{2}}(x)\right\}^2,$$

$$\quad (4.1.24)$$

where $x = \left(\dfrac{\rho z_0}{\sqrt{2}\,zw}\right)^2$

The Fraunhofer diffraction pattern is formed in the focal plane of a spherical lens.

Consider the original function of the form:

$$f'_n(r,\theta) = \exp\left(-\frac{r^2}{w^2} + in\theta\right) \quad (4.1.25)$$

where w is the radius of the Gaussian beam waist. Then, the complex amplitude of the Fraunhofer diffraction of the Gaussian beam in the waist on the SPP will be described by the expression:

$$F'_n(\rho,\varphi) = \frac{(-i)^{n+1}k}{f}\exp(in\varphi)\int_0^{\infty}\exp\left(-\frac{r^2}{w^2}\right)J_n\left(\frac{k}{f}r\rho\right)rdr \quad (4.1.26)$$

We know the reference integral [4]:

$$\int_0^\infty \exp(-px^2) J_n(cx) x\, dx = \frac{c\sqrt{\pi}}{8p^{3/2}} \exp\left(-\frac{c^2}{8p}\right)\left[I_{(n-1)/2}\left(\frac{c^2}{8p}\right) - I_{(n+1)/2}\left(\frac{c^2}{8p}\right)\right],$$

(4.1.27)

where $I_v(x)$ is the modified Bessel function or the Bessel function of the second kind. In view of (4.1.27), the expression (4.1.26) has the form:

$$F'_n(\rho,\varphi) = (-i)^{n+1} \exp(in\varphi)\left(\frac{kw^2}{4f}\right)\sqrt{2\pi x}\exp(-x)\left[I_{(n-1)/2}(x) - I_{(n+1)/2}(x)\right],$$

(4.1.28)

where $x = \dfrac{1}{2}\left(\dfrac{kw\rho}{2f}\right)^2$.

The function of the intensity of the Fraunhofer diffraction pattern of the Gaussian beam on the SPP is:

$$\overline{I}_n(\rho) = |F'_n(\rho,\varphi)|^2 = 2\pi\left(\frac{kw^2}{4f}\right)^2 x\exp(-2x)\left[I_{(n-1)/2}(x) - I_{(n+1)/2}(x)\right]^2.$$

(4.1.29)

Equation (4.1.29) shows that at $x = 0$ at the centre of the Fourier plane the intensity is zero $(n \neq 0)$: $\overline{I}'_n(0) = 0$. The factors $x\exp(-2x)$ in the equation (4.1.29) show that in the far field the intensity distribution is annular. The radius of the ring can be found from the equation:

$$(n-4x)I_{(n-1)/2}(x) + (n+4x)I_{(n+1)/2}(x) = 0$$

(4.1.30)

Let us find the form of the function of intensity on the outside of the ring when $\rho \to \infty$ (or $x \to \infty$).

For this we use the asymptotic behaviour of the Bessel function [5]:

$$I_v(x) \approx \frac{\exp(x)}{\sqrt{2\pi x}}\left(1 - \frac{4v^2 - 1}{8x}\right), \quad x \gg 1$$

(4.1.31)

Then, instead of (4.1.29) when $x \to \infty$ we get:

$$\overline{I}'_n(\rho) \approx \left(\frac{nf}{k\rho^2}\right)^2$$

(4.1.32)

Note that the expression (4.1.32) can be obtained from the equation (4.1.29), letting the Gaussian beam radius $w \to \infty$ to infinity for fixed ρ.

We find the type of function of intensity on the inner side of the ring. When ρ tends to zero (for fixed w), the argument of the Bessel function x also tends to zero, and we can use the first terms of the expansion of the cylindrical functions:

$$I_v(x) \approx \left(\frac{x}{2}\right)^v \Gamma^{-1}(v+1), \quad x \ll 1 \tag{4.1.33}$$

where $\Gamma(x)$ is the the gamma function. Then, instead of (4.1.29) when $\rho \to 0$ we get:

$$\overline{I}'_n(\rho) \approx \pi \Gamma^{-2}\left(\frac{n+1}{2}\right)\left(\frac{kw^2}{f}\right)\left(\frac{kw\rho}{4f}\right)^{2n} \tag{4.1.34}$$

Equation (4.1.34) that the intensity near the centre of the Fourier plane increases as the $2n$ degree of the radial coordinate:

$$\overline{I}'_n(\rho) \approx (w\rho)^{2n}, \quad \rho \ll 1 \cdot \tag{4.1.35}$$

If tending ρ zero is accompanied by tensing the Gaussian beam radius w to infinity so that their product $w\rho$ remains constant, then the equation (4.1.34) shows that the intensity near the centre of the Fourier plane will tend to infinity in proportion to the square of the radius of the waist:

$$\overline{I}'_n(\rho \to 0, w \to \infty) \approx w^2, \quad \rho w = \text{const} \tag{4.1.36}$$

but in the most central point at $\rho = 0$ the intensity will be zero, $\overline{I}'_n(\rho = 0) = 0$ if any w.

4.1.2. Theoretical analysis of the diffraction of plane and Gaussian waves on a spiral axicon

The Fraunhofer diffraction of a plane wave on a finite radius spiral axicon with the transmission function $\text{circl}(r/R)\exp(i\alpha r)$ is described as follows:

$$F(\rho) = \int_0^R \exp(i\alpha r) J_n\left(\frac{k}{f}\overline{\rho}r\right) r \, dr \tag{4.1.37}$$

Consider the integral:

$$I = \int_0^R \exp(i\alpha r) J_n(\rho r) r \, dr, \quad \rho = \frac{k}{f} \bar{\rho} \qquad (4.1.38)$$

Using the integral representation of the Bessel function

$$J_n(x) = \frac{(-i)^n}{2\pi} \int_0^{2\pi} \exp(in\varphi) \exp(ix\cos\varphi) \, d\varphi \qquad (4.1.39)$$

We obtain [8]:

$$I = \frac{(-i)^{n+2}}{2\pi} \frac{\partial}{\partial \alpha} \left[\exp(i\alpha R) \int_0^{2\pi} \exp(in\varphi) \frac{\exp(iR\rho\cos\varphi)}{\alpha + \rho\cos\varphi} \, d\varphi - \int_0^{2\pi} \frac{\exp(in\varphi) \, d\varphi}{\alpha + \rho\cos\varphi} \right].$$

$$(4.1.40)$$

Using the relation

$$\exp(ix\cos\varphi) = \sum_{m=-\infty}^{+\infty} i^m \exp(-im\varphi) J_m(x), \qquad (4.1.41)$$

instead of the integral (4.1.40) we obtain an expression for the diffraction in the form of a series:

$$I = \frac{(-i)^n}{2\pi} \frac{\partial I_1^n}{\partial \alpha} - \frac{\exp(i\alpha R)}{2\pi} \sum_{m=-\infty}^{+\infty} i^m \left(iRI_1^m + \frac{\partial I_1^m}{\partial \alpha} \right) J_{m+n}(R\rho) \quad (4.1.42)$$

where

$$I_1^n = \int_0^{2\pi} \frac{\exp(in\varphi) \, d\varphi}{\alpha + \rho\cos\varphi}. \qquad (4.1.43)$$

The integrals (1.4.43) and their derivatives are calculated using the theory of residues. Expressions for the integrals I_1^n and $\partial I_1^n / \partial \alpha$ are listed below.

Case 1. $0 < \rho < |\alpha|$

$$I_1^n = \frac{2\pi \, \mathrm{sgn}\,\alpha}{\sqrt{\alpha^2 - \rho^2}} \left(\frac{-\alpha + \mathrm{sgn}\,\alpha \sqrt{\alpha^2 - \rho^2}}{\rho} \right)^{|n|} \qquad (4.1.44)$$

$$\frac{\partial I_1^n}{\partial \alpha} = -2\pi \, \mathrm{sgn}\,\alpha \left(\frac{-\alpha + \mathrm{sgn}\,\alpha \sqrt{\alpha^2 - \rho^2}}{\rho} \right)^{|n|} \frac{\alpha + \mathrm{sgn}\,\alpha |n| \sqrt{\alpha^2 - \rho^2}}{(\alpha^2 - \rho^2)^{\frac{3}{2}}} \qquad (4.1.45)$$

Case 2: $\rho > |\alpha|$

$$I_1^n = \frac{\pi i}{\sqrt{\rho^2 - \alpha^2}}\left[\left(\frac{-\alpha - i\sqrt{\rho^2 - \alpha^2}}{\rho}\right)^{|n|} - \left(\frac{-\alpha + i\sqrt{\rho^2 - \alpha^2}}{\rho}\right)^{|n|}\right] = \pi i \frac{\chi^{*|n|} - \chi^{|n|}}{\sqrt{\rho^2 - \alpha^2}}$$

(4.1.46)

where $\chi = \left[-\alpha + i\left(\rho^2 - \alpha^2\right)^{1/2}\right]/\rho$.

$$\frac{\partial I_1^n}{\partial \alpha} = \pi i \left[\alpha\left(\rho^2 - \alpha^2\right)^{-\frac{3}{2}}\left(\chi^{*|n|} - \chi^{|n|}\right) - i\left(\rho^2 - \alpha^2\right)^{-1}|n|\left(\chi^{*|n|} + \chi^{|n|}\right)\right].$$ (4.1.47)

Consider the scalar paraxial diffraction of a collimated Gaussian beam with the complex amplitude

$$E_0(r) = \exp\left(-\frac{r^2}{w^2}\right)$$

(4.1.48)

where w is the radius of the Gaussian beam waist.

The spiral axicon to which the approximation of a thin transparant is described by the transmission function of the type

$$\tau_n(r, \varphi) = \exp(i\alpha r + in\varphi)$$

(4.1.49)

(r, φ) are the polar coordinates in the plane of the spiral axicon when $z = 0$, z is the optical axis, α is the axicon parameter, $n = 0, \pm1, \pm2, \ldots$ is the number of the SPP.

Then the paraxial diffraction of the waves (4.1.48) on the spiral axison (4.1.49) is described by the Fresnel transform:

$$F_n(\rho, \theta, z) = -\frac{ik}{2\pi z}\exp\left(ikz + \frac{ik\rho^2}{2z}\right) \times$$

$$\times \int_0^R \int_0^{2\pi} \exp\left[-\frac{r^2}{w^2} + i\alpha r + in\varphi + \frac{ikr^2}{2z} - \frac{ik}{z}\rho r\cos(\varphi - \theta)\right] r\,dr\,d\varphi,$$

(4.1.50)

wherein (ρ, θ) are the polar coordinates in the plane z (z is the optical axis), $k = 2\pi/\lambda$ is the wavenumber. With the reference integral [4]

$$\int_0^\infty x^{\lambda+1} \exp\left(-px^2\right) J_\nu(cx)\,dx =$$

$$= \frac{c^\nu p^{-(\nu+\lambda+2)/2}}{2^{\nu+1}\nu!}\Gamma\left(\frac{\nu + \lambda + 2}{2}\right){}_1F_1\left[\frac{\nu + \lambda + 2}{2}, \nu + 1, -\left(\frac{c}{2\sqrt{p}}\right)^2\right],$$ (4.1.51)

instead of (4.1.50), we obtain:

$$F_n(\rho,\theta,z) = \frac{(-i)^{n+1} k}{z} \exp\left[in\theta + ikz + \frac{ik\rho^2}{2z} \right] \times$$

$$\times \left(\frac{k\rho}{2z}\right)^n \frac{\gamma^{-(n+2)/2}}{2^{n+1} n!} \sum_{m=0}^{\infty} \frac{(i\alpha)^m \gamma^{-m/2}}{m!} \Gamma\left(\frac{m+n+2}{2}\right) {}_1F_1\left[\frac{m+n+2}{2}, n+1, -\left(\frac{k\rho}{2z\sqrt{\gamma}}\right)^2\right],$$

$$(4.1.52)$$

where $\gamma = 1/w^2 - ik/(2z)$, ${}_1F_1(a,b,x)$ is the degenerate or confluent hypergeometric function:

$$_1F_1(a,b,x) = \sum_{m=0}^{\infty} \frac{(a)_m x^m}{(b)_m m!}, \qquad (4.1.53)$$

$(a)_m = \Gamma(a+m)/\Gamma(a)$, $(a)_0 = 1$, and $\Gamma(x)$ is the gamma function.

From (4.1.52) it follows that the diffraction pattern is a set of concentric rings. At $\rho = 0$ in the centre of the diffraction pattern at any $n \neq 0$ the intensity would be zero. Since the complex amplitude (4.1.52) depends on a combination of variables $k\rho/(2z\sqrt{\gamma})$, the radii ρ_l of the local maxima and minima of the diffraction pattern should satisfy the expression:

$$\rho_l = \frac{wza_l}{z_0}\left(1 + \frac{z_0^2}{z^2}\right)^{1/4} \qquad (4.1.54)$$

where a_l are the constants depending only on the number of the ring $l = 1, 2,...$ the diffraction pattern and the parameter α, $z_0 = kw^2/2$ is the Rayleigh length.

When $\alpha = 0$ (i.e. no axicon), from (1.4.52) we obtain the ratio for the complex amplitude of Fresnel diffraction of the Gaussian beam on the SPP:

$$F_n(\rho,\theta,z,\alpha=0) = \frac{(-i)^{n+1} k}{z} \exp\left[i(n\theta + kz) + \frac{ik\rho^2}{2z} \right]\left(\frac{k\rho}{2z}\right)^n \times$$

$$\times \frac{\gamma^{-(n+2)/2}}{2^{n+1} n!} \Gamma\left(\frac{n+2}{2}\right) {}_1F_1\left[\frac{n+2}{2}, n+1, -\left(\frac{k\rho}{2z\sqrt{\gamma}}\right)^2\right]. \qquad (4.1.55)$$

Given the connection between the hypergeometric and Bessel functions:

$$J_{(n-1)/2}(x) = \exp(-ix)\left(\frac{x}{2}\right)^{(n-1)/2} \Gamma^{-1}\left(\frac{n-1}{2}\right) {}_1F_1\left(\frac{n}{2}, n; 2ix\right) \quad (4.1.56)$$

and a recurrence relation for the hypergeometric functions

$$ {}_1F_1\left(\frac{n}{2}, n+1; 2ix\right) = \left(i\frac{d}{dx} + 2\right) {}_1F_1\left(\frac{n}{2}, n; 2ix\right) \quad (4.1.57)$$

we can instead of (4.1.55) get the well-known expression for the Fresnel diffraction of the Gaussian beam on the SPP:

$$E_n(\rho,\theta,z,\alpha=0) = \frac{(-i)^{n+1}\sqrt{\pi}}{2}\left(\frac{z_0}{z}\right)^2 \left(\frac{\rho}{w}\right)\left[1+\left(\frac{z_0}{z}\right)^2\right]^{-3/4} \times$$

$$\times \exp\left[i\frac{3}{2}\tan^{-1}\left(\frac{z_0}{z}\right) - i\frac{k\rho^2}{2R_0(z)} + i\frac{k\rho^2}{2z} - \frac{\rho^2}{w^2(z)} + in\theta + ikz\right] \times$$

$$\times \left\{ I_{\frac{n-1}{2}}\left[\rho^2\left(\frac{1}{w^2(z)} + \frac{ik}{2R_0(z)}\right)\right] - I_{\frac{n+1}{2}}\left[\rho^2\left(\frac{1}{w^2(z)} + \frac{ik}{2R_0(z)}\right)\right]\right\},$$

$$(4.1.58)$$

where $w^2(z) = 2w^2[1+(z/z_0)^2]$, $R_0(z) = 2z[1+(z/z_0)^2]$, $I_\nu(x)$ is the Bessel function of the second kind and the νth order.

When $z \to \infty$ ($z \gg z_0$) from the expression (1.4.52) we obtain the relation for the complex amplitude of the Fraunhofer diffraction of the Gaussian beam on the spiral axicon ($\gamma = 1/w^2$):

$$F_n(\rho,\theta,z\to\infty) = \frac{(-i)^{n+1}}{2^n n!z}z_0\exp\left(in\theta + ikz + \frac{ik\rho^2}{2z}\right)\left(\frac{z_0\rho}{zw}\right)^n \times$$

$$\times \sum_{m=0}^{\infty}\frac{(i\alpha w)^m}{m!}\Gamma\left(\frac{m+n+2}{2}\right){}_1F_1\left[\frac{m+n+2}{2}, n+1, -\left(\frac{z_0\rho}{zw}\right)^2\right].$$

$$(4.1.59)$$

When $\alpha = 0$ (i.e. no axicon) and $z\to\infty$ ($z \gg z_0$) from (1.4.52) follows the expression for the complex amplitude of the Fraunhofer diffraction of the Gaussian beam on the SPP:

$$F_n(\rho,\theta,z\to\infty,\alpha=0) =$$

$$= \frac{(-i)^{n+1}}{2^n n!z}z_0\exp\left(in\theta + ikz + \frac{ik\rho^2}{2z}\right)\left(\frac{z_0\rho}{zw}\right)^n\Gamma\left(\frac{n+2}{2}\right){}_1F_1\left[\frac{n+2}{2}, n+1, -\left(\frac{z_0\rho}{zw}\right)^2\right].$$

$$(4.1.60)$$

It is interesting to compare the expression (4.1.60) with the complex amplitude of the Fraunhofer diffraction of the limited flat wave of radius R on the SPP, when the focal length of the spherical lens is f:

$$E_n(\rho,\theta) = \frac{(-i)^{n+1}\exp(in\theta + ikz)}{(n+2)n!}\left(\frac{kR^2}{f}\right)\left(\frac{kR\rho}{2f}\right)^n {}_1F_2\left[\frac{n+2}{2},\frac{n+4}{2},n+1,-\left(\frac{kR\rho}{2f}\right)^2\right],$$

(4.1.61)

where ${}_1F_2(a,b,c,x)$ is the hypergeometric function:

$$_1F_2(a,b,c,x) = \sum_{m=0}^{\infty}\frac{(a)_m x^m}{(b)_m (c)_m m!}.$$

(4.1.62)

4.1.3. Numerical modeling of the diffraction of different beams on elements of singular optics

Tables 4.1.1 and 4.1.2 show the results of paraxial simulation of diffraction of different beams on the elements of singular optics using the Fourier and Fresnel transforms. The following parameters of the illuminating beams were chosen: wavelength of incident light $\lambda = 0.63 \cdot 10^{-6}$ m, the radius of the circular aperture $R = 1$ mm, the Gaussian beam radius $\sigma = 0.7234$ mm.

Table 4.1.1. Diffraction of different beams on the SPP of the second order in the Fresnel and Fraunhofer zones

Distance from element	Bounded flat beam	Gaussian beam

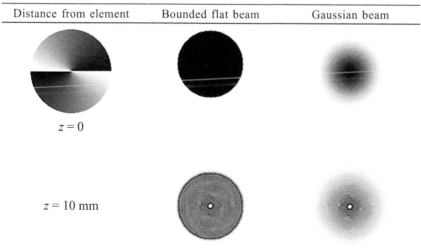

$z = 0$

$z = 10$ mm

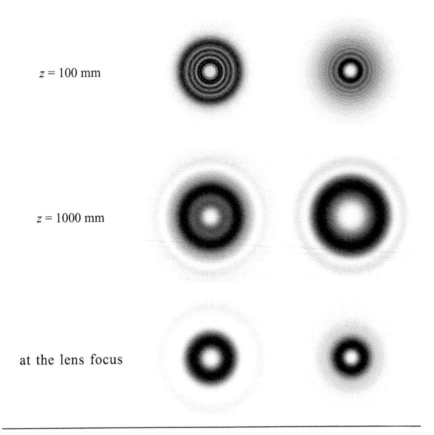

$z = 100$ mm

$z = 1000$ mm

at the lens focus

Table. 4.1.2. Diffraction beams on the various elements $T(\varphi) = \arg\left[\sum_p \exp(in_p\varphi)\right]$, in a zone of Fresnel and Fraunhofer.

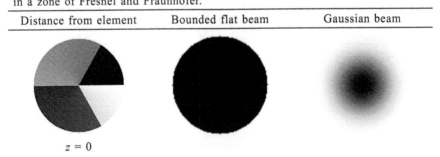

Distance from element	Bounded flat beam	Gaussian beam

$z = 0$

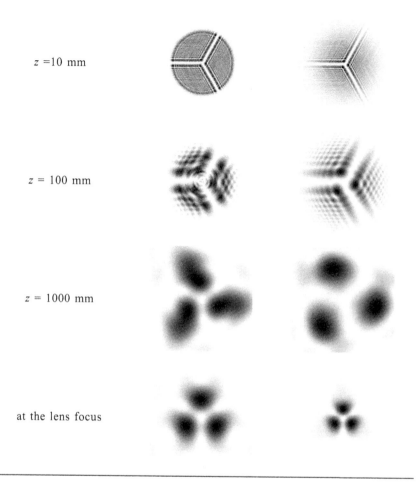

The nature of diffraction for the limited flat beam and the Gaussian beam varies in the Fresnel zone, but the distributions of the beams in the far-field are very similar.

The chapter assumes the use of the singular optics elements for apodization of the transmission function of the lens. In this case, it is desirable to know the distribution of the beam in the near-field diffraction. In the scalar case, we can use a diffraction Rayleigh-Sommerfeld integral of the first type:

$$E(u,v,z) = -\frac{z}{2\pi} \iint_{\Sigma_0} E_0(x,y) \frac{e^{ik\ell}}{\ell^2} \left(ik - \frac{1}{\ell} \right) dxdy \qquad (4.1.63)$$

where $E_0(x, y)$ is the input field $\ell = \sqrt{(u-x)^2 + (v-y)^2 + z^2}$, Σ_0 is an area in which input field is given, $k = 2\pi/\lambda$ is the wave number, λ is wavelength.

We can also use an integral propagation operator, based on the expansion in plane waves:

$$E(u,v,z) = \frac{1}{\lambda^2} \times$$

$$\times \iint_{\Sigma_0} E_0(x,y) \left\{ \int_{\Sigma_s} \int \exp\left(ikz\sqrt{1-\xi^2-\eta^2}\right) \exp\left(ik[\xi(u-x)+\eta(v-y)]\right) d\xi \, d\eta \right\} dx \, dy,$$

$$(4.1.64)$$

where Σ_S : $\sigma_1 \le \sqrt{\xi^2 + \eta^2} \le \sigma_2$ is the region of the considered spatial frequencies. If $\sigma_1 = 0$, $\sigma_2 = 1$ we consider only propagating waves, and at $\sigma_2 > 1$ evanescent waves are also taken into account. The expression (4.1.64) can be calculated through the fast Fourier transform algorithm.

Furthermore, if the input field is presented in the radial-vortex type:

$$E_0(r, \varphi) = E_0(r) \exp(im\,\varphi) \qquad (4.1.65)$$

expression (4.1.64) can be simplified [9, 10]

$$E(\rho,\theta,z) = -i^{n+1}k \exp(in\theta) \times$$

$$\times \int_0^{\sigma_0} \exp\left(ikz\sqrt{1-\sigma^2}\right) \left(\int_0^{r_0} E_0(r)J_n(k\sigma r)r dr \right) J_n(k\sigma\rho)\sigma d\sigma, \qquad (4.1.66)$$

where r_0 is the radius of the limiting aperture, σ_0 is the range of the considered spatial frequencies, n is the order of the vortex phase singularity.

Tables 4.1.3 and 4.1.4 show the results of simulation of diffraction in the near field of a bounded plane wave on the SPP and the vortex axicon by the expression (4.1.66).

The calculation parameters: $r_0 = 10\lambda$, the sampling step $\Delta r = 0.01\lambda$, $\sigma_0 = 3$, sample rate $\Delta\sigma = 0.01$.

Table 4.1.3. Diffraction of a restricted plane wave on the first-order SPP in the near field

Phase in the aperture with radius 10λ ($z = 0$)	Intensity distribution in the range $z \in [0.1\lambda, 10\lambda]$ $x \in [0.1\lambda, 10\lambda]$	The distribution in the plane $z = 0.3\lambda$, $x \in [0.1\lambda, 10\lambda]$	
		Intensity	Phase

The modeling shows that the action elements of singular optics effect even at a distance of less than a wavelength, i.e. very near the plane of the optical element.

The spiral axicon is described by:

$$E_0(r,\varphi) = \exp\left(-ik\alpha_0 r + im\varphi\right) \qquad (4.1.67)$$

where α_0 is the conical wave parameter that determines the angle at which rays intersect the optical axis.

The modeling results show that increasing parameter of the conical wave α_0 for the axisymmetric case ($n = 0$) decreases the size of the central light spot. When $\alpha_0 = 0.9$ the scalar theory predicts overcoming the diffraction limit at half intensity FWHM = 0.39λ (limit for the lens 0.51λ). In the presence of the vortex component the intensity at the optical axis is zero and a compact light ring forms.

4.1.4. Conclusions

In this section the scalar diffraction theory was used to discuss the formation of vortex beams with diffraction elements of singular optics.

Table 4.1.4. Diffraction of a limited plane wave on axially symmetric and vortex axicons in the near field

	$\alpha_0 = 0.5$		$\alpha_0 = 0.9$	
	n=0	n=1	n=0	n=1
Phase function				
Spectral distribution				
Axial distribution				
Intensity at the maximum, rad.=2λ	$z=11\lambda$, $\rho = 2\lambda$ FWHM = 0.66λ	$z=11\lambda$	$z=2.5\lambda$, FWHM = 0.39λ	$z=2.5\lambda$

The paraxial scalar theory was used to consider different beams on the spiral phase plate. The complex amplitude of the 'pure' optical vortex, formed in the diffraction of the unlimited plane wave, is proportional to the difference of two Bessel functions of the first kind of the $[(n+1)/2]$- and $[(n-1)/2]$-th orders. In diffraction of the Gaussian beam the complex amplitude in the far field zone is proportional to the difference of two Bessel functions of the second kind of the same orders.

Also, analytical expressions were derived for the diffraction of the limited plane wave and a Gaussian beam on the spiral axicon. In the first case, the complex amplitude is proportional to a series of Bessel functions, in the second – to a series of Kummer functions.

Numerical simulation using paraxial operators of propagation showed that the character of diffraction of spiral elements for a limited flat beam and a Gaussian beam varies in the Fresnel zone, but in the far-field zone the distribution of beams is practically the same.

For the diffraction pattern in the near field we used a scalar integral operator of propagation, based on the expansion in plane waves. The simulation showed that the effect of the elements of singular optics operates even at a distance less than the wavelength, i.e. very near the plane of the optical element. Thus, their use for apodization of the transmission function of the focusing lens is fully justified.

However, in the near-field diffraction we can not neglect the vector nature of the electromagnetic field, so we will continue to take into account also the influence of the polarization of the laser beam.

4.2. Vector representation of the field in the focal region for the vortex transmission function

Recently, special attention has been paid to the possibility of reducing the size of the focal spot and/or deepening the focus of the high-aperture focusing system [10–18]. Some particular type of polarization was always considered, especially radial [19–21], because it produces the highest power of the longitudinal component, useful for many applications [22].

Many intracavity [23–26] and extra-cavity schemes [27–29] have been developed to generate radially polarized beams. Moreover, there are ways to generate almost arbitrarily (non-uniform) polarized beams [30–33], but they are quite complex to implement, or require expensive devices.

Thus, on the one hand, one must consider the possibility of subwavelength localization of light and deepening of the focus for a more general type of polarization, on the other hand it is useful to determine the possibility for the most common and easiest to implement polarizations – linear and circular.

Shapr focusing of the laser beams with vortex singularity is an urgent task in the field of optical manipulation of micro- and nano-objects [34–40], in lithography [41,42] and microscopy [43,44]. Therefore, the focusing of vortex laser beams by using a high-aperture optical system has been studied in many works, especially recently [45–49].

The question of the mutual influence of phase and polarization optical vortices, their transition into each other, as well as compensation or amplification in this case of the orbital angular momentum has a long history [50–54]. This paper, based on the compensation of polarization vortices using a vortex transmission phase function, shows the possibility of formation of subwavelength structures in both the individual components of the vector field and the total intensity.

We consider a high-aperture aplanatic focusing optical system in which the focal region is located at a distance from the aperture significantly larger than the wavelength. Then the electric field vector in a homogeneous dielectric medium close to the focus may be described by the Debye approximation [45, 55]

$$\mathbf{E}(\rho,\varphi,z) = -\frac{if}{\lambda} \times$$

$$\times \int_0^\alpha \int_0^{2\pi} B(\theta,\phi)T(\theta)\mathbf{P}(\theta,\phi)\exp\left[ik(\rho\sin\theta\cos(\phi-\varphi)+z\cos\theta)\right]\sin\theta d\theta d\phi,$$

$$(4.2.1)$$

where (ρ,φ,z) are the cylindrical coordinates of the focal area (θ,ϕ) are the spherical angular coordinates of the exit pupil of the focusing system, $B(\theta,\phi)$ is the transmission function, $T(\theta)$ is the apodization function of the pupil, $\mathbf{P}(\theta,\phi)$ is the polarization matrix, $n\sin\alpha = NA$ is the numerical aperture, n is the refractive index of the medium, $k = 2\pi/\lambda$ is the wave number, λ is the wavelength, f is the focal length.

The pupil apodization function $T(\theta)$ [22]:

$$T(\theta) = \begin{cases} \sqrt{\cos\theta}, & \text{Sine}: r = f\sin\theta \\ 1, & \text{Herschel}: r = 2f\sin(\theta/2) \\ \sqrt{\theta/\sin\theta}, & \text{Lagrange}: r = f\theta \\ (1/\cos\theta)^{3/2}, & \text{Helmholtz}: r = f\tan\theta \end{cases} \qquad (4.2.2)$$

depends on the geometry of the focusing system. In particular, the sine rule used for microscopic lenses and aplanatic systems, the Herschel rule – for uniform focusing, and focusing is also carried out using the Lagrange function, and the Helmholtz function can be used to describe the planar diffractive lenses. There are other functions, including those for a parabolic mirror [15].

In Fig. 4.2.1 it is clear that the effect of the apodization function $T(\theta)$ is almost negligible for the small numerical aperture and greatly differs at $\theta = \pi/2$. In particular, when using the Helmholtz condition the peripheral contribution increases markedly. In the remainder of this paper we will consider only the sine rule of apodization.

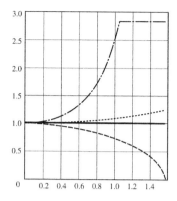

Fig. 4.2.1. The dependence of the apodization function $T(\theta)$ on the angle θ for the sine conditions (dashed line), the Herschel condition (solid line), the Lagrange condition (dotted line) and the Helmholtz condition (dot-dash line).

The polarization matrix of the focusing system $\mathbf{P}(\theta,\phi)$ has the following form [16]:

$$
\mathbf{P}(\theta,\phi) = \begin{bmatrix} 1+\cos^2\phi(\cos\theta-1) & \sin\phi\cos\phi(\cos\theta-1) & \cos\phi\sin\theta \\ \sin\phi\cos\phi(\cos\theta-1) & 1+\sin^2\phi(\cos\theta-1) & \sin\phi\sin\theta \\ -\sin\theta\cos\phi & -\sin\theta\sin\phi & \cos\theta \end{bmatrix} \begin{bmatrix} a(\phi,\theta) \\ b(\phi,\theta) \\ c(\phi,\theta) \end{bmatrix}
$$

$$(4.2.3)$$

where $a(\theta,\phi)$, $b(\theta,\phi)$, $c(\theta,\phi)$ are the polarization functions for x-, y- and z-components of the incident beam.

For commonly used types of polarization, these functions have a simple form and do not depend on θ.

Then, if the transmission function $B(\theta,\phi)$ can be written as a radial portion and the superposition of angular harmonics [57, 58]:

$$B(\theta,\phi) = R(\theta)\Omega_B(\phi) \tag{4.2.4}$$

where

$$\Omega_B(\phi) = \sum_{m=-M_1}^{M_2} d_m \exp(im\phi) \tag{4.2.5}$$

the integral over ϕ in (4.2.1) can be taken analytically for special cases of polarization.

In particular, if the coefficients of the input polarization are also represented as a superposition of angular harmonics (4.2.5): $a(\phi) = \Omega_a(\phi)$, $b(\phi) = \Omega_b(\phi)$, $c(\phi) = \Omega_c(\phi)$, the integrals over ϕ in (4.2.1) will be expressed through the corresponding sum of the Bessel functions of the first kind of different orders [56]:

$$\int_0^{2\pi} \exp(it\cos(\phi-\varphi))\Omega_p(\phi)\Omega_B(\phi)\mathrm{d}\phi =$$

$$= \int_0^{2\pi} \exp(it\cos(\phi-\varphi))\sum_{l,m} p_l d_m \exp\left[i(l+m)\phi\right]\mathrm{d}\phi = \qquad (4.2.6)$$

$$= 2\pi \sum_{l,m} p_l d_m i^{l+m} \exp\left[i(l+m)\varphi\right]J_{l+m}(t),$$

where $t = k\rho \sin\theta$.

Then the vector electric field in the focal region is calculated by the one-dimensional integration:

$$\mathbf{E}(\rho,\varphi,z) = -ikf\int_0^\alpha \mathbf{Q}(\rho,\varphi,\theta)q(\theta)\mathrm{d}\theta \qquad (4.2.7)$$

where

$$q(\theta) = R(\theta)T(\theta)\sin\theta \exp(ikz\cos\theta) \qquad (4.2.8)$$

and the form of $\mathbf{Q}(\rho,\varphi,\theta)$ is dependent on the polarization of the input field.

The function $q(\theta)$ for small values of the numerical aperture (α is small) is close to zero because of the factor $\sin\theta$. The effect of the apodization function $T(\theta)$ in this case (close to unity) is negligible.

Therefore, a high degree of focusing can be achieved at small numerical apertures only due to the amplitude distribution (which leads to energy losses) or of a special type $R(\theta)$ (e.g. the use of radially polarized beams of a high order [20, 23, 24]).

Considering the function $R(\theta)$ as a purely phase one, we can write for the common types of polarization of the input field – linear, circular, radial and azimuthal – explicit expressions for the field intensity in the focal plane and to analyze the possibility of subwavelength localization of light in the focal region on the basis of the expression (4.2.7) of the high-aperture focusing systems using the singular phase transmission function (4.2.5).

4.2.1. Uniformly polarized beams (linear and circular polarization)

4.2.1.1. Linear x-polarization

For linear x-polarization of radiation incident on a high-aperture aplanatic focusing system the input polarization coefficients are written as $a(\phi) = 1$, $b(\phi) = 0$, $c(\phi) = 0$, and the polarization vector of the system has the form:

$$\mathbf{P}(\theta,\phi) = \begin{bmatrix} 1 + \cos^2\phi(\cos\theta - 1) \\ \sin\phi\cos\phi(\cos\theta - 1) \\ -\cos\phi\sin\theta \end{bmatrix} \qquad (4.2.9)$$

Then the vector $\mathbf{Q}(\rho,\varphi,\theta)$ in (4.2.7) for the vortex transmission function (4.2.5) can be expressed by:

$$\mathbf{Q}(\rho,\varphi,\theta) = i^m \exp(im\varphi) \begin{bmatrix} J_m(t) + C2_m(t)(\cos\theta - 1) \\ SC_m(t)(\cos\theta - 1) \\ -C_m(t)\sin\theta \end{bmatrix}, \ t = k\rho \sin\theta (4.2.10)$$

where

$$C2_m(t) = \frac{1}{4}\left[2J_m(t) - e^{i2\varphi}J_{m+2}(t) - e^{-i2\varphi}J_{m-2}(t)\right],$$

$$SC_m(t) = \frac{i}{4}\left[e^{i2\varphi}J_{m+2}(t) - e^{-i2\varphi}J_{m-2}(t)\right],$$

$$C_m(t) = \frac{i}{2}\left[e^{i\varphi}J_{m+1}(t) - e^{-i\varphi}J_{m-1}(t)\right].$$

Consider the axial (at $r = 0$) field intensity distribution (4.2.7). The vortex transmission vector in this case will contain non-zero components only for the orders of the vortex singularity: $|m| \leq 2$

$$\mathbf{Q}_{m=0}(0,0,\theta) = \frac{1}{2}(1 + \cos\theta)\begin{bmatrix} 1 \\ 0 \\ 0 \end{bmatrix} \qquad (4.2.11)$$

$$\mathbf{Q}_{m=1}(0,0,\theta) = -\frac{1}{2}\sin\theta\begin{bmatrix} 0 \\ 0 \\ 1 \end{bmatrix} \qquad (4.2.12)$$

$$\mathbf{Q}_{m=2}(0,0,\theta) = \frac{1}{4}(\cos\theta - 1)\begin{bmatrix} 1 \\ i \\ 0 \end{bmatrix}$$ (4.2.13)

As can be seen from equations (4.2.11)–(4.2.13), in each case the axial distribution has a certain polarization: initial linear x-polarization at $m = 0$, longitudinal at the vortex singularity of the first order, and circular at the vortex singularity of the second order.

Table 1.1 shows the distribution of the field in the focal region for the linearly polarized full-aperture vortex transmission function:

$$B(\theta,\phi) = circ(\theta,\alpha)\exp(im\phi) = \exp(im\phi)\begin{cases} 1, \ \theta \in [0,\alpha] \\ 0, \ \theta > \alpha \end{cases}, \ |m| \leq 2$$

(4.2.14)

for an aplanatic system with focus $f = 101\lambda$ and $\alpha = 82°$. The transverse area is bounded by the dimensions $3\lambda \times 3\lambda$, and the longitudinal region varies with the range of five wavelengths $z \in [-2.5\lambda, 2.5\lambda]$, the cross section is shown in focus ($z = 0$).

Topological pictures in the table are for visualizing the structure of various 'optical bottles' (the dark areas, surrounded by light) [59] formed in the focal region when using the vortex transmission function. The size of these hollow capsules is usually (see the Figure) even smaller than the subwavelength of the focal spots, but in this study they are not considered.

From Table 4.2.1 it is seen that the vortex field of the corresponding order provides the subwavelength localization of light in the individual components, but the influence of the other components at a constant distribution over the aperture $R(\theta)$ (4.2.14) eliminates this effect in the total field intensity.

It is known [14] that when using a narrow annular aperture with the maximum radius for the focusing system we can distinguish the longitudinal component of radial polarization. Consider in a more general case the effect of the narrow annular aperture of arbitrary radius $R(\theta) = \delta(\theta - \theta_0)$.

With linear x-polarization expression (4.2.7) takes the following form:

Table 4.2.1. Longitudinal $z \in [-2.5\lambda, 2.5\lambda]$ and tranverse $z = 0$, $3\lambda \times 3\lambda$ distributions for the linearly polarized full-aperture transmission function (4.2.14)

	$|E|^2$, topology*	$|E_x|^2$, arg E_x	$|E_y|^2$, arg E_y	$|E_z|^2$, arg E_z
$m=0$	FWHM ($|$): 0.47λ, (—): 0.74λ			
$m=1$			FWHM (—): 0.36λ	
$m=2$		FWHM ($|$): 0.22λ	FWHM ($|$): 0.42λ	

*) Topology – the structure of the field with the normalization only in sections, excluding changes in energy along the optical axis.

$$\mathbf{E}(\rho,\varphi,z) = -ikf i^m \exp(im\varphi) \begin{bmatrix} J_m(t_0) + C2_m(t_0)(\cos\theta_0 - 1) \\ SC_m(t_0)(\cos\theta_0 - 1) \\ -C_m(t_0)\sin\theta_0 \end{bmatrix} \sqrt{\cos\theta_0} \sin\theta_0 \exp(ikz\cos\theta_0)$$

(4.2.15)

where $t_0 = k\rho \sin\theta_0$.

Obviously, the main contribution to the central light spot comes from the terms containing $J_0(k\rho \sin\theta_0)$. It is also understood that the smallest dimension is reached at $\theta_0 \to 90°$.

However, in this case, the apodization function $\sqrt{\cos\theta_0} \to 0$, so the maximum value on the axis for the defined components should be found separately.

In particular, at $m = 0$ using a narrow annular aperture the axial intensity of the x-axis component is determined by the expression:

$$\left| E^x_{m=0}(0,0,z) \right|^2 = \frac{(kf)^2}{4} \left| \exp(ikz\cos\theta_0)(1+\cos\theta_0)\sin\theta_0\sqrt{\cos\theta_0} \right|^2$$

$$(4.2.16)$$

which no longer depends on the defocusing z.

Figure 4.2.2 shows the function $\tau(\theta) = (1+\cos\theta)^2 \sin^2\theta\cos\theta$ of angle $\theta \in [0, \pi/2]$. The graph shows that the peak is reached at $\theta_0 = 0.8$ rad.

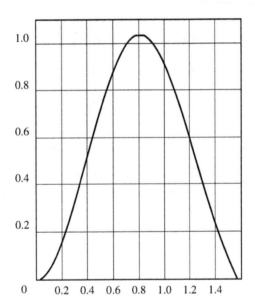

Fig. 4.2.2. The dependence of the axial intensity of the x-component on the angle θ for linear x-polarization for $m = 0$.

Table 4.2.2 shows the transverse distribution of the total intensity and the intensity of the x-component in the focal plane (and also sections) when using a narrow annular aperture of the maximum radius of the considered focusing system $r_0 = 100\lambda$ and the optimal radius $r_0 = f\sin(\theta_0) \approx 73.5\lambda$. Using the aperture with the optimal radius provides the most compact uniform transverse distribution, and the aperture with the maximum radius leads to the minimum achievable size of the focal spot with a loss of symmetry.

When $|m| = 1$ using a narrow annular aperture leads to the following form of the axial intensity of the longitudinal component:

$$\left|E_{m=1}^{z}(0,0,z)\right|^{2} = \frac{\left(kf\right)^{2}}{4}\left|\sin^{2}\theta_{0}\sqrt{\cos\theta_{0}}\right|^{2} \qquad (4.2.17)$$

When $|m|=2$ both transverse components contributed to the axial intensity (see (4.2.13)), and the intensity can be maximized by selecting the radius $R(\theta) = \delta(\theta-\theta_{0})$:

$$\left|E_{m=1}^{x,y}(0,0,z)\right|^{2} = \frac{\left(kf\right)^{2}}{16}\left|\left(\cos\theta_{0}-1\right)\sin\theta_{0}\sqrt{\cos\theta_{0}}\right|^{2}. \qquad (4.2.18)$$

Figure 1.3a and 1.3b show the plots $\tau(\theta)=\sin^{4}\theta_{0}\cos\theta_{0}$ and $\tau(\theta)=\left(\cos\theta-1\right)^{2}\sin^{2}\theta\cos\theta$ in relation to the angle $\theta\in[0,\pi/2]$, respectively. It can be seen that the maxima, moving to the edge of the aperture, are taking smaller and smaller values.

Table 4.2.2. The transverse ($z = 0$, $3\lambda\times3\lambda$) distributions (negative) when using a narrow annular aperture of maximum and optimum radii for linear x-polarization with $m = 0$.

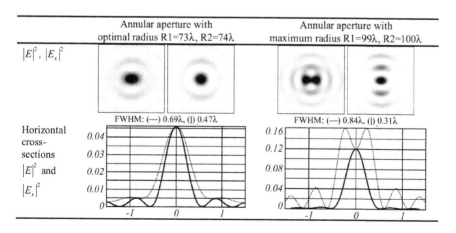

$\left|E\right|^{2}$, $\left|E_x\right|^{2}$	Annular aperture with optimal radius R1=73λ, R2=74λ	Annular aperture with maximum radius R1=99λ, R2=100λ
Horizontal cross-sections $\left|E\right|^{2}$ and $\left|E_x\right|^{2}$	FWHM: (—) 0.69λ, (l) 0.47λ	FWHM: (—) 0.84λ, (l) 0.31λ

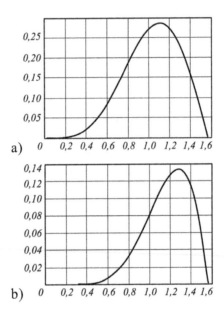

Fig. 4.2.3. The dependence of the axial intensity on the angle θ for linear x-polarization for the longitudinal component at $|m| = 1$ (a) and for transverse components at $|m| = 2$ (b).

Of course, it is physically not possible and unfavourable from the energy viewpoint to create an infinitely narrow slit, so the width of the slit has a certain size (in our calculations it was equal to the wavelength), and the above calculations are not quite correct.

Table 4.2.3 shows a comparison of the longitudinal and transverse field intensity distributions for the linearly polarized input field for $m = 0$ limited by an annular aperture; the inner and outer radii of this aperture are designated as R1 and R2. The transverse distribution of the size $3\lambda \times 3\lambda$ ($z = 0$) is shown in focus, and the longitudinal area is considered in the range $z \in [-50\lambda, 50\lambda]$, to show the degree of independence of the resultant distribution on defocusing.

As shown in Table 4.2.3, the use of an annular aperture with its average radius close to the maximum, results in a reduction of the range of invariance to defocusing and the energy concentration in the focal region. Thus, if the task is to ensure the most compact energy concentration in a small area rather than the optimum compromise between the depth of the field and its lateral compactness, it is necessary to choose the maximum possible radius of the annular aperture. In this case the numerical aperture of the focusing optical system is the largest.

Table 4.2.3. Longitudinal and transverse intensity distribution of the field for the linearly polarized field for $m = 0$, limited by the annular aperture

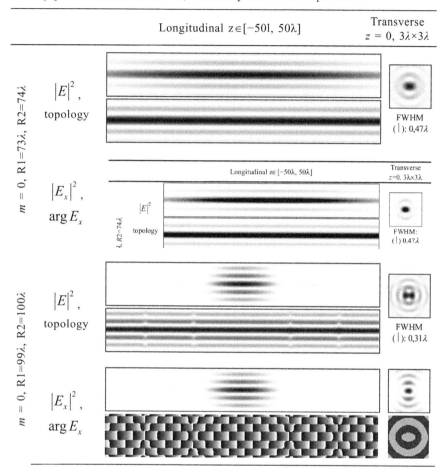

Table 4.2.4 shows similar results for the vortex transmission function $0 < |m| \leq 2$. Maximizing the contribution of the axial component allows to improve the overall outcome for linear polarization only when $m = 0$ and achieve subwavelength localization in only one direction.

Table 4.2.4 shows that when using the vortex field of high orders in the individual components of the vector field we can achieve significant subwavelength localization (FWHM = 0.28λ) of radiation in one of the directions.

This may be important in the interaction of laser radiation with matter, selectively sensitive to various components of the vector electromagnetic field [60]. For example, in [61] the authors used different field components for the excitation of different orientations of fluorescing molecules. Also, in [62] on the contrary the excitation of separate fluorescing molecules can be used to separate the components of the vector field in the focal plane.

Table 4.2.4. Longitudinal $z \in [-50\lambda, 50\lambda]$ and transverse ($z = 0$, $3\lambda \times 3\lambda$) field intensity distribution for the linearly polarized field with the vortex transmission function ($0 < |m| \leq 2$) limited by an annular aperture

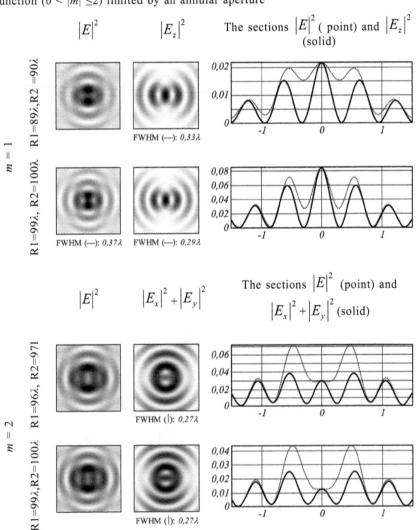

4.2.1.2. Circular polarization

For the circular polarization of the input field the coefficients with the normalization taken into account are equal to

$a(\phi) = 1/\sqrt{2}$, $b(\phi) = \pm 1/\sqrt{2}\,i$, $c(\phi) = 0$ and the vector of polarization:

$$\mathbf{P}(\theta,\phi) = \frac{1}{2} \begin{bmatrix} \left[1 + \cos^2\phi(\cos\theta - 1)\right] \pm i\left[\sin\phi\cos\phi(\cos\theta - 1)\right] \\ \left[\sin\phi\cos\phi(\cos\theta - 1)\right] \pm i\left[1 + \sin^2\phi(\cos\theta - 1)\right] \\ -\sin\theta\left[\cos\phi \pm i\sin\phi\right] \end{bmatrix}. \qquad (4.2.19)$$

The vector for the vortex field in (4.2.7):

$$\mathbf{Q}(\rho,\varphi,\theta) = \frac{1}{2} i^m \exp(im\varphi) \begin{bmatrix} J_m(t) + \frac{1}{2}\left[J_m(t) + E2_m(t)\right](\cos\theta - 1) \\ \mathrm{sgn}(p)i\left\{J_m(t) + \frac{1}{2}\left(J_m(t) - E2_m(t)\right)(\cos\theta - 1)\right\} \\ -E1_m(t)\sin\theta \end{bmatrix}$$

$$(4.2.20)$$

where $E1_m(t) = \mathrm{sgn}(p)ie^{\mathrm{sgn}(p)i\varphi}J_{m+\mathrm{sgn}(p)1}(t)$, $E2_m(t) = -e^{\mathrm{sgn}(p)i2\varphi}J_{m+\mathrm{sgn}(p)2}(t)$
and $t = k\rho\sin\theta$, $\mathrm{sgn}(p)$ is the sign of the input polarization.

The behaviour on the axis with the values of the vortex index $|m| \le 2$:

$$\mathbf{Q}_{m=0}(0,0,\theta) = \frac{1}{2}(1 + \cos\theta)\begin{bmatrix} 1 \\ \mathrm{sgn}(p)i \\ 0 \end{bmatrix} \qquad (4.2.21)$$

– the circular polarization of the same sign as the input field.

$$\mathbf{Q}_{m=1}(0,0,\theta) = \frac{1}{2}\begin{bmatrix} 0 \\ 0 \\ -\sin\theta \end{bmatrix} \qquad (4.2.22)$$

there is only the longitudinal component at the sign at the input polarization '–' (a similar result when $m = -1$ and the sign of the input polarization '+').

$$\mathbf{Q}_{m=2}(\theta) = \frac{1}{4}(1-\cos\theta)\begin{bmatrix} -1 \\ i \\ 0 \end{bmatrix}$$

(4.2.23)

– the circular polarization of the same sign as the input field (similar result with $m = -2$ and the sign of input polarization '+').

The situation is similar to the case of linear polarization, but involves both transverse components.

With a further increase of the index m all components on the axis are zero.

Table 4.2.5 shows the distribution of the field in the focal region the circular '+'-polarization of the full-aperture vortex transmission function at $| m | \le 2$ for the aplanatic system with focus $f = 101\lambda$ and $\alpha = 82°$.

Table 4.2.6 shows the results for comparison of the effect of the annular aperture, the average radius of which is close to optimum, and with the maximum radius.

In the first case, i.e. for the optimum aperture at $m = 0$, the total intensity provides a deep focus (more 100λ) at the focal spot (FWHM: 0.56λ) slightly smaller than the scalar limit for the lens (FWHM: 0.61λ), but at other values of m the subwavelength localization occurs only in the individual components.

In the second case (for the maximum aperture radius) in the total intensity we can achieve the same subwavelength localization (FWHM: 0.56λ) but at $|m| = 1$. Thus, there is a reduction in the depth of focus a few times and the increase of the intensity in the focal plane as a result of this. For other values of m significant subwavelength localization (FWHM 0.36λ) occurs only in the individual components.

4.2.2. Cylindrical vector beams (radial and azimuthal polarization)

4.2.2.1. The radial polarization

When the radial polarization of the radiation incident on high-aperture aplanatic focusing system, the coefficients are written as input polarization $a(\phi) = \cos\phi$, $b(\phi) = \sin\phi$, $c(\phi) = 0$ and polarization matrix system has the form:

$$\mathbf{P}(\theta,\phi) = \begin{bmatrix} \cos\phi\cos\theta \\ \sin\phi\cos\theta \\ -\sin\theta \end{bmatrix}$$

(4.2.24)

Table 4.2.5. Longitudinal $z \in [-2.5\lambda, 2.5\lambda]$ and transverse $z = 0$, $3\lambda \times 3\lambda$ distribution for the plane-polarized vortex full-aperture transmission function

$|E|^2$, topology *	$|E_x|^2$, $\arg E_x$	$|E_y|^2$, $\arg E_y$	$|E_z|^2$, $\arg E_z$

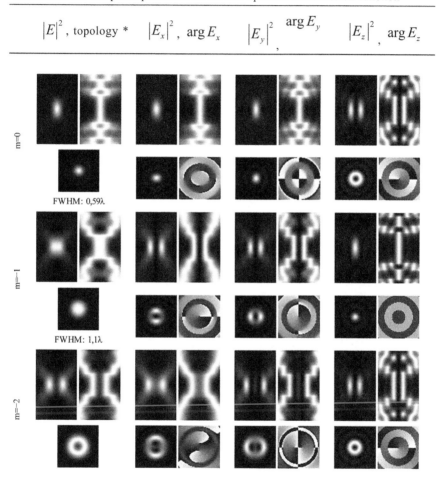

Then the matrix $\mathbf{Q}(\rho,\varphi,\theta)$ in (4.2.7) for the vortex transmission function (4.2.5) can be expressed by:

$$\mathbf{Q}(\rho,\varphi,\theta) = i^m \exp(im\varphi) \begin{bmatrix} C_m(t)\cos\theta \\ S_m(t)\cos\theta \\ -J_m(t)\sin\theta \end{bmatrix} \qquad (4.2.25)$$

where $C_m(t) = \dfrac{i}{2}\left[e^{i\varphi}J_{m+1}(t) - e^{-i\varphi}J_{m-1}(t)\right]$,

$S_m(t) = \dfrac{1}{2}\left[e^{i\varphi}J_{m+1}(t) + e^{-i\varphi}J_{m-1}(t)\right]$, $t = k\rho\,\sin\theta$.

Table 4.2.6. The transverse ($z = 0$, $3\lambda \times 3\lambda$) field intensity distribution for the vortex transmission function with circular polarization ($|m| \le 2$) limited by a narrow annular aperture

The vortex transmission vector at the axis (at $\rho = 0$) will contain non-zero components only for the orders of the vortex singularity $|m| \le 1$:

$$\mathbf{Q}_{m=0}(0,0,\theta) = -\sin\theta \begin{bmatrix} 0 \\ 0 \\ 1 \end{bmatrix} \qquad (4.2.26)$$

$$\mathbf{Q}_{m=1}(0,0,\theta) = \frac{1}{2}\cos\theta \begin{bmatrix} 1 \\ i \\ 0 \end{bmatrix} \qquad (4.2.27)$$

When $m = 0$ only the longitudinal component forms in the focal region on the axis, and at $|m| = 1$ only the transverse components will be present. We note that the situation is similar to the linear polarization with the vortex of the first and second order, but in this case the intensity is 4 times higher. Table 4.2.7 shows the distribution of the field in the focal region for the radially polarized full-aperture vortex transmission functions $|m| \leq 1$ for the aplanatic system with the focus $f = 101\lambda$ and $\alpha = 82°$. The longitudinal and transverse dimensions are the same as in Table 4.2.1. From Table 4.2.7 it is clear that the full-aperture field with the uniform amplitude does not allow subwavelength localization of light in the total intensity.

With radial polarization and the use of the narrow annular aperture expression (4.2.7) takes the following form:

$$\mathbf{E}(\rho,\varphi,z) = -ikf i^m \exp(im\varphi) \begin{bmatrix} C_m(t_0)\cos\theta_0 \\ S_m(t_0)\cos\theta_0 \\ -J_m(t_0)\sin\theta_0 \end{bmatrix} \sqrt{\cos\theta_0}\,\sin\theta_0\,\exp(ikz\cos\theta_0)$$

$$(4.2.28)$$

where $t_0 = k\rho\,\sin\theta_0$.

Obviously, for $m = 0$ the optimum value θ_0 is obtained as in the previous section with the vortex of the first order, and for $|m| = 1$ only the transverse components contribute to the axial intensity:

$$\left|E^{x,y}_{m=1}(0,0,z)\right|^2 = \frac{(kf)^2}{4}\sin^2\theta_0\cos^3\theta_0 \qquad (4.2.29)$$

maximization of which, as seen in Fig. 4.2.4, occurs at $\theta_0 = 0.7$ radians. Table 4.2.8 compares the action of annular apertures with different average radii.

Table 4.2.7. Longitudinal $z \in [-2.5\lambda, 2.5\lambda]$ and transverse ($z = 0$, $3\lambda \times 3\lambda$) distributions for radially polarized full-aperture vortex transmission function $|m| \leq 1$

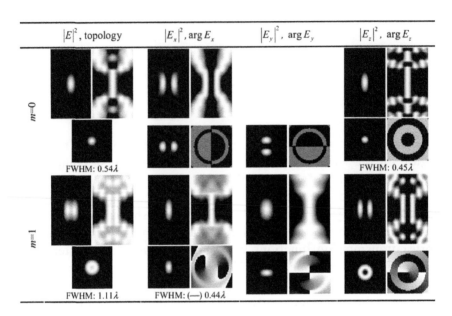

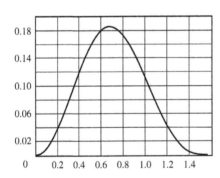

Fig. 4.2.4. The dependence of the axial intensity of the transverse components on angle θ for radial polarization with $m=1$.

The above results of numerical simulations show that using the optimal radius of the annular aperture for $m = 0$ we can reach a compromise between the depth of focus, which exceeds 60λ, and the compactness of the focal spot – it has the size FWHM 0.44λ, which is less than the scalar limit of the lens.

Increasing the radius of the annular aperture reduces the FWHM to 0.37λ, but due to a significant (almost 3 times) reduction in the depth of focus.

Table 4.2.8. Longitudinal and transverse distribution of the field intensity for the radially polarized vortex transmission function ($|m|\leq1$) limited by a narrow annular aperture

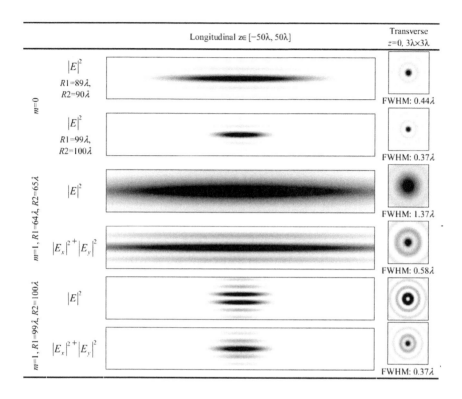

Since in radial polarization the contribution of the transverse components is proportional to $\cos\theta$, then for the high-aperture this contribution decreases. Therefore, when $m = 1$, the annular aperture with a maximum radius, as seen from Table 4.2.6, actually leaves only the longitudinal component. To keep the transverse components in the common field, we need to use an aperture with the radius optimum for these components, which leads to a broadening of the focal spot.

4.2.2.2. The azimuthal polarization

In the case of the azimuthal polarization of the input field the coefficients are written as $a(\phi) = -\sin\phi$, $b(\phi) = \cos\phi$, $c(\phi) = 0$ and the polarization matrix of the system has the form:

$$\mathbf{P}(\theta,\phi) = \begin{bmatrix} -\sin\phi \\ \cos\phi \\ 0 \end{bmatrix}. \qquad (4.2.30)$$

The matrix for the vortex field $\mathbf{Q}(\rho,\varphi,\theta)$ in (4.2.7) is:

$$\mathbf{Q}(\rho,\varphi,\theta) = i^m \exp(im\varphi) \begin{bmatrix} -S_m(t) \\ C_m(t) \\ 0 \end{bmatrix} \qquad (4.2.31)$$

where $S_m(t) = \dfrac{1}{2}\left[e^{i\varphi} J_{m+1}(t) + e^{-i\varphi} J_{m-1}(t) \right]$,

$C_m(t) = \dfrac{i}{2}\left[e^{i\varphi} J_{m+1}(t) - e^{-i\varphi} J_{m-1}(t) \right]$, $t = k\rho \sin\theta$.

As follows from (4.2.31) the longitudinal component is always missing.

Table 4.2.9 shows the distribution of the field in the focal region for the azimuthally polarized full-aperture vortex transmission function $|m| \leq 1$ for an aplanatic system with focus $f = 101\lambda$ and $\alpha = 82°$.

It is worth noting that for $m = 1$ the vortex singularity in the centre of the focal region is missing, since the polarization vortex is compensated by the scalar vortex of the transmission function.

As shown in Table 4.2.9, only the transverse components contributed to the total intensity, but in order to achieve subwavelength localization we must cut off the rays corresponding to the low numerical aperture of the focusing system.

In imposing a narrow annular aperture expression (4.2.7) in the case of azimuthal polarization takes the following form:

$$\mathbf{E}(\rho,\varphi,z) = -ikf i^m \exp(im\varphi) \begin{bmatrix} S_m(t_0) \\ -C_m(t_0) \\ 0 \end{bmatrix} \sqrt{\cos\theta_0}\, \sin\theta_0 \exp(ikz\cos\theta_0)$$

$$t_0 = k\rho \sin\theta_0 \qquad (4.2.32)$$

On the optical axis (at $\rho = 0$), the vector (4.2.31) will contain non-zero components only for one singularity order $|m| = 1$:

$$\mathbf{Q}_{m=\pm 1}(0,0,\theta) = \frac{\pm i}{2}\cos\theta \begin{bmatrix} 1 \\ i \\ 0 \end{bmatrix} \qquad (4.2.33)$$

Table 4.2.9. Longitudinal $z \in [-2.5\lambda, 2.5\lambda]$ and transverse $z = 0$, $3\lambda \times 3\lambda$ distributions for azimuthally polarized full-aperture vortex transmission function $|m| \leq 1$

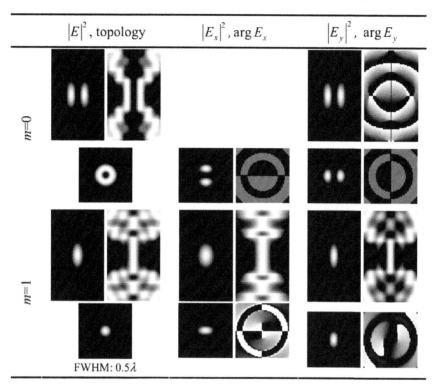

which corresponds to circular polarization and similarly to radial polarization with the vortex of the first order (see (4.2.27)), with the important difference that the longitudinal component is absent and does not distort the total intensity.

For all other values $|m| \neq 1$ at the axial points we obtain the absolute zero of intensity because all components are zero.

Table 4.2.10 shows the comparison of action of a narrow annular gap of different radius, and it is clear that the results obtained for the total intensity correspond in the radial polarization only to the transverse components. When choosing the average radius of the annular aperture close to the optimum we obtain a compromise between depth of focus (more than 100 wavelengths) and the size of the focal spot (FWHM: $0.58\lambda < 0.61\lambda$ – the scalar limit for the lens). When we select the maximum radius there is a substantial subwavelength localization (FWHM: 0.37λ) with a simultaneous decrease of the depth of focus (30 wavelengths).

Table 4.2.10. Longitudinal and transverse field intensity distribution for the azimuthally polarized vortex transmission function ($|m| = 1$) limited by a narrow annular aperture

4.2.3. Generalized vortex polarization

From the analysis of the previous sections it is clear that if we do not consider the interaction of electromagnetic radiation with matter, selectively sensitive to the transverse and longitudinal components [60–62], it is sometimes useful to 'drive' with the input vortex field the transverse (or longitudinal) component from the centre to ensure that various components are not superposed and the total intensity does not broaden. Thus, the higher the number of the vortex, the larger the radius of the zero intensity zone. However, the above types of polarization include vortex singularities no higher than the second order.

Therefore, this section discusses the generalized vortex polarization, i.e. the polarization whose coefficients contain vortex singularities of arbitrary order and their superpositions. This polarization can be implementd by one of the methods of generating random-polarized beams [30–33].

Equation (4.2.7) is valid when the input polarization coefficients $a(\phi)$, $b(\phi)$, $c(\phi)$ can be represented as (4.2.5).

In the particular case of

$$a(\phi) = a\exp(il\phi), \quad b(\phi) = b\exp(ip\phi), \quad c(\phi) = c\exp(is\phi) \quad (4.2.34)$$

the polarization vortex:

$$P(\theta,\phi) = \begin{vmatrix} a\exp(il\phi)\left[1+\cos^2\phi(\cos\theta-1)\right]+b\sin\phi\cos\phi\exp(ip\phi)(\cos\theta-1) \\ +c\cos\phi\exp(is\phi)\sin\theta\, a\sin\phi\cos\phi\exp(il\phi)(\cos\theta-1) \\ +b\exp(ip\phi)\left[1+\sin^2\phi(\cos\theta-1)\right]+c\sin\phi\exp(is\phi)\sin\theta \\ -\left[a\exp(il\phi)\cos\phi+b\exp(ip\phi)\sin\phi\right]\sin\theta \\ +c\exp(is\phi)\cos\theta \end{vmatrix}$$

(4.2.35)

and the vortex transmission vector:

$$Q(\rho,\theta)=i^m\exp(im\varphi)\times$$

$$\times\begin{bmatrix} ai^l\,e^{il\varphi}\left[J_{m+l}(t)+C2_{m+l}(t)(\cos\theta-1)\right]+bi^p\,e^{ip\varphi}\,SC_{m+p}(t)(\cos\theta-1) \\ +ci^s\,e^{is\varphi}\,C_{m+s}(t)\sin\theta\, ai^l\,e^{il\varphi}\,SC_{m+l}(t)(\cos\theta-1)+bi^p\,e^{ip\varphi} \\ \left[J_{m+p}(t)+S2_{m+p}(t)(\cos\theta-1)\right]+ci^s\,e^{is\varphi}\,S_{m+s}(t)\sin\theta \\ -\left[ai^l\,e^{il\varphi}\,C_{m+l}(t)+bi^p\,e^{ip\varphi}\,S_{m+p}(t)\right]\sin\theta+ci^s\,e^{is\varphi}\,J_{m+s}(t)\cos\theta \end{bmatrix},$$

(4.2.36)

where

$$C_m(t)=\frac{i}{2}\left[e^{i\varphi}J_{m+1}(t)-e^{-i\varphi}J_{m-1}(t)\right],\quad S_m(t)=\frac{1}{2}\left[e^{i\varphi}J_{m+1}(t)+e^{-i\varphi}J_{m-1}(t)\right],$$

$$SC_m(t)=\frac{i}{4}\left[e^{i2\varphi}J_{m+2}(t)-e^{-i2\varphi}J_{m-2}(t)\right],$$

$$C2_m(t)=\frac{1}{4}\left[2J_m(t)-e^{i2\varphi}J_{m+2}(t)-e^{-i2\varphi}J_{m-2}(t)\right],$$

$$S2_m(t)=\frac{1}{4}\left[2J_m(t)+e^{i2\varphi}J_{m+2}(t)+e^{-i2\varphi}J_{m-2}(t)\right],\quad t=k\rho\sin\theta.$$

Note that a non-zero longitudinal component can be formed in any of the above examined variants (except azimuthal) of sharp focusing, with $s=0$, and $c\neq0$.

Consider the possibility to maximize the axial intensity of the longitudinal components when using a narrow annular aperture:

$$E_z(\rho,\varphi,z)=-ikfi^m\,e^{im\varphi}\left\{\begin{array}{l}\left[ai^l\,e^{il\varphi}\,C_{m+l}(t_0)+bi^p\,e^{ip\varphi}\,S_{m+p}(t_0)\right]\sin\theta_0+ \\ ci^s\,e^{is\varphi}\,J_{m+s}(t_0)\cos\theta_0\end{array}\right\}\times$$

$$\times\sqrt{\cos\theta_0}\,\sin\theta_0\,\exp\left(ikz\cos\theta_0\right)$$

(4.2.37)

where $t_0 = k\rho \sin\theta_0$.

Obviously, a 'well-selected' subwavelength central spot in the longitudinal component can be obtained at $m = -s$ (including $s = 0$) and $|l|, |p| \gg |s|$. However, in this case, the spot size will decrease at $\theta_0 \to 90°$ which also leads to a decrease in the intensity of this component, because $\cos\theta_0 \to 0$ (see Table 4.2.11).

Therefore, it is better to consider the case of the transverse incident field. At the zero input longitudinal component $c(\phi) = 0$ expression (4.2.37) will include terms proportional to $J_{m+l\pm1}(t_0)$ and $J_{m+p\pm1}(t_0)$ and consequently become non-zero at the centre point when the optical vortex is applied to the input $m = -(l\pm1)$ or $m = -(p\pm1)$ (see Table 4.2.12).

If $l = p$ and $b = ia$, i.e. at the input there is circular polarization of the l-th order, then:

$$|E_z(\rho,\varphi,z)|^2 = (kf)^2 \left| aJ_{m+l+1}(k\rho\sin\theta_0)\sqrt{\cos\theta_0}\sin^2\theta_0 \right|^2 \quad (4.2.38)$$

will be non-zero at $\rho = 0$ in the case of a vortex transmission function with $m = -(l+1)$.

But the situation in this case for all the components will be reduced to the usual circular polarization, because the transverse components receive a similar compensation (see Table 4.2.11).

Therefore, to form substantially subwavelength structures as in Table 4.2.12, it is necessary that $l \neq p$, even if the $|l| = |p|$.

For example, if $l = -p$, i.e. there the vortex polarization of l-th order at the input, then:

Table 4.2.11. The transverse ($z = 0$, $3\lambda \times 3\lambda$) field intensity distribution for the vortex polarization with coefficients $a(\phi) = \exp(i9\phi)$, $b(\phi) = \exp(i9\phi)$, $c(\phi) = 1$

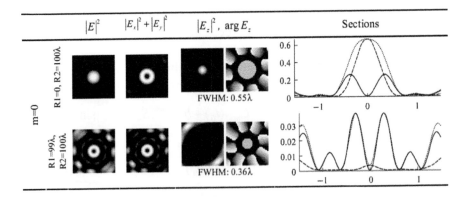

Table 4.2.12. The transverse ($z = 0$, $3\lambda \times 3\lambda$) field intensity distribution for the vortex polarization with coefficients $a(\phi) = \exp(2i\phi)$, $b(\phi) = \exp(-i3\phi)$, $c(\phi) = 0$, using a narrow annular aperture with radii $R1 = 99\lambda$, $R2 = 100\lambda$

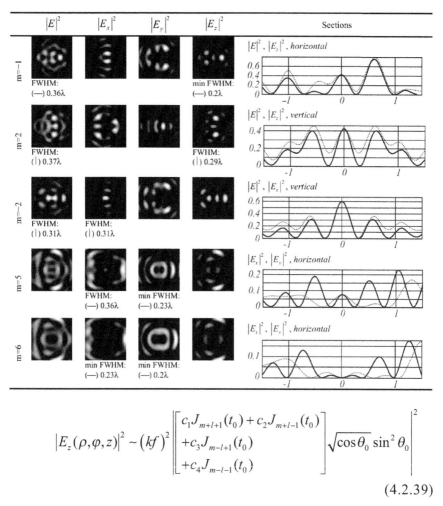

$$|E_z(\rho,\varphi,z)|^2 \sim (kf)^2 \left| \begin{bmatrix} c_1 J_{m+l+1}(t_0) + c_2 J_{m+l-1}(t_0) \\ +c_3 J_{m-l+1}(t_0) \\ +c_4 J_{m-l-1}(t_0) \end{bmatrix} \sqrt{\cos\theta_0} \sin^2\theta_0 \right|^2$$

(4.2.39)

will be non-zero at $\rho = 0$ in the same cases as for arbitrary $l \neq p$ (see Table 4.2.12).

4.2.4. Conclusions

In this section, the analytical and numerical analysis was carried out of the possibility of subwavelength light localization and deepening of the focus of the high-aperture focusing system using the vortex phase transmission functions for different types of input polarization.

As a generalized case we reviewed the polarization the components of which can be represented as a superposition of angular harmonics.

It is shown that in sharp focusing of the laser beams with the vortex singularity the subwavelength localization of light (and shadow areas) in the individual components of the vector field is possible with all types of polarization. This moment is useful to consider in the interaction of laser radiation with matter, selectively sensitive to the transverse and longitudinal components of the electromagnetic field.

In order to form a subwavelength structure in the total field intensity we need certain types of polarization and additional apodization of the function of the pupil of the focusing system, such as a narrow annular slit. The optimal choice of the radius of such a slit, maximizing the contribution of the specific components, allows to reach a compromise between the depth of focus and the size of the focal spot.

For example, for linear x-polarization by maximizing the transverse component we can obtain in the total intensity an elongated focal spot with FWHM 0.47λ for the narrow side invariant to defocusing over more than 50 wavelengths. When FWHM 0.31λ sensitivity to defocusing is much higher – at a distance of 15 wavelengths from the focal plane.

In the case of the azimuthal polarization a similar result will be symmetrical and a circular focal spot having no longitudinal component, with FWHM 0.37λ and the total length of about 30 wavelengths can be formed.

Also, a symmetrical result, but with a lower degree of the subwavelength localization is obtained using circular polarization. At the same time, as in linear polarization, changing the order of the vortex singularity we can control the contribution to the longitudinal and transverse components in the axial distribution of the total intensity.

In addition, the longitudinal component is amplified by a narrow annular aperture with the maximum radius and has the greatest power in the radial polarization.

Using the generalized vortex polarization coupled with the vortex beams limited by a narrow annular aperture, allows to create in the focal region complex subwavelength details in both total and component's intensity distributions. In the above examples components with FWHM less than 0.22λ were obtained.

4.3. Application of axicons in high-aperture focusing systems

In [63] using the paraxial scalar model it was shown that a lensacon [64] consisting of two low-aperture elements (lens and axicon), provides a central light spot, whose size corresponds to the non-linear increase in the numerical aperture of the tandem.

The presence in the tandem at least one high-aperture element requires a more rigorous theory. This section discusses the linear and vortex axicons as a supplement to the high-aperture lens in the model of the sharply focusing system in the Debye approximation.

As shown in the previous section, reducing the transverse dimension of the focal spot and/or increasing the longitudinal extent of the focal field of the high-aperture focusing system is achieved most efficiently in the radial polarization of the incident light. This effect is due to the fact that in the radial polarization the best energy redistribution in the z-component of the electric field is obtained in the focus.

However, even in this case we cannot achieve the theoretically predicted limit according to the size of the zero-order Bessel beam (FWHM = 0.36λ) [65]. The point is that, despite the strong longitudinal component, the transverse components also provide a significant contribution to the focal intensity increasing the general size of the central spot. To minimize the contribution of the transverse components of the focal area we need to make additional efforts. For example, as shown in the previous case, it is possible to use a narrow annular aperture allowing the passage throught the lens of only peripheral rays with the greatest angle to the optical axis. But this simple to implement method leads to a considerable loss of efficiency.

To ensure the full-aperture gain of the longitudinal components it is promising a parabolic mirror and a diffraction lens [15], as well as a microaxicon [21]. All these elements must be of the high-aperture type, and hence in the case of the diffractive elements the microrelief the suwavalength zones that are quite complicated to manufacture.

Also known are studies in which zone plates were used [18, 22, 65]; the radii of the zones in these plates were calculated using a special procedure according to the parameters of the focusing system, and increasing the number of zones to more than three is undesirable due to the loss of efficiency [18].

In this section, it is proposed to use phase axicons as similar full-aperture optical elements, removing the transverse components from the focal region in sharp focusing of the radially polarized beam. When the high-aperture lens is supplemented by even a 'weak' convergent axicon, the focal region looks like a cone, the tip of which has a smaller transverse dimension than the focal spot of a separate lens. This effect is due to the fact that the axicon increases the numerical aperture of the central rays of the lens and allows to redirect them from the focal plane closer to the lens plane. Depending on the parameters of the axicon we vary the length and 'sharpness'of the cone formed.

In addition to the longitudinal component, the transverse components of the electric field are also useful in various applications. In particular, during the passage through the hollow metal waveguides the radially polarized beams exhibit large energy losses on the walls of the waveguide, and azimuthally-polarized beams, respectively, the minimum losses [66]. Also, the various components of the vector electromagnetic field vector can be for three-dimensionally oriented excitation of fluorescent molecules [61].

This section shows that in the linear polarization, which occurs in the majority of lasers, the application of the high-aperture lensacon a light spot in the focus having in the transverse component the area at half intensity HMA = 0.139λ instead of 0.237λ for an individual lens. However, due to the side lobes, consisting mainly of the longitudinal component, the total electric field (the sum of the intensities of all components) in focus looks flattened in one direction. The size of the light spot in this area can be reduced to 0.32λ.

The effectiveness of using vortex axicons [67, 68] in the task of forming the long narrow distributions for circular and azimuthal polarizations has also been shown. In this case the compensation of polarization singularity [69] present in the circular polarization takes place, which allows to obtain axially symmetric distributions. A similar result is obtained in azimuthal polarization with the only difference being that in this case the light spot is more compact due to the absence of the longitudinal component.

The possibility of achieving subwavelength localization of light in the various components of the electric vector field using the vortex function has been thoroughly discussed in the previous section [70]. This section shows that the introduction of the vortex component in the axicon with circular polarization can control the contribution of various components of the vector field in the central portion, and hence the focal cone tip that may be useful in the interaction of electromagnetic radiation with the substances having selective sensitivity to the longitudinal or the transverse component of the vector field [60, 61].

4.3.1. Adding the axicon to the lens: the paraxial scalar model

When illuminating a spherical lens with a focal length f with a flat beam limited by a circular aperture with radius R, a pattern forms in the focal plane of the lens the transverse amplitude of which is proportional to $\sim J_1(k\rho R f^{-1})\rho^{-1}$, $k = 2\pi/\lambda$ [71]. The size of the central spot of this distribution is determined by the first root of the Bessel function of the first order $J_1(\gamma_{11}) = 0$, $\gamma_{11} = 3.83$.

Thus, the minimum achievable radius of the central spots, formed by the lens with the circular aperture, is defined by the expression:

$$\rho_{lens} = \frac{3.83\lambda f}{2\pi R} = 1.22\frac{\lambda f}{2R} \qquad (4.3.1)$$

We use the expression for the numerical aperture of aplanatic lens in a medium with refractive index n:

$$NA_{lens} = n\sin\theta \approx n\frac{R}{f} \qquad (4.3.2)$$

where θ is the so-called aperture angle of the lens (Fig. 4.3.1).

Then it is possible to estimate the minimum radius of the focal spot in the free space as follows:

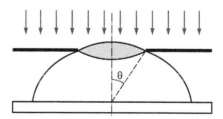

Fig. 4.3.1. Aperture angle of aplanatic lens.

$$\rho_{lens} = \frac{0.61}{NA}\lambda \geq 0.61\lambda \qquad (4.3.3)$$

Also, the size of the focal spot is determined by the full width at half maximum of the intensity (Fig. 4.3.2), in the case of the lens it is:

$$\text{FWHM}_{\min}^{\text{lens}} = 0.515\lambda \qquad (4.3.4)$$

and the minimum area of the focal spot, which is determined by the area at half-decay of the intensity will be:

$$\text{HMA}_{\min}^{\text{lens}} = \pi \cdot \left(\text{FWHM}_{\min}^{\text{lens}} / 2\right)^2 \approx 0.21\lambda^2 \qquad (4.3.5)$$

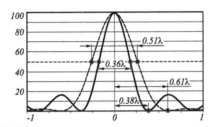

Fig. 4.3.2. The size of the focal spot for the lens (solid) and the axicon (dotted line).

The longitudinal size of the focal spot is associated with the concept of the depth of focus (DOF), which for the diffraction-limited systems is estimated by the expression:

$$\text{DOF}_{\text{lens}} = \frac{\lambda}{2\sin^2\theta} \geq 0.5\lambda \qquad (4.3.6)$$

from which it is evident that by increasing the numerical aperture the depth of focus decreases and tends to limit at half wavelength.

It is also known [72] that when using a lens with a narrow annular aperture a pattern is formed in the focal plane the amplitude of which is proportional to the zero-order Bessel function $-J_0(k\alpha_0\rho)$, with the first root of this function having a smaller value: $J_0(\gamma_{01}) = 0$, $\gamma_{01}=2.405$, hence the minimum achievable size becomes smaller.

However, such a scheme has low efficiency due to the loss of most of the energy shielded by the aperture.

The light field with the intensity proportional to the zero-order Bessel function $\sim |J_0(k\alpha_0\rho)|^2$ can be formed by using an optical element with better energy parameters – refractive axicon [73], as well as a linear diffraction axicon or a binary diffractive kinoform [74].

For the diffractive axicon the complex transmission function is described by:

$$\tau_{\text{ax}}(r) = \exp(-ik\alpha_0 r) \qquad (4.3.7)$$

where the parameter α_0 determines the angle of convergence θ of the rays from the axicon to the optical axis:

$$\alpha_0 = \sin\theta \qquad (4.3.8)$$

and in fact is equal to the numerical aperture of the axicon.

The radius of the central spot in free space:

$$\rho_{ax} = \frac{2.405}{k\alpha_0} = \frac{0.38}{\sin\theta}\lambda \geq 0.38\lambda \qquad (4.3.9)$$

The size of the central light spot defined by the full width at half-decay of intensity (Fig. 4.3.2) for the axicon is:

$$\text{FWHM}^{ax}_{min} = 0.357\lambda \qquad (4.3.10)$$

and the minimum area of the focal spot, which is determined by the area at half-decay of intensity will be:

$$\text{HMA}^{ax}_{min} = \pi \cdot \left(\text{FWHM}^{ax}_{min}/2\right)^2 = 0.1\lambda^2 \qquad (4.3.11)$$

The length of the focal region (maximum distance of the conservation of diffraction-free beam propagation):

$$z^{ax}_{max} = \frac{R}{\alpha_0} \qquad (4.3.12)$$

and in the limiting case at $\sin\theta \to 1$ when the DOF of the axicon is equal to the radius of the optical element R.

To achieve the minimum diffraction limit in the free space, all the elements listed above must have the maximum numerical aperture.

However, if we consider a tandem of two optical elements, in particular such as the lensacon [64], then the focal spot close to the diffraction limit cna be formed using 'weaker' optics [65].

If we use a collecting axicon, the complex transmission function of the lensacon is:

$$\tau_{lx}(r) = \exp\left[-ik\left(\alpha_0 r + \frac{r^2}{2f}\right)\right] \qquad (4.3.13)$$

and in this case, a Bessel beam with a decreasing scale forms [65]:

$$I_{lx}(\rho,z) \sim \frac{2\pi A^2}{z\alpha_0}\left(\frac{\alpha_0 fz}{f-z}\right)^3 J_0^2\left(\frac{k\alpha_0 f}{f-z}\rho\right) = \eta(z)J_0^2[\beta(z)\rho],$$

$$(4.3.14)$$

up to the distance

$$z_{max}^{lx} = \frac{R}{\alpha_0 + R/f} \le z_{max}^{ax} \qquad (4.3.15)$$

The minimum radius of the central light spot is achieved at the end of the segment of self-reproduction:

$$\rho_{min}^{lx} = \frac{2.405}{2\pi(\sin\theta + R/f)}\lambda = \frac{0.38}{(\sin\theta + R/f)}\lambda \qquad (4.3.16)$$

and, in general, it will be smaller than the focal spot formed by each of the elements of the tandem individually.

Depending on the ratio α_0 and the numerical aperture of the lens the decrease in the central light spot when the axicon is added to the lens (or vice versa, the lens to the axicon) can be very significant.

In particular at $\alpha_0 = 0.5$, ($\rho_{ax} = 0.76\lambda$) and $R/f = 0.5$ ($\rho_{lens} = 1.22\lambda$) we should obtained the minimum attainable limit for the axicon $\rho_{min}^{lx} = 0.38\lambda$. When $R = 2000\lambda$ (about 1 mm for optical wavelengths) $z_{min}^{ax} = 2R \approx 2$ mm, and $z_{max}^{lx} = R \approx 1$ mm.

For the numerical simulation of the action of the lensacon in the paraxial case we can use the Fresnel transform for the axially symmetric fields:

$$F(\rho,z) = \frac{k}{iz}\exp(ikz)\exp\left(\frac{ik\rho^2}{2z}\right)\int_0^\infty \tau(r)\exp\left(\frac{ikr^2}{2z}\right)J_0\left(\frac{kr\rho}{z}\right)r\,dr. \qquad (4.3.17)$$

Table 2.3.1 shows the results of calculations with the following parameters: $\lambda = 0.5$ μm, $R = 1$ mm, $f = 2$ mm (numerical aperture $NA = R/f = 0.5$), $\alpha_0 = 0.5$, 20 pixels per wavelength were used for integrands, and the axial region of the radius 2λ (1 μm) had 50 pixels per wavelength. The longitudinal intensity distribution was calculated in the range $z \in [0.5$ mm, 2.5 mm].

The results of numerical simulation agree well with the above reasoning. Longitudinal distribution patterns demonstrate how the energy is redistributed along the optical axis.

Joining the lens with the axicon, on the one hand, halves the segment of the self-reproduction of the diffraction-free beam in comparisn with the axicon, but, on the other hand, a more compact pattern is formed, and the intensity in the central light spot is increased by almost an order of magnitude. As compared with the lens, the intensity of the focal spot is almost 5 times less, but the depth of focus increases by 40 times.

Table 4.3.1. Average aperture lensacon in the paraxial model

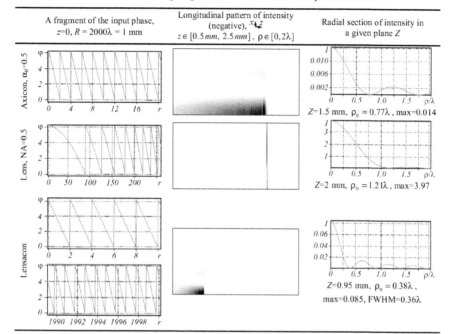

A fragment of the input phase, $z=0$, $R = 2000\lambda = 1$ mm	Longitudinal pattern of intensity (negative), $^x\!\!\downarrow\!z$ $z \in [0.5\,mm,\ 2.5\,mm]$, $\rho \in [0,2\lambda]$	Radial section of intensity in a given plane Z
Axicon, $\alpha_0=0.5$		$Z=1.5$ mm, $\rho_0 = 0.77\lambda$, max=0.014
Lens, NA=0.5		$Z=2$ mm, $\rho_0 = 1.21\lambda$, max=3.97
Lensacon		$Z=0.95$ mm, $\rho_0 = 0.38\lambda$, max=0.085, FWHM=0.36λ

4.3.2. Apodization of the short-focus lens by the axicon: the non-paraxial vector model in the Debye approximation

Consider a high-aperture aplanatic focusing optical system (4.2.1) in the radial vortex representation (4.2.7).

Linear x-polarization

Most modern lasers have a linear polarization. Figure 4.3.3 shows the axial intensity distribution pattern formed by a lensacon with radius $R = 1$ mm, composed of a lens with a focal length $f = 2$ mm ($NA = 0.5$) and an axicon with different values of the parameter α_0 (from 0.1 to 0.5) using paraxial (Fig. 4.3.3 a) and non-paraxial vector (Fig. 4.3.3 b) models. In the latter (non-paraxial) case, the incident wave is linearly polarized, and Fig. 4.3.3 b shows the total intensity of the electric field vector. The area in question has the dimensions $z \in [500\lambda, 4000\lambda]$ with wavelength $\lambda = 0.5$ μm, and in the non-paraxial model the distance is counted from the focal plane.

The results shown in Fig. 4.3.3 indicate that when using a weak axicon there is good agreement between the results of the paraxial and non-paraxial models, but with an increase in the numerical aperture of the axicon the mismatch is becoming stronger, and at $\alpha_0 = 0.5$ the Debye approximation actually stops working because the considered the area is too far from the plane of focus.

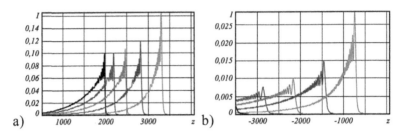

a) b)

Fig. 4.3.3. The axial intensity distribution generated by the lensacon in the paraxial (a) and non-paraxial (b) cases for a lens with $NA = 0.5$ ($R = 2000\lambda$, $f = 4000\lambda$) and axicons with α_0: 0.1 (blue line), 0.2 (red line), 0.3 (green line), 0.4 (brown line), 0.5 (black line).

This situation can be avoided by raising the numerical aperture of the focusing lens.

Figure 4.3.4 gives the results for the tandem with the radius $R = 1$ mm, consisting of a lens with a focal length $f = 1.05$ mm ($NA = 0.95$) and the axicon with different values of the parameter α_0 (0.1 to 0.95); the considered axial segment $z \in [600\lambda, 2150\lambda]$. The application of the paraxial model when using high-aperture optical elements is not correct, so the results for the two models differ significantly.

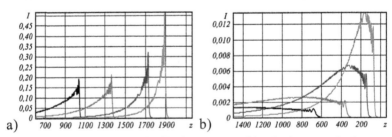

a) b)

Fig. 4.3.4. The axial intensity distribution generated by the lensacon in the paraxial (a) and non-paraxial (b) cases for a lens with $NA = 0.95$ ($R = 2000\lambda$, $f = 2100\lambda$) and axicons with α_0: 0.1 (blue line), 0.2 (red line), 0.5 (green line). and 0.95 (black line).

Next we will consider a including a high-aperture lens with focus $f = 101\lambda$ and the radius $R = 100\lambda$. For the infrared wavelength range (for example, $\lambda = 10$ μm) such an element can not be attributed to micro-optical elements, because for a hundred wavelengths the radius is already $R = 1$ mm, which is quite convenient for the experiments.

The correctness of the results in the Debye approximation is provided at a high Fresnel number

$$N_F = \frac{R^2}{\lambda f}$$

(4.3.18)

In this case $N_F \approx 100$ is sufficiently large [55].

The results of numerical simulation for a single lens with focus $f = 101\lambda$, and the radius $R = 100\lambda$ and a lensacon with different axicons are shown in Table 4.3.2. The number of pixels in the azimuthal angle is equal to 200λ, the transverse dimension of the focal region is $4\lambda \times 4\lambda$.

As can be seen from Table. 4.3.2 for sharp focusing the behaviour of the beam formed by a lensacon with a weak axicon is close to that predicted by the paraxial model.

However, due to the contribution of the longitudinal component the cross section ceases to be symmetrical and looks elongated. At the same time, on the vertical axis it is possible to overcome the limit reached not only by the lens, but also the axicon: FWHM = 0.32λ. Importantly, in the tandem with a strong lens, even when using a weak axicon (in particular with a numerical aperture of 0.1) a limit can be achieved in one direction generated separately by the high-aperture axicon (FWHM = 0.36λ).

Table 4.3.2. Longitudinal and transverse distribution of the linear x-polarization

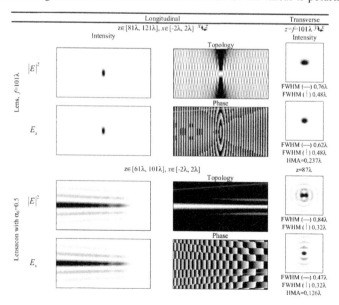

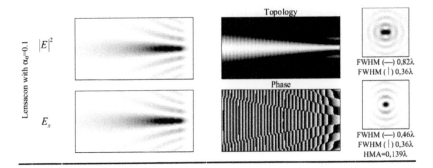

*) Topology – the distribution without normalization of intensity to the maximum value in the considered range of values of z.

Figure 4.3.5 shows the results of the comparison of the paraxial and non-paraxial models for a tandem with radius $R = 100\lambda$ for a lens with a focal length $f = 101\lambda$ mm ($NA = 0.95$) and axicons with different values of parameter α_0: from 0 (no axicon) to 0.95. The considered axial segment $z \in [20\lambda, 120\lambda]$, in the transverse plane $\rho \in [0, 0.6\lambda]$. In the non-paraxial vector case the pattern of the transverse distribution ceases to be symmetric, so Fig. 4.3.5 d shows a vertical section when the total intensity is equal to the intensity of the x-component.

As can be seen from Fig. 4.3.5 b for the high-aperture lens and the axicon the paraxial model predicts a decrease in the radius of the central light spot to 0.2λ. At the same time, when using the non-paraxial vector model the narrowing can be esimated only to 0.33λ in one coordinate.

The main contribution to the total intensity of the axial distribution is provided by the x-component:

$$E_x(\rho,z) = -\frac{kf}{2}\int_0^\alpha B(\theta)\exp(ikz\cos\theta)\sqrt{\cos\theta}\,(1+\cos\theta)J_0(k\rho\sin\theta)\sin\theta\,d\theta.$$

$$(4.3.19)$$

Imposition of an annular aperture on the lens: $B(\theta) = \delta(\theta - \theta_0)$ leads to the formation of the field, which is proportional to the zero-order Bessel function, which is independent of defocusing z:

$$|E_x(\rho,z)|^2 = \frac{(kf)^2}{4}J_0^2(k\rho\sin\theta_0)\left|\cos\theta_0(1+\cos\theta_0)^2\sin^2\theta_0\right|. \quad (4.3.20)$$

The radius of the central spot is associated with an azimuth angle θ_0 and can not be less than

$$\rho^\delta_{min} = 0.38\lambda \qquad (4.3.21)$$

When the axicon is used as $B(\theta)$, the distribution on the axis (at $\rho = 0$) is described by the expression:

$$E_x(0,z) = \frac{kf}{2} \int_0^\alpha \exp\left[ik(z\cos\theta - \alpha_0 f \sin\theta)\right]\sqrt{\cos\theta}\,(1+\cos\theta)\sin\theta\,d\theta,$$

$$(4.3.22)$$

which shows that for small values of the numerical aperture, i.e. small angles θ, the effect of the axicon is insignificant, but if the numerical aperture increases the effect is enhanced, and at $\cos(\theta) \rightarrow 0$ the dependence on the distance z disappears.

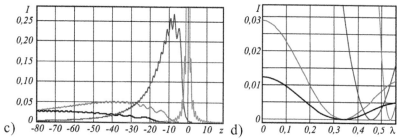

c)   -80 -70 -60 -50 -40 -30 -20 -10 0   10 z d)   0   0,1 0,2 0,3 0,4 0,5 λ

Fig. 4.3.5. Axial (a), (c) and transverse vertical (b), (d) intensity distributions formed by the lensacon in the paraxial (a), (b) and non-paraxial (c), (d) case for a lens with $NA = 0, 95$ ($R = 100\lambda$, $f = 101\lambda$) and axicons with α_0: 0 (no axicon, blue line), 0.1 (red line), 0.5 (green line) and 0.95 (black line).

Radial polarization

In this case, using the high-aperture focusing optical system the main contribution to the overall axial intensity distribution is provided by the longitudinal component [75]:

$$E_z(\rho,z) = -kf \int_0^\alpha B(\theta)\exp(ikz\cos\theta)\sqrt{\cos\theta}\,J_0(k\rho\sin\theta)\sin^2\theta\,d\theta.$$

$$(4.3.23)$$

Superposition of a narrow annular aperture on the lens also leads to the formation of a field proportional to the zeroth order Bessel function:

$$|E_z(\rho,z)|^2 = (kf)^2\, J_0^2(k\rho\sin\theta_0)\left|\cos\theta_0\sin^4\theta_0\right|, \qquad (4.3.24)$$

but the intensity (4.3.24) as $\theta_0 \rightarrow 90°$ will be significantly higher than (4.3.20). In particular, at $\theta_0 = 82°$ the increase will be equal to a factor of 3.

However, inserting a narrow annular aperture to the optical system is associated with the loss of energy efficiency, and it is therefore desirable to perform the phase (not amplitude) apodization of the pupil function.

If the axicon is used as $B(\theta)$, the intensity on the axis (at $\rho = 0$) is described by an expression similar to (4.3.22):

$$E_z(\rho,z) = -kf \int_0^\alpha \exp\left[ik(z\cos\theta - \alpha_0 f \sin\theta)\right]\sqrt{\cos\theta}\,\sin^2\theta\,d\theta.$$
$$(4.3.25)$$

We determine the distance at which the axial value does not vary greatly from the condition:

$$z_{max} \le \frac{\varepsilon}{\cos\theta} + \alpha_0 f\,\mathrm{tg}\,\theta \qquad (4.3.26)$$

from which it is clear that at $\theta \to 90°$ $z_{max} \to \infty$ (similar to introducing a narrow annular aperture), while for small ρ we should have $z_{max} \le \varepsilon + \alpha_0 f \theta$, i.e. increasing the range of the weak variation of the function (4.3.25) is due to the introduction of the term $\alpha_0 f \theta$ associated with the phase function of the axicon.

Thus, the addition of an axicon to the lens increases the numerical aperture of the lens by a constant value which for the high-aperture lenses affects only the rays passing through the central part (Fig. 4.3.6 a).

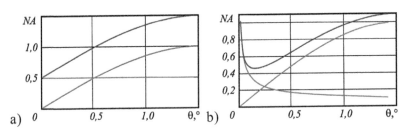

Fig. 4.3.6. The graph for the numerical aperture NA for (a) the lensacon (blue corresponds to high-aperture lens, green – axicon with $\alpha_0 = 0.5$, and red – the total numerical aperture) and (b) the lens, supplemented by the fraxicon $\exp\left(-i\sqrt{r}\right)$.

Obviously, the optimum addition in terms of the alignment of the numerical aperture with respect to the lens is the phase element the numerical aperture of which has a dependence: $1-\sin\theta$. This tandem will operate as a 'strong' linear axicon.

Figure 4.3.6 *b* shows an example of adding a fracxicon [42] of the type $\exp\left(-i\alpha\sqrt{r}\right)$ with $\alpha = 1$ to the high-aperture lens. In this case, the numerical aperture almost never surpasses the limit for the free space and significantly increases in the central part.

Table 4.3.3 gives the results of numerical simulations of focusing a plane wave limited by an annular aperture with radii $R_1 = 99\lambda$ and $R_2 = 100\lambda$, a lens with a focal length of 101λ, as well as the focusing of a plane wave limited by a circular aperture of radius $R = 100\lambda$, with a lensacon with the axicon parameter $\alpha_0 = 0.95$ ($z^{ax}_{max} \approx 105\lambda$) with radial polarization. The number of pixels for the azimuthal angle is equal to 200λ. The transverse dimension $-4\lambda \times 4\lambda$.

As can be seen from Table 4.3.3, in the case of radial polarization the distribution of the total intensity has radial symmetry. The radius of the central light spot reaches the minimum predicted for the zeroth order Bessel function and never decreases.

The contribution of the transverse components for the lens with a narrow (width λ) annular aperture is virtually absent in the focal region. The energy loss in inserting a diaphragm is proportional to the area of the blocked cenral part of the lens.

A lensacon with the full aperture, and thus no loss of input energy, is characterized by the formation of a conical focal region and there is no contribution of the transverse components to the tip of this region. By increasing the numerical aperture of the axicon the focal cone becomes more elongated along the optical axis (see Table 4.3.4) with an almost uniformly distributed intensity (Fig. 4.3.7). Note that the core and the sharp edge of the cone consist only of the longitudinal component, and the 'tail' is complemented by the transverse component (Fig. 4.3.8).

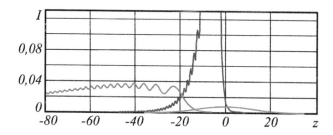

Fig. 4.3.7. The axial distribution of the total intensity, formed by a lens with a narrow annular aperture (blue) and the full-aperture lensacon with $\alpha_0 = 0.1$ (red) and the lensacon with $\alpha_0 = 0.95$ (green).

Table 4.3.3. Longitudinal and transverse distributions in radial polarization for the lens with a narrow annular aperture and a lensacon with a high-aperture axicon

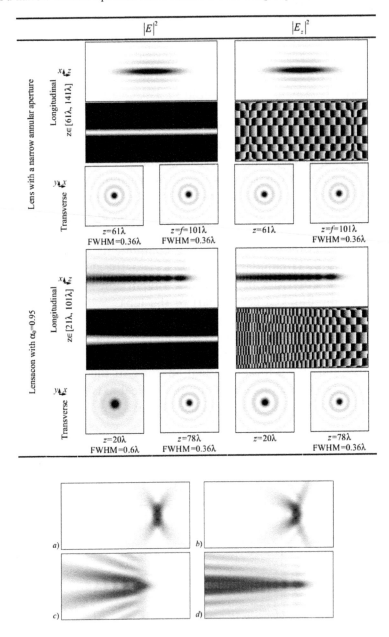

Fig. 4.3.8. The longitudinal distribution of the transverse component (red) and the longitudinal component (blue) for (a) a separate lens and a lensacon with (b) $\alpha_0 = 0.01$, (c) $\alpha_0 = 0.05$, and (d) $\alpha_0 = 0.95$.

Table 4.3.4. Longitudinal distribution in radial polarization for the full-aperture lens and lensacon with low-aperture axicons, transverse dimension $4\lambda \times 4\lambda$

	Transmission function	Intensity (negative) $z \in [86\lambda, 106\lambda], x \in [-2\lambda, 2\lambda]$	Topology
Lens			
Lensacon with $\alpha_0 = 0.01$			
Lensacon with $\alpha_0 = 0.05$			
Lensacon with $\alpha_0 = 0.1$			

4.3.3. Using vortex axicons for spatial redistribution of components of the electric field in the focal region

In [75] the authors demonstrated the possibility of subwavelength localization of radiation in sharp focusing of the vortex field. The central light spot for the azimuthal and circular polarizations is symmetrical.

In azimuthal polarization the effect of subwavelength localization is only possible for the transverse component of the vector electric field, and in circular polarization – also for the longitudinal component.

Azimuthal polarization

In the second section it was shown that for the azimuthal polarization the presence of the vortex component in the transmission function (4.2.5) leads to the field in the focal region in the form of (4.2.31). As follows from (4.2.31), the longitudinal component is always missing.

Table 4.3.5. Longitudinal and transverse ($4\lambda \times 4\lambda$) distributions of the total intensity of the field for azimuthally polarized vortex transmission function ($|m|=1$) and vortex lensacon

Transmission function	Longitudinal $z \in [-30\lambda, 5\lambda]$, $x \in [-2\lambda, 2\lambda]$ $^{x} \!\!\perp_{\!z}$	Tranverse $^{y} \!\!\perp_{\!x}$
Lens with a vortex phase function		$z=f=101\lambda$ FWHM: 0.5λ
Vortex lensacon with $\alpha_0=0.1$		$z=96\lambda$ FWHM: 0.41λ

On the optical axis (at $\rho = 0$) the focal field will contain non-zero components for only one order of the vortex singularity $|m| = 1$, which are expressed through $J_0(t)$.

Circular polarization

For circular polarization the presence of the vortex component in the transmission function allows one to get the field of the form (4.2.30) in the focal region. On the optical axis (at $\rho = 0$) the focal field will contain non-zero-valued components for the vortex singularity orders $|m| \leq 2$, which are also expressed in terms of the zero-order Bessel functions.

The use of vortex filters allows to select various components of the vector field (longitudinal or transverse) in the axial distribution: vortex phase singularity compensates polarization singularities [69, 75], and the various components include the non-zero axial terms proportional to $J_0(t)$ that determines the minimum achievable size of the central spot of the Bessel beam of the zero order, i.e. FWHM = 0.36λ.

If weak vortex axicons are used, as shown in the Table 4.3.7, the intensity in the axial segment varies considerably and the formation of a compact spot light occurs only at the end of this segment. However, the intensity at the edge of the focal cone in this case is much larger (comparison is shown in Fig. 4.3.5c), which may be useful in some problems.

Table 4.3.6. Longitudinal and transverse ($4\lambda \times 4\lambda$) distributions of the total intensity of the field and components for the vortex lensacon with $\alpha_0 = 0.95$ ($|m| \leq 2$) and high-aperture lens ($NA = 0.95$) with circular polarization

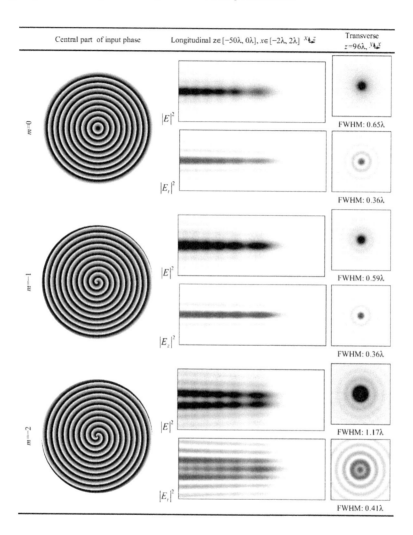

Thus, the results of numerical simulation in the Debye approximation show that it is possible to form a compact light spot when using the high-aperture lens and the low-aperture axicon in the tandem. The increase in the numerical aperture of the axicon will lead to a longer focal region with the nearly uniform intensity distribution along this region, which will lead to a decrease in the intensity values in each cross section.

Table 4.3.7. Longitudinal and transverse ($4\lambda \times 4\lambda$) distributions of the total intensity of the field and components for the vortex lensacon with $\alpha_0 = 0.1$ ($|m| \leq 2$) and high-aperture lens ($NA = 0.95$) with circular polarization

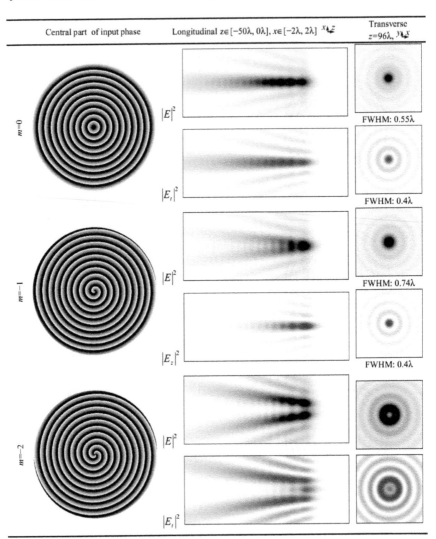

4.3.4. Conclusions

The section deals with the application of axicons in high-aperture focusing systems with different polarization in the Debye model.

The studies confirmed the possibility of reducing the size of the focal spot formed by the high-aperture lens by supplementing the lens with a diffractive axicon (the effect is observed even when using the low-aperture axicon). This approach provides an alternative and energy-efficient concept as compared to the annular aperture.

The focal area of the use of such lensacon looks like a cone, which is consistent with the results of paraxial modelling [65]. However, in contrast to the paraxial model, the non-paraxial assessments determine as the minimum achievable limit the size of the central spot of the zero-order Bessel beam (FWHM = 0.36λ). Depending on the 'strength' of the diffraction axicon the focusing in the range of these subwavelength dimensions it occurs on a more or less extended segment.

However, in case of linear polarization, this limit can be overcome (FWHM = 0.32λ) in one of the transverse directions.

For the circular polarization, the focal spot size reduction while maintaining its axial symmetry can be achieved using vortex axicons. In this case the compensation of polarization singularities takes place.

Similar compensation is obtained for the azimuthal polarization with the difference that in this case a more compact light spot is formed in the absence of the longitudinal component.

In addition, we have shown that the introduction of the vortex component (first and second order) to the axicon in the case of circular polarization allows one to manage the contribution of different components of the vector field to the core and edge of the focal cone. In this case, each component can achieve subwavelength localization of light (FWHM = 0.41λ), which can be useful in the interaction of electromagnetic radiation with the substances having selective sensitivity to the various components of the vector field.

4.4. Control of contributions of components of the vector electric field at the focus of the high-aperture lens using binary phase structures

In the previous section we demonstrated the possibility of isolating individual components of the electric field when using axicons and vortex axicons as an additional transmission function of the lens. However, the production of multilayer diffractive optical elements, allowing to implement such transmission function, is associated with certain difficulties. On the other hand, the superposition of optical vortices with opposite signs can be produced using a binary phase function [76, 77].

In this section, in order to reduce the transverse dimension of the central light spot of the focusing system with a high numerical aperture we consider changing the contribution of various components of the vector electric field in the focal region through supplementing this system with binary phase diffractive optical elements.

4.4.1. Maximizing the longitudinal component for linearly polarized radiation

Since most modern lasers emit linearly polarized light, and the use of polarization converters leads to considerable complication of the optical circuit, the search for simple ways to reduce the size of the focal spot in linear polarization remains relevant. Without the loss of generality, we consider polarization along one of the transverse axes.

With linear x-polarization and the transmission function (4.2.4) without the vortex component $\Omega_B(\phi)=1$ the optical axis different from zero has only the x-component of the vector (4.2.7):

$$E_x(0,0,z) = -\frac{ikf}{2} \int_0^\alpha R(\theta)T(\theta)\sin\theta(\cos\theta+1)\exp(ikz\cos\theta)d\theta. \qquad (4.4.1)$$

If, however $\Omega_B(\phi)=\cos\phi$, then values on the optical axis only the z-component will differ from zero:

$$E_z(0,0,z) = \frac{ikf}{2} \int_0^\alpha R(\theta)T(\theta)\sin^2\theta\exp(ikz\cos\theta)d\theta, \qquad (4.4.2)$$

and when $\Omega_B(\phi)=\sin\phi$ we have the zero in the total intensity of the electric vector $|E|^2 = |E_x|^2 + |E_y|^2 + |E_z|^2$.

If, however $\Omega_B(\phi)=\sin 2\phi$, then only the y-component on the optical axis is different from zero:

$$E_y(0,0,z) = -\frac{ikf}{2} \int_0^\alpha R(\theta)T(\theta)\sin\theta(1-\cos\theta)\exp(ikz\cos\theta)d\theta. \qquad (4.4.3)$$

The expressions in (4.4.1)–(4.4.3), with other conditions being equal, differ in the integrands: $1+\cos\theta$ for the x-component, $\sin\theta$ for the z-component, and $1-\cos\theta$ for the y-component, respectively. Obviously, in the range of angles $\theta \in [0, \pi/2]$, the total contribution to the x-component will be greater than to the y-component.

Figure 4.4.1 shows the path of the rays in the sharp focusing of the linearly polarized field. In presence of the binary phase changing the direction of vectors to half-rays to the opposite, the longitudinal components will add up, and the transverse ones – deducted.

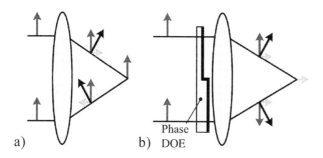

a)   b) Phase DOE

Fig. 4.4.1. Action of the binary phase transmission function of the high-aperture lens with linear polarization.

Table 4.4.1 shows the results of numerical modelling for an aplanatic (free of spherical aberration and coma) lens with a numerical aperture $NA_{obj} \approx 0.99$. In this case, the pupil apodization function $T(\theta) = \sqrt{\cos\theta}$ is considered. Red and green colours correspond to the transverse components (x and y), and blue – the longitudinal component. The table also shows the values of full width at half maximum of the intensity in the horizontal direction FWHM (–), the half maximum intensity area HMA and the intensity value at the central point of the focal plane I (0,0,0).

As can be seen from Table. 4.4.1 the introduction of the linear phase singularity leads to exclusion in the centre of the focal region of the transverse components and the appearance of the longitudinal component (except the 3rd line), and the maximum value is achieved at the perpendicular arrangement of the singular line and the direction of polarization.

For a diffractive lens we used the Helmholtz rule and the pupil apodization function $T(\theta) = (1/\cos\theta)^{3/2}$ [15]. Based on a comparison of the pupil apodization function for the lens and a diffractive lens (see Fig. 4.2.1), in [15] it was assumed that using a parabolic mirror or a diffractive lens we can achieve a more compact focusing.

The numerical aperture of the diffractive lens is determined by the formula:

Table 4.4.1. Simulation results for the aplanatic lens with $NA = 0.99$ at linear x-polarization, $R(\theta) = 1$

Phase transmission function	Longitudinal horizontal distribution in the plane $y=0$ $(^x\!\llcorner\!^z)$ $z \in [-3\lambda, 3\lambda]$, $x \in [-1,5\lambda, 1,5\lambda]$			Transverse distribution in focus	FWHM(λ), HMA, I(0,0,0)
	$\|E_x\|^2$	$\|E_z\|^2$	$\|E\|^2$	$\|E\|^2$ $(^x\!\llcorner\!^y)$	
					0.75λ 0.28λ² 3.32
					1.52λ 0.64λ² 1.58
					0.42λ - 0.07

$$NA_{dl} = \sin\left[\arctg\left(\frac{R}{f}\right) \right] n \qquad (4.4.4)$$

where R is the radius of the lens, f is the focal distance, n is the refractive index of the optical medium.

Table 4.4.2 shows the comparative results of the numerical simulation for the aplanatic lens and diffractive lenses with a high numerical aperture, having a transmission function $B(\theta,\phi) = R(\theta)\Omega_B(\phi)$, where $R(\theta) = 1$ and $\Omega_B(\phi) = 1$ (first line), $\Omega_B(\phi) = \arg(\cos\phi)$ (second row), and $\Omega_B(\phi) = \arg(\sin 2\phi)$ (third line). It shows the longitudinal horizontal section of the intensity $|E|^2$ in the region $z \in [-3\lambda, 3\lambda]$, $x \in [-1.5\lambda, 1.5\lambda]$ cross section in the focal plane $x, y \in [-1.5\lambda, 1.5\lambda]$.

From Table. 4.4.2 it is clear that using the diffractive lens with a numerical aperture $NA = 0.99$ focusing can be produced in a smaller area of the focal spot than with the aplanatic lens with the same numerical aperture. Also when adding the transmission function with a phase jump to the focusing system the transverse components in the central part are suppressed more efficiently and, therefore, the longitudinal component is increased on the axis. This is due to a sharp increase in the pupil apodization function at high angles θ for the diffractive lens, as shown in the second section in Fig. 4.2.1.

However, to sufficiently suppress the transverse components (red and green) near the axis to obtain a compact central light spot, it is necessary to increase the numerical aperture of the diffractive lens

Table 4.4.2. Simulation results for an aplanatic lens and a diffractive lens with a numerical aperture $NA = 0.99$ linear x-polarization, $R(\theta) = 1$

Phase transmission function	Aplanatic lens			Diffractive lens	
	Longitudinal ($^x\llcorner^z$, y=0) and transverse ($^x\llcorner^y$, z=0) distributions $|E|^2$		FWHM($|$), HMA, I(0,0,0)	Longitudinal ($^x\llcorner^z$, y=0) and transverse ($^x\llcorner^y$, z=0) distributions $|E|^2$	FWHM($|$), HMA, I(0,0,0)

(Table rows with plotted distributions)

Row 1: 0.75λ, 0.28λ², 3.32 (aplanatic); 0.82λ, 0.27λ², 1.34 (diffractive)

Row 2: 1.52λ, 0.64λ², 1.58 (aplanatic); 1.44λ, 0.45λ², 0.9 (diffractive)

Row 3: 0.41λ, -, 0.07 (aplanatic); 0.38λ, -, 0.11 (diffractive)

to $NA = 0.9987$ (Fig. 4.4.2). Thus, it is possible to overcome the diffraction limit (HMA = 0.1312) for the lens (HMA = 0.212), but not for the Bessel beam (HMA = 0.112). The value of the intensity in the central spot in this case decreases almost 8 times compared with the aplanatic lens due to the lengthening of the longitudinal size of the focal region and the emergence of side lobes. However, compared with the same diffractive lens not having any additions, there are no energy losses there.

The 'staggered' binary phase allows to select y-component in the plane $y = 0$ (on the horizontal FWHM (–) = 0.381), but its energy is initially too low in comparison with other components, so the intensity value at the central point is 30 times less than for the focal spot of the lens.

However, as will be shown below, when using a narrow annular slit the focal plane retains a substantially lower amount of energy.

It is known that at the radial polarization of the incident light the contribution of the transverse component to the focal region can be reduced and, therefore, the transverse dimension of the focal spot can also be reduced either by introducing a narrow annular aperture transmitting only the peripheral rays, or by means of annular structures, such as amplitude-phase and purely phase.

In the latter cases the energy in the central zone of the lens is not blocked and can be redirected from the focal region to another part of the space so as to form a predetermined three-dimensional distribution in the region around the focus.

As shown in Fig. 4.4.3 due to the phase of the ring structure in the central portion of the lens can increase the numerical aperture and change the inclination of the central rays, thereby redistributing the input vector field component along the optical axis.

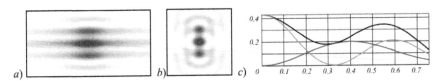

a)   b)   c)

Fig. 4.4.2. The simulation results for a diffractive lens with a numerical aperture $NA = 0.9987$ with linear x-polarization: $B(\theta,\phi)=\arg(\cos\phi)$: longitudinal ($y = 0$) (a) and transverse ($z = 0$) (b) distribution $|E|^2$, and (c) the section in the focal plane along the x-axis.

A similar approach was reviewed in the scalar case in [63], where the tandem of a lens and an ax Phase as presented in the form of diffractive optical elements whose phase function is proportional to the radial coordinate in fractional degrees.

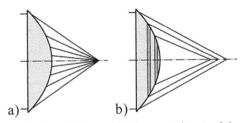

a)   b)

Fig. 4.4.3. The change in the of the rays in the central part of the lens by an annular phase structure.

Table 4.4.3 shows the comparative results of the simulation ($z \in [-20\lambda, 20\lambda]$, $x,y \in [-1.5\lambda, 1.5\lambda]$) for the aplanatic lens and the diffractive lens with $NA = 0.99$ with a narrow annular aperture approximating the function:

$$B_\delta (\theta,\phi) = \delta(\theta - \pi/2) \qquad (4.4.5)$$

and

$$B_{\delta c}(\theta,\phi) = \delta(\theta - \pi/2)\cos\phi \qquad (4.4.6)$$

As can be seen from Table 4.3.3, the imposition of a narrow annular aperture leads to 'pulling' of the focal spot along the optical axis, which corresponds to the formation of the zero-order Bessel beam for the x-component. Additional introduction of the phase jump allows one to get such distribution for the z-component.

In this case, the energy is distributed almost uniformly on a fairly extensive area, so only the appropriate fraction remains in the focal plane – the intensity in the central spot in 1600 times lower than for a lens without an aperture. When using the diffractive lenses we obtain about the same results, with the difference that the depth of focus (length of the focal region) is significantly greater and the proportion of energy in each plane is reduced proportionally.

Table 4.4.4 shows the simulation results for a high-aperture focusing system having a 'semi-annular' transmission function:

$$B1(\theta,\phi) = \arg[R1(\theta)\cos\phi] = \arg[\cos(0.01kf\sin\theta)\cos\phi], \quad (4.4.7)$$

$$B2(\theta,\phi) = \arg[R2(\theta)\cos\phi] = \arg[GL_{3,0}((0.03kf\sin\theta)^2)\cos\phi], \quad (4.4.8)$$

where $R_2(\theta) = GL_{3,0}(t)$ is the radial part of the Laguerre–Gaussian mode [5].

The amplitude–phase distributions, proportional to the radial-symmetric Laguerre–Gaussian modes, have been used in [19] for radial polarization.

As can be seen from Table 4.4.4, using binary phase apodization we can achieve a significant reduction of the size of the focal spot

Table 4.4.3. Simulation results for an aplanatic lens and a diffractive lens with a numerical aperture $NA = 0.99$ at linear x-polarization $z \in [-20\lambda, 20\lambda]$, $x,y \in [-1.5\lambda, 1.5\lambda]$

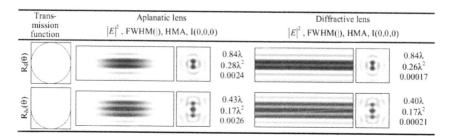

Trans-mission function	Aplanatic lens $|E|^2$, FWHM(|), HMA, I(0,0,0)				Diffractive lens $|E|^2$, FWHM(|), HMA, I(0,0,0)		
$R_\delta(\theta)$				0.84λ 0.28λ² 0.0024			0.84λ 0.26λ² 0.00017
$R_{\delta c}(\theta)$				0.43λ 0.17λ² 0.0026			0.40λ 0.17λ² 0.00021

Table 4.4.4. Simulation results for an aplanatic lens with $NA = 0.99$ supplemeted with a semi-annular structure at linear x-polarization $z \in [-3\lambda, 3\lambda]$, $x,y \in [-1.5\lambda, 1.5\lambda]$

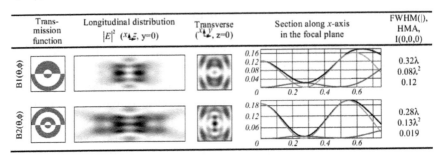

Trans-mission function	Longitudinal distribution $\|E\|^2$ (x-z, $y=0$)	Transverse (x-y, $z=0$)	Section along x-axis in the focal plane	FWHM(|), HMA, $I(0,0,0)$
B1(θ,ϕ)			0.16 / 0.12 / 0.08 / 0.04 / 0 0.2 0.4 0.6	0.32λ 0.08λ^2 0.12
B2(θ,ϕ)			0.18 / 0.12 / 0.06 0 0.2 0.4 0.6	0.28λ 0.13λ^2 0.019

(HMA = 0.08l2) without such a significant loss of energy as when using a narrow annular aperture, although compared to conventional sharp focusing the intensity in the central light spot decreases 27 times. This situation is in full accordance with the theory proposed by Toraldo de France [78], when the size of the central spot is reduced due to the loss of efficiency and growth of side lobes. Nevertheless, in some optical applications such losses are not significant, particularly when compared with the narrow diaphragm.

For example, the scanning microscopes can be operated with only a few photons which is half a dozen orders of magnitude less than the power of a conventional laser.

Circular polarization

Circular polarization is also the common and easy to implement type of polarization. The circular polarization can be produced from the linear by using the quarter-wave plate [79]. As shown above, by using the vortex transmission functions it is possible to compensate 'polarization singularity' present in the circular polarization, which allows to obtain axisymmetric distributions. The binary transmission function corresponds to the superposition of two vortex functions with opposite signs, one compensates the polarization singularity, and the second makes a phase singularity in the corresponding components. Therefore, the intensity minimum forms in the focal region when using the binary transmission functions.

4.4.2. Increasing efficiency by focusing the radially polarized radiation

In the radial polarization of the radiation incident on the high-aperture focusing system only the z-component of the vector (4.2.7) will differ from zero on the optical axis; this component is fully compatible with expression (4.4.2) but with a factor of 2, i.e. the intensity of the longitudinal component will be 4 times higher than for linear polarization.

However, only using the radial polarization we can not overcome the diffraction limit, in this case, FWHM = 0.54l.

Using a narrow annular aperture allows to form in the focal region an extended Bessel beam with the expected FWHM = 0.37l and low intensity (375 times lower than in the focal spot). Application of the full-aperture binary phase annular structures can significantly reduce the size of the central light spot with a much smaller energy loss – FWHM = 0.33l with the energy decreasing about 8 times compared with the conventional lens.

The bilens (lens with a linear phase jump), respectively, will increase in the central part the contribution of the transverse component, but the azimuthal polarization can be used more effectively in this case.

As can be seen from Table 4.4.5, according to the findings of [80], using axicons and other annular phase structures, such as the radial part of expressions (4.4.7) and (4.4.8), as the transmission function can increase the numerical aperture of the lens in the

Table 4.4.5. Simulation results for the aplanatic lens with NA = 0.99 with radial polarization

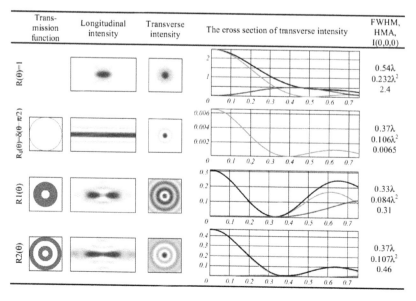

Transmission function	Longitudinal intensity	Transverse intensity	The cross section of transverse intensity	FWHM, HMA, I(0,0,0)
$R(\theta)=1$				0.54λ 0.232λ² 2.4
$R_\delta(\theta)=\delta(\theta-\pi/2)$				0.37λ 0.106λ² 0.0065
$R1(\theta)$				0.33λ 0.084λ² 0.31
$R2(\theta)$				0.37λ 0.107λ² 0.46

central parts and send the central rays to the extrafocal area. Thus, the focus will retain only the longitudinal component and the central rays are not simply blocked, but can be deflected so as to form a predetermined distribution of intensity around the focus (for example, 'optical bottles').

Note that the central ray blocking allows only highlight the longitudinal component and get the intensity distribution proportional to the zero-order Bessel function (FWHM = 0.36λ). Using the additional annular phase structure we can receive a smaller central light spot (FWHM = 0.33λ), although, as in the linear polarization, by a corresponding reduction of the energy in this spot.

4.4.3. Overcoming of the diffraction limit for azimuthal polarization by the transverse components of the electric field

In the case of the azimuthal polarization of the input field the longitudinal component is always missing. Vertical and horizontal binary asymmetric structures will generate on the optical axis respectively the x- or y-component with the highest attainable value of the intensity. The action of the bilens in azimuthal polarization is fundamentally different from the result of linear polarization. In the present case (see Table 4.4.6) the rotation of the bilens leads only to the rotation of the total intensity pattern (although the contribution of the component in the axis point changes at the same time).

Supplementing the lens with the functions of the form (4.4.7) and (4.4.8) gives rise on the axis to non-zero values of the transverse components (longitudinal component in the azimuthal polarization is absent), and subwavelength localization is minimum of all the considered polarization variants – HMA = $0.054λ^2$. The intensity in the central spot in this case will be 40 times less than the intensity of the focal spot formed by introducing a phase jump.

A positive aspect for axially symmetric polarizations (radial and azimuthal) is the independence of the distribution of the total intensity on the rotation of the asymmetric binary structure.

4.4.4. Conclusions

In the section based on the vector model of the high-aperture lens in the Debye approximation we showed the possibility of using binary phase structures for the management of the contribution of components of the vector electric field on the optical axis at different polarizations of the incident radiation on the focusing system.

Table 4.4.6. Simulation results for the aplanatic objective with $NA = 0.99$ at azimuthal polarization

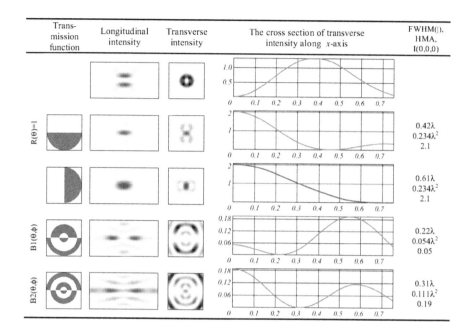

	Transmission function	Longitudinal intensity	Transverse intensity	The cross section of transverse intensity along x-axis	FWHM(), HMA, I(0,0,0)
$R(\theta)=1$					
					0.42λ 0.234λ² 2.1
					0.61λ 0.234λ² 2.1
B1(θ,φ)					0.22λ 0.054λ² 0.05
B2(θ,φ)					0.31λ 0.111λ² 0.19

In particular, the introduction of a linear phase jump perpendicular to the direction of linear polarization leads to exclusion of the transverse components in the centre of the focal region and the appearance of the longitudinal component.

This control allows one to create in the focal region a central light spot much smaller than the diffraction limit, equal to HMA = $0.2\lambda^2$.

Application of the phase annular structures reduces the area of the central light spot to FWHM = 0.33λ (HMA = $0.084\lambda^2$) with an efficiency of 13% for the radial polarization.

Asymmetric binary phase structures can be used to reduce the area of the central light spot for linear polarization – HMA = $0.08\lambda^2$ with an efficiency of 3.6% and azimuthal polarization – HMA = $0.054\lambda^2$ with an efficiency of 2.4%.

Thus, the size of the central spot is reduced due to the loss of efficiency and the growth of side lobes. However, these losses are much less than when using narrow annular apertures. In addition, the energy of the central zone of the lens can be used to form a predetermined three-dimensional distribution in a region close to the focus, or sufficiently far away. In the latter case it is more convenient to use diffractive lenses.

4.5. Minimizing the size of the light or shadow focal spot with controlled growth of the side lobes

This section examines not just the control of the contribution of different components of the electric field in the focal region [70], but also minimizing the size of the focal spot (light or shadow) due to optimal selection of the transmission function of the pupil.

In [19] to reduce the size of the focal spot, formed by the high-aperture focusing system, it is proposed to use radially polarized Laguerre–Gaussian modes of higher radial order, without the vortex phase.

The vortex phase transmission function as a factor that reduces the size of the light spot in the individual components of the sharply focused electric field at different polarizations was considered in [75], and the ability to reduce the focal spot in the total intensity distribution due to additional variations of the radius was shown in [70] . The use of Zernike polynomials as transmission functions, including those with the vortex phase dependence, was discussed in [81]. This enabled to introduce simultaneously the vortex phase dependence and amplitude changes in the radius. It has been shown that it is possible to reduce the size of not only the light spot but also the shadow area.

In the field of optical trapping and manipulation [82] and STED-microscopy [83] and shadow microscopy [84] there is also a problem of focusing in a shady spot or a light ring with a compactly localized zero intensity in the centre. Formation of such a focal distribution is carried out in the introduction of the vortex phase singularity to the focused beam.

This section explores the minimization of light and shadow focal spot at different types of polarization due to the additional apodization of the focusing system by an optical element with the vortex phase dependence on the angle and the polynomial amplitude dependence on the radius. The order and direction of the vortex phase component are selected depending on the type of polarization of the beam incident on the focusing system [85], while the coefficients in a low degree radial polynomial are optimized considering fulfilling certain conditions, in particular, conservation of energy efficiency and ensuring a given level of side lobes. Search for the coefficients is performed based on the minimization of the functional using Brent's method [86].

4.5.1. Scalar diffraction limit: theoretical analysis

The result of focusing a limited plane wave with wavelength λ and wave number $k = 2\pi/\lambda$ by a lens with radius R and a focal length f in the scalar case corresponds to the expression:

$$E_0(\rho,\varphi) = \frac{k}{if} \exp(ikf) \int_0^R J_0\left(\frac{kr\rho}{f}\right) r\, dr = \exp(ikf)\frac{R}{i\rho} J_1\left(\frac{k\rho R}{f}\right), \quad (4.5.1)$$

where (ρ,φ) are the polar coordinates in the focal plane, $Jn(x)$ is the Bessel function of n-th order.

The radius of the focal spot can be determined by the first zero of the Bessel function of the first order:

$$\rho_l = \frac{3.83 f}{kR} \approx \frac{0.61\lambda}{NA} \qquad (4.5.2)$$

where NA is the numerical aperture of the lens.

A well-known method of reducing the size of the focal spot is not use the full aperture but use a narrow peripheral ring with width δ. In this case, we obtain a range associated with the zero-order Bessel function (for small δ we can take the integrand at the mid-point):

$$E_0(\rho,\varphi) = \frac{k}{if} \exp(ikf) \int_{R-\delta}^R J_0\left(\frac{kr\rho}{f}\right) r\, dr \approx \exp(ikf)\frac{kR\delta}{if} J_0\left(\frac{k\rho R}{f}\right),$$

$$(4.5.3)$$

where the radius of the focal spot can be determined by the first zero of the Bessel function of zero order:

$$\rho_{l\delta} = \frac{2.4 f}{kR} \approx \frac{0.38\lambda}{NA} \qquad (4.5.4)$$

As can be seen from a comparison of formulas (4.5.1) and (4.5.3), reducing the transverse dimension of the focal spot is achieved at the expense of decrease of the intensity in the focus ($\rho = 0$). When $\delta \sim \lambda$ the fall is proportional to the square of the focal length.

Thus, the scalar diffraction limit in the free space ($NA = 1$) for the focal spot light using the aberration-free full-aperture focusing system in the radius is $\rho_l \approx 0.61\lambda$ (at half intensity FWHMI$_i \approx 0.51\lambda$), and using a narrow annular aperture $\rho_{l\delta} \approx 0.38\lambda$ (FWHMI$_{l\delta} \approx 0.36\lambda$).

Also relevant is the problem of focusing in a shady spot or a light ring with the compactly localized zero intensity in the centre. Formation of such a focal distribution is carried out in the introduction of a vortex phase singularity to the focused beam.

The result of focusing a limited plane wave with the introduced vortex phase singularity of first order in the scalar case corresponds to the expression:

$$E_1(\rho,\varphi) = \frac{k}{if} \exp(ikf) \exp(i\varphi) \int_0^R J_1\left(\frac{kr\rho}{f}\right) r\,dr. \qquad (4.5.5)$$

The integral in (4.5.5) is usually written by the superposition of Bessel and Struve functions [5]:

$$\int_0^R J_1(\alpha r) r\,dr = \frac{\pi R}{2\alpha}\left[J_1(\alpha R)H_0(\alpha R) - J_0(\alpha R)H_1(\alpha R)\right], \qquad (4.5.6)$$

which is inconvenient for analysis. Therefore, we consider the integral in (4.5.5) via the hypergeometric function:

$$\int_0^R J_1(\alpha r) r\,dr = \frac{\alpha R^3}{6} {}_1F_2\left(\frac{3}{2};\frac{5}{2},2;-\frac{\alpha^2 R^2}{4}\right) \qquad (4.5.7)$$

where ${}_1F_2(x;y_1,y_2;z) = \sum_{n=0}^{\infty} \dfrac{(x)_n}{(y_1)_n \cdot (y_2)_n} \cdot \dfrac{z^n}{n!}$,

$(a)_0 = 1,\ (a)_n = a\cdot(a+1)\cdot\ldots\cdot(a+n-1)$

Thus, equation (4.5.5) can be written as:

$$E_1(\rho,\varphi) = \exp(ikf)\exp(i\varphi)\frac{k^2 R^3 \rho}{6if^2} {}_1F_2\left(\frac{3}{2};\frac{5}{2},2;-\frac{(k\rho R)^2}{4f^2}\right). \qquad (4.5.8)$$

If we retain in the right side of (4.5.8) only the first two terms of the series and equate the derivative to zero, we get an estimate for the radius of the focal ring:

$$\hat{\rho} = \frac{2\sqrt{10}f}{3Rk} \approx \frac{0.336\lambda}{NA}.$$

When calculating the exact hypergeometric function (e.g. in Matlab) or calculating the integral in (4.5.8) by the rectangle method we obtain a higher value (Fig. 4.5.1):

$$\rho_d = \frac{0.39\lambda}{NA}. \qquad (4.5.9)$$

The size of the light ring can also be reduced by superposing a narrow annular aperture which transmits only the peripheral rays:

$$E_1(\rho,\varphi) = \frac{k}{if}\exp(ikf)\exp(i\varphi)\int\limits_{R-\delta}^{R} J_1\left(\frac{kr\rho}{f}\right)r\,dr \approx \exp(ikf)$$

$$\exp(i\varphi)\frac{kR\delta}{if}J_1\left(\frac{k\rho R}{f}\right). \qquad (4.5.10)$$

In this case, the radius of the light ring is determined by the maximum value of the Bessel function of the first order:

$$\rho_{d\delta} = \frac{1.84 f}{kR} \approx \frac{0.293\lambda}{NA}. \qquad (4.5.11)$$

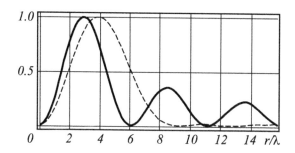

Fig. 4.5.1. The results of the calculation of the normalized intensity for the expression (4.5.8) (dashed line) and the expression (4.5.10) (solid line) with $NA = 0.1$.

Size reduction is achieved due to the loss of energy in the central ring, and hence by reducing the intensity of the light barrier. However, the compactness of the light funnel may be a more important feature than its depth.

From the above calculations it follows that the scalar diffraction limit in the free space ($NA = 1$) for the light ring using an aberration-free full-aperture focusing system is equal in the radius to $\rho_d \approx 0.39\lambda$ (the inner diameter of the ring at half intensity is also equal to this value), and using a narrow annular aperture $\rho_{d\delta} \approx 0.29\lambda$.

Thus, the diffraction limit for the light ring is smaller than for the light spot.

To reach the scalar limit when using focusing systems with a high numerical aperture is problematic due to the contribution to the focal region of various components of the electric field. This problem is solved using polarizations of special types, allowing to maximize the contribution of one of the components that actually leads to the scalar case. For the full-aperture apodization it is most simple to add phase optical elements to the focusing system and the focusing of laser modes of high orders [19, 81]. In the latter case, the distribution in the exit pupil of the system is presented in the polynomial form. Next we consider the possibility of minimizing the transverse focal size using low-order polynomials when fulfilling certain conditions, including the conservation of energy efficiency or ensuring a given level of side lobes.

4.5.2. Optimization of the transmission function of the focusing system

Consider the sharp focusing of the laser beams with different polarization when using the superposition of the following type as the transmission function $B(\theta,\phi)$ [81]:

$$B_m(\theta,\phi) = \exp(im\phi)\sum_{s=0}^{S} c_s \sin^s \theta \qquad (4.5.12)$$

where c_s are the coefficients of the polynomial at a fixed value of the vortex order m.

This type of illuminating beam for common types of polarization allows to use the expression of the form (4.2.7), where

$$q(\theta) = \exp(ikz\cos\theta)\sqrt{\cos\theta}\sum_{s=0}^{S} c_s \sin^{s+1}\theta \qquad (4.5.13)$$

After selecting a certain type of polarization and the order of the vortex phase dependence of m the optimization of the coefficients in the superposition (4.5.13) can be made on the basis of minimizing the functional binding the output distribution intensity and the target function $T(\rho,\varphi)$ in the field of focus:

$$\Phi\left\{\left|E_m\left(\rho,\varphi,z=0\right)\right|^2 ; T(\rho,\varphi)\right\} \xrightarrow[c_s]{} \min \qquad (4.5.14)$$

The target function is a superposition of the functions expressing the conditions which must be satisifed by the intensity distribution in the focal plane. In particular, the conditions of the intensity concentration in a particular area or a 'fine' for exceeding the values of the normalized intensity of a given threshold in some range.

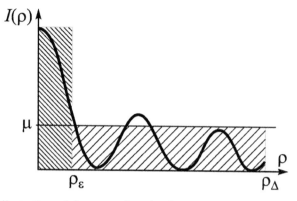

Fig. 4.5.2. Illustration of the target function in the focal region at a fixed value of the polar angle φ.

If we consider the functional (4.5.14) for a fixed value of the polar angle φ, then the functions participating in it will depend only on the radius and we can demonstrate the imposition of the conditions on the obtained intensity distribution (see Fig. 4.5.2)

To achieve a certain compromise between reducing the size of the focal spot and suppressing the side lobes it is necessary to impose the condition of concentration of energy in the focal plane in a circle having a radius ρ_ε and providing in the ring with the radii $[\rho_\varepsilon, \rho_\Delta]$ a predetermined level μ of the sidelobes. In the area of a radius greater than ρ_Δ the intensity distribution is not controlled.

The minimization of the functional (4.5.14) for a fixed value of the polar angle φ was performed by Brent [86]. A software module that implements this method was included in the open resource ALGLIB [87].

4.5.3. Minimization of the light spot for radially polarized radiation

It is known that radial polarization is characterized by the smallest transverse dimension of the focal spot associated with maximizing the contribution of the longitudinal component to the total intensity on the optical axis.

In radial polarization in the absence of the phase vortices the contribution of the longitudinal component to the total intensity on the optical axis is maximum, however, the size of the focal spot formed is also affected by the transverse components which contribute to the vicinity of the optical axis (Table 4.5.1, line 1).

When using a narrow annular aperture with the radius close to the radius of the focusing lens, and the relative width δ (normalized to the radius of the aperture), the contribution of the transverse component is minimized and the intensity in the focal region is determined solely by the longitudinal component [81]:

$$\left| E_0^z (\rho,\varphi,z=0) \right|^2 \sim \delta^{3/2} J_0^2 (k\rho) \qquad (4.5.15)$$

which implies that the radius of the central light spot is $\rho_\delta \approx 0.38\lambda$, which corresponds to the scalar limit (4.5.4). The intensity in it will be very small, since the relative width of the annular aperture $\delta \ll 1$ (Table 4.5.1, line 2).

If we enter a phase jump of π radians with respect to the average radius of a narrow annular aperture (which corresponds to the introduction of superposition of two destructively interfering beams):

$$\tau(\theta) = \begin{cases} 0, & \theta < \Theta - \delta, \\ 1, & \Theta - \delta < \theta < \Theta - \delta / 2, \\ \exp(i\pi), & \theta \geq \Theta - \delta / 2, \end{cases} \qquad (4.5.16)$$

ρ_d

we get the following distribution in tl   )cal plane for the longitudinal (main) component ($NA = 1$):

$$E_0^z(\rho,\varphi,z=0) = ikf \int_0^{\Theta} \tau(\theta) J_m (k\rho\sin\theta)\sin^2\theta\sqrt{\cos\theta}\, d\theta$$

$$= ikf \int_0^{\arcsin\Theta} \tau(x)x^2 \left(1-x^2\right)^{-1/4} J_0 (k\rho x)\, dx \approx$$

$$\approx 2^{1/4}\delta^{3/4}\left[3^{-1/4} J_0\left(k\rho\left(1-\frac{3\delta}{4}\right)\right) - J_0\left(k\rho\left(1-\frac{\delta}{4}\right)\right)\right].$$

$$(4.5.17)$$

Equation (4.5.17) shows that in the centre of the focal region the value differs from zero. The radius (before the first zero) of this weak spot light can be estimated, using an approximation of Bessel functions:

$$\rho_{\delta ph} \approx \frac{\lambda}{\pi}\sqrt{\frac{0.288 + 0.72\delta}{0.288 + 1.488\delta}} \qquad (4.5.18)$$

By reducing the width of the annular aperture we obtain the limiting value, less the scalar limit (4.5.4) $\rho_{\delta ph} \xrightarrow{\delta \to 0} = \lambda/\pi \approx 0.318\lambda$. Note that formula (4.5.18) predicts an even more size when increasing

Table 4.5.1. The results of numerical simulation for a beam with radial polarization without vortex components ($m = 0$)

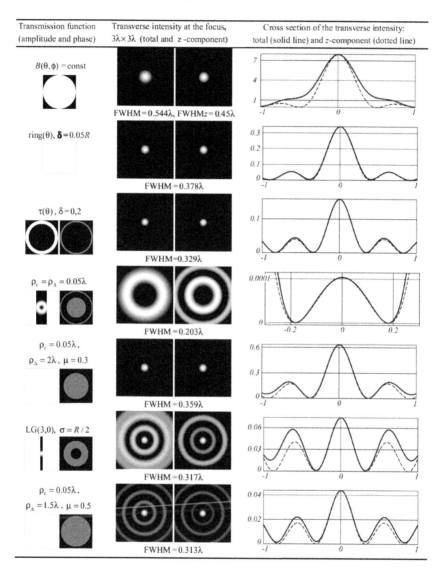

Transmission function (amplitude and phase)	Transverse intensity at the focus, $3\lambda \times 3\lambda$ (total and z-component)	Cross section of the transverse intensity: total (solid line) and z-component (dotted line)
$B(\theta,\phi) = \mathrm{const}$	FWHM = 0.544λ, FWHMz = 0.45λ	
ring(θ), $\delta = 0.05R$	FWHM = 0.378λ	
$\tau(\theta)$, $\delta = 0.2$	FWHM = 0.329λ	
$\rho_\varepsilon = \rho_\Delta = 0.05\lambda$	FWHM = 0.203λ	
$\rho_\varepsilon = 0.05\lambda$, $\rho_\Delta = 2\lambda$, $\mu = 0.3$	FWHM = 0.359λ	
LG(3,0), $\sigma = R/2$	FWHM = 0.317λ	
$\rho_\varepsilon = 0.05\lambda$, $\rho_\Delta = 1.5\lambda$, $\mu = 0.5$	FWHM = 0.313λ	

the width of the aperture, but in this case the approximation in which the expression is obtained stops working. Table 4.5.1 (row 3) shows the results of using the aperture ring with the width $\delta = 0.2$ with the introduction of a phase jump by π radians along the mean radius of the aperture. It is seen that the light spot size is reduced to FWHM = 0.329λ for some growth of the sidelobes the intensity level of which, however, is less than 30% of the main peak. These sidelobes may be effectively filtered out [89, 90]. Note also that the effect of the sidelobes, if their intensity is below a certain level, may be offset in the non-linear interaction of light with the recording medium. [91].

In the case of full-aperture transmission function having a power-law dependence on the radial coordinate, the longitudinal component will be as follows:

$$E_0^z(\rho,\varphi,z=0) = ikf \int_0^{\pi/2} (\sin\theta)^{s+2} J_0(k\rho\sin\theta)(\cos\theta)^{1/2} \, d\theta. \quad (4.5.19)$$

Then the intensity near the optical axis is approximately described by the expression [81]:

$$I_0^z(\rho,\varphi,z=0) \approx \left\{\frac{kf}{2}\frac{\Gamma(3/4)\Gamma((s+3)/2)}{\Gamma((s+3)/2+3/4)}\left[1 - \frac{(s+3)}{(s+3)+3/2}\left(\frac{k\rho}{2}\right)^2\right]\right\}^2,$$

$$(4.5.20)$$

which implies that the radius of the central light spot is determined by the degree s:

$$\rho_s = \frac{\lambda}{\pi}\sqrt{1 + \frac{3}{2(s+3)}} \quad (4.5.21)$$

With increasing degree s the radius of the central light spot will tend to value $\rho_s \xrightarrow[s\to\infty]{} \lambda/\pi \approx 0.32\lambda$.

Thus, the use of the full-aperture power function will reduce the size of the light spot more than introducing a narrow annular aperture ($\rho_\delta \approx 0.38\lambda$), reaching virtually the same spot size as in the introduction of the annular phase jump ($\rho_{\delta ph} \approx 0.38\lambda$).

However, this effect is achieved by increasing the degree s, which actually results in the concentration of energy of the illuminating beam in the peripheral portion of the lens. Thus, the approximation of the power function of the narrow annular aperture is carried out.

Consider another approach to reducing the central spot – at the expense of destructive interference. In this case, in the superposition (4.5.13) select the coefficients in a special way [81].

When using superposition (4.5.13) the field in the focal plane takes the following form:

$$\mathbf{E}_0\left(\rho,\varphi,z=0\right)=kf\begin{bmatrix}\left[\Gamma(1/4)/16\right]k\rho\cos\varphi\sum_{s=0}^{S}c_s\Phi_s^x(k\rho)\\[2mm]\left[\Gamma(1/4)/16\right]k\rho\sin\varphi\sum_{s=0}^{S}c_s\Phi_s^y(k\rho)\\[2mm]i\left[\Gamma(3/4)/2\right]\sum_{s=0}^{S}c_s\Phi_s^z(k\rho)\end{bmatrix}\qquad(4.5.22)$$

where

$$\Phi_s^x(t)=\Phi_s^y(t)=\frac{\Gamma\left((s+3)/2\right)}{\Gamma\left((2s+11)/4\right)}\,_1F_2\left(\frac{s+3}{2};\frac{2s+11}{4},2;-\frac{t^2}{4}\right),$$

$$\Phi_s^z(t)=\frac{\Gamma\left((s+3)/2\right)}{\Gamma\left((2s+9)/4\right)}\,_1F_2\left(\frac{s+3}{2};\frac{2s+9}{4},1;-\frac{t^2}{4}\right).$$

If we write the first few terms with respect to s in (4.5.22), such as the transverse components:

$$E_0^{x,y}\left(\rho,\varphi,z=0\right)=k^2\rho f\frac{\Gamma(1/4)}{16}\begin{bmatrix}\cos\varphi\\\sin\varphi\end{bmatrix}\left\{\frac{c_0\Gamma(3/2)}{\Gamma(11/4)}\,_1F_2\left(\frac{3}{2};\frac{11}{4},2;-\frac{k^2\rho^2}{4}\right)+\right.$$

$$+\frac{c_1\Gamma(2)}{\Gamma(13/4)}\,_1F_2\left(2;\frac{13}{4},2;-\frac{k^2\rho^2}{4}\right)+$$

$$\left.+\frac{c_2\Gamma(5/2)}{\Gamma(15/4)}\,_1F_2\left(\frac{5}{2};\frac{15}{4},2;-\frac{k^2\rho^2}{4}\right)+\frac{c_3\Gamma(3)}{\Gamma(17/4)}\,_1F_2\left(3;\frac{17}{4},2;-\frac{k^2\rho^2}{4}\right)+...\right\},$$

it becomes clear that there is a relationship for the terms of the same parity. In this case, the Zernike polynomials are simpler [81].

In the presence of only the zero and second degree of the radial polynomial radial near the optical axis we can write:

$$E_0^{x,y}\left(\rho\to0,\varphi,z=0\right)\approx k^2f\rho\begin{bmatrix}\cos\varphi\\\sin\varphi\end{bmatrix}\frac{c_0\Gamma(1/4)}{16}\frac{\Gamma(3/2)}{\Gamma(11/4)}\times$$

$$\times\left\{\left(1+\frac{6}{11}\frac{c_2}{c_0}\right)-\frac{3}{22}\left(1+\frac{2}{3}\frac{c_2}{c_0}\right)\left(\frac{k^2\rho^2}{2}\right)+\frac{1}{66}\left(1+\frac{14}{19}\frac{c_2}{c_0}\right)\left(\frac{k^2\rho^2}{2}\right)^2\right\},$$

$$E_0^z\left(\rho \to 0, \varphi, z = 0\right) \approx ikf \frac{c_0 \Gamma(3/4)}{2} \frac{\Gamma(3/2)}{\Gamma(9/4)} \times$$

$$\times \left\{ \left(1 + \frac{2c_2}{3c_0}\right) - \frac{1}{3}\left(1 + \frac{15 \cdot 2c_2}{13 \cdot 3c_0}\right)\left(\frac{k^2 \rho^2}{2}\right) + \frac{5}{13 \cdot 24}\left(1 + \frac{21 \cdot 2c_2}{17 \cdot 3c_0}\right)\left(\frac{k^2 \rho^2}{2}\right)^2 \right\}.$$

(4.5.23)

From (4.5.23) it is clear that at the focus ($\rho = 0$) with $c_2 = -3c_0/2$ the contribution of the longitudinal component is reset. The contribution of the transverse component is reset when $c_2 = -11c_0/6$. Thus, the Zernike polynomial (2, 0), in which $c_2 = -2c_0$, is nearly optimal for maximizing the contribution of the longitudinal component [81].

Analytical calculations for a large number of terms are time-consuming, so it is easier to use the numerical optimization procedure for the coefficients with certain conditions satisfied.

In particular, the dimensions of the focal spot can be arbitrarily reduced due to the repeated increase of side lobes and departure of almost all the energy out of focus [78] (see Table 4.5.1, line 4). In order to achieve a compromise between the reduction of the size of the focal spot and holding the energy at the centre we impose a condition of concentration of energy in the focal plane in a circle of radius ρ_ε and ensuring in the ring with radii $[\rho_\varepsilon, \rho_\Delta]$ the given level μ of the sidelobes (Fig. 4.5.2) .

Outside the radius ρ_Δ the intensity level is not controlled and can substantially exceed the value of the centre. Increasing the control region, usually together with a decrease in the loss of energy, leads to the enlargement of the size of the focal spot.

Line 5 in Table. 4.5.1 shows the results of the use of full-aperture transmitting functions in the form of a third-order polynomial with coefficients chosen so as to concentrate energy in a circle of radius ρ_ε $\rho_\varepsilon = 0.05\lambda = 0.05$ and hold the sidelobe level in the ring with radii $[0.05\lambda, 2\lambda]$ not greater than $\mu = 0.3$ of the value in the focal spot. In this case, the size of the focal spot is smaller than with the narrow annular aperture, and the intensity at the focus is two-fold higher.

In [19] it was proposed to reduce the size of the focal spot by using transmission functions proportional to the Laguerre-Gaussian (LG) modes. The results of focus for the LG modes depend on the waist radius and the radial mode number. The increase in the mode number does not necessarily lead to the reduction in the size of the focal spot. In particular, the following results were obtained for $\sigma = R/2$: FWHM $= 0.437\lambda$ (radial number of the LG mode $n = 1$), FWHM $= 0.875\lambda$ ($n = 2$), FWHM $= 0.317\lambda$ ($n = 3$ – Table 4.5.1, line 6), FWHM $= 0.384\lambda$ ($n = 4$), FWHM $= 0.475\lambda$ ($n = 5$).

In this case also there is a significant increase in the side lobes which in the total intensity approaches the level of $\mu = 0.86$ of the maximum (Table 4.5.1, line 6).

The result for the LG modes can improve in the level of the side lobes to $\mu = 0.5$ while maintaining the same size of the focal spot (FWHM $= 0.313l$) using the optimization algorithm to the superposition (4.5.13) – Table 4.5.1, line 7.

In the presence of the first order vortex components ($|m| = 1$) at radial polarization the transverse components will already have non-zero values on the optical axis, and the longitudinal component, on the contrary, will have the annular distribution [81]. The contribution of the longitudinal component to the total intensity is comparable to the contribution of the transverse components. This is especially important to consider in the interaction of radiation with the media, selectively sensitive to various components of the electric vector.

Note that in this case the presence of the transverse components in the central part of the focal region prevents obtaining a 'clean' light ring. The azimuthal polarization is better suited for this purpose.

4.5.4. Formation of light rings of the subwavelength radius azimuthal polarization of laser radiation

In azimuthal polarization the field in the centre of the focal plane ($z = 0$) contains non-zero components only for one singularity order $|m| = 1$. For all other values of $|m| \neq 1$ the absolute intensity zero will form at the axial points because all components are zero. Therefore, this polarization can be used more efficiently for the formation of light rings (Table 4.5.2, line 1).

At $m = 0$ taking (4.5.13) into account, the expression in the focal plane is reduced to a compact form:

$$\mathbf{E}_0\left(\rho,\varphi,z=0\right)=kf\begin{bmatrix}-\sin\varphi\\\cos\varphi\\0\end{bmatrix}\int_0^{\pi/2}\sum_{s=0}^{S}c_s\cdot\left(\sin\theta\right)^{s+1}J_1\left(k\rho\sin\theta\right)\left(\cos\theta\right)^{1/2}\,d\theta.$$

$$(4.5.24)$$

The integrals in (5.4.24) can be taken using the hypergeometric functions and analysis of these functions [81] shows that the radius

Table 4.5.2. The results of numerical simulation for the beam with azimuthal polarizationand with no vortex components ($m = 0$)

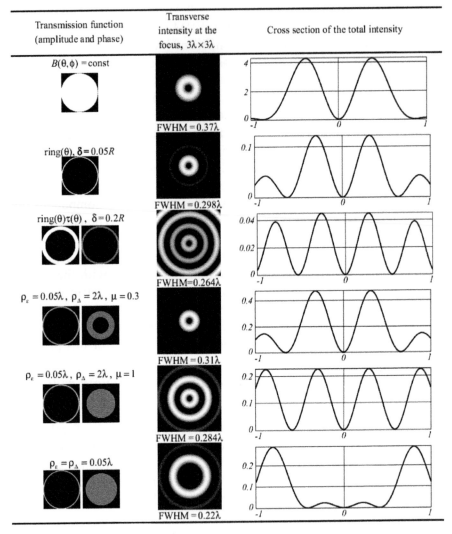

Transmission function (amplitude and phase)	Transverse intensity at the focus, $3\lambda \times 3\lambda$	Cross section of the total intensity
$B(\theta, \phi) = \text{const}$	FWHM $= 0.37\lambda$	
$\text{ring}(\theta), \delta = 0.05R$	FWHM $= 0.298\lambda$	
$\text{ring}(\theta)\tau(\theta), \delta = 0.2R$	FWHM $= 0.264\lambda$	
$\rho_\varepsilon = 0.05\lambda, \rho_\Delta = 2\lambda, \mu = 0.3$	FWHM $= 0.31\lambda$	
$\rho_\varepsilon = 0.05\lambda, \rho_\Delta = 2\lambda, \mu = 1$	FWHM $= 0.284\lambda$	
$\rho_\varepsilon = \rho_\Delta = 0.05\lambda$	FWHM $= 0.22\lambda$	

of the light ring depends on the degree s, but has a limiting value $\rho_{min} \approx 0.26\lambda$ (lower bound).

In superposing a narrow annular aperture the integral in (4.5.24) takes the simple form that is independent on polynomials:

$$\int_{\pi/2-\delta}^{\pi/2} J_1(k\rho\sin\theta)\sin\theta(\cos\theta)^{1/2}\,d\theta \sim \delta^{3/4}J_1(k\rho) \qquad (5.4.25)$$

i.e. the radius of the light ring is determined by the first maximum of the Bessel function of the first order $\rho_\delta \approx 0.293\lambda$, which corresponds to the scalar limit (4.5.9) (see Table 4.5.2, line 2). The intensity of the focus ring will fall with a decrease in the width of the annular aperture δ.

Also, the introduction of the additional annular phase jump by π radians along the mean radius of the aperture can reduce the size of the light ring to FWHM $= 0.264\lambda$ due to a further reduction in the intensity of the ring (see Table 4.5.2, line 3).

When using superposition (4.5.13) we can try to reach a compromise between the reduction of the size of the light ring and loss of energy in the central part. For this purpose, in particular, we can retain the low intensity of the sidelobes relative to the values in the central light ring in a sufficiently wide range (Table 4.5.2, line 4).

If the side lobes are allowed to increase to the intensity level of the central ring (Table 4.5.2, line 5), then we can overcome the diffraction limit of the central ring twice as as than when using the narrow annular aperture.

The central light ring can be reduced even more only due to a significant loss of energy (see Table 4.5.2, line 6).

Circular polarization

This type of polarization is attractive by the simplicity of obtaining from linear polarization which is characteristic of most laser sources. It does not require complex or costly deviced, as in the previous two cases of radial and azimuthal polarization.

In the case of circular polarization the sign of polarization and the sign of the vortex order are inter-related. For values of the vortex order $|m| \leq 2$ with the sign opposite to that of polarization there are components with non-zero values on the optical axis ($\rho = 0$). This fact will have a negative impact on attempts to get the annular distribution in the total intensity.

Consider the case of circular '+'-polarization. Then, at $m = 0$, $m = -2$ the transverse components will have non-zero values on the optical axis and in the longitudinal component there will be a light ring. When $m = -2$ the energy of the transverse components at the axis is less than at $m = 0$. When $m = -1$ the situation is reversed, but for high-aperture focusing systems usually the contribution to the longitudinal component is large. Therefore, if it is necessary to form a shady spot it is more convenient to use the option with $m = -2$.

When using a narrow annular aperture the contributions of the longitudinal and transverse components are comparable and will not interfere with each other both in the formation of a compact light spot and a light ring.

Thus, for the more common and easiest to implement types of polarization (linear or circular) we can obtain optimal results only for the individual components of the electric field – longitudinal or transverse. The effect of optimization in the total intensity is deteriorating due to the joint contributions of various components. In this case, the contribution of the unwanted components can be levelled out using the materials selectively sensitive to various components of the electric field, as well as in a system of two colliding beams [89].

4.5.5. Conclusions

The section discusses the possibility of minimizing the light and shadow focal spots at different types of polarization with the controlled growth of sidelobes due to optimal apodization of the focusing system. The considered apodization involves the introduction of a polynomial amplitude dependence on the radius. This optimization of the coefficients in the polynomial dependence can be performed by different methods in order to maintain energy efficiency and provide a specified level of sidelobes.

Studies have shown the effectiveness of radial polarization to form a light spot of the minimum size and azimuthal polarization – for the light rings.

In this case, the limit predicted by scalar theory, is achieved simply by introducing a narrow annular aperture, blocking the central part of the focusing system. To increase the efficiency we can use a wider annular aperture with a radial phase jump. Such an optical element can also produce a smaller focal spot than predicted by the scalar diffraction theory.

The light or shadow focal spot can be greatly reduced only by transferring the energy to the sidelobes. In this case, to control the growth of the intensity of the sidelobes in a region near the focal spot we can use the amplitude radial–polynomial distribution (which is optimized, for example, by the Brent method).

Reaching a compromise between the reduction of the size of the focal spot and holding the energy at the centre depends on the size of the focal region controlled. Outside the controlled region the intensity level can greatly exceed the value in the centre. However, increasing the control region, usually together with a decrease of the energy loss leads to enlargement of the focal spot size. Thus, the results are entirely consistent with the Toraldo Di Francia's theory [78].

For the more common and easiest to implement types of polarization, in particular, circular or linear, similar results can be obtained only for individual components of the electric field – longitudinal or transverse. The effect of optimization in the total intensity deteriorates due to the joint contributions of the various components. In this case, the contribution of the undesirable components may be decreased in the interaction of radiation with the media selectively sensitive to various components of the electric field, as well as a system of two colliding beams [89].

Closing remarks

The most effective way of formation of vortex fields is the use of phase diffractive optical elements, the simplest of which is the spiral phase plate and the helical axicon. The chapter discussed in detail the use of singular optics elements in sharp focusing, when the combination of the polarization and phase characteristics leads to a variety of effects, including overcoming the diffraction limit in the far field diffraction.

It was shown analytically and numerically that the subwavelength localization in the individual components of the vector field is possible with all types of polarization and it is important to take this into account in the interaction of laser radiation with matter, selectively sensitive to various components of the vector electromagnetic field.

However, for the formation of substantially subwavelength details in the full intensity of the electric field it is necessary to have certain types of polarization and the additional pupil apodization function, such as a narrow annular gap. This apodization is a simple but energy-inefficient way. Therefore, the chapter examines an alternative and more energy-efficient way of supplementing the focusing lens with a diffractive axicon. Adding such phase apodization allows one to manage the distribution along the optical axis for the various components of the electric field. In particular, in this case the contribution of the transverse component in the focal plane is reduced by changing the inclination angle of the central rays. For the radially polarized radiation the focal area will look like a cone, the tip of which contains only the longitudinal component of the electric field and has the transverse dimension smaller than the diffraction limit (FWHM = 0.41λ).

The chapter also shows the possibility of using binary phase structures for the management of the contribution of components of the vector electric field on the optical axis at different polarizations of the incident light on focusing system. In particular, the introduction of a linear phase jump perpendicular to the direction of linear polarization leads to exclusion in the centre of the focal region of the transverse components and the appearance of the longitudinal component. The use of binary phase ring structures makes it possible to reduce the size of the central light spot for radial polarization to FWHM = 0.33λ with an efficiency of 13%. The asymmetric binary phase structures have been used to reduce the area of the central light spot for linear polarization – HMA = $0.08\lambda^2$ with an efficiency of 3.6%, and for azimuthal polarization – HMA = $0.054\lambda^2$ with an efficiency of 2.4%.

Thus, the size of the central spot can be reduced by the loss of efficiency and growth of side lobes that are fully consistent with the Toraldo Di Francia's theory. Although these losses aree much less than when using the narrow annular apertures, the sidelobe growth higher than 30% of the maximum is not desirable in the imaging systems. In this regard, the chapter examined the algorithm of achieving a certain compromise between the reduction of the size of the focal spot and the growth of the side lobes and diffraction efficiency. The compromise is achieved by optimizing the apodization function represented by a radial polynomial of low degree. The search for the polynomial coefficients was performed by the Brent method based on minimization of the functional, which includes several conditions, such as reducing the size of the focal spot at providing a given level of the sidelobes.

References

1. Bereznyi A.E., et al., DAN SSSR, 234 (4), 802–805 (1984).
2. Khonina S.N., Kotlyar V.V., Shinkarev M.V., Soifer V.A., Uspleniev G.V. The rotor phase filter, J. Mod. Opt., 1992. Vol. 39. No. 5. P. 1147-1154.
3. Diffractibe computer optics (ed. V.A. Soifer), Moscow, Fizmatlit, 2007.
4. Prudnikov A.P., et al., Integrals and series. Special functions. Moscow, Nauka, 1983.
5. Handbook of mathematical function, ed by M. Abramovitz, I.A. Stegun - NBS, Appl. Math. Ser. 55, 1964.
6. Kotlyar V.V., Almazov A.A., Khonina S. N., Soifer V.A., Elfstrom H. and Turunen J., Generation of phase singularity through diffracting a plane or Gaussian beam by

a spiral phase plate, J. Opt. Soc. Am. A, Vol. 22, No. 5, 849-861 (2005)

7.   Kotlyar V.V., Khonina S.N., Kovalev A.A., Soifer V.A., Elfstrom H., and Turunen J., Diffraction of a plane, finite-radius wave by a spiral phase plate, Opt. Lett. 31, 1597-1599 (2006).

8.   Kotlyar V.V., Kovalev A.A., Khonina S.N., Skidanov R.V., Soifer V.A., Elfstrom H., Tossavainen N., and Turunen J., Diffraction of conic and Gaussian beams by a spiral phase plate, Appl. Opt. Vol. 45, No.12, 2656-2665 (2006).

9.   Khonina S.N., et al., Komp. Optika, 34 (3), 317-332 (2010).

10.   Khonina S.N., et al.,Izv. Samars. Nauch. Tsentra RAN, 12(4), 15-25 (2010).

11.   Karman, G.P. Airy pattern reorganization and subwavelength structure in a focus, G. P. Karman, M. W. Beijersbergen, A. van Duijl, D. Bouwmeester and J. P. Woerdman, J. Opt. Soc. Am. A. – 1998. – Vol. 15, No. 4. – P. 884-899.

12.   Quabis, S. Focusing light to a tighter spot, S. Quabis, R. Dorn, M. Eberler, O. Glockl and G. Leuchs, Opt. Commun. – 2000. – V.179. – P.1–7.

13.   Kant, R. Superresolution and increased depth of focus: an inverse problem of vector diffraction, J. Mod. Opt. – 2000. –Vol. 47, N. 5. – P. 905-916.

14.   Dorn, R. Sharper focus for a radially polarized light beam, R. Dorn, S. Quabis and G. Leuchs, Phys. Rev. Lett. – 2003. – V.91. – P.233901.

15.   Davidson, N. High-numerical-aperture focusing of radially polarized doughnut beams with a para-bolic mirror and a flat diffractive lens, N. Davidson, N. Bokor, Opt. Lett. – 2004. –Vol. 29, No. 12. – P. 1318-1320.

16.   Sheppard, C.J.R. Annular pupils, radial polarization, and superresolution, C.J.R. Sheppard and A. Choudhury, Appl. Opt. – 2004. – Vol. 43, No. 22. – P. 4322-4327.

17.   Pereira, S.F. Superresolution by means of polarisation, phase and amplitude pupil masks, S.F. Pereira, A.S. van de Nes, Opt. Commun. – 2004. – Vol. 234. – P.119 124.

18.   Wang, H. Creation of a needle of longitudinally polarized light in vacuum using binary optics, H. Wang, L. Shi, B. Lukyanchuk, C. Sheppard and C.T. Chong, Nature Photonics. – 2008. –Vol. 2. – P. 501-505.

19.   Kozawa, Y. Sharper focal spot formed by higher-order radially polarized laser beams, Y. Kozawa and S. Sato, J. Opt. Soc. Am. A. – 2007. – V.24. – P.1793-1798.

20.   Lerman, G.M. Effect of radial polarization and apodization on spot size under tight focusing conditions, G.M. Lerman and U. Levy, Opt. Express. – 2008. – Vol. 16, No. 7. – P. 4567-4581.

21.   Kotlyar V.V., et al., Komp. Optika, 2009. – V.33, No.1. – 52-60

22.   Zhan, Q. Cylindrical vector beams: from mathematical concepts to applications, Advances in Optics and Photonics. – 2009. – Vol. 1. – P.1–57.

23.   Kozawa, Y. Generation of a radially polarized laser beam by use of a conical Brewster prism, Yuichi Kozawa and Shunichi Sato, Opt. Lett. – 2005. – V.30(22). – P.3063-3065.

24.   Niz'ev V.G., et al., Kvant. ELektronika.– 2009. – No. 39(6). – P.505-514.

25.   Bomzon, Z. Radially and azimuthally polarized beams generated by space-variant dielectric subwavelength gratings, Z. Bomzon, G. Biener, V. Kleiner, and E. Hasman, Opt. Lett. – 2002. – V.27(5). – P.285-287.

26.   Yonezawa, K. Compact Laser with Radial Polariza-tion Using Birefringent Laser Medium, Jpn., K. Yonezawa, Y. Kozawa, and S. Sato, J. Appl. Phys. – 2007. – V.46(8A). – P.5160–5163.

27.   Tidwell, S.C. Generating radially polarized beams interferometrically, S.C. Tidwell, D.H. Ford, and W.D. Kimura, Applied Optics. – 1990. – V.29. – P.2234–2239.

28.   Passilly, N. Simple interferometric technique for generation of a radially polarized light beam, Nicolas Passilly, Renaud de Saint Denis, and Kamel Aït-Ameur, Fran-

çois Treussart, Rolland Hierle, and Jean-François Roch, J. Opt. Soc. Am. A. – 2005. – V.22(5). – P.984-991.

29. Volpe, G. Generation of cylindrical vector beams with few-mode fibers excited by Laguerre–Gaussian beams, G. Volpe, D. Petrov, Opt. Comm. – 2004. – V.237. – P.89–95.

30. Davis, J.A. Two dimensional polarization encoding with a phase only liquid-crystal spatial light modulator, J.A. Davis, D.E. McNamara, D.M. Cottrell, and T. Sonehara, Appl. Opt. – 2000. – V.39. – P.1549–15541.

31. Neil, M.A.A. Method for the generation of arbitrary complex vec-tor wave fronts, Mark A.A. Neil, Farnaz Massoumian, Rimvydas Juskaitis, and Tony Wilson, Opt. Lett. – 2002. – V.27(21). – P.1929-1931.

32. Iglesias, I. Polarization structuring for focal volume shaping in high-resolution microscopy, Ignacio Iglesias, Brian Vohnsen, Opt. Commun. – 2007. – Vol. 271. – P.40–47.

33. Khonina, S.N. Polarization converter for higher-order laser beams using a single bi-nary diffractive optical element as beam splitter, Khonina S.N., Karpeev S.V., Alferov S.V., Opt. Lett. – 2012. - Vol. 37, No. 12. – P. 2385-2387

34. Cicchitelli, L. Longitudinal components for laser beams in vacuum, L. Cicchitelli, H. Hora, and R. Postle, Phys. Rev. A. – 1990. – Vol. 41. – P. 3727–3732.

35. Simpson, N.B. Optical tweezers and optical spanners with Laguerre-Gaussian modes, N.B. Simp-son, L. Allen, M.J. Padgett, J. Mod. Opt. – 1996. – Vol. 43(12). – P. 2485-2491.

36. Heckenberg, N.R. Trapping microscopic particles with singular beams, N.R. Heckenberg, T.A. Nieminen, M.E. J Friese, H. Rubinsztein-Dunlop, Proc. SPIE. – 1998. –Vol. 3487. – P. 46-53.

37. Helseth, L.E. Mesoscopic orbitals in strongly focused light, Opt. Commun. – 2003. – Vol. 224. – P. 255–261.

38. Soifer V.A. et al., Fizika Elementarnykh Chastits i Atomnogo Yadra. – 2004. – № 35(6). – C. 1368-1432.

39. Skidanov R.V., Komp. Optika. – 2007. – No. 31(1). – P. 14-21.

40. Franke-Arnold, S. Advances in optical angular momentum, S. Franke-Arnold, L. Allen, and M. Padgett, Laser Photonics Rev. – 2008. – Vol. 2. – P. 299–313.

41. Levenson, M.D. Optical vortex masks for via levels, M.D. Levenson, T. Ebihara, G. Dai, Y. Morikawa, N. Hayashi, S.M. Tan, J. Microlith. Microfab. Microsys. – 2004. – Vol. 3(2). – P. 293-304.

42. Unno, Y. Impact of mask errors and lens aberrations on the image formation by a vortex mask, Y. Unno, T. Ebihara, M.D. Levenson, J. Microlith. Microfab. Microsys. – 2005. – Vol. 4(2). – P. 023006.

43. Willig, K.I. STED microscopy resolves nanoparticle assemblies, K.I. Willig, J. Keller, M. Bossi, S.W. Hell, New J. Phys. – 2006. – Vol. 8. – P. 106.

44. Torok P. The use of Gauss-Laguerre vector beams in STED microscopy, P. Torok and P.R.T. Munro, Opt. Express. – 2004. – Vol. 12, No. 15. – P. 3605-3617.

45. Helseth, L.E. Optical vortices in focal regions, Opt. Commun. – 2004. – Vol. 229. – P. 85–91.

46. Liu, P. Phase singularities of the transverse field component of high numerical aperture dark-hollow Gaussian beams in the focal region, P. Liu, B. Lu, Opt. Commun. – 2007. – Vol. 272. – P. 1–8.

47. Singh, R. K. Focusing of linearly-, and circularly polarized Gaussian background vortex beams by a high numerical aperture system afflicted with third-order astigmatism, Rakesh Kumar Singh, P. Senthilkumaran, Kehar Singh, Opt. Commun. –

2008. – Vol. 281. – P. 5939–5948.

48. Chen, B. Tight focusing of elliptically polarized vortex beams,, Baosuan Chen and Jixiong Pu, Appl. Opt. – 2009. – Vol. 48, No. 7. – Pp. 1288-1294.

49. Rao, L. Focus shaping of cylindrically po-larized vortex beams by a high numerical-aperture lens, Lianzhou Rao, Jixiong Pu, Zhiyang Chen, Pu Yei, Opt. & Las. Techn. – 2009. – Vol. 41. – P. 241–246.

50. Beth, R.A. Mechanical detection and measurement of the angular momentum of light, Phys. Rev. – 1936. – Vol. 50. – P. 115–125.

51. Holbourn, A.H.S. Angular momentum of circularly polarized light, Nature (London). – 1936. – Vol. 137. – P. 31.

52. Allen, L. Orbital angular momentum of light and the transformation of Laguerre–Gaussian laser modes, L. Allen, M. W. Beijersbergen, R. J. C. Spreeuw, and J. P. Woerdman, Phys. Rev. A. – 1992. – Vol. 45. – P. 8185–8189.

53. Barnett, S.M. Orbital angular-momentum and non-paraxial light-beams, S.M. Barnett and L. Allen, Opt. Commun. – 1994. – Vol. 110. – P. 670–678.

54. Soskin, M.S. Topological charge and angular momentum of light beams carrying optical vortices, M.S. Soskin, V.N. Gorshkov, M.V. Vasnetsov, J.T. Malos, and N.R. Heckenberg, Phys. Rev. A. – 1997. – Vol. 56. – P. 4064–4075.

55. Richards, B. Electromagnetic diffraction in optical systems. II. Structure of the image field in an aplanatic system, B. Richards and E. Wolf, Proc. Royal Soc. A. – 1959. – Vol. 253. – P. 358–379.

56. Prudnikov Yu.A., et al., Integrals and series. Special functions. Moscow, Nauka, 1983.

57. Gori, F. Polarization basis for vortex beams, J. Opt. Soc. Am. A. – 2001. – Vol. 18, No. 7. – P. 1612-1617.

58. Schwartz, C. Backscattered polarization patterns, optical vortices, and the angular momentum of light, Chaim Schwartz and Aristide Dogariu, Opt. Lett. – 2006. – Vol. 31, No. 8. – P. 1121-1123.

59. Bokor, N. A three dimensional dark focal spot uniformly surrounded by light, Nandor Bokor, Nir Davidson, Opt. Commun. – 2007. – Vol. 279. – P. 229–234.

60. Grosjean, T. Photopolymers as vectorial sensors of the electric field, T. Grosjean and D. Courjon, Opt. Express. – 2006. –Vol. 14, No. 6. – P. 2203-2210.

61. Xie, X.S. Probing single molecule dynamics, X.S. Xie and R.C. Dunn, Science. – 1994. – Vol. 265. – P. 361–364.

62. Beversluis, M.R. Programmable vector point-spread function engineering, M.R. Beversluis, L. Novotny, and S.J. Stranick, Opt. Express. – 2006. – Vol. 14. – P. 2650-2656.

63. Khonina S.N., et al., Komp. Optika. – 2009. – V.33, No. 4. – C. 401-411

64. Koronkevich, V.P. Lensacon, V.P. Koronkevich, I.A. Mikhaltsova, E.G. Churin, and Yu.I. Yurlov, Appl. Opt. – 1993. – V. 34(25). – P. 5761-5772.

65. Kalosha, V.P. Toward the subdiffraction focusing limit of optical superresolution, V.P. Kalosha, I.Golub, Opt. Lett. – 2007. – V. 32. – P. 3540-3542.

66. § Yirmiyahu, Y. Excitation of a single hollow waveguide mode using inhomogeneous anisotropic subwavelength structures, Yaniv Yirmiyahu, Avi Niv, Gabriel Biener, Vladimir Kleiner, and Erez Hasman, Opt. Express. – 2007. – V. 15(20). – P. 13404-13414.

67. Khonina, S.N. Trochoson, S.N. Khonina, V.V. Kotlyar, V.A. Soifer, M.V. Shinkaryev, G.V. Uspleniev, Optics Communications. – 1992. – V. 91 (3-4). – P. 158-162.

68. Khonina, S.N. Rotation of microparticles with Bessel beams generated by diffractive elements, S.N. Khonina, V.V. Kotlyar, R.V. Skidanov, V.A. Soifer, K. Jefimovs,

J. Simonen, J. Turunen, Journal of Modern Optics. – 2004. – V. 51(14). – P. 2167–2184.

69. Zhan, Q. Properties of circularly polarized vortex beams, Opt. Lett. – 2006. – Vol. 31, No. 7. – P. 867-869.

70. Khonina, S.N. Controlling the contribution of the electric field components to the focus of a high-aperture lens using binary phase structures, Khonina S.N., Volotovsky S.G., J. Opt. Soc. Am. A. – 2010. - Vol.27, No.10. – P. 2188-2197)

71. Goodman G. Introduction into Fourier Optics. Moscow, Mir, 1970.

72. Durnin, J. Exact solutions for non-diffracting beams. I. The scalar theory, J. Opt. Soc. Am. A. – 1987. – V. 4, N. 4. – P. 651–654

73. McLeod, J.H. The axicon: a new type of optical element, J. Opt. Soc. Am. – 1954. – V. 44. – P. 592–597.

74. Fedotovsky, A. Optical filter design for annular imaging, Fedotovsky, A., Lehovec, H., Appl. Opt. – 1974. – Vol. 13(12). – P. 2919–2923.

75. Khonina, S.N. Vortex phase transmission function as a factor to reduce the focal spot of high-aperture focusing system, Khonina S. N., Kazanskiy N. L., Volotovsky S. G., Journal of Modern Optics. – 2011. – V. 58(9). – P. 748–760

76. Khonina, S.N. Generation of rotating Gauss-Laguerre modes with binary-phase diffractive optics, S.N. Khonina, V.V. Kotlyar, V.A. Soifer, M. Honkanen, J. Lautanen, J. Turunen, Journal of Modern Optics. – 1999. – V. 46(2). – P. 227-238.

77. Khonina, S.N. Encoded binary diffractive element to form hyper-geometric laser beams, S.N. Khonina, S.A. Balalayev, R.V. Skidanov, V.V. Kotlyar, B. Paivanranta, J. Turunen, J. Opt. A: Pure Appl. Opt. – 2009. – V. 11. – P. 065702-065709.

78. Toraldo di Francia, G. Supergain antennas and optical resolving power, G. Toraldo di Francia, Nuovo Cimento, Suppl. – 1952. – Vol. 9. – P. 426.

79. Landberg G.S. Optics. Textbook, Moscow, Fizmatlit, 2003.

80. Khonina S.N., Komp. Optika – 2010. – V. 34(1). – 35-51

81. Khonina S.N., et al. Komp. Optika – 2011. – V. 35, No. 2. – C. 203-219.

82. Artl, J. Generation of a beam with a dark focus surrounded by regions of higher intensity: the optical bottle beam, J. Artl and M.J. Padgett, Opt. Lett. – 2000. – V. 25. – P. 191-193.

83. Hell, S.W. Breaking the diffraction resolution limit by stimulated-emission-depletion fluorescence microscopy, S.W. Hell and J. Wichmann, Opt. Lett. – 1994. – V. 19. – P. 780-782.

84. Biss, D.P. et al., Appl. Opt. – 2006. – V. 45. – P. 470-479.

85. Khonina, S.N. Influence of vortex transmission phase function on intensity distribution in the focal area of high-aperture focusing system, S.N. Khonina, N.L. Kazanskiy and S.G. Volotovsky, Optical Memory and Neural Networks (Allerton Press). – 2011. – V. 20(1). – P. 23-42.

86. Brent, R.P. Algorithms for Minimization Without derivatives, R.P. Brent. – Prentice-Hall, 1973. – 195 p.

87. http://alglib.sources.ru/

88. Sales, T.R.M. Diffractive superresolution elements, T.R.M. Sales and G.M. Morris, J. Opt. Soc. Am. A. – 1997. – Vol. 14. – P. 1637.

89. Bewersdorf, J. 4pi-confocal microscopy is coming of age, Joerg Bewersdorf, Alexander Egner, Stefan W. Hell, G.I.T. Imaging & Microscopy. – 2004. – Vol. 4. – P. 24-25.

90. Helseth, L.E. Breaking the diffraction limit in nonlinear materials, L.E. Helseth, Opt. Commun. – 2005. – Vol. 256. – P. 435.

91. Helseth, L.E. Smallest focal hole, Opt. Commun. – 2006. – V. 257. – P. 1-8.

5
Optical trapping and manipulation of micro- and nano-objects

The optical trapping and rotation of micro-objects are based on the well-known phenomenon of light pressure. After developing lasers it became possible to generate the radiation pressure force sufficient for acceleration, deceleration, deflection, direction and stable trapping of microscopic objects, whose dimensions range from tens of nanometers to tens of micrometers. If the refractive index is greater than the refractive index of the medium, the force resulting from changes in the direction of the light acts on the micro-object so that it moves into the region of the highest light intensity.

The first experiments with the trapping and acceleration of micro-objects, suspended in liquid and gas, are described in [1]. In 1977, changes were detected in the force of radiation pressure on the transparent dielectric spherical objects, depending on the wavelength and size [2].

If in the first studies it was shown that a micro-object can be trapped and moved linearly, then subsequent studies considered the possibility of rotating and orienting micro-objects in space. Optical rotation allows non-contact drives for micromechanical systems [3], also has many applications in biology [4].

There are three main ways of rotation of microscopic objects.

– Due to the spin angular momentum, which exists in the fields with circular polarization. In this case only the birefringent micro-objects, such as micro-objects made of Iceland spar, rotate [5, 6]. The main drawback of this method is the restriction on the material from which the micro-object is made.

– Due to the orbital angular momentum, which arises due to the spiral shape of the wave front, such as Laguerre–Gaussian (LG) and Bessel beams of higher orders. Transfer of the orbital angular momentum is due to partial absorption of light in the micro-object. This method is presented in [3, 7, 8]. In these studies, Bessel and LG beams formed with the use of amplitude holograms, which is unprofitable from the standpoint of energy efficiency. It is much more efficient to use pure phase DOEs, for example to create Bessel beams (BB) [9]. There are works in which the micro-objects move along paths other than the circle, for example, a light triangle, square, spiral [10, 11–14].

– By changing the phase shift in the interference pattern (in trapping of a microscopic object in the interference pattern) between the beam having a helical wavefront (i.e., LG beam) and the Gaussian beam. This pattern is rotated by changing the optical path length of one of the beams. This method is described in [15]. The main drawback of this method is the need to use a fairly complex optical circuit. In this case it is also easier to use the DOE that forms a superposition of Bessel of LG modes [16]. Rotating BB or LG which in propagation along the optical axis is accompanied by the rotation of the intensity distribution in the beam cross section can be used to rotate microscopic objects with variable speed by using the linear displacement of the radiation source or a focusing lens. In this case, the optical system is reduced, in fact, to one DOE.

2D-arrays of traps (micro-objects are pressed to the table of the microscope) have potential application for building elements of micro-optomechanical systems [17, 18], the formation of different micro-configurations [60], the sorting of biological cells [54] and other applications that do not require manipulation of longitudinal objects.

A system of two traps was realized with the help of a beam splitter and refractive optics [21, 22]. However, this approach is very complicated if a larger number of traps is required. An alternative and more promising approach is the separation and guiding of the laser beam with the DOE [7, 23–26].

In [27] the authors proposed to supplement a dynamic diffractive element, which is a matrix of $N \times N$ programmable phase gratings, with a $N \times N$ matrix of microlenses. In [7] the iterative method was used for calculating the phase DOE for creating 2D and 3D arrays of optical traps. In the experiments, the matrix was formed from eight Gaussian beams. The main disadvantage of spatial light modulators based on liquid crystals remain low diffraction efficiency (strong diffraction noise due to the high discreteness of the modulators) and the insufficient resolution of the matrix of pixels to handle the complex phase distributions. Also, the final pixel size limits the maximum variation of the diffraction orders (high carrier spatial frequencies are accompanied by the binarization of the phase profile and diffraction efficiency decreases).

Measurements have shown [28] that 15% of the energy of the incident beam remains after the liquid crystal modulator. Energy losses are due to several reasons:

1)   the opaque part of the panel (core loss, up to 65%),

2)   the structure of the liquid crystal modulator is similar to the grating, generating high orders (54% loss),

3)   the inability to concentrate all the energy in a useful manner, because the modulator has a maximum phase shift of less than 2π (maximum ratio achieved between the first and zero order of 2:1) [28],

4)   the discrepancy between the square aperture of the panel and the round profile of the incident beam (8%). Thus, the use of DOE for the formation of multiorder light beams for micro-object rotation problems if no dynamics is required, is preferred to the use of dynamic light modulators.

There are many studies in which solutions with separable variables for the Helmholtz and Schrödinger equations are used in optics. These studies examined multimode Bessel beams [29], multiorder LG beams [30], non-paraxial light beams that retain their structure during propagation [33], parabolic beams, Gauss–Helmholtz waves, paraxial light beams that retain their structure up to scale, Ince–Gaussian modes [33], elegant Ince–Gaussian beams [33], Hermite–Laguerre-Gaussian modes [32], optical vortices [33]. Some of these beams have been realized with laser resonators [33], liquid crystal displays [34, 35], phase DOEs [29, 36, 37]. These beams can provide additional new features in the problem of 'optical tweezers'.

There is a considerable number of works [38–48] concerned with the calculation of forces acting on the micro-object. In the well-known papers on the calculation of forces acting on the micro-object using the geometric optics approach, restrictions are imposed on the shape of the micro-object and the shape of the beam, and, as a rule, the motion parameters of the micro-object are not considered. For example, in [39] the authors considered only spherical micro-objects in a Gaussian beam. In [42], the force was calculated for the non-spherical micro-objects, but the authors consider the case of a

Gaussian beam. In [38] the spherical and elliptical micro-objects in Gaussian and LG beams were considered.

5.1. Calculation of the force acting on the micro-object by a focused laser beam

This section describes the derivation of the expressions for the force acting on a two-dimensional dielectric cylindrical object from a monochromatic electromagnetic wave.

5.1.1. Electromagnetic force for the three-dimensional case

In [1] a formula is derived which expresses the conservation of the total momentum of the system of the electromagnetic field plus the object V, bounded by the surface S:

$$\frac{\partial}{\partial t}\int_{V_1} P_i dV + \frac{\partial}{\partial t} P_{0i} = -\oint_{S_1} \sigma_{ik} n_k dS, \qquad (5.1)$$

where P_i are the coordinates of the vector of the momentum of the electromagnetic field (V_1 and S_1 are the volume and the surface restricting it, which include an object $V \in V_1$) that is associated with the Umov–Poynting vector by the relation:

$$\mathbf{P} = \frac{\mathbf{S}}{c} = \frac{1}{4\pi c}\left[\mathbf{E}\times\mathbf{H}\right], \qquad (5.2)$$

P_{0i} are the coordinates of the momentum vector of the object, $\delta P_{0i}/\delta t$ are the coordinates of the force vector of the light on the object ($\mu = 1$):

$$\sigma_{ik} = \frac{1}{4\pi}\left(\frac{|\mathbf{E}|^2 + |\mathbf{H}|^2}{2}\delta_{ik} - E_i E_k - H_i H_k\right); \qquad (5.3)$$

σ_{ik} is the Maxwell stress tensor of the electromagnetic field ($\sigma_{ik} = \sigma_{ki}$); $\mathbf{E}$, $\mathbf{H}$ are the vectors of the stress of the electric and magnetic fields in a vacuum.

After averaging over the time period of $T = 2\pi/\omega$ of the monochromatic light:

$$\mathbf{E}(\mathbf{x},t) = \mathrm{Re}\left\{\mathbf{E}(\mathbf{x})e^{i\omega t}\right\}, \quad \mathbf{H}(\mathbf{x},t) = \mathrm{Re}\left\{\mathbf{H}(\mathbf{x})e^{i\omega t}\right\} \qquad (5.4)$$

instead of equation (8.1) we obtain:

$$F_i = \left\langle \frac{\partial P_{0i}}{\partial t} \right\rangle = -\oint \langle \sigma_{ik} \rangle n_k dS, \tag{5.5}$$

as

$$\left\langle \frac{\partial}{\partial t} \int_V P_i dV \right\rangle = \int_V \left\langle \frac{\partial}{\partial t} P_i \right\rangle dV = 0. \tag{5.6}$$

It can be shown that:

$$
\left\langle \frac{\partial P_x}{\partial t} \right\rangle = \frac{1}{4\pi c} \left\{ \left\langle \mathrm{Re}\left(i\omega E_y(\vec{x})e^{i\omega t}\right)\mathrm{Re}\left(H_z(\vec{x})e^{i\omega t}\right) \right\rangle + \right.
$$
$$
+ \left\langle \mathrm{Re}\left(E_y(\vec{x})e^{i\omega t}\right)\mathrm{Re}\left(i\omega H_z(\vec{x})e^{i\omega t}\right) \right\rangle -
$$
$$
- \left\langle \mathrm{Re}\left(i\omega E_z(\vec{x})e^{i\omega t}\right)\mathrm{Re}\left(H_y(\vec{x})e^{i\omega t}\right) \right\rangle -
$$
$$
\left. - \left\langle \mathrm{Re}\left(E_z(\vec{x})e^{i\omega t}\right)\mathrm{Re}\left(i\omega H_y(\vec{x})e^{i\omega t}\right) \right\rangle \right\} = 0, \tag{5.7}
$$

where Re (...) is the real part of complex number $\langle f(t) \rangle = \dfrac{1}{T}\displaystyle\int_0^T f(t)dt$.
Similarly to (5.7) for the other projections of the momentum vector
of the electric field it can be shown that $\left\langle \dfrac{\partial P_y}{\partial t} \right\rangle = \left\langle \dfrac{\partial P_z}{\partial t} \right\rangle = 0$.

To obtain expressions for the time-averaged stress tensor (8.3) we
take into account that

$$\left\langle \mathrm{Re}\left(E_i(\vec{x})e^{i\omega t}\right)\mathrm{Re}\left(E_j(\vec{x})e^{i\omega t}\right) \right\rangle == \frac{1}{2}\mathrm{Re}\left[E_i(\vec{x})E_j^*(\vec{x})\right]. \tag{5.8}$$

Then instead of (5.5) we obtain (ε_2 is the dielectric constant of
the medium):

$$
F_x = \frac{1}{8\pi}\oint_s \left\{ \frac{1}{2}\left[\varepsilon_2 |E_x|^2 + |H_x|^2 - \varepsilon_2 |E_y|^2 - \right.\right.
$$
$$
\left. - |H_y|^2 - \varepsilon_2 |E_z|^2 - |H_z|^2 \right] dS_x +
$$
$$
\left. + \mathrm{Re}\left(\varepsilon_2 E_x E_y^* + H_x H_y^*\right) dS_y + \mathrm{Re}\left(\varepsilon_2 E_x E_z^* + H_x H_z^*\right) dS_z \right\},
$$

$$F_y = \frac{1}{8\pi} \oint_S \left\{ \frac{1}{2} \left[\varepsilon_2 |E_y|^2 + |H_y|^2 - \varepsilon_2 |E_x|^2 - \right. \right.$$

$$\left. - |H_x|^2 - \varepsilon_2 |E_z|^2 - |H_z|^2 \right] dS_y +$$

$$\left. + \mathrm{Re}\left(\varepsilon_2 E_y E_z^* + H_y H_z^* \right) dS_z + \mathrm{Re}\left(\varepsilon_2 E_y E_x^* + H_y H_x^* \right) dS_x \right\},$$

$$F_z = \frac{1}{8\pi} \oint_S \left\{ \frac{1}{2} \left[\varepsilon_2 |E_z|^2 + |H_z|^2 - \varepsilon_2 |E_x|^2 - \right. \right.$$

$$\left. - |H_x|^2 - \varepsilon_2 |E_y|^2 - |H_y|^2 \right] dS_z$$

$$\left. + \mathrm{Re}\left(\varepsilon_2 E_z E_x^* + H_z H_x^* \right) dS_x + \mathrm{Re}\left(\varepsilon_2 E_z E_y^* + H_z H_y^* \right) dS_y \right\}, \qquad (5.9)$$

where $dS_x = -\dfrac{\partial z}{\partial x} dxdy$, $dS_y = \dfrac{\partial z}{\partial y} dxdy$, $dS_z = dxdy$, $E_1 = E_x$, $E_2 = E_y$, $E_3 = E_z$

(and similarly for H_i and F_i).

5.1.2. Electromagnetic force for the two-dimensional case

We rewrite the expression (5.9) for the force of the action of light on the micro-object in the 2D case. For the TE-polarization ($H_x = E_y = E_z = 0$) the electric field is directed along the axis X:, $E_x \neq 0$, Z is the optical axis, the 2D-object has the form of a cylinder with the arbitrary cross-sectional shape and has an infinite length along the axis X. The plane YOZ is the plane of incidence. In this case the relation (5.9) takes the form:

$$F_x = 0,$$

$$F_y = \frac{1}{8\pi} \oint_{S_1} \left\{ \frac{1}{2} \left[|H_y|^2 - \varepsilon_2 |E_x|^2 - |H_z|^2 \right] dS_y + \mathrm{Re}\left(H_y H_z^* \right) dS_z \right\}, \qquad (5.10)$$

$$F_z = \frac{1}{8\pi} \oint_{S_1} \left\{ \frac{1}{2} \left[|H_z|^2 - \varepsilon_2 |E_x|^2 - |H_y|^2 \right] dS_z + \mathrm{Re}\left(H_z H_y^* \right) dS_y \right\},$$

Here S_1 is already a contour enclosing a section of the object in the plane YOZ. Force F_z is directed along the optical axis and is analogous to the scattering force for the Rayleigh particles [2], and

F_y is directed across the optical axis and is analogous to the gradient force [2]. The relationship between the projections H_y, H_z and E_x (TE-polarization) follows from Maxwell's equations:

$$H_y = \frac{i}{k_0 \mu} \frac{\partial E_x}{\partial z}, H_z = \frac{1}{i k_0 \mu} \frac{\partial E_x}{\partial y}, \tag{5.11}$$

and between the projections E_y, E_z and H_x (TM-polarization):

$$E_y = \frac{1}{i k_0 \varepsilon} \frac{\partial H_x}{\partial z}, E_z = \frac{i}{k_0 \varepsilon} \frac{\partial H_x}{\partial y}, \tag{5.12}$$

where $k_0 = 2\pi/\lambda$ is the wave number of light with a wavelength λ, ε is the dielectric constant of the medium, μ is the magnetic permeability of the medium. Similar to (5.10), the force of light pressure with TM-polarization for the 2D object will have the following projections ($E_x = H_y = H_z = 0$):

$$F_x = 0,$$

$$F_y = \frac{1}{8\pi} \oint_{S_1} \left\{ \frac{1}{2} \left[\varepsilon_2 |E_y|^2 - \varepsilon_2 |E_z|^2 - |H_x|^2 \right] dS_y + \varepsilon_2 \, \mathrm{Re}\left(E_y E_z^*\right) dS_z \right\},$$

$$\tag{5.13}$$

$$F_z = \frac{1}{8\pi} \oint_{S_1} \left\{ \frac{1}{2} \left[\varepsilon_2 |E_z|^2 - \varepsilon_2 |E_y|^2 - |H_x|^2 \right] dS_z + \varepsilon_2 \, \mathrm{Re}\left(E_z E_y^*\right) dS_y \right\},$$

where (as in equation (5.10)) $dS_y = n_y dl = \sin \phi \, dl = dz$ and $dS_z = n_z dl = \cos \phi \, dl = dy$ and dl is the element of the arc.

5.1.3. Calculation of force for a plane wave

To calculate the force exerted by the light field on a cylindrical object, we must calculate the integral over the contour within which the object resides. As follows from the formulas for calculating the force projections (5.10), (5.13), the force should not change when the radius of integration R_i changes, if the object is completely enclosed in the integration contour: $R_i > R$.

We calculate the iterative algorithm of the diffraction field of a plane wave on a cylindrical object, and we also calculate the force acting on it at various radii of integration. Simulation parameters: the incident wave is flat, the entire calculated diffraction field 10×10 μm, the wavelength 1 μm. The object is a cylinder with a circular

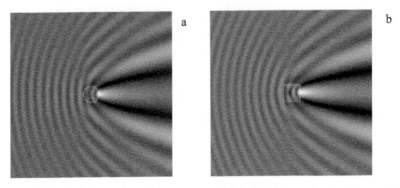

Fig. 5.1. The diffraction field $|E_x|$ of a TE-polarized plane wave on a) a cylinder with a circular cross section and b) a cylinder with a square cross-sectional shape.

Table 5.1. Dependence of the projection of force from the radius of integration

	Cylindrical object with a circular cross-section			
R_i, μm	5	3.75	2.5	1
$F_z \cdot 10^{-10}$ N/m	0.33176	0.32036	0.33213	0.31781

	Cylindrical object with a square cross-section			
R_i, μm	5	3.75	2.5	1
$F_z \cdot 10^{-10}$ N/m	0.32137	0.31588	0.31792	0.31688

cross-section, a diameter of 1 μm, or a square with 1 μm side. The refractive index of the cylinder n_1 = 1.4 (ε_1 = 1.96). The density of the light energy flux is 100 mW/m over the entire diffraction field.

Figure 5.1 shows the amplitude of the diffraction field ($|E_x|$, TE-polarization) of a plane wave on a cylindrical object with the above parameters.

Table 5.1 shows the dependence of the projection of the force F_z acting on the cylinder along the axis of light propagation Z, on the radius of integration R_i. Since the incident wave is flat and extends along the axis Z, the projection of force along the Y axis must be zero. Simulation shows that the projection of force on the Y-axis in this case is three orders of magnitude less than the projection of the force along the axis Z. For example, at the radius of integration R_i = 5 μm for a cylinder with a circular cross-section the projection of force F_z = 0.33176 · 10^{-10} N/m and projection F_y = 0.0007617· 10^{-10} N/m.

As shown in Table 5.1, the fluctuations of the results of calculation of force are less than 5%. The number of samples over the entire diffraction field 256×256. This result proves that the force is calculated correctly using formulas (5.10) (to within 5%).

Let us consider the dependence of the force, calculated by formulas (5.10), on the resolution of the diffraction field.

Table 5.2 shows the dependence of the projection of force F_z on the Z axis on the number of counts in the entire diffraction field for the above parameters. The force was calculated for an integration radius of 2.5 μm. All of the diffraction field was 5×5 μm in size.

Table 5.2 shows that when the number of counts in the entire diffraction field is 64×64, calculation of the force in the case of a circular cylinder is less accurate due to an error in the description of the boundary of the circular cylinder by a broken line. This does not apply to the last three values of force for a cylinder with a circular cross-section of 2% and 1.5% for the square.

Table 5.3 shows the projection of force F_z in the Z-axis under the same conditions on the number of samples K, taken for the diameter of a cylinder with a circular cross section, at a fixed resolution of the diffraction field – 256×256 pixels.

Table 5.3 shows that at low resolution of the object (in this case 12 samples per diameter of the circular cylinder), the value of the projection of the force acting on the cylinder is considered to be inaccurate. The difference for the last three values of the forces in Table 5.3 is 8%. When taking less than 12 samples per wavelength the iterative algorithm ceases to converge for the given parameters.

Table 5.2. Dependence of the projection of force F_z on the number of counts in the entire diffraction field

$K \times K$	64 × 64	128 × 128	256 × 256	512 × 512
	Cylindrical object with a circular cross-section			
$F_z \cdot 10^{-10}$ N/m	0.4058	0.3523	0.3479	0.3454
	Cylindrical object with a square cross-section			
$F_z \cdot 10^{-10}$ N/m	0.3220	0.3259	0.3216	0.3324

Table 5.3. Dependence of the projection of force F_z on the number of samples K taken at the diameter of the circular cylinder

K	12	25	50	100
$F_z \cdot 10^{-10}$ N/m	0.4046	0.3594	0.3497	0.3327

5.1.4. Calculation of force for a non-paraxial Gaussian beam

In this section we calculate the projection of force by the formulas (5.10) acting from the non-paraxial Gaussian beam on a dielectric cylinder with a circular cross section, depending on the displacement L of the centre of the cylinder from the centre of the beam waist.

Projections of the force, calculated by formula (5.10), acting on a cylinder with a circular cross-section in the case of a TE-polarized wave are shown in Fig. 5.2. The parameters of the experiment:

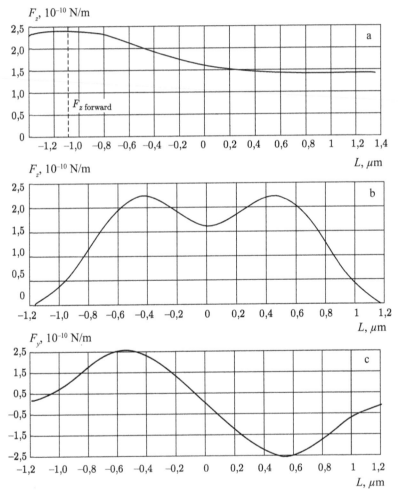

Fig. 5.2. TE-polarization: a) the dependence of the projection of force F_z on the displacement L of the object along the Z axis through the centre of the waist ($Y = 0$), the dependence of the projections of the forces F_z (b) and F_y (c) on displacement L of the object along the Y axis through the centre of the waist ($Z = 0$).

$D = \lambda = 2\omega_0 = 1$ μm, $\varepsilon_2 = 1$ (centre), $\varepsilon_1 = 2$, (object), the power of incident radiation per unit length is $P = 0.1$ W/m. The offset from the centre of the waist L has the dimension of a μm.

Similar projections of force in the case of TM-polarization, calculated by the formulas (5.13), are shown in Fig. 5.3.

Figures 5.2b and 5.3b shows that at the transverse displacement of the cylinder along the Y axis there is a projection of force F_y tending to return the cylinder to the centre of the waist. Moreover, the maximum projection of the force F_y and F_z is obtained for the

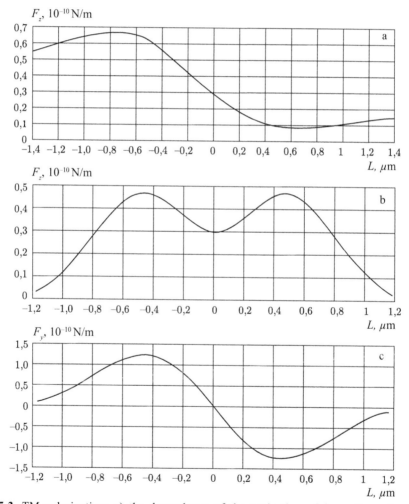

Fig. 5.3. TM-polarization: a) the dependence of the projection of force F_z on the displacement L of the object along the Z axis through the centre of the waist ($Y = 0$), the dependence of the projections of the forces F_z (b) and F_y (c) on displacement L of the object along the Y axis through the centre of the waist ($Z = 0$).

transverse displacement of the cylinder L approximately equal to the radius of the waist of the Gaussian beam: $L \approx \omega_0$.

In [3] the results are presented of numerical simulation of the force acting on a Kerr microsphere in the 3D case. Simulation parameters: the refractive index of the sphere $n_1 = 1.4$, refractive index $n_2 = 1.33$, sphere diameter $D = 2$ μm, wavelength $\lambda = 1.06$ μm, relative aperture (the ratio of the aperture of the lens to the focal length) NA = 1.4, the shift from the focus along the axis Z $L = 1$ μm. The force acting perpendicular to the propagation of light when a subject moves from the centre in a plane perpendicular to the propagation of radiation at the given parameters, $F = 0.3 \cdot 10^{-10}$ N. Figures 5.2b and 5.3b show that the projection of force is of the same order of magnitude per unit length of the cylinder $(0.5-1) \cdot 10^{-10}$ N/m.

Figure 5.4 shows the interference pattern of two Gaussian beams directed against each other with a waist at the origin, creating a standing wave. Figure 5.4a shows the amplitude of the total field $|E_x|$ (TE-polarization), Fig. 5.4b is the modulus of the projection of the Umov–Poynting vector on the axis of light propagation Z. The first Gaussian beam is directed along the axis Z, the second beam in the opposite direction of the axis Z. For the first Gaussian beam the wavelength is 1 μm, the radiation power of 50 mW/m, the Gaussian beam waist is at the origin, its diameter is 1 μm. The radiation power of the second beam 50 mW/m, the wavelength is also equal to 1 μm and the diameter of the waist is 1.5 μm. If a dielectric object with the size of the order of the wavelength is placed in such a field, then this field will be a trap for it: the object is drawn into the intensity maxima of the field. Figure 5.5 is a plot of the dependence of the

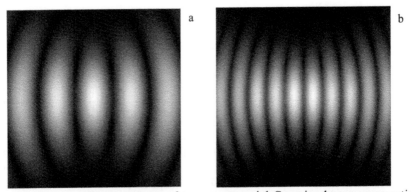

Fig. 5.4. The interference pattern of two non-paraxial Gaussian beams propagating in opposite directions along the axis Z: a) the total amplitude of the electrical field vector $|E_x|$, and b) the projection on the Z axis of the Umov–Poynting vector $|S_z|$.

F_z, 10^{-10} N/m

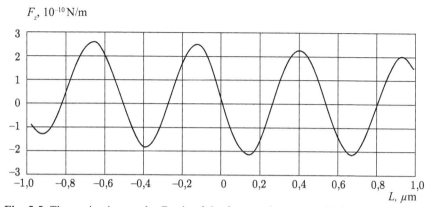

Fig. 5.5. The projection on the Z axis of the force acting on a cylinder with a circular cross section with $\varepsilon_1 = 2$, depending on the displacement of the centre circle of the cylinder along the axis Z.

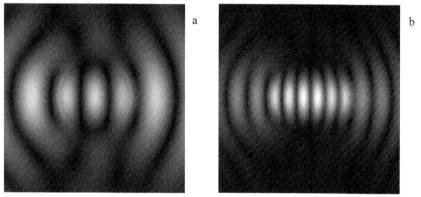

Fig. 5.6 Same as in Fig. 8.4, but in the presence of a cylinder with a circular cross section in the centre of the waist.

projection of force F_z directed along the Z axis on the displacement L from the axis Z. The object is a cylinder with a circular cross section with a diameter of 1 μm, dielectric constant $\varepsilon_1 = 2$. The diffraction field has a size of 2.5×2.5 μm. Figure 5.5 shows that near the waist along the Z axis almost periodically over a distance of about 0.25 μm there are points at which the force is zero. If the centre of the cylinder coincides with these points, then the cylinder will be in a stable or unstable equilibrium. The points of stable and unstable equilibria alternate, that is, approximately every 0.5 μm the cylinder will be in the 'optical trap' (at the point of stable equilibrium).

Figure 5.6 shows the diffraction of Gaussian beams directed against each other, shown in Fig. 5.4, on a cylinder with a circular

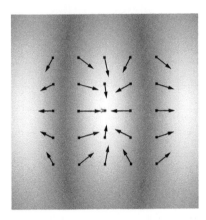

Fig. 5.7. The field vectors of the forces acting on the part of two colliding Gaussian beams on a cylinder with a circular cross section, which centre is located at different points in the interference pattern: the centre circle of the cylinder coincides with the beginning of each hand, and the length of each arrow is proportional to the modulus of strength at this point.

cross section, as described above. Figure 5.6a represents the strength of the electric field $|E_x|$ (TE-polarization), Fig. 5.6b – a projection of the Umov–Poynting vector on the axis Z. The object is located in the centre of the waist ($z = 0$). For visualization, the object in Fig. 5.6a is slightly obscured itself.

Figure 5.7 shows the central part of the diffraction pattern in Fig. 5.6a with the size of 0.31×0.31 μm. The arrows displayed the direction of the force acting on this cylinder by radiation, with the object placed in each specific point in space. One can see that the object is 'drawn' into the maxima of the interference pattern. The length of the arrows is proportional to the absolute force value.

If the refractive index is less than the refractive index of the particles, under certain conditions one can observe the 'trapping' of the particle along the Z axis, not only in the case of two colliding beams, but also in the case of a focused Gaussian beam.

Figure 5.8 shows a graph of the projection of force F_z in the displacement of the cylinder over distance L along the axis Z. The parameters of the experiment: the wavelength 1 μm, the diameter of the Gaussian beam waist $2\omega_0 = 1$ μm, the dielectric constant of the particles $\varepsilon_1 = 1.2$, the medium $\varepsilon_2 = 1$, the particle diameter $D = 2$ μm. From the graph we can see the trapping mechanism: the projection of force F_z in front of the focus is positive and directed towards the focus, behind the focus it is negative and pushes the particle back into focus. From numerical experiments it was

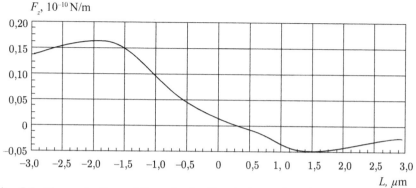

Fig. 5.8. The projection of force in the Z axis for a Gaussian beam acting on a cylinder with a circular cross section with $\varepsilon_1 = 1.2$ (medium $\varepsilon_2 = 1$).

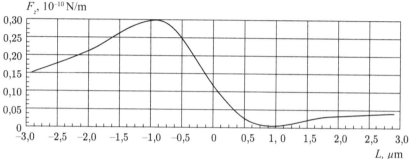

Fig. 5.9. The boundary of 'trapping': the projection of force in the Z axis for a non-paraxial Gaussian beam and a cylinder with a circular cross section with $\varepsilon_1 = 1.35$.

determined that the ability to trapping depends on the dielectric constant of the particle. For the given parameters 'trapping' occurs when $1 < \varepsilon_1 < 1.35$. A plot of the force F_z under these parameters and the dielectric constant of the particle $\varepsilon_1 = 1.35$ is shown in Fig. 5.5.

Figures 5.8 and 5.9 show that in the displacement of the cylinder along the optical axis at a distance $L \approx 0.4$ μm the force exerted on it by the light beam goes to zero: $F_y = F_x = 0$. The existence of such an equilibrium point for the cylinder can be explained in terms of the two forces (scattering and gradient) acting on the cylinder near the waist of the Gaussian beam. Indeed, when the centre of the cylinder is exactly in the centre of the beam waist, then it is subjected only to the scattering force (photons push the cylinder forward), which is proportional to the intensity $|E_x|^2$. At the offset from the centre of the cylinder along the optical axis a gradient force arises due to the presence of the gradient of intensity $\Delta|E_x|^2$, which is aimed at the centre of the beam. At displacement $L \approx 0.4$ μm those forces are equivalent and the cylinder is in equilibrium.

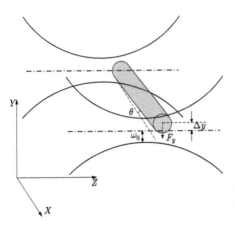

Fig. 5.10. Defining of the slope of a three-dimensional finite cylinder in the two-dimensional model.

A real cylindrical object has a finite length. But the two-dimensional approximation, which we consider here, can be applied to the description of the real situation if the length of the cylinder will be much larger than the diameter of its cross section. Indeed, consider the case where a three-dimensional dielectric cylinder of finite length is located near the waist of a cylindrical Gaussian beam (see Fig. 5.10).

Let the cylinder axis tilted at an angle θ, in the plane XY. Then the maximum deviation from the stable equilibrium point of the cylinder cross-section in the YZ plane will be equal to $\Delta y = l \cdot \mathrm{tg}(\theta)$, $y/a \ll 1$, where l is the length of the cylinder, a is the radius of its cross section. This results in a projection force F_y directed to the point of maximum intensity on the beam axis. That is, small rotations and displacements of the three-dimensional finite-length cylinder near a stable equilibrium of the cylindrical Gaussian beam waist will give rise to forces seeking to return the cylinder to the 'optical trapping' position.

5.1.5. Calculation of forces for the refractive index of the object smaller than the refractive index of the medium

It is interesting to calculate the light force and field, acting on a dielectric 2D object whose refractive index is smaller than that of the medium.

Figure 5.11 shows the diffraction pattern of a plane wave in a medium with a refractive index $n_2 = 1.33$ (water) on a cylindrical object with a circular cross section with a refractive index $n_1 = 1$ (cylindrical air bubble). The diameter of the cylinder is equal to the

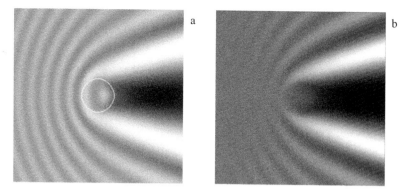

Fig. 5.11. Modulus of the strength of the electric field $|E_x|$ (a) and modulus of the projection of the Umov–Poynting vector on the axis Z $|S_z|$ (b) for diffraction of a plane wave on an air cylinder with a circular cross section in water.

wavelength and is equal to 1 μm. Figure 5.11a is the modulus of the strength of the electric field $|E_x|$ (TE-polarization), Fig. 5.11b is a projection of the Umov–Poynting vector on the axis Z. The energy does not propagate behind the 'air bubble' which can be clearly seen on the sections shown in Fig. 5.12, taken along the Z axis through the point $Y = 0$.

Figure 5.12a shows the value of the amplitude $|E_x|$, Fig. 5.12b – the projection of the Umov–Poynting vector on the axis Z.

If such an object is placed near the focus of a Gaussian beam, it will be pushed out of it, as illustrated in the graphs in Fig. 5.13.

Figure 5.13a is a plot of the dependence of the force F_z along the Z axis on the displacement L along the axis Z, Fig. 5.13b – the dependence of force F_y on the displacement L along the Y axis through the focus. The Gaussian beam has a wavelength of 1 μm, the diameter of the waist is 1 μm, the radiation power 100 mW/m. It is seen that in deviation in either direction from the focus in the transverse direction the force, directed toward the deflection, increases, which leads to the movement of the focus in this direction. In deviation along the axis Z of light the force exerted on an object in front of the focus is less in the absolute value than behind the focus.

5.2. Methods for calculating the torque acting on a microobject by a focused laser beam

In this section, we calculate the torque acting on a cylindrical micro-

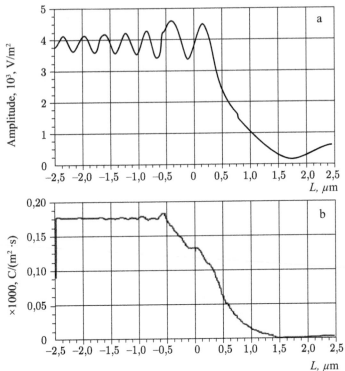

Fig. 5.12. Section of Fig. 8.11a (a) and 8.11b (b) along the Z axis through the point $Y = 0$.

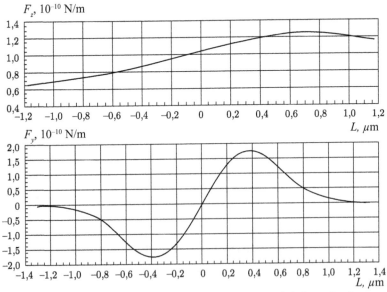

Fig. 5.13. The projections of the force of the non-paraxial Gaussian beam on 'a cylindrical air bubble' in the water: on the longitudinal axis F_z (a) and transverse axis F_y (b).

object with an elliptical cross-section from the side of the focused non-paraxial Gaussian beam. Calculation of the moment of the force was conducted depending on the size and shape of the integration region encompassing the micro-object under study. We consider the torque on the micro-object, located at the point where the force exerted by light is zero. We also consider a torque acting on the micro-object with an elliptical section in the standing wave.

5.2.1. The orbital angular momentum in cylindrical microparticles

Figure 5.14 shows the scheme of the problem.

Light propagates along the Z axis in a medium with permittivity ε_1 and falls on an object with a dielectric constant ε_2. Then the torque **M** at any point A can be calculated using the formula [4]:

$$\mathbf{M} = \oint_S \left[\mathbf{r} \times \left(\tilde{T} \cdot \mathbf{n} \right) \right] dS, \tag{5.14}$$

where **n** is the normal to the surface S, covering the object in question, A is the point at which we calculated torque **M**, **r** is the radius vector from point A to the integration surface S, $\tilde{T}$ is the Maxwell's stress tensor of the electromagnetic field.

The product of the Maxwell stress tensor on the normal can be written as:

$$\left(\tilde{T} \cdot \mathbf{n} \right) = \begin{bmatrix} T_{ii} & T_{ij} & T_{ik} \\ T_{ji} & T_{jj} & T_{jk} \\ T_{ki} & T_{kj} & T_{kk} \end{bmatrix} \begin{pmatrix} n_x \\ n_y \\ n_z \end{pmatrix} = $$

$$= \begin{bmatrix} T_{ii}n_x + T_{ij}n_y + T_{ik}n_z \\ T_{ji}n_x + T_{jj}n_y + T_{jk}n_z \\ T_{ki}n_x + T_{kj}n_y + T_{kk}n_z \end{bmatrix} = \begin{pmatrix} t_x \\ t_y \\ t_z \end{pmatrix}. \tag{5.15}$$

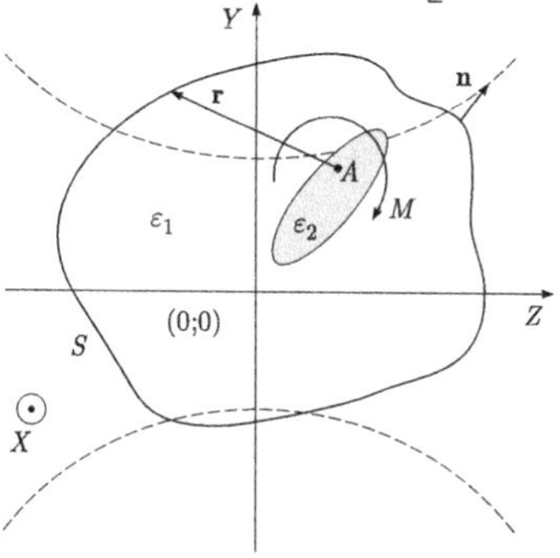

Fig. 5.14. Scheme of the problem.

Then

$$\mathbf{r} \times \left(\tilde{T} \cdot \mathbf{n} \right) = \begin{vmatrix} i & j & k \\ r_x & r_y & r_z \\ t_x & t_y & t_z \end{vmatrix} = \mathbf{i}(r_y t_z - r_z t_y) - \tag{5.16}$$

$$-\mathbf{j}\left(r_x t_z - r_z t_y \right) + \mathbf{k}\left(r_x t_y - r_y t_x \right).$$

For the two-dimensional case (cylindrical object) this makes sense only in relation to the axis X, i.e.

$$M_x = \oint_S \left(r_y t_z - r_z t_y \right) dS. \tag{5.17}$$

As $E_y = E_z = H_x = 0$, for TE-polarization, we finally obtain:

$$M_x = \frac{1}{4} \oint_S \left[\varepsilon_0 \varepsilon_1 \left| E_x \right|^2 + \mu\mu_0 \left| H_y \right|^2 - \mu\mu_0 \left| H_z \right|^2 \right] \times$$

$$\times r_y dy - \frac{1}{2} \oint_S \mu\mu_0 \operatorname{Re}\left(H_z H_y^* \right) r_y dz -$$

$$-\frac{1}{4} \oint_S \left[\varepsilon_0 \varepsilon_1 \left| E_x \right|^2 - \mu\mu_0 \left| H_y \right|^2 + \mu\mu_0 \left| H_z \right|^2 \right] r_z dz - \tag{5.18}$$

$$-\frac{1}{2} \oint_S \mu\mu_0 \operatorname{Re}\left(H_y H_z^* \right) r_z dy.$$

Calculation of the final formula for the moment for TM-polarization differs by the substitution in (5.17) $E_x = H_y = H_z = 0$.

5.2.2. The results of numerical simulation of the torque

To verify the correctness of the formula (5.18), we calculated the torque acting on an elliptical microparticle by a non-paraxial Gaussian beam with different sizes of square integration contour S. Figure 5.15b shows a plot of the changes of torque M_x on the integration contour, whose parameters are shown in Fig. 5.15a. The waist of the Gaussian beam is located in the centre of coordinates, the light propagates in the positive direction along the axis Z.

The diffraction field is calculated by the method described in the previous section. The integration was carried out on the square contour S, the number of counts in Fig. 5.15a 254×254. This is convenient because it is not necessary to calculate the normal $\mathbf{n}$ at each point of the circuit – the contour integral (5.18) splits into a sum of integrals, some of which are taken on the contour sides parallel

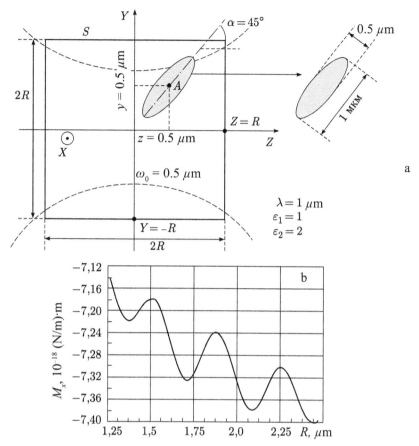

Fig. 5.15. a) The calculation of the torque and b) the results of the calculation of torque M_x for different R.

to the axis Y, and another part on the sides of the contour parallel to the axis Z. The power of the incident radiation was equal to $P = 100$ mW/m, the wavelength $\lambda = 1$ µm, the dielectric constant of the medium $\varepsilon_1 = 1$, the dielectric constant of the particle $\varepsilon_2 = 2$. Point A, in relation to which the moment is calculated, coincides with the centre of an elliptical particle.

Note that the torque **M** is measured in units [N ·m]. However, in this case the dimension of M_x is [(N/m)·m] because the force in the two dimensional case is expressed, due to an infinitely long cylindrical object along the X axis, in units N/m and is unit force per unit length of the micro-object.

Figure 5.15b shows that the fluctuations in the value of M_x with a change in R are less than 4%.

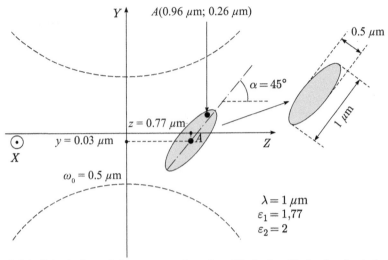

Fig. 5.16. Calculation of the torque when the elliptical cylinder is situated at the equilibrium point (the force of the light to an object is equal to zero).

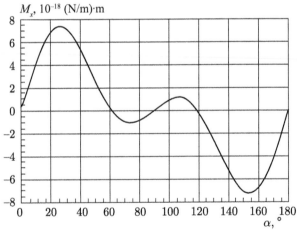

Fig. 5.17. Dependence of the projection on the X axis of the torque acting on the elliptical cylinder in the focus of a Gaussian beam, on angle α.

Note that the magnitude of the torque **M** depends on the location of point A, relative to which it is calculated, if the force acting on the particle is not zero. If we calculate the torque M_x for the scheme in Fig. 5.15a at the same location of the cylinder, but when the point A has coordinates (0.25 μm, 0.25 μm), the value will be $M_x = 5.8 \cdot 10^{-17}$ (N/m)·m.

Figure 5.16 shows the scheme of calculating the moment of the force M_x for the case when the force of light acting on a particle is zero. An elliptical cylinder with dielectric constant $\varepsilon_2 = 2$ is in the water with $\varepsilon_1 = 1.77$.

The torque in finding the elliptical cylinder at the focus with point A, located in the centre of the particle, is $M_x = 7.11 \cdot 10^{-8}$ (N/m)·m. If we move the observation point A to a point with the coordinates (0.96 μm, 0.26 μm), then the moment will be equal to $M_x = 7.11 \cdot 10^{-8}$ (N/m)·m. Thus, we see that the torque acting on the particle is almost independent of the point relative to which it is calculated, if the particle is at the point where the force exerted by the light field is zero.

Figure 5.17 shows a plot of the dependence of the torque on angle α. The simulation parameters are the same as in Fig. 5.15a, but the object is located in the centre of the waist, point A is located in the centre of the object, the major diameter of the ellipse is 1.2 μm, minor 0.3 μm, the dielectric constant of the object $\varepsilon_1 = 2.25$. The angle α was determined as shown in Figure 5.16 (measured from the Z-axis counter-clockwise).

Figure 5.17 shows that the positions of the elliptical particle in the focus when its long axis lies along and perpendicular to the optical axis, are stable equilibria ($M_x > 0$ for a particle rotating in the clockwise direction). When α is approximately equal to 60° and 120°, there are two points of unstable equilibrium, when the moment vanishes. This agrees well with the data in [4], where the same plot of the dependence the torque on the rotation angle of the particle α was obtained.

Figure 5.18 is a plot of the dependence of the torque M_x on the displacement L of a circular particle with a radius of 0.25 μm on the Y axis through the centre of the waist. Point A is located in the centre

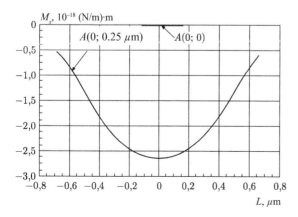

Fig. 5.18. The dependence of the torque M_x on the displacement L of a spherical particle along the Y axis for the two positions of point A relative to the centre of the particle.

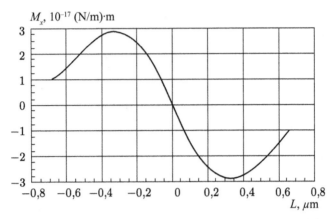

Fig. 5.19. The dependence of the torque on circular displacement L of the absorbing particles along the Y axis with respect to point A (0, 0).

of particle (0; 0) and 0.25 μm above the centre of the Y axis. The other parameters are the same as in Fig. 5.15.

As can be seen from Fig. 5.18 the torque acting on a circular non-absorbing particle in relation to the centre is equal to zero when it is displaced along the Y axis, and varies with the displacement of point A up by 0.25 μm from the centre of the particle along the Y axis. If to the function, describing the dielectric permittivity of the particles, we add an imaginary component (the absorbing particle), the torque on the circular particle in relation to the centre will be different from zero at the displacement of the particle identical to that in Fig. 5.18. Figure 5.19 shows a plot of the dependence of the torque acting on a circular particle with $\varepsilon_2 = 2 + 1i$ with respect to the centre of the particle. The other parameters are the same as in the calculation of the graph in Fig. 5.18.

As can be seen from Fig. 5.19, the torque acting on a circular absorbing particles relative to the centre is not equal to zero when the particle is displaced from the optical axis.

Figure 5.20 is a plot of the dependence of the torque M_x acting on an elliptical dielectric ($\varepsilon_2 = 2 + 0i$) and absorbing microparticles ($\varepsilon_2 = 2 + 0i$), on the angle of rotation of the particle α. The particle is located in the coordinates (0, 0.25) with respect to the beam waist, the moment was calculated relative to the centre of the ellipse. The other simulation parameters are the same as in Fig. 5.15a.

As can be seen from Fig. 5.20, the addition of the imaginary part to the function of dielectric permittivity of the particle increases the

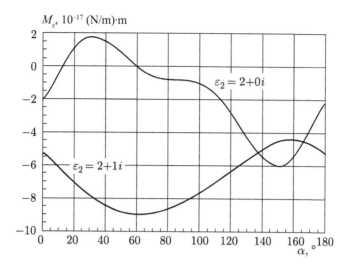

Fig. 5.20. The dependence of torque M_x on the elliptical particle on the rotation angle α of the particle.

scattering strength, due to which at any angle α the torque is non-zero and is directed counterclockwise ($M_x < 0$).

Figure 5.21a shows the results of the calculation of the dependence of the torque M_x acting on the elliptical particle, located at the centre of the Gaussian beam, on the angle of rotation α. All parameters are the same as in Fig. 5.15a, the particle is taken only half the size (the smaller diameter of the ellipse 0.25 μm, larger diameter 0.5 μm), the dielectric constant and the particle is taken. Figure 5.21b shows the same chart, but for the two Gaussian beams, the first waist radius $\sigma_1 = 0.5$ μm, second $\sigma_2 = 0.6$ μm, the intensity of the two Gaussian beams was the same and equal to 50 mW/m. Diffraction of the oppositely directed beams on a particle at $\alpha = 45°$ is shown in Fig. 5.21c.

As can be seen from Fig. 5.21, by adding the imaginary part of dielectric permittivity if the cylinder is located in the centre of the waist of a Gaussian beam, the torque is strongly attenuated and has several zero points. It is also clear that the position of stable equilibrium when adding a second oppositely directed Gaussian beam will be observed at the location of the major axis of an elliptical particle along the axis Y, i.e. the elliptical particle tends to settle along the line of maximum intensity.

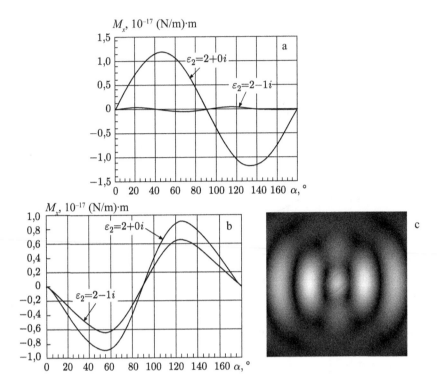

Fig. 5.21. a) The dependence of the torque M_x on the angle α for a single Gaussian beam b) the dependence of torque M_x on angle α for the two oppositely directed Gaussian beams, c) the amplitude of the field $|E_x|$ of two oppositely directed Gaussian beams on an elliptical particle.

5.1.7. A geometrical optics method for calculating the force acting by light on a microscopic object

As can be seen in the previous section in the problem of calculating the electromagnetic forces in the approach is very capacious as regards the volume of calculations, as well as in the problems of calculating the forces acting on the micro-object in the light beam it is generally required to calculate the forces in some areas; consequently, the amount of computation grows as the square of the size of the area. In this case, for example, the calculation of forces to simulate the motion of the micro-object can take dozens of hours on a PC. In this regard, we use a simple geometrical optics approach, which gives approximate results. The geometrical optics approach has been used in the calculation of the simplest optical traps [32]. However, this and other studies generally considered some of the simpler cases in which restrictions are imposed either on the form of the micro-object [32, 86], or the shape of the beam [78]. We consider the method of calculating the forces acting on the micro-object of arbitrary shape in a light beam with a given distribution of intensity and phase [15].

Consider the micro-object of arbitrary shape in the light beam. We assume that the observed number of conditions:

1.   The light beam is given by functions of intensity $I(x, y)$ and phase $\varphi(x, y)$.

2.   The micro-object is bounded by two surfaces: the top, which is given by the function $f_1(x, y)$ and the bottom, which is given by the function $f_2(x, y)$ (Fig. 5. 22). The functions $f_1(x, y)$ and $f_2(x, y)$ are unique.

3.   The micro-object is moving in the plane xy (however, this method is easy to calculate the force extends to three-dimensional motion).

4.   The light beam is incident on the micro-object vertically from top to bottom.

The unit vectors $\vec{a}(a_x, a_y, a_z)$, $\vec{b}(b_x, b_y, b_z)$, $\vec{c}(c_x, c_y, c_z)$ define the direction of the incident and refracted rays. The vector $\vec{a}(a_x, a_y, a_z)$ is determined by the function $\varphi(x, y)$. This vector must always be perpendicular to the wavefront.

The components of the force **F** of a single beam on the micro-object are determined by the formula

$$F_x = \frac{N}{c}(a_x - c_x),$$

$$F_y = \frac{N}{c}(a_y - c_y),$$

$$\text{(5.19)}$$

where N is the power of the beam, c is the velocity of light.

For the whole beam, this formula takes the form

$$F_x = \frac{1}{c}\iint_\Omega I(x,y)(a_x - c_x)\,dx\,dy,$$

$$F_y = \frac{1}{c}\iint_\Omega I(x,y)(a_y - c_y)\,dx\,dy,$$

$$\text{(5.20)}$$

where Ω is the region of maximum cross-sectional area of a microscopic object in a plane perpendicular to the direction of

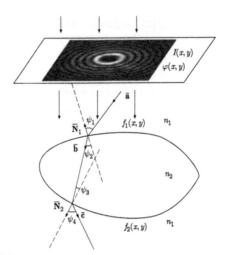

Fig. 5.22. Scheme of refraction of light rays on a micro-object.

propagation of the light beam.

It should be borne in mind that the direction vector for the refracted output beam depends on the direction vector of the incident beam

$$F_x = \frac{1}{c} \iint_\Omega I(x,y)\left(a_x - c_x(\vec{\mathbf{a}})\right) dx\, dy,$$

$$F_y = \frac{1}{c} \iint_\Omega I(x,y)\left(a_y - c_y(\vec{\mathbf{a}})\right) dx\, dy. \tag{5.21}$$

The dependence $\vec{\mathbf{c}}(\vec{\mathbf{a}})$, i.e. direction of the beam after the micro-object, depending on the initial beam direction, can be determined from the following relations

$$\left(\vec{\mathbf{N}}_1, -\vec{\mathbf{a}}\right) = \cos(\psi_1),$$

$$\left(\vec{\mathbf{N}}_1, -\vec{\mathbf{b}}\right) = \cos(\psi_2),$$

$$\left(\vec{\mathbf{a}}, \vec{\mathbf{b}}\right) = \cos(\psi_2 - \psi_1),$$

$$n_1 \sin\psi_1 = n_2 \sin\psi_2,$$

$$\left(\vec{\mathbf{N}}_2, \vec{\mathbf{b}}\right) = \cos(\psi_3), \tag{5.22}$$

$$\left(\vec{\mathbf{N}}_2, \vec{\mathbf{c}}\right) = \cos(\psi_4),$$

$$\left(\vec{\mathbf{b}}, \vec{\mathbf{c}}\right) = \cos(\psi_3 - \psi_4),$$

$$n_2 \sin\psi_3 = n_1 \sin\psi_4,$$

where n_1, n_2 are the refractive indices of the medium and micro-object, $\vec{N}_1$ and $\vec{N}_2$ are the normal vectors to the surfaces $f_1(x, y)$ and $f_2(x, y)$, ψ_1, ψ_2, ψ_3, ψ_4 are the angles of incidence and refraction at the surfaces. By simple transformations we obtain from (5.22)

$$b_x = \frac{A_1^2}{A_1^2 + A_2^2}\left[\left(\frac{A_3 A_2}{A_1} + K_1 K_2\right) + \sqrt{\left(\frac{A_3 A_2}{A_1^2} + K_1 K_2\right)^2 - \left(1 + \frac{A_2^2}{A_1^2} + K_1^2\right)\left(\frac{A_3^2}{A_1^2} + K_1^2 - 1\right)}\right].$$

(5.23)

Here we have introduced a number of intermediate symbols which can significantly simplify the writing of the formula:

$$A_1 = a_y - \frac{N_{1y} a_z}{N_{1z}}, \qquad A_2 = a_x - \frac{N_{1x} a_z}{N_{1z}},$$

$$A_3 = \frac{a_z \cos\psi_2}{N_{1z}} + \cos(\psi_1 + \psi_2),$$

$$K_1 = \frac{\cos\psi_2}{N_{1z}} - \frac{N_{1y} A_3}{A_1}, \qquad K_2 = \frac{N_{1x}}{N_{1z}} - \frac{N_{1y} A_2}{A_1}.$$

The same calculation procedure is used to determine other components of the vectors $\vec{b}(b_x, b_y, b_z)$ and $\vec{c}(c_x, c_y, c_z)$. Since the cumulative record of these formulas is very large and is similar to the formula (5.23), it will not be presented here.

To determine the components of the direction vector $\vec{a}(a_x, a_y, a_z)$ we need to use the phase $\varphi(x, y)$ of the light beam. As mentioned above, the direction vector must always be perpendicular to the wavefront. Then the direction vector is described by the following relation

$$\vec{a} = \frac{\operatorname{grad}\varphi(x, y)}{|\operatorname{grad}\varphi(x, y)|}.$$

(5.24)

For normal vectors we can write the relation

$$\vec{N}_1 = \left(\frac{\dfrac{\partial f_1}{\partial x}}{\sqrt{\left(\dfrac{\partial f_1}{\partial x}\right)^2 + \left(\dfrac{\partial f_1}{\partial y}\right)^2 + 1}}, \frac{\dfrac{\partial f_1}{\partial y}}{\sqrt{\left(\dfrac{\partial f_1}{\partial x}\right)^2 + \left(\dfrac{\partial f_1}{\partial y}\right)^2 + 1}}, -\frac{1}{\sqrt{\left(\dfrac{\partial f_1}{\partial x}\right)^2 + \left(\dfrac{\partial f_1}{\partial y}\right)^2 + 1}} \right),$$

$$\vec{N}_2 = \left(\frac{\dfrac{\partial f_2}{\partial x}}{\sqrt{\left(\dfrac{\partial f_2}{\partial x}\right)^2 + \left(\dfrac{\partial f_2}{\partial y}\right)^2 + 1}}, \frac{\dfrac{\partial f_2}{\partial y}}{\sqrt{\left(\dfrac{\partial f_2}{\partial x}\right)^2 + \left(\dfrac{\partial f_2}{\partial y}\right)^2 + 1}}, -\frac{1}{\sqrt{\left(\dfrac{\partial f_2}{\partial x}\right)^2 + \left(\dfrac{\partial f_2}{\partial y}\right)^2 + 1}} \right).$$

$$(5.25)$$

Similarly, we determine the parameters of the reflected light rays (Fresnel reflection). Using (5.21)–(5.25), we can calculate the force acting on an arbitrarily shaped micro-object from an arbitrary light beam. Knowing the mechanical properties of the medium in which the micro-object is located, we can also simulate the motion of a microscopic object. In modelling the motion of a microscopic object in the light beam was solved by a system of equations of motion.

$$\begin{cases} m\dfrac{d\mathbf{v}}{dt} = \mathbf{F}_l + \mathbf{F}_f, \\ \dfrac{d\mathbf{r}}{dt} = \mathbf{v}, \end{cases}$$

where $\mathbf{F}_l$ is the force acting on the micro-object from the light beam, $\mathbf{F}_f$ is the force of viscous friction. Based on the above proposed method software has been developed that allows one not only to calculate the force of the light on the microscopic object in a given light field, but also to simulate the motion of a microscopic object in a given environment. Figure 5.23 shows the interface of the software.

However, the geometrical optics approach for the micro-objects comparable in size to the wavelength usually gives very inaccurate results where we need to calculate, for example, the intensity behind a microscopic object. It is necessary to verify how we apply this approach to calculate the force of light. To do this, we compare the values of force obtained by two methods: geometrical optics, which has been described in this section, and the electromagnetic method, which was described in Section 5.1.

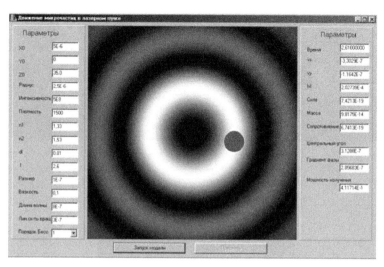

Fig. 5.23. Interface software for the simulation of the motion of microscopic objects in the light field.

5.1.8. Comparison of results of calculations by geometrical optics and electromagnetic methods

The method of calculating the force (5.19)–(5.25) allows us to calculate the effect of light also for the cylindrical micro-object [52]. A non-paraxial Gaussian beam was approximated by a system of rays with the parameters on the same beam shown in Fig. 5.24. A cylindrical object with a diameter of 1 μm and a refractive index of 1.41 was shifted across the axis of the Gaussian beam, with the values of the forces F_x and F_y calculated in each position. Figure 5.24 shows approximation of a Gaussian beam by a system of rays and refraction of the rays by this system at different positions of the cylindrical micro-object with respect to the optical axis.

For a cylindrical micro-object we calculated the force of light in the Gaussian beam. Figure 5.25a is a plot of the dependence of the projections of the force F_z on the displacement with respect to the waist along the propagation axis of the beam, superimposed on the graph presented in Section 5.1. Figure 5.25b is a plot of the dependence of projections of force F_y on the displacement relative to the axis of beam propagation, superimposed on the graph presented in Section 5.1 [53–55].

For the longitudinal force (Fig. 5.25a) there is an area where the standard deviation of force, obtained in the geometrical optics

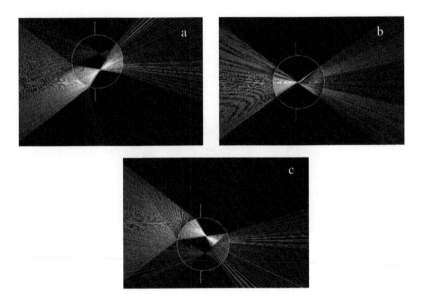

Fig. 5.24. Refraction of beams on a cylindrical micro-object 1 μm in diameter in different positions relative to the beam axis, (a) the displacement by 0.5 μm upwards, (b) exactly on the beam axis, (c) the displacement by 0.5 μm downward.

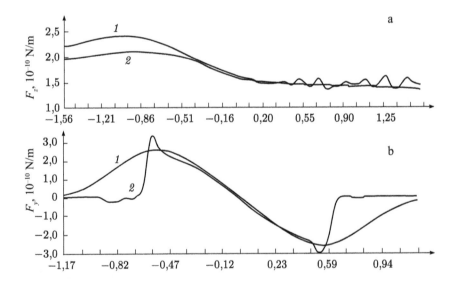

Fig. 5.25. a) The dependence of force F_z on the displacement L of the object along the Z axis through the centre of the waist $(Y = 0)$, b) dependence of force F_y on the displacement L of the object along the Y axis through the centre of the waist $(Z = 0)$ (1 – exact calculation, 2 – calculation by geometrical optics .)

approximation, from the force calculated in the framework of the electromagnetic approach is no more than 0.1. As seen in Fig. 5.25b, for the shear force there is also a region in which the standard deviation is not more than 0.1.

To determine the accuracy, a standard formula was used for calculating the standard deviation (RMS):

$$\sigma = \frac{\sqrt{\dfrac{1}{N}\sum_{i=1}^{N}\left(f_{1i}-f_{2i}\right)^{2}}}{\langle f_{2}\rangle}, \tag{5.26}$$

where f_{1i} is the sampling point of the compared function, f_{2i} is the discretization point of the reference function, N is the number of discretization points of the function, $\langle f_{2}\rangle$ is the mean value of the reference function.

These results suggest that the geometrical optics method for calculating the forces acting on the micro-object from the light beam has the error of no more than 0.1, except in cases of no practical interest, such as displacement of more than 0.5 μm from the beam axis in Fig. 5.25b. In this case, the method has a much lower computational complexity than the exact methods of calculation, which allows not only to calculate the forces acting on the micro-object in the light field, but also to simulate the motion of the micro-object.

5.2. Rotation of micro-objects in a Bessel beam
5.2.1. Rotation of micro-objects in Bessel beams

In optical systems for the rotation of micro-objects with the help of a Bessel beam (BB) [56,57,58,61] the light energy is concentrated using a ring BB 'squeezed' through a spherical lens. However, it

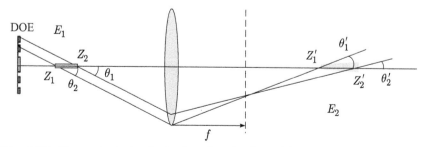

Fig. 5.26. The optical circuit for imaging the Bessel beam, used to manipulate microscopic objects.

appears that BB has the property to maintain its diameter near the axicon or DOE, loses this property in imaging using a spherical lens and begins to disperse.

We show that the image of the diffractionless BB produced using with a spherical lens leads to a divergent BB. Figure 5.26 shows the optical scheme.

As an initial function we choose the zero-order Bessel beam:

$$\Psi_0(r) = J_0(\alpha r). \tag{5.27}$$

To find how the function Ψ_0 is transformed by a lens, we need to simulate the propagation of the beam in free space over distance a by Fresnel transform, then multiply by the complex transmission function of the lens with a focal length f, and again apply the Fresnel transform at distance z:

$$\Psi(u,v,z) = \left(\frac{k}{2\pi i}\right)^2 \frac{1}{az}$$

$$\times \int_{-\infty}^{\infty}\int_{-\infty}^{\infty}\int_{-\infty}^{\infty}\int_{-\infty}^{\infty} \Psi_0(r)\exp\left\{\frac{ik}{2}\left[\frac{(x-\xi)^2}{a} + \frac{(y-\eta)^2}{a} - \frac{\xi^2+\eta^2}{f} + \frac{(\xi-y)^2}{z} + \frac{(\eta-v)^2}{z}\right]\right\} dxdyd\xi d\eta. \tag{5.28}$$

The distance z is related to the distance a by the lens formula. Equation (5.28) uses the transmission function of the paraxial lens in the form of:

$$\tau(\xi,\eta) = \exp\left[-\frac{ik}{2f}\left(\xi^2+\eta^2\right)\right]. \tag{5.29}$$

The evaluation of the integral (5.28) gives:

$$\Psi(\rho,z) = \frac{f}{(z-f)} J_0\left(\frac{\alpha f\rho}{f-z}\right)\exp\left[-i\frac{\alpha^2}{2k}\left(a+\frac{fz}{f-z}\right)+i\frac{k\rho^2}{2(f-z)}\right], \tag{5.30}$$

where $\rho^2 = u^2 + v^2$.

Equation (5.30) shows that the BB (5.27) like the non-paraxial beam diverges for $z > f$. This is due to the fact that the lens produces a divergent parabolic wave front in the BB. On the contrary, the beam diverges before the focal plane.

Figure 5.27 shows the results of the experiment compared with the theoretical results for the Bessel function of the fifth order; the experiments were caried out used a lens with a focal length $f = 50$ mm, the distance from the DOE to the lens 200 mm, a He–Ne laser with a wavelength of 0.633 μm.

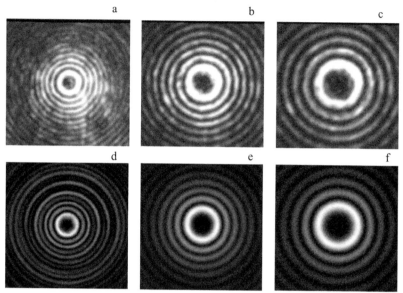

Fig. 5.27. The divergent paraxial Bessel beam at distances of 100 mm, 125 mm, 150 mm, respectively (a), (b), (c) (experiment), (d), (e), (f) (theory).

Figure 5.27 shows a qualitative agreement between theory and experiment.

It may seem that when imaging with a lens, the diffractionless BB (5.27) becomes a divergent BB (5.30) due to the fact that it was calculated using the paraxial Fresnel transformation. But it is not so. It can be shown that the use of Fresnel transformation for BB (5.27) retains its diffractionless nature. Indeed, we choose as the initial light field the BB in the form:

$$\Psi_0(r,\varphi,z=0) = J_n(\alpha r)\exp(in\varphi),\tag{5.31}$$

Then at a distance z from the plane $z = 0$, we obtain:

$$\Psi(\xi,\eta,z) = \frac{k}{2\pi i z}\int_0^\infty\int_0^{2\pi} J_n(\alpha r)\exp(in\varphi)\exp\left[\frac{ik}{2z}\left(r^2 + \rho^2\right)\right]\exp\left[-\frac{ik}{z}r\rho\cos(\theta-\varphi)\right]r\,dr\,d\varphi,$$

$$\tag{5.32}$$

where $\rho^2 = \xi^2 + \eta^2$, $= \operatorname{arctg}(\eta/\xi)$.

Replacing the integral with respect to φ (5.32) by the Bessel function of the n-th order, instead of (5.32) we get:

$$\Psi(\rho,\theta,z) = \frac{(-i)^{n+1}k}{z}\exp\left(\frac{ik}{2z}\rho^2\right)\exp(in\theta)\int_0^\infty J_n(\alpha r)J_n\left(\frac{kr\rho}{z}\right)\exp\left(\frac{ik}{2z}r^2\right)r\,dr.$$

$$(5.33)$$

The integral in (5.33) can be calculated, then instead of (5.33) we get:

$$\Psi(\rho,\theta,z) = \exp\left(-i\frac{z\alpha^2}{2k}\right)J_n(\alpha\rho)\exp(in\theta). \tag{5.34}$$

Equation (5.34) shows that the Frenel transform preserves the original paraxial non-diverging BB (with the accuracy up to a phase factor) [60, 61]:

$$\Psi(\rho,\theta,z) = \exp\left(-i\frac{z\alpha^2}{2k}\right)\Psi_0(\rho,\theta). \tag{5.35}$$

In the two-dimensional case we obtain a simpler relation between the scalar and vector BBs. In the two-dimensional case, the expansion of a complex function satisfying the Helmholtz equation in plane waves has the form [75]:

$$\Psi(x,z) = \int_{-\infty}^{\infty} \Psi_0(t)\exp\left[ik\left(xt + z\sqrt{1-t^2}\right)\right]dt, \tag{5.36}$$

where z is optical beam axis (on the y-axis there are no changes $\partial/\partial y = 0$).

If we select $\Psi_0(t)$ in the form of:

$$\Psi_0(t) = \frac{(-i)^n\exp\left[in\arccos\left(\frac{t}{a}\right)\right]}{2\pi\sqrt{a^2 - t^2}}\mathrm{rect}\left(\frac{t}{a}\right), \tag{5.37}$$

then substituting (5.37) into (5.36) we obtain an expression for the non-paraxial two-dimensional beam which at $z = 0$ coincides with the Bessel beams:

$$\Psi(x,z) = \frac{(-i)^n}{2\pi}\int_{-a}^{a}\frac{\exp\left[in\arccos\left(\frac{t}{a}\right)\right]}{\sqrt{a^2 - t^2}}\exp\left[ik\left(xt + z\sqrt{1-t^2}\right)\right]dt. \tag{5.38}$$

From Eq. (5.38) at $z = 0$ and after the change $t = a\cos\varphi$, we obtain

$$\Psi(x,z=0) = \frac{(-i)^n}{2\pi}\int_0^\pi \exp(in\varphi)\exp(ika\cos\varphi)d\varphi = J_n(kax). \qquad (5.39)$$

To get a compact notation for the Bessel beam at any z we can write Eq. (5.36) as

$$\Psi(x,z) = \int_{-\pi}^{\pi} \Psi_0(\theta)\exp\left[ik(x\cos\theta + z\sin\theta)\right]d\theta. \qquad (5.40)$$

Then, for

$$\Psi_0(\theta) = \frac{(-i)^n}{2\pi}\exp(in\theta), \qquad (5.41)$$

instead of (5.38) we obtain

$$\Psi(x,z) = \frac{(-i)^n}{2\pi}\int_{-\pi}^{\pi}\exp(in\theta)\exp\left[ik(x\cos\theta + z\sin\theta)\right]d\theta = J_n(kr)\exp(in\varphi), \qquad (5.42)$$

where $x = r\sin\varphi$, $y = r\cos\varphi$.

At $z = 0$ from Eq. (5.42) we get:

$$\Psi(x,z=0) = J_n(kr)(i\,\mathrm{sgn}\,x)^n. \qquad (5.43)$$

The scalar two-dimensional Bessel beam (5.42) can be regarded as a vector beam, assuming that $\Psi(x, z)$ is the projection on the y-axis of the vector of the electric field $E_y(x, z) = \Psi(x, z)$ for the TE-polarized monochromatic electromagnetic wave. This field is described by three quantities E_y, H_x, H_z, where H_x and H_z are the projections on the x and z axes of the vector of the strength of the magnetic field of the wave. Projections of the magnetic vector can be found through E_y:

$$H_x = \frac{1}{ik}\frac{\partial E_y}{\partial z},$$
$$\qquad\qquad\qquad\qquad\qquad (5.44)$$
$$H_z = \frac{i}{k}\frac{\partial E_y}{\partial x}.$$

With the help of equations (5.42) and (5.44) we can find an expression for the Umov–Pointing vector of the two-dimensional Bessel beam. Indeed, the Umov–Poynting vector is defined for complex vector fields in the form:

$$\mathbf{S} = \frac{c}{4\pi} \operatorname{Re}\left[\mathbf{E} \times \mathbf{H}^*\right], \tag{5.45}$$

where c is the speed of light.

In two dimensions, taking into account (5.42), instead of (5.44) we get:

$$S_x = \frac{ic}{4\pi k}\left(E_y \frac{\partial E_y^*}{\partial x} - E_y^* \frac{\partial E_y}{\partial x} \right) = \frac{c}{4\pi k} \operatorname{Im}\left(E_y \frac{\partial E_y^*}{\partial x} \right), \tag{5.46}$$

$$S_z = \frac{ic}{4\pi k}\left(E_y \frac{\partial E_y^*}{\partial z} - E_y^* \frac{\partial E_y}{\partial z} \right) = \frac{c}{4\pi k} \operatorname{Im}\left(E_y \frac{\partial E_y^*}{\partial z} \right). \tag{5.47}$$

Substituting (5.42) into (5.46) and (5.47), we obtain the projection of the Umov–Poynting vector for the two-dimensional Bessel beam with TE-polarization

$$S_x(x,z) = \frac{cnz}{4\pi k\, r^2} J_n^2(kr), \tag{5.48}$$

$$S_z(x,z) = \frac{-cnx}{4\pi k\, r^2} J_n^2(kr). \tag{5.49}$$

From equations (5.48) and (5.49) it follows that at $z = 0$ $S_x (x, z = 0) = 0$ and

$$S_z(x, z = 0) = \frac{-cn}{4\pi\, kx} J_n^2(kx), \tag{5.50}$$

and at $x = 0$, $S_z (x = 0, z) = 0$ and

$$S_x(x = 0, z) = \frac{cn}{4\pi\, kz} J_n^2(kz). \tag{5.51}$$

Arrows in Fig. 5.28 arrows indicate the direction of the Umov–Poynting vector, which follows from equations (5.48)–(5.51).

Let a monochromatic Bessel beam be linearly polarized along axis x:

$$\mathbf{E} = \mathbf{e}_x U(x,y,z) = \mathbf{e}_x J_n(\alpha r)\exp\left[i\left(\beta z + n\varphi\right)\right], \tag{5.52}$$

where $\alpha = k \sin\theta$, $\beta = k \cos\theta$, θ is the angle of the conical wave to the optical axis z, (r, φ) are the polar coordinates. From Maxwell's equations

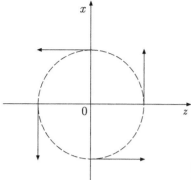

Fig. 5.28. The arrows indicate the direction of the Umov–Poynting vector at $r = $ const for a 2D Bessel beam of the n-th order.

$$\begin{cases} \text{rot } \mathbf{E} = ik\mathbf{H}, \\ \text{rot } \mathbf{H} = -ik\varepsilon\mathbf{E}. \end{cases} \qquad (5.53)$$

We find the rest of the projection of the electric and magnetic vectors:

$$\begin{vmatrix} E_x = U, \\ E_y = \dfrac{1}{k^2\varepsilon}\dfrac{\partial^2 U}{\partial y\partial x}, \\ E_z = \dfrac{1}{k^2\varepsilon}\dfrac{\partial^2 U}{\partial z\partial x}, \end{vmatrix} \quad \begin{vmatrix} H_x = 0, \\ H_y = \dfrac{1}{ik}\dfrac{\partial U}{\partial z}, \\ H_z = \dfrac{i}{k}\dfrac{\partial U}{\partial y}. \end{vmatrix} \qquad (5.54)$$

It is seen that the projection H_y and H_z of the order k^{-1}, and E_y and E_z of the order k^{-2}, i.e. small compared with E_x.

The Umov–Poynting vector is defined by equation (5.45) with (5.54) takes the form:

$$S_z = \frac{c}{4\pi k}\,\text{Im}\left[U^*\frac{\partial U}{\partial z}\right], \qquad (5.55)$$

$$S_y = \frac{c}{4\pi k}\,\text{Im}\left[U^*\frac{\partial U}{\partial y}\right], \qquad (5.56)$$

$$S_x = \frac{c}{4\pi\varepsilon k}\,\text{Im}\left[\frac{\partial^2 U}{\partial y\partial x}\frac{\partial U^*}{\partial y} + \frac{\partial^2 U}{\partial z\partial x}\frac{\partial U^*}{\partial y}\right]. \qquad (5.57)$$

We substitute the expression for U from (5.52) into (5.55)–(5.57) and obtain the projection of the Umov–Poynting vector of the 3D vector paraxial Bessel beam:

$$S_z = \frac{c\beta}{4\pi k} J_n^2(\alpha r), \tag{5.58}$$

$$S_y = \frac{cnx}{4\pi kr^2} J_n^2(\alpha r), \quad x = r\cos\varphi, \tag{5.59}$$

$$
\begin{aligned}
S_x = \frac{-cn}{4\pi k\varepsilon} \Biggl\{ & \left(\frac{\beta^2 y}{k^2 r^2} + \frac{n^2 x^2 y}{k^2 r^6} \right) J_n^2(\alpha r) - \\
& - \left(\frac{\alpha y}{k^2 r^3} - \frac{2\alpha x^2 y}{k^2 r^5} + \frac{\alpha x y^2}{k^2 r^5} \right) J_n(\alpha r) \frac{\partial J_n(t)}{\partial t} + \\
& + \left(\frac{\alpha^2 y^3}{k^2 r^4} - \frac{\alpha^2 x^2 y}{k^2 r^4} \right) \left(\frac{\partial J_n(t)}{\partial t} \right)^2 + \frac{\alpha^2 x^2 y}{k^2 r^4} J_n(\alpha r) \frac{\partial^2 J_n(t)}{\partial t^2} \Biggr\}.
\end{aligned} \tag{5.60}
$$

Note that equation (5.58) and (5.59) are similar and almost identical with the equations (5.49) and (5.48), respectively.

If in (5.60) we leave only the terms proportional to k^{-1}, and the terms with k^{-2} and k^{-3} ignored, instead of (5.60), we obtain a simple expression:

$$S_x = \frac{-cn}{4\pi k\varepsilon} \left\{ \frac{\beta^2 y}{k^2 r^2} J_n^2(\alpha r) + \left(\frac{\alpha^2 y^3}{k^2 r^4} - \frac{\alpha^2 x^2 y}{k^2 r^4} \right) \left(\frac{\partial J_n(t)}{\partial t} \right)^2 + \frac{\alpha^2 x^2 y}{k^2 r^4} J_n(\alpha r) \frac{\partial^2 J_n(t)}{\partial t^2} \right\}. \tag{5.61}$$

From Eq. (5.58)–(5.60) we see that at $x = 0$:

$$
\begin{cases}
S_y = 0, \\
S_x = \dfrac{-cn}{4\pi k\varepsilon} \left[\dfrac{\beta^2 y}{k^2 r^2} J_n^2(\alpha r) + \dfrac{\alpha^2 y^3}{k^2 r^4} \left(\dfrac{\partial J_n(t)}{\partial t} \right)^2 \right].
\end{cases} \tag{5.62}
$$

The sign S_x is determined by the product ny, with $n > 0$ and at $r = $ const projection S_x is aimed at in Fig. 5.29 (z-axis is directed toward the observer).

At $y = 0$:

$$
\begin{cases}
S_x = 0, \\
S_y = \dfrac{cnx}{4\pi kr^2} J_n^2(\alpha r).
\end{cases} \tag{5.63}
$$

The sign S_y is determined by the product nx and for $n > 0$ the projection is shown in Fig. 5.29.

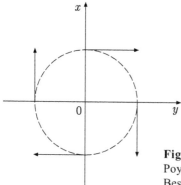

Fig. 5.29 The arrows indicate the direction of the Poynting vector in the cross section (x, y) 3D paraxial Bessel beam at r = const. The light is linearly polarized along the axis x.

The orbital angular momentum of the electromagnetic field is given by [27]:

$$\mathbf{M} = [\mathbf{r} \times \mathbf{S}] = \left[\mathbf{r} \times \left\{ \frac{c}{4\pi} \mathrm{Re}[\mathbf{E} \times \mathbf{H}] \right\} \right]. \tag{5.64}$$

The projection on the optical axis of the orbital angular momentum for a linearly polarized electromagnetic field calculated in the paraxial approximation has the form:

$$M_z = \frac{1}{4\pi kc} \left[y \, \mathrm{Im} \left(E \frac{\partial E^*}{\partial x} \right) - x \, \mathrm{Im} \left(E \frac{\partial E^*}{\partial y} \right) \right]. \tag{5.65}$$

For a linearly polarized Bessel beam

$$E_x = J_n(\alpha r) \exp(in\varphi) \exp(i\beta z), \quad \alpha^2 + \beta^2 = k^2 \tag{5.66}$$

projection on the z axis of the orbital angular momentum will be:

$$M_z = \frac{n J_n^2(\alpha r)}{4\pi kc}. \tag{5.67}$$

The expression (5.67) up to a constant coincides with the first term in the equation obtained in [93].

Diffractive optical elements can generate Bessel beams which retain the modal nature at a great distance along the propagation axis. Based on geometrical considerations, the distance at which the single-modal nature of Bessel light fields J_n (αr) exp ($in\varphi$), is estimated by the following formula [45]:

$$z_{max} = R \left[\left(\frac{2\pi}{\alpha \lambda} \right)^2 - 1 \right]^{1/2}, \tag{5.68}$$

where R is the radius of the DOE, α is the parameter of the Bessel function.

In [92] it is shown that the formation of BBs by holographic optical elements, the maximum distance over which they maintain the character of their mode increases by about two times compared with the method of forming a BB with a narrow gap [44]. However, we need some distance from the plane of the holographic optical element so that the beam can form. Thus, the segment of the optical axis in which BB, formed by the final phase DOE, retains its modal nature, begins at some z_{min}, required for a beam to form, and ends at z_{max}, defined by the DOE radius R and the BB parameter α.

In [81] it is proposed to form a Bessel mode beam with a spiral zone plate, the transmission function of which is a function of:

$$\tau(r,\varphi) = \text{sgn}\left(J_n\left(\alpha r\right)\right)\exp\left(in\varphi\right). \tag{5.69}$$

A helical DOE with transmittance (69) effectively forms a light field whose amplitude is proportional to the Bessel functions J_n (αr) exp $(in\varphi)$, near the optical axis in the interval $0 < z < Rk/\alpha$ [81]. At the same time, the DOE with transmittance (5.69) forms a light ring in the Fourier plane with a maximum intensity [68].

In calculating the phase of the DOE, for the formation of BB of the 5-th order we used the following parameters: $R = 3$ mm, $\lambda = 633$ nm, $\alpha = 44.5$ mm^{-1}. Figure 5.30a shows a template (600 × 600 samples), used at the University of Joensuu (Finland) to made a 16-gradation DOE (discretization step 10 μm). Figure 5.30b shows the central part of the DOE microrelief at a magnification of 50 (top view), and Fig. 5.30c – at a magnification of 200 (oblique vies). Pictures of the microrelief are obtained with an interferometer NEWVIEW 5000 of the firm Zygo (USA).

The results of comparing the experimental formation of a Bessel beam of the 5-th order and the numerical simulation based on the Fresnel integral transform are shown in Fig. 5.31. The fabricated phase DOE was illuminated by a collimated beam of an He–Ne laser. The resulting intensity distribution at different distances behind the DOE was recorded by a CCD-camera. Figure 5.31a–d (top row) shows the experimentally recorded intensity distributions at the following distances from the plane of the DOE: 300 mm (a), 400 mm (b), 500 mm (c) 600 mm (d). Figure 5.31e–h (bottom row) shows the corresponding patterns of numerical simulation [60].

A comparison of the patterns in Fig. 5.31 shows a good agreement between theory and experiment.

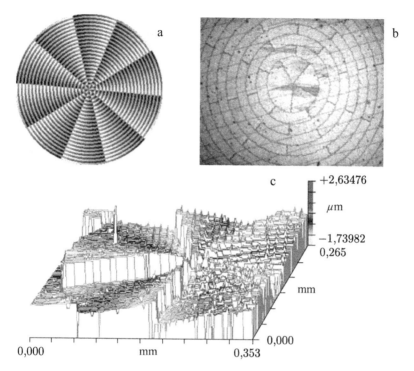

Fig. 5.30. The phase DOE that forms a Bessel beam of the fifth-order: phase pattern (a) and the form of the central part of the microrelief at a magnification of 50 times (b) and 200 times (s).

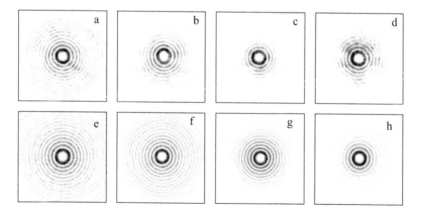

Fig. 5.31. Experimentally recorded intensity distribution (negative) in the cross-section at the following distances from the plane of the DOE: 300 mm (a), 400 mm (b), 500 mm (c) 600 mm (d), and the corresponding results of numerical simulation (e–h).

Experiments were carried out by the rotation of microscopic objects in the optical setup [14, 19] shown in Fig. 5.32a. The basis of the set is a modified microscope Biolam–M. A standard optical circuit was used for generating laser radiation. The appearance of the installation is shown in Fig. 5.32b.

In the development of the optical setup it was necessary to satisfy several conflicting requirements: firstly, for the most efficient focusing it was necessary to use a microscope objective with high magnification, and secondly, the size of the DOE determined the size of the beam incident on the microscope objective, and for example, for a 90× microscope the beam size was significantly greater than

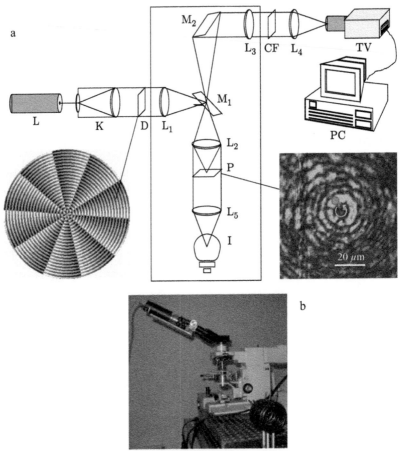

Fig. 5.32. Optical diagram of the experimental setup (a): L – argon laser, K – collimator, D – DOE, L_1 – corrective lens, M_1 – a semitransparent mirror of the microscope, M_2 – rotating mirror, L_2 – microscope objective, P – cell with microobjects, L_3 – the eyepiece of the microscope, CF – a red filter, TV – TV camera, L_4 – camera lens, L_5 – condenser of the illuminator, I – light fixture, picture of the experimental setup (b).

the entrance aperture, which inevitably leads to a decrease in beam energy. In addition, the use of a microscope for focusing and image formation leads to the need to combine the focal and working planes of the microscope objective. Both of these problems were successfully solved with the help of corrective lenses L_1. To determine the minimum required beam power, the developed method for calculating the forces was used to define the minimum intensity of 2×10^8 W/ m^2 at which motion is possible of a micro-object with a diameter of 5 μm, with a refractive index of 1.5 in the Bessel beam of the 5-th order. When using the ×16 microscope objective, the power of the beam in the working plane is 90 mW. Given that the losses by reflection from the refractive surfaces of the focusing system are 55–60% (experimentally obtained value), it follows that the power of the beam at the output of the laser should be about 200 mW. At the same time it is taken into account that the central ring of BB receives no more than 30% of the energy (experimentally determined value).

The work of the installation will now be described. The argon laser beam travels from collimator K to the DOE D, which forms the fifth-order Bessel beam. The correction lens L_1 then forms the final beam, which then enters the optical system of the microscope (lenses L_1, L_2). The generated beam is imaged by a decrease in the cell with an aqueous suspension of micro-objects. Background illumination is provided by the lamp I, through the lens L_5. Lens L_2 (microscope objective 16×, 20×, 90×) is used for focusing and at the same time to form an image of the cell. Yeast cells were used as micro-objects. Figure 5.33 shows the different stages of movement of yeast cells trapped by the first light ring of the Bessel beam. Light filter CF in the experiment was chosen so that the micro-object could be seen but the beam not. The micro-object made a total of eight revolutions, and then was stuck to the bottom. The parameters of the experiment:

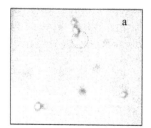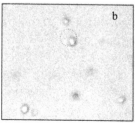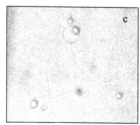

Fig. 5.33. Yeast cells are trapped by a light Bessel beams and make 8 turns around the ring with a diameter 17 μm (the first ring of the Bessel beam), a, b, c – stages of movement by 0.5 s. The trajectory is shown by the contour.

the cell size 4.5×7 μm, the Bessel beam power 150–200 mW, the diameter of the trajectory 17 μm [19].

As an object for experiments with rotation in the light beam the yeast cells have two significant drawbacks:

1.   The preparation of these micro-objects for experiment takes several hours;

2.   It is impossible to accurately determine the size and shape of yeast cells so it is also difficult to simulate their motion in light beams.

These deficiencies are not found in polystyrene balls, which are manufactured for use in chromatography. In addition, these balls are often used in the work of other experimenters, which facilitates the comparison of experimental results. Experiments were carried out with polystyrene balls with a diameter of 5 μm. They are made with good accuracy of ±0.1 μm, so are a good target for simulation. It should be noted that the diameter of the balls of 5 μm is very large for experiments with rotation, usually other experimenters used balls with a diameter 1 μm (mass 125 times smaller), but we you think about the future practical use of the experimental results (e.g. in micromechanics), this size is the most suitable, because it is comparable with the characteristic dimensions of the micro-mechanical devices. In the next experiment, BB was focused so that the size of the first ring was 3 μm, which is smaller than the size of the microspheres, and such a beam could carry out the seizure of a micro-object and move it to 30 μm to the side. The micro-object was moved by displacement of the beam by turning the mirror by 1°. Stages of motion of the micro-object are shown in Fig. 5.34 (trapped micro-object is highlighted by the outline). The beam power was about 200 mW, and a microscope objective with a magnification of ×90 was used [19].

It was interesting to experiment with the combined motion of a microscopic object. That is, moving the beam to ensure that the micro-object is also rotated. Such an experiment has been done. The phases of the motion of a pair of bonded microspheres are presented in Fig. 5.35. The parameters of this experiment: beam power 250 mW, microscope objective ×20. During the displacement by 50 μm the spheres made 4 turns, rotating as a whole [19].

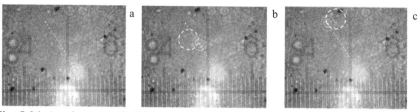

Fig. 5.34. A polystyrene ball is trapped by BB (diameter of the first light ring 3 μm) and moved by 30 μm in a straight line, a, b, c – movement through the stages by 2 s.

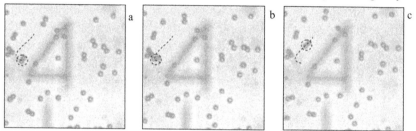

Fig. 5.35. Polystyrene beads are trapped by a Bessel beam and linearly moved by 50 μm, while rotating around the centre of the beam (4 turns), a–c – stages of movement after 1 s.

5.2.2. Optical rotation using a multiorder spiral phase plate

An experiment with the rotation of microscopic objects [6, 18, 63, 64, 88, 90] was conducted using the optical system shown in Fig. 5.36. The laser light travelled through the collimator to the DOE D, which forms the laser beam with a set of optical vortices. Then, using the optical microscope system (lens L_6 and L_2), the laser beam can be focused into a cell containing an aqueous suspension of polystyrene microspheres. Background light is generated by the lamp I, with the use of lenses L_5. Lens L_2 (microscope objective 16×, 20×) focuses the illumination light and at the same time forms an image of the workspace. The laser beam is focused by the microobjective L6.

The scheme presented in Fig. 5.36 differs from the circuit in Fig. 5.28a by the fact that focusing and observation are carried out through different microscopes. At the same time focusing the laser beam is conducted from the bottom to minimize the friction force of the microscopic object on the bottom of the cell, but, unfortunately, imposes limitations on the power of the light beam (at a specific power the micro-objects are squeezed up and leave the working plane). Experiments with rotation of the micro-objects were carried out using a DOE, forming four optical vortices with the numbers of orders (±3, ±7). The phase of this DOE is shown in Fig. 5.37a. The central part of the relief of this DOE is presented in Fig. 5.37b. Figure 5.37c shows the intensity distribution of the element in the zone of Fraunhofer diffraction.

Figure 5.38 shows the various stages of the movement of polystyrene microspheres trapped by a laser beam in an optical ring (optical vortex of seventh-order). The diameter of the orbit was 12 μm. Microspheres were in the water. Focusing was carried out with a ×20 microscope.

The light beam with the optical vortex simultaneously trapped and rotated a group of micro-objects. It should be noted that the light beams with optical vortices of high orders rotate the micro-objects more efficiently. As can be clearly seen in Fig. 5.38, the microsphere is trapped in a light beam with an optical vortex of order 3 but does not move, while a group of microspheres in the light beam from the optical vortex of seventh order rotates at an appreciable rate. Sophisticated experiments in which micro-objects were rotated in different diffraction orders were carried out.

Figure 5.39 shows the various stages of the movement of polystyrene beads, trapped by several optical vortices. Light beams with optical vortices trapping and at the same time the group of microspheres in different orders. Four microspheres were trapped in optical vortices of the 3rd and −3rd order, four microspheres in an optical vortex of −7th order, and five microspheres in an optical

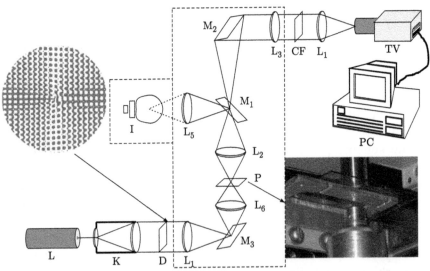

Fig. 5.36. The experimental optical system: L – argon laser, K – collimator, D = DOE, L_1 – corrective lenses, M_1 – semitransparent mirror of the microscope, M_2 – rotating mirror, L2 – microscope objective, P – cell with microspheres, L_3 – the eyepiece of the microscope, CF – red filter, TV – CCD-camera, L_5 – condenser illuminator, and I – light fixture.

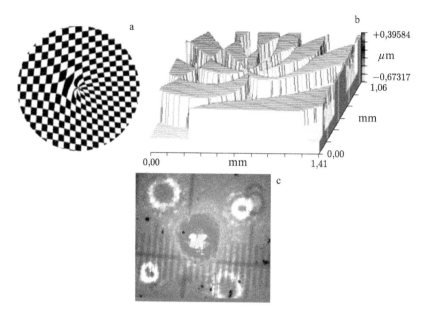

Fig. 5.37. Phase of the binary DOE to create optical vortices with −7, −3, 3, 7 orders of magnitude (a), the central part of the microrelief (b), the intensity distribution in the diffraction pattern (c).

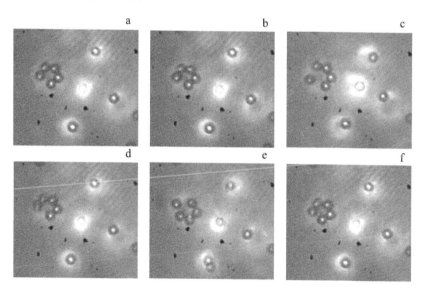

Fig. 5.38. The trapping and rotation of microscopic objects in the optical vortex of order 7, stage movement are shown in 2.5 s.

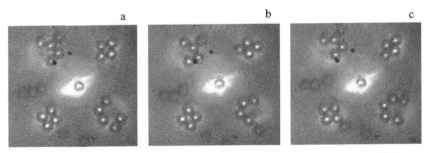

Fig. 5.39. The trapping and rotation of microspheres in optical vortices of 3rd, –3rd and 7th-order, stage of the movement are shown at intervals of 5 s.

vortex of 7th order were trapped. In the optical vortex of 7th order microspheres did not rotate, apparently due to the fact that some of the microspheres adhered to the bottom of the cell.

Groups of microspheres rotate in different orders of the light beam with optical vortices of the 3rd and 7th orders. Moreover, Fig. 5.39 shows that the microspheres in the optical vortices of opposite sign rotate in opposite directions.

5.2.3. Rotation of microscopic objects in a vortex light ring formed by an axicon

Lithography technology allows us to produce binary DOEs [55]. However, the helical axicon phase function is not binary. Therefore, it is necessary to use a simple method of converting a grayscale function to a binary one, which is based on use of the carrier frequency. In this case, the transmission function of the binary phase axicon has the form:

$$\tau_{n\beta}(r,\varphi) = \text{sign}\left[\cos\left(\alpha r + n\varphi + \beta r \cos\varphi\right)\right]\text{rect}\left(\frac{r}{R}\right), \qquad (5.70)$$

where $\beta = 2\pi/T$ and T is the period of the carrier spatial frequency [69]. As is well known, the binary DOE creates two identical diffraction orders, each of them has an efficiency of about 41% [91]. For the spatial separation of the orders it is necessary to satisfy the condition $\beta > \alpha$. Therefore, the radii of the rings in the Fraunhofer diffraction pattern for the vortex axicon (spiral or helical) are approximated by the expression $\alpha f/k$. Figure 5.40a shows a binary phase DOEs, forming two identical rings with the same numbers $n = 10$, but with different signs. Figure 5.40b shows the calculated Fraunhofer diffraction pattern for the DOE shown in Fig. 5.40a. DOE

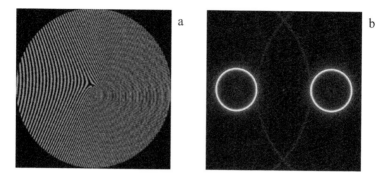

Fig. 8.40. a) the binary phase helical axicon of 10th order with a spatial carrier frequency, b) calculated diffraction pattern of a plane wave on the DOE shown in Fig. 8.40a.

has a radius $R = 2$ mm, the wavelength $\lambda = 532$ nm.

The axicon parameters $\alpha = 50$ mm^{-1}, the spatial carrier frequency $\beta = 100$ mm^{-1}, the focal length of the spherical lens $f = 420$ mm.

The phase in Fig. 5.40a was used in the manufacture of an amplitude photomask with a resolution of 3 μm using a circular laser writing station CLWS-200. Then, a DOE was produced by etching on a glass substrate 2.5 mm thick, with a refractive index of 1.5.

When illuminating the DOE with a radius $R = 2$ mm (Fig. 5.40) with a plane wave with wavelength $\lambda = 532$ nm, two identical light rings (each with an efficiency of about 41%) form in the focal plane of the lens $f = 460$ mm. The radial section of the ring, shown in Fig. 5.41, was measured with a CCD-camera. The produced DOE was used to trap and rotate polystyrene microspheres with a diameter of 5 μm. The DOE was illuminated by a collimated light beam of a solid-state neodymium laser with a wavelength of 532 nm and 500 mW power. A bright ring of radius 37.5 μm was formed in the focal plane of the microscope objective ($\times$40). Figure 5.42 shows two successive shots of microspheres, separated by a time interval of ten seconds, ten polystyrene microspheres, moving on a light ring, can clearly seen.

5.2.4. Optical rotation in a double light ring

The DOEs used to produce a double ring were manufactured in three different ways: by electron-beam lithography with electron beam direct writing on the resist, the technology of optical lithography using a binary photomask and wet etching of a glass substrate, and

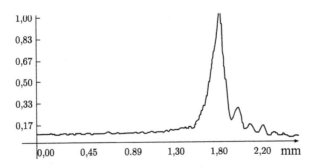

Fig. 5.41. Experimental radial section of the light ring in the focal plane of a lens (f = 460 mm).

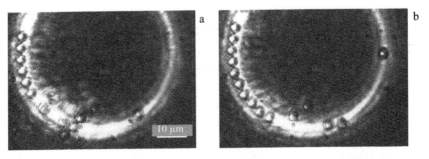

Fig. 5.42. Ten polystyrene beads with a diameter of 5 μm move along the bright ring with a radius of 37.5 μm, formed by a helical axicon with the number n = 10, with an average speed of about 4 μm/s. Figures (a) and (b) are separated by an interval of 10 s.

using the liquid crystal display or a dynamic spatial light modulator (SLM) [79].

The diffraction element was fabricated on a substrate of fused silica with a diameter 1 inch and a thickness of 3 mm. The stages of manufacture are shown in Fig. 5.43 First, a PMMA-resist with a thickness of 200 nm was deposited on the upper surface of the SiO$_2$ substrate. A 15 nm layer of Cu was then deposited on the substrate (Fig. 5.43a) for ensuring electrical conductivity. A cathode-ray machine Leica Lion LV1 was used for exposure. After exposure, the conductive layer was removed. Then, the resulting relief was coated with a 50 nm layer of Cr (Fig. 5.43c). Etching of the SiO$_2$ substrate was carried out step by step through reactive ion etching in the atmosphere CHF$_3$/Ar (Fig. 5.43d). After etching the Cr mask was removed by wet etching (Fig. 5.43f). The measured depth of the relief of the DOE was 578 nm.

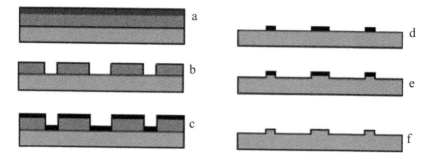

Figure 5.43. Stages of production of binary DOEs with electron-beam lithography and reactive ion etching.

Fig. 5.44. The central part of the relief of the DOE, size 260 × 350 μm.

The synthesized image of the phase of the DOE was used to produce a photomask on a glass substrate with a deposited layer of chromium. Recording on the photomask was done at the laser writing station CLWS-200, with a positioning accuracy of 50 nm and a resolution of 0.6 μm. The DOE was produced by standard methods of photolithography on a glass substrate, thickness 2.5 mm. The depth of etching was 0.5 μm. The error of etching in the height was about 50 nm. Figure 5.44 shows a profile of a binary central part of the DOE with the size of 260 × 350 μm, measured with an NewView 5000 Zygo interferometer.

The cross section for one period of modulation of the surface relief is shown in Fig. 5.45. Figure 5.45 shows that for the wavelength of laser light of 532 nm the depth of etching of glass was 0.5 μm. The trapezoidal single step did not exceed 25% of its width (approximately 20 μm).

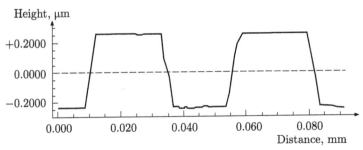

Fig. 5.45. Section of the central part of the DOE microrelief shown in Fig. 8.44.

5.2.5 Formation of the DOE with a liquid-crystal display

CRL OPTO SXGA SLM (spatial light modulator) with the active-region of 1316 × 1024 (1280 × 1024, excluding the boundary region) pixels (the size of one pixel is 15 μm) was used to display the image of the phase of the DOE with the 512 × 512. The microdisplay generates the phase image due to reflection of polarized laser light from the different planes: in the closed state of an individual element of the resolution of the microdisplay light is reflected from the outer surface of the thin film, in the open state – from the inner surface. The result is a binary DOE with a diameter of 6.5 mm. The image formed on the microdisplay is updated with a frequency of 63Hz. At the same time, switching of individual pixels does not exceed 10 μs.

Figure 5.46 shows the distribution of light in a double ring formed in the focal plane of a spherical lens with a focal length $f = 138$ mm using a binary DOE (5.70), the phase of which is shown in Fig. 5.40a, which was made by a variety of ways: electron lithography (a); optical lithography (b) and with SLM (c). Only the minus first diffraction

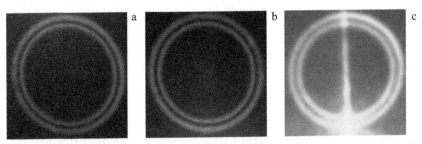

Fig. 5.46. The measured intensity in the Fraunhofer diffraction patterns (double ring) formed using a binary DOE implemented in different ways: by electron-beam lithography (a); optical lithography (b) and with liquid-crystal display (c).

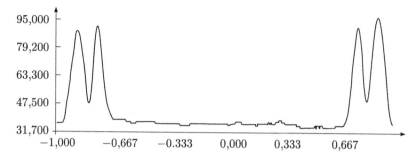

Fig. 5.47. Radial section of the intensity of the Fraunhofer diffraction pattern (Fig. 8.46b) in relative units. The horizontal axis is in millimeters.

order is shown. The size of diffraction patterns in Fig. 5.46 is 2×2 mm.

Figure 5.47 shows the radial cross section of a double ring of the Fraunhofer diffraction pattern (Fig. 5.46b). It is seen that the ring radius is about 0.8 mm, and width is about 0.3 mm.

|The experiments, the optical scheme of which is shown in Fig. 5.48, used a solid-state laser with a wavelength of 532 nm and a power of 500 mW [20]. In order to minimize power losses in reflections on the refractive surfaces the beam is not expended by

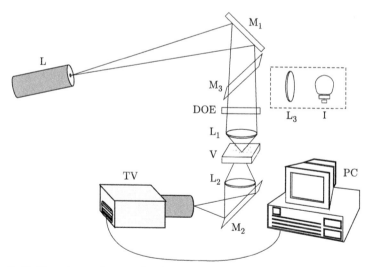

Fig. 5.48. Experimental setup for the rotation of microscopic objects. L – solid-state laser, M_1 – the first mirror, M_3 – semi-transparent mirror to illuminate the work area, M_2 – second rotating mirror, L_1 – focusing microscope objective (40×), L_2 – microscope objective, constructs the image of the working area (20×), L_3 – lens of the illuminator, I – light fixture, V – cell with microobjects, TV – TV camera, PC – a computer.

the collimator and the desired size is achieved by increasing the distance between the laser and the first turning mirror. Rotating mirror M_1 directs the light beam to the DOE, then the microscope objective L_1 (40×, water immersion, the focal length 4.3 mm) focuses the beam in the work area inside the cell V with micro-objects, microscope objective L_2 (20×) forms an image of the workspace, and the mirror M2 turns the light beam in the horizontal direction of the camera. Micro-objects are polystyrene beads with a diameter of 5 μm.

Polystyrene microspheres were trapped by light rings and they move along therm with an approximately constant speed. Different stages of the movement of microspheres at intervals of 2.5 s in a double ring of light are shown in Fig. 5.48.

The radius of the inner ring was 37 μm, the radius of the outer ring 48 μm. As can be seen from Fig. 5.49, there is a steady movement of microspheres along the inner ring of light with an average speed of about 3–4 μm/s and the movement of microspheres along the outer ring of light at a speed of 0.5–0.7 μm/s. This difference in speed is caused by different intensities of the rings. The difference of the intensities of the rings is due to the fact that the DOE is illuminated with a Gaussian beam with the radius smaller than the radius of DOE (to reduce losses during focusing).

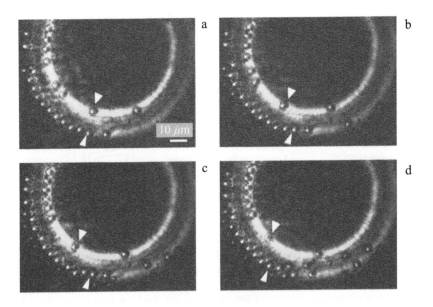

Fig. 5.49. The various stages of movement of microspheres in a double ring of light formed by a compound axicon.

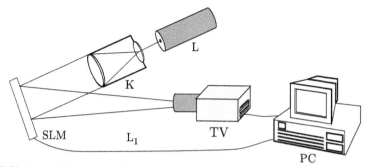

Fig. 5.50. Optical layout for the experiment. L – solid-state laser with a wavelength of 532 nm and a power of 500 mW, K – collimator, TV – TV camera, PC – personal computer.

5.2.5. Rotation of micro-objects by means of hypergeometric beams and beams that do not have the orbital angular momentum using the spatial light modulator

Spatial light modulators allow one to generate in real-time the phase DOEs, including for the problems of optical rotation [5, 34, 71, 87, 94]. Unfortunately, the main disadvantage of dynamic modulators, as working for as transmission and reflective, is low diffraction efficiency. This places increased demands on the beams formed. In particular, the need to minimize the number of orders, but it is also clear that it is most efficient to use light beams that have modal properties.

Experimentally the hypergeometric modes were formed with a liquid-crystal micro-display with an optical arrangement shown in Fig. 5.50 [10].

The laser beam is expanded by a collimator and is incident on the SML of light at an angle close to 90°, and is reflected towards the camera. The DOE is formed on the dynamic CRL OPTO modulator with a resolution of 1316 × 1024, and the physical size of 6.5 mm. The modulator is illuminated by a plane beam from a solid-state laser with a wavelength of 532 nm and a power of 500 mW. At the same time, the binary phase, obtained in the modulator, was encoded in two ways.

1. Binary phase $S(r, \varphi)$ by adding a linear carrier, satisfies the equation:

$$S(r,\varphi) = \text{sgn}\left\{\cos\left[\gamma \ln\frac{r}{w} + n\varphi + \alpha x\right]\right\}, \qquad (5.71)$$

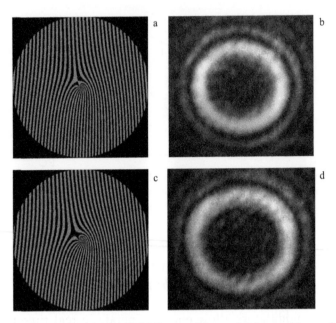

Fig. 5.51. Phase formed in the SLM for $n = 10$, $\gamma = 5$ (a), $n = 10$, $\gamma = 10$ (c) and $n = 5$, $\gamma = 10$ (d) and the corresponding intensity distributions at a distance of 2000 mm from the modulator (b), (d) and (e).

where sgn (x) is the sign function, α is the spatial carrier frequency, x is the Cartesian coordinate.

The coded element forms two symmetric orders with hypergeometric modes in the Fraunhofer diffraction plane.

2. With a quadratic radial encoding the binary phase of the DOEs is calculated by the formula

$$S(r,\varphi) = \text{sgn}\left\{\cos\left[\gamma\ln\frac{r}{w} + n\varphi + \frac{kr^2}{2f}\right]\right\},\tag{5.72}$$

where f is the focal length of the spherical lens.

This means that because of the actual addition of the lens, the hypergeometric modes are formed in a convergent beam.

Figure 5.51 shows the phases of the linearly coded phase DOEs and the intensity distributions at a distance of 2000 mm from the SLM for different n and γ.

Since the energy in linear encoding is divided between two orders, and the diffraction efficiency of DOEs, formed on the SLM is low, the resulting images have insufficient contrast. To get rid of this shortcoming, we used a quadratic coding (5.72). Figure 5.52 presents the phases of these types of elements (central parts), forming

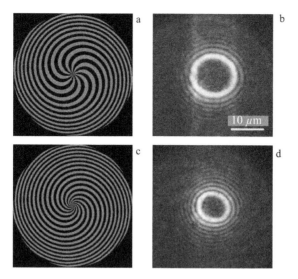

Fig. 5.52. (a) the phase of the DOE for the hypergeometric mode with parameters $n = 10$, $\gamma = 1$, (b) the intensity distribution in the Fresnel diffraction zone, (c) the phase of the DOE for the hypergeometric mode with parameters $n = 10$, $\gamma = 10$, (d) intensity distribution in the Fresnel diffraction zone.

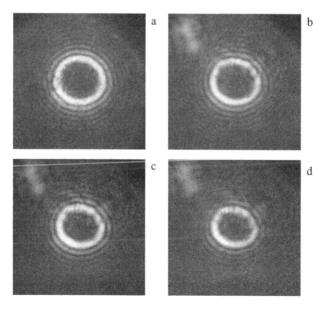

Fig. 5.53. The spread of the beam $n = 10$, $\gamma = 1$ (a) 700 mm, (b) 725 mm (c), 750 mm, (d) 775 mm.

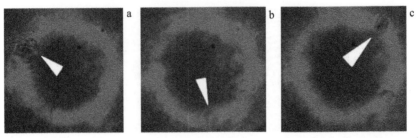

Fig. 8.54. The various stages of movement of a microsphere in the hypergeometric mode with parameters $n = 10$, $\gamma = 10$, formed by a binary DOE, the phase of which is shown in Fig. 8.52c.

hypergeometric modes, and the intensity distribution at a distance of 700 mm from the DOE.

Figure 5.53 shows the distribution of intensity at different distances from the DOE.

Reducing the size of the hypergeometric mode is due to the fact that the image was shot in the converging beam. As can be seen from Fig. 5.53, the structure of the distribution is preserved, which proves the modal nature of the light beam. Also, it was experimentally determined that the brightest central ring receives 35–40% of the energy beam, which is somewhat higher than in BB. Unfortunately, the low diffraction efficiency of the SLM in combination with relatively low laser power (500 mW) does not allow the hypergeometric modes to be used for the rotation of microscopic objects (in most studies the laser power was greater than 1W). A group of microscopic objects was trapped. Experiments with the rotation of microscopic objects were carried out with the phase DOE shown in Fig. 5.52c. The optical system shown in Fig. 5.48 was used in an experiment with the rotation of polystyrene beads with a diameter of 5 μm in the hypergeometric mode with parameters $n = 10$, $\gamma = 10$. Figure 5.54 presents the successive stages of movement (with an interval of 15 s) of trapped polystyrene beads along the brightest light ring of the hypergeometric mode.

The experiment proves the possibility of using the hypergeometric modes in problems of optical trapping and rotation of micro-objects. The presence of an additional parameter γ allows to adjust the radius of the brightest ring without changing the parameters of the optical system.

Any paraxial optical field, described by a complex amplitude E (x, y) at $z = 0$ can be decomposed into a number of LG modes in the basis:

$$E(x,y) = \sum_{n,m} C(n,m)\Psi_{nm}(x,y),\qquad(5.73)$$

where $C(n, m)$ are the complex coefficients with indices n and m, where m is the azimuthal index,

$$\Psi_{nm}(x,y) = a^{-1}\sqrt{\frac{n!}{(n+|m|)!}}\left(\frac{r}{a}\right)^{|m|}L_n^{|m|}\left(\frac{r^2}{a^2}\right)\exp\left[-\frac{r^2}{2a^2}+im\varphi\right],\qquad(5.74)$$

where $a = \omega_0/\sqrt{2}$, ω_0 is the Gaussian beam waist radius, (r, ϕ) are the polar coordinates, $L_n^m(x)$ are the associated Laguerre polynomials. In [58] a condition is defined for the number of modes in equation (5.73) in which the intensity in the cross section of a multimode LG beam will rotate as it propagates along the axis z:

$$B = \frac{2(n-n')+|m|-|m'|}{m-m'} = \text{const},\qquad(5.75)$$

where (n, m) and (n', m') are numbers of any two numbers of the linear combination (5.73). Constant $B/4$ is equal to the number of revolutions performed by the multi-mode LG beam $z = 0$ to $z = \infty$. Half of these rotations the beam performs in the distance from $z = 0$ and $z = z_0$, where $z_0 = k\omega_0^2$ is the Rayleigh length, $k = 2\pi/\lambda$ is the wave number of light. In [66] an equation was derived for the projection on the z axis of the linear density of the orbital angular momentum of a linearly polarized laser beam at unit power unit consisting of a superposition of LG modes (5.73):

$$wJ_z = \frac{\sum\limits_{n,m} m|C(n,m)|^2}{\sum\limits_{n,m}|C(n,m)|^2},\qquad(5.76)$$

where w is the angular frequency of light. From equations (5.73)–(5.76) it follows that:

1) the phases of the coefficients $C(n, m)$ do not affect the values of B and J_z, but affect the kind of intensity of the light field from equation (5.73) $I(x, y, z) = |E(x, y, z)|^2$;

2) the number of revolutions according to equation (5.75) during the rotation of the beam (5.73) depends only on the combination

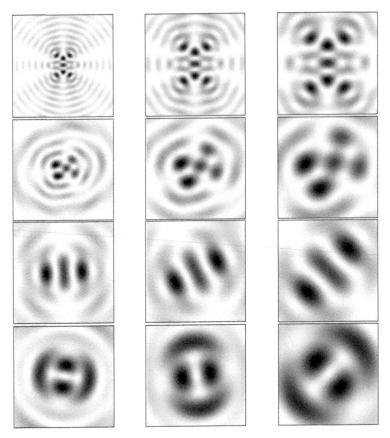

Fig. 5.55. Calculated intensity distributions (negative) in the cross-section of multimode LG beams, calculated at a distance $z = 30$ mm (column 1), $z = 40$ mm (column 2), $z = 50$ mm (column 3) for: 4-mode (row 1), 5-mode (row 2), 3-mode (row 3) and 2-mode (row 4).

of numbers of modes (n, m) and does not depend on the choice of coefficients $C(n, m)$;

3)   the orbital angular momentum (5.76) is determined only by the azimuthal numbers m of LG modes and the values of moduli of the coefficients $|C(n, m)|$ and is independent of the number n. Therefore, using a suitable choice of a combination of numbers (n, m) and the moduli of the coefficients $|C(n, m)|$ can be realized by different variants of combinations of values of B and J_z.

Figure 5.55 shows examples. The first row shows the intensity distributions of a four-mode GL beam with coefficients $C(8,0) = 1$, $C(11,2) = 1$, $C(10, -4) = i$, $C(9, 6) = 1$. Such a beam is not rotating $(B = 0)$, but has a positive orbital angular momentum $(wJ_z = 1)$. In the second row of Fig. 5.55 there are intensity distributions of a 5-mode LG beam with coefficients $C(2, 2) = i$, $C(3,1) = 1$, $C(4,0) = -1$, $C(4, -2) = 1$, $C(4, -4) = 1$. Such a beam is rotated counter-clockwise ($B = -1$) and has a negative orbital angular momentum ($wJ_z = -3/5$). The third row of Fig. 5.55 shows the cross-section of intensity of the three-mode GL beam with coefficients $C(10, -2) = 1$, $C(8,0) = 1$, $C(4,2) = 1$. This beam is rotated counter-clockwise ($B = -3$), but has no orbital angular momentum ($wJ_z = 0$). The fourth row of Fig. 5.55 shows a two-mode LG beam with coefficients $C(1, -1) = 1$, $C(9,1) = 1$. This beam is rotated in a clockwise direction ($B = 8$) and also has an orbital angular momentum ($wJ_z = 0$). Simulation parameters: wavelength $\lambda = 633$ nm, the waist radius of the fundamental LG mode $\omega_0 = 0.1$ mm; the size of each image in Fig. 5.55 1×1 mm, the distance at which the intensities in Fig. 5.55 were calculated (from left to right) $z = 30$ mm, $z = 40$ mm, $z = 50$ mm.

For a superposition of Bessel modes (BM) (5.73) we write:

$$\Psi_{nm}(x,y) = \left[\sqrt{\pi}R J'_m(\gamma_n)\right]^{-1} J_m(k\alpha_n r)\exp(im\varphi), \qquad (5.77)$$

where $\alpha_n = \cos\theta_n = \gamma_n/kR$, θ_n is the angle of inclination to the z-axis of the conical wave, $J_m(x)$, $J'_m(x)$ is the Bessel function and its derivative, γ_n is the root of the Bessel functions. BMs (5.77) are normalized to unity in the circle of radius R. The laser beam (5.77), consisting of BM, will rotate [33] at a finite distance from the reference plane ($z = 0$), provided that the number of modes (n, m), occurring in the superposition (5.77), will satisfy the condition:

$$B_1 = \frac{\alpha_n^2 - \alpha_{n'}^2}{m - m'} = \text{const}. \qquad (5.78)$$

The number $B_1/2$ is the number of rotations performed by the intensity in the beam cross section at a distance equal to one wavelength λ.

The projection on the z axis of the linear density of the orbital angular momentum of the laser beam at a power unit consisting of a superposition of BM is calculated using equation (5.76). From equations (5.76) and (5.78) it follows that by selecting the numbers (n, m) we can generate Bessel beams with rotation of intensity in the cross section ($B_1 \neq 0$), but with zero orbital angular momentum ($J_z = 0$). Figure 5.56 shows the intensity distribution of a two-mode Bessel beam with the coefficients $C(\alpha_5, 3) = 1$ and $C(\alpha_{10}, -3) =$

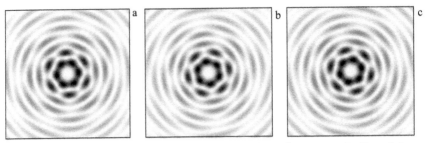

Fig. 5.56. Intensity distribution in the cross section of a two-mode Bessel beam calculated at different distances z from the initial plane: 1 m (a), 2 m (b) and 3 m (c).

1. Such a beam is rotated counterclockwise ($B_1 = -12.5 \cdot 10^{-8}$) and has no orbital angular momentum ($J_z = 0$). The calculation parameters: $\lambda = 633$ nm, $\alpha_5 = 5 \cdot 10^{-4}$, $\alpha_{10} = 10 \cdot 10^{-4}$, the size of diffraction patterns in Fig. 5.56 is 5×5 mm.

The rotating beam can be produced by the hypergeometric mode. Like Bessel modes, the hypergeometric modes have infinite energy, and therefore, in practice, they can be produced with the help of an amplitude-phase filter or a digital hologram with a circular aperture. Therefore, the hypergeometric modes will keep their modal properties only at a finite distance along the optical axis.

For the superposition of hypergeometric modes (5.73) instead of (5.74) and (5.77) we write ($R \gg 1$):

$$\Psi_{nm}(x,y) = \left(2\pi r^2\right)^{-\frac{1}{2}} \exp\left(i\frac{\pi n}{\ln R}\ln r + im\varphi\right), \quad n = 0, \pm 1, \pm 2, \ldots \qquad (5.79)$$

These modes are orthonormal in the ring with radii R and R^{-1}:

$$\int_0^R \int_0^{2\pi} \Psi_{nm}(r,\varphi)\Psi_{n'm'}^*(r,\varphi)\, r\, dr\, d\varphi = \delta_{nn'}\delta_{mm'}, \qquad (5.80)$$

where $\delta_{nn'}$ is the Kronecker symbol. In the Fresnel diffraction zone the mode (5.79) has the form ($R \to \infty$, $x = kr^2/2z$):

$$\Psi_{nm}(r,\varphi,z) = \frac{1}{2\pi|m|!}\left(\frac{2z}{k}\right)^{\frac{i\gamma-1}{2}} \exp\left[\frac{i\pi}{4}(-|m|+i\gamma-1)+ix+im\varphi\right] \times$$

$$\times x^{\frac{|m|}{2}}\Gamma\left(\frac{|m|+i\gamma+1}{2}\right) {}_1F_1\left(\frac{|m|+i\gamma+1}{2}, |m|+1; -ix\right), \qquad (5.81)$$

where $\gamma = \pi n$, ${}_1F_1(a, b, x)$ is the confluent hypergeometric function,

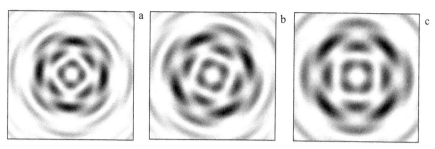

Fig. 5.57. The distribution of intensity in the cross section of the two-mode hypergeometric beam, calculated at different distances z from the initial plane: 1.5 m (a), 2 m (b) and 2.5 m (c).

$\Gamma(x)$ is the gamma function.

The condition for the rotation of the beam is multimode hypergeometric form:

$$B_2 = \frac{n - n'}{m - m'} = \text{const}, \tag{5.82}$$

where $B_2/4$ is the number of rotation which the beam (5.73) with the modes (5.79) makes in the interval from $z = 1$ to $z = R$. Note that in the interval from $z = R^{-1}$ to $z = 1$ the hypergeometric beam makes $B_2/4$ turns in the opposite direction.

The projection on the z axis of the linear density of the orbital angular momentum of the laser beam (5.73) with the modes (5.79) per unit power is described as previously by (5.76). Therefore, using the rotation condition (5.82) and the expression for the orbital angular momentum (5.76) we can generate a laser beam, which, for example, will rotate the transverse distribution of intensity, and there will be a zero orbital angular momentum. Figure 5.57 shows the intensity distribution of a two-mode hypergeometric beam with coefficients C (0, 2) = 1 and C (3, –2) = 1. Such a beam is rotated counterclockwise ($B_2 = -0.75$) and has no orbital angular momentum ($J_z = 0$). The calculation parameters: $\lambda = 633$ nm, $\gamma_0 = 0$, $\gamma_3 = 13.597$, the size of diffraction patterns in Fig. 5.57 is 4×4 mm.

Note that in [66] the authors studied a special case of hypergeometric modes at $\gamma = -i$, which are formed with a spiral phase plate with the transmission exp ($im\varphi$). These modes have the same phase velocities, and therefore their linear combination (5.73) can not rotate during propagation. It also follows from the rotation condition (5.82) at $n = n' = $ const ($B_2 = 0$).

In conclusion, we present some experimental results. The experiments were carrieds out using a binary liquid-crystal SLM

Fig. 5.58 Binary phase pattern formed on the microdisplay.

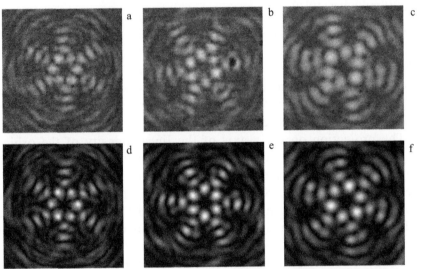

Fig. 5.59. Diffraction pattern of a rotating two-mode Bessel beam at different distances from the microdisplay (a, b, c – experiment, d, e, f – theory): $z = 720$ mm (a, d); $z = 735$ mm (b, e); $z = 765$ mm (c, f).

CRL Opto SXGA H1 1280 × 1024. Figure 5.58 shows a binary phase intended to generate a light field representing a superposition of two Bessel modes with numbers C $(\alpha_1, 3) =$ C $(\alpha_2, -3) = 1$ $(\alpha_1 = 1.4 \times 10^{-4}, \alpha_2 = 7 \times 10^{-3})$. The size of the formed phases is 7 × 7 mm.

The intensity distribution in the cross section of one of the two beams formed, measured at different distances from the microdisplay with a CCD camera, is shown in Fig. 5.59 [12].

As can be seen from Fig. 5.59, there is qualitative agreement between the experimental and the theoretical data. The experiments

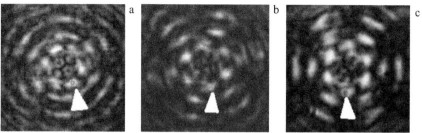

Fig. 5.60. The rotating beam with trapped polystyrene ball with a diameter 1 mm.

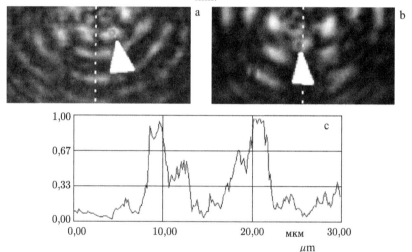

Fig. 5.61 The rotating beam with trapped polystyrene microspheres with a diameter 1 mm, the interval between frames (a) and (b) 10 s, (c) section along the dotted line of the beam (b).

were conducted using the optical scheme shown in Fig. 5.48. In this scheme, the mirror M1 was replaced by a spatial light modulator. A rotating multimode BB was formed with $C(\alpha_1, 3) = C(\alpha_2, -3) = 1$, $\alpha_1 = 1.4 \times 10^{-4}$, $\alpha_2 = 0.7 \times 10^{-4}$. Beam power was approximately 5 mW, $\lambda = 0.532$ μm.

Figure 5.60 shows the different positions of a rotating beam with zero orbital angular momentum with a trapped polystyrene sphere with a diameter of about 1 μm. Pictures a, b, c were taken at different shifts of the focusing microscope objective (16×) from the initial plane: 0 mm (a), 0.1 mm (b), 0.2 mm (c).

Since the displacement of the micro-object is small we consider separately Fig. 5.60a and c. The dotted line indicates the middle of the beam and was used to construct the cross section of the beam in Fig. 5.61b (Fig. 5.61c).

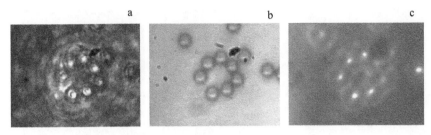

Fig. 5.62. The experimental image with a clearly visible Bessel beam of the 5th order (a), an image with the beam removed and clearly visible microobjects (b), and the image of the correlation peaks at the location of micro-objects.

As can be seen from Fig. 5.61, the microsphere trapped at the maximum intensity is rotated following the rotation of the beam. The beam cross section shows that the maxima in Fig. 5.61b are oriented vertically. This experiment shows that using the DOE and a very simple optical arrangement it is possible to control the rotation of the micro-object together with the beam. This effect is usually achieved by using rather complex interferometers.

5.2.6. Quantitative investigation of rotation of micro-objects in light beams with an orbital angular momentum

The motion of micro-objects in different light beams can be compared most conveniently using the average speed. To determine the average speed, special software was developed allowing processing and separating micro-objects in the image sequence.

To automatically determine the speed the correlation function with one of the images of the micro-objects is calculated. Figure 5.62 shows the different stages of processing the experimental images in Fig. 5.62a, in Fig. 5.62b the effect of the light beam due to the separation of colours is completely removed, in Fig. 5.62c correlation peaks are clearly visible on the site of the micro-objects.

After this, the coordinates of the micro-object were determined from the coordinates of the maximum of the correlation peaks. The average speed was defined with respect to both time and the ensemble of micro-objects. The first stage included the determination of the average linear velocity of each micro-object $\langle v_i \rangle$ separately as follows:

$$\langle v_i \rangle = \frac{1}{t} \int_0^t |\mathbf{v}_i(t)| dt, \tag{5.83}$$

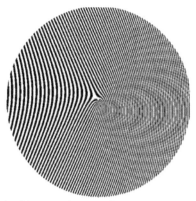

Fig. 5.63. Phase of the binary axicon to form Bessel beam of the tenth order.

where t is the time of observation, $v_i(t)$ is the velocity of the micro-object versus time. We then determine the average velocity V of the ensemble of micro-objects:

$$V = \frac{1}{N}\sum_{i=1}^{N}\langle v_i\rangle. \tag{5.84}$$

The Bessel beam of the fifth-order was formed using a DOE [14] the phase of which is shown in Fig. 5.30a, and the transmission function is given by (5.69). The Bessel beam of the tenth order was produced using a binary helical axicon whose phase is shown in Fig. 5.63, and the transmission function has the form [89]:

$$E_2(r,\varphi) = \text{sgn}\{\exp(in\varphi + iar + iyr\cos\varphi)\}, \tag{5.85}$$

where y is the carrier spatial frequency, a is the parameter of the axicon, $n = 10$ is the order of the helical axicon.

The determining factor for the speed of the micro-objects is the presence of dry and viscous friction forces. But if the force of viscous friction is quite easy to define, and it depends only on the properties of the liquid (as they are the same throughout the volume of the cell) and on the form of a microscopic object, the force of dry friction can greatly vary depending on the location of the micro-object. To minimize the influence of dry friction force on the bottom of the cell the Bessel beam was positioned at the same place with an error of no more than 2 µm, so the nature of the friction of one cell should not be changed by changing the beams. The strength of viscous friction is proportional to the velocity of the micro-object, consequently, the velocity of the micro-object can indicate the magnitude of force of

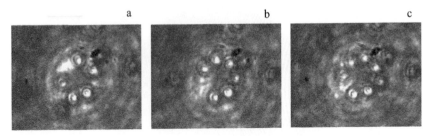

Fig. 8.64. The various stages of the movement of polystyrene beads in a light ring of the Bessel beam formed by the DOE (helical axicon of the 5th order).

the light beam acting on the micro-object. Computational experiments were carried out prior to each full-scale experiment (using the developed method) to estimate the velocity of microscopic objects in a particular light beam for the given parameters. So simulation was carried out for the fifth-order BB for a $16\times$ focusing microscope objective and the beam power 230 mW (at the output of the laser). The average velocity of the spheres with a diameter of 5 μm was 8 μm/s, which gave reason to believe the success of full-scale experiment with the same parameters.

Initially, the circuit was fitted with a DOE the phase of which is shown in Fig. 5.30a. The DOE formed a fifth-order BB in which the very bright ring trapped as a result seven microspheres. Different stages of the movement of these microspheres at intervals of seven seconds are shown in Fig. 5.64. The diameter of the bright ring of the BB was about 18 μm.

Figure 5.65 presents the processed images with clearly visible micro-objects, and the almost completely 'removed' beam.

In this experiment the average velocity was determined using seven micro-objects with the observation time of 29 s. A total of 116 images with the successive stages of movement were processed. The average velocity was 1.3 ± 0.1 μm/s.

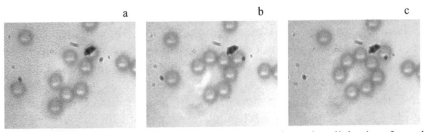

Fig. 5.65. Different stages of the movement of microspheres in a light ring, formed by a helical axicon of the 5th order, after processing to determine the average velocity.

Table 5.4. Comparative experimental parameters

Beam	Beam power in the working plane (mW) (including losses)	Average intensity of the bright ring (W/m²)	Diameter of the bright ring (μm)	Average speed of micro-objects (μm/s)	Average speed excluding stopped micro-objects (μm/s)
5th order BB	230	$27 \cdot 10^7$	18	1.3 ± 0.1	3.4 ± 0.4
10th order BB	100	$8 \cdot 10^7$	37	2.9 ± 0.3	3.1 ± 0.4

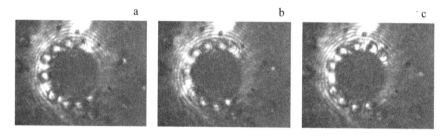

a   b   c

Fig. 5.66. The various stages of the movement of micro-objects in a light ring formed by the binary axicon of the 10th order (Fig. 8.63).

For comparison, a similar experiment was carried out with the BB of the 10th order. Different stages of movements of the micro-objects at intervals of one second in the BB of the 10th order are shown in Fig. 5.66.

In this experiment the average velocity was determined using nine microparticles with the observation time of 22 s. A total of 88 images with successive stages of movement were processed. The average speed was 2.9 ± 0.3 μm/s. It should be borne in mind that the use of the binary axicon (Fig. 5.63a) reduces by half the beam energy, as the energy is divided between the two (plus and minus first) orders. Table 5.4 presents the parameters of both experiments for comparison.

As shown in Table 5.4, using the BB of the 10th order the velocity of the micro-objects is more than doubled, with half the beam energy. If we assume that the force of viscous friction is proportional to the velocity of micro-objects, then at the same beam energy the force directed along the ring in the BB of the 10th order will be superior to the same effect for the BB of the 5th order four times. It should however be noted that it is difficult to take into account the effect of friction of micro-objects on the bottom of the cell, which increases with the beam power (due to the pressure of light as a result of Fresnel reflection from the micro-object). In particular, because of this force in both experiments there was a complete arrest of movement of some micro-objects (for a short time) in a number of stages. To minimize the influence of friction forces in the same experiments, the average velocity was measured in two stages. In the first stage the overall average velocity was calculated from (5.83) and (5.84), in the second stage we determined micro-objects and time intervals, during which their actual velocity was less than half the average speed defined in the first stage. This was followed again by determination of the average velocity from (5.83) and (5.84), but the above-mentioned micro-objects were not considered. The resulting average velocity is indicated in the rightmost column of Table 5.4.

In all the above, as a rule special attention was paid to the formation of an optical vortex beam without regard to the effectiveness of this beam in rotation tasks. At the same time, it is obvious that if we consider the task of efficient transmission of the torque to micromechanical systems, it is necessary to investigate how changes of the order number of the optical vortex will change the amount of energy transferred from the beam to a microscopic object. At a qualitative level it was determined that the velocity of the micro-objects increases with increasing numbers of the order of an optical vortex [39, 40]. However, quantification of this relationship was not carried out. To perform such a study, further experiments were carried out by the rotation of polystyrene beads in light beams with the angular harmonics of the 30th and 31st order [17]. The experimental setup for optical rotation is shown in Fig. 5.48.

The experiment used a solid-state laser with a wavelength of 532 nm and a power of 500 mW. The beam was not collimated in order to minimize power losses in reflections on the refractive surfaces, and the desired size is achieved by increasing the distance between the laser and the first rotating mirror. Polystyrene microspheres with a diameter of 5 μm were used as the micro-objects.

In order to form a set of 4 optical vortices (numbers of orders –31, –30, 30, 31) experiments were carried out with a DOE, the binary phase of which is shown in Fig. 5.67a [16, 17]. Figure 5.67b shows the central part of the microrelief. Figure 5.67c shows the distribution of intensity for the DOE in the area of Fraunhofer diffraction.

The scheme included an element, the phase of which is shown in Fig. 5.67a. The optical vortex of 30th orderr trapped 14 micro-objects as a result.

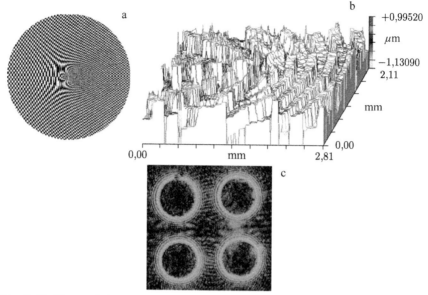

Fig. 5.67. Phase of the DOE for the formation of optical vortices of –31, –30, 30, 31 orders (a), the central part of the DOE microrelief (b), the intensity distribution in the diffraction pattern (c).

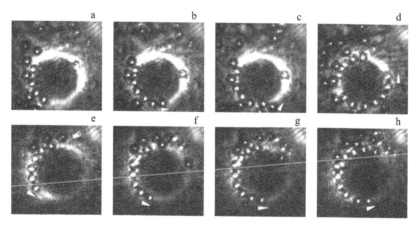

Fig. 5.68. The various stages of motion of micro-objects in the vortex beams, formed by the DOE: a–d) optical vortex of order 30, (e–h) optical vortex of order 31.

Different stages of the movement of these micro-objects the interval of seven seconds are shown in Fig. 5.68 (a–d). Exactly the same experiment was performed for an optical vortex of order 31, the stages of its movement are shown in Fig. 5.68 (e–h).

The motion of micro-objects in different light beams can be compared most conveniently using the average speed. Experimental

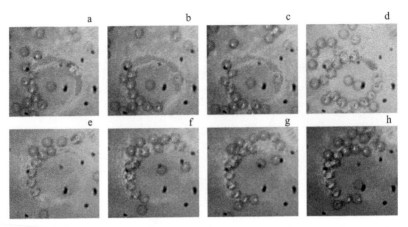

Fig. 5.69. Various stages of motion of micro-objects in the vortex beams, formed by the DOE: (a–d) optical vortex of order 30; (e–h) optical vortex of order 31, after computer processing of images.

images were processed to determine the average velocity by the method described in the previous section. Figure 5.69 presents the processed images with clearly visible micro-objects, and an almost invisible beam. The different brightness of the images is due to the change in background illumination, almost invisible on the original images.

In this experiment, the average velocity was determined using fourteen and eleven microspheres (for the 30th and 31st orders, respectively) during 19 s. 76 images of consecutive stages of movement for an optical vortex of the 30th order were processed. The determination of the average velocity did not take into account the moments of time during which the microspheres stayed under the influence of friction. The average velocity was 14 ± 3 μm/s.

175 images were processed for an optical vortex of the 31st order, i.e. the total duration of the experiment was 44 s. The average speed was 11 ± 3 μm/s.

Figure 5.70 shows the images processed to determine the average velocity of various stages of micro-objects in optical vortices in the 3rd and 7th orders.

In this experiment, the average velocity was determined using eight microspheres (for the optical vortex of the 3rd order) during 12 s, Only 48 images with the successive stages of movement were processed. The average velocity was 4 ± 2 μm/s. The average velocity in the micro-objects in the optical vortex of 7th order was determined using the results of several experiments (not only in Fig. 5.70). Taken together, 16 microspheres

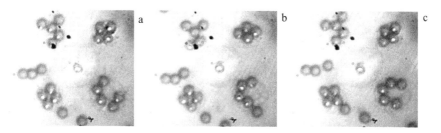

Fig. 5.70. The various stages of motion of micro-objects in the vortex beams, formed by a binary DOE the phase of which is shown in Fig. 8.37a.

Table 5.5. Comparative experimental parameters

The order of optical vortex	Beam power in the working plane (mW) (including losses)	Aaverage intensity of the bright ring (W/m²) × 10⁸	The diameter of the ring (μm)	Average speed of micro-objects (μm/s)
3	50	3.2	9	4 ± 2
7	50	2.1	13	6 ± 2
30	40	0.9	27	14 ± 3
31	40	0.9	28	11 ± 3

were used and 203 images were processed (total time of four experiments 51 s). The average speed was 6 ± 2 μm/s. These data were used to compile Table 5.5.

As shown in Table 5.5, at increasing numbers of the order the velocity of the micro-objects is initially almost doubled and in the further growth of the number of the order does not change so much (though with reduced intensity). If we assume that the force of viscous friction is proportional to the velocity of micro-objects, then at the same beam energy the force directed along the optical vortex ring should increase with the number of the order. It should be noted that this is difficult to take into account the effect of friction of micro-objects on the bottom of the cell, which increases with the beam power (due to the pressure of light from the Fresnel reflection from the micro-objects). In addition, as shown in Fig. 5.69 and Fig. 5.70, the motion of micro-objects is very uneven (there are short stops), indicating heterogeneities in the bottom of the cell. It is also extremely difficult to accurately determine the power of the specific beam, as available devices allow one to define an integrated beam power (i.e. all four rings simultaneously with the zero-order). This power is then to be divided in proportion to the brightness of each image.

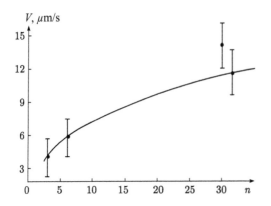

Fig. 5.71. Theoretical dependence of the velocity of polystyrene beads with a diameter of 5 μm on the number of the singularity of the light field (curve), and experimental data on the dependence of the velocity of polystyrene beads with a diameter of 5 μm on the number of the singularity of the light field (points).

According to [76], the orbital angular momentum of the light field transmitted to the micro-object can be expressed by the formula

$$M = \frac{\lambda n P}{2 \pi c} \eta_{abs},\qquad(5.86)$$

where M is the transmittted moment, λ is wavelength, n is the order (number) of the singularity, P is beam power, η_{abs} is the absorption coefficient of micro-objects. Using this formula and assuming that there is complete coincidence for one of the experimentally obtained points, a theoretical curve of the dependence of the velocity of the micro-objects on the number of singularity was constructed. Figure 5.71 shows this curve with the superimposed experimental points.

As can be seen from the graph in Fig. 5.71, the experimental data are in good agreement with the theoretical ones within the experimental error.

5.3. Rotating microparticles in light beams designed and formed to obtain the maximum torque
5.3.1. Formation of the light beams, consisting of several light rings

To increase the efficiency of the transmission of the torque from the vortex beam to some micro-object, such as a microturbine, it is necessary to satisfy several conditions:

1. The uniform distribution of light on the surface of a rotating micro-object.
2. The incidence of light at a certain optimum angle at every point of the micro-object.

Unfortunately, with the increasing order of the vortex beam the width of the light ring in the intensity distribution of this beam relative to the beam diameter is reduced, which leads to an increase in the intensity of light on a small part of the surface of the microscopic object, which may lead to the destruction of the micro-object, if it is necessary to rotate it at high speed. The easiest way to avoid it is to use a vortex beam with an intensity distribution in the form of several light rings.

For vortex beams the intensity distribution in the form of multiple rings consider DOEs similar to those shown in Fig. 5.72. ???The phase function of the DOE is a collection of rings and in each ring. the vortex phase is rotated by 180° relative to the previous ring(correction to the phase function π/2). Diffraction on this DOE results in the formation of a light beam consisting of several rings.

The width of the rings in the DOE in Fig. 5.72 is different due to the fact that for the same width of the rings in the DOE the formed beam has rings of different intensity, which is undesirable (Fig. 5.73b).

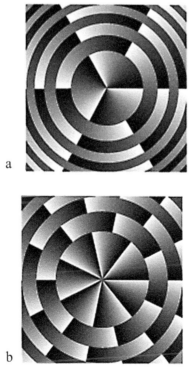

a

b

Fig. 5.72. DOE to form a vortex beam of several light rings: (a) 3rd order, (b) 7th order.

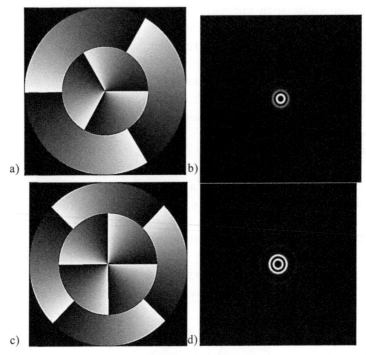

Fig. 5.73. DOE and the corresponding light rings with equal energies (a, b) and equal brightnesses (c, d).

Unfortunately, in all rings of this beam the order of the helical phase is the same which means that the slope of the wavefront in the rings differs because of the different diameter of the rings (Fig. 5.74).

When creating a microturbine with straight blades (i.e., what is permitted by technology) the transmission efficiency of the torque from the different rings is different. To increase this efficiency, the slope of the wavefront for the different light rings should be similar (second condition). Unfortunately, due to interference effects it is not possible to create an element that would allow us to form ring-shaped vortex beams with different numbers of singularity for each ring. But the annular shape of the vortex beam is not essential for the problem of rotation of microturbines. To transfer the torque from the vortex beam efficiently to the microturbine we must satisfy the condition of equality of the forces acting on the blade as a whole in the microturbine at any of its orientation.

$$I_s(\phi) = \int_0^R I(r,\phi)\,dr = const \qquad (5.87)$$

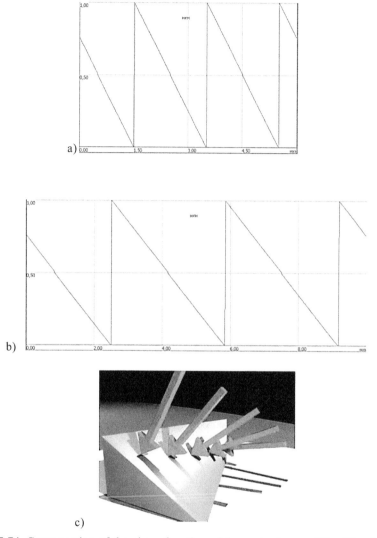

Fig. 5.74. Cross-section of the phase function of the vortex beam of the 3rd order of 1 μm radius (a), the cross section of the phase function of the vortex beam of the 3rd order with a radius of 1.5 μm (b), three-dimensional model of the phase surface with the image of the direction of propagation of light in the form of arrows (c).

where $I(r, \varphi)$ is the intensity of the vortex beam in polar coordinates centered on the axis of the beam, I_s is the total intensity of the beam on the selected sector. Essentially this means that the total intensity on a line drawn from the centre of the beam should not changed in any of its rotation. Consider whether it is possible to generate light beams that satisfy this condition. To do this, we make an element similar to that shown in Fig. 5.72, but with different values for the orders of the singularity of the rings. The phase function of an element is determined by the formula

$$\Phi(r,\phi) = \begin{cases} \exp(in\phi), r < R \\ \exp(im\phi) r \geq R \end{cases} \qquad (5.88)$$

where r, ϕ are polar coordinates, R is the boundary radius picked empirically.

Consider one of these diffractive optical elements (Fig. 5.75a). The boundary area of the element is a phase screw of the 1st order, the central region is the phase screw of the 2nd order. Figure 5.75b shows the distribution of the intensity in the beam formed in this element.

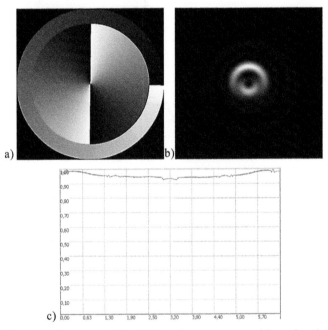

Fig. 5.75. The phase function of the DOE to form a superposition of optical vortices of the 1st and 2nd order (a), the intensity distribution in the superposition of optical vortices of the 1st and 2nd order (b) a plot of I_s in the superposition of the optical vortices of the 1st and 2nd order (c).

As seen from the graph in Fig. 5.75c, the resulting beam satisfies (5.87) (I_s value is almost unchanged), but the beam is asymmetric, leading to problems with the alignment of microturbines. To resolve this asymmetry we will try to increase the order of singularity of optical vortices in the superposition, while maintaining the difference of the order numbers. Figure 5.76a shows the phase function of the diffractive optical element forming a superposition of optical vortices in the 4th and 5th order.

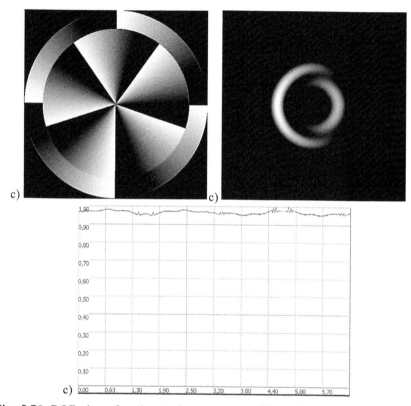

Fig. 5.76. DOE phase function to form a superposition of optical vortices of the 4th and 5th order (a), the intensity distribution in a superposition of optical vortices of the 4th and 5th order (b), and a plot of I_s in the superposition of optical vortices of the 4th and 5th order (s).

As can be seen from Fig. 5.76a simple increase in the order of the optical vortices included in the superposition does not lead to significant improvement. Another way is to increase the difference in the numbers of orders in superposition. Figures 5.77 a, c, d shows the phase functions of the diffractive optical elements forming the superposition with the difference in the numbers in 2, 3, 4, respectively.

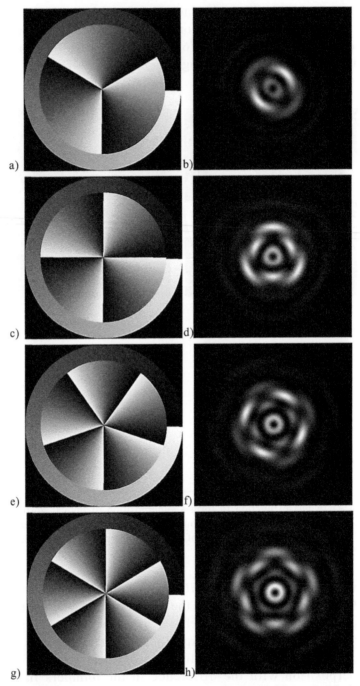

Fig. 5.77. DOE phase function for the formation of superpositions of optical vortices with the difference of the numbers of orders 2, 3, 4, 5 respectively (a, c, e, g), and the intensity distribution in the beams formed by them (b, d, f, h).

As can be seen from Fig. 5.77, starting with the differences of numbers 2, there are quite acceptable intensity distributions by the criterion of symmetry. Also on the basis of Fig. 5.77 it can be concluded that an increase in the difference of the numbers of the orders leads to a broadening of the vortex curve around the inner ring. Significant expansion of this curve is undesirable because in this case the wave front in the edge regions of the curve will have a non-optimal angle, resulting in a loss of efficiency in the transmission of torque. In Fig. 5.78 the intensity distribution in the beam includes circles at the radius of the maximum torque transmission efficiency.

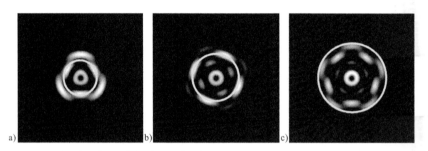

Fig. 5.78. Distribution of the intensity in superposition of vortices of the 1st and 4th order (a), 1st and 5th order (b), 1st and 6th order (c) with the superimposed circles of the optimal radius.

As can be seen from Fig. 5.78, the circle of the maximum transmission efficiency of the torque for the superposition of vortex beams of the 1st and 4th order runs almost along the inner edge of the vortex curve. For superposition of the vortex beams of the 1st and 5th order it runs around the centre, and for the superposition of vortex beams of the 1st and 6th order along the external border. Based on this, it can be assumed that the torque is transmitted most efficiently in superposition of the vortex beams of the 1st and 5th order. This hypothesis was tested using a mathematical model that can estimate the magnitude of the torque transmitted from the light beam to the microturbine of the desired shape. Figure 5.79 represented different sections of the turbine blade.

So in the case in Fig. 5.79a the blade shape such that a vortex beam, designed for it, exerts no pressure in the vertical direction, i.e. there is no z-component of the force, as the vertical momentum of the beam is not changed. But in this case the projection of the force on the y-axis is not possible, this maximum is attained in the case of the blade shape shown in Fig. 5.79b, but in this case a non-

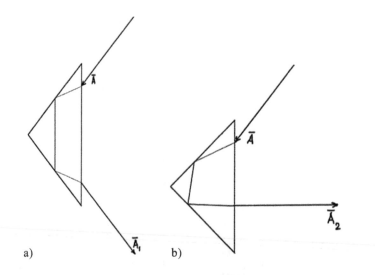

Fig. 5.79. A – the incident beam; A_1 – the beam coming out of the blades with a smaller angle at the base; A_2 – the beam caming out of the blades with a large angle at the base.

Fig. 5.80. Three-dimensional models of microturbines to receive the torque from the vortex light beams.

zero component of force appears along the axis z which gives rise to frictional forces in micromechanical systems. For this form of the blade is better to use the system to focus the beam from the bottom, when the force of light pressure compensates for the force of gravity. These cross-sections were used to build three-dimensional models of microturbines (Fig. 5.80).

Figure 5.80 shows three-dimensional models of microturbines for the case of no z-component of the force, and for the maximum force along the coordinate y.

Calculation of the forces acting on the microturbines, was conducted by the method described in [52]. A special simulation program was developed to calculate the forces and torque on the basis of the method described in [52]. A window of this program is shown in Fig. 5.81.

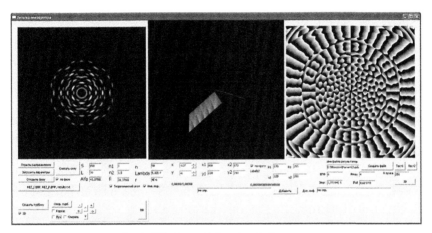

Fig. 5.81. Window of the simulation program.

The program allows one to set both the parameters of the vortex beam and the shape of the blade. Using the program, we calculated the torque acting on a microturbine with a diameter of 20 μm with the shape of the blades, eliminating the z-component of the force in the beams shown in Fig. 5.77. The calculation results are shown in Table 5.6.

Table 5.6. Torque of microturbines in the superposition of vortex beams

The beam	Superposition of the 1st and 4th-order	Superposition of the 1st and 5th order	Superposition of the 1st and 6th orders
Torque (N·m)	$1.8 \cdot 10^{-15}$	$2.3 \cdot 10^{-15}$	$1.9 \cdot 10^{-15}$
Relative torque	1	1,27	1,05

As shown in Table 5.6 the highest torque is obtained for superposition of optical vortices in the 1st and 5th order. When focusing with a ×90 microlens, the diameter of the focused beam from the superposition of optical vortices of the 1st and 5th order 12 μm. Accordingly, it is the maximum size of the microturbine to which this beam transmits torque with maximum efficiency. If it is required to rotate a larger microturbine and the possibilities of technology are such that the larger microturbine will be more efficient, it is necessary to form a wider beam.

5.3.2. Formation of the superposition of high-order optical vortices

For the broadening of vortex beams of superposition optical vortices have two options:

1. Form a superposition of two optical vortex of a higher order;
2. To form a beam from a superposition of three or more optical vortices.

In the formation of the superposition of two vortex beams the structure of the diffractive optical element remains the same: the boundary region is responsible for the formation of the inner ring and the inner region is responsible for the formation of the outer curve. To select the beam we must determine the geometric parameters of the microturbine. Based on the possibility of manufacturing technology and strength properties of materials the diameter of the microturbine must be at least 20 μm, but no more 30 μm (for larger diameter of the turbines the required intensity of the beam is very high). The diffractive optical elements calculated for these sizes are shown in Fig. 5.82.

The beam in Fig. 5.82b is a superposition of optical vortices in the 5th and 15th orders. The beam diameter of the outer boundary of the vortex curve 22 diameter of the inner ring 8 μm. Figure 5.82d shows the beam produced by the superposition of the vortex beams of 1st, 11th and 21st order. The diameter of the beam at the outer boundary of the vortex curve 29 μm, inner ring 1.5 μm. Unfortunately, this configuration is inconvenient in terms of manufacturing microturbines. Take advantage of the central ring in this case is not possible because it will have to limit thin blade mount to the central axis (up 0.5 μm), which is unacceptable for reasons of safety. To look at a final decision Is a graph of the polar angle (Fig. 5.83).

As can be seen from Fig. 5.83 the I_s value changes by no more than 4% of the maximum value. The change is stochastic in nature, ie, probably due to the sampling error of the image. This means that the gradient force will not prevent the rotation of the turbine and hence a vortex beam can be used to rotate the microturbine.

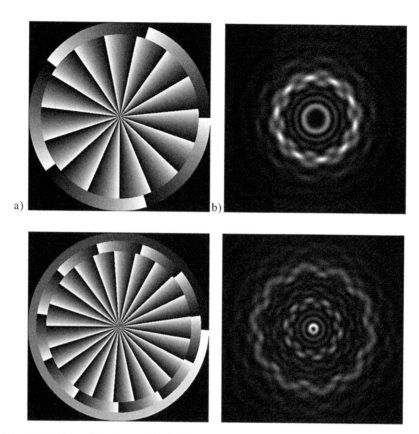

Fig. 5.82. Phase function of the DOE to form superpositions of optical vortices in of the 5th and 15th orders, respectively, (a), the superposition of the 1st, 11th and 21st order (c), the intensity distribution in the formed beams (b, d).

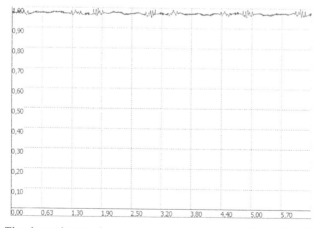

Fig. 5.83. The dependence of I_s on the polar angle for the intensity distribution in Fig. 5.82b.

5.3.3. Experimental formation of vortex beams superpositions

Several DOEs were produced for the experiments with the formation of superpositions of the vortex beams. As the technology enables to produce only binary DOEs, the encoded analogues of the halftone DOEs were calculated for the experiments. The DOEs were made for both the formation of the beams with multiple rings and the same number of singularity and the DOEs to form a superposition of vortex beams with different numbers of the singularity of the DOE. Figure 5.84 is a view of the central part of the microrelief of the DOE form the formation of a beam consisting of optical vortices of the 6th and 48th orders.

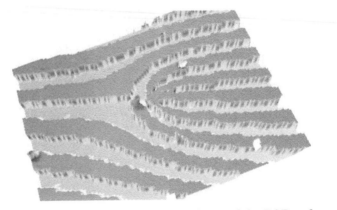

Fig. 5.84. View of the microrelief of the central part of the DOE to form a beam consisting of optical vortices of the 6th and 48th order.

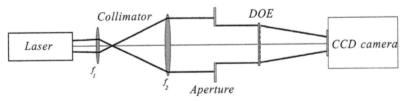

Fig. 5.85. The experimental optical scheme.

The optical scheme of the experiment is shown in Fig. 5.85. We used a solid-state laser with a wavelength of 532 nm and a beam divergence of 1.2 mrad. Two lenses forming a collimator were used for beam expansion. The aperture cut the beam to the desired diameter. The camera was used to capture the resultant intensity distributions at a distance of 1000 mm from the DOE.

Figure 5.86 shows vortex beams with two or four rings, as well as the beam of superposition of the of the first order and the vortex of the sixth order.

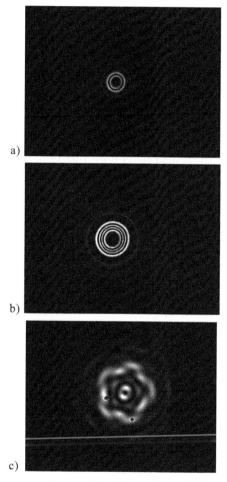

a)

b)

c)

Fig. 5.86. Multi-ring vortex beams (2) and two rings, (b) four rings, superposition of the vortices of the first and sixth orders.

As can be seen from Fig. 5.86 there is qualitative agreement between the simulation results and the generated beams.

5.4. Light beams specially formed for linear displacement and positioning of micro-objects

This section describes calculations of the transmission functions of the DOE forming light traps for objects whose refractive index is higher than that of the environment. The calculated traps have a certain amplitude–phase distribution. The distribution of intensity defines the capture area and the distribution phase allows the automatic movement of the captured objects due to the gradient of the phase [52, 93]. To calculate the DOE forming such complex distribution of light fields and having the phase transmission function, we use the methods of encoding the amplitude [94–97].

Calculated focusators can be divided into several classes, depending on the tasks for which they are intended. Thus, attention will be given to the focusators forming light traps used for precise positioning of micro-objects; focusators forming light fields with a given direction of the gradient of the phase for the automatic movement of the captured micro-objects from the initial position to the end; focusators forming light fields for sorting objects of a certain size in the total flow of particles in a liquid medium; focusators, generating sets of light traps for filtering of individual objects from the general flow of particles.

Modelling of the produced elements will be carried out by the software which uses the fast Fourier transform (FFT) algorithm to calculate the distribution of the light field in the focal plane.

5.4.1. Encoding of amplitude by the local phase jump method

Methods of coding the amplitude commonly used when it is required to get the phase function, forming an arbitrary given distribution of the light field. Earlier, the coding of the amplitude was used for the calculation of binary phase DOEs forming Gauss–Hermite modes [96], hypergeometric modes [98] and laser Airy beams [99]; spatial filters for mode selection [96]; spatial matched filters [96]; composite DOEs to form several images at the same time [97]. The most common methods of implementing the coding of the amplitude are the local phase jump method [96, 98] and the random phase mask method [94, 95, 97, 100, 101].

The method of encoding the amplitude by means of the local phase jump is based on replacing each count of the amplitude A_n of the initial distribution by the section of the relief of the transparent DOE with the width Δx, which has a stepped phase jump $\Delta\varphi_n$. The value of Δx is constant and determined by the specific dimensions of the calculated elements. The phase jump has two free parameters (Fig. 5.87): the width of the phase steps δx_n and its height $\Delta\varphi_n$. For the calculations we have chosen a method based on the variation in the width of the phase jump. In [96] it was found that from a technological point of view δx_n can be calculated efficiently using the following expression:

$$\delta x_n = \frac{\Delta x}{2}(1 - A_n)$$

(5.89)

which implies that the value of the width of the phase step is proportional to the normalized amplitude $0 \le A_n \le 1$.

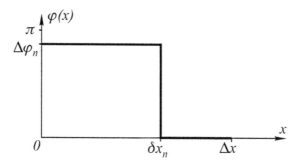

Fig. 5.87. Type of local phase jump in encoding the amplitude.

By changing the amplitude level above which the encoding operation is not applied, this method can be used to vary widely efficiencies and formation errors – we can achieve the high quality formation of the specified field, but sacrifice the efficiency of its formation, and vice versa. Thus, selecting the coding level we can achieve the suitable ratio of the error of formation and the effectiveness of forming the predetermined light field for each specific task.

Thus, the calculation of the phase function of the DOE will take place in several steps: by the analytical method or with the help of digital images we specify the amplitude–phase distribution, which must be formed in the plane of focus:

$$A(\xi,\eta) = \sqrt{I_{out}(\xi,\eta)} \exp\left(i\Phi_{out}(\xi,\eta)\right)$$

(5.90)

where I_{out} (ξ,η) is the given intensity distribution, and Φ_{out} (ξ, η) is the given phase distribution in the plane of focus. To calculate the amplitude and phase distribution forming in the focal plane the given light field $A(\xi, \eta)$ with the help of the FFT algorithm we perform the Fourier integral transform:

$$T(x,y) = \int\limits_{-\infty}^{\infty} \int A(\xi,\eta) \exp\left[\frac{ik}{f}(\xi x + \eta y)\right] d\xi d\eta = \mathfrak{I}^{-1}\{A(\xi,\eta)\} \quad (5.91)$$

where x, y are the Cartesian coordinates in the plane of the element, f is the focal length of the optical system, $k = 2\pi/\lambda$ is the wavenumber.

For the resulting distribution $T(x,y)$ of the transmission function of the element we carry out the operation of the encoding of the amplitude by the method of the local phase jump (5.89) and also carry out the reverse transformation (5.91) to observe the generated amplitude–phase distribution in the plane of focus $F(\xi,\eta) = \mathfrak{I}\{T(x,y)\}$.

Next, to calculate the efficiency of formation ε and the standard deviation of intensity (RMS) we use the following formulas:

$$\varepsilon = \frac{\sum\limits_{\xi,\eta \in D} F^2(\xi,\eta)}{\sum\limits_{x,y \in G} I_{in}(x,y)} \quad (5.92)$$

$$\delta^2 = \frac{\sum\limits_{\xi,\eta \in D} \left[F^2(\xi,\eta) - I_{out}(\xi,\eta)\right]^2}{\sum\limits_{\xi,\eta \in D} I_{out}^2(\xi,\eta)} \quad (5.93)$$

where G is the area that defines the boundaries of the DOE, D is the area defined by the boundaries of the trap.

According to (5.89), the maximum width of the phase step is $\Delta x/2$ wherein the initial amplitude at a point must be equal to 0. Thus, such a step will deflect half of the incident light on it, and half of the incident light will also be sent to the same point as in the case of absence of steps, thus distorting the picture formed by the light field. In some cases, when the spectrum calculated according to (5.91) is dominated by low frequencies, this leads to the fact that for any choice of the amplitude level above which the encoding operation is not used, high values of the error of formation of the light distribution are obtained. To solve this problem we carry out the operation of adding zeros to the distribution $A(\xi, \eta)$ which from the physical point of view is equivalent to a decrease in the radius of the incident light, so that it overlaps only the central areas of the DOE, whose contribution to the formation of the light field in this case is maximal.

5.4.2. Modification of the method of encoding the amplitude for the calculation of light fields with a complex structure

Methods of encoding the amplitude give good results when the resultant distribution of the light field has a simple structure, for example, in the case of formation of simple geometric shapes of [96, 102] or single-mode beams [96, 98]. When the structure of the formed fields is more complicated, for example, when the reference distributions represent light grids or arrays divided in the plane of geometric figures, there is an increase in the error of formation of light fields and a reduction in the effectiveness of their formation. The increase in the formation error occurs due to the appearance of significant intensity differences between the different portions of the images formed with the change in the amplitude level above which the encoding operation is not applied, does not completely correct this problem. To resolve this issue we can use an approach which requires one more iteration of the encoding of the amplitude, but with a modified reference distribution of the light field.

This approach can be described by the following sequence of steps: according to (5.90) from the given distributions of intensity and phase we obtain the complex output amplitude $A(\xi, \eta)$; the amplitude encoding is performed according to the previously described scheme; the Fourier transformation is carried out according to (5.91), and the resultant distribution of the light field $F(\xi, \eta)$ is analysed. If the result is recognized as unsatisfactory (i.e. the error of formation is above some acceptable level), then we replace the modulus of the original complex amplitude $A(\xi, \eta)$ taking the resultant distribution $F(\xi, \eta)$ into account:

$$|A'(\xi,\eta)| = ||A(\xi,\eta)| + \mu(|A(\xi,\eta)| - |F(\xi,\eta)|)|, \qquad (5.94)$$

where $0 \le \mu \le 1$; the encoding operation is performed for the calculated output amplitude distribution $A'(\xi, \eta)$:

$$A'(\xi,\eta) = |A'(\xi,\eta)| \exp(i\Phi_{out}(\xi,\eta)).$$

In the examples below the use of this approach for encoding the amplitude improves the quality and effectiveness of forming predetermined light fields when the encoding of the amplitudes as described in Section 5.4.1 scheme does not yield acceptable results.

5.4.3. Positioning of micro-objects using focusators

Positioning of the particles is of great interest for solving the problem of assembly of microsystems, as it is clear that for this it is necessary to position the component parts in the points with specified coordinates. To solve this problem, the light fields are commonly used which represent a set of light peaks in the points with the defined coordinates [23, 24, 27, 103]. In this case, to capture a specific micro-object the region of the maximum light intensity must include some part of the object and this requires placing the formed trap close to the object. This considerably complicates the automatic positioning of micro-objects, and in the absence of external control of the arrangement of the traps significantly reduces the probability of their capture in the given coordinates. To solve this problem, we can be used a focusator in a light cross with the phase gradient along each of its constituent beams. This form of the light trap allows to increase the area of the capture of particles moving in the general stream and due to the presence of the phase gradient the captured particle will move along the beam to the centre of the cross. Thus, the stability of the capture of a micro-object in the centre of the cross will be achieved due to the presence of the phase gradient for each of the generated rays.

The reference amplitude–phase distribution of the light field to calculate this focusator is shown in Fig. 5.88. The phase varies along the rays forming the cross from 0 radians to 2π at the edges in the cross centre. The dimension of the input images is 128×128 pixels.

The parameters of the encoding operation are as follows: width of the portion of the DOE encoding one count of the input amplitude distribution, $\Delta x = 8$; the amplitude level above which the encoding operation is not applicable $A_{enc} = 0.165$; the initial distribution of the complex amplitude in the calculation was supplemented with zeros to the size of 1024×1024 counts.

The calculation was made according to the method of encoding the amplitude by the scheme described in section 5.4.1. The dimension of the calculated element was 1024×1024 pixels. The phase function $\varphi(x, y)$ after encoding the amplitude was of the multilevel type and its binary form was produced by encoding the phase by superposition

Fig. 5.88. The distribution of intensity (negative) (a) and phase (b) of the reference image to calculate the focusator in a cross with a phase gradient.

of the grating with the carrier frequency β: as a result, the phase function shown in Fig. 5.89 was calculated for the DOE. The phase beyond the calculation contour is not defined.

$$\phi_b(x,y) = \arg\left[\phi^*(x,y) \cdot e^{-i\beta x} + \phi(x,y) \cdot e^{i\beta x} \right] \qquad (5.95)$$

After such encoding the calculated binary element formed a predetermined light distribution in two diffraction orders +1 and −1. The carrier frequency in encoding was 27000 lines/m. The DOE diameter was 4 mm.

As a result, the phase function shown in Fig. 5.89 was calculated for the DOE. The phase beyond the calculation contour was not defined.

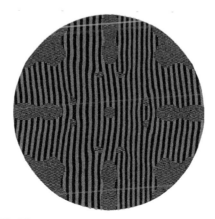

Fig. 5.89. The phase function of a focusator in a cross.

The resulting distribution of the light field in the focal plane, calculated using the FFT algorithm, is shown in Fig. 5.90. The total efficiency of the formation of a given complex distribution in one of the orders, calculated using the formula (5.93), for this case is equal to 43%. RMS is equal to 39%.

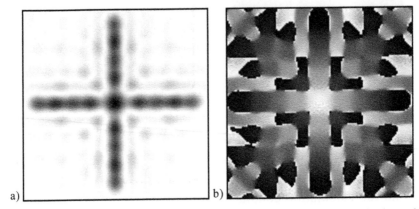

a)   b)

Fig. 5.90. The light field distribution formed in the focusing plane: the intensity (negative) (a) and the phase (b).

On the basis of this element we calculated a number of light fields as a set of light traps and light gratings. Individual light crosses can be formed at the points at which the individual micro-objects should be captured and thus it is possible to produce automatic precise positioning of a set of micro-objects. Examples of reference distributions for the calculation of a number of focusators, forming such sets of light traps, are shown in Figs. 5.91–5.93.

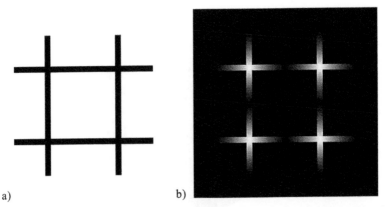

a)   b)

Fig. 5.91. The distribution of intensity (negative) (a) and the phase (b) of the reference image to calculate the focusator in a grating with four nodes.

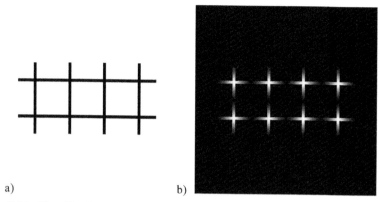

Fig. 5.92. The distribution of intensity (negative) (a) and the phase (b) of the reference image to calculate a focusator in a grating with eight nodes.

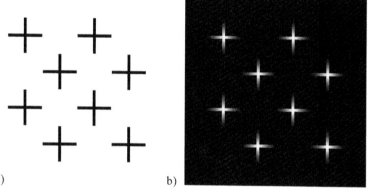

Fig. 5.93. The distribution of intensity (negative) (a) and the phase (b) of the reference image to calculate a focusator in a set of eight crosses.

The coding parameters for the case of the formation of the grating with four nodes: DOE diameter 4 mm; the dimension of the input images 256×256 pixels; the initial distribution of the complex amplitude in the calculation was supplemented with zeros to the size of 1024×1024 counts; the width of the portion of the DOE encoding one count of the input amplitude distribution, $\Delta x = 4$; the amplitude level above which the encoding operation is not applicable $A_{enc} = 0.1$; the carrier frequency when encoding 27000 lines/m.

The calculation was made according to the simple encoding of the amplitude using the scheme described in section 5.4.1. The calculated phase function of the DOE, forming in the focal plane the image of the grating with four nodes, is shown in Fig. 5.94. The phase beyond the calculation contour is not defined.

Fig. 5.94. The calculated phase function of the focusator to a grating with four nodes.

The results of the Fraunhofer diffraction of the plane wave on the DOE are shown in Fig. 5.95. The efficiency of forming the given complex distribution in one of the orders is 26%. The RMS is equal to 40%.

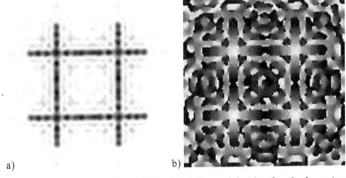

a)   b)

Fig. 5.95. The distribution of the light field formed in the focal plane (one of the orders), intensity (negative) (a) and the phase (b) for the focusator in the grating with four nodes.

The coding parameters for the case of the formation of a grating with eight nodes: the DOE diameter 4 mm; the dimension of the input images 256×256 pixels; the initial distribution of the complex amplitude in the calculation was supplemented with zeros to the size of 1024×1024 counts; the width of the portion of the DOE encoding one count of the input amplitude distribution, $\Delta x = 4$; the amplitude level above which the encoding operation is not applicable $A_{enc} = 0.06$; the carrier frequency when encoding – 55000 lines/m.

The calculation was carried out by the modified method of coding the amplitude using the scheme described in section 5.4.2. The calculated phase function of the DOE, forming in the focal plane the image of the lattice with eight nodes, is shown in Fig. 5.96. The phase beyond the calculation range is not defined.

Fig. 5.96. The calculated phase function of the focusator to a grating with eight nodes.

The results of Fraunhofer diffraction of the plane wave on the DOE are shown in Fig. 5.97. The efficiency of forming the given complex allocation in one of the orders is 18%. RMS is equal to 44%.

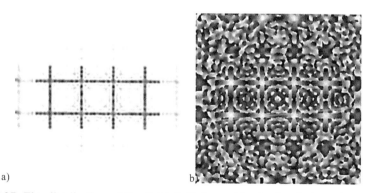

a)   b)

Fig. 5.97. The distribution of the light field formed in the focal plane (one of the orders): intensity (negative) (a) and the phase (b) for a focusator to a grating with eight nodes.

The coding parameters for the case of formation of a set of eight crosses: DOE diameter 4 mm; the dimension of the input images 256×256 pixels; the initial distribution of the complex amplitude in the calculation was supplemented with zeros to the size of 1024×1024 counts; width of the portion of the DOE encoding one count of the input amplitude distribution, $\Delta x = 4$; the amplitude level above which the encoding operation is not applicable $A_{enc} = 0.06$; the carrier frequency when encoding 55000 lin/m.

The calculation was made by the modified method of encoding the amplitude using the scheme described in section 5.4.2. The calculated phase function of the DOE, forming in the focal plane a set of eight crosses, is shown in Fig. 5.95. The phase beyond the calculation range is not defined.

Fig. 5.98. The calculated phase function of a focusator to a grating with eight nodes.

The results of Fraunhofer diffraction of the plane wave on the DOES are shown in Fig. 5.99. The efficiency of forming the given complex distribution in one of the orders is 15%. RMS is 38%.

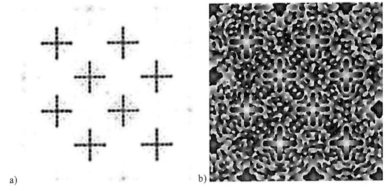

a)   b)

Fig. 5.99. The distribution of the light field formed in the focal plane (one of the orders): intensity (negative) (a) and phase (b) for a focusator to a set of eight crosses.

It is seen that the forming efficiency of these light traps differs and is determined by the complexity of the structures formed.

5.4.4. Focusators to move micro-objects along predetermined paths

Moving micro-objects along a predetermined path is one of the most common tasks encountered when manipulating micro-objects. As mentioned in the introduction, this operation can be done in two ways – either by the light modulator with its display showing a sequence of holograms defining the new position of the formed trap [81, 104, 105], or via the formation of specific amplitude and phase distributions when the phase distribution defines the direction of movement of the captured object [93, 106]. Methods based on the movement of the captured micro-objects by the use of formed amplitude and phase distributions are more favorable for the implementation of the automatic movement, because they do not require an operator that would calibrate the manipulation system when the external experimental conditions change. The method using PMS is disadvantageous in terms of the efficiency of formation of a given field and requires a high positioning accuracy of the optical elements of the scheme. In [106] the vortex nature of the generated field imposed restrictions on the resultant distribution of the resultant light field. So, it is impossible to form light fields without the vortex nature or with the random phase distribution, that is, for example, we can not form a light curve along which the phase first increases and then decreases. Also, we can not arbitrarily set the order for changing the phase of the same curve, which would change the velocity of the trapped particles without changing the power of the laser used. Coding of the amplitude allows one to remove these restrictions and to calculate the elements the efficiency of which exceeds the values obtained with the PMS. It should be noted that all focusators, forming a predetermined amplitude and phase distribution, may be used to move micro-objects along a predetermined path.

Consider the case of calculation of the focusators forming the letter 'Γ', a light triangle and a square with the gradient of the phase. Figure 5.100 shows the reference distribution of intensity and phase in the case of forming the letter 'Γ'. The operation of encoding amplitude and phase were carried out with the following parameters: the dimension of the input images 256×256 pixels; width of the portion of the DOE encoding one count of the input amplitude distribution, $\Delta x = 4$; the amplitude level above which the encoding operation is not applicable $A_{enc} = 0.075$; the initial distribution of the complex amplitude is supplemented with zeros to the size of

1024×1024 counts; DOE diameter 4 mm; the carrier frequency in encoding the phase $\beta = 55\ 000$ lines/m.

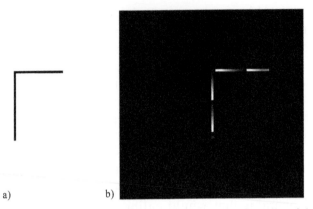

a)   b)

Fig. 5.100. The distribution of intensity (negative) (a) and phase (b) of the reference image to calculate a focusator to the letter 'Γ'.

The calculation was made by the method of encoding the amplitude using the scheme described in section 5.4.1. As a result, the phase function was calculated for the DOE and in shown in Fig. 5.101. The phase beyond the calculation range is not defined.

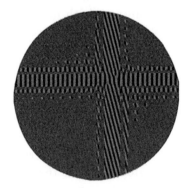

Fig. 5.101. The calculated phase function of the focusator to the letter 'Γ'.

The dimension of the calculated element was 1024×1024 pixels. The results of Fraunhofer diffraction of the plane wave on the DOE are shown in Fig. 5.102. The efficiency of forming the given complex distribution in one of the orders is 13%. The RMS is equal to 51%.

The reference intensity and phase distribution for the case of formation of a light distribution in the form of an equilateral triangle contour is shown in Fig. 5.103.

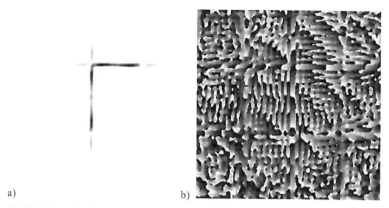

a)   b)

Fig. 5.102. The distribution of the light field formed in the focal plane (one of the orders): intensity (negative) (a) and phase (b) for a focusator to the letter 'Γ'.

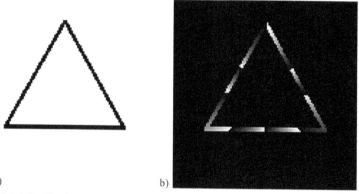

a)   b)

Fig. 5.103. The distribution of intensity (negative) (a) and phase (b) of the reference image to calculate a focusator in an equilateral triangle.

The operation of encoding the amplitude and the phase was performed with the following parameters: the dimension of the input images 256×256 pixels; width of the portion of the DOE encoding one count of the input amplitude distribution, $\Delta x = 4$; amplitude level above which the encoding operation is not applicable $A_{enc} = 0.075$; the initial distribution of the complex amplitude is supplemented with zeros to the size of 1024×1024 counts; DOE diameter 4 mm; the carrier frequency in encoding the phase $\beta = 55$ 000 lines/m.

The calculation was made by the method of encoding the amplitude using the scheme described in section 5.4.1. The phase function calculated for the DOE is shown in Fig. 5.104. The phase

Fig. 5.104. The calculated phase function of the focusator in the contour of an equilateral triangle.

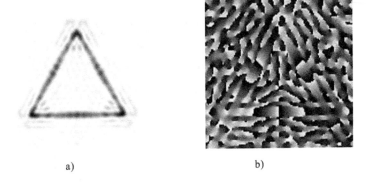

a)   b)

Fig. 5.105. The distribution of the light field (one of the orders) formed in the focal plane: intensity)negative (a) and the phase (b) of the reference image for calculating a focusator in the contour of the square.

beyond calculation range is not defined. The dimension of the calculated element was 1024×1024 pixels.

The results of Fraunhofer diffraction of the plane wave on the DOE are shown in Fig. 5.105. The efficiency of forming the given complex distribution in one of the orders is 35%. the RMS is 50%.

For the case of formation of a light distribution in the form of a square contour the reference intensity and phase distribution is shown in Fig. 5.106.

The operation of encoding the amplitude and the phase was performed with the following parameters: the dimension of the input images 256×256 pixels; width of the portion of the DOE encoding one count of the input amplitude distribution $\Delta x = 4$; the amplitude

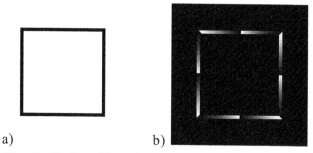

a)　　　　　　　　　　b)

Fig. 5.106. The distribution of intensity (film) (a) and phase (b) of the reference image to calculate a focusator in the square contour.

level above which the encoding operation is not applicable $A_{enc} = 0.05$; the initial distribution of the complex amplitude was supplemented with zeros to the size of 1024×1024 counts; DOE diameter 4 mm; the carrier frequency in encoding the phase $\beta = 55000$ lines/m.

The calculation was made by the method of encoding the amplitude using the scheme described in section 5.4.1. As a result, the phase function shown in Fig. 5.107 was calculated for the DOE. The phase beyond the calculation range was not defined. The dimension of the calculated element was 1024×1024 pixels.

The results of Fraunhofer diffraction of the plane wave on the DOE are shown in Fig. 5.105. The efficiency of forming the given complex distribution in one of the orders is 34%. RMS is equal to 49%.

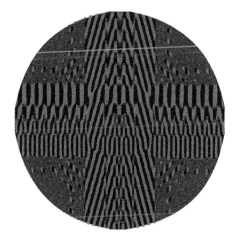

Fig. 5.107. The calculated phase function focusator in the loop of a square.

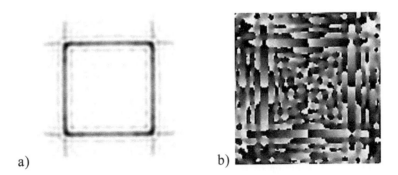

a)   b)

Fig. 5.108. The distribution of the light field generated in the focal plane (one of the orders): intensity (negative) (a) and the phase (b) for a focusator in the contour of a square.

5.4.5. Focusators for sorting and filtering micro-objects

The task of sorting micro-objects arises when it is necessary to choose from the general stream the particles whose parameters differ from the parameters of the other particles. These parameters may include: the size of micro-objects under study [103, 107–112], their color [113], the refractive index [111, 113]. For the implementation of the various types of sorting it was necessary to use different approaches, such as the use of arrays of optical vortices with fractional topological charges [103], the use of arrays of traps [111], the use of traps in the form of a light line with an asymmetric distribution of the intensity [107], the use of damped waves [108]; (optical traps on a chip), formed on the basis of the Talbot effect [109].

If we generate across the flow several light lines with different parameters, and each of them will have a phase gradient, we can achieve the simultaneous sorting of two or more different species of micro-objects in the total flow. Since the gripping force exerted on the micro-object depends on its size, sorting can be carried out by forming several light lines with different parameters, such as the intensity or magnitude of the differentialof the phase at the ends of the light segment. Also, light fields in the form of several light lines with the same parameters can be used to improve the efficiency of sorting, when a portion of the particles not sorted by one of the light lines and passed on, will deviate from the total flow under the effect of the following line. Encoding allows the calculation of the amplitude of the phase function of the DOE, which can generate such a set of light lines.

Consider the case of forming two light lines with different values of the phase difference at the ends of these segments. The reference distributions of the intensity and the phase of the light field is shown in Fig. 5.109.

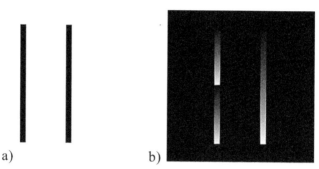

a)   b)

Fig. 5.109. The distribution of intensity (negative) (a) and phase (b) of the reference image to calculate a focusator in two parallel lines.

The operation of encoding the amplitude and the phase was performed with the following parameters: the dimension of the input images of 128×128 pixels; width of the portion of the DOE encoding one count of the input amplitude distribution, $\Delta x = 8$; the amplitude level above which the encoding operation is not applicable $A_{enc} = 0.2$; the initial distribution of the complex amplitude was supplemented with zeros to the size of 2048×2048 counts; DOE diameter 4 mm; the carrier frequency in encoding the phase $\beta = 55\ 000$ lines/m.

The calculation was made according by the method of encoding the amplitude using the scheme described in section 5.4.1. As a result, the phase function shown in Fig. 5.110 was calculated for the DOE. The phase beyond the calculation range was not defined. The dimension of the calculated element was 1024×1024 pixels.

The results of the Fraunhofer diffraction of the plane wave on the DOE are shown in Fig. 5.111. The efficiency of forming the given complex distribution in one of the orders is 14%. RMS is equal to 47%.

In this case, the carrier frequency when encoding can be chosen in such a way that all the four generated light lines will be at the same distance from each other (Fig. 5.112). That is, during the sorting of micro-objects we can use all four light lines formed that should improve its efficiency and speed. In this case, the efficiency of formation of the light distribution is doubled and equal to 28%.

Fig. 5.110. The calculated phase function of the focusator in two parallel lines.

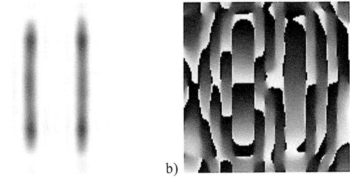

a)   b)

Fig. 5.111. The distribution of the light field formed in the focal plane of the field (one of the orders): intensity (negative) (a) and phase (b) for a focusator in two parallel lines.

5.4.6. Focusators to filter certain micro-objects

The task of filtering the individual micro-objects [114] can also be attributed to the problem of sorting micro-objects. Filtration implies a process of dividing the particles in the stream where some of them continues to move downstream, and a portion remains in the filter system. Then, the filtered particles can be removed from the general flow and manipulated separately. To perform filtering we can use

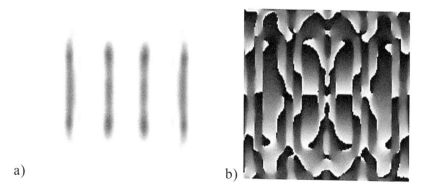

Fig. 5.112. The distribution of the light field generated in the focal plane (one of orders): intensity (negative) (a) and phase (b) for a focusator in four parallel lines.

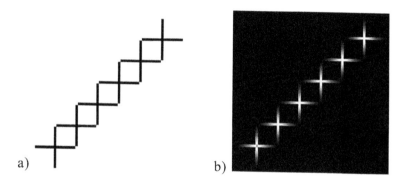

Fig. 5.113. The distribution of intensity (negative) (a) and phase (b) in the reference image to calculate a focusator in six crosses, arranged diagonally.

the previously described light focusators in the light cross and the gradient of the phase. Such traps can be formed at an angle to the flow by increasing the area of possible capture, thereby increasing the filtration efficiency.

Consider the case of forming a set of six light crosses arranged diagonally [115]. In the case of such traps the particles to be filtered will be captured in the light cross region and due to the phase gradient will move in their centres. The reference intensity and phase distributions of such light field are shown in Fig. 5.113.

The operation encoding the amplitude and phase was performed with the following parameters: the dimension of the input images 256×256 pixels; width of the portion of the DOE encoding one count of the input amplitude distribution $\Delta x = 4$; the amplitude level above which the encoding operation is not applicable A_{enc} = 0.07; the initial distribution of the complex amplitude was supplemented with zeros to the size of 1024×1024 counts; DOE diameter 4 mm; the carrier frequency in encoding the phase β = 55000 lines/m.

The calculation was carried out using the modified method of coding the amplitude using the scheme described in section 5.4.2. As a result, the phase function calculated for the DOE is shown in Fig. 5.114.

The dimension of the calculated element was 1024×1024 pixels. The phase beyond the calculation range was not defined.

Fig. 5.114. The calculated phase function of a focusator in six crosses, arranged diagonally.

The results of Fraunhofer diffraction of the plane wave on the DOE are shown in Fig. 5.115. The efficiency of forming the complex distribution in one of the orders is 15%. RMS is 43%.

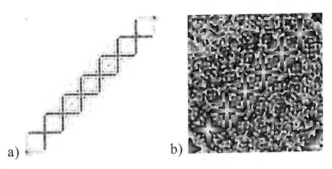

a)   b)

Fig. 5.115. The distribution of the light field generated in the focal plane (one of orders): intensity (negative) (a) and phase (b) for a focusator in six crosses arranged diagonally.

5.4.7. *Experimental study of the motion of micro-objects in light fields with a given amplitude–phase distribution generated by binary focusators*

For the experiments with manipulation the binary focusators whose phase functions were calculated in the first chapter were fabricated by photolithography on a glass substrate with a resolution of 2 μm. The diameter of the produced elements was 4 mm.

The experimentally defined amplitude–phase distributions were formed by binary focusators using the optical system shown in Fig. 5.116.

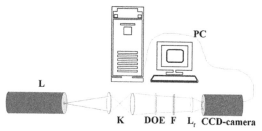

Fig. 5.116. The experimental optical scheme: L – laser, K – collimator, D – DOE, F – filters, L_f – focusing lens.

Two convex lenses forming the collimator K were used to expand the laser beam which then falls on a binary DOE D. Next in the beam path are darkening light filters F, which serve to attenuate the radiation, and the collecting lens L_f, which serves to focus the laser beam on the CCD-camera matrix VS-CTT-252. The experiment used a neodymium solid-state laser doped with rare earth metals L, emitting at a wavelength of 532 nm. The radiated output power was about 500 mW.

Figure 5.117 shows three-dimensional images of different parts of the microrelief of the produced diffractive optical elements whose phase functions were shown previously 1, in Figs. 5.99 and 5.110, respectively.

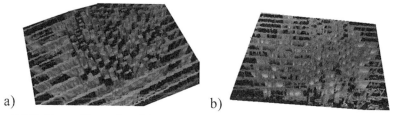

a)   b)

Fig. 5.117 Three-dimensional image of an area of the microrelief of produced diffractive optical elements.

Figure 5.118 shows the intensity distribution in the focus of the lens formed by the binary focusator in a light cross with the gradient of the phase whose phase function is shown in Fig. 5.89.

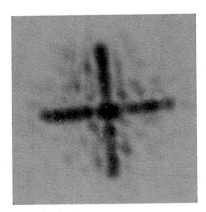

Fig. 5.118. The intensity distribution (negative) formed by a binary focusator in a cross.

Figure 5.119 shows the intensity distribution in the focus of the lens formed by a a binary focusator in two parallel lines with the phase gradient whose phase function is shown in Fig. 5.110.

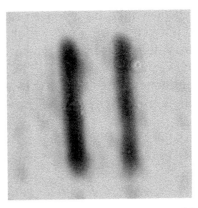

Fig. 5.119. The intensity distribution (negative) generated by a binary focusator in two parallel lines.

As seen from the drawings, the structure of the generated fields agrees well with the simulation results presented earlier.

Some experiments on the formation of the defined amplitude and phase distributions were conducted using dynamic spatial light modulators using the optical circuit shown in Fig. 5.120. The laser beam, expanded by means of two convex lenses forming the collimator K, is directed through the attenuating light filters F to the display of the dynamic light modulator (SLM), so that the angle of incidence is as small as possible. The laser beam reflected from the display of the modulator is directed at the focusing lens L_f, which projects the image received in its focus on the matrix of the CCD-camera. The modulator was illuminated with the laser light of wavelength of 457 nm and 532 nm. The phase function of the DOE formed on the binary dynamic modulator CRL OPTO with a resolution of 1316×1024 pixels with a resolution (the size of one pixel is 15 μm).

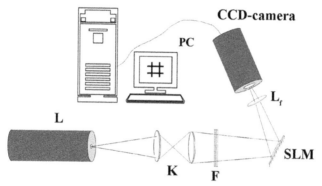

Fig. 5.120. The experimental optical scheme: L – laser, K – collimator, D – DOE, F – filters, Lf – focusing lens, SLM – spatial light modulator.

Figure 5.121 shows the intensity distribution in the focus of the lens in the form of a grating with four nodes generated by a binary focusator; its phase function is shown in Fig. 5.94.

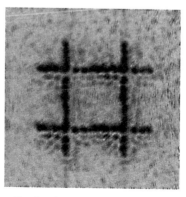

Fig. 5.121. The intensity distribution (negative) in the form of a grating with four nodes generated by a binary focusator.

Figure 5.122 shows the intensity distribution in the focus of the lens in the form of a grating with eight nodes formed by s binary focusator; its phase function is shown in Fig. 5.96.

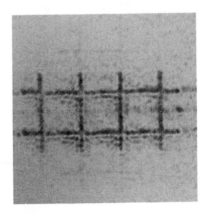

Fig. 5.122. The intensity distribution (negative) in the form of a grating with eight nodes formed by a binary focusator.

Figure 5.123 shows the intensity distribution in the focus of the lens as a set of eight crosses formed by a binary focusator whose phase function is shown in Fig. 5.98.

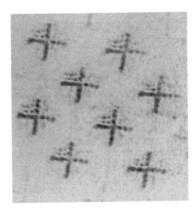

Fig. 5.123. The intensity distribution (negative) as a set of eight crosses formed by binary focusators.

Figure 5.124 shows the intensity distribution in the focus of the lens as a set of six crosses arranged diagonally and formed by a binary focusator with the phase function shown in Fig. 5.114.

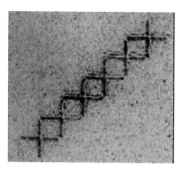

Fig. 5.124. The distribution of intensity (negative) as a set of six crosses arranged diagonally and formed by a binary focusator.

Figure 5.125 shows the intensity distribution in the focus of the lens formed by a binary focusator in an equilateral triangle; its phase function is shown in Fig. 5.104.

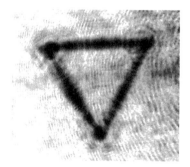

Fig. 5.125. The intensity distribution (negative), formed in a binary focusator in an equilateral triangle.

Figure 5.126 shows the intensity distribution in the focus of the lens formed by a binary focusator in a square contour, eeith the phase function shown in Fig. 5.107.

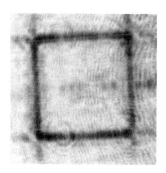

Fig. 5.126. The intensity distribution (negative) formed by a binary focusator in a square contour.

As seen from the drawings, the binary light modulator can also be used to form light traps, which agrees well with the simulation results presented earlier.

The experiments with positioning of micro-objects [102] were carried out using the optical system shown in Fig. 5.127. The laser radiation source was a neodymium solid-state laser doped with rare earth metals emitting at a wavelength of 532 nm. The radiated output power was about 500 mW. In order to minimize power losses in the reflection on the refracting surfaces of the lens the desired size of the laser beam to illuminate the DOE is not achieved using a collimator, and by increasing the distance between the laser output aperture and the first deflecting mirror M_1. The laser beam rebounded from the M_1 is directed to the DOE. Microlens L_1 (×20) focuses the beam generated by the DOE on a glass substrate with a suspension of microparticles in a water solution. The manipulation surface was illuminated with an illuminator system consisting of a lamp and the lens L_3. Microlens L_2 (×16) is used to construct the image of the manipulation plane on a matrix on the CCD-camera.

The experiments with automatic positioning of micro-objects were carried out using transparent polystyrene microspheres with a diameter of 5 µm.

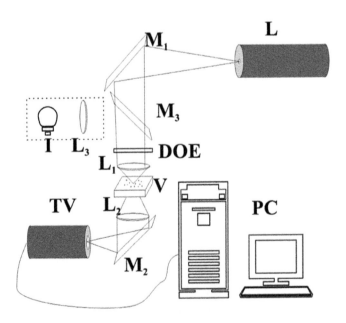

Fig. 5.127. The experimental optical scheme for manipulating microscopic objects.

Figure 5.128 shows stages of movement of the microspheres captured in the light cross region. The time interval between frames 0.75 s. As seen from the images, the captured microspheres move from one of the edges of the cross to the centre, which is fully consistent with the calculations. The measured velocity of the particles was 3.84 ± 0.60 µm/s.

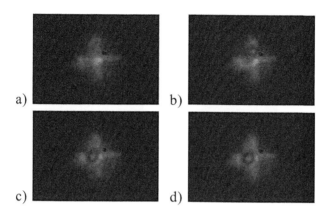

Fig. 5.128. Stages of movement of a polystyrene microsphere in the light beam in the form of a cross. The interval between the frames 0.75 s.

The experiment to capture and deflect the transparent micro-objects from the general flow using two light lines with a phase gradient was carried out using the optical scheme shown in Fig. 5.127. Manipulation was conducted using the same laser as in the experiments with the positioning of the individual particles. Manipulated objects were transparent polystyrene microspheres with a diameter of 5 µm.

Figures 5.129 and 5.130 show the stages of motion of microparticles captured in the light line region. They were formed using a DOE with the phase function of this DOES is shown in Fig. 5.110. The time interval between the frames in Fig. 5.129 was 1.25 s, in Fig. 5.130 1.5 s.

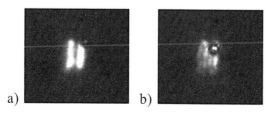

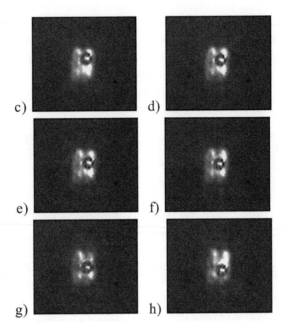

Fig. 5.129. Stages of movement of a polystyrene microsphere captured in one of the light lines. The interval 1.25 s.

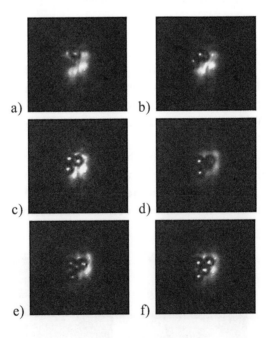

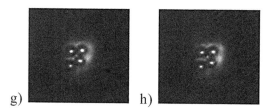

g)   h)

Fig. 5.130. Stages of movement of polystyrene microspheres trapped in light lines. The interval between frames 1.5 s.

As can be seen if the polystyrene microsphere enters the area of the light it begins to move in the direction perpendicular to the general flow; this is fully consistent with the calculations. Thus, the particles are deflected from the general flow, which can be considered as sorting. The velocity of the microspheres for the case in Fig. 5.129 was 0.82 ± 0.06 μm/s. According to the simulation results shown in Fig. 5.130, the phase along one of the lines is growing twice as fast. Therefore, the velocity of movement of the captured polystyrene microsphere along each of the lines for the case in Fig. 5.130 will be different: for example, for the line with the greater phase difference at its ends the particle velocity is 1.45 ± 0.4 μm/s, 1.77 ± 0.4 μm/s; for a line with a smaller gradient it is respectively 1.01 ± 0.12 μm/s.

5.4.8. The linear movement of micro-objects in superposition of two vortex beams

Consider the method of calculating the DOE forming the optical field for linear movement of a micro-object on the basis of two vortex beams. Instead of calculating a special focusator, the required element is calculated based on a linear encoded DOE to form an optical 'vortex' and generates a light beam in the form of strips with the phase gradient directed along it.

Calculation of the element is based on a linearly coded optical 'vortex'. The coding of the phase function of the optical 'vortex' to reduce it to the binary form is carried by the formula:

$$\varphi_b(x,y) = \arg\left[\varphi^*(x,y) \cdot e^{-i\beta x} + \varphi(x,y) \cdot e^{i\beta x}\right] \qquad (5.96)$$

where α is the carrier frequency, φ is the initial phase, φ_b is the new binary phase.

With linear encoding, to form a binary element, the resulting image shows two instead of one ring and the direction of phase changes to the opposite. If we change the parameter such as the linear encoding frequency this will change the distance between the rings (Fig. 5.131), and at a certain frequency we can obtain an image in which the rings are in contact, and a strip will form at the point of contact. At the same time, the fact that the beams in the +1 and −1 orders are equal in magnitude but opposite in sign, provides a strong phase gradient directed along the formed light strip, as shown in Fig. 5.131.

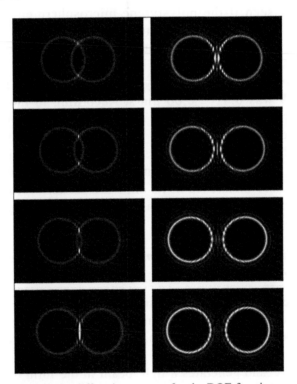

Fig. 5.131. Changes in the diffraction pattern for the DOE forming an optical vortex with $n = 60$ as the encoding frequency changes from 18500 m^{-1} to 36000 m^{-1} in increments of 2500 m^{-1} (frequency increases from top to bottom).

Fig. 5.132. Directions of increasing phase along the rings for DOE forming two vortex beams of the 60th order.

It should also be noted that only the vortex beams with the even order are suitable for the formation of the required strips as for the odd order two arcs form instead of one straight strip in the area of contact of the rings (Fig. 5.133). Though such a beam may also be used, for example, for capture and linear movement of opaque micro-objects.

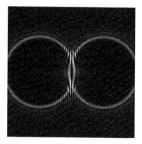

Fig. 5.133. The pattern formed at odd order of singularity ($n = 101$).

An increase in the order of the beam increases the diameter of the rings, as well as the length of the strip (Fig. 5.134), but each order has its own the optimal frequency of coding the phase for the formation of the strip.

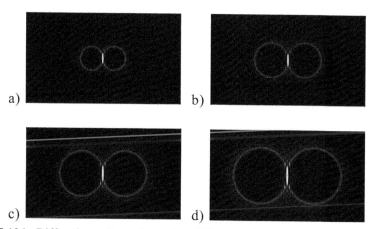

Fig. 5.134. Diffraction patterns formed at different orders of optical vortex beams: $n = 40$ (a), $n = 60$ (b), $n = 80$ (c), $n = 100$ (d).

The calculations were carried out using a modelling program that lets one calculate the DOE with the given coding parameters and the order of singularity as well as simulate diffraction patterns formed by this element. For easier calculation the program included the facility to calculate several DOEs simultaneously with different coding frequency at a predetermined pitch, obtain from them ring images and measure the length and the diffraction efficiency of the strip. The diffraction efficiency in this case is the ratio of the energy within the strip to the energy used for formation of the whole diffraction pattern. From the results of the calculations of the program it was possible to select several DOEs forming optical vortex beams with different order. These DOEs have the optimal coding frequency of the phase at which the longest and most effective strip. The parameters are shown in Table 5.7.

Table 5.7. Dependence of the efficiency and length of the strip on the singularity at the optimum coding frequency

Order	Encoding frequency, m^{-1}	Efficiency, %	Length of the strip, pixel
20	9060	18,244	17
40	17760	16,004	24
60	26260	11,386	29
80	34320	8,4575	38
100	42360	6,5517	46

From Table 5.7 it can be noted that in addition to increasing the length of the strip, increasing the singularity decreases the effectiveness of the strip, however, depending on the conditions of the problem we must choose what is more important: the length of the strip or its effectiveness.

All calculated elements were formed on a CRL OPTO spatial light modulator. With the help of computer the binary image of the DOE was displayed on the screen of the modulator, which created the appropriate phase delay. Figure 5.135 is a general view of an optical circuit for obtaining a diffraction pattern by a modulator.

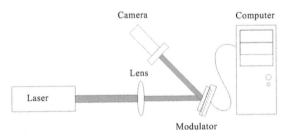

Fig. 5.135. General view of the optical circuit to obtain a diffraction pattern with a modulator.

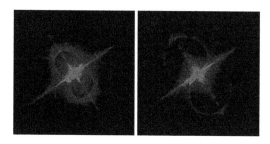

Fig. 5.136. Examples of images obtained by the modulator.

Figure 5.136 shows that the strip does form, but it also shows that a lot of noise is generated by the central peak which is due to the fact that the quality of imaging by the modulator is much lower compared to the quality of the DOE made on the optical substrate.

Further, all five DOEs calculated above were made on glass and examined. The optical circuit shown in Fig. 5.137 was assembled on the table for this purpose.

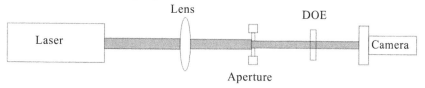

Fig. 5.137. General view of the optical circuit to obtain a diffraction pattern using DOE fabricated on glass.

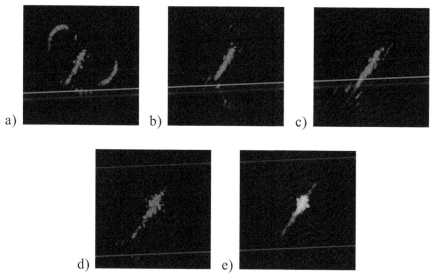

Fig. 5.138. Images obtained using DOE made on glass and forming optical vortex beams with $n = 20$ (a), $n = 40$ (b), $n = 60$ (a), $n = 80$ (c), $n = 100$ (d).

As can be seen from Fig. 5.138, the DOE data also do not give the highest quality image, and in the middle of the strip the intensity also increases. This may be due to a small error of the height of the relief of the DOE. The sufficiently high diffraction efficiency of producing the DOE enabled experiments with manipulating them using microparticles.

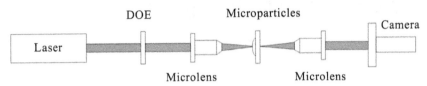

Fig. 5.139. General view of the optical circuit for micromanipulation.

A special scheme (Fig. 5.139) was constructed for this purpose on the optical table. DOE was located at a distance from the laser so that the output beam expanded and covered the entire item. After the DOE the laser beam is focused by the microlens to a size suitable for the manipulation of particles with the size of 5 μm. In the focus of the microscope objective there was a glass substrate coated with a drop of a suspension of polystyrene microparticles with a diameter of 5 μm and behind it was another microlens producing images. Further, the laser beam was redirected by the mirror to the camera in front of which there was a set of light filters allowing attenuation of the beam to the extent at which it was possible to obtain images of the particles and the light beam.

The microparticles were manipulated using a DOE forming a vortex beam of the 20th order. It was possible to trap and move the particle along the strip, but due to the increased intensity in the centre when reaching the centre of the strip, the particle was firmly fixed in this intensity peak. Stages of capture and movement are shown in Fig. 5.140.

Figure 5.140 shows how the particle is captured on the left in the region of increased intensity and moves right along the strip approximately to its middle.

5.5. Formation of arrays of light 'bottles' with the DOE

We consider the method for calculating transmission functions of the DOE forming light traps for objects whose refractive index is lower than that of the environment, and their arrays. Such traps include light 'bottles' and hollow light beams. The ultimate goal is the algo-

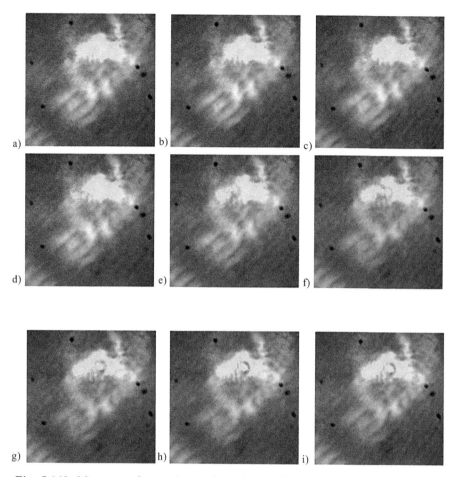

Fig. 5.140. Moments of capturing and moving particles with an interval between frames of 0.25s.

rithm that allows one to calculate the phase functions of the DOE forming both individual traps which have the form of simple geometric shapes – circle, square loop, equilateral triangle, etc., and a set of traps for manipulating multiple micro-objects.

The developed method is based on the superposition of Bessel beams of the 0th order, therefore, first we consider the properties of these beams and highlight the ones that will be used by us in the future to solve the problem. Next we describe an algorithm for calculating the complex transmission functions of the DOE, and identify differences in its application to the case of forming the light 'bottle' and the hollow light beams. In order to improve the

uniformity of the generated distributions we consider a modification
of this algorithm, which uses the gradient method calculation.

Modeling of the elements will be carried by the software, which
calculates the distribution of the light field in different planes using
the Fresnel transform, implemented on the basis of the algorithm of
the fast Fourier transform (FFT).

5.5.1. Bessel light beams and their remarkable properties

The Single-mode Bessel beam is described by the complex
amplitude

$$\Psi_{nm}(r,\phi,z) = J_m(k\rho_n r)\exp\left(ik\sqrt{1-\rho_n^2}\,z\right)\exp(im\phi) \quad (5.97)$$

where $J_m(x)$ is the Bessel function of the mth order, $k = 2\pi/\lambda$ is the
wave number of the light with a wavelength λ, (r, f) are the polar
coordinates in a plane perpendicular to the axis of propagation of
the beam z, $\rho_n = \sin\theta_n$, θ_n is the angle inclination to the z axis of
the conical wave generating the Bessel beam.

The Bessel beams possess a number of remarkable properties:
applied to the final segment of the axis without diffraction [66, 116]
they may form a light 'pipe' on the optical axis [59], self-replication
at a certain distance from an obstacle located on the optical axis
[117, 118], may possess an orbital angular momentum [60, 64, 119,
120], the longitudinal intervals [121], as well as to rotate during
their propagation [60, 122].

The diffractive optical elements allow to form the Bessel beams,
preserving the character of the mode at a great distance along the
propagation axis. Based on geometrical considerations the distance at
which the mode nature of of the single-mode Bessel light field $J_n(\alpha r)$
$\exp(in\varphi)$ is preserved is estimated by the following formula [65]:

$$z_{\max} = R\left[\left(\frac{2\pi}{\alpha\lambda}\right)^2 - 1\right]^{1/2} \quad (5.98)$$

where R is the radius of the DOE, α is the parameter of the Bessel
function.

To form the diffraction-free Bessel beams we can used diffractive
optical elements; their transmission function is as follows [67, 68]:

$$\tau(r,\phi) = \mathrm{sgn}\left[J_m(\alpha r)\right]\exp(im\phi) \qquad (5.99)$$

where $\tau(r,\varphi)$ are the polar coordinates in the plane of the element, $\alpha = kp_\alpha$ is the parameter of the Bessel function.

The helical DOE with the transmittance (5.99) effectively forms the light field with the amplitude proportional to the Bessel function $Jn(\alpha r)\exp(im\varphi)$, near the optical axis to some $z_{max} = Rk/\alpha$ [67].

It is worth noting that the use of the final phase DOE the Bessel beam begins to take shape only at a distance from the plane of the element. Thus, a segment in which the Bessel beam retains its modal character starts with some z_{min}, necessary for the formation of the beam, and ends at z_{max}.

Bessel beams are widely used in the field of laser manipulation. Thus, due to the fact that such beams propagate without distortion on some segment of the optical axis, in [64] it was possible to create a vertical chain of 16 silica beads with a diameter of 5 μm, to move it as a whole and tilt by 5°. The ability to restore the structure of the beam after passing through the obstacles allows the simultaneous manipulation in several planes [123]. The longitudinal frequency of the Bessel beams can be used to create three-dimensional light traps.

5.5.2. Multimode Bessel beams

Multimode Bessel beams have the same properties as the single-mode ones [84, 121, 124, 125, 126, 127, 128, 129, 130], however they are also widely used in laser manipulation tasks.

Based on the formula (5.99) we can device the following formula to calculate the complex amplitude of the element forming the superposition of N Bessel beams:

$$T(x,y) = \sum_{p=1} C_p \cdot \mathrm{sgn}\left(J_0(\alpha_p \vec{r})\right) \times \exp\left[i(xu + yv)\right], \quad (5.100)$$

where C_p are complex factors responsible for the contribution of individual beams in the overall distribution of the light field formed; u, v are the parameters corresponding to the axial displacement of the centre of the Bessel beam; x, y are the Cartesian coordinates, $r^2 = x^2 + y^2$. The squared modulus of the coefficient C_p sets the intensity of a single light beam, and the phase of this factor – the phase shift of the beam.

The properties of the multimode beams, such as the longitudinal periodicity, forward rotation during propagation, Fourier invariance and the invariance to truncation depending on the ratio of the parameters of the individual single-mode beams have been investigated in [69, 121, 131, 132]. Thus, choosing the parameters of beams in superposition we can ensure that the generated beam is a superposition, having properties that can be applied to create three-dimensional complex-shaped traps.

As is known, the single-mode Bessel beam of the 0th order is a light axial segment in which the radially symmetric intensity distribution in the cross section is described by the Bessel function of the corresponding order (Fig. 5.141).

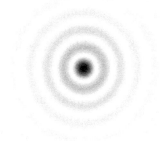

Fig. 5.141. The intensity distribution in the cross section of the Bessel beam of the zeroth order (negative).

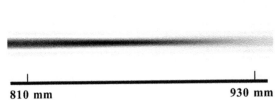

Fig. 5.142. Intensity distribution (negative) of the superposition of Bessel beams of 0th order with the parameters $\alpha_1 = 21.85 \cdot 10^3$ m^{-1}, $\alpha_2 = 17.08 \cdot 10^3$ m^{-1}, $\alpha_3 = 10.31 \cdot 10^3$ m^{-1} along the optical axis.

810 mm   930 mm

The longitudinal interference of several concentric single-mode Bessel beams can lead to the fact that the resultant intensity distribution will differ from that shown in Fig. 5.141. So, if we consider the superposition of three Bessel beams of 0th order with the parameters $\alpha_1 = 21.85 \cdot 10^3$ m^{-1}, $\alpha_2 = 17.08 \cdot 10^3$ m^{-1}, $\alpha_3 = 10.31 \cdot 10^3$ m^{-1}, we can see that the intensity distribution of the light field in a limited area of the beam propagation axis represents the light axial length almost devoid of additional rings, the presence of which is characteristic of a single-mode Bessel beam of 0th order; at the edges of this segment the broadens (Fig. 5.142).

5.5.3. Formation of light 'bottles' through the use of composite Bessel beams of 0th order

The superposition of Bessel beams of 0th th order with the parameters $\alpha_1 = 21.85 \cdot 10^3$ m^{-1}, $\alpha_2 = 17.08 \cdot 10^3$ m^{-1}, $\alpha_3 = 10.31 \cdot 10^3$ m^{-1} along the optical axis, presented in the previous section, can be used for the further construction of three-dimensional light traps of the light 'bottle' type and hollow light beams. For this we can use the approach considered in [97]. It is proposed to form a complex distribution of light fields in the plane of focus by their construction from individual light spots. Thus, it is possible to control not only the shape of the calculated light trapbut also the amplitude–phase distribution in its individual portions.

First, consider how to use the beam which is a superposition of three coaxial Bessel beams of 0th order with the parameters $\alpha_1 = 21.5 \cdot 10^3$ m^{-1}, $\alpha_2 = 17.08 \cdot 10^3$ m^{-1}, $\alpha_3 = 10.31 \cdot 10^3$ m^{-1} to form a simple light bottle. As described above, this beam allows to form the light axial segment without additional light rings, wherein in the edges of the central portion the light spot expands. Consider two such complex Bessel beams located at such a distance from each other where they will interfere very little. We assume that each of these beams is formed by an element with a radius of 3 mm, the transmission function of which is calculated by formula (5.100) at the wavelength of the incident light $\lambda = 532$ nm. One can assume that when the distance between them is reduced because of the low intensity of the rings surrounding the central peak these rings will provide an additional little impact on the interference pattern. Thus, the two light segment will extend along the axis without exerting any detectable effect on each other, as long as the distance between them is such that they can be viewed not as two separate light segments, but as one, but elongated in the direction of one of the transverse coordinates. As a result, we obtain the distribution of the light field the sections of which are shown in Fig. 5.143.

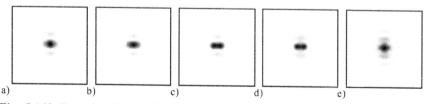

a)   b)   c)   d)   e)

Fig. 5.143. Intensity distribution (negative) of the superposition of two separated Bessel beams in space at a distance of 800 mm (a), 825 mm (b), 850 mm (c), 875 mm (d), 900 mm (e) from the plane of the element.

If we reduce the distance between the light spots further, the section will be shortened in the transverse direction and as a result one spot will form when they are combined (Fig. 5.144).

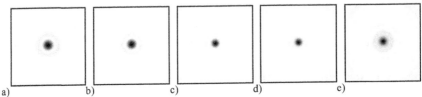

a)   b)   c)   d)   e)

Fig. 5.144. Intensity distributions (negative) of the superposition of two Bessel beams combined at distances of 800 mm (a), 825 mm (b), 850 mm (c), 875 mm (d) and 900 mm (d) from the plane element.

If we consider not two but a larger number of beams arranged in a line at a certain distance from each other, it is possible to form a distribution of the light field which will be a light rectangle parallel to the axis of beam spreading. At the same time changing the distance between the individual interfering complex Bessel beams we can change the transverse dimension of the light box.

For the simplest form of the light 'bottle' the layout of the generated beams is shown in Fig. 5.145.

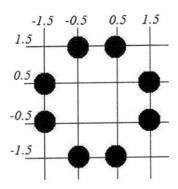

Fig. 5.145. Arrangement of the Bessel beams for the case of forming a light 'bottle'.

As seen, this circuit comprises eight points located symmetrically about the origin of the coordinates at the the circumferential diameter of 3 units. The grid size is proportional to the size of the minimum diffraction spot formed by an element with a radius of 3 mm at a distance z at which the light 'bottle' forms at an emission wavelength of $\lambda = 532$ nm. All coefficients C_p will be considered equal to 1. The phase component of these factors is equated to zero, so that in the light 'bottle' region there is no phase gradient, the presence of which may cause the movement of the trapped object [60]. The radius of the calculated element is as described previously $R = $ 3 mm. Calculating the transmission function of the DOE using Eq. (2.4), we obtain the amplitude–phase distribution shown in Fig. 5.146. As can be seen, the phase of the cell has a binary form due to the symmetrical arrangement of the formed beams. If we ignore the amplitude component of this complex field, it is possible to obtain a pure phase transmission function.

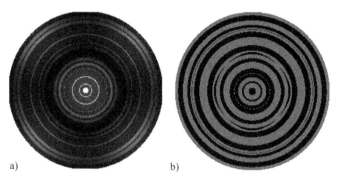

a)   b)

Fig. 5.146. Amplitude (A) and phase (b) transmission function of the DOE, forming a single light 'bottle'.

Modelling using the Fresnel integral

$$F(u,v,z) = -\frac{ik}{2\pi z} \int\limits_{-\infty}^{\infty} \int T(x,y) \exp\left[\frac{ik}{2z}\left((x-u)^2 + (y-v)^2\right)\right] dx\,dy$$
(5.145)

where x, y are the coordinates of the input plane, u, v are the coordinates in the output plane at distance z, $T(x, y)$ is the transmission function of the DOE, $k = 2\pi/\lambda$ is the wavenumber. Passage through such a phase DOE of a plane light wave with a wavelength of 532 nm and a radius of 2.1 mm gives the distribution of the light field shown in Fig. 5.147 (longitudinal distribution) and 5.148 (cross-sections).

The analysis of the figures shows that the light 'bottle' is formed in the area of the axis of 820 to 875 mm.

If we carefully consider the intensity distriobutions presented in Fig. 5.148, we can see that the minimum intensity is quite small and thus the ratio of the minimum intensity in the centre of the ring to the maximum at its perimeter is equal to 0.26 to 1.

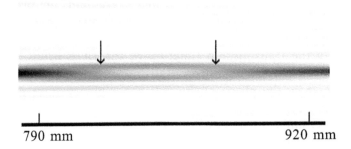

790 mm   920 mm

Fig. 5.147. Intensity distributions (negative) for the formed light 'bottle' along the optical axis. Arrows indicated the region of formation of the light 'bottle'

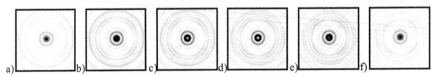

a)   b)   c)   d)   e)   f)

Fig. 5.148. Intensity distributions (negative) for the generated light 'bottles' at a distance of 790 mm (a), 810 mm (b), 830 mm (c), 850 mm (d), 870 mm (e) and 890 mm (f) from plane of the element.

The size of this area can be significantly increased, and the ratio of the minimum intensity to maximum reduced if to the above layout of the beams we add one more compound Bessel beam (Fig. 5.149), with a coefficient of expansion C_p of 2.4 assumed to be equal to $1 \cdot e^{i\pi}$.

The introduction of such a beam does not change the nature of a binary phase distribution (Fig. 5.150), but will lead to destructive interference.

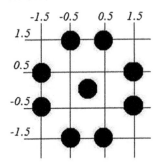

Fig. 5.149. Arrangement of the superposition of Bessel beams in the case of forming a light 'bottle'

The result is an increase in the size of the area of low intensity (Figs. 5.151 and 5.152) and due to this the ratio of the minimum to maximum intensity decreases from 0 to 1.

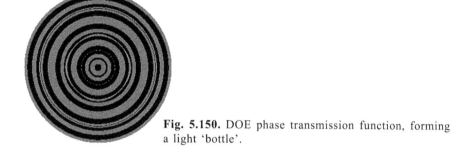

Fig. 5.150. DOE phase transmission function, forming a light 'bottle'.

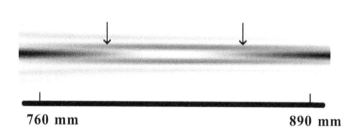

760 mm   890 mm

Fig. 5.151. Intensity distribution (negative) for the produced light 'bottle' along the optical axis. Arrows indicate the region of formation of the light 'bottle'.

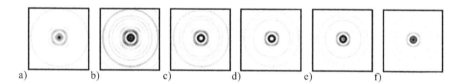

a)   b)   c)   d)   e)   f)

Fig. 5.152. Intensity distributions (negative) for the generated light 'bottle' at a distance of 770 mm (a), 790 mm (b), 810 mm (c), 830 mm (d), 850 mm (e) and 870 mm (f) from plane element.

In some cases, which will be discussed further, the introduction of such additional beams with a phase shift allows to remove the intensity peaks, which arise in the central part of traps due to interference between the beams forming the outer boundary of the light 'bottles'.

The efficiency of formation of traps using the calculated elements is calculated using the formula (5.92). Then for the different planes remote from the element at a distance of 770 mm, 790 mm, 810 mm, 830 mm, 850 mm, 870 mm we obtain efficiencies of 8%, 9%,

11%, 12%, 13%, 15%, respectively, the average efficiency is equal to 11.33%.

The same procedure can be used to calculate the phases of the elements forming the single light 'bottle', whose cross section will have a shape different from circle. In these cases, the location point of formed beams in respective circuits may be symmetrical or asymmetrical with respect to the origin of the coordinates. If in the first of these cases the transmission function of the phase distribution will have a binary form, then in the second phase it will be multi-level. To demonstrate this, we consider two more cases of the formation of single light 'bottles' – in the first of them the external border of the trap in the section will have the form of the square, in the second – an equilateral triangle. Note that due to the lack of radial symmetry in these figures the coefficients C_p in the expansion (5.4) are different with each other. Therefore, to calculate them it is convenient to use one of many currently existing iterative algorithms. Next to this, we consider the developed algorithm based on the gradient method.

5.5.4. Algorithm for increasing the uniformity of light traps formed on the basis of the gradient procedure

The gradient methods have repeatedly been used in several studies to calculate the self-replicating multimode laser beams [130], multiorder gratings [133–137], composite DOE [138], and the quantized DOEs [139]. Thus, it can be concluded that these methods are widely used for the calculation of the diffraction elements. The essence of the gradient methods is to minimize the functional, which is usually calculated as an error in determining the distribution of the light field in any plane.

Let $I_0(x, y)$ be the reference intensity distribution of the formed light 'bottles' in a plane at a distance z from the element. This distribution can be set both analytically and with the help of a graphic file. Let the number of points that make up this image is N. The coordinates of these points will determine the axial displacement of the superposition of three Bessel beams of 0th order with parameters $a_1 = 21.85 \cdot 10^3$ m^{-1}, $a_2 = 17.08 \cdot 10^3$ m^{-1}, $a_3 = 10.31 \cdot 10^3$ m^{-1}, which form the outline of the trap. As noted above, such a superposition allows to generate on the axis of the beam a light spot by removing almost entirely the concentric circles due to interference. To calculate the light field formed by the DOE at a distance z from it we can use the Fresnel integral transform (5.145).

The task of finding the phase $T(x, y)$ and respective C_p arguments in the expression (5.144) for the superposition we can formulate as a problem of minimizing the following quadratic criterion:

$$K = \frac{\iint\limits_D \left[\left| B(u,v) \right| - A_0(u,v) \right]^2 dudv}{\iint\limits_D A_0^2(u,v) dudv} \qquad (5.146)$$

where $A_0(u,v) = \sqrt{I_0(u,v)}$, D is the limited area in which the the error is calculated.

To minimize this criterion we can use the gradient method. In the first iteration, all coefficients C_p^1 are equal to 1. Next, the formula (5.145) is used for the Fresnel conversion of the complex DOES transmittance function calculated by the formula (5.144) to get the distribution of the light field formed by this element at a distance z from the input plane.

On the basis of the given and received distributions of the light fields we calculate the error K. New coefficients $\omega_p^{j+1} = \left| C_p^{j+1} \right|$ are calculated as follows:

$$\omega_p^{(j+1)} = \omega_p^{(j)} + \mu^{(j)} \frac{\partial K}{\partial \omega_p^{(j)}} \qquad (5.147)$$

where j is the number of iterations.

To calculate the gradient $\dfrac{\partial K}{\partial \omega_p^{(j)}}$ we used the following numerical procedure: value of the pth coefficient $\omega_p^{(j)}$ was increased by a small amount Δ, while all other factors remain unchanged. Then, based on this vector of the coefficients $\omega_p^{(j)}$ we calculated the Fresnel transform of the resultant complex transmission function $T(x, y)$ of the DOE. The result is a new estimate of the error $K_p^{(j)}$ and is then used to calculate the right-hand derivative, i.e.,

$$\frac{\partial K}{\partial \omega_p^{(j)}} = \frac{K_p^{(j)} - K^{(j)}}{\Delta} \qquad (5.148)$$

In order to make sure that the values of the coefficients $\omega_p^{(j+1)}$ are not negative as a result of calculation by the above formula, in

(5.147) we impose restrictions on the step $\mu^{(j)}$ which was selected at each iteration as follows:

$$\mu^{(j)} = \frac{\min_p \left\{ \dfrac{\omega_p^{(j)}}{\partial K \Big/ \partial \omega_p^{(j)}} \right\}}{\gamma \cdot \eta^{j-1}} \tag{5.149}$$

Here the parameters γ and η are responsible for the rate of convergence of the algorithm. It is recommended to choose the values of the following periods: $2 \leq \gamma \leq 3$ and $1.1 \leq \eta \leq 2$.

Further, based on the calculated values of the coefficients $C_p^{(j+1)}$ according to (2.4) we calculate the new estimnate $T^{(j+1)}(x, y)$ of the complex function of the DOE and a new iteration of the algorithm starts.

Note that when using arbitrary number as initial values of coefficients C_p^i due to the fact that the gradient methods are sensitive to the initial approximation and because of the specificity of the calculation step $\mu^{(j)}$, we were unable to obtain satisfactory results. Thus, as an initial evaluation of the coefficients C_p^i it is recommended to select constant values.

With this algorithm, we calculated some elements that form the light 'bottle' with the outer boundary of the 'bottle' the form of a square, an equilateral triangle and a rectangle with an aspect ratio of 2 to 1. To analyze the intensity distributions generated by these traps, we used in the following quantities: the effectiveness of the formation of the light 'bottle', calculated according to the formula (5.92), and the uniformity of the intensity distribution along the contour:

$$U = 1 - \frac{\sum\limits_{m=1}^{M} \left| I_m - \overline{I} \right|}{M \cdot \overline{I}}, \tag{5.150}$$

where I_m is the intensity at the point m, M is the number of measurement points $\overline{I} = \dfrac{1}{M} \sum\limits_{m=1}^{M} I_m$, is the average value of the intensity of the contour.

Figure 5.153 shows the layout of complex beams to form a light 'bottle' with the section of the external border of the 'bottle' in the form of a square contour. The total number N of single-mode beams involved in the calculation of the transmission function according to (5.144) is equal to 48.

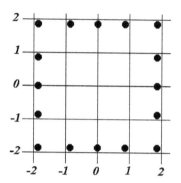

Fig. 5.153. Arrangement of the superposition of Bessel beams to form a light 'bottle', the outer boundary of which is in the form of a square.

The initial values of coefficients C_p^1 for $1 \le p \le N$ is chosen equal to 1. The radius of the DOE calculated – 4mm. Distribution of a light field formed in the first iteration shown in Fig. 5.154.

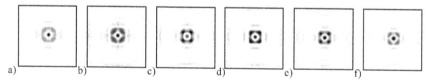

Fig. 5.154. Intensity distributions (negative) formed on the first iteration, at distances of 760 mm (a), 780 mm (b), 800 mm (c), 820 mm (d), 840 mm (e) and 860 mm (f) from the plane of the element in the case of the calculation of the light 'bottle' for the scheme in Fig. 5.153.

As a result of six iterations of the above algorithm we derived a complex transmission function (Fig. 5.155).

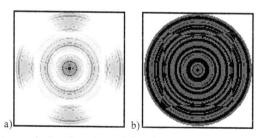

Fig. 5.155. The transmission function of the element forming the light 'bottle', the outer boundary of which is in the form of a square loop: the amplitude (negative) (a) and phase (b).

The results of Fresnel diffraction of a plane wave with a wavelength of 532 nm on the DOE are shown in Figs. 5.156 and 5.157. As can be seen, a single light 'bottle' of good quality forms. The value of the uniformity of the intensity distribution for the cross section at a distance of 800 mm from the plane of the element, calculated by the formula (5.150), is equal to 0.89. The average efficiency of the formation of traps is 9%.

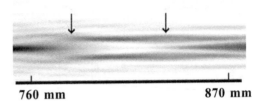

760 mm   870 mm

Fig. 5.156. The intensity distribution (negative) for the formed light 'bottle', the outer boundary of which has the shape of the contour of the square, along the optical axis. Arrows indicated the region of formation of the light 'bottle'.

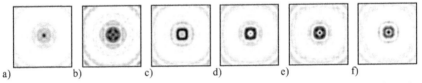

a)   b)   c)   d)   e)   f)

Fig. 5.157. Intensity distribution (negative) for the generated light 'bottle', the outer boundary of which has the shape of a square, at distances of 760 mm (a). 780 mm (b), 800 mm (c), 820 mm (d), 840 mm (e) and 860 mm (f) from the plane of the element.

Similar traps have been calculated the outer boundary of which had the form of an equilateral triangle and a rectangle. Arrangement of the superposition of Bessel beams to form a light 'bottle', the outer boundary of which has the shape of a triangle is shown in Fig. 5.158. The total number N of single-mode beams involved in the calculation of the transmission function according to (5.144) is equal to 45.

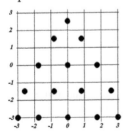

Fig. 5.158. Arrangement of the superposition of Bessel beams of light to form a 'bottle', the outer boundary of which has the shape of an equilateral triangle.

The initial values of coefficients C_p^1 for $1 \leq p \leq N$ are chosen equal to 1. The radius of the calculated DOE was 3 mm. 16 iterations were performed in the calculations. The corresponding distribution of the light field in the first iteration, the calculated transmission function and the resulting distribution of the light field in the last iteration are represented, respectively, in Figs. 5.159, 5.160, 5.161, 5.162. The value of the uniformity of the intensity distribution at a distance of 820 mm was equal to 0.83. The average value of the efficiency was 11%.

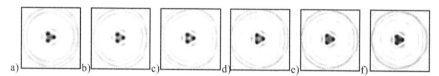

a)   b)   c)   d)   e)   f)

Fig. 5.159. Intensity distributions (negative) formed in the first iteration, at distances of 760 mm (a), 780 mm (b), 800 mm (c), 820 mm (d), 840 mm (e) and 860 mm (f) from the plane element in the case of the calculation of the light 'bottle' for the scheme in Fig. 5.158.

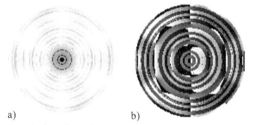

a)   b)

Fig. 5.160. The transmission function of the element forming the light 'bottle', the outer boundary of which has the shape of an equilateral triangle: the amplitude (negative) (a) and phase (b).

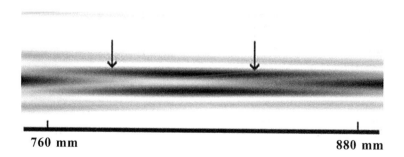

760 mm   880 mm

Fig. 5.161. Intensity distributions (negative) for the formed light 'bottle', the outer boundary of which has the shape of an equilateral triangle, along the optical axis. Arrows indicated the region of formation of the light 'bottle'.

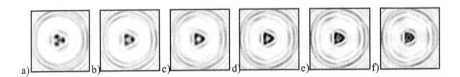

Fig. 5.162. Intensity distributions (negative) for the formed light 'bottle', the outer boundary of which has the shape of an equilateral triangle, at distances of 750 mm (a), 780 mm (b), 800 mm (c), 820 mm (d), 840 mm (e) and 860 mm (f) from the plane of the element.

In the case of forming the light 'bottle', which has a cross section in the form of a rectangle with an aspect ratio of 2 to 1, corresponding to the arrangement of beams, the transmission function and the intensity distribution of the formed light fields are shown in Figs. 5.163–5.167, respectively. 5 iterations of the above procedure were performed in the calculations. For a given trap the value of the uniformity of the intensity distribution at a distance of 830 mm is equal to 0.87, the average efficiency is 7%.

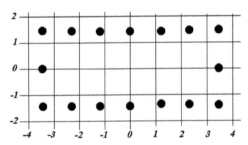

Fig. 5.163. Arrangement of the superposition of Bessel beams to form a light 'bottle', the outer boundary of which has the shape of the contour of a rectangle.

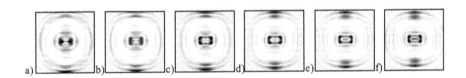

Fig. 5.164. Intensity distribution (negative) formed in the first iteration, at distances of 760 mm (a), 780 mm (b), 800 mm (c), 820 mm (d), 840 mm (e) and 860 mm (f) from the plane element in the case of the calculation of the light 'bottle' for the scheme in Fig. 5.163.

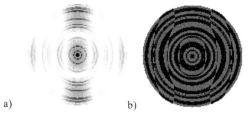

a)   b)

Fig. 5.165. The transmission function of the element forming the light 'bottle', the outer boundary of which has the shape of the contour of a rectangle: the amplitude (negative) (a) and phase (b).

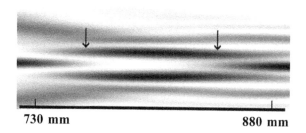

730 mm   **880 mm**

Fig. 5.166. Intensity distributions (negative) for the formed light 'bottle', the outer boundary of which has the shape of the contour of a rectangle along the optical axis. Arrows indicated the region of formation of the light 'bottle'.

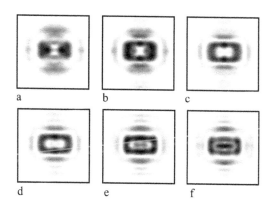

a   b   c

d   e   f

Fig. 5.167. Intensity distributions (negative) for the formed light 'bottle', the outer boundary of which has the shape of the contour of a rectangle, at distances of 770 mm (a), 790 mm (b), 810 mm (c), 830 mm (d), 850 mm (e) and 870 mm (f) from the plane of the element.

From a comparison of the intensity distributions shown in Figs. 5.156, 5.161 and 5.166 it can be seen that light 'bottles' of different shapes are formed at slightly different distances from the element; this can be explained by different transverse dimensions of the traps so that the light 'bottles' of different shapes will form at slightly different distances.

5.5.5. Formation of arrays of light 'bottles'

The above method of forming the light 'bottles' through the use of superpositions of the Bessel beams can be used not only for the formation of light 'bottles' with the external border in the section being different from the ring, but also for the formation of arrays of such traps. Earlier, the formation of arrays of light 'bottles' was seen only in a small number of works, so in [140] a generalized method of phase contrast was used for this purpose, and [141] composite holograms forming several self-replicating Bessel beams remote from each other were formed. In the first of these works the light field was formed by a spatial light modulator.

Figures 5.168 and 5.169 show the layout of the superposition of Bessel beams of 0th order with the parameters $\alpha_1 = 21.85 \cdot 10^3$ m^{-1}, $\alpha_2 = 17.08 \cdot 10^3$ m^{-1}, $\alpha_3 = 10.31 \cdot 10^3$ m^{-1} in the case of formation of two or three light 'bottles' with their boundaries touching in a plane in which the transverse dimension of the region with the minimum intensity is maximal. The radius of the calculated elements was 3 mm, the wavelength of the incident radiation 532 nm.

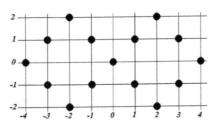

Fig. 5.168. Arrangement of the superposition of Bessel beams to form two light 'bottles'.

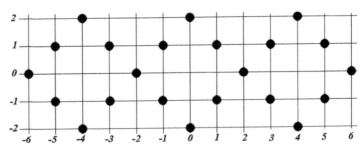

Fig. 5.169. Arrangement of the superposition of Bessel beams to form three light 'bottles'.

As can be seen from Fig. 5.168 two light bottles were formed using the circuit with 15 compound Bessel beams, thus, the total number of single-mode beams participating in the construction of traps is 45. Table 5.8 shows the calculated coefficient values C_p (x, y) for each of the individual compound Bessel beams.

Table 5.8. The values of coefficients C_p in the case of the formation of two light 'bottles'

The coordinates of the centre of the beam	Coefficient C_p	Coordinates of the centre of the beam	Coefficient C_p
(1;1)	1	(–2;–2)	0.73
(1;–1)	1	(–2;2)	0.73
(–1;1)	1	(2;–2)	0.73
(–1;–1)	1	(2;2)	0.73
(0;0)	0.82	(3;–1)	0.73
(–4;0)	0.73	(3;1)	0.73
(–3;–1)	0.73	(4;0)	0.73
(–3;1)	0.73		

As a result of the symmetric about the origin location of the beams, the phase calculated by the formula (5.144) of the complex transmission function takes the binary form (Fig. 5.170 b).

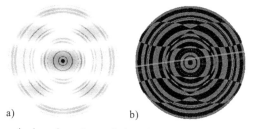

a)   b)

Fig. 5.170. Transmission function of the element forming two light 'bottles': amplitude (negative) (a) and phase (b).

If you ignore the amplitude component, such a phase element will form the light field, the distribution of which will have the form shown in Fig. 5.171. As can be seen form the light, 'bottle' with the help of such an element is not obtained.

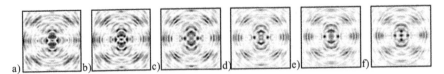

Fig. 5.171. Intensity distributions (negative) generated by the phase function in Fig. 5.170 b at distances of 790 mm (a), 810 mm (b), 830 mm (c), 850 mm (d), 870 mm (e) and 890 mm (f) from the plane of the element.

The method for coding the amplitude, as described in the previous chapter, allows to transfer from the amplitude–phase transmission function of the DOE to only the phase function. Figure 5.172 shows the phase of the encoded element. The amplitude level above which the encoding operation is not performed is equal to 0.1. Encoding is performed on both spatial coordinates.

Fig. 5.172. Coded transmission function of the phase element, forming two light 'bottles'.

The distribution of the light field formed by such an element in the space is shown in Figs. 5.173 and 5.174. It can be seen that the two light 'bottles' form. The average efficiency is equal to 8.5%. The value of the uniformity of the intensity distribution at a distance of 840 mm is equal to 0.80.

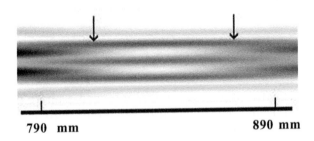

790 mm   890 mm

Fig. 5.173. Intensity distributions (negative) for the two generated light 'bottles' along the optical axis. Arrows indicated the region of formation of light 'bottles'.

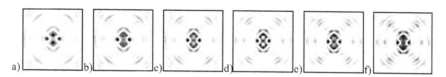

a) b) c) d) e) f)

Fig. 5.174. Intensity distributions (negative) for the formed light 'bottle', the outer boundary of which is triangular in shape, at distances of 780 mm (a), 800 mm (b), 820 mm (c), 840 mm (d), 860 mm (e) and 880 mm (f) from the plane of the element.

Table 5.9 shows the calculated coefficient values C_p (x, y) in the case of forming three light 'bottles'.

Table 5.9. The values of coefficients C_p in the case of the formation of the three light 'bottles'

The coordinates of the centre of the beam	Coefficient C_p	Coordinates of the centre of the beam	Coefficient C_p
(-6;0)	0,6	(0;2)	1,0
(-5;-1)	0.7	(1;-1)	1.0
(-5;1)	0.7	(1;1)	1.0
(-4;-2)	0.8	(2;0)	0.9
(-4;2)	0.8	(3;-1)	0.9
(-3;-1)	0.9	(3;1)	0.9
(-3;1)	0.9	(4;-2)	0.8
(-2;0)	0.9	(4;2)	0.8
(-1;-1)	1.0	(5;-1)	0.7
(-1;1)	1.0	(5;1)	0.7
(0;-2)	1.0	(6;0)	0.6

In this case due to the symmetrical about the origin arrangement of the beams the phase calculated by the formula (5.144) the transmission function of the DOE has a binary form. The coding of the amplitude at the level of 0.035 (carried out by one of the spatial coordinates) resulted in the phase distribution for the element showsn in Fig. 5.175.

Fig. 5.175. Coded transmission function of the phase element, forming three light 'bottles'.

The distribution of the light field generated by such an element of is shown in Figs. 5.176 and 5.177. The average value of the efficiency is 10%, the values obtained for the uniformity of the intensity distribution at a distance of 840 mm is 0.77.

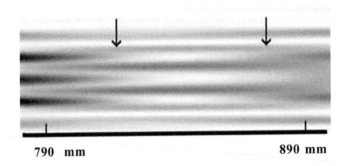

790 mm   890 mm

Fig. 5.176. The intensity distributions (negative) for the three-formed light 'bottles' along the optical axis. Arrows indicated the region of formation of light 'bottles'.

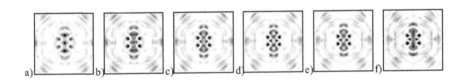

Fig. 5.177. Intensity distribution (negative) for the three-formed light 'bottles' at distances of 780 mm (a), 800 mm (b), 820 mm (c), 840 mm (d), 860 mm (e) and 890 mm (f) from the plane element.

With a further increase in the number of traps using this approach it is difficult to achieve a good combination of the diffraction efficiency and the quality of the formation of light 'bottles'. Increasing the value of the efficiency of formation of traps and the quality of their formation at their number $M > 3$ can be achieved if they are not placed in a row but in a few rows. We consider this approach in more detail on the example of simple light 'bottles' with a ring-shaped cross section.

To create a single light 'bottle', we used a Bessel beam of the 0th order, which was formed by an axicon with parameters $R = 3$ mm, $\alpha = 21.85 \cdot 10^3$ m^{-1}, $\lambda = 532$ nm; its the transmission function is calculated according to the formula (5.3). If we consider only the phase distribution, the transmission function of such DOE will have the form shown in Fig. 5.178.

Fig. 5.178. Phase transmission function of the element forming the light 'bottle'.

In illuminating this element with plane wave with a wavelength $\lambda = 532$ nm and a radius of 3 mm self-replicating light 'bottles' will form along the axis of the beam. The distribution of the light field shown in Figs. 5.179 and 5.180 shows that one of the 'bottles' is formed on the portion of the optical axis between 500 and 524 mm.

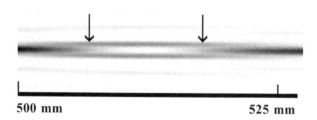

500 mm   **525 mm**

Fig. 5.179. The intensity distributions (negative) for the formed light 'bottle' along the optical axis. Arrows indicated the region of formation of the light 'bottle'.

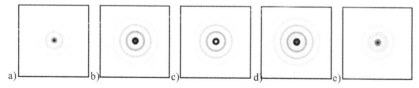

a)   b)   c)   d)   e)

Fig. 5.180. Intensity distributions (negative) for the light 'bottle' formed at distances of 502 mm (a), 507 mm (b), 512 mm (c), 517 mm (d) and 522 mm (e) from the plane of the element.

If several such light beams are produced simultaneously, then with the proviso that the adjacent beams will have a phase shift equal to π, we can produce an array of light 'bottles' with the distance between them smaller than the traps themselves. This approach is similar to that reviewed in [141], wherein an array of light 'bottles' was formed using composite holograms, but the distance between the traps formed was several times greater than the transverse dimensions of traps, which limits the use of such complicated traps.

The transmission function of the elements forming several such beams was calculated using the formula (5.144). To produce phase elements the amplitude component in the case of the number of generated traps $M < 3$ was simply ignored; when $M \geq 3$ the amplitude was encoded by the previously described partial encoding. In the case of the three traps the amplitude level above which the encoding operation was not applied was 0.2; in the case of the five traps it was 0.27. As a result, we obtained the following distributions of the phase functions of the elements forming two, three and, respectively, five light bottles (Fig. 5.181). The binary form of the phase distributions as well as in the above example was due to symmetrical arrangement of the formed beams relative to the axis.

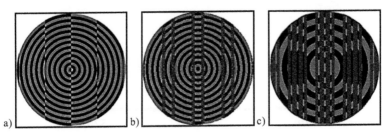

Fig. 5.181. Phase transmission functions of the elements forming the two (a), three (b) and five (c) light 'bottles'.

The distributions of light fields obtained by simulation of the passage of a plane wave with a wavelength λ through the DOE, forming two or three light bottles, are shown in Figs. 5.182–5.185, respectively.

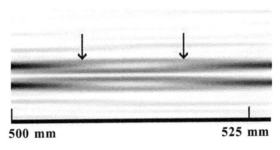

500 mm   **525 mm**

Fig. 5.182. The intensity distributions (negative) for the two generated light 'bottles' along the optical axis. Arrows indicated the region of formation of light 'bottles'.

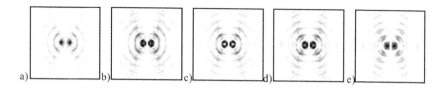

Fig. 5.183. Intensity distributions (negative) for two light 'bottles' formed at distances of 502 mm (a), 507 mm (b), 512 mm (c), 517 mm (d) and 522 mm (e) from the plane of the element.

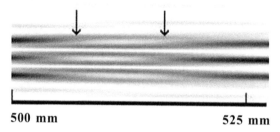

500 mm   525 mm

Fig. 5.184. Intensity distributions (negative) for the three-formed light 'bottles' along the optical axis. Arrows indicated the region of formation of light 'bottles'.

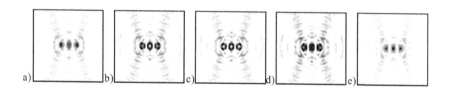

Fig. 5.185. Intensity distributions (negative) for three light 'bottles' formed 502 mm (a), 507 mm (b), 512 mm (c), 517 mm (d) and 522 mm (e) from the plane of the element.

The average efficiency of the formation of the corresponding fields, calculated according to the formula (5.92), is equal to: in the case of two light 'bottles' 6.0%, 5.5% for the three bottles. In the case of the formation of five light 'bottles' by this procedure the traps in the area from 500 to 525 mm are distorted, but due to the fact that we are considering the self-replicating beams they then form in the segment of the beam axis from 880 mm to 960 mm, as shown in Figs. 5.186 and 5.187. The effectiveness of their formation is equal to 8%.

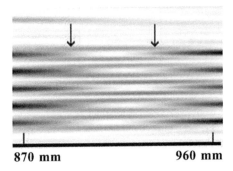

870 mm   960 mm

Fig. 5.186. The intensity distributions (negative) for the five-formed light 'bottles' along the optical axis. Arrows indicated the region of formation of light 'bottles'.

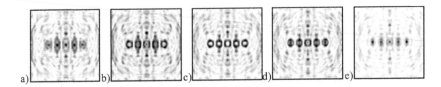

Fig. 5.187. Intensity distributions (negative) for five formed light 'bottles' at distances of 880 mm (a), 900 mm (b), 920 mm (c), 940 mm (d) and 960 mm (e) from the plane of the element.

Now consider how the effectiveness is influenced by the fact that the light 'bottles' will be formed in two rows. For this we consider the element forming six traps in the same plane, arranged in two rows, the spacing between rows will be commensurate with the distance between the traps in each of them. When forming such a trap in a single row, in order to ensure that the formed traps have the same energy characteristics, it is necessary to perform coding of the amplitude with a sufficiently high amplitude level above which the encoding operation is not applied, which reduces the efficiency of their formation. The transmission function of the element has the form shown in Fig. 5.188. The encoding amplitude level is equal to 0.3.

Fig. 5.188. Phase transmission function of the element forming six light 'bottles' in two rows.

The generated light field distributions are shown in Figs. 5.189 and 5.190. The efficiency calculated according to the formula (5.92) is 12.5%.

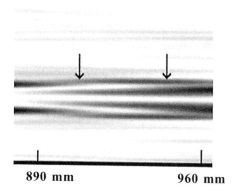

890 mm   960 mm

Fig. 5.189. Intensity distributions (negative) for six light 'bottles' formed in two rows along the optical axis.

Arrows indicated the region of formation of light 'bottle'

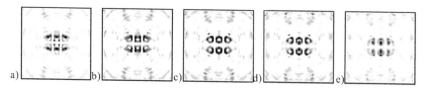

Fig. 5.190. Intensity distributions (negative) for six light 'bottles' formed in two rows at distances of 880 mm (a), 900 mm (b), 920 mm (c), 940 mm (d) and 960 mm (e) from the plane of the element.

Thus, it was possible to increase the efficiency by 4.5% compared with the case of forming five light 'bottles' disposed in a row by 4.5%. It is worth noting that in this case some of the traps are distorted in one of the directions along the optical axis. The presence of these distortions leads to the fact that the light 'bottle' is not closed completely, but due to the fact that the size of the rings forming the trap is reduced, we can expect a stable capture of particles into such a trap.

5.5.6. *Formation of arrays of hollow light beams*

As noted earlier, the phases of the complex coefficients C_p in the expression (5.144) are responsible for the phase shift of the formed beams with respect to each other. Thus, if we produce N_{mp} composite Bessel beams of 0th order with the parameters $a_1 = 21.85 \cdot 10^3$ m^{-1}, $a_2 = 17.08 \cdot 10^3$ m^{-1}, $a_3 = 10.31 \cdot 10^3$ m^{-1}, and the values of the phases of the corresponding coefficients C_p will increase linearly, we can produce arrays of not light 'bottles' but of hollow light beams. Such beams may be regarded as universal traps for both transparent and non-transparent particles.

Consider the formation of two contiguous hollow light beams by means of the above approach to generate light 'bottle' and their arrays. As the layout of the individual components of Bessel beams will choose the scheme shown in Fig. 5.191.

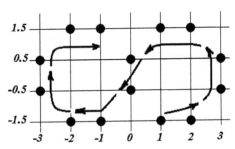

Fig. 5.191. Arrangement of the superposition of Bessel beams to form two hollow light beams.

The total number of single-mode beams participating in the formation of the light field is 42. The dotted arrow indicates the direction of the linear growth of the phases of the coefficients C_p for the respective beams. The phases of the coefficients C_p increases from 0 rad for beams with coordinates (1; −1.5) to the value of $52/14\pi$ for beams with coordinates (−1, 1.5). Unlike the case of the formation of arrays of light 'bottles', in the case of forming the hollow light beams we do not need to calculate the amplitudes of the coefficient C_p, in calculating all of them were chosen equal to 1. The radius of the DOE was 3 mm. As a result of the calculation of the transmission function according to (5.144), and the subsequent operations of coding the amplitude above 0.2 and binarization of the phase we obtained the DOE phase function shown in Fig. 5.192.

Fig. 5.192. transmission function member forming two hollow light beam.

The results of Fresnel diffraction of the plane light wave with a wavelength $\lambda = 532$ nm and a radius of DOE of 2 mm are shown in Fig. 5.193 and 5.194. The effectiveness of the formation of light traps, calculated in accordance with (5.146), is 34%.

700 mm   900 mm

Fig. 5.193. Intensity distributions (negative) for the two hollow light beams.

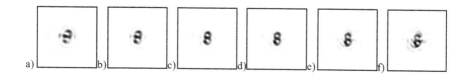

a)   b)   c)   d)   e)   f)

Fig. 5.194. Intensity distributions (negative) for the two formed hollow light beams at a distance of 700 mm (a), 740 mm (b), 780 mm (c), 820 mm (d), 860 mm (e) and 900 mm (f) from the plane of the element.

A similar procedure can be used to calculate the phase function elements forming a larger number of contiguous hollow light beams. If the coefficients C_p of the beams, which are located in symmetric about the origin orders, will be complex-conjugate, the phase distribution of the transmission function according to (5.144) is binary. For the cases of forming any even number of hollow light beams we can calculate the phase function of the binary type.

The schemes for the formation of three or more hollow light beams are formed from the scheme for the formation of the two beams by adding an additional six points for each new beam. Thus, in calculating the transmission function according to (5.144) for the case of three hollow beams the total number N of the used Bessel beams of 0th order is 60, in the case of four – 78, and so forth. The appropriate schemes are shown in Fig. 5.195.

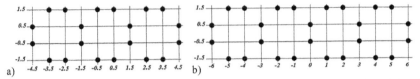

Fig. 5.195. Arrangement of the superposition of Bessel beams to form three (a) and four (b) hollow light beams.

As noted above, in the case of an odd number of the formed hollow light beams the phase distribution of the transmission function is not binary. To bring it to the binary form we need to use the encoding process of the phase by the grating with a carrier according to (5.95).

According to the scheme for the formation of hollow three light beams (Fig. 5.195 a), calculations by (5.144) and subsequent encoding operations of the amplitude above the 0.2 level and coding of the phase by the grating according to the carrier frequency $\beta =$ 10 000 lines/m gives a binary phase distribution (Fig. 5.196).

Fig. 5.196. Transmission function of the element forming three hollow light beams.

The radius of the calculated DOE was 3 mm. The result of Fresnel diffraction of the plane light wave with a wavelength $\lambda = 532$ nm and a DOE radius of 2 mm is shown in Figs. 5.197 and 5.198. Since the phase encoding is performed according to (5.95), two diffraction orders are formed, with each containing three hollow light beams. Calculations according to (5.92) for the total efficiency of both orders give a value of 31.5%.

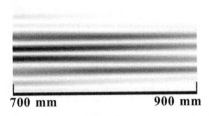

700 mm   900 mm

Fig. 5.197. Intensity distributions (negative) along the optical axis for the three hollow light beams.

Fig. 5.198. Intensity distributions (negative) for the three hollow light beams formed at a distance of 700 mm (a), 740 mm (b), 780 mm (c), 820 mm (d), 860 mm (e) and 900 mm (f) from the plane of the element.

As mentioned above, in the case of the formation of four hollow light beams the coefficients C_p can be chosen in such a way as to the phase distribution of the transmission function of the DOE, calculated according to (5.144), was binary form (Fig. 5.199 b). The radius of the DOE was 3 mm.

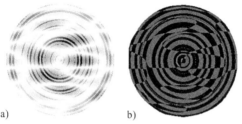

a)   b)

Fig. 5.199. The transmission function of the element forming four hollow light beams: amplitude (negative) (a) and phase (b).

The results of the Fresnel diffraction of the plane light wave with a wavelength $\lambda = 532$ nm and a radius of 2 mm on the amplitude-phase element is shown in Figs. 5.200 and 5.201. As can be seen, the hollow light beams have discontinuities.

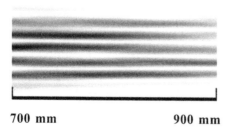

700 mm   **900 mm**

Fig. 5.200. Distributions of the intensity (negative) along the optical axis for the four hollow light beams formed.

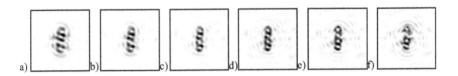

Fig. 5.201. Intensity distributions (negative) for the four hollow light beams formed at a distance of 700 mm (a), 740 mm (b), 780 mm (c), 820 mm (d), 860 mm (e) and 900 mm (f) from the plane of the element.

The quality of the beams generated in this case can be improved by changing the coefficients C_p. If we simply increase the phases of the coefficients C_p of the respective beams when moving on the contour of the scheme of their arrangement, the calculated distribution of the phase transmission function will not have a binary form (Fig. 5.202). The radius of the DOE is 3 mm.

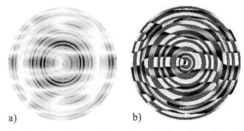

a)   b)

Fig. 5.202. Transmission function of the element forming four hollow light beams: amplitude (negative) (a) and phase (b).

As a result of encoding the amplitude above 0.2 and the subsequent binarization we obtain the phase transmission function of the DOE shown in Fig. 5.203.

Fig. 5.203. Phase transmission function of the element forming four hollow light beams.

The results of Fresnel diffraction of a plane light wave with a wavelength $\lambda = 532$ nm and a DOE radius of 2 mm are shown in Figs. 5.204 and 5.205.

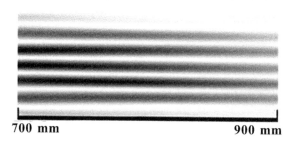

700 mm   **900 mm**

Fig. 5.204. Intensity distributions (negative) along the optical axis for the four light hollow beams formed.

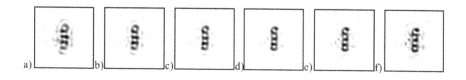

Fig. 5.205. Intensity distributions (negative) for the four hollow light beams formed at a distance of 700 mm (a), 740 mm (b), 780 mm (c), 820 mm (d), 860 mm (e) and 900 mm (f) from the plane of the element

In this case, the resultant intensity distribution is more uniform and has no discontinuities. Calculations according to (5.92) to give the value of the average efficiency of 8%. Thus, as in the case of forming the light 'bottles' by increasing the number of hollow light beams formed the efficiency of their formation is reduced.

5.5.7. Experimental formation of arrays of light 'bottles' with the binary DOE

The binary DOEs whose phase functions were previously calculated and are presented in Figs. 5.150, 5.172, 5.175, 5.178, 5.181, 5.196, 5.203, were fabricated by photolithography. The resolution in recording the template was 1 mm, the depth of etching the element 530 nm. The diameters of the produced elements were 4 and 4.4 mm. The appropriate three-dimensional images of different parts of the produced diffractive optical elements are shown in Fig. 5.206.

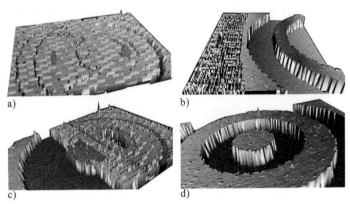

Fig. 5.206. Three-dimensional image of an area of the microrelief of the produced diffractive optical elements.

To assess the strength of capturing a microscopic object we use Stokes' law

$$F = 6\pi r \eta \upsilon \qquad (5.151)$$

where F is the force of viscous friction, which is equal to the force of optical trapping, r is the radius of the microsphere, η is viscosity, υ is the velocity of the entrapped microsphere. Since the micro-objects are manipulated in an aqueous solution, then η is equal 0.9 mPa·s.

For the experimental formation of arrays of light 'bottles' the authors of [142] used the optical scheme shown in Fig. 5.207. The diaphragm cut the beam to the desired size so that the diameter of the beam and the DOE matched. The laser wavelength was 532 nm.

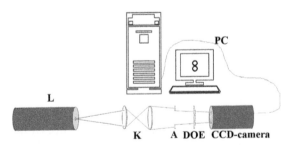

Figure 5.207. The experimental optical scheme for the formation of arrays of light 'bottles'.

Figure 5.208–5.210 shows the results of forming one, two and, respectively, three light 'bottles' using binary elements; their phase functions are shown in Figs. 5.150, 5.172, 5.175.

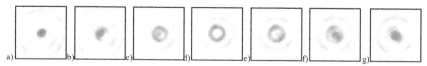

Fig. 5.208. Intensity distributions (negative) at different distances from the plane of the element forming a light 'bottle': a) 625 mm, b) 650 mm, c) 675 mm, d) 700 mm, e) 725 mm, f) 750 mm, g) 775 mm.

Fig. 5.209. Intensity distributions (negative) at different distances from the plane of the element, forming two light 'bottles': a) 775 mm, b) 825 mm, c) 850 mm, d) 875 mm, e) 900 mm, f) 925 mm, g) 975 mm.

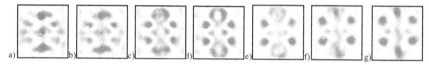

Fig. 5.210. Intensity distributions (negative) at different distances from the plane of the element forming three light 'bottles': a) 800 mm, b) 825 mm, c) 850 mm, d) 875 mm, e) 925 mm, f) 950 mm, g) 975 mm.

The results of Fresnel diffraction of a plane light wave on the corresponding elements are shown in Figs. 5.152, 5.174, 5.177. As can be seen, the formed images are in good agreement with the simulation results. Minor discrepancies in the distance at which the light 'bottles' form is due to the divergence of the laser beam.

Figure 5.211–5.215 show the results of forming one, two, three, five and , respectively, six light 'bottles' by the binary elements; their phase functions are shown in Figs. 5.178, 5.181.

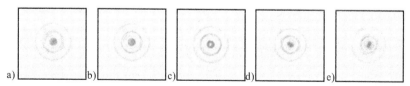

Fig. 5.211. Intensity distributions (negative) at different distances from the plane of the element forming a light 'bottle': a) 195 mm, b) 200 mm, c) 205 mm, d) 210 mm, e) 215 mm.

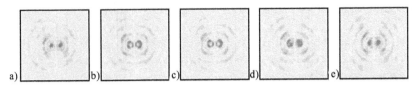

Fig. 5.212. Intensity distributions (negative) at different distances from the plane of the element forming two light 'bottles': a) 215 mm, b) 218 mm, c) 220 mm, d) 223 mm, e) 230 mm.

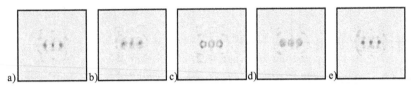

Fig. 5.213. Intensity distributions (negative) at different distances from the plane of the element forming three light 'bottles': a) 210 mm, b) 215 mm, c) 220 mm, d) 225 mm, e) 230 mm.

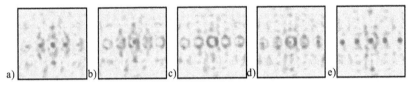

Fig. 5.214. Intensity distributions (negative) at different distances from the plane of the element forming five light 'bottles': a) 410 mm, b) 420 mm, c) 430 mm, d) 440 mm, e) 455 mm.

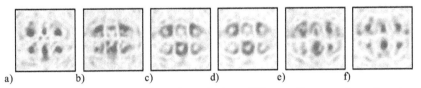

Fig. 5.215. Intensity distributions (negative) at different distances from the plane of the element forming six light 'bottles': a) 395 mm, b) 415 mm, c) 425 mm, d) 435 mm, e) 445 mm, f) 460.

The results of the Fresnel diffraction of a plane light wave on the elements with the given phase functions are shown in Figs. 5.180, 5.183, 5.185, 5.187, 5.190. The mismatch in the distances at which the trap is formed is due to the fact that in the calculation the diameter of the elements is taken as 6 mm, and the experiment were carried out with elements with a diameter of 4 mm. This was due to the need to simplify the input of the beam generated by the DOE in the focusing microlens. On the basis of the presented images it can be concluded that the quality of the light 'bottles' is high.

5.5.8. Capture of transparent micro-objects in the system of light 'bottle'

The light 'bottles' are intended to capture the opaque particles (more generally, for the capture of particles whose refractive index is lower than that of the environment), but as shown by the calculations in [143] using the method proposed in [52] they can also be used to manipulate transparent microscopic objects with the refractive index is higher than that of the environment.

As is known, the total momentum of the beam

$$\vec{P} = \frac{E}{c}\vec{A} \qquad (5.152)$$

where E is the energy in the light beam, c is the velocity of light, $\vec{A}$ is the unit vector defining the direction of light propagation.

If we need to calculate the force, the energy is divided by the duration t:

$$F = \frac{E}{tc}(1+k) = \frac{W}{c}[1+k] \qquad (5.153)$$

where W is the total power of the light beam incident on the surface.

On the basis of (5.152), the pressure of light at normal incidence on the material object is described by a simple formula

$$p = \frac{I}{c}(1+\chi) \qquad (5.154)$$

where I is intensity, χ is the reflectance of the surface.

If the light beam hits the refractive object of complex shape, force calculations are slightly more complicated (Fig. 5.216).

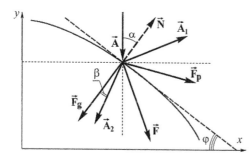

Fig. 5.216. Refraction of the beam on the surface with Fresnel reflection taken into account.

Assume, for simplicity, without the loss of generality, that originally the refractive surface is hit vertically with a single light beam with power P. Two beams form as a result: the reflected beam with power P_1 and the refracted one with power P_2. The reflected beam is the cause of the light pressure force, which can be determined by the formula

$$\vec{F}_p = -2\chi \frac{W}{c} \cos\alpha \vec{N} \qquad (5.155)$$

where the angle α is the angle of incidence, W is the power of the incident beam, $\vec{N}$ is the normal vector to the surface at the point of beam incidence, $\chi = \dfrac{\sin^2(\alpha - \beta)}{\sin^2(\alpha + \beta)}$ is the Fresnel reflection coefficient in the case of s-polarized light. s-polarization is not chosen at random: all capture effects, which will be described below, are most evident in the case of s-polarization, although they are also present for p-polarization. Equation (5.155) is obtained from (5.153) according to the law of conservation of momentum.

The refracted beam gives the gradient force (the law of conservation of momentum)

$$\vec{F}_g = 2(1-\chi)\frac{W}{c}\sin\frac{\alpha - \beta}{2} \cdot \frac{\vec{A} - \vec{A}_2}{\left|\vec{A} - \vec{A}_2\right|} \qquad (5.156)$$

where $\vec{A}$ is the unit vector in the direction of propagation of the reflected beam, $\vec{A}_2$ is the unit vector in the direction of propagation of the refracted beam, β is the angle of refraction.

The total force $\vec{F}$ is obtained by vector adding the light pressure force and gradient force $\vec{F}_g$

$$\vec{F} = \vec{F}_p + \vec{F}_g. \qquad (5.157)$$

As can be seen from Fig. 5.216, the projections of forces $\vec{F}_p$ and $\vec{F}_g$ on the x-axis can be pointing in different directions. This means that the projection of the total force $\vec{F}$ may be directed in the one or the other direction. For small values of φ – the angle of the tangent to the surface at the point of incidence to the axis x, the projection $\vec{F}_p$ on the x-axis is smaller than the projection $\vec{F}_g$ on the x-axis, and the projection of the total force on the x-axis is positive. But the increase of φ increases the reflection coefficient and from a certain angle, the sign of the projection of force $\vec{F}$ on the x-axis changes, i.e. the power of light pressure becomes dominant and displaces the particle from the beam. Consequently, the beams of the 'light bottle' type can be used to capture the transparent microparticles. When the beam radius is properly, the spherical transparent microparticle will be captured at the centre of the light 'bottle'.

Consider the case of the spherical microparticles analytically. As can be seen from Fig. 5.217, incidence angle α coincides with the angle between the vertical direction and at the entry point of the beam in the microparticle. Φ is the angle, in this case coincides with the angle φ in Fig. 5.217. Thus, the formula (5.155) is transformed into

$$\vec{F}_p = 2\chi \frac{W}{c} \cos\varphi \left(-\vec{N}\right) \tag{5.158}$$

where $\vec{N} = (\sin\varphi, \cos\varphi)$.

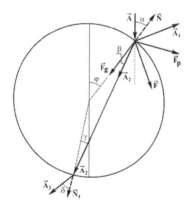

Fig. 5.217. The refraction of the light beam on a spherical microparticle.

For simplicity, we consider only the projection of forces on the x-axis

$$F_{px} = -2\chi \frac{W}{c} \cos\phi \sin\phi = -\chi \frac{W}{c} \sin 2\phi. \tag{5.159}$$

For the force $\vec{F}_g$ the projection on the x-axis can be written on the basis of vectors $\vec{A}_2 = (-\sin(\phi-\beta), -\cos(\phi-\beta))$ and the ratio

$$n_1 \sin\phi = n_2 \sin\beta, \tag{5.160}$$

wherein n_1 – refractive index, n_2 – the refractive index of the microparticle.

After substitution we obtain

$$F_{gx} = 2(1-k)\frac{P}{c}\sin\frac{\phi-\beta}{2} \cdot \frac{\sin(\phi-\beta)}{\sqrt{\sin^2(\phi-\beta)+(-1+\cos(\phi-\beta))^2}} =$$

$$= 2(1-k)\frac{P}{c}\sin\frac{\phi-\beta}{2} \frac{\sin(\phi-\beta)}{\sqrt{2(1-\cos(\phi-\beta))}} =$$

$$= 2(1-\kappa)\frac{P}{c}\sin\frac{\left(\phi-\arcsin\left(\dfrac{n_1}{n_2}\sin\phi\right)\right)}{2}\frac{\sin\left(\phi-\arcsin\left(\dfrac{n_1}{n_2}\sin\phi\right)\right)}{\sqrt{2\left(1-\cos\left(\phi-\arcsin\left(\dfrac{n_1}{n_2}\sin\phi\right)\right)\right)}}.$$

$$\text{(5.161)}$$

Finally, substituting the value of the Fresnel reflection coefficient, we obtain a formula for the projection of the gradient force on the *x*-axis

$$F_{gx} = 2\left\{1-\frac{\sin^2\left[\phi-\arcsin\left(\dfrac{n_1}{n_2}\sin\phi\right)\right]}{\sin^2\left[\phi+\arcsin\left(\dfrac{n_1}{n_2}\sin\phi\right)\right]}\right\}\frac{P}{c}\sin\frac{\phi-\arcsin\left(\dfrac{n_1}{n_2}\sin\phi\right)}{2}\cdot$$

$$\cdot\frac{\sin\left(\phi-\arcsin\left(\dfrac{n_1}{n_2}\sin\phi\right)\right)}{\sqrt{2\left(1-\cos\left(\phi-\arcsin\left(\dfrac{n_1}{n_2}\sin\phi\right)\right)\right)}}.$$

$$\text{(5.162)}$$

However, formula (5.162) calculates the gradient force only in view of the refractive index, while the refraction is effected on an additional surface (Fig. 5.217).

Of course, except for the second refractive index, there is also a reflection on the second border of the microsphere, but due to the specific geometry of the problem the angles of incidence and reflection are very close to the vertical and the account of the second reflection at a high complexity of adds nothing to the balance of forces. Taking into account the two refractions, the formula (5.156) is transformed into

$$\vec{F_g} = 2(1-k)\frac{P}{c}\sin\frac{\pm-\delta}{2}\cdot\frac{\vec{A}-\vec{A_3}}{\left|\vec{A}-\vec{A_3}\right|}. \qquad (5.163)$$

After transformations we obtain the final formula for the gradient force

$$F_{lx} = 2\left\{1 - \frac{\sin^2\left[\phi - \arcsin\left(\dfrac{n_1}{n_2}\sin\phi\right)\right]}{\sin^2\left[\phi + \arcsin\left(\dfrac{n_1}{n_2}\sin\phi\right)\right]}\right\}\frac{P}{c}\sin\frac{\phi - \arcsin\left(\dfrac{n_1}{n_2}\sin\phi\right)}{2}\cdot$$

$$\cdot\frac{\sin\left[2\left(\phi - \arcsin\left(\dfrac{n_1}{n_2}\sin\phi\right)\right)\right]}{\sqrt{2\left(1 - \cos\left[2\left(\phi - \arcsin\left(\dfrac{n_1}{n_2}\sin\phi\right)\right)\right]\right)}}.$$

For the calculation, consider the following parameters: $n_1 = 1.33$, $n_2 = 1.56$. These are refractive indices of the water and polystyrene, respectively. We sum up the projection of forces $\vec{F_g}$ and $\vec{F_p}$ on the x-axis, normalize the sum for the maximum value and construct a chart for it (Fig. 5.218, solid line).

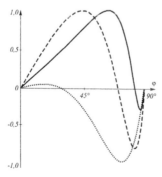

Fig. 5.218. Projection of the total force exerted on a spherical microparticle, depending on the value of the central angle ϕ of the beam entry point: the solid line for the refractive indices $n_1 = 1.33$, $n_2 = 1.56$, the dotted line — for the refractive indices $n_1 = 1$, $n_2 = 1.56$, dashed line — for the refractive indices $n_1 = 1.0$, $n_2 = 2.4$.

As can be seen from Fig. 5.218, the projection of force changes its sign if the central angle φ is approximately 83°, and reaches a maximum when the angle φ is approximately 87°. This means that whenthe diameter of the annular light is beam slightly smaller than the diameter of the spherical microparticles it will hold the transparent microparticle in its geometric centre. That is, if we correctly choose the size of the beam we obtain an universal trap for any given size of micro-objects. The radius of the beam at maximum beam intensity is related to the angle φ by the formula

$$R = r \, sin\phi \qquad\qquad (5.165)$$

where r – radius of the microparticles, R – the radius of the beam at maximum intensity. With the above parameters – the angle of the maximum capture φ = 87° and the radius of the microparticle 2.5 µm – the radius of the beam should be 2.49 µm, i.e., the beam and microparticle radii are virtually identical. The maximum value of the holding force with approximately 4 times less than the gradient holding force (Fig. 5.218, solid line).

It should be noted that an increase in the refractive index difference the value of this force increases while the angle φ, at which the force changes sign conversely decreases. Thus, when a polystyrene microparticle is in the air n_1 = 1.0, n_2 = 1.56, the maximum value is only 1.5 times smaller than the maximum value of the gradient force (Fig. 5.218, dotted line). And if we consider, for example, a diamond particle in air n_1 = 1.0, n_2 = 2.4, the capture force of light pressure will exceed the gradient force almost 10 times (Fig. 5.218, the dashed line). Each force in Fig. 5.218 is normalized to its maximum value.

The transparent particles are handled using the optical scheme shown in Fig. 5.127. The wavelength of the laser light used – 532 nm, the maximum average laser power – 2000 MW.

Figure 5.219 presents stages of the movement of a capture polystyrene microsphere in a single light 'bottle'. The time intervals between individual frames are 5 s. The average speed of the microparticles was 0.60 ± 0.04 µm/s. The capture force, calculated in accordance with (5.151), was 0.025 ± 0.002 pN.

Figure 5.220 shows stages of the movement of two captured polystyrene microspheres in the two light 'bottles'. The time interval between individual frames was 5 s. The average speed of the microparticles was 1.00 ± 0.04 µms. The capture force, calculated in accordance with (5.151), was 0.042 ± 0.002 pN.

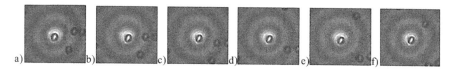

Fig. 5.219. Stages of movement of a polystyrene microsphere trapped in a solitary light 'bottle'. The time interval between frames 5 s.

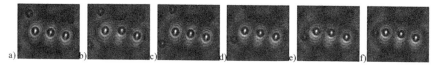

Fig. 5.220. Stages of movement of two polystyrene microspheres trapped in two light 'bottles'. The time interval between frames 5 s.

Figure 5.221 presents stages of movement of three captured polystyrene microspheres in three light 'bottles'. The time interval between individual frames was 5 s. The average speed of the microparticles was 0.7 ± 0.04 μm/s. The capture, calculated in accordance with (5.151), was 0.030 ± 0.002 pN.

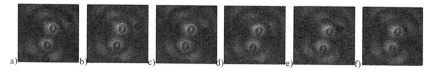

Fig. 5.221 Stages of movements of three polystyrene microspheres trapped in the three light 'bottles'. The time interval between frames 5 s.

Figure 5.222 presents the results of capture of polystyrene microspheres in a system of six light 'bottles'.

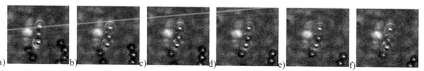

Fig. 5.222. Stages of movement of three polystyrene microspheres trapped in the optical field of six light 'bottles'. The time interval between frames 4 s.

It was not possible to capture six microspheres at the same time: some the microparticles fell out of traps, because of the division of radiation energy between the six traps the absolute value of capture strength was particularly low for transparent micro-objects. Therefore, Fig. 5.222 presents stages of movement of only three polystyrene microspheres. The time interval between individual frames was 4 s. The average speed of the microparticles was 0.50 ± 0.04 μm/. The capture force, calculated in accordance with (5.151), was 0.021 ± 0.002 pN.

Thus, it is clear that the light 'bottles' can actually be used to capture and move the transparent micro-objects.

5.5.9. Capture and displacement of metallic tin microparticles with the shape close to spherical

Commercially available metal powders contain very irregularly shaped microparticles. Optical trapping of such microparticles is associated with difficulties of purely technical nature. Tin microparticles were produced for the experiments to capture the microparticles in the opaque light 'bottles' by creating an electric arc between two tin electrodes near the surface. During their fall the microparticles of the molten metal gained nearly completely spherical shapes and froze as if splashed in the water. Thus, it was possible to obtain a water slurry of spherical microparticles of a size of 1 to 20 μm. The optical trapping and movement of the tin microparticles were carried out using the optical circuit shown in Fig. 5.127. The beam was focused to produce a little 'bottle' of the correct size using a ×20 microscope objective. The power of the beam was about 1 watt. the wavelength of laser light used 532 nm.

Unfortunately, the above-described method for producing metal microparticles provides only a relatively low concentration of the microparticles in the slurry and a large variation in the parameters, so it was difficult to capture two approximately equal-sized microparticles. Figure 5.223 presented stages movement at intervals of 2 st of one of the tin microparticles with a diameter of about 3 μm in a double light 'bottle'. The microparticle was alternately captured and moved in each of the two areas of low intensity, which proved the efficiency of both traps in the beam.

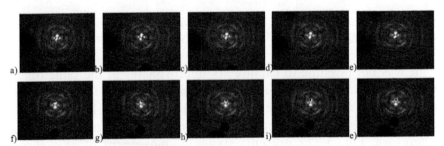

Fig. 5.223. Phases of displacement of tin microparticles of a diameter of 3 μm in a double light 'bottle'. The time interval between frames 2 s.

The average travel speed was 1.5 μm/s. Assuming that the particle has a shape close to spherical, and, using the formula (5.151), we obtained the value of the capture force of 0.038 pN.

5.5.10. The deposition and positioning of micro-objects using arrays of hollow light beams

If the manipulation of micro-objects is carried out in a cell containing a suspension of microparticles, at the bottom of the cell the concentration of the microparticles is relatively small, whereas in the bottom layer their number is an order of magnitude higher. To create from the microparticles structures at the bottom of the cell one must use the hollow light beams retaining the structure during propagation in free space [69]. Arrays of such beams allow to solve the problem of simultaneous and parallel positioning of a large number of micro-objects.

Hollow light beams look like some light tubes only when we consider their free circulation after the DOE. In focusing these light beams becomes a kind of light funnel that is very convenient fpr precipitating the microparticles from the suspension to the bottom of the substrate (Fig. 5.224).

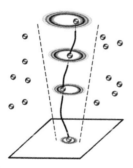

Fig. 5.224. Deposition of microparticles from an aqueous suspension on the optical substrate.

The use of several adjacent hollow beams can also place microparticles in the correct order simultaneously with their deposition.

To carry out experiments with precipitation and positioning of transparent micro-objected, we used the optical scheme shown in Fig. 5.127. L_1 – focusing microlens (×8), L_2 – imaging microscope objective (×20). Manipulation objects were selected transparent polystyrene microspheres with a diameter of 5 μm. The laser wavelength – 532 nm, the maximum average output power of the laser – 2000 MW.

Figure 5.225 shows the precipitation and positioning stages of polystyrene microspheres in a triple hollow beam. It is evident that with the help of the hollow beams the process of alignment of the microparticles in a line is fast. Thus, there is no need for large displacements of the trapping beam to a specific side, as microparticles themselves fall into the trap.

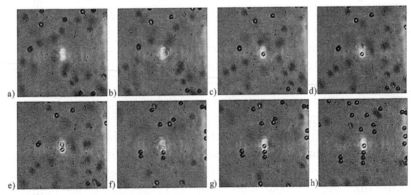

Fig. 5.225. Stages of the deposition and alignment into a line of a group of polystyrene microparticles: the initial state of the substrate (a), the beginning of the deposition of the first microparticle (b), the end of the deposition of the first microparticle (c), the beginning of the deposition of the second microparticle (d), the end of the deposition of the second microparticle (de), the beginning of the deposition of the third microparticle (f), the end of the deposition of the third microparticle (g), depositing the fourth microparticle (h). Moments of time of frames: a) 0 s; b) 4 s; c) 12 s; d) 16 s; d) 20 s; e) 50 s; g) 58 s; h) 88 s.

5.6. The light beams generated by the DOE for non-damaging capture of biological micro-objects

Manipulation of the individual cells plays an important role in problems in vitro fertilization, intercellular interactions, cell fusion, embryology, microbiology, study of stem cells and regenerative medicine. Studies of the behaviour of individual cells provide information about the chemical interactions of intracellular mechanical properties of the cells allow the study of the electromagnetic properties of the material of cells.

Despite the huge number of advantages over the contact techniques, optical trapping, however, can cause damage to the captured biomaterial [144], which in some cases is a highly undesirable effect.

Damage to the biological micro-objects remains the greatest problem in the development of technology of the optical tweezers. The main cause of damage also remains not fully clarified. Possible causes are called multiphoton absorption [145], the formation of free oxygen atoms [146] and the heat absorption by the solution surrounding the laser focus [147].

Research by Mirsaidov and his colleagues [148] examines in detail the problem of photodamage of the captured biomaterial in bacteria *E. coli*. Comparing multiple traps with time-division and multiple traps on the basis of 'holographic optical tweezers' HOT (similar techniques are used widely enough, see, e.g., [149], [150]), the authors conclude that in both cases the main role is played by the total energy absorbed by the cell. The results of this study show that critical to the *E. coli* cells is the amount of energy 5 J.

Summarizing various studies and damage factors discussed in them, it can be said that to minimize this damage we should select the correct laser wavelength – to avoid excessive absorption, watching the exposure time of the biological material within the optical trap and possibly limiting the radiant power dissipated in the object. In addition, the main risk of damage is not so much destruction and cell death (usually the power and wavelength of the exciting laser are such that cell death occurs only after tens of minutes), but the genetic damage, and also the distribution of the radiation inside the cell. For example, damage to the cell nucleus will affect its viability much stronger than the damage to less significant organelles.

Various modifications of the laser beam used for capture are subject to relentless interest from researchers. The existing literature on this subject can be classified according to two criteria: *object of* modification of the beam and *dynamics* the DOE used. There two main target areas of the modifications: the creation of systems of optical traps allowing to capture and manipulate multiple particles at the same time and giving some special properties to the beam. The DOE used here can be either dynamic (so-called HOT, or static – microrelief on a transparent plate or fiber (see, e.g., [151]). Of course, the purpose of the modification often determines whether the given DOE should be dynamic. Dynamic DOEs are usually controlled from the computer and can be easily changed in real time, that is their great advantage. The process of making static DOEs, at the same time, it is very complicated and in most of its stages is irreversible. On the other hand, the energy efficiency of dynamic DOEs – modulators – is small and is about 15%, so the static DOEs are preferred unless there are no dynamic changes of the beam profile.

A good representative of a series of studies using optical trapping with optical modulators are studies under the direction of Moradi [152], which shows the ability to create multiple traps with dynamic DOEs with control of the capture force in each unit trapped. Similar work was carried out a little earlier by Emiliani and his colleagues [153].

Special features of the trapped beam also have different purposes. Firstly, such modifications are made to achieve a more sustainable capture or fixed orientation of the captured objects. In this case, the beam cross section is usually comparable with the size of the captured object. Second, distributions different from Gaussian are created for sorting or conveying movement of the captured particles in space. This allows the optical manipulation with a stationary beam.

The first modification of the above type is often represented by different superposition modes of free space. For example, various orders of Laguerre–Gaussian modes are used for controlling the orientation and rotation of the captured particles, as shown for example in [154] and [155]. Of particular interest are the so-called 'optical bottle' – beams, the transverse intensity distribution in which has large areas of zero intensity. Study [156] presented a new type of beams – hollow hypersinus-Gaussian beams forming a clearly marked ring at the waist. Olson and his colleagues have developed a method for obtaining 'the optical tunnels' using genetic algorithms to find the coefficients of the spherical functions giving the necessary distribution in superposition [157].

In [158] the authors studied the properties of the beam which was a superposition of two axicons with opposite orders. It was noted that the beam can be used to capture objects of a certain size, since the capture is possible only starting from a certain size – dependent though on the order of the selected axicons.

The modes can be combined not only in the focal plane, but also in a more complicated matter. Developed in [159], a composition with a small offset of the Gaussian beam and Gauss–Laguerre modes (1.1) forms in the longitudinal section the intensity distribution similar to a horseshoe.

Sometimes the existing distributions are subject to small modifications with the help of phase DOEs. For example, in [160] the authors the Lorentz–Gauss beams and evaluated their possible use for optical trapping. It was shown that the Lorentz–Gaussian beams in the capture of the object in front of the focal plane provide a more stable capture.

Finally, a very important application of static DOEs are various optical conveyors – intensity distributions which move all the trapped particles in a certain direction, and the filters – distribution of intensity, trapping particles of only one type (one form, density, size, etc.) . In such schemes, the energy efficiency of optical components often plays a big role. As an example we mention [161] – note, however, that the modulators – DOEs in this study were used to create optical traps combinations allowing to control their particle collection and their further spatial distribution. It is also possible to use as an example eddy light fields to move micro-objects along a predetermined path without displacement of the beam [11, 162, 163].

Finally, some modifications of the beam do not require any complex modulators nor DOEs expensive to manufacture. Study [164] presented the elegant in its simplicity solution to change the rigidity of the trap by simply changing the exit pupil of the lens, projecting a beam in the object plane. It is shown that the profile of the potential well close to the maximum beam intensity is strongly dependent on the shape of the aperture.

This study deals with the modification of the optical beam used for capture so as to reduce the input power of the beam and at the same time ensure that the intensity distribution in the focal plane is such that most of the radiation fell on the periphery of the captured biological object. The genetic information whose damage should be avoided most of all is situated in the kernel, which in turn is often located near the geometric centre of the nucleus. For the object chosen in these experiments – *Saccharomyces cerevisiae yeast* – that is true in most cases, as can be seen from the study [165].

In this work, the intensity distribution in the laser beam is formed using a static micro-relief on a quartz plate, and for the theoretical estimation of the dynamics in the optical trap we used multibeam geometrical optics.

5.6.1. Modification of the Gaussian beam to optimize the power characteristics of the optical trap

We simulate the forces acting on a spherical object of radius 5 μm, placed in a Gaussian beam. The theoretical basis for this calculation are presented in [52] and [166], and there are confirmed by experiment. Simulation was conducted using a software package described in [167], which allows to calculate the forces acting in the light beam on an inhomogeneous light beam ellipsoidal micro-object. The simulation is based on the laws of geometrical optics which, although giving a fairly significant error in calculations, provides the computational efficiency to rapidly assess the efficiency of the studied beam.

During the simulation the beam is represented as a large number of rays, the intensity and direction of which are determined by the characteristics of the phase and amplitude characteristics of the beam. This is followed by tracing the path of each beam through the cell model and the force of the beam at the interfaces of the media is calculated (3).

The simulation results show that the force of the optical capture is maximum if the object is somewhat off-centre in relation to the beam (point (0,0) in Fig. 5.226).

Obviously, it is possible to find a form of the light beam, which at the same radiation power will provide a greater value of the force in the direction of one of axes in the plane of the cross section of the beam (e.g., along the axis) than the Gaussian beam.

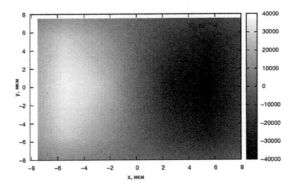

Fig. 5.226. Field of the x-component of the force acting on a spherical object with a radius of 5 μm from the Gaussian beam.

To do this, we deform the beam in a special way.

Suppose that the initial Gaussian beam has radius a. We modify the beam so that the equi-illuminated curves of the beam are the closed curves formed of two arcs of circles of different radii disposed with a slight offset (see Fig. 5.227).

As a parameter that determines the amount of bending we can selected $d = t_1 \times a$ – the coordinate of the intersection of the arc with the axis y. The arc of the auxiliary circle can be given by the equation $y = c(x)$. This amplitude distribution will take the form of a crescent, and can be described by the formulas:

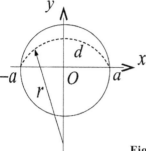

Fig. 5.227. Modification of the Gaussian beam.

$$A(x,y) = \exp\left(\frac{-x^2}{a^2}\right) \times \exp\left(-\frac{\left(y - c(x) - t_3 \times a\right)^2}{\left(t_2 \times a\right)^2}\right); c(x) = \sqrt{r^2 - x^2} - r + d;$$

$$r = \frac{a}{2} \times \left(\frac{1 + t_1^2}{t_1}\right). \tag{5.166}$$

The parameters t_1, t_2 and t_3 determine the particular shape of the beam. Parameter t_1 determines the ratio of the inner and outer semi-circle, confining the crescent. Parameter t_2 determines how quickly the intensity of the light field falls to the edges of the beam. Parameter t_3 defines a fixed width of the crescent independent of x.

The resulting distribution is now used to calculate the forces acting on an object illuminated by the beam of such shape. We use the method of the steepest gradient descent to optimize the shape of the beam for the parameters t_1, t_2 and t_3 and we find that the maximum power is reached at $t_1 \to 1$, t_2, $t_3 \to 0$. In this case, the intensity distribution collapses to a point, i.e., becomes a delta function. Thus, in the plane of capturing an object there is a particular point such that to reach the maximum value of force in a predetermined direction it is most advantageous to put all the energy of the beam at this point. The location of the point is determined by the shape of the object.

By virtue of diffraction limitations, however, we cannot produce the light beams with the intensity distribution in the form of delta functions. To get as close as possible to the optimal distribution, we stop optimization at the stage when the diffraction limit has not yet been crossed. When using an aqueous medium ($n \sim 1.33$) and a laser wavelength of $\lambda \sim 500$ nm, diffraction the limit will be equal to

$$d_{min} = \frac{\lambda}{2n} \sim 0.2 \ \mu m. \tag{5.167}$$

To meet the diffraction limitations, we impose a lower bound on the parameter $t_2 \geq 0.04$ that will correspond to the distance $a-d = 0.2 \ \mu m$ (characteristic width of the crescent).

For optimized distribution parameters in this case we obtain the following values: $t_1 = 0.94$, $t_2 = 0.04$, $t_3 = 0.0$. The intensity distribution of the light beam with these parameters shown in Fig. 5.225 ??????

Simulating illumination of an ellipsoid with such a beam and taking the contribution of the force of each beam as

$$F_i = \frac{\Delta P}{\Delta t} = \frac{-1}{c_0}(-I_r \vec{a}_r - I_d \vec{a}_d + I_e \vec{a}_e) \tag{5.168}$$

where I_e, I_r, I_d are respecively the intensity of the incident, reflected and transmitted beams, and $\vec{a}_e, \vec{a}_r, \vec{a}_d$ are their directions, we obtain a numerical estimate [167].

The calculated values of the forces acting on a spherical object with radius $R = 5 \ \mu m$ illuminated with a Gaussian beam, the beam–crescen and the beam–delta function are shown in Table 5.10.

Fig. 5.228. The optimized distribution of the intensity.

For ease of comparison, all results are normalized to the maximum value corresponding to the delta function.

Table 5.10. Values of force in illumination of an object with radius $R = 5$ μm with light beams with a power 300 mW with different intensity distributions

Beam type	The absolute value of the force, pN	The normalized value of the force
Gaussian beam	103	0.31
The optimized beam – crescent	293	0.88
The delta function	333	1

The table shows that the optimized distribution in the form of a crescent provides 0.88 of the theoretically attainable maximum and the Gaussian distribution only 0.31.

To capture the ellipsoidal particles, the beam–crescent was somewhat transformed: by simple substitution $x \rightarrow x_{mod} = kx$ the outer circular arc is transformed into an elliptical one, where k is the ratio of the major and minor axes of the ellipse. For *Saccharomyces cerevisiae* $k \sim 1.5$.

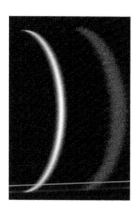

Fig. 5.229. Theoretical (left) and experimental (right), intensity distribution in the beam-crescent.

5.6.2 Measurements of the energy efficiency of the DOE forming beams–crescents

To form the above-described intensity distribution, a binary DOE was calculated using the algorithm described in [96] and manufactured by photolithography. The resolution of the element was 1 μm, the depth of etching 480 ± 50 nm.

Figure 5.230 shows the phase function of the produced DOE, and Fig. 5.231 a part of its microrelief.

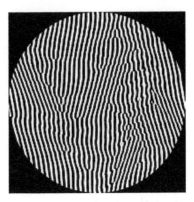

Fig. 5.230. DOE phase function, forming the 'crescent' distribution.

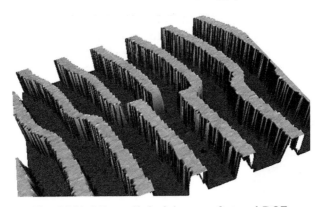

Fig. 5.231. Microrelief of the manufactured DOE.

Experiments were carried out by measuring the effectiveness of DOE fabrication. These experiments served two purposes. First, the measurements were taken of the diffraction efficiency of the phase plate, which is one of the indicators of quality of both the calculation of the phase distribution and the manufacturing quality of the DOE. Second, comparison with the Gaussian beam, which has passed the same optical system allows further comparison of the efficacy of the intensity distribution in the form of a crescent: knowing which part of the incoming light has reached the beam focus in the case of the Gaussian beam and the beam–crescent, and measuring the maximum capture force generated by each trap we can draw some conclusions about the power efficiency of the trap.

Figure 5.232 is a diagram used to measure the energy efficiency of the crescent-shaped beams.

The beam of a solid-state laser with a wavelength of $\lambda = 532$ nm was directed through a collimator to the DOE, after which it was focused by a lens at the receiving element of the light meter. The aperture cut off the mirrored order and higher diffraction orders the images. The collimator and the focuser were two lenses with a focal length $f = 400$ mm.

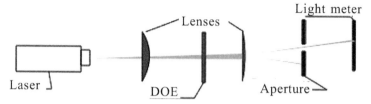

Fig. 5.232. Optical circuit for measuring the energy efficiency of the beam–crescent.

The energy efficiency was assessed by comparing the luminance achieved with the crescent beam with the luminance achieved by the Gaussian beam. To measure the latter, we used the same scheme, but without the DOE. The aperture was opened to the same radius of the hole.

Table 5.11 shows the measurement results.

Table 5.11. Results of measuring the energy efficiency of the crescent-shaped beams

Type of beam	Illumination, lux	Energy efficiency
Gaussian beam	950 ± 100	1
Beam-crescent	280 ± 50	0.29 ± 0.05

5.6.3. Experiments with optical manipulation of cells of Saccharomyces cerevisiae

For the series of experiments, the *Saccharomyces cerevisiae* culture was held in a glucose solution for 10 hours. For the capture experiments a small amount of culture was dissolved in distilled water to provide the particle concentration suitable for capture.

The optical diagram of the experimental setup is shown in Fig. 5.233. The laser radiation source was a laser with the wavelength $\lambda = 532$ nm, and the variable power (200–2000 mW). The subject plane was illuminated with white light from an incandescent lamp. The image in the object plane was recorded with a CCD-camera TMDSIPCAM8127J3 (30 frames/s, 1080 p). The CCD camera with a filter cut off most of the laser light to avoid illumination of the the image of the captured particles.

Using rotary mirrors, the laser beam was directed at the DOE and then focused inside droplets of the *Saccharomyces cerevisiae* yeast suspension using an immersion lens with a 40-fold magnification. Then the ×8 lens projected an image of what is happening in the object plane which with the help of another deflecting mirror travelled through the filter of green to the CCD-camera. The image from the camera in real-time was displayed on the PC screen, allowing to manually control the capture process using microscrews of the specimen table, with visual inspection.

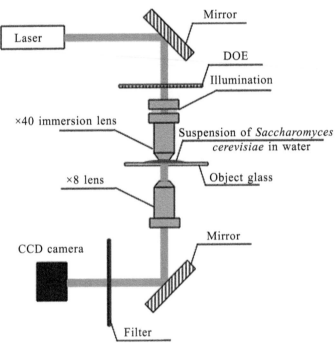

Fig. 5.233. Optical scheme of the experiment.

The purpose of the series of experiments was to determine the critical speed at which the captured object flies out from the optical trap formed by the crescent beam and, for comparison, a Gaussian beam. In this case the advantage of the proximity of the form of captured objects to spherical form was utilized, with a sufficient degree of accuracy we can estimate the maximum holding force of the trap, assuming its linear dependence on the speed. In [167] it was previously shown that the equilibrium between the force of resistance of the medium and the forces of radiation pressure occurs almost instantaneously, while the coefficient of the resistance force dependence on the speed can be set equal to $6\pi r\eta$. Simply put, we can ignore the relaxation time of the liquid–cell system regards the light pressure force at the moment of separation equal to the force of viscous friction. Consequently, from Stokes' formula we have

$$F_{max} = v_{max}\ 6\pi r\eta \qquad (5.169)$$

Thus, to evaluate the forces use the method described in [168].

Attempts to capture yeast cells by the crescent-shaped beams proved to be very successful.

The initial assumption was fully justified: in capture and movement the main share of radiation falls on the periphery of the cell, as shown in Figure 5.236. In addition, the shape of the beam provides additional rigidity of the trap in the lateral movement.

For comparison, similar experiments were performed with the Gaussian beam.

In capture with the Gaussian beam the cells occasionally flew out of the trap due to transverse movement of microflows.

Table 5.12. Comparison of the numerical results of experiments

Beam type	Input power of the laser beam, mW	Beam efficiency, rel. units	Radiation power at the beam focus	Maximum capture force, pN
Crescent	1300	0.29	377	1200
Gaussian 250	1	250	400	

Experimental results show that at comparable radiation power supplied to the beam focus the capture force strength of the crescent beam is several times greater than the force that can be obtained using the Gaussian beam.

Figure 5.234 and 5.235 show one episode each from a series of experiments with the crescent and Gaussian beams, respectively. Both episodes depict speed changes depending on the time of capture of the cell by the beam until its departure from the beam.

The uneven speed changes in the experiment were due to the imperfection of the experimental setup – the stage is driven by microscrews, rotated manually.

However, it is seen that the output from the micro-object from the crescent beam occurs at a higher speed and, as seen in Fig. 5.236, the crescent beam captured simultaneously two cells, thus ensuring greater resistance force of the medium.

We can also note that the experimental data agree well with the calculated data given in Table 5.10: if we take the following ratio as metrics

$$\tilde{A} = \frac{F}{I}, \qquad (5.170)$$

where F – grip strength, and I – the power of the radiation beam in focus, then the theoretical calculations have

$$\frac{\sigma_{cresc}}{\sigma_{gauss}} = \frac{0.88\sigma_\Delta}{0.31\sigma_\omega} = 2.8 \qquad (5.171)$$

where σ_Δ is the ratio of the capture force of the beam with an intensity distribution tending to the delta function to the radiation power, while the experiment gives the following values:

$$\frac{\sigma_{cresc}}{\sigma_{gauss}} = \frac{3.18 \cdot 10^{-9} \frac{s}{m}}{1.6 \cdot 10^{-9} \frac{s}{m}} \approx 2 \qquad (5.172)$$

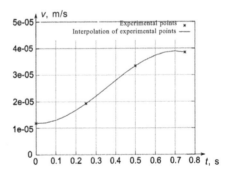

Fig. 5.234. The time dependence of the velocity of the particles in one of the experiments with the crescent-shaped beam.

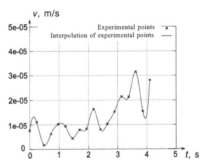

Fig. 5.235. The time dependence of the velocity of the particles in one of the experiments with the Gaussian beam.

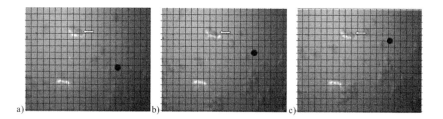

Fig. 5.236. a, b, c – the capture of yeast cells by the crescent-shaped beam. The black dot marks the position of the cell stationary relative to the fixed table. White arrow – captured particles. The time interval between frames 0.55 s.

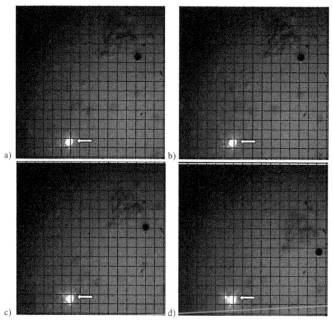

Fig. 5.237. a, b, c – the capture of yeast cells by the Gaussian beam. The black dot marks the position of the cell stationary relative to the fixed table. White arrow – captured particles. The time interval between frames 0.55 s.

The significant increase in the speed of the micro-object in relation to the power of the working beam, shows the possibility of power reduction when using the crescent-shaped beams to move the biological micro-objects. Unfortunately, due to technical reasons no test was performed of the viability of the cells after the move, but even without this test, it is clear that the use of beams of the shaped matched with the moving biological micro-object at least allows to use beams of lesser power, which in any case should decrease the likelihood of damage to the biological object.

References

1.   Ashkin, A. Acceleration and trapping of particles by radiation pressure, A. Ashkin, Phys. Rev. Lett. - 1970. - Vol. 24(4). - P. 156-159.

2.   Ashkin, A. Observation of resonances in the radiation pressure on dielectric spheres, A. Ashkin, J. M. Dziedzic, Phys. Rev. Lett. - 1977. - Vol. 38(23). - P. 1351-1354.

3.   He, H. Direct observation of transfer of angular momentum to absorptive particles from a laser beam with a phase singularity, H. He, M. E. J. Friese, N. R. Heckenberg, H. Rubinsztein-Dunlop, Phys. Rev. Lett. - 1995. - Vol. 75(5). - P. 826-829.

4.   Sato, S. Optical trapping and rotational manipulation of microscopic particles and biological cells using higher-order mode Nd:YAG laser beams, S. Sato, M. Ishigure, H. Inaba, Electron. Lett. - 1991. - Vol. 27(20). - P. 1831-1832.

5.   Bretenaker, F. Energy exchange between a rotating retardation plate and a laser beam, F. Bretenaker, A. Le Floch, Phys. Rev. Lett. - 1990. - Vol. 65(18). - P. 2316.

6.   Chen, C. Analysis of vector Gaussian beam propagation and the validity of paraxial and spherical approximations, C. Chen, P. Konkola, J. Ferrera, R. Keilmann, M. Schaffenberg, Journal of Optical Society of America A. - 2002. - Vol.19(2). - P. 404-412.

7.   Cojoc, D. Design and fabrication of diffractive optical elements for optical tweezer arrays by means of e-beam lithography, D. Cojoc, E. Di Fabrizio, L.Businaro, S. Cabrini, F. Romanato, L. Vaccari, M. Altissimo, Microelectronic Engineering. - 2002. - Vol. 61–62. - P. 963–969.

8.   Friese, M. E. J. Optical angular-momentum transfer to trapped absorbing particles, M. E. Friese, J. Enger, H. Rubinsztein-Dunlop, N. R. Heckenberg, Phys. Rev. A. - 1996. - Vol. 54(2). - P. 1593-1596.

9.   Khonina, S. N. The phase rotor filter, S. N. Khonina, V. V. Kotlyar, M. V. Shinkaryev, G. V. Uspleniev, J. Modern Optics. - 1992. -Vol. 39(5). - P. 1147-1154.

10.   Volostnikov V.G., et al., Izv. Samarsk. Nauchn. Tsentra RAN. - 2000. - T.2, № 1. - C. 48-52.

11.   Abramochkin, E. G. Microobject manipulations using laser beams with nonzero orbital angular momentum, E. G. Abramochkin, S. P. Kotova, A. V. Korobtsov, N. N. Losevsky, A. M. Mayorova, M. A. Rakhmatulin, V. G. Volostnikov, Laser physics – 2006. - Vol. 16(5). – P. 842-848.

12.   Abramochkin, E. G. Application of spiral laser beams for beam shaping problem, E. G. Abramochkin, E. V. Razueva, V. G. Volostnikov, Proc. of LFNM. - 2006. - 29 June - 1 July 2006, Kharkiv, Ukraine. - P. 275 - 278.

13.   Abramochkin, E. G. Fourier invariant singular wave-fields and beam shaping problem, E. G. Abramochkin, E. V. Razueva, V. G. Volostnikov, Proc. of LFNM. - 2006. - 29 June - 1 July 2006, Kharkiv, Ukraine. - P. 370 - 373.

14.   Abramochkin, E. G. Gaussian beams: new aspects and applications, E. G. Abramochkin, V. G. Volostnikov, Proc. of LFNM. – 2006. - 29 June - 1 July 2006, Kharkiv, Ukraine. - P. 267 - 274.

15.   Paterson, L. Controlled rotation of optically trapped microscopic particles, L. Paterson, M. P. MacDonald, J. Arlt, W. Sibbett, P. E. Bryant, K. Dholakia, Science. - 2001. - Vol. 292(5518). - P. 912-914.

16.   Kotlyar V.V., et al., Zh. Teor. Fiz., - 1997. - V. 23, No. 17. – 1-6.

17.   Friese, M. E. J. Optical alignment and spinning of laser-trapped microscopic particles, M. E. J. Friese, T. A. Nieminen, N. R. Heckerberg, H. Rubinsztein-Dunlop, Nature. - 1998. - Vol. 394(6691). - P. 348-350.

18.   Higurashi, E. Optically induced angular alignment of trapped birefringent microob-

jects by linear polarization, E. Higurashi, R. Sawada, T. Ito, Appl. Phys. Lett. - 1998. - Vol. 73(21). - P. 3034-3036.

19. Holmlin, R. E. Light-driven microfabrication: Assembly of multicomponent, three-dimensional structures by using optical tweezers, R. E. Holmlin, M. Schiavoni, C. Y. Chen, S. P. Smith, M. G. Prentiss, G. M. Whitesides, Angew. Chem. Int. Ed. Engl. - 2000. - Vol. 39(19). - P.3503-3506.

20. Grover, S. Automated single-cell sorting system based on optical trapping, S. C. Grover, A. G. Skirtach, R. C. Gauthier, C. P. Grover, J. Biomed. Opt. - 2001. - Vol. 6(1). - P. 14-22.

21. Fällman, E. Design for fully steerable dual-trap optical tweezers, E. Fällman, O. Axner, Appl. Opt. - 1997. - Vol. 36(10). - P. 2107-2113.

22. Sasaki, K. Pattern formation and flow control of fine particles by laser-scanning micromanipulation, K. Sasaki, M. Kosioka, H. Misawa, N. Kitamura, H. Masuhara, Opt. Lett. - 1991. - Vol. 16(19). - P. 1463-1465.

23. Dufresne, E. R. Optical tweezer arrays and optical substrates created with diffractive optical elements, E. R. Dufresne, D. G. Grier, Rev. Sci. Instr. - 1998. - Vol. 69(5). - P. 1974-1977.

24. Dufresne, E. R. Computer-generated holographic optical tweezer arrays, E. R. Dufresne, G. C. Spalding, M. T. Dearing, S. A. Sheets, D. G. Grier, Rev. Sci. Instrum. - 2001. - Vol. 72(3). - P. 1810-1816.

25. Grier D. G., Dufresne E. R. Apparatus for applying optical gradient forces. US Patent 6,055,106. The University of Chicago. 2000.

26. Hahn, J. Real-time digital holographic beam-shaping system with a genetic feedback tuning loop, J. Hahn, H. Kim, K. Choi, B. Lee, Appl. Opt. - 2006. - Vol. 45(5). - P. 915-924.

27. Curtis, J. E. Dynamic holographic optical tweezers, J. E. Curtis, B. A. Koss, D. G. Grier, Optics Communications. - 2002. - Vol. 207(1-6). - P. 169–175.

28. Reicherter, M. Optical particle trapping with computer-generated holograms written on a liquid-crystal display, M. Reicherter, T. Haist, E. U. Wagemann, H. J. Tiziani, Opt. Lett. - 1999. - Vol. 24 (9). - P. 608-610.

29. Harris, M. Optical helices and spiral interference fringes, M. Harris, C.A. Hill, J. M. Vaughan, Optics Communications. - 1994. - Vol. 106(4-6). - P. 161-166.

30. Khonina S.N., et al., Komp. Optika – 1999. - Вып.19. – C.107-111.

31. Bandres, M. A. Parabolic nondiffracting optical wave fields, M. A. Bandres, J. C. Gutierrez-Vega, S. Chavez-Cedra, Opt. Lett. - 2004. - Vol. 29(1). - P. 44-46.

32. Abramochkin, E.G. Generalized Gaussian beams, E. G. Abramochkin, V. G. Volostnikov, J. Opt. A: Pure and Appl. Opt. – 2004. - Vol. 6(5). - P. S157-S161.

33. Kotlyar, V. V. Generation of phase singularity through diffracting a plane or Gaussian beam by a spiral phase plate, V. V. Kotlyar, A. A. Almazov, S. N. Khonina, V. A. Soifer, H. Elfstrom, J. Turunen, J. Opt. Soc. Am. A. – 2005. - Vol. 22(5). - P. 849-861.

34. Bentley, J. B. Generation of helical Ince-Gaussian beams with a liquid-crystal display, J. B. Bentley, J. A Devis, M. A. Bandres, J. C. Gutierrez-Vega, Opt. Lett. – 2006. - Vol. 31(5). - P. 649-651.

35. Chattrapiban, N. Generation of nondiffracting Bessel beams by use of a spatial light modulator, N. Chattrapaban, E. A. Rogers, D. Cofield, W. T. Hill, R. Roy, Optics Let. - 2003. - Vol. 28(22). - P. 2183-2185.

36. Kotlyar, V. V. Calculation of phase formers of non-diffracting images and a set of concentric rings, V. V. Kotlyar, S. N. Khonina, V. A. Soifer, Optik. - 1996. - Vol. 102(2). - P. 45-50.

37.   Kotlyar, V. V. Rotation of multimodal Gauss-Laguerre light beans in free space and in a fiber, V. V. Kotlyar, V. A. Soifer, S. N. Khonina, Optics and Lasers in Engineering. - 1998. - Vol. 29(4-5). P. 343-350.

38.   Rakhmatulin M.P., Development of methods of handling micro-objects by laser radiation. Dissertation, Samara, 2003.

39.   Ashkin, A. Observation of a single-beam gradient force optical trap for dielectric particles, A. Ashkin, J. M. Dziedzic, J. E. Bjorkholm, S. Chu, Optics Letters. - 1986. - Vol. 11(5). - P. 288-290.

40.   Ashkin, A. Optical levitation by radiation pressure, A. Ashkin, J. M. Dziedzic, Appl. Phys. Lett. - 1971. - Vol. 19(8). - P. 283-285.

41.   Ashkin, A. Forces of a single-beam gradient laser trap on a dielectric sphere in the ray optics regime, A. Ashkin, Biophys J. - 1992. - Vol.61(2). - P. 569-582.

42.   Ganic, D. Exact radiation trapping force calculation based on vectorial diffraction theory, D. Ganic, X. Gan, M. Gu, Opt. Express. - 2004. - Vol. 12(12). - P. 2670-2675.

43.   Lemire, T. Coupled-multipole formulation for the threatment of electromagnetic scattery by a small dielectric particles of arbitrary sphere, T. Lemire, J.Opt.Soc.Am. A. -1997. - Vol. 14(2). - P. 470-474.

44.   Marston, P. L. Radiation torque on a sphere caused by circulalarly-polarized electromagnetic wave, P. L. Marston, J. H. Chrichton, Physical Review A. - 1984. - Vol. 30(5). - P. 2508-2516.

45.   Navade, Y. Radiation forces on a dielectric sphere in the Rayleigh scatterry regime, Y. Navade, T. Asakure, Opt. Commun. –1996. - Vol. 124 (5-6). - P. 529-541.

46.   Nieminen, T. A. Calculation and optical measurement of laser trapping forces on non-spherical particles, T. A. Nieminen, H. Rubinsztein-Dunlop, N. R. Heckenberg, Journal of Quantitative Spectroscopy & Radiative Transfer. - 2001. - Vol. 70(4-6).- P. 627–637.

47.   Rohrbach, A. Optical trapping of dielectric particles in arbitrary fields, A. Rohrbach, E. H. Stelzer, J. Opt. Soc. Am. A. – 2001. - Vol. 18(4). - P. 813-835.

48.   Pobre, R. Radiation force on a nonlinear microsphere by a lightly focused Gaussian beam, R. Pobre, C. Salome, Appl. Opt. – 2002. - Vol. 41(36). – P. 7694-7701.

49.   Debye, P. Der Lichtdruck and Kugeln von beliebige Material, P. Debye, Ann. Phys, 1909. – V. 335(11). – P. 57-136.

50.   Rockstuhl, C. Calculation of the torque on dielectric elliptical cylinders, C. Rockstuhl, H. P. Herzig, Opt. Soc. Am. A. - 2005. – V. 22(1). – P. 109-116.

51.   Shaohui, Y. Transverse trapping forces of focused Gaussian beam on ellipsoidal particles, Y. Shaohui, Y. Baoli, J. Opt. Soc. Am. B. – 2007. – Vol. 24(7). - P. 1596-1602.

52.   Skidanov R.V., Komp. Optika – 2005. - No. 28. –18-21.

53.   Kotlyar V.V., et al., Komp. Optika. – 2005. - No. 27, – 105-111.

54.   Kotlyar V.V., et al., Komp. Optika – 2003. - Вып. 25. – 24-28.

55.   Kotlyar V.V., et al., Opticheskii Zhurnal. – 2005. - V. 72, No. 5. - С.55-61.

56.   Arlt, J. Optical micromanipulation using a Bessel light beams, J. Arlt, V. Garces-Chavez, W. Sibbett, K. Dholakia, Opt. Comm. - 2001. - Vol. 197. - P. 239-245.

57.   Garces-Chavez, V. Simultaneous micromanipulation in multiple planes using a self-reconstructing light beam, V. Garces-Chavez, D. McGloin, H. Melville, W. Sibbett, K. Dholakia, Nature. - 2002. - Vol. 419(6903). -P.145-147.

58.   MacDonald, M. P. Creation and manipulation of three-dimensional optically trapped structures, M. P. MacDonald, L. Paterson, K. Volke-Sepulveda, J. Arlt, W. Sibbett, K. Dholakia, Science. - 2002. - Vol. 296(5570). - P. 1101-1103.

59.   Turunen, J. Holographic generation of diffraction-free beams, J. Turunen, A. Vasara, A. T. Friberg, Applied Optics. – 1988. – Vol. 27(19). - P. 3959-3962.

60. Soifer V.A., et al., Komp. Optika. – 2005. -Вып.28. - С.5-17.

61. Khonina, S.N. Rotating microobjects using a DOE-generated laser Bessel beam, S. N. Khonina, R. V. Skidanov, V. V. Kotlyar, V. A. Soifer, Proceedings of SPIE. - 2004. - Vol. 5456. - P. 244-255.

62. Miller, W. Symmetry and separation of variables, W. Miller. – Cambridge University Press., 1977. – 318 p.

63. Allen, L. Orbital angular momentum of light and the transformation of Laguerre-Gaussian laser modes, L. Allen, M. W. Beijersbergen, R. J. C. Spreeuw, J. P. Woerdman, Phys. Rev. A. - 1992. - Vol. 45(11). - P. 8185-8189.

64. Volke-Sepulveda, K. Orbital angular momentum of a high-order Bessel light beam, K. Volke-Sepulveda, V. Garces-Chavez, S. Chavez-Cerda, J. Arlt, K. Dholakia, J. Opt. B: Quantum Semiclass. Opt. - 2002. - Vol. 4(2). - P. S82–S89.

65. Durnin, J. Exact solution for nondiffracting beams. I. The scalar theory, J. Durnin, J. Opt. Soc. Am. A. – 1987. - Vol. 4(4). - P.651-654.

66. Durnin, J. Diffraction-free beams, J. Durnin, J. J. Miceli, J. H. Eberly, Phys. Rev. Lett. - 1987. - Vol. 58(15). - P. 1499–1501.

67. Paterson, C. Higher-order Bessel waves produced by axicon-type computer-generated holograms, C. Paterson, R. Smith, Optics Comm. - 1996. - Vol. 124(1-2). - P. 121-130.

68. Fedotowsky, A. Optimal filter design for annular imaging, A. Fedotowsky, K. Lehovec, Appl. Opt. - 1974. - Vol. 13(12). - P. 2919-2923.

69. Skidanov R.V., Komp. Optika. – 2006. -No. 29. - 4-23.

70. Kotlyar V.V., et al., Komp. Optika. – 2005. - No.. 28. - 29-36.

71. Soifer V.A., et al. Proc. Second International Symposium Golografiya EKSPO-2005. – 69-70.

72. Khonina, S.N. DOE-generated laser beams with given orbital angular moment: application for micromanipulation, S. N. Khonina, R. V. Skidanov, V. V. Kotlyar, V. A. Soifer, J. Turunen, Proc. SPIE Int. Soc. Opt. Eng. -2005. – Vol. 5962. – P. 59622W.

73. Khonina, S. N. Optical micromanipulation using DOEs matched with optical vorticies, S. N. Khonina, R. V. Skidanov, V. V. Kotlyar, A. A. Kovalev, V. A. Soifer, Proc. SPIE. –2006. - Vol. 6187. -P. 61871F.

74. Skidanov, R. V. Optical microparticle trapping and rotating using multi-order DOE, R. V. Skidanov, S. N. Khonina, V. V. Kotlyar, V. A. Soifer, Proc. Of the ICO Topical Meeting on Optoinformatic/Information Photonics. – 2006. - 4-7 Sep. 2006, Saint-Peterburg, Russia. - P. 466-469.

75. Soifer, V.A. Remarkable laser beams formed by computer-generated optical elements: properties and applications, V. A. Soifer, V. V. Kotlyar, S. N. Khonina, R. V. Skidanov, Proc. SPIE. -2006. - Vol. 6252. - P. 62521B.

76. Guo, C.-S. Optimal annulus structures of optical vortices, C.–S. Guo, X. Liu, J.-L. He, H. – T. Wang, Opt. Express. – 2004. - Vol. 12(19). - P. 4625-4634.

77. Kotlyar, V.V. Diffraction of conic and Gaussian beams by a spiral phase plate, V. V. Kotlyar, A. A. Kovalev, S. N. Khonina, R. V. Skidano, V. A. Soifer, H. Elfstrom, N. Tossavainen, J. Turunen, Appl. Opt. – 2006. - Vol. 45(12). - P.2656-2665.

78. Soifer, V.A. Optical data processing using DOEs, V. A. Soifer, V. V. Kotlyar, S. N. Khonina, R. V. Skidanov, Methods for Computer Design of Diffractive Optical Elements, Wiley-Interscience Publication John Wiley & Sons, Inc, 2002. - Chapter 10. - P. 673-754.

79. Kotlyar, V. V. Diffraction of a finite-radius plane wave and a Gaussian beam by a helical axicon and a spiral phase plate, V. V. Kotlyar, A. A. Kovalev, R. V. Skidanov, O. Yu. Moiseev, V. A. Soifer, J. Opt. Soc. Am. A. – 2007. - Vol. 24(7). – P. 1955-1964.

80. Soifer V.A., et al., Proc. All-Russian Semina: Yu.I. Desinyuk – founder of optical holography. FTI RAN, St. Petersburg. 2007 – 116-123.

81. Leach, J. Interactive approach to optical tweezers control, J. Leach, K. Wulff, G. Sinclair, P. Jordan, J. Courtial, L. Thomson, G. Gibson, K. Karunwi, J. Cooper, Z. J. Laczik, M. Padgett, Appl. Opt. – 2006. - Vol. 45(5). - P. 897–903.

82. Xun, X. D. Phase calibration of spatially nonuniform spatial light modulators, X. D. Xun, R. W. Cohn, Appl. Opt. – 2004. - Vol. 43(35). - P. 6400–6406.

83. Kotlyar V.V., et al., Komp. Optika. – 2006. - No..30. - 16-22.

84. Kotlyar V.V., et al., Komp. Optika. – 2007. - V. 31, No. 1. - 35-38.

85. Skidanov, R. V. Micromanipulation in Higher-Order Bessel Beams, R. V. Skidanov, V. V. Kotlyar, V. V. Khonina, A. V. Volkov, V. A. Soifer, Optical Memory and Neural Networks (Information Optics). – 2007. - Vol. 16(2). - P. 91-98.

86. Davis, J.A. Diffraction-free Beam Generated with Programmable Spatial Light Modulators, J. A. Davis, J. Guertin, D. M. Cottrell, Appl. Opt. – 1993. - Vol.32(31). - P. 6368-6370.

87. Davis, J. A. Intensity and phase measurements of nondiffracting beams generated with the magneto-optic spatial light modulator, J. A. Davis, E. Carcole, D. M. Cottrell, Appl. Opt. – 1996. - Vol. 35(4). - P. 593-598.

88. Skidanov V.V., et al., Komp. Optika. – 2007. V. 31, No. 1. - 14-21.

89. Skidanov V.V., et al., Izv. SNTs RAN. –2006. – V. 8, No. 4. - 1200-1203.

90. Gao, M Generation and application of the twisted beam with orbital angular momentum, M. Gao, C. Gao, Z. L, Chinesse optics letters. – 2007. - Vol. 5(2). – P. 89-92.

91. Farafonov, V.S. A new solution of the light scattering problem for axisymmetric particles, V. S. Farafonov, U. B. Ilin, T. A. Henning, Journal of Quantitative Spectroscopy and Radiative Transfer. - 1999. - Vol. 63(2-6). - P. 205-215.

92. Hong Du, Hao Zhang Ultra high precision Mie scattering calculations, 2002.

93. Jesacher, A. Fully phase and amplitude control of holographic optical tweezers with high efficiency, A. Jesacher, C. Maurer, A. Schwaighofer, S. Bernet, M. Ritsch-Marte, Optics Express. - 2008. - Vol. 16(7). - P.4479-4486.

94. Montes-Usategui, M. Fast generation of holographic optical tweezers by random mask encoding of Fourier components, M. Montes-Usategui, E. Pleguezuelos, J, Andilla, E. Martin-Badosa, Optics Express. - 2006. - Vol.14(6). - P.2101-2107.

95. Montes-Usategui, M. Adding functionalities to precomputed holograms with random mask multiplexing in holographic optical tweezers, M. Montes-Usategui, J. Mas, M. S. Roth, E. Martin-Badosa, Applied Optics. - 2011. - Vol.50(10). - P. 1417-1424

96. Kotlyar V.V., et al., Komp. Optika. – 1999. – No. 19. – 54-64.

97. Kotlyar V.V., et al., Komp. Optika. – 2001. – No. 21. – 36-39.

98. Khonina, S.N. Encoded binary diffractive element to form hyper-geometric laser beams, S.N. Khonina, S. A. Balalayev, R. V. Skidanov, V. V. Kotlyar, B. Palvanranta, J. Tutunen, J. Opt. A: Pure Appl. Opt. – 2009. – Vol. 11(6). – P. 065702-065708.

99. Khonina S.N., et al., Komp. Optika. – 2009. - V. 33, No. 2. – 138-146.

100. Cohn, R.W. Approximating fully complex spatial modulation with pseudorandom phase-only modulation, R. W. Cohn, M. Liang, Applied Optics. – 1994. - Vol. 33(20). - P. 4406-4415.

101. Hassenbrook, L.G. Random phase encoding of composite fully complex filters, L. G. Hassenbrook, M. E. Lhamon, R. C. Daley, R. W. Cohn, M. Liang, Optics Letters. - 1996. - Vol.21(4). - P. 272-274.

102. Skidanov R.V., et al., Komp. Optika. - 2010. - V.34, No. 2. - 214-218.

103. Guo, C.–S. Optical sorting using an array of optical vortices with fractional topo-

logical charge, C.-S. Guo, Y.-N. Yu, Zh. Hong, Optics Communications. - 2010. - Vol.283(9). - P.1889-1893.

104. Sinclair, G. Assembly of 3-dimdensional structures using programmable holographic optical tweezers, G. Sinclair, P. Jordan, J. Courtial, M. Padgett, J. Cooper, Z. Laczik, Optics Express. - 2004. - Vol. 12(22). - P.5475-5480.

105. Polin, M. Optimized holographic optical traps, M. Polin, K. Ladavac, S.-H. Lee, Y. Roichman, D. G. Grier, Optics Express. - 2005. - Vol.13(15). - P. 5831-5845.

106. Abramochkin E.G., et al., Prod. 27th School for Coherent Optics and holography. Irkutsk – 2005. 203-207.

107. Cheong, F. C. Optical travelator: transport and dynamic sorting of colloidal microspheres with an asymmetrical line optical tweezers, F. C. Cheong, C. H. Sow, A. T. S. Wee, P. Shao, A. A. Bettiol, J. A. van Kan, F. Watt, Applied Physics B. - 2006. - Vol.83(1). - P.121-125.

108. Marchington, R.F. Optical deflection and sorting of microparticles in a near-field optical geometry, R. F. Marchington, M. Mazilu, S. Kuriakose, V. Garces-Chavez, P. J. Reece, T. F. Krauss, M. Gu, K. Dholakia, Optics Express. - 2008. - Vol. 16(6). - P. 3712-3726.

109. Sun, Y. Y. Large-scale optical traps on a chip for optical sorting, Y. Y. Sun, X.-C. Yuan, L. S. Ong, J. Bu, S. W. Zhu, R. Liu, Applied Physics Letters. - 2007. - Vol. 90(3). - P. 031107-1 - 031107-3.

110. Rodrigo, P. J. Interactive light-driven and parallel manipulation of inhomogeneous particles, P. J. Rodrigo, R. L. Eriksen, V. R. Daria, J. Gluckstad, Optics Express. - 2002. - Vol. 10(26). - P. 1550-1556.

111. Dasgupta, R. Microfluidic sorting with a moving array of optical traps, R. Dasgupta, S. Ahlawat, P. K. Gupta, Applied Optics. - 2012. - Vol. 51(19). - P. 4377-4387.

112. Rodrigo, P. J. Dynamic formation of optically trapped microstructure arrays for biosensor applications, P. J. Rodrigo, R. L. Eriksen, V. R. Daria, J. Gluckstad, Biosensors and Bioelectronics. - 2004. - Vol. 19(11). - P. 1439-1444.

113. Rodrigo, P. J. Real-time interactive optical micromanipulation of a mixture of high-and low-index particles, P. J. Rodrigo, V. R. Daria, J. Gluckstad, Optics Express. - 2004. - Vol.12(7) - P. 1417-1425.

114. Prentice, P. A. Manipulation and filtration of low-index particles with holographic Laguerre-Gaussian optical trap arrays, P. A. Prentice, M. P. MacDonald, T. G. Frank, A. Cuschieri, G. C. Spalding, W. Sibbett, P. A. Campbell, K. Dholakia, Optics Express. - 2004. - Vol.12(4). - P. 593-600.

115. Porfir'ev A.P., Vest. Samarsk. Gosud. Aerokosmich. Univ., - 2010. -No. 4(24). - 269-275.

116. Balalaev S.A., et al., Komp. Optika. - 2007. - V.31, No. 4. - 23-29.

117. Herman, R.M. Hollow beams of simple polarization for trapping and storing atoms, R. M. Herman, T. A. Wiggins, J. Opt. Soc. Am. - 2002. - Vol. 19(1). – P. 116-121.

118. McQueen, C. A. An experiment to study a "nondiffracting" light beam, C. A. McQueen, J. Arlt, K. Dholakia, Am. J. Phys. - 1999. - Vol. 67(10). - P. 912–915.

119. Khonina, S.N. An analysis of the angular momentum of a light field in terms of angular harmonics, S. N. Khonina, V. V. Kotlyar, V. A. Soifer, P. Paakkonen, J. Simonen, J. Turunen, Journal of Modern optics. –2 001. - Vol. 48(10). - P. 1543-1557.

120. Khonina, S. N. Rotation of microparticles with Bessel beams generated by diffractive elements, S. N. Khonina, V. V. Kotlyar, R. V. Skidanov, V. A. Soifer, K. Jefimovs, J. Simonen, J. Turunen, Journal of Modern Optics. - 2004. - Vol. 51(14). - P.2167-2184.

121. Kotlyar, V.V. An algorithm for the generation of laser beams with longitudinal pe-

riodicity: rotating images, V. V. Kotlyar, S. N. Khonina, V. A. Soifer, Journal of Modern Optics. – 1997. - Vol. 44(7). - P. 1409-1416.

122. Paakkonen, P. Rotating optical fields: experimental demonstration with diffractive optics, P. Paakkonen, J. Lautanen, M. Honkanen, M. Kuittinen, J. Turunen, S. N. Khonina, V. V. Kotlyar, V. A. Soifer, A. T. Friberg, Journal of Modern Optics. - 1998. - Vol. 45(11). - P. 2355-2369.

123. Garces-Chavez, V. Simultaneous micromanipulation in multiple planes using a self-reconstructing light beam, V. Garces-Chavez, D. McGloin, H. Melville, W. Sibbett, K. Dholakia, Nature. - 2002. - Vol. 419(6903). -P.145-147.

124. Porfirev, A. P. The generation of the hollow optical beams array by the superpositions of Bessel beams, A. P. Porfirev, Asia-Pacific Conference on Fundamental Problems of Opto- and Microelectronics (Russia, Moscow-Samara, 4-8 July). - LPI, 2011. - 1 DVD-ROM, SAMP15.

125. Khonina, S. N. Astigmatic Bessel laser beams, S. N. Khonina, V. V. Kotlyar, V. A. Soifer, K. Jefimovs, P. Paakkonen, J. Turunen, Journal of Modern Optics. - 2004. - Vol. 51(5). - P.677-686.

126. Chavez-Cerda, S. Interference of traveling nondiffracting beams, S. Chavez-Cerda, M. A. Meneses-Nava, J. M. Hickmann, Optics Letters. - 1998. - Vol. 23(24). - P.1871-1873.

127. Dudley, A. Superpositions of higher-order Bessel beams and nondiffracting speckle fields, A. Dudley , R. Vasilyeu, A. Forbes, N. Khilo, P. Ropot, V. Belyi, N. Kazak, Proc. of SPIE. - 2009. - Vol. 7430. - P.74300A-1 - 74300A-7.

128. McGloin, D. Interfering Bessel beams for optical micromanipulation, D. McGloin, V. Garces-Chavez, K. Dholakia, Optics Letters. - 2003. - Vol. 28(8). - P.657-659.

129. Vasilyeu, R. Generating superpositions of higher–order Bessel beams, R. Vasilyeu, A. Dudley, N. Khilo, A. Forbes,Optics Express. - 2009. - Vol. 17(26). - P. 23389-23395.

130. Diffractive computer optics. V.A. Soifer (editor), Moscow, Fizmatlit, 2007.

131. Khonina, S. N. Generating a couple of rotating nondiffarcting beams using a binary-phase DOE, S. N. Khonina, V. V. Kotlyar, V. A. Soifer, J. Lautanen, M. Honkanen, J. Turunen, Optik. - 1999. - Vol. 110(3). - P. 137-144.

132. Kotlyar V.V., et al., Avtometriya. - 1997. - No. 5. - 46-54.

133. Doskolovich, L. L. Analytical initial approximation for multiorder binary gratings design, L. L. Doskolovich, V. A. Soifer, G. Alessandretti, P. Perlo, M. Repetto, Pure and Appl.Opt.- 1994. - Vol.3(6). - P.921-930.

134. Krackhardt, U. Upper bound on the diffraction efficiency of phase-only fanout elements, U. Krackhardt, J. N. Mait, N. Streibl, Applied Optics. - 1992. - Vol. 31(1). - P.27-37.

135. Zhou, C. Numerical study of Dammann array illuminators, C. Zhou, L. Linen, Applied Optics. - 1995. - Vol. 34(26). - P.5961-5969.

136. Kotlyar V.V., et al., Komp. Optika. -1999. - No. 19. - 74-79.

137. Doskolovich L.L., et al., Komp. Optika. - 1996. - No.16. - 31-35.

138. Khonina, S. N. A method for design of composite DOEs for the generation of letter image, S. N. Khonina, V. V. Kotlyar, V. V. Lushpin, V. A. Soifer, Optical Memory and Neural Networks. - 1997. - Vol. 6(3). - P.213-220.

139. Kotlyar V.V., et al., Komp. Optika. - 1996. - No. 16. - 50-53.

140. Gluckstad, J. Dynamic array of dark optical traps, J. Gluckstad, V. R. Daria, P. J. Rodrigo, Applied Physics Letters. - 2004. - Vol. 84(3). - P. 323-325.

141. Yuan, X.-C. The generation of an array of nondiffracting beams by a single composite computer generated hologram, X.-C.Yuan, S. H. Tao, B. S. Ahluwalia, J.Opt.A.:

Pure Appl.Opt. - 2005. - Vol.7(1). - P.40-46.

142. Skidanov R.V., et al., Komp. Optika. - 2012. - V.36, No. 1. - 80-90.

143. Skidanov R.V., et al., Komp. Optika. - 2012. - V.36, No. 2. - 211-218.

144. Leitz, G. Stress response in Caenorhabditis elegans caused by optical tweezers: wavelength, power, and time dependence, G. Leitz, E. Fällman, S. Tuck, O. Axner, Biophys. J. - 2002 - Vol. 82(4). - P. 2224-2231.

145. König, K. Cell damage in near-infrared multimode optical traps as a result of multiphoton absorption, K. König, H. Liang, M.W. Berns, B.J. Tromberg, Opt Lett. - 1996. - Vol. 21(14). - P. 1090–1092.

146. Thanh, S.D. Photogenerated singlet oxygen damages cells in optical traps, S.D. Thanh, N.C. Zakharov, arXiv:0911.4651 - 2009.

147. Peterman, E.J.G. Laser-Induced Heating in Optical Traps, E.J.G. Peterman, F. Gittes, C.F. Schmidt, Biophysical Journal. - 2003. - Vol. 84(2). - P. 1308–1316.

148. Mirsaidov, U. Optimal optical trap for bacterial viability, U. Mirsaidov, W. Timp, K. Timp, M. Mir, P. Matsudaira, G. Timp, Phys Rev E Stat Nonlin Soft Matter Phys. - 2008. - Vol. 78(2 Pt 1). - P. 021910.

149. van der Horst, A. Calibration of dynamic holographic optical tweezers for force measurements on biomaterials, A. van der Horst, N.R. Forde, Opt Express. - 2008. - Vol. 16(25). - P. 20987-21003.

150. Emiliani, V. Wave front engineering for microscopy of living cells, V. Emiliani, D. Cojoc, E. Ferrari, V. Garbin, C. Durieux, M. Coppey-Moisan, E. Di Fabrizio, Opt Express. - 2005. - Vol. 13(5). - P. 1395-1405.

151. Xin, H. Targeted delivery and controllable release of nanoparticles using a defect-decorated optical nanofiber, H. Xin, B. Li, Opt Express. - 2011. - Vol. 19(14). - P. 13285-13290.

152. Moradi, A. Strength control in multiple optical traps generated by means of diffractive optical elements, A. Moradi, E. Ferrari, V. Garbin, E. Di Fabrizio, D. Cojoc, Optoelectronics and Advanced Materials - Rapid Communications. - 2007. - Vol. 1(4). - P. 158-161.

153. Emiliani, V. Multi force optical tweezers to generate gradients of forces, V. Emiliani, D. Sanvitto, M. Zahid, F. Gerbal, M. Coppey-Moisan, Opt Express. – 2004. - Vol. 12(17). - P. 3906-3910.

154. Dasgupta, R. Optical orientation and rotation of trapped red blood cells with Laguerre-Gaussian mode, R. Dasgupta, S. Ahlawat, R.S. Verma, P.K. Gupta, Opt Express. – 2011. - Vol. 19(8). - P. 7680-7685.

155. Daria, V.R. Simultaneous transfer of linear and orbital angular momentum to multiple low-index particles, V.R. Daria, M.A. Go, H.-A. Bachor, Journal of Optics. – 2011. - Vol. 13(4). - P. 044004.

156. Sun, Q. Hollow sinh-Gaussian beams and their paraxial properties, Q. Sun, K. Zhou, G. Fang, G. Zhang, Z. Liu, S. Liu, Opt Express. – 2012. . Vol. 20(9). - P. 9682-9691.

157. Olson, C.C. Tailored optical force fields using evolutionary algorithms, C.C. Olson, R.T. Schermer, F. Bucholtz, Opt Express. – 2011. - Vol. 19(19). - P. 18543-18557.

158. Jesacher, A. Size selective trapping with optical "cogwheel" tweezers, A. Jesacher, S. Fürhapter, S. Bernet, M. Ritsch-Marte, Opt Express. – 2004. - Vol. 12(17). - P. 4129-4135.

159. Brijesh, P. Spatially shaping the longitudinal focal distribution into a horseshoe-shaped profile, P. Brijesh, T. J. Kessler, J. D. Zuegel, D. D. Meyerhofer, Quantum Electronics and Laser Science Conference. - 2007. QELS '07 (2007). - P. 1 –2.

160. Jiang, Y. Radiation force of highly focused Lorentz-Gauss beams on a Rayleigh particle, Y. Jiang, K. Huang, X. Lu, Opt Express. – 2011. - Vol. 19(10). - P. 9708–9713.

161. Steuernagel, O. Coherent Transport and Concentration of Particles in Optical Traps using Varying Transverse Beam Profiles, O. Steuernagel, arXiv:physics/0502023 – 2005.

162. Volostnikov V.G., et al., Kvant. Elektronika. - 2002. - V. 32, No. 7. - 565-566.

163. Abramochkin E.G., et al., Izv. RAN. Ser. Fiz. - 2008. - V. 72, No. 1. - C. 76-79.

164. Bowman, R. Particle tracking stereomicroscopy in optical tweezers: Control of trap shape, R. Bowman, G. Gibson, M. Padgett, Opt. Express. – 2010. - Vol. 18(11). - P. 11785-11790.

165. Saito, T.L. SCMD: Saccharomyces cerevisiae Morphological Database, T.L. Saito, M. Ohtani, H. Sawai, F. Sano, A. Saka, D. Watanabe, M. Yukawa, Y. Ohya, S. Morishita, Nucleic Acids Res. - 2004 - Vol. 32 (Database issue) - P. D319-322.

166. Simpson, S.H. Computational study of the optical trapping of ellipsoidal particles, S.H. Simpson, S. Hanna, Phys. Rev. A. – 2011. - Vol. 84(5). - P. 053805.

167. Skidanov R.V., et al., Komp. Optika. – 2010. - V. 34, No. 3. - 308-314.

168. Svoboda, K. Biological Applications of Optical Forces, K. Svoboda, S. M. Block, Annual Review of Biophysics and Biomolecular Structure. - 1994. - Vol. 23. - P. 247-285.

Conclusion

A number of important directions which, according to the authors, should be followed in future investigations are also listed.

Regardless of the possibility of localisation of the energy of the plasmon mode in the subwave size range, a significant shortcoming of the plasmon structures, examined in the first chapter, are the high losses through absorption in the metal, directly proportional to the degree of localisation of the mode energy. At the same time, there are surface electromagnetic waves which can propagate in the completely dielectric structures. These waves are referred to as Bloch surface waves (BSW) or surface states of the photonic crystals and, as indicated by the name, propagate along the interface between the homogeneous medium and the photonic crystal or between two different photonic crystals. When all the materials of a structure are dielectrics, the losses through absorption during propagation of such a surface wave can be almost completely eliminated. The absence of the absorption losses makes the BSW promising in different applications, in particular, in integrated optics and optical transmission system and information processing on the nanoscale.

The results described in the first chapter can also be used as a basis for solving new fundamental problems of controlling the application of the BSW. These tasks include the calculation and development of a new class of elements of 'two-dimensional' optics designed for fulfilling the given transformations of the BSW (refraction, reflection and focusing), the calculation of the diffractive structures for the effective excitation of the BSW and formation of high-frequency interference patterns of the BSW.

An interesting aspect of the problem of calculating the diffractive gratings with the resonance properties for differentiation and integration of the optical power source is, according to the authors, the development of a method for calculating and producing planar structures for the given transformations of the time optical signals designed for information processing in the chip. In this case, the time signals may be represented by the plasmon pulses, and the operations of transformation of the pulse (differentiation and integration) will be performed by the microstructures located directly on the surface.

The photonic crystals, discussed in the second chapter, are promising components for information technologies. Using the subwavelength holes calculated by the appropriate procedure in planar waveguides, it is possible to construct not only the PC-waveguides, PC-collimators and PC lenses, investigated in the second chapter, but also PC-prisms, PC-mirrors, PC-filters, PC-resonators, etc. The planar and compact nature (the size does not exceed tens of a micron) of these components in nanophotonics make them promising for the construction of multi-component devices in telecommunication systems, planar sensors, and planar microparticle manipulating devices.

The material in the third chapter can be used as a basis for further investigations in the problems of sharp focusing of laser light, including:

– sharp focusing of the light of femtosecond pulses taking into account the linear and non-linear dispersion of the material of the diffractive lens;

– modelling of sharp focusing of light by means of nanostructured waveguide planar and 3-D micro-optics components. It is well-known that the nanoslits in the planar waveguides are capable of retaining the light inside the slit, and the intensity inside the nanoslit may be an order of magnitude larger than the intensity inside the waveguide;

– optimisation of the parameters of micro-optics components, forming 'photonic' nanojets, focuses with a large focal depth, including those with the bent trajectories. There are light fields propagating along the parabolic trajectories, for example, Airy laser beams. The diffraction of a plane wave on the angular phase step also results in the formation of a hyperbolic 'photonic jet'. It is important to calculate the micro-optics components, capable of focusing the light in a line deflected from the optical axis in the required manner.

The methods of focusing singular laser beams, developed in the fourth chapter, require additional relatively complicated investigations. These investigations in the fourth chapter are planned to continue taking into account the presence of wave aberrations in the focusing system. The majority of investigations reported the negative effect of aberrations on the structure in the focal region. However, the authors plan to investigate the positive aspect of apodization the transmission function of the lens by the distribution, corresponding to the Zernike polynomials. The primary optical aberrations, such as shift, tilting, spherical aberration, astigmatism, are described by the Zernike polynomials of the low orders and can be used in particular for decreasing the transverse size of the focal spot below the diffraction limit. Another important task for focusing system is the problem of decreasing the longitudinal size of the light spot. The main methods of solving this problem are based on the interference of several coherent beams, mostly propagating in the opposite directions. The application of vortex beams for this purpose opens different possibilities in the formation of the three-dimensional distribution in the focal region.

The further continuation of the studies of optical micro-manipulation should take place by carrying out joint investigations and development with experts in the area of nanotechnology, micromechanics and microbiology in order to develop efficient means for contactless micro-manipulation of opaque and transparent objects with the size from several nanometres to several micrometres, situated in a liquid or free space.

Index

A

anomalies
  Rayleigh–Wood's anomalies 2
  Wood's anomalies 2, 79, 80, 81, 90
axicon 286, 293, 294, 295, 298, 299, 301, 302, 303, 304, 305, 306, 307,
        311, 312, 313, 314, 315, 316, 317, 318, 319, 320, 321, 322, 323, 324,
        325, 326, 327, 328, 329, 330, 331, 417, 418, 419, 423, 425, 435,
        437, 438, 440, 444, 445, 446, 561, 577, 472, 578, 473, 474, 476,
        477, 478, 479, 480, 481, 483, 484, 485, 486, 487, 488, 489, 493,
        494, 499, 502, 521, 526, 561, 572, 573, 574, 577, 578, 579, 583,
        481, 589, 486, 596, 493, 598, 597, 493

B

bandgaps 182, 184, 220, 221, 224
beams
  Gaussian beam 529, 530, 531, 537, 539, 541, 542, 543, 544, 545, 546,
        547, 549, 552, 553, 558, 583, 588
  LG beam 529, 588, 590
  non-paraxial Gaussian beam 537, 545, 546, 547, 558
  vortex laser beams 423, 425, 426, 447

C

coefficient
  Fourier coefficient 38
constant
  dielectric constant 19, 38, 67, 79, 88, 96, 97, 99, 100, 101, 103, 108, 112,
        118, 122, 124, 131, 147, 148, 226, 228, 294, 356, 367, 532, 534,
        540, 541, 542, 546, 548, 550, 552
  propagation constant 26, 53, 74, 75, 96, 99, 100, 108, 109, 113, 114, 115,
        116, 120, 123, 147, 152, 156, 222, 225, 246, 254, 260, 261, 361
coronagraph 424

D

dielectric step 3, 91, 107, 108, 110, 122
diffraction
  Fraunhofer diffraction 575, 577, 581, 582, 585, 599

E

efficiency
  diffraction efficiency 529, 530, 584, 585, 587
equation
  Helmholtz equation 42, 48, 191, 224, 226, 230, 361, 363, 364, 563
  Maxwell's equations 1, 7, 10, 16, 35, 37, 48, 182, 195, 202, 214, 224,
    262, 268, 270, 281, 286, 291, 406, 534, 565
  Richards–Wolf 285, 286, 291, 295
  Schrödinger equation 190, 192, 306
equatorial geometry of magnetization 20

F

fibres
  Photonic crystal fibres 219
FIMMWAVE 259, 260, 261
FullWAVE 6.0 185, 205, 206, 207, 214
function
  Fadeyeva's function 85
  Heaviside function 81, 85

G

gratings
  Bragg gratings 3, 79, 83, 131, 132, 179, 269

I

integral
  Fresnel integral 569

L

laser
  He–Ne laser 294, 304, 561, 569
  InGaN/GaN semiconductor laser 150
  Nd:YAG laser 297, 710
Laurent rule 15, 16, 38
lens
  aplanatic lens 285, 286, 287, 289, 295, 296, 297, 332, 475, 496, 497, 498,
    499, 500
  Mikaelian lens 189, 190, 201, 202, 217, 264, 368, 371, 372, 374, 379,

384, 421
two-dimensional photonic crystal gradient lenses 193
Veselago lens 201, 263
Vesselago superlens 199
lithography
  electron-beam lithography 578, 580, 581
  photolithography 580

M

matrix
  Halley's matrix 65
  scattering matrix 1, 2, 26, 27, 29, 30, 53, 56, 57, 58, 60, 61, 62, 63, 64,
      65, 66, 67, 68, 70, 71, 73, 75, 76, 80, 84, 86, 88, 89, 165, 166,
      176, 177
  Toeplitz matrix 12, 13, 15, 38, 51
method
  BOR-FDTD 268, 269, 270, 280, 298, 311, 324, 326, 327, 330, 332, 333,
      334, 343, 344, 406, 408, 409, 415
  FDTD method 185, 190, 203, 205, 206, 268, 269, 298, 311, 324, 326,
      327, 330, 332, 334, 335, 336, 343, 406
  finite difference method 224, 266
  finite-difference time-domain method 190, 261, 415
  Fourier modal method 1, 2, 3, 7, 33, 40, 41, 48, 49, 50, 57, 67, 73, 88,
      107, 119, 131, 137, 167, 175, 176, 177, 180
  Halley's method 61, 70, 71
  Householder method 60, 65
  iterative method 529
  Krylov method 225, 238, 246, 247, 248, 249, 250, 251
  method of matched sinusoidal modes 224, 225, 249, 259
  method of matched sinusoidal modes (MSM method) 224, 225
  method of the scattering matrix 1, 26
  MSM method 224, 225, 250, 251, 253, 256, 259, 260, 261
  multipole method 222
  Newton's method 60, 61, 65, 69, 70, 71, 72, 234
microaxicon 294, 298, 301, 302, 303, 304, 311, 331
model 2, 58, 67, 88, 89, 90, 91, 107, 119, 185, 191, 195, 223, 251, 252,
      253, 255, 256, 259, 260, 286, 298, 349, 425, 472, 474, 479, 480,
      483, 493, 504, 543, 606, 610, 481, 700
  Drude–Lorentz model 67, 88
modes
  Laguerre–Gaussian mode 296
  TE- 534
  transverse–magnetic modes 226

N

nanorods 188
number
  Bloch's wave number 184

P

Padé approximant 86, 87, 88
plasmon–polaritons 2, 91, 92, 95, 101, 102, 107, 118, 122, 131, 168
polarization
  TE-polarization 7, 19, 47, 48, 135, 138, 158, 167, 168, 172, 173, 174, 175, 191, 195, 203, 211, 243, 313, 372, 385, 390, 533, 535, 539, 541, 544, 547, 565, 401
  TM-polarization 7, 15, 47, 133, 136, 141, 158, 168, 169, 170, 173, 213, 214, 246, 368, 390, 534, 538, 547

R

Rayleigh frequency 84

S

scanning near-field optical microscopy 269, 292, 337
scheme
  Kretschmann scheme 100, 106
singular optics 423, 425, 426, 440, 444, 446
superlens 199, 200, 201, 205, 263, 358, 359, 360, 367, 420

T

TE- 534
theorem
  Floquet–Bloch theorem 10
Toeplitz matrix 12, 13, 15, 38, 51
transform
  Fresnel transform 561

U

Umov–Poynting vector 7, 32, 54, 55, 97, 98, 99, 100, 531, 539, 541, 544, 564, 565, 566

V

vector
  Umov–Poynting vector 531, 539, 541, 544, 564, 565, 566

W

waveguide
  plasmon waveguide 104, 119, 121, 122
waves
  damped electromagnetic waves (DEW) 132
  E-waves 7, 24, 43
  H-waves 5, 6, 7, 9, 24, 32, 44, 45